The History of Mathematics: A Source-Based Approach Volume 2

Frontispiece to Euler's *Introductio in Analysin Infinitorum*, 1748

AMS/MAA | TEXTBOOKS

VOL 61

The History of Mathematics: A Source-Based Approach Volume 2

June Barrow-Green
Jeremy Gray
Robin Wilson

MAA PRESS
An Imprint of the American Mathematical Society
Providence, Rhode Island

2020 *Mathematics Subject Classification*. Primary 01-01, 01A05;
Secondary 01A45, 01A50, 01A55.

For additional information and updates on this book, visit
www.ams.org/bookpages/text-61

The ISBN numbers for this series of books includes

ISBN 978-1-4704-4382-5 (number 2)
ISBN 978-1-4704-4352-8 (number 1)

Library of Congress Cataloging-in-Publication Data

The first volume was catalogued as follows:

Names: Barrow-Green, June, 1953– author. | Gray, Jeremy, 1947– author. | Wilson, Robin J., author. Title: The history of mathematics : a source-based approach / June Barrow-Green, Jeremy Gray, Robin Wilson.

Description: Providence, Rhode Island : MAA Press, an imprint of the American Mathematical Society, [2018]- | Series: AMS/MAA textbooks ; volume 45 | Includes bibliographical references and index. Identifiers: LCCN 2018034323 | ISBN 9781470443825 | 9781470456931 (ebook) |Softcover ISBN 9781470472993

Subjects: LCSH: Mathematics–History. | Mathematics–Study and teaching. | AMS: History and biography – Instructional exposition (textbooks, tutorial papers, etc.). | History and biography – History of mathematics and mathematicians – General histories, source books. | History and biography – History of mathematics and mathematicians – Indigenous European cultures (pre-Greek, etc.). | History and biog-raphy – History of mathematics and mathematicians – Egyptian. | History and biography – History of mathematics and mathematicians – Babylonian. | History and biography – History of mathematics and mathematicians – Greek, Roman. | History and biography – History of mathematics and mathematicians – China. | History and biography – History of mathematics and mathematicians – India. | History and biography – History of mathematics and mathematicians – Medieval. | History and biography – History of mathematics and mathematicians – 17th century.

Classification: LCC QA21 .B24 2018 | DDC 510.9–dc23 LC record available at https://lccn.loc.gov/2018034323

∞ The paper used in this book is acid-free and falls within the guidelines
established to ensure permanence and durability.
Visit the AMS home page at `https://www.ams.org/`

11 10 9 8 7 6 5 4 3 2   27 26 25 24 23 22

Contents

Acknowledgments

We would like to thank the following people for their help: Tom Archibald, Henk Bos, Joseph Dauben, Caitanya Vaghela Durley, José Ferreirós, Raymond Flood, Catherine Goldstein, Niccolò Guicciardini, Victor Katz, Snezana Lawrence, Jesper Lützen, Peter Neumann, Jeanne Peiffer, Steve Russ, and the late Jackie Stedall. We thank Commander D.G. Turnbull for generously allowing us to reproduce letters from the volumes of the Newton Correspondence edited by his father. We also thank the family of Dirk Struik for their generosity in allowing us to reprint some passages from his *A Source Book in Mathematics: 1200–1800*. We also wish to thank our present and former colleagues at the Open University who have helped to keep the history of mathematics alive, especially Gloria Baldi, Mick Bromilow, Rebecca Browne, Giles Clark, Derek Goldrei, Sara Griffin, Tracy Johns, and Derek Richards and, for their help with TEX, Camilla Jordan and John Trapp. We also thank Gresham College, London, for a grant towards the production costs of this book.

John Fauvel was the driving force behind the Open University course upon which this book is based. Not only did he possess a wide range of knowledge about the history of mathematics, he was able to balance great complexity of material with an exquisite sensitivity to the needs and aspirations of students, to the point of making them joint explorers in the study of the subject. His emphasis on the patient reading of primary sources in translation was empowering, and his influence through the Open University, the British Society for the History of Mathematics, and in many other ways led to a remarkable growth in the appreciation of what the history of mathematics can offer. His untimely death in 2001 deprived us all of his wisdom, energy, and enthusiasm, and we hope that this book will spread his influence further.

We also thank these people at the MAA and the AMS for their support: Donald J. Albers, Stephen Kennedy, Beverly Ruedi, Chris Thieverge, Michael Haggett, and Jennifer Wright Sharp.

Conventions. We have used the following conventions in this book:

- Book titles are given in italics, and if the title is not in English we have followed it with a translation, which is in italics if it is the title of an English translation of the original book and in roman otherwise.
- A reference of the form '(Boyer 1959, 110)' is to page 110 of the item listed under Boyer in the Bibliography as published in 1959.
- A reference of the form 'Boyer (1959, 110) wrote' is to be read as an abbreviation for 'Boyer, on p. 110 of his book (or article) of 1959, wrote'.

Many references are given to sources in English in the form F&G 15.D4, which stands for the entry D4 in Chapter 15 of Fauvel and Gray, *The History of Mathematics: a Reader*.

Permissions & Acknowledgments

The American Mathematical Society gratefully thanks the following people and institutions for permission to reproduce figures and extracts.

Collections École Polytechnique

Figure 13.1, p. 337. Photogragh of the original premises of the École Polytechnique. © Collections École polytechnique (Palaiseau, France), PHX 800001.

T. Dénés

Figure 14.12, p. 390. Photograph of bust of János Bolyai. Source: Downloaded from `www.titoktan.hu/Bolyai_a.htm`. Courtesy of T. Dénés.

European Mathematical Society, Zürich

Figure 19.3, p. 555. A manuscript page of Évariste Galois. Source: Peter M. Neumann, *The mathematical writings of Évariste Galois*, Heritage of European Mathematics, European Mathematical Society, Zürich, 2011.
Excerpt, pp. 552–554.

Georg-August-Universität, Göttingen

Figure 15.22, p. 448. Group portrait of The Mathematics Club of Göttingen, 1902 (Karl Schwartzchild Nachtclass). SUB Göttingen, Cod. Ms. K. Schwarzchild, 23 : 1,16. Courtesy of Georg-August-Universität, Göttingen.

Dirk Struik

Excerpts, pp. 105–113, 115–116, 116–118, 164–166, 255–256, and 277–279. Source: Dirk Struik, *A Source Book in Mathematics 1200–1800*, Harvard University Press, Cambridge, Massachusetts, 1969. Used with permission of the Struik family.

Interdisciplinary Scientific Center J.-V. Poncelet (ISCP)

Figure 15.1, p. 414. Portrait of Jean-Victor Poncelet. Courtesy of the Interdisciplinary Scientific Center J.-V. Poncelet, Moscow, Russia.

LibraryThing

Figure 20.7, p. 588. Portrait of Gustav Lejeune Dirichlet. Public domain, downloaded from `www.librarything.com`.

Mathematisches Forschungsinstitut Oberwolfach

Figure 21.7, p. 605. Photograph of Gösta Mittag-Leffler. Source: Konrad Jacobs, Erlangen, Mathematisches Forschungsinstitut Oberwolfach gGmbH (MFO): used under Creative Commons License Attribution–ShareAlike 2.0 Germany (CC by-SA 2.0) license.

Conservapedia

Figure 20.4, p. 578. Portrait of William Thomson (Lord Kelvin). Public domain, downloaded from `www.conservapedia.com`.

Queen's College, Oxford

Frontispiece, p. ii. Frontispiece to Euler's *Introductio in Analysin Infinitorum*, 1748. Courtesy of the Provost and Fellows of The Queen's College, Oxford.

Figure 5.3, p. 129. Drawing of the great comet of 1680, from Newton's *Principia*, Book III. Courtesy of the Provost and Fellows of The Queen's College, Oxford.

Figure 5.4, p. 133. The title page of Newton's *Principia*. Courtesy of the Provost and Fellows of The Queen's College, Oxford.

Figure 5.6, p. 140. Drawing of parallelograms approximating the area under a curve, from Newton's *Principia*. Courtesy of the Provost and Fellows of The Queen's College, Oxford.

Figure 5.8, p. 142. Drawing of a particle obeying a central force as it sweeps out equal areas in equal times, from Newton's *Principia*, Book I. Courtesy of the Provost and Fellows of The Queen's College, Oxford.

Figure 7.2, p. 195. Euler's *Mechanica* (1736). Courtesy of the Provost and Fellows of The Queen's College, Oxford.

Figure 7.3, p. 197. Euler's *Introductio in Analysin Infinitorum* (1748). Courtesy of the Provost and Fellows of The Queen's College, Oxford.

Figure 7.4, p. 200. Euler's *Lettres à une princesse d'Allemagne* (1768). Courtesy of the Provost and Fellows of The Queen's College, Oxford.

Figure 9.3, p. 254. MacLaurin's explanation of why he wrote his *Treatise of Fluxions*. Courtesy of the Provost and Fellows of The Queen's College, Oxford.

Royal Museums Greenwich Picture Library

Figure 5.2, p. 128. The Royal Observatory, Greenwich.

Cdr DG Turnbull

Excerpts, pp. 87–88 and 88–94. Source: HW Turnbull, *The mathematical correspondence of Isaac Newton*, Cambridge, 1960. Used with permission from Cdr DG Turnbull.

University of California Press Books

Excerpts, pp. 101–102, 129–130, and 137–138. Source: Isaac Newton, *The Principia: The Authoritative Translation and Guide: Mathematical Principles of Natural Philisophy* (Julia Budenz, author; I. Bernard Cohen and Anne Whitman, translators), Berkeley, California, 2016.

University of St. Andrews Special Collections

Figure 3.5, p. 62. The opening page of Newton's *De Analysi* (1711). Courtesy of University of St. Andrews Special Collections.

Figure 6.18, p. 181. Title page of Émilie du Châtelet's *Principles Mathématiques de la Philosophie Naturelle* (the French translation of Newton's *Principia*). Courtesy of University of St. Andrews Special Collections.

Figure 9.7, p. 272. Euler's *Institutiones Calculi Differentialis* (1755). Courtesy of University of St. Andrews Special Collections.

Figure 9.8, p. 273. Euler's *Institutiones Calculi Integralis*, Vol. 1 (1768). Courtesy of University of St. Andrews Special Collections.

Wellcome Collection

Figure 2.16, p. 37 D. Loggan's line engraving of John Wallis. Downloaded from `Wellcomecollection.org`, used under Creative Commons Attribution 4.0 International (CC by 4.0) license.

Figure 6.14, p. 179. Portrait of Pierre-Louis Moreau de Maupertuis. Downloaded from `Wellcomecollection.org`, used under Creative Commons Attribution 4.0 International (CC by 4.0) license.

Figure 11.1, p. 311. Portrait of Pierre-Simon Laplace. Downloaded from `Wellcomecollection.org`, used under Creative Commons Attribution 4.0 International (CC by 4.0) license.

Figure 18.8, p. 535. Portrait of William Rowan Hamilton. Downloaded from `Wellcomecollection.org`, used under Creative Commons Attribution 4.0 International (CC by 4.0) license.

Wikimedia Commons

Figure 1.1, p. 8. Frontispiece to Thomas Sprat's *History of the Royal Society* (1667).

Figure 2.5, p. 21. Portrait of René Descartes by Franz Hals.

Figure 3.1, p. 52. Portrait of Isaac Newton.

Figure 3.2, p. 53. Drawing showing Newton's rooms in Trinity College, Cambridge.

Figure 3.9, p. 68. Portrait of Gottfried Wilhelm Leibniz.

Figure 4.8, p. 106. The opening page of Leibniz's first article on the calculus.

Figure 5.9, p. 149. Portrait of Pierre Varignon.

Figure 6.2, p. 156. Portrait of Jakob Bernoulli.

Figure 6.3, p. 157. Portrait of Johann Bernoulli.

Figure 6.4, p. 158. Portrait of Brook Taylor.

Figure 6.5, p. 158. Portrait of Marquis de l'Hôpital.

Figure 6.15, p. 180. Portrait of Voltaire.

Figure 6.16, p. 180. Portrait of Émilie de Breteuil, Marquise du Châtelet.

Figure 6.17, p. 181. Frontispiece to Voltaire's *Elements of the Philosophy of Newton*.

Figure 6.20, p. 184. Portrait of Alexis-Claude Clairaut.

Figure 7.1, p. 193. Portrait of Leonhard Euler.

Figure 7.5, p. 204. Portrait of Jean le Rond D'Alembert.

Figure 7.6, p. 204. Portrait of Joseph-Louis Lagrange.

Figure 8.2, p. 231. A map of the city of Königsberg.

Figure 8.5, p. 233. Drawing of a new system of bridges.

Figure 9.2, p. 253. Portrait of Colin MacLaurin.

Figure 9.10, p. 277. The title page of Lagrange's *Théorie des Fonctions Analytiques.*

Figure 11.2, p. 313. Portrait of Mary Somerville.

Figure 13.2, p. 338. Portrait of Gaspard Monge.

Figure 13.4, p. 347. Portrait of Niels Henrik Abel.

Figure 13.5, p. 348. Portrait of Joseph Fourier.

Figure 13.6, p. 351. Portrait of Carl Friedrich Gauss.

Figure 13.9, p. 360. Portrait of Joseph Diaz Gergonne.

Figure 13.10, p. 360. The title page of the first issue of Gergonne's *Annales.*

Figure 13.11, p. 361. Portrait of August Leopold Crelle.

Figure 13.12, p. 361. The title page of the first issue of Crelle's *Journal.*

Figure 13.13, p. 366. Portrait of Joseph Liouville.

Figure 13.14, p. 366. The title page of the first issue of Liouville's *Journal.*

Figure 14.7, p. 382. Portrait of Johann Heinrich Lambert.

Figure 14.11, p. 390. Portrait of Nikolai Ivanovich Lobachevskii.

Figure 14.21, p. 407. Photograph of Bernhard Riemann.

Figure 14.22, p. 410. Photograph of Eugenio Beltrami.

Figure 15.13, p. 426. Photograph of August Ferdinand Möbius.

Figure 15.18, p. 433. The 27 straight lines on a cubic surface. Used under Creative Commons Attribution–ShareAlike Unported 3.0 (CC by-SA 3.0) license.

Figure 15.19, p. 436. Photograph of Felix Klein.

Figure 16.1, p. 452. Portrait of Bernard Bolzano.

Figure 16.6, p. 458. The title page of Cauchy's *Cours d'Analyse* (1821).

Figure 16.8, p. 462. Portrait of Augustin-Louis Cauchy.

Figure 17.1, p. 476. Photograph of Richard Dedekind.

Figure 17.2, p. 482. Photograph of Georg Cantor.

Figure 17.4, p. 494. Photograph of Giuseppe Peano.

Figure 17.6, p. 496. Photograph of Gottlob Frege.

Figure 17.8, p. 499. Photograph of Bertrand Russell.

Figure 18.1, p. 512. Portrait of Adrien-Marie Legendre.

Figure 18.9, p. 540. Portrait of Hermann Günther Grassmann.

Figure 18.10, p. 542. Photograph of James Clerk Maxwell.

Figure 18.11, p. 542. Photograph of Josiah Willard Gibbs.

Figure 19.1, p. 549. Portrait of Paolo Ruffini.

Figure 19.2, p. 552. Drawing of Évariste Galois.

Figure 19.4, p. 556. The 1830 Revolution, 'Prise de l'Hôtel de Ville: le Pont d'Arcole', by Amédée Bourgeois.

Figure 19.6, p. 566. Photograph of Camille Jordan.

Figure 20.1, p. 570. Title page of Fourier's *Théorie Analytique de la Chaleur*. Used under Attribution-partage dans les mêmes conditions 3.0 non transposé (CC by-SA 3.0) license.

Figure 20.5, p. 580. Drawing of Thomson's machine for predicting the tides.

Figure 21.2, p. 598. Photograph of George William Hill.

Figure 21.3, p. 599. Photograph of Jules Henri Poincaré.

Figure 21.5, p. 603. Photograph of King Oscar II.

Figure 21.6, p. 605. Photograph of Sonya Kovalevskaya.

Figure 21.8, p. 608. Photograph of Charles Hermite.

Figure 21.9, p. 608. Photograph of Karl Weierstrass.

Figure 21.10, p. 608. Photograph of Lars Edvard Phragmén.

Figure 22.1, p. 624. Photograph of Eliakim Hastings Moore.

Figure 22.3, p. 624. Photograph of Heinrich Maschke.

Figure 22.5, p. 629. Photograph of David Hilbert.

Figure 22.6, p. 629. Photograph of Hermann Minkowski.

Introduction

This book is a history of some of the principal development in mathematics from about 1650 to the start of the 20th century. The topics selected are mainstream, but we hope that our treatment is fresh.

The presentation is divided into two parts. The first begins with the independent discoveries of the calculus by Isaac Newton and Gottfried Wilhelm Leibniz, and traces its advances through to the end of the 18th century. Unlike some older histories of the subject, we have emphasised some of the applications of the subject by including the history of differential equations in our account. Little mathematical detail is required to show how the new calculus transformed not only the study of curves but also the theory of mechanics; it enabled mathematicians to ask a whole raft of new questions and frequently to answer them by simultaneously refining the tools and methods that they had.

We also look at one of the great works of mathematical physics and arguably its founding text: Newton's *Principia Mathematica* (1687). This was not itself a calculus text — although there are some intriguing connections — and one of Leonhard Euler's achievements fifty years later was to recast the subject in the language of the calculus, and thus to make it much more accessible. Here we trace a journey from Newton to Laplace that stands as one of the most influential examples of the power of mathematics.

The 18th century — the age of Euler, D'Alembert, and Lagrange — saw many other breakthroughs, from the formulation and solution of the wave equation, which profoundly influenced mathematicians' ideas about what functions could be, to Euler's revival of the theory of numbers that was carried forward by Lagrange and then by Gauss. Moreover, and in part by the sheer force of Euler's example, mathematics was transformed into a public enterprise that involved publications in journals and books if prestige was to be gained.

The second part of the book is about mathematics in the 19th century. We look at the discovery of new geometries, and new ways of thinking about geometry: the non-Euclidean geometryof János Bolyai in Hungary and Nikolai Ivanovich Lobachevskii in Russia; the rise of projective geometry in France and Germany; and, more briefly, differential geometry in the hands of Gauss and Riemann, and axiomatic geometry in the work of Hilbert and others.

Then we turn to the rigorisation of the calculus. This was largely the achievement of Cauchy, although he left much for others to do, and we trace this story as far as the introduction of set theory as a possible foundation for mathematics. We look, therefore, at Cantor's introduction of a theory of infinite sets, at Dedekind's foundational ideas, and at Frege and Russell's hopes of deriving all of mathematics from logic.

Algebra also changed greatly in the 19th century, going from the largely symbolic and manipulative form that it had (and school algebra still has) to a more structural form. We consider such topics as the Fundamental Theorem of Algebra, the nature of complex numbers, developments in number theory due to Gauss, and the introduction of vectors. We discuss investigations (from Lagrange via Galois to Jordan) into the question of whether there are algebraic solutions (formulas) for polynomial equations of any degree as a key source of what may be called 'structural algebra'. This can be thought of as the creation of a family of concepts that are adequate to explain what can otherwise seem to be the result of blind calculation, and it ushered in what was to become a new branch of mathematics: group theory.

The applications of mathematics flourished in this period, and we look at a few examples: Fourier's ideas about the flow of heat and the representation of arbitrary functions by infinite sums of sines and cosines (Fourier series); the origins of potential theory in the work of Green and Gauss and the first attempts by Dirichlet to make a rigorous subject out of it; the wave equation, the invention of the telegraph, and the work of Thomson and Heaviside. A final chapter takes up Poincaré's study of celestial mechanics and its implications for both pure and applied mathematics. A short coda, describing the state of mathematics on the brink of the 20th century, concludes this history.

Studying the history of mathematics is by no means the same as studying mathematics itself, although some familiarity with mathematics is advisable. The questions that this book addresses are questions about the *history* of mathematics, not mathematical questions with a historical flavour (exciting though that can be, too). We are interested in understanding who did the mathematics, and why? Were they teachers — and if so, who were their students and why were they there? Was there a cultural or philosophical dimension to their mathematical work? What does it mean to *discover* something in mathematics? How was mathematical knowledge disseminated? Surprisingly rich answers to questions such as these can be obtained without one having to master the accompanying mathematics. What was done is interesting, but why it was done is interesting too. What was the context that made the work important in its day? Why is it still of interest to study it today? These are the central questions that flow through this book.

To answer these questions, historians of mathematics rely on what we might call the 'facts', which can be drawn from many sources: documents, written texts, and also various artefacts. The historian's task is to make these mute objects speak again — but inevitably they do so with the historian's voice. This is ultimately because the big questions raised here do not have simple answers, and so studying the history of mathematics involves a certain amount of disagreement. Historians produce arguments, based on their selections of the evidence; their conclusions are not so much facts as opinions. We can ask that their opinions are well argued and well informed, but opinions they remain, and other historians, perhaps bringing forth new facts, can disagree. This gives the history of mathematics a necessarily provisional character, but it opens the way to new and important work in the subject, and you will find much recent historical work reflected in these pages.

We have deliberately made use of extensive quotations from original sources (more can be found through the footnotes), because we believe that grappling with these texts is essential in the study of the history of mathematics. Understanding these extracts is not always easy, but it is rewarding. Those we present provide the best way in to understanding what people did and why, and thus to understanding how and why mathematics developed in the way it did. They generate questions, many of which we pursue in some detail, and they speak to us across the centuries. They are best appreciated with pen in hand, as you follow the calculations and draw the diagrams.

This is the second and final volume of our history of mathematics. It follows the previous volume, which we refer to as *Volume* 1, but it can be read independently. Like that volume, it also contains suggested exercises, which — unlike the exercises in most textbooks on the subject — are firmly historical, rather than primarily mathematical. They will be found at the end of the book, along with suggestions for how they can be approached.

Part I

The 17th and 18th centuries

1

Introduction: The 17th and 18th centuries

The 17th and 18th centuries in European history are marked by a number of important events, ending with the French Revolution, the transformation whose continuing repercussions have done so much to shape the Europe of today. So we might expect that the decades before 1789 were very different, and indeed they were.

Mathematics at the time of Descartes was in an intriguingly mixed state. René Descartes himself had moved to the new, and Protestant, country of the Netherlands, where he felt less likely to be prosecuted by the Catholic Church that had recently and harshly disciplined Galileo for seeming to disagree with Church doctrine about the rotation of the Earth. There he found people able to read his work, some of whom, like him, were independent scholars only loosely attached to the university world. In France a few mathematicians had university positions, while others, including Marin Mersenne, Pierre de Fermat, and Blaise Pascal, did not. In England and Scotland, too, innovative work was also done by people in various walks of life, from surveyors and astronomers to the professionals attached to Gresham College in London.

One factor that changed things in England was the English Civil War of 1642–1651. By the time it was over and the monarchy had been restored in 1660 under a constitution that now limited its powers, an influential group of scholars based in Oxford and London had come together, and in 1662 they persuaded the new King, Charles II, to establish the world's first strictly scientific society, the Royal Society. Its significance lay in its attitude to knowledge, summed up in its motto 'Nullius in verba' ('On the word of no-one'). Claims of any kind were to be accepted only on the basis of evidence, not on rumour, nor even on the word of a gentleman. Robert Hooke, Christopher Wren, Edmond Halley, William Brouncker, and others applied themselves to observation and argument, and the King, who paid for their work, expected that the nation would in some way benefit from their labours — for example, by finding reliable ways to determine longitude at sea and by assisting in navigation, trade, and the expansion of British naval power.

To help bring this about, the Royal Society created a journal called the *Philosophical Transactions*, the first volume of which was published in 1665. This mirrored a trend being felt elsewhere in Europe: the first issue of the first academic journal to be

Figure 1.1. Frontispiece to Thomas Sprat's *History of the Royal Society* (1667)

founded there, the Parisian *Journal des Sçavans*, had appeared a few months earlier, on 5 January 1665.[1] Journals did not mark the end of personal correspondence — indeed, many of the articles they published took the form of open letters and responses to previous articles. But whereas in an earlier period such communications would have been private, or given a very limited circulation, now, if the author wished, they could be distributed among the subscribers to a journal, read more widely, and draw unknown others into the debate. Thus was created what historians now call the 'Republic of Letters', an informal international group of people who were interested in this or that

[1] This journal was later renamed the *Journal des Savants*.

aspect of science, or antiquity, or any form of scholarship. Relationships in that world might be close or hostile, amicable, or competitive, but the dynamic was conducive to the growth of enquiry, and often transcended national boundaries. Even during times of war, gentlemen scholars from opposing countries could travel freely and call upon the aid of other like-minded folk.

Although it may have been unclear at the time, and only slowly became clearer in England, we also see here the origins of another significant innovation in the life of mathematics and science: the learned academy, often with royal support. In the 18th century influential mathematicians increasingly had positions in the Academies of Sciences in Paris, Berlin, or St Petersburg. At the highest levels the place for gifted amateurs with their own money was shrinking.

Notably, few of these dominant mathematicians were attached to universities. Research was not then the primary task of any university, and one can wonder on occasion whether teaching was either. The Bernoulli brothers, Jakob and Johann, were professors in the Netherlands and Switzerland at the start of the 18th century, and Joseph-Louis Lagrange was briefly a professor in Turin at the start of his career before moving to the Academy in Berlin, but Leonhard Euler never held a university position and moved between the Academies in St Petersburg and Berlin.

So when Isaac Newton went up to the University of Cambridge in 1661, he was not going to a modern high-powered research institution but to a place where young men of varying degrees of wealth and influence were educated, as much or as little as they saw fit, so that they could pursue careers in the Church or at Court. In Isaac Barrow Cambridge possessed a mathematician of real ability, but he left in 1669 to seek a better life at Court, and Newton succeeded him as the Lucasian Professor at Cambridge. Newton then had to make his way with the assistance of members of the Royal Society.

The origins of the Lucasian Chair are revealing. It had been founded in 1663 with the bequest of Henry Lucas, who had been the Member of Parliament for Cambridge University and who left four thousand books and an endowment of one hundred pounds to pay for a professor who was not to be active in the Church; Charles II signed the Chair formally into existence in January 1664. Such bequests are a sure sign that reform was wanted, and the Chair became one for mathematical sciences, not least because of its first two occupants.

In due course, Newton left Cambridge for London, where he eventually became Master of the Royal Mint and President of the Royal Society. The first edition of his hugely influential *Philosophiæ Naturalis Principia Mathematica* (The Mathematical Principles of Natural Philosophy) had been published by the Royal Society in 1687, with the active assistance of Halley. The second and third editions were prepared with the assistance of other Cambridge mathematicians, such as Roger Cotes and Brook Taylor, but after that, British involvement in mathematics and many branches of science declined, for reasons that are not well understood, and by 1800 even the Royal Society had become something of a gentlemen's club where social status eclipsed scientific merit.

The only intellectual figure who can bear comparison to Newton during his lifetime was Gottfried Wilhelm Leibniz. Just as Newton's interests extended beyond mathematics to embrace (indeed, redefine) many areas of physics, and also theology, and even alchemy, Leibniz was also occupied with philosophy, linguistics, logic, law, and (to a lesser extent than Newton) physics. But he was unlucky with his employment, and wound up spending almost forty years as a rather minor Court Councillor in Hannover. This prevented him from applying his prodigious skills to the tasks that posterity would have preferred. Even so, not only did he create a version of the calculus that was as good as Newton's, he had the time and the opportunity to foster the next generation of mathematicians, and so his influence on mathematics was arguably the greater of the two. In some measure he had this opportunity because, unlike Newton, he was strongly in favour of publishing. The mathematical and scientific journal that he and the German philosopher Otto Mencke helped to create in 1682, the *Acta Eruditorum Lipsiensium* (Acts of the Scholars of Leipzig), was well received and it carried his two influential presentations of the calculus in 1684 and 1686 that we describe in Section 4.4. This was the first time that the new subject was put into print, and they established Leibniz as a major original mathematician.

Mathematics in France also had its vicissitudes. The country that had produced Descartes, Fermat, and Pascal at the start of the 17th century was unable to match them at the end. Instead, mathematics was briefly dominated by the feuding Bernoulli brothers from Switzerland and their children. The younger brother, Johann, began the 18th-century enterprise of mastering, extending, and applying the calculus, a development only hinted at by Newton in his *Principia* but which Newton's followers felt was almost forced upon them if they were to understand, let alone extend, his work.

The Bernoullis had learned the calculus by corresponding with Leibniz about it, and it fell to Johann to teach the man who was to dominate the middle decades of the 18th century, his fellow Swiss, Leonhard Euler. Euler transformed almost every branch of mathematics he touched, and created many others. He revived the theory of numbers that Fermat had begun a century earlier; he rewrote the calculus in the language of functions; he wrote on celestial mechanics and optics, fluid flow, and differential equations. Above all, he published, in the journals of the learned Academies, as well as his many books, and by his example — he was a lucid and an exceptionally prolific writer — he established that publication should henceforth be the rule. He also won prizes from the Academies for his researches into many specific topics and in turn organised prize competitions; these competitions were a characteristic feature of mathematical life in the 18th century and provided a way, not always successful, of promoting research in specific areas.

If Euler had a rival it was the French mathematician Jean le Rond D'Alembert, who was ten years younger than Euler and a major figure in the Parisian Académie des Sciences. In D'Alembert's view, the 18th century was the century of mechanics, and he was almost wholly focused on creating mathematics that would deepen our understanding of the physical world. He was, apparently, a gifted conversationalist and a popular figure in social circles and the salons of the time, but he was a poor writer, and much of his influence was to be exerted through the work of his protégé, Joseph-Louis Lagrange. That said, D'Alembert was the first to extend the calculus effectively to functions of several variables, and to describe the motion of a vibrating string correctly.

Lagrange was a more austere figure — an algebraist by temperament, who first came to people's notice when he took up and enriched two of Euler's subject areas, the calculus of variations (which we do not discuss) and the theory of numbers. He later wrote a very important paper on the algebraic solution of polynomial equations, as well as a thorough, if unsuccessful, attempt to base the calculus on solely algebraic considerations, and most extensively on celestial mechanics. Too shy ever to meet Euler, he nonetheless succeeded him in Berlin before moving to the Academy in Paris in 1787.

Celestial mechanics at that time may be summed up as the attempt to see whether Newton's law of gravity was capable of describing all the delicate and complicated behaviour of the motion of the planets, their satellites, and the Moon. As we shall discuss, the motion of even three bodies (such as the Sun, the Earth, and the Moon) under their mutual gravitational attraction is exceedingly difficult to understand, and throughout the 18th century numerous figures contributed to its analysis. Their work was consummated by Pierre-Simon Laplace, whose five-volume treatise on celestial mechanics showed that all the known complexities of the various orbits can indeed be explained satisfactorily on Newtonian grounds, although long-term predictions were to remain impossible.

In less than two hundred years, mathematics was rewritten around the calculus, which replaced geometry as the relevant core discipline in the subject: indeed, much of geometry was rewritten using the terms and methods of the calculus. But this was not the modern rigorous calculus that we know — that was to be a creation of the 19th century. This was a calculus with powerful new methods and uncertain foundations, and was largely cast in a formal, algebraic language. It discussed functions (generally expressed as infinite series), their derivatives, and their integrals, and above all, as in celestial mechanics, differential equations of many kinds. Its legacy to succeeding generations was both its astonishing efficacy and a growing need to re-establish mathematics on the rigorous foundations that had been associated, however generously, with Euclid's *Elements*.

2

The Invention of the Calculus

Introduction

It is impossible to tell the story of the invention of the calculus with a surprise at the end: you will already know that it was invented by Newton and Leibniz. But they did not create exactly the same calculus; each chose to emphasise different aspects of the contemporary mathematical scene, and to develop their work in different ways. So by looking in a little detail at what each individual did, we have an opportunity not only to savour one of the truly dramatic moments in the history of mathematics, but also to learn about it in a more satisfying way. Yet it is also legitimate to speak of the *creation of the calculus* by Newton and Leibniz, because what they did in common so greatly transformed what anyone achieved before. Each produced

- a systematic method for finding tangents
- a systematic method for finding areas
- a profound connection between tangency and area problems.

These three together provide a reasonable characterisation of the family of ideas that one now calls 'the calculus'.

Our story is a 17th-century one that enables us to see how these ideas were brought together. While the opening two sections of this chapter help us to establish the sources from which the calculus acquired its characteristics, we do not mean to imply that history had 'in mind' the creation of the calculus, or that the work of Newton and Leibniz was 'merely' the 'inevitable' culmination of all that went before it. Section 2.3 documents how matters stood in the late 1650s: even as late as this no-one was looking for the calculus. Only in Chapter 3 do we describe what Newton and Leibniz did.

One generality in particular is striking, and to modern eyes unexpected. The problems that these mathematicians discussed, and the language in which they formulated their answers, are always geometrical in character. As with Descartes's work, upon which Newton and Leibniz built, so here also the new ideas entered as methods.[1] Various techniques, including those of the calculus, arose as means for providing geometrical problems with geometrical answers. The means themselves were not necessarily

[1] Descartes's ideas about geometry were discussed in *Volume* 1, Chapter 12.

geometrical — that was part of their novelty — and in later chapters we will see how this tension between geometry and the new solution methods led to a wholesale transformation of people's conception of mathematics itself. For the moment, we merely emphasise the role of new mathematical reasoning: it is in the methods, and not in the questions or the answers.

2.1 Tangents, maxima, and minima

The problem of finding tangents to curves arose in classical mathematics, but the Greek study of tangents was quite lopsidedly developed.[2] Difficult questions had been raised about circles touching circles, and about conics and their tangents, but apart from the spiral of Archimedes and some results on conics by Apollonius, few other curves had been successfully treated.

The Greek concept of a tangent did not make the problem an easy one. Insofar as it emerged from various specific occurrences, a line t (as in Figure 2.1) was taken to be a tangent to a curve C at a point P if two conditions were met: t passes through P, and the part of the curve near P lies entirely on one side of t and so does not cross it.

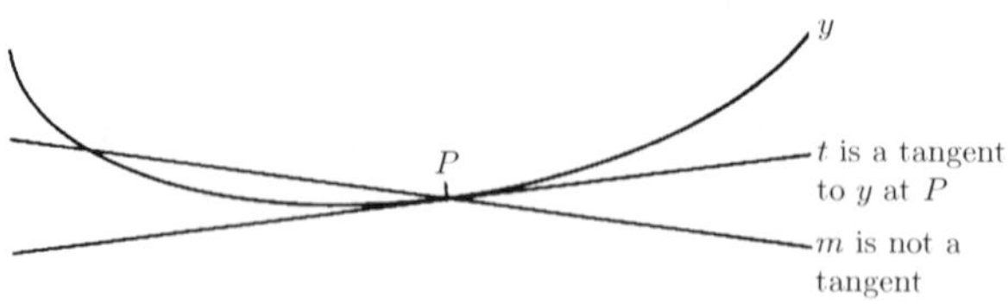

Figure 2.1. A tangent to a curve

In the specific case where the curve C is a circle, Euclid had shown (in his *Elements*, Book III, Prop. 16) that these conditions imply that any other line through P crosses C.

Other classical texts used this definition implicitly, but none of them addressed the general question of how to find a tangent to a given curve. This was to be a problem that mathematicians in the 17th century were to encounter over and over again: the classical examples provided instances of rigorous proofs, but not of discovery methods. When it came to finding out new results, you were very much on your own.

In fact, there are two kinds of problem about tangents. In the first, we are given a curve and a specified point on the curve and asked to find the tangent to the curve at the specified point. In the second, we are given a curve and a specified point not on the curve and asked to find tangents to the curve through the specified point. We might expect the answer to the first problem to be a unique tangent, but the answer to the second is not: for example, there are two tangents to a circle from any specified point outside it, and no tangents at all from a point inside it. The mathematicians we discuss in this chapter were mostly interested in the first of these problems.

Another problem concerned maxima and minima: typically, one would want to find the maximum value of a given expression that represents a curve. That this problem came to be seen as closely related to that of finding tangents is part of our story, but at first the problems were distinct. Maxima and minima had been studied in Greek

[2] As we described in *Volume* 1, Chapter 4.

times. Apollonius, for instance, had investigated the maximum and minimum straight-line segments between a point and a conic, Heron of Alexandria had considered the minimum path-length of light,[3] and Pappus had noticed that the figure enclosing the maximum area for a given perimeter is a circle.[4] But these were random insights that were not united into any coherent theory.

Figure 2.2. Pierre de Fermat (1601?–1665)

Fermat's adequality methods. The first successful and potentially general approach to the problem of finding maxima and minima was proposed by Pierre de Fermat in 1636. It was not published until 1679, but it circulated in manuscript and became quite well known (Mersenne sent a copy to Descartes in January 1638).[5] Because it will lead us to a good illustration of the way in which algebraic methods entered into geometrical problems, we look at it in detail.

---

[3] See Heron, *Catoptrics*, 4–5, in (Cohen and Drabkin 1948, 62–66) and F&G 5.A3(c).
[4] See Pappus, *Mathematical Collection* V, in (Thomas 1939, 615–621) and F&G 5.B5.
[5] See Fermat, *Oeuvres* 1, 133–135, Struik, *A Source Book*, 223–224, and F&G 11.C1.

Fermat on maxima and minima.

> The whole theory of evaluation of maxima and minima presupposes two unknown quantities and the following rule:
>
> Let a be any unknown of the problem (which is in one, two, or three dimensions, depending on the formulation of the problem). Let us indicate the maximum or minimum by a in terms which could be of any degree. We shall now replace the original unknown a by $a+e$ and we shall express thus the maximum or minimum quantity in terms of a and e involving any degree. We shall adequate, to use Diophantus' term, the two expressions of the maximum or minimum quantity and we shall take out their common terms. Now it turns out that both sides will contain terms in e or its powers. We shall divide all terms by e, or by a higher power of e, so that e will be completely removed from at least one of the terms. We suppress then all the terms in which e or one of its powers will still appear, and we shall equate the others; or, if one of the expressions vanishes, we shall equate, which is the same thing, the positive and negative terms. The solution of this last equation will yield the value of a, which will lead to the maximum or minimum, by using again the original expression.
>
> Here is an example.[6]
>
> *To divide the segment AC at E so that $AE \times EC$ may be a maximum.*
>
> We write $AC = b$; let a be one of the segments, so that the other will be $b - a$, and the product, the maximum of which is to be found, will be $ba - a^2$. Let now $a + e$ be the first segment of b; the second will be $b - a - e$, and the product of the segments, $ba - a^2 + be - 2ae - e^2$; this must be adequated with the preceding: $ba-a^2$. Suppressing common terms: $be \sim 2ae + e^2$. Suppressing e: $b = 2a$. To solve the problem we must consequently take the half of b.
>
> We can hardly expect a more general method.

To understand better what Fermat had in mind, we note that he took the idea of 'adequality' from Diophantus's *Arithmetica*, a book written in the 3rd century AD. There it has the sense of something like 'approximately equal', so let us presume that it had that sense for Fermat. The behaviour of the quantity e seems to derive from a perception by Fermat that the value of the expression concerned does not change much at a maximum or minimum, so

$$(b-a)\times a \text{ is much the same as } ((b-(a+e))\times(a+e),$$

and for this to be the case, it can be verified that

$$be \text{ is approximately equal to } 2ae + e^2.$$

So we must take e as something sufficiently small for this to be true — that is, as something small enough to throw away once we have divided through by e. So (using our symbol $\sim$ for adequality)

$$b \sim 2a + e,$$

and thus $b \sim 2a$ (after throwing away e).

[6] It is instructive to draw the figure that Fermat described.

Fermat's account, as given in the slightly modernised form of his recent editors, raises a number of problems. How convincing would its first readers have found it? What is to 'adequate'? What is *e*? Is it some very small quantity? Is it finite? Why can it simply be suppressed? As often happens with hard mathematics, the example is probably easier to understand than the generalities that precede it, and the method certainly gives the right answer in this case: the line should be divided halfway along. Whatever its theoretical justification might be, the method at least seems easy to use. Overall, we may judge that a simple and efficacious — albeit mysterious — technique is no less valuable in mathematics than in medicine; we should not expect it to have been rejected.

This method is certainly effective, while the logical justification given for these steps is perhaps less than illuminating, but Fermat added no more words of explanation. Nor did he talk of 'indivisibles' or 'infinitesimals' as his contemporaries sometimes did. Still less did he talk of a quantity tending to 0, as a modern text might do. He presented exactly what you have seen: a simple, algebraic technique based on an insight that he conveyed only in this obscure way.

It so happens, however, that Fermat wrote at least one other account of his method, perhaps around 1638, which sheds a little more light on how his mind was working.[7] Indeed, it may well indicate how he developed his method.

He began by remarking that he had been led to the study of problems about maxima and minima by studying some old work by Viète,[8] 'and thus for resolving easily all the difficulties concerning limiting conditions which have caused so many problems for ancient and modern geometers'.[9] Then he went on:

Fermat on maxima and minima, continued.

> Maxima and minima are in effect unique and singular, as Pappus said and as the ancients already knew, although Commandino claimed not to know what the term 'singular' signified in Pappus. It follows from this that on one side and the other of the point constituting the limit one can take an ambiguous equation, and that the two ambiguous equations thus obtained are accordingly correlative, equal and similar.
>
> For example, let it be proposed to divide the line b in such a way that the product of the segments shall be a maximum. The point answering this question is evidently the middle of the given line, and the maximum product is equal to $b^2/4$; no other division of this line gives a product equal to $b^2/4$.
>
> But if one proposes to divide the same line b in such a way that the product of the segments shall equal z'' (this area being besides supposed to be less than $b^2/4$) there will be two points answering the question, and they will be found situated on one side and the other of the

[7] See Fermat, *Oeuvres* 1, 133–135, F&G 11.C2, and (Stedall 2008, 72–73).

[8] François Viète (1540–1603) was a French mathematician who lived earlier than Fermat and whose work greatly influenced him. See *Volume* 1, Chapter 9.

[9] Fermat wrote in Latin, but left the term 'limiting conditions' in Greek: 'diorismoi'. In Greek mathematics a diorism was a 'condition of possibility' or, as we might say, a necessary condition for a problem to have solutions.

> point corresponding to the maximum product. In fact let a be one of the segments of the line b, one will have $ba - a^2 = z''$; an ambiguous equation, since for the segment a one can take each of the two roots.

Figure 2.3 may help with your understanding (Fermat did not leave us a diagram of his own). What Fermat had noticed is that, on dividing the segment b into segments of lengths a and $b - a$, the only value of z'' for which the equation $a(b - a) = z''$ has a unique root is the value $z'' = b^2/4$. There are no solutions if z'' is greater than $b^2/4$, and there are two if z'' is less.

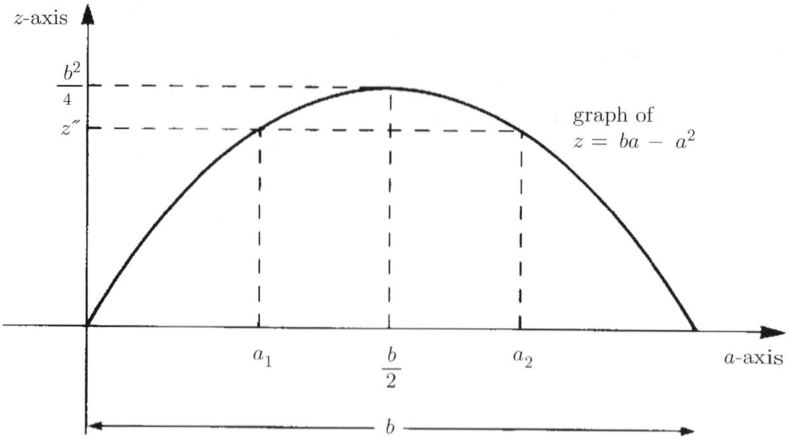

Figure 2.3. Fermat's method for finding maxima and minima

He then argued that if the unequal roots of $a(b - a) = z''$ are a and e — he used the same symbol e that he had used before, but now it is finite and not one that he can suppress — then they get closer and closer as z'' tends to the value $b^2/4$. As he put it:

> If, in place of the area z'', one takes another greater value, although always less than $b^2/4$, the segments a and e will differ less from each other than the previous ones, the points of division approaching closer to the point constituting the maximum of the product. The more the product increases the more on the contrary diminishes the difference between a and e until it will vanish exactly at the division corresponding to the maximum product; in this case there will only be a unique and singular solution, the two quantities a and e becoming equal.

So the special (unique or singular) equation, one might say, is also the one giving the special (maximum) value.

Fermat then gave another example, to show that the method works even when the answer is not obvious. We can conclude that, in Fermat's opinion, finding a maximum or a minimum was a question of finding repeated roots of an equation. We see from our graph that the maximum point is distinguished among all nearby points because it meets a horizontal line in two coincident points, whereas nearby horizontal lines meet the curve in pairs of points whose separation can be made arbitrarily small. While this is not without difficulties, it certainly makes more sense than his first account, for now the algebra is coupled with some intuitively plausible geometry.

This concludes our discussion of Fermat's method for finding maxima and minima.

We now turn to his method for finding tangents.[10] The phrase 'a method of tangents' is used here to mean a systematic way of finding a tangent to a given curve at a given point on the curve. So there might be various kinds of tangent method, depending on how the curve has been presented. If the definition tells you how to draw the curve dynamically (like the quadratrix), then you might look for a way of capturing the direction of motion at the given point.[11] If the definition is static, you may feel more inclined to look for a property of lines through the given point which characterises the tangent. If the curve is given by means of an equation in two variables, you might look for a more formal, symbolic method. Any of these ways of proceeding, if successful and systematic, can be called a *tangent method.*

Fermat set himself the task of finding the tangent to a parabola at an arbitrary point on it. As in Figure 2.4, he took the parabola BDN with vertex D and axis DC, and let B be an arbitrary point on the parabola and BE the tangent to it, meeting the axis at E. He let BC be the perpendicular through B to the axis.[12]

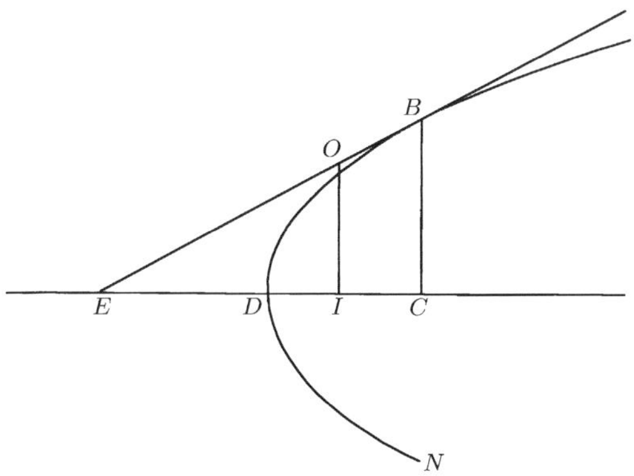

Figure 2.4. A tangent to a parabola

He now let O be a point on the tangent other than B, and let OI be the perpendicular through O to the axis. The crucial property of the tangent that Fermat used is that it lies outside the parabola, so $OI > JI$ because O lies outside the parabola.

He knew that

$$\frac{CD}{ID} = \frac{BC^2}{OJ^2},$$

[10]See Fermat, *Oeuvres* I, 133–135, extract in Struik, *A Source Book*, 223–224, and F&G 11.C1.

[11]The quadratrix is defined as the curve that marks the points common to a horizontal line that descends with uniform velocity and a straight line that rotates about a fixed point with uniform angular velocity; it was discussed in *Volume* 1, Section 3.3.

[12]We have reversed Fermat's original figure to make it easier to follow his argument. In his figure, the axis DIC pointed to the left, which reminds us that coordinate axes were novel at the time and far from standard. For a discussion of the way that Descartes introduced coordinate methods in his *La Géométrie*, see *Volume* 1, Section 13.4.

because the points J and O lie on the parabola. This gave

$$\frac{CD}{ID} > \frac{BC^2}{OI^2}.$$

Next he observed that the triangles BCE and OIE are similar, so

$$\frac{BC}{OI} = \frac{CE}{IE},$$

and so

$$\frac{BC^2}{OI^2} = \frac{CE^2}{IE^2},$$

from which he deduced that

$$\frac{CD}{ID} > \frac{CE^2}{IE^2}.$$

He next introduced a simpler set of letters. He let $CD = d$, $CE = a$, and $CI = e$, so this inequality can be written as

$$\frac{d}{d-e} > \frac{a^2}{a^2+e^2-2ae}.$$

He then concluded his argument as follows:

> Removing the fractions,
>
> $$da^2 + de^2 - 2dae > da^2 - a^2e.$$
>
> Let us then adequate, following the preceding method; by taking out the common terms we find
>
> $$de^2 - 2dae \sim -a^2e,$$
>
> or, which is the same,
>
> $$de^2 + a^2e \sim 2dae.$$
>
> Let us divide all terms by e:
>
> $$de + a^2 \sim 2da.$$
>
> On taking out de, there remains $a^2 = 2da$, consequently $a = 2d$.
>
> Thus we have proved that CE is the double of CD — which is the result.

It is the result because the tangent at an arbitrary point B on the parabola is now known to be the line through B that meets the axis at E, where $CE = 2 \times CD$. Fermat then claimed that the method never fails, that it can be extended to a number of beautiful problems, and that he had used it to find centres of gravity of figures.

The connection between the two methods emerges as the similarity of the algebraic techniques used to solve them both. The adequality argument is the same as before and, in keeping with its place in Fermat's writing, is bereft of any explanation. But nothing is being maximised or minimised, even though you could argue that something (the equation of a line through E) is being shown to have coincident roots.

Fermat's contemporaries found this method attractive, if obscure, although Descartes was scathing at first and pointed out that there was no quantity that was being maximised.[13] But once his professional pride was assuaged by an admission to that effect from Fermat, he accepted the method. Others wanted a proof that the method always works, for Fermat presented it only in cases where the answer was already known.

[13]See Descartes's letter to Mersenne in Fermat, *Oeuvres*, 2, 126–132.

Fermat did not succeed in giving his method the security of a proof, but he did make it more plausible. In any case, his contemporaries were willing to rely on methods of their own which had no better guarantee than that they too were successful, when they could be checked. He also succeeded in extending his method to curves defined in various ways. He found tangents to the cycloid (defined below, see Figure 2.9) by means of it, which showed it off to good advantage. Later mathematicians were willing to use his method even if, as did Christiaan Huygens, they regretted its obscurities.

Descartes's method of normals. Fermat's great rival, René Descartes, also had a method of tangents — or, rather, a method of *normals* — a way of finding a straight line that meets a given curve at right angles. Once a normal to a given curve at a given point is obtained the tangent at that point can immediately be found (because the tangent and the normal are perpendicular). Descartes, in fact, does not seem to have been so interested in tangents, but he regarded the problem of finding a normal to a given curve at a specified point on the curve as 'not only the most useful and most general problem in geometry that I know, but even, that I have ever desired to know', because it enabled him to describe the angle at which two curves intersect.[14]

Figure 2.5. René Descartes (1596–1650)

Others were of the same opinion — we shall look at several contemporary attempts at this problem — and because Descartes published his approach in Book II of *La Géométrie*, it rapidly came to exercise a greater influence than Fermat's.

Descartes's motivation for studying the problem of normals derived from his interest in the path of light through lenses. The refraction of light is studied with respect to the normals to a lens, rather than the tangents, so his interest in finding normals had substantial practical roots.

Although constructing a normal is no simpler than constructing a tangent, Descartes had a subtle geometrical perception up his sleeve, which he found by the classical mode of analysing the problem. If the normal has been found, what can we then say about it? Descartes noticed (see Figure 2.6) that if any point on the normal to the curve S at the point P, say, were taken as the centre of a circle C passing through P, then P would be the only point of the curve that the circle went through in the region

[14]See Descartes, *La Géométrie*, p. 95, extracts in F&G 11.A9, and (Stedall 2008, 74–78).

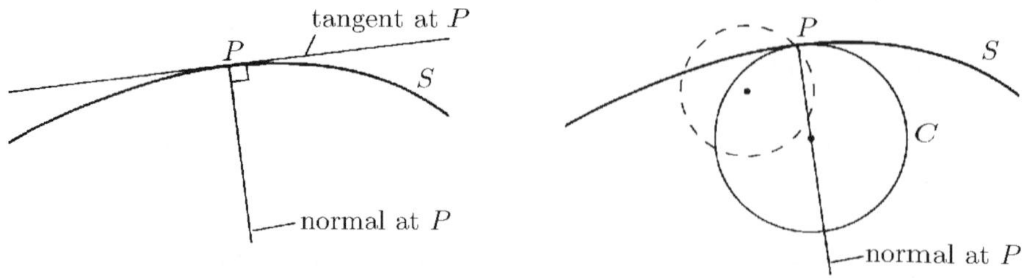

Figure 2.6. The circle C and the curve S have a common tangent at P

of P, whereas other circles through P (whose centres were not on the normal) would have to cut the curve somewhere nearby as well.

In Figure 2.6 the dashed circle does not have its centre on the normal; it cuts the curve S at the point P and at another point nearby. The circle C does have its centre on the normal; it touches the curve S at the point P.

It was this geometrical insight that was readily handled by his algebraic techniques. If we can somehow obtain an equation expressing the relation between the curve and these circles, then the roots of the equation tell us where the curve and the circle intersect. The presence of two equal roots indicates a circle that touches the curve (that is, it meets it at two identical points), and so the centre of such a circle lies on the normal. And only one point on the normal needs to be established in order to specify the normal completely, because the straight line through that point and P is the normal being sought. Descartes always looked for the point where the normal meets a coordinate axis, as this ensures that the other coordinate of the required point is 0.

Having moved rather rapidly through Descartes's argument, let us go through it again in more detail, following Descartes's own presentation.[15] This is a demanding text, and you may find it helpful to look at our comments at the end.[16]

Descartes's method of normals.

> [1] The angle formed by two intersecting curves can be as easily measured as the angle between two straight lines, provided that a straight line can be drawn making right angles with one of these curves at its point of intersection with the other. This is my reason for believing that I shall have given here a sufficient introduction to the study of curves when I have given a general method of drawing a straight line making right angles with a curve at an arbitrarily chosen point upon it. And I dare say that this is not only the most useful and most general problem in geometry that I know, but even that I have ever desired to know.
>
> [2] Let CE be the curved line. It is desired to draw a straight line at right angles to it, through the point C. I suppose the problem to have been solved, and that the sought-for line is CP, which I prolong to the point P where it meets the straight line GA. [This was a common method

[15]See Descartes, *La Géométrie* 95–111, and F&G 11.A9.

[16]We have introduced the numbering of the paragraphs, and have followed the standard editorial convention of setting our additions to a text in square brackets, as we do elsewhere in this book.

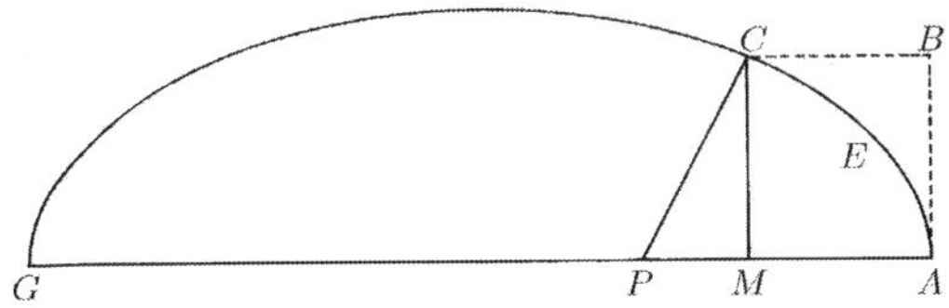

Figure 2.7. Descartes's method of normals — the analysis

at the time for locating a point as the intersection of two lines.] (GA is the line to whose points all those of CE are referred; so that putting MA or CB equal to y, and CM or BA equal to x, I have some equation showing the relation between x and y.)

Then I put $PC = s$, and $PA = v$, whence $PM = v - y$. Since the triangle PMC is right-angled, the square on the hypotenuse s^2 is equal to $x^2 + v^2 - 2vy + y^2$, the sum of the squares on the two sides. That is to say, $x = \sqrt{s^2 - v^2 + 2vy - y^2}$, or equally $y = v + \sqrt{s^2 - x^2}$. By this means I can get rid of one of the two unknown quantities x or y from the equation relating the points of the curve CE to those of the straight line GA. This is easily done by putting throughout $\sqrt{s^2 - v^2 + 2vy - y^2}$ in place of x, the square of this in the place of x^2, its cube in place of x^3, and so on. That is if it is x I want to get rid of; or if it is y, I put in its place $v + \sqrt{s^2 - x^2}$, and its square or cube, etc., in place of y^2, y^3 etc. After this process there always remains an equation in only one unknown quantity, x or y.

[3] For example, if CE is an Ellipse, MA the segment of its diameter on which CM is ordinate, and which has r for its latus rectum and q its major axis then by Book I Proposition 13 of Apollonius we have $x^2 = ry - ry^2/q$. Getting rid of x^2 from this gives

$$s^2 - v^2 + 2vy - y^2 = ry - ry^2/q,$$

or

$$y^2 + \frac{qry - 2qvy + qv^2 - qs^2}{q - r}$$

equals nothing. For it is better here to consider the whole together in this way, than as one part equal to the other. ...

[4] Such an equation having been found it is to be used, not to determine x, y, or z, which are known, since the point C is given, but to find v or s, which determine the required point P. With this in view, observe that if the point P fulfils the required conditions, the circle about P as centre and passing through the point C will touch but not cut the curve CE; but if this point P be ever so little nearer to or farther from A than it should be, this circle must cut the curve not only at C but also in another point. Now if this circle cuts CE, the equation involving x and y as unknown quantities (supposing PA and PC known) must have two unequal roots.

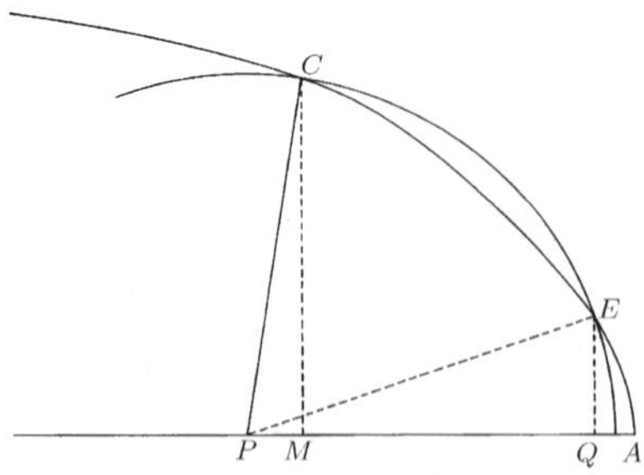

Figure 2.8. Descartes's method of normals — finding P

Suppose, for example, that the circle cuts the curve in the points C and E. Draw EQ parallel to CM. Then x and y may be used to represent EQ and QA respectively in just the same way as they were used to represent CM and MA; since PE is equal to PC (being radii of the same circle), if we seek EQ and QA (supposing PE and PA given) we shall get the same equation that we should obtain by seeking CM and MA (supposing PC and PA given). It follows that the value of x, or y, or any other such quantity, will be two-fold in this equation — that is, the equation will have two unequal roots. If the value of x be required, one of these roots will be CM and the other EQ; while if y be required, one root will be MA and the other QA. It is true that if E is not on the same side of the curve as C, only one of these will be a true root, the other being drawn in the opposite direction, or less than nothing. The nearer together the points C and E are taken however, the less difference there is between the roots; and when the points coincide, the roots are exactly equal, that is to say, the circle through C will touch the curve CE at the point C without cutting it.

[5] Furthermore, it is to be observed that when an equation has two equal roots, its left-hand member must be similar in form to the expression obtained by multiplying by itself the difference between the unknown quantity and a known quantity equal to it; and then, if the resulting expression is not of as high a degree as the original equation, multiplying it by another expression which will make it of the same degree. This last step makes the two expressions correspond term by term.

[6] For example, I say that the first equation found in the present discussion, namely

$$y^2 + \frac{qry - 2qvy + qv^2 - qs^2}{q - r}$$

must be of the same form as the expression obtained by making $e = y$ and multiplying $y - e$ by itself, that is, as $y^2 - 2ey + e^2$. We may then compare the two expressions term by term, thus: Since the first term,

y^2, is the same in each, the second term,

$$\frac{qry - 2qvy}{q - r},$$

of the first is equal to $-2ey$, the second term of the second; whence, solving for v, or PA, we have $v = e - \frac{r}{q}e + \frac{1}{2}r$; or, since we have assumed e equal to y, $v = y - \frac{r}{q}y + \frac{1}{2}r$. In the same way, we can find s from the third term,

$$e^2 = \frac{qv^2 - qs^2}{q - r},$$

but since v completely determines P, which is all that is required, it is not necessary to go further.

When trying to follow the train of thought in a source such as this, it is useful to draw the diagram for yourself, constructing and labelling it as the discussion proceeds. Paragraph [2] sets up the analysis. Descartes started with the curve CE, whose normal at C is what is ultimately required. Points on this curve are specified in relation to the straight line AG, and to the point A in particular. So the point C is at a distance y along from A, and x up. For the purposes of analysis, he supposed the problem solved, and that the normal cuts the axis AG at P, to which he assigned the distances v from A and s from C. Then he constructed a right-angled triangle PMC and applied the Pythagorean theorem to write down the algebraic relationship between the lengths x, y, s, and v. Because he knew the relationship between x and y in the case of any particular curve, either x or y can be eliminated from the equation of the curve and the equation in x, y, s, and v. This left him with just one of the original 'unknowns' x or y, at the cost of having introduced two further unknowns v and s, which are indeed what are required for the normal to become known in the end.

At this stage you might be feeling a little at sea, so it is perhaps worth restating the problem in other terms.

What do we know? We are given a curve, defined in terms of the relationship between x and y (measured as above) for any point C on it.

What do we want? Given some specified point C on the curve — that is, some specified pair of values for x and y — we want to be able to say just where the point P is (that is, say what v or s are). This will uniquely define the normal.

How far have we got? We have eliminated one of x or y, at our choice, so as to leave just one curve variable to have to deal with, and it still needs specification.

Descartes now broke off from his description of the general method (which he resumed in paragraph [4]) to give three examples of how particular curves are treated up to this point. We look at just one of these, the case of the ellipse (paragraph [5]). He needed the relationship between x and y for points on an ellipse, which he obtained from Apollonius's *Conics*. (If you chase up the proposition referred to you may be able to see why Descartes's contemporaries were so impressed by his algebraic reformulations![17]) Then Descartes eliminated x from the equation of the curve, by the method described in the previous paragraph.

In paragraph [4] he brought in the argument about the circles that was outlined earlier. So much for the geometry.

[17]See Apollonius, *Conics*, Book I, Proposition 13, and F&G 4.D4(c).

Box 1. Descartes's method of finding normals to the ellipse.
For the ellipse, Descartes had to determine P by saying that the equation

$$y^2 + \frac{qry - 2qvy + qv^2 - qs^2}{q - r} = 0$$

must be of the form $y^2 - 2ey + e^2 = 0$. He first computed what e must be, and then expressed v in terms of e. Finally, he found the position of P, which is v units away from the origin, and he obtained

$$y^2 + \frac{qr - 2qv}{q - r}y + \frac{qv^2 - qs^2}{q - r} = 0,$$

which he compared term by term with $y^2 - 2ey + e^2 = 0$. In this way he found that

$$e = -\frac{qr - 2qv}{q - r},$$

which he rewrote as

$$v = e - \frac{re}{q} + \frac{r}{2}.$$

Similarly, looking at the third term he found that

$$e^2 = \frac{qv^2 - qs^2}{q - r}.$$

In this case he did not need the result, for he could now set $e = y$ (which says that the roots coincide) to get

$$v = y - \frac{ry}{q} + \frac{r}{2},$$

thus finding the point P.

Then in paragraph [5] Descartes sprang his final algebraic insights. He was rather proud of these, referring to them a few pages later as 'not the least important feature of my general method'. His point was that if we know that an equation has two equal roots (or we want it to have them, which comes to the same thing here), then it must be of the form $(y-e)^2$ if the equation is quadratic, or of the form $(y-e)^2$ times something else if the equation is more than quadratic, where in each case e is the repeated root. So comparing this form term by term with the equation produced from the method in paragraph [2] enables us to write v and s in terms of y or x, and thus to construct the normal. In this way, Descartes converted his geometrical insight into an algebraic form.

After this impressive display of his originality, Descartes went through the three particular curves he gave as examples earlier. Box 1 discusses the example of the ellipse.

What is to be made of all this? Is it a solution worthy of the problem? The answer must be 'yes'. The method is quite general; it should apply at least to all the curves that Descartes admitted into geometry. But it is imperilled by complexity: even for simple curves we may not be able to eliminate either x or y or solve the final equation for v. So it is interesting to see how it was received.

His method met with just these criticisms. He had illustrated it with only three particular curves (the ellipse, the curve sometimes called a Cartesian parabola,[18] and one other), curves for which the crucial algebraic step — that of determining v so that the circle meets the curve in a single (repeated) point — is easy. But in general this determination is not easy, and in some cases it can be virtually impossible. Moreover, there are curves (such as the quadratrix) to which the method simply does not apply. But Descartes's published algebraic method is remarkable for entirely avoiding arguments about very small quantities, and perhaps partly for that reason it was soon taken up despite its attendant computational problems.

The publication of his method in the second Latin edition of *La Géométrie* (1659–1661) was accompanied by a plethora of commentaries. The most important of these was written by Jan Hudde and published in the form of a letter to Frans van Schooten, the editor of this edition.[19] Hudde was a Dutch mathematician, a pupil of van Schooten, who later gave up mathematics to become the Burgomaster of Amsterdam. Hudde directly attacked the question of determining when an equation has repeated roots. Hudde's rule, as it became called, was a flexible procedure for finding the repeated roots of any equation that has them, and as such it was fairly easy to apply.[20]

Kinematic methods. The last tangent method we discuss is a kinematic one, which was used to great effect by Gilles Personne de Roberval and Evangelista Torricelli. Because of its generality and intuitive simplicity it was of immense importance, although we shall treat it only briefly. It is again best approached by an example, in this case a cycloid (see Figure 2.9).

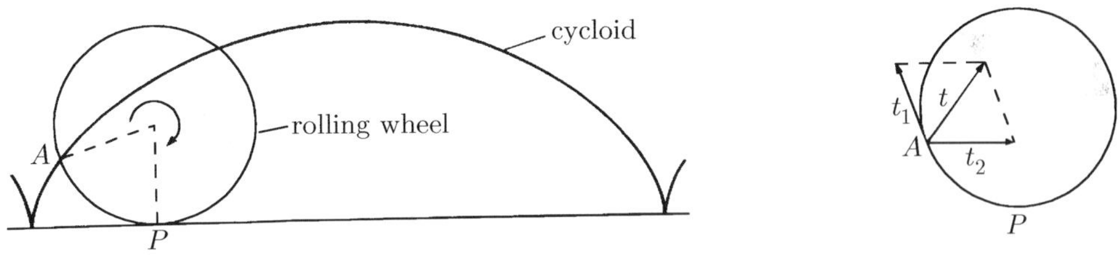

Figure 2.9. The generation of a cycloid

A cycloid is the curve traced out by a point A on the rim of a circular wheel rolling along a straight line. At any instant its motion is composed of the horizontal motion of the wheel along the line and the circular motion of the point A around the centre of the wheel. Roberval took the direction of the tangent at A to be the direction of the motion of A, which in turn he took to be the composition of these two separate motions.[21]

---

[18]This is a curve with equation $y^3 - by^2 - cdy + bcd + dxy = 0$ that is 'generated by the motion of a parabola', *La Géométrie*, 96.

[19]See Descartes, *Geometria*, 2nd edn., p. 507, in F&G 11.B5.

[20]It lies beyond the scope of this book to explain Hudde's rule, but you will find a brief passage by Hudde in F&G 11.B5.

[21]Roberval may have done this work by 1636, and he taught it between 1639 and 1644, but it was published only posthumously. See (Roberval 1693, 193) and F&G 11.E1.

> To find the tangent to the figure at a given point, I draw a tangent to the circle which passes through the said point, for each point of the circle moves along the tangent to the circle. I then consider the movement which we have given to our point by carrying it along the diameter moving parallel to itself. Drawing through the same point the curve of this movement, if I obtain a parallelogram ... and if at the same point I draw the diagonal, I have the tangent to the figure which has these two movements as its components, i.e. the circular and the direct. That is how one proceeds when one supposes the movements are equal. If one had supposed that, instead of being equal, the movements had been in some other ratio, the parallelogram would have been constructed with its sides in that ratio.

In the notation of Figure 2.9, he found that t_1 and t_2 are equal in magnitude, and so t was readily obtained. Indeed, if t_1 represents the motion of A around the centre of the wheel, and t_2 the motion of A due to sideways motion of the wheel, then t represents the actual motion of the point A due to t_1 and t_2, which Roberval obtained as the diagonal of the parallelogram with sides t_1 and t_2.

Roberval's career tells us something about the mathematical life of his times. Alone of the French mathematicians we have met so far, he occupied a professional position: the Chair of mathematics at the Collège de France established by a legacy from Peter Ramus.[22] Paradoxical as it may seem, this is precisely why Roberval so seldom published his findings — the Chair was awarded only after a public examination and kept for three years. Like a modern-day chess champion, Roberval had to balance releasing his discoveries (which was good for esteem) with concealing them and so keeping his competitive edge; in fact, Roberval kept the Chair for the rest of his life, from the age of 31 to his death at 75. His letters in the Mersenne correspondence were one way that he was willing to spread his findings, and it was Mersenne who published Roberval's method of tangents in 1644. But Roberval's secrecy, and his inclination to claim results without circulating proofs, seem to have made him unpopular; Descartes, in particular, had a strong dislike of him.

Despite the habitual secrecy with which Roberval surrounded his work, kinematic ideas spread through Europe. Torricelli took them up; he may well have discovered them independently, as Barrow subsequently did in England. Unlike other infinitesimal ideas, the concept of 'motion at an instant' has survived in mathematics to this day, providing for many people a crucial pedagogical resting point for the calculus. It occurs again in Barrow's work, and still later in Newton's. It is a precalculus idea, both historically and intuitively, that the calculus had to struggle to make precise. We can look at its lively existence in applied mathematics and wonder whether, even if calculus captures the minds of mathematicians, motion does not live in their hearts.

In view of the profusion of tangent methods developed between 1635 and 1660, it is hard to judge how the problem of finding tangents was regarded. The profusion of methods suggests that no-one was completely happy with any one method, so the problem was far from being completely resolved; but three methods were being pursued, so presumably no-one thought that the general problem was insoluble. Also, nobody seems to have found any method sufficiently general to drive out the others. Even so, the progress made had taken everyone well beyond their classical inheritance.

---

[22]Ramus was a controversial educational reformer in the 16th century in France. We described his work briefly in *Volume* 1, Chapter 9.

2.2 Area and volume problems

Just as with problems about tangents, so problems about areas and volumes were, in essence, problems about curves. They were geometrical problems inviting geometrical solutions: find the areas enclosed by curves or underneath segments of curves; find the volumes of solids obtained when curves are rotated around an axis.

Sometimes these problems were studied for practical reasons, sometimes for their own sake. Simon Stevin, for example, the leading Dutch mathematician of the second half of the 16th century, wrote a treatise on mechanics in 1586. It contains many ingenious geometrical arguments that were to be taken up, or perhaps independently rediscovered, by many people forced for practical reasons to deal with curves that Archimedes had left untouched. It also marks a deliberate retreat from full Archimedean rigour. In contrast, the Italian priest Bonaventura Cavalieri, who wrote fifty years later, seems to have needed no such motivation for a lifetime's investigation of area and volume.

To understand what the problems were, mathematicians again considered what had come down from the Greeks. One classical author stood out: Archimedes, who alone of the major Greek mathematicians had confronted these problems in any generality. Accordingly, to do their work, practical mathematicians like Stevin read Archimedes diligently.[23] However, the problems solved by Archimedes were few in nature. Worse, as with tangents, the Ancients' area methods were not discovery methods: they neither suggested how the theorems were discovered, nor did they look likely to suggest new theorems. It was clear that the results were correct — they had proofs of a logical force that no one disputed — but no way was opened to new results. The indirect proof method of 'reductio ad absurdum', used twice to justify a conjectured area, with its attendant laborious exhaustion arguments, was difficult to apply.

The classical legacy did not help them. Johannes Kepler, for instance, found himself 'repelled by the thorny reading' of Archimedes.[24] So mathematicians started to devise their own techniques. In 1615, Kepler wrote his influential *Nova Stereometria Doliorum Vinariorum* (New Measurement of Large Wine Casks). Here he considered solids of revolution.

To obtain a solid of revolution, draw a curve in the (x, y) plane as shown in Figure 2.10 and rotate it in space once around the y-axis. The surface so generated defines a solid of revolution.

Kepler calculated the volumes of no fewer than ninety solids by regarding a solid as being composed of infinitely many infinitesimal pieces whose volumes he could determine. For example, he regarded a sphere as made up of infinitely many cones, each with its vertex at the centre and its base on the surface of the sphere (see Figure 2.11).

Each cone has a volume that is given by the formula for a cone with a flat base:

$$\text{volume} = \tfrac{1}{3}(\text{base} \times \text{height}).$$

[23]The *Editio Princeps*, or first edition of Archimedes' works, entitled *Archimedis Opera quae quidem Exstant Omnia ...*, was edited by Thomas Geschauff Venatorius and published in Basel in 1544 in Greek with a Latin translation. Some of Archimedes' works were also translated into Latin by Commandino in an edition published in Venice in 1558 under the title *Archimedis Opere Nonnulla ...*. There was also a version of the *Editio Princeps* published by D. Rivault in Paris in 1615. See Heath, *The Works of Archimedes*, xxix.

[24]Kepler, *Nova Stereometria Doliorum Vinariorum*, preface. We discussed aspects of Kepler's life and work in *Volume* 1, Chapter 10.

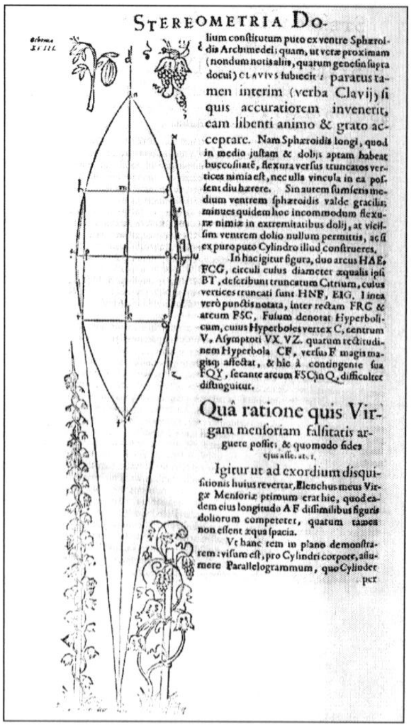

STEREOMETRIA DO-
lium constitutum puto ex ventre Sphæroidis Archimedei; quam, ut veræ proximam (nondum notis aliis, quarum genesin supra docui) CLAVIUS subiecit: paratus tamen interim (verba Clavij) si quis accuratiorem invenerit, eam libenti animo & grato acceptare. Nam Sphæroidis longi, quod in medio justam & dolijs aptam habeat buccositatē, flexura versus truncatos vertices nimia est, nec ulla vincula in ea possent diu hærere. Sin autem sumseris medium ventrem sphæroidis valde gracilis; minues quidem hoc incommodum flexuræ nimiæ in extremitatibus dolij, at vicissim ventrem dolio nullum permittis, ac si ex puro puto Cylindro illud construeres.

In hac igitur figura, duo arcus HAE, FCG, circuli cuius diameter æqualis ipsi BT, describunt truncatum Citrium, cuius vertices truncati sunt HNF, EIG. Linea verò punctis notata, inter rectam FRC & arcum FSC, Fusum denotat Hyperbolicum, cuius Hyperboles vertex C, centrum V, Asymptoti VX VZ. quarum rectitudinem Hyperbola CF, versus F magis magisq; affectat, & hic à contingente sua FQY, secante arcum FSC in Q, difficulter distinguitur.

Quâ ratione quis Virgam mensoriam falsitatis arguere possit; & quomodo fides ejus asse. at. r.

Igitur ut ad exordium disquisitionis huius revertar, Elenchus meus Virgæ Mensoriæ primum erat hic, quod eadem eius longitudo AF dissimilibus figuris doliorum competeret, quarum tamen non essent æqua spacia.

Vt hanc rem in plano demonstrarem: visum est, pro Cylindri corpore, assumere Parallelogrammum, quo Cylinder per

Figure 2.10. The volume measurement of what Kepler called a 'lemon' in his *Nova Stereometria Doliorum Vinariorum* (1615)

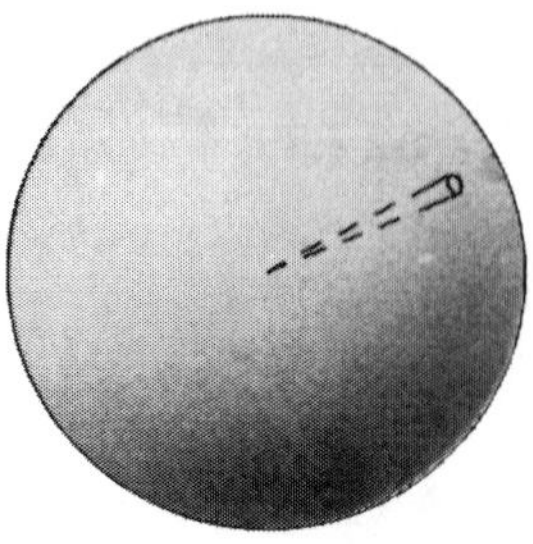

Figure 2.11. A sphere with an infinitesimal cone inside it

Kepler took this as the volume of the infinitesimal cone, and argued that each cone has the same height r, the radius of the sphere, so the volume of the sphere, which is the sum of the volumes of all the cones, is $r/3$ times the sum of the areas of the bases. That sum is the surface area of the sphere, which is $4\pi r^2$, so the volume of the sphere is $\frac{4}{3}\pi r^3$.

In this way Kepler rederived Archimedes' formula for the volume of a sphere, but in a flexible and heuristic way that, while not conforming to Greek standards of proof, promised to be capable of quite general application.

In calculating the volumes of his solids of revolution, he not only raised and solved questions that Archimedes had not tackled but also went some way to demonstrating the fecundity of his new techniques. This point was not lost on his contemporaries, and his book has been called the source of the inspiration for all later computations of volume.[25]

The mathematicians who came of age in the 1620s and 1630s certainly felt themselves to be living in a vibrant mathematical culture. Just as they produced several tangent methods, so too they produced several methods for finding areas and volumes. We shall need to distinguish four of them, although many mathematicians used various mixtures of these as they felt circumstances dictated:

1. methods relying on the idea of *indivisible* elements
2. methods relying on *infinitesimal* elements of area or volume
3. arithmetical summation methods
4. methods relying on finite elements of volume and some sort of approximation argument.

The first of these, although it looks deceptively like the second, is different. This is because in the first method an area (say) is taken to be made up of lines, and lines have no area. The second method, the method of infinitesimals exemplified by Kepler's approach, invokes objects that do have a certain area, albeit infinitesimal. The other methods use objects of finite area to approximate the sought-for area and then somehow infer the correct answer from these approximations. In the third, cunning choices of approximating areas lead to efficient ways of calculating their total area. In the fourth and more general method, mathematicians worked out the total area in a variety of ways, but by and large they avoided the use of exhaustion and double reductio ad absurdum, while retaining the idea of making successively closer approximations.

We now turn to examples of these methods so as to make clear what is meant by each of them.

The method of indivisibles. The leading exponent of the first kind of reasoning (indivisibles) was Bonaventura Cavalieri, who was a member of an Italian monastic order and a frequent correspondent of Galileo's. Unfortunately, his work (*Geometria Indivisibilibus*, 1635), which he published when he was 37, is not easy to read. This is because he had to confront a rather obvious paradox or potential source of error, which he described in a letter to Evangelista Torricelli — namely, the following 'proof' that triangles ABC_1 and ABC_2 have the same area (see Figure 2.12). Every vertical line that occurs in $\triangle ABC_1$ also occurs in $\triangle ABC_2$, and vice versa, so if the triangles are made up of lines then they must have the same area! This is false, obviously, but why?

Cavalieri's answer was that an accurate use of his method depends on how the figures are decomposed into lines, and the first version of his theory of indivisibles (which occupies six-sevenths of his book) bears the burden of attempting to formulate a theory of proportion in the manner of Eudoxus — not for area but for collections of lines.[26] The ratio of the areas of two figures is shown to be the ratio of 'all the lines' of the first figure to 'all the lines' of the second. This is interesting because, by summoning up the ghost of Eudoxus, Cavalieri seems to be trying to maintain classical standards

[25]This was the opinion of the distinguished 19th-century German historian of mathematics Moritz Cantor in his *Vorlesungen über die Geschichte der Mathematik* (Lectures on the History of Mathematics), Vol. II, p. 750, cited in (Boyer 1959, 110).

[26]Eudoxus is believed to be the originator of similar ideas in Books V and VI of Euclid's *Elements*.

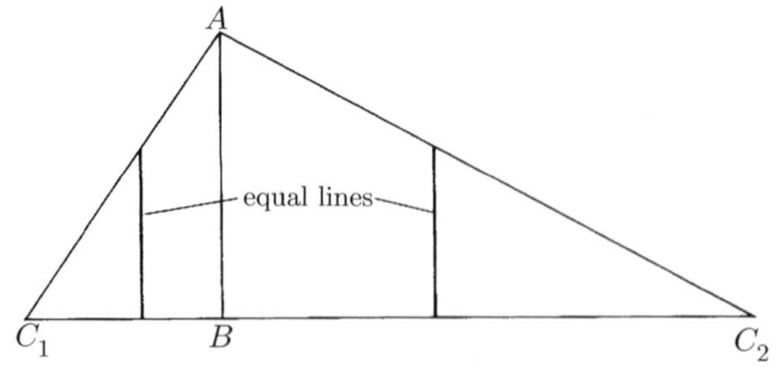

Figure 2.12. Torricelli's paradox — do the triangles ABC_1 and ABC_2 have equal areas?

of rigour. It is also interesting because, by introducing the concept of 'all the lines' of a figure, and then the principle named after him, he was trying to introduce a flexible new tool. Indeed, in conformity with the generalisation suggested at the start of this chapter, the methods that Cavalieri used to compute ratios of 'all the lines' were quite algebraic, but their complexities will prevent us from looking at them in detail.

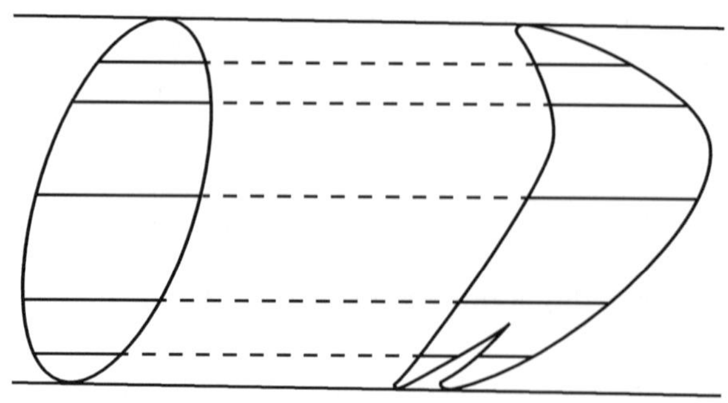

Figure 2.13. Two figures of equal area

Cavalieri's principle asserts that two plane figures have the same area if they lie between the same parallels, and any line drawn parallel to the two given parallel lines cuts off equal chords in each figure (see Figure 2.13).[27]

Because Cavalieri felt that the resulting theory was too difficult for most readers, as he wrote to Galileo in 1634, he added a seventh part of his book, in which individual lines can be compared according to certain rules designed to eliminate the paradox above.[28] This made the seventh part of his book a source for a second theory using indivisibles. Either way, the work was criticised because it was not clear how an area could be made up of lines, because lines have no thickness.

What prospered better was Cavalieri's second version of his theory, in which his principle was re-interpreted so as to permit him to deal with individual lines. In this new approach each line is moved on its own, so to speak, from one figure to the next. In order to get a sense of how it goes, we look at one of its most beautiful applications: Roberval's computation of the area under the cycloid, which he included in his *Traité*

[27]For the relevant excerpt from Cavalieri's *Geometria Indivisibilibus*, Book II, see (Stedall 2008, 62–65).
[28]See Galileo, *Le Opere*, 16, 113.

des Indivisibles (begun in 1634, but first published in 1693).[29] Roberval, by the way, claimed that he had invented the method independently.

To follow the text, note that Roberval first described (not altogether clearly) the construction of the cycloid. Then he considered chords in the semicircle and using them constructed the companion curve to the cycloid (see Figure 2.14). Then he deduced the theorem he wanted.

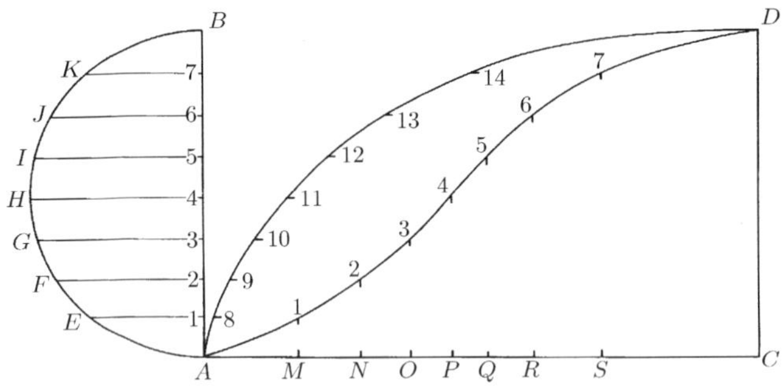

Figure 2.14. The cycloid, its companion, and its generating circle

Roberval on the area under a cycloid.

We suppose that the diameter AB of the circle $AEFGB$ moves parallel to itself, as if carried by some other body, until it has arrived at CD and turned through a semi-circle. While it travelled, the point A at the extremity of the said diameter moves round the circumference of the circle $AEFGB$, and moves as far as the diameter, in such a way that when the diameter is at CD the point A has reached B, and the line AC is equal to the circumference $AGHB$. Now, the path of the diameter is divided into infinitely many parts equal amongst themselves and to each part of the circumference AGE which is also divided into infinitely many parts all equal to themselves and to the parts of AC run through by the diameter, as was said. And then I consider the path of the said point A carried by the two movements: the one of the diameter forwards, the other its proper motion along the circumference.

To find the said path, I see that when it has reached E it has risen above its initial position which it has left; the height is marked by drawing the sine $E1$ through the point E to the diameter AB, and the versed sine $A1$ is the height of the said A when it has reached E.[30] Likewise when it has reached F, I draw the sine $F2$ through the point F to AB, and $A1$ will be the height of A when it has reached G; and doing this at all the places on the circumference that A runs through I find all the heights and elevations above the end of the diameter A, which are $A1\,A2\,A3\,A4\,A5\,A6\,A7$; therefore to find the places which the

[29] See the extract in F&G 11.E1.

[30] The versed sine of a is $1 - \cos a$.

> said point A passes through, and to know the curve which is drawn by the two movements, I carry each of the heights on each of the diameters $M\,N\,O\,P\,Q\,R\,S\,T$ and I find that $M1\,N2\,O3\,P4\,Q5\,R6\,S7$ are the same as those taken on AB. Then I take the same sines $E1\,F2\,G3$ etc. and I carry them at the height found on this diameter and draw them towards the circle, and the ends of these sines form two curves of which one is $A\,8\,9\,10\,11\,12\,13\;\,14D$ and the other $A\,1\,2\,3\,4\,5\,6\,7\,D$.
>
> I know how the curve $A\,8\,9\,D$ is drawn, but to know which movements produce the other I say that while AB has run along the line AC the point A has risen up the line AB and marked all the points $1\,2\,3\,4\,5\,6\,7$, — the first space while AB reached M, the second while AB reached N, and so always the one space is equal to the other until the diameter has arrived at CD, when the point A has risen to B. That is how the curve $A\,1\,2\,3\,D$ is drawn. Now, these two curves enclose a space, being separated one from the other by all the sines and joining together again at the ends $4, A, D$. Now, each part contained between these two curves is equal to each part of the area of the circle AEB contained in that circumference, for the ones and the others are made of equal lines, i.e. of height $A1, A2$ etc. and of sines $E1, F2$ etc., which are the same as those of the diameters $M\,N\,O$ etc., and thus the figure $A\,4\,D\,12$ is equal to the semi-circle AHB. Now, the curve $A\,1\,2\,3\,D$ divides the parallelogram $ABCD$ in two equally, because the lines of one half are equal to the lines of the other half, and the line AC to the line BD, and consequently, according to Archimedes, the half is equal to the circle, to which on adding the semi-circle, i.e. the space between the two curves, one will have a circle and a half for the space $A\,8\,9\,D\,C$; and doing the same for the other half, the whole figure of the cycloid will make three times the circle.

The effort needed to understand the passage is amply repaid by the vivid sense that it gives not only of the power and elegance of the method but also of Roberval's skill as a mathematician. The text is not too lucid, for Roberval wrote as if the diameter of the rolling wheel that generates the cycloid is not rotating — but his expository task was not easy, and the best policy for the reader is to stay with the sense. The point A has two motions as before, one around the centre of the wheel and the other with the wheel along the ground, and Roberval always discussed each motion separately before discussing how they were to be composed. When you have read it, ask yourself: Why is the area between the cycloid and its companion curve equal to the area of the semi-circle? And why does the companion curve divide the area of the rectangle into two equal pieces?

In fact, the first two areas are equal by Cavalieri's principle. If you shade them in horizontally, joining E to 1 in the semi-circle and 8 to 1 under the cycloid, and so on, the lines at each level are of the same length: $E1 = 81$, $F2 = 92$, and so on; in Roberval's phrase, they are 'made of equal lines'. The second pair of areas are equal because, wonderfully, the companion curve is symmetric about the midpoint of the rectangle $ABDC$. So Roberval's approach was to regard an area as composed of indivisible lines, which can be moved around according to intuitively sensible rules to make other equal areas. To see it is to believe it!

This method of indivisibles became all the rage after Torricelli's exposition of his version of Cavalieri's methods in his *Opera Geometrica* (1644). Torricelli had succeeded Galileo as the mathematician and philosopher at the Court of Duke Ferdinando II of Tuscany in 1642 at the age of 33, when he moved to the Medici palace in Florence. There he wrote his *Opera Geometrica*, which was published at the Duke's expense.[31] He died, most probably of typhoid, in 1647.

Torricelli had initially been doubtful about the method of indivisibles, because it seemed so easy to prove obviously false results with it, but he changed his mind and presented a theory shorn of Cavalieri's careful attention to the foundations. He simply presented an area as made up of lines, and supplied rigour by using the method of exhaustion, but he referred to this approach as Cavalieri's method, and in the form of Torricelli's far more readable book it travelled to northern Europe. One noteworthy result in the book, tucked away as an Appendix after a long and unremarkable account of Archimedes' squaring of the parabola, is that the area under a cycloid is three times the area of the generating circle; Roberval's name is not mentioned and Torricelli may well not have known of his work. Another is his striking discovery of a curve that, when rotated about its axis, generates a surface of infinite area that bounds a finite volume.[32]

Cavalieri may have had reservations about the presentation of his method, but he found Torricelli an inspiring correspondent, and the two men became close friends. Outside Italy, Cavalieri's method enjoyed a vogue in the 1640s — despite a wholesale attack on it launched by the Jesuit Paul Guldin from Austria, who alleged that the method was not only worthless but had been stolen from Kepler and others.[33] He may have been motivated by an earlier criticism that Cavalieri had levelled against one of his own arguments, but in any case Cavalieri easily defended himself against this charge of plagiarism, arguing correctly that his method was one of true indivisibles, whereas Kepler's was a method of infinitesimals. Guldin's claim was never accepted by others. (Nothing if not even-handed, Guldin also criticised Kepler for lack of rigour.)

As for the validity of the method, this too was generally acknowledged. To take an example, in 1650 van Schooten defended Cavalieri's ideas against the reservations of Huygens. Yet another who was impressed by indivisible methods was the Scotsman James Gregory, who travelled to Italy in the 1660s to learn them from a pupil of Torricelli.

That so many mathematicians can be mentioned in this connection shows that Cavalieri's original idea was widely, if not universally, regarded as powerful and useful. The method of indivisibles never became accepted as the only way to tackle area problems, however, nor was it ever thought to be unproblematic. Even in its day, it existed alongside the steadily more sophisticated attempts to use approximate arguments. It is to those that we must now turn.

The method of infinitesimals. An alternative to the method of indivisibles, which at least recognises the obvious problem, is to consider that the lines making up a figure have a certain width. The problem now is that if this width is finite and the number of

---

[31] Torricelli is best remembered for his experiments in the 1640s that established the principle of the barometer.

[32] Torricelli, *Opera Geometrica*, 115–116, in (Stedall 2008, 100–101).

[33] The Jesuit order is often thought of as the most intellectual branch of the Roman Catholic Church. It ran a network of schools and colleges; among its pupils were Christoph Clavius and René Descartes.

lines making up a figure is still taken to be infinite, then the area of the figure would seem to be infinite. The solution proposed was to say that the width is not finite — rather, it is less than any finite non-zero quantity — but it is not zero either. Such a quantity came to be called an *infinitesimal*. The defining character of such a quantity is that infinitely many of them taken together can amount to a finite quantity.

Galileo appears to have been one of the first to contemplate such objects, when he discussed the 'paradox of the wheel' in the first day of discussions in his *Two New Sciences*.[34] In this paradox, Galileo considers two concentric regular hexagons, one twice the size of the other.

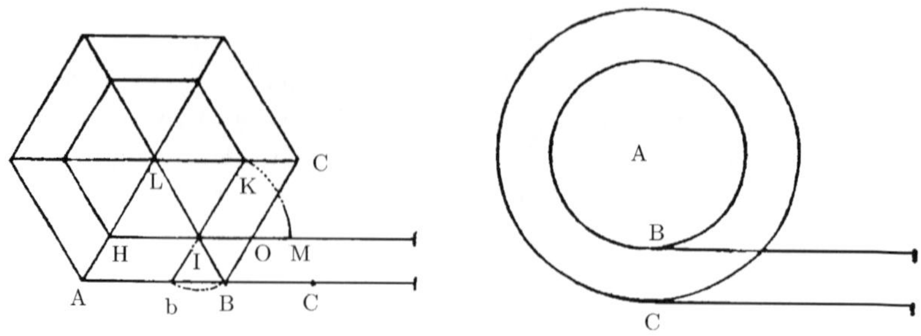

Figure 2.15. Galileo's paradox of the wheel, from *Two New Sciences*, p. 50

The larger hexagon has one edge resting on the straight line *AB*, and the smaller one has one edge resting on the straight line *HI*.

When the larger one is made to roll along the line *AB* the smaller one also rotates, but successive copies of its sides are separated by gaps (which are also the sizes of the sides). In this way it is clear that when the hexagons have both made one complete revolution, and their centres have therefore moved through a distance six times the length of a side of the larger hexagon, each has traversed the same distance along the lines *AB* and *HI*.

Galileo now supposed the hexagons to be replaced by circles, which he regarded as 'polygons having an infinitude of sides' (p. 24) — the above argument does not depend on the polygons being hexagons: they could have any number of sides. How now does one explain the apparent paradox that the smaller circle has rolled exactly as far as the larger one, and so must have the same circumference?

Sagredo (the intelligent foil for Salviati, who speaks for Galileo in the dialogue) tries to argue that the points of the smaller circle slide along the line and in this way contribute at each point a bit of length. Salviati objects that there would have to be infinitely many of these slips (skid marks, as it were) and so their total contribution would be infinite. Instead, says Salviati, the circle contains infinitely many points that must be thought of as empty spaces, and these empty spaces have the property that infinitely many of them in total make only a finite amount. Although Galileo did not use the word, these would be infinitesimal magnitudes. Note that every circle must have them, so it would seem that infinitesimals can also be of different sizes.

Recently, the historian Tiziana Bascelli has argued that Torricelli, who knew Galileo in the last months of his life and copied out the dialogue for the fifth day of the *Two*

[34]Galileo wrote his *Two New Sciences* in the form of a dialogue. We also discussed it in *Volume* 1, Chapter 10.

New Sciences, was inspired by this discussion of empty spaces to contemplate the idea of infinitesimal lines of different widths, as he did in some of his unpublished work.[35]

It seems, however, that infinitesimal methods were not much used in area and volume questions, and came into their own only for questions concerning maxima, minima, and tangents to curves. We shall see several uses of infinitesimals in this setting as this book proceeds, and in this naive sense they survive to this day when people find it helpful to talk of infinitesimal (instantaneous) changes of position, speed, or direction at an instant.

The arithmetical method of Wallis. One mathematician who can be said to have applied infinitesimals to the study of areas, albeit in a highly arithmetical way, was John Wallis. By his own account, Wallis learned only a little mathematics when he went up to Cambridge in 1631 at the age of 15, and discovered his mathematical abilities during the English Civil War only when he became a code-breaker for Parliament. He became the Savilian Professor of Geometry in Oxford in 1649, in no small part, apparently, for his services to the winning side, and set about remedying the gaps in his education. He began with the English mathematician William Oughtred's *Clavis Mathematicae* (The Key of the Mathematics, 1647), but the books that truly educated him were Frans van Schooten's Latin edition of Descartes's *La Géométrie* of 1649 and Torricelli's *Opera Geometrica*. He published his first book, an algebraic treatment of conic sections, in 1655; in this book the symbol ∞ for infinity appeared in print for the first time.

Figure 2.16. John Wallis (1616–1703)

He then began writing his much more remarkable book on the subject of areas and volumes. We can let him introduce his fundamental idea in his own, somewhat confusing, words, taken from the opening pages of his *Arithmetica Infinitorum* (*The Arithmetic of Infinitesimals*) of 1656:

[35]See (Bascelli 2014).

Wallis on infinitesimals.

> If there is proposed a series, of quantities in arithmetic proportion (or as the natural sequence of numbers) continually increasing, beginning from a point or 0 (that is, nought, or nothing), thus as 0, 1, 2, 3, 4, etc., let it be proposed to inquire what is the ratio of the sum of all of them, to the sum of the same number of terms equal to the greatest.
>
> The simplest method of investigation, in this and various problems that follow, is to exhibit the thing to a certain extent, and to observe the ratios produced and to compare them to each other; so that at length a general proposition may become known by induction.
>
> It is therefore the case, for example, that:
>
> $$\frac{0+1}{1+1} = \frac{1}{2} \qquad \frac{0+1+2=3}{2+2+2=6} = \frac{1}{2}$$
>
> $$\frac{0+1+2+3=6}{3+3+3+3=12} = \frac{1}{2} \qquad \frac{0+1+2+3+4=10}{4+4+4+4+4=20} = \frac{1}{2}$$
>
> $$\frac{0+1+2+3+4+5=15}{5+5+5+5+5+5=30} = \frac{1}{2} \qquad \frac{0+1+2+3+4+5+6=21}{6+6+6+6+6+6+6=42} = \frac{1}{2}$$
>
> Proposition 2. Theorem: If there is taken a series, of quantities in arithmetic proportion (or as the natural sequence of numbers) continually increasing, beginning from a point or 0, either finite or infinite in number (for there will be no reason to distinguish), it will be to a series of the same number of terms equal to the greatest, as 1 to 2.
>
> That is, if the first term, is 0, the second 1 (for otherwise some adjustment must be applied), and the last is l, the sum will be $\frac{l+1}{2}l$ (for in this case the number of terms will be $l + 1$). Or (putting m for the number of terms, whatever the second term) $\frac{1}{2}ml$.

We can deduce a number of things about Wallis's mathematics from this extract alone. Certainly, his examples are clearer than his explanations. You may well feel that it was only after the examples that you became clear about what he was trying to say. Also, his reasoning in the proposition is decidedly loose: what is the 'greatest' term? The historian Jacqueline Stedall commented, in her edition and translation of Wallis's book, that

> Wallis's reasoning seems to break down immediately at this point, because if his series contains an infinite number of terms increasing indefinitely it can have no greatest term. What he is really thinking of, however, though he does not yet make it clear, is a series with a finite greatest term l, arrived at by m steps of size d, thus $0, d, 2d, 3d, \ldots, md = l$. When m is finite it is clear that the sum of terms is $\frac{1}{2}(m+1)l$, or, to $(m+1)l$ as 1 to 2. Wallis allowed the number of steps m to become infinitely large, by making d arbitrarily small, indeed infinitesimally small, but in such a way that md remains always equal to l and is therefore finite. In that case, Wallis argued ('by induction') that the same ratio of 1 to 2 would still hold.[36]

Now that we have been rescued by Stedall, we can also distinguish the characteristic feature of an argument involving infinitesimals as opposed to indivisibles. With infinitesimals there is some quantity which is non-zero but arbitrarily small, or smaller

[36] See (Stedall 2004, 14).

than any finite quantity, but such that infinitely many copies of it make a finite quantity. In the case at hand, this is the step size d.

The method was not invented by him, but no-one else carried it to such lengths and obtained so many new results by its means. Surely no-one, then or since, has so magnificently disdained rigour to such good effect. As another Wallis scholar, Kirsti Pedersen, said of Wallis's *Arithmetica Infinitorum*, it 'is not burdened with proofs, for he relied boldly and confidently on his really astounding intuition as to the correlation between the sums of different series'.[37]

Wallis first applied his method to a result about the area contained in a spiral, a topic previously treated by Archimedes. Then, in Proposition 19, he turned to determine the area under the parabola between $x = 0$ and $x = 1$; this is also not a new result, but Wallis was building up a list of known results in a way that he hoped would lead him to a generalisation. That is the exciting part.

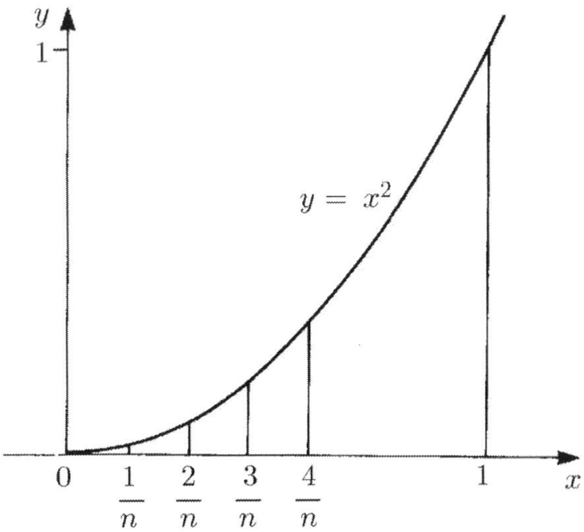

Figure 2.17. An example of Wallis's method for finding areas

He began, unexceptionably enough, by dividing the interval from 0 to 1 at the points $1/n, 2/n, \ldots, (n-1)/n$ (see Figure 2.17) and writing

$$\alpha = \frac{0^2 + 1^2 + 2^2 + \cdots + n^2}{n^2 + n^2 + \cdots + n^2} = \frac{\frac{1}{6}n(n+1)(2n+1)}{(n+1)n^2} .$$

The numerator is correct, although Wallis simply assumed it to be so because it is correct when n is small. The right-hand side simplifies to $\frac{1}{3} + \frac{1}{6n}$ so, said Wallis, the quotient of the sums taken to infinity yields the value 1/3. So what? Well, the height of the vertical line above $x = k/n$ is $y = k^2/n^2$. So in true Cavalierian (strictly, Torricellian) spirit, the area under $y = x^2$ is related to the sum of the lines — that is, to

$$\frac{0^2 + 1^2 + 2^2 + \cdots + n^2}{n^2 + n^2 + \cdots + n^2} = \alpha .$$

[37]See (Pedersen 1980, 38).

Moreover, the relationship is exact (and the area is the sum of all the lines), when the area is covered by the lines, which presumably occurs when n is infinite. But this yields the value for the area, which is correct. Now because it is correct, and because the method also gives the correct answer when applied to $y = x^3$, $y = x^4$, and so on, Wallis felt justified in extending it to other cases, like $y^p = x^q$, where the answer was not known. In this way he found the area under all curves $y = x^{q/p}$, provided only that q/p does not equal -1, which is the case of the hyperbola. He found that the area under $y = x^{q/p}$ between $x = 0$ and $x = 1$ is $\frac{p}{p+q}$, which is a considerable generalisation of his earlier result, even if in Wallis's hands it rested on mere analogy.

Nor did Wallis stop there. He passionately wanted to evaluate the area of a circle in a way that would allow him to evaluate π (if we may be permitted to introduce the symbol introduced only in 1706, by William Jones). This involved Wallis in finding the area of the quarter-circle bounded by

$$x^2 + y^2 = 1, \text{ or } y = (1 - x^2)^{1/2},$$

and the lines $x = 0$ and $y = 0$. This is the area of a quarter-circle of radius 1, which is $\frac{1}{4}\pi$. Finding that his present methods did not enable him to do this directly, he stepped back from that problem and went pattern hunting — recall that Wallis had been a code-breaker.

His earlier results enabled him to evaluate the area $A_{k,m}$ under the curve

$$y = (1 - x^k)^m, \text{when } m \text{ is an integer,}$$

for he could then write $(1 - x^k)^m$ as a polynomial in x^k. So he listed his results in a table for all integer values of m between 1 and 10 and for $k = 1, \frac{1}{2}, \frac{1}{3}, \ldots, \frac{1}{10}$. The case he wanted is $k = 2$ and $m = \frac{1}{2}$. He found that there was a pattern in the table that enabled him to calculate or guess an entry exactly, or to express it in terms of the area under the still-unknown value for $y = (1 - x^2)^{1/2}$. Moreover, he could see how the entries grow as k decreases and m increases, and this enabled him to compare the entries he could find exactly with those involving the one he wanted, and indeed to estimate the ones he wanted to any level of accuracy. The result, which is given here in the more intelligible form into which Newton cast it in 1665, was

$$\frac{\pi}{4} = \frac{1}{2} \times \frac{2 \times 2 \times 4 \times 4 \times 6 \times 6 \times etc.}{1 \times 3 \times 3 \times 5 \times 5 \times etc.}.$$

Wallis's methods were surely amazing. At once splendidly unrigorous and yet accurate, they challenged other mathematicians to do as well by any means, fair or foul, and made it plain that Wallis had no time for the old, classical methods.[38]

The method of approximation. If an area cannot easily be said to be made up of indivisible lines (which have no thickness), nor of infinitesimal lines (which have a mysterious thickness), perhaps it might be possible to regard an area as made up, approximately, of a finite number of strips of finite width. The method of computing an area by means of approximations with finite strips is perhaps best explained by an example: Fermat's computation of the area under $y = k/x^2$, lying to the right of $x = a$, and bounded by the x-axis.[39] It is a remarkable result, because it shows that although

[38] Extract in (Stedall 2008, 89–95).

[39] Fermat, *Oeuvres* I, 255–259, with French translation 219–221, F&G 11.C5, (Stedall 2008, 78–83).

the figure has an infinite perimeter, its area is *finite*. Fermat wrote it down only in 1658, but the ideas in it go back to the 1640s, if not earlier.

A geometric progression is a sequence of the form

$$a, ar, ar^2, \dots, ar^n, \dots.$$

The sum of the first $n+1$ terms is

$$a(1 + r + r^2 + \cdots + r^n) = a\frac{1 - r^{n+1}}{1 - r},$$

and so, if $0 < r < 1$, the sum of the entire progression is $\dfrac{a}{1-r}$.

Fermat interpreted that result in this way:

> The entire method is based on a well-known property of the geometric progression, namely the following theorem:
> *Given a geometric progression the terms of which decrease indefinitely, the difference between two consecutive terms of this progression is to the smaller of them as the greater one is to the sum of all following terms.*

That is, in a geometric progression the ratios

$$\frac{ar^n - ar^{n+1}}{ar^{n+1}} \quad \text{and} \quad \frac{ar^n}{\sum_{j=n+1}^{\infty} ar^j}$$

are equal because

$$\sum_{j=n+1}^{\infty} ar^j = \frac{ar^{n+1}}{1-r}.$$

Fermat's argument for the curve with equation $x^2y = K$ (see Figure 2.18), which can also be written as $y = K/x^2$, may be modernised (by using Cartesian coordinates instead of Fermat's Euclidean formulation) as follows. With the origin at A, choose points $G, H, O, M, \dots$ on the x-axis so that

$$AG = a, \quad AH = a/r, \quad AO = a/r^2, \quad AM = a/r^3, \quad \dots.$$

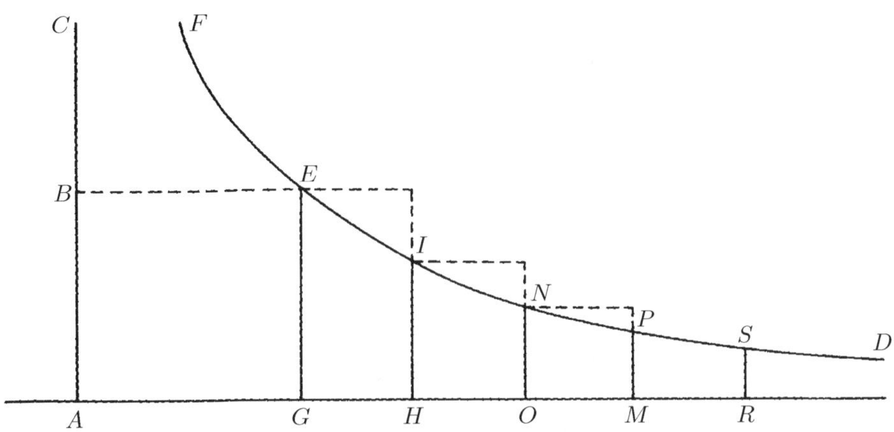

Figure 2.18. The area under $y = K/x^2$

Therefore

$$GH = (a/r) - a = a(1/r - 1), \quad HO = (a/r^2) - (a/r) = (a/r)(1/r - 1),$$

and so on. The corresponding points intervals on the curve satisfy

$$GE = K/a^2, \ \ HI = Kr^2/a^2, \ \ ON = Kr^4/a^2, \ \ MP = Kr^6/a^2, \ldots.$$

This means, on considering the first two rectangles, that

$$GH \times GE = a(1/r - 1) \times K/a^2 = K(1/r - 1)/a,$$

and

$$HO \times HI = a(1/r - 1) \times Kr^2/ra^2 = rK(1/r - 1)/a,$$

and so

$$(GH \times GE)/(HO \times HI) = 1/r\,.$$

Similarly, we can see that the ratio $(HO{\times}HI)/(OM{\times}ON)$ is also $1/r$. So the areas of successive rectangles form a geometric progression, and so, although infinite in number, they have a finite sum:

$$K\frac{(1/r) - 1}{a}\frac{1}{1 - r} = \frac{K}{ar}.$$

Today, a mathematician would conclude the argument by saying that as r approaches 1 the width of the strips becomes smaller and smaller and collectively the strips approximate the area under the curve and to the right of GE better and better, so the area under the curve is K/a, which is the correct result.

Fermat argued:

> But the lines AO, AH, AG, which form the ratios of the parallelograms, define by their construction a geometric progression; hence the infinitely many parallelograms $EG \times GH, HI \times HO, NO \times OM$, etc., will form a geometric progression, the ratio of which will be AH/AG. Consequently, according to the basic theorem of our method, GH, the difference of two consecutive terms, will be to the smaller term AG as the first term of the progression, namely, the parallelogram $GE{\times}GH$, to the sum of all the other parallelograms in infinite number. According to the adequation of Archimedes, this sum is the infinite figure bounded by HI, the asymptote HR, and the infinitely extended curve IND.
>
> Now if we multiply the two terms by EG we obtain
>
> $$GH/AG = (EG \times GH)/(EG \times AG);$$
>
> here $EG{\times}GH$ is to the infinite area the base of which is HI as $EG{\times}GH$ is to $EG{\times}AG$. Therefore, the parallelogram $EG{\times}AG$, which is a given rectilinear area, is adequated to the said figure; if we add on both sides the parallelogram $EG \times GH$, which, because of infinite subdivisions, will vanish and will be reduced to nothing, we reach a conclusion that would be easy to confirm by a more lengthy proof carried out in the manner of Archimedes, namely, that for this kind of hyperbola the parallelogram AE is equivalent to the area bounded by the base EG, the asymptote GR, and the curve ED infinitely extended.

Fermat's explanation of the adequality argument is that for the approximation to be good, GH must be small, and for this it is enough that $1/r$ be greater than but very close to 1, because the width of GH is $a(1/r - 1)$. The geometric progression is

$$GH \times GE,\ GH \times GE \times r,\ GH \times GE \times r^2, \ldots,$$

which has the finite sum $GH \times GE \times 1/(1-r)$ provided that $r < 1$, so $1/r > 1$. The two conditions Fermat required are therefore compatible.

The merit of Fermat's method is that the strips are made up of comprehensible rectilinear figures of known area; its potential drawback is the need for an approximation or adequality argument at the end. Its peculiar feature in this case is the cunning choice of widths — it gradually became more usual with this style of argument to use strips of constant width — but the choice here was made to deal with the really interesting aspect of the problem at hand, namely that the area is finite but its base is infinite. Consequently, an approximation to the area with infinitely many strips would be a natural one to look for, and their total area would be a sum containing infinitely many terms. The paradigm case of a summation in which infinitely many terms are added and their total value is finite is the geometric progression, and this may be why Fermat proposed the argument he did. Unfortunately, we do not know how Fermat came to suspect that the area is finite. In going beyond Euclid and Archimedes by completing the infinite sum, Fermat was in line with contemporary practice. The first to defend this point of view in print was Gregory of Saint-Vincent, in his *Opus Geometricum* of 1647, but others, like Torricelli, were also using it.

Conclusions. What lessons can be drawn from what we have seen? Certainly, there was a great deal going on in the central decades of the 17th century. There were several methods in use for finding areas, which made up in heuristic power what they may have lacked in rigour. All contained a large measure of algebra. None stood especially commended, but that does not mean that they were not productive. On the contrary, they are a measure of how differently things stood then than now. To the mathematicians of the 1650s these methods represented a rich new arsenal of techniques which they could look forward to refining and extending in their own research. The opinion of Pascal is worth taking seriously here, for he is in the first rank of mathematicians and philosophers. The trenchant expression of his views that we give below, taken from a letter that he wrote in 1658, speaks for itself, and although it concerns the method of indivisibles specifically, it can bear a more general interpretation and may serve as a conclusion to this section.[40]

Pascal on the method of indivisibles.

> I wanted to write this note to show that everything which is proved by the true rules of indivisibles will also be proved with the rigour and the manner of the ancients, and that therefore the methods differ, the one from the other, only in the way they are expressed: which cannot hurt reasonable people once one has alerted them to what that means.
>
> And that is why I do not find any difficulty in what follows in using the language of indivisibles, the sum of lines or the sum of planes; and thus when for example I consider the diameter of a semi-circle

[40]See Pascal, *Oeuvres* VIII, 352, and F&G 11.E2.

divided into an indefinite number of equal parts at the points Z, from which ordinates ZM are taken, I shall find no difficulty in using this expression, the sum of the ordinates, which seems not to be geometric to those who do not understand the doctrine of indivisibles and who imagine that it is to sin against geometry to express a plane by an indefinite number of lines; which only shows their lack of intelligence, for one understands nothing other by that than the sum of an indefinite number of rectangles made on each ordinate with each of the equal portions of the diameter, whose sum is certainly a plane which only differs from the space of the semicircle by a quantity less than any given quantity.

2.3 The situation mid-century

1658 and 1659 were momentous years. René-François de Sluse, Huygens, and Wallis began a productive discussion of the curve known as the cissoid; Pascal, who had deserted mathematics for the religious life, briefly returned and provoked a storm with a competition to investigate the cycloid; van Schooten published his massive second Latin edition of Descartes's *La Géométrie*; and certain curves were rectified, thereby refuting an opinion held by Descartes. Moreover, these years saw the last flicker for a while of French interest in mathematics: the geometer Girard Desargues died in 1661, Pascal in 1662, and Fermat in 1665. Descartes himself had died in 1650, and Mersenne even earlier, in 1648.

By the 1660s, the active centre of mathematics had shifted to the Netherlands. The leading figure there was that of Christiaan Huygens, but there were also van Schooten, Sluse, Hudde, and van Heuraet.[41] A community about to become active was the rather more dispersed British one: Barrow and Newton in Cambridge, James Gregory in Edinburgh, Wallis in Oxford, and in London Christopher Wren and John Collins (a minor mathematician but a useful correspondent). Our task is to bring into focus the tumultuous developments of the previous forty years by elaborating on what these mathematicians regarded as the most important problems in mathematics, and what they hoped the most productive methods for dealing with them might be.

The cissoid. The cissoid of Diocles is one of the antique curves.[42] Here it is enough to observe that its equation can be written in the form $y^2(2-x) = x^3$, in which form it has a cusp at the origin and the vertical line $x = 2$ is an asymptote (see Figure 2.19). It had already been studied by contemporary mathematicians before 1658; for example, Huygens found tangents to it by Descartes's method in 1653.

Then in May 1658 Sluse wrote to Huygens that if the cissoid is rotated about its asymptote then the surface so obtained has two surprising properties:

- the volume of the enclosed solid can be computed and the calculation reduces to finding the area of a circle
- consequently, although the axis of the solid is infinitely long, the volume of the enclosed solid is finite

[41]Christiaan Huygens was the most important mathematically inclined natural philosopher between Galileo and Newton; for a study of his work and influence, see (Yoder 2004).

[42]It is discussed in *Volume* 1, Chapter 4.

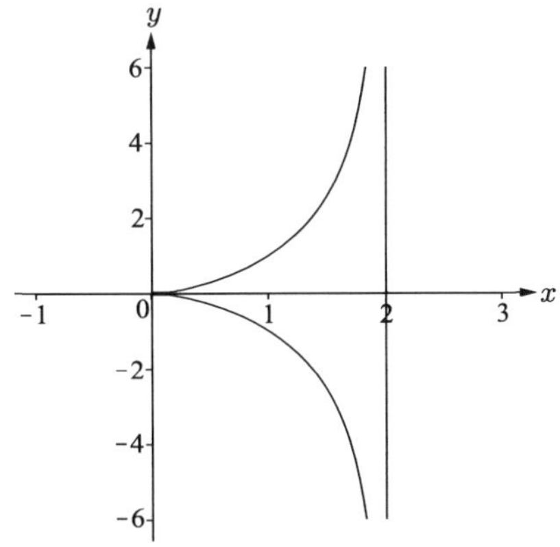

Figure 2.19. The cissoid of Diocles

He also remarked that the area between the cissoid and its asymptote is also finite.

Thus inspired, Huygens gave a very elegant computation of the area, which Sluse had been unable to do, and wrote to Wallis at the end of the year, challenging him, in effect, to find the area by the methods of his *Arithmetica Infinitorum*. The letter took its time to reach Wallis, but he then succeeded within a week and included his solution in his *Two Tracts on the Cycloid and the Cissoid*, which he was just then compiling.

Huygens' method is thoroughly Archimedean in spirit, and its discovery required Huygens' great visual imagination. Wallis's method similarly drew on his extraordinary skills at interpolating values in formulas. So both methods were ad hoc, and required the special skills of their proponents, even for a familiar curve. Although Huygens appreciated the strength of Wallis's approach, he deliberately sought to stick close to classical methods in order to be as rigorous as possible. Wallis saw the justice of Huygens' point of view and in his letter to him he even promised that he would eventually give a strictly geometrical demonstration — but he never did. So it seems that in their different ways Huygens and Wallis could both feel happy with their own studies of the cissoid. They had obtained new results, and their techniques stood confirmed as a consequence.

The cycloid. The story is similar when we turn to the cycloid. Pascal proposed several problems on the cycloid that he had been able to resolve while fighting off an attack of toothache. He did so under the curious pseudonym of Amos Dettonville, which is an anagram of Lovis (= Louis) de Montalte, the pen-name under which he had published his classic theological polemic, *Lettres Provinciales* (1656–1657). The problems concerned the area and the centre of gravity of a segment of the cycloid and the volume and centroid of the solid obtained by rotating the cycloid about its base. Unfortunately, the challenge was not widely distributed, and by the time the closing date came around (1 October 1658) only three answers had been received: from Wallis, and the Jesuits Honoré Fabri and Antoine de Lalouvère.

Pascal pronounced none of these entries satisfactory, and instead published his own work. Needless to say, the contestants were annoyed by this, and went on to publish their own versions. Pascal, who was strongly opposed to the Jesuits on theological grounds, then aggravated matters by publishing the false claim that Lalouvère had plagiarised his solution from Roberval. But it is Pascal's work that is the most interesting, because it contains a list of transformations converting one area calculation to another, one volume calculation to another, and so forth. Pascal justified these transformations by exhibiting alternative dissections of an area or volume into infinitesimal pieces; by compiling such a list, he exhibited a set of connections between different problems. This in turn suggested that a more general technique might exist for dealing with these problems, for which it would be profitable to look. It also suggests to us that the state of the art was still ad hoc.

Arc lengths. Another surprising development in the late 1650s was the computation of arc-length along various curves — technically, their 'rectification' or straightening out (see Figure 2.20).

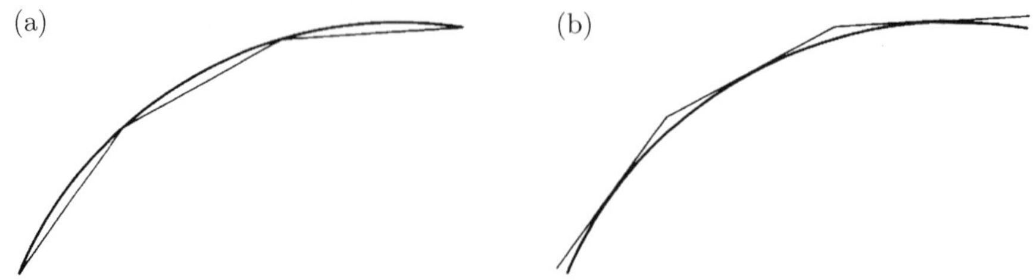

Figure 2.20. Rectification of a curve: (a) by chords; (b) by tangents

The first to accomplish this was William Neile, who was a student at Oxford when he rectified the semi-cubical parabola.[43] When Wallis published Neile's discovery in his own *Two Tracts etc.*, he made the shrewd observation that the method is general and depends on regarding an infinitesimal piece of arc as equivalent to the hypotenuse of a right-angled triangle (see Figure 2.21).

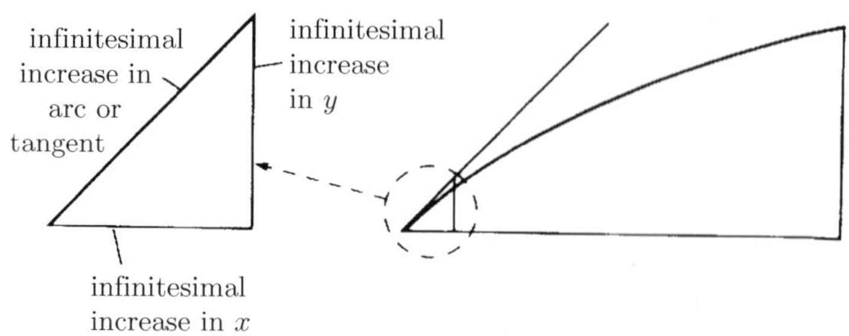

Figure 2.21. Rectification of a curve: infinitesimal tangents

However, generality of method is much clearer in van Heuraet's approach.[44] It was published as a four-page commentary in van Schooten's 2nd edition of *La Géométrie*

[43] The semi-cubical parabola is a curve with equation $y^2 = x^3$. For an extract from Wallis's Latin letter to Huygens of 1659 in which he described Neile's method, and an English translation, see (Stedall 2008, 102–104).

[44] See Descartes, *Geometria*, 2nd edn., 517–520, transl. in (Grootendorst and van Maanen 1982, 101–105), and F&G 11.B4.

of 1659, where van Heuraet pointed out that it applies to all the curves with equations of the form $ky^m = x^{m+1}$. Van Heuraet's method, which he seems to have discovered independently of Neile, is marginally the simpler of the two, and makes more obvious the appeal to Pascal's characteristic triangle (defined below, in Section 3.3).[45]

However, the most dramatic consequence of the work of Neile and van Heuraet was that it refuted Descartes's dictum that 'the ratios between straight and curved lines are not known, and I believe cannot be discovered by human minds'.[46] This claim was intended to apply only to geometrical (or algebraic) curves, for Descartes knew perfectly well that some mechanical (or transcendental) curves could be rectified — the quadratrix, for example. By shattering this belief, Neile and van Heuraet did much to confirm peoples' faith in the indivisible or infinitesimal methods currently employed; now even Descartes's work was being left behind. Rectifications of other curves, albeit transcendental ones, soon followed: Wren and Fermat rectified the cycloid in 1659 and 1660 respectively, and Huygens the cissoid. In 1668, James Gregory published his *Geometriae Pars Universalis* (The Universal Part of Geometry), which contains a general treatment of the problem of rectification, so it seems that the problem was indeed one that appealed to mathematicians.

Areas and tangents. No-one in 1660 would have predicted the invention or discovery of the calculus within twenty-five years. In retrospect, perhaps the largest missing piece of the puzzle was the connection between area problems and tangency problems, still largely unsuspected in 1660. Its history is also surprising, so we shall now work towards it.

As we have seen, tangent methods were many and varied. In 1659 Hudde published what became his celebrated rules. These rules were greatly to excite the young Newton in 1664, not least because they gave promise of being general and easy to use. An extension of them was published by Sluse in 1672 — but by then Newton had found this extension independently, one of his many unpublished discoveries.

All these rules were firmly algebraic, although the old Greek concept of a tangent still exerted its influence. But whatever approach was adopted, finding tangents to curves was always more of a routine process than finding areas. This is so today (as any student of the calculus will testify) and it was true in 1659: there were Hudde's rules for tangents, which dealt with algebraic curves, but no rules for areas. It is therefore interesting to see that during the 1660s certain mathematicians began to find that some area questions were closely connected to tangency questions.

Isaac Barrow. The most successful proponent of this view was Isaac Barrow, who presented it in his *Lectiones Geometriae* of 1670, although Gregory's book of 1668 had clearly hinted at the idea. Barrow's presentation is clearly an exercise in geometry: we shall soon see how different this is from the Fundamental Theorem of the Calculus. From a question about one curve, ZGE, and areas, Barrow (in a perfectly general way) produced a new curve VIF and a question about tangents (see Figure 2.22), as we shall now see.[47]

---

[45] This is not to be confused with Pascal's triangle of binomial coefficients.

[46] Descartes, *La Géométrie*, 91, and F&G 11.A8.

[47] See Barrow, *Lectiones Geometricae*, in Struik, *A Source Book*, 255–256, and F&G 11.E3.

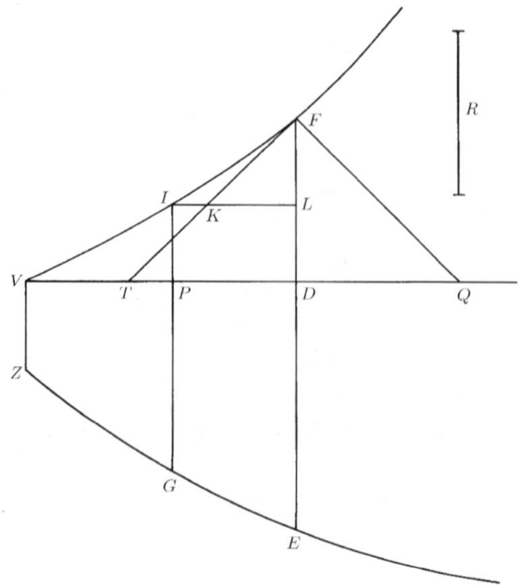

Figure 2.22. A fundamental theorem of Barrow

Barrow on tangents and areas.

> Let ZGE be any curve of which the axis is VD and let there be perpendicular ordinates to this axis (VZ, PG, DE) continually increasing from the initial ordinate VZ; also let VIF be a line such that, if any straight line EDF is drawn perpendicular to VD, cutting the curves in the points E, F, and VD in D, the rectangle contained by DF and a given length R is equal to the intercepted space $VDEZ$; also let $DE : DF = R : DT$, and join [T and F]. Then TF will touch the curve VIF. For, if any point I is taken in the line VIF (first on the side of F towards V), and if through it IG is drawn parallel to VZ, and IL is parallel to VD, cutting the given lines as shown in the figure; then $LF : LK = DF : DT = DE : R$, or $R \times LF = LK \times DE$.
>
> But, from the stated nature of the lines DF, LK, we have $R \times LF =$ *area* $PDEG$: therefore $LK{\times}DE =$ *area* $PDEG < DP{\times}DE$; hence $LK < DP < LI$. Again, if the point I is taken on the other side of F, and the same construction is made as before, plainly it can be easily shown that $LK > DP > LI$, from which it is quite clear that the whole of the line TKF lies within or below the curve VIF. Other things remaining the same, if the ordinates, VZ, PG, DE, continually decrease, the same conclusion is attained by similar argument; only one distinction occurs, namely, in this case, contrary to the other, the curve VIF is concave to the axis VD.

We see that Barrow was using a Greek definition of tangent, where the line meets the curve just once. He established that the line TF is a tangent to the curve at F by proving that the line always lies below the curve, except at F. The best way of regarding his theorem is to think of the line FDE as moving, so $VDEZ$ is the area swept out so far by the line DE on its journey from VZ. This is actually quite a significant step to take, for Barrow is thinking of areas like these where the right-hand boundary can (and does) move. The point F is then fixed by the rule $DF \times R = area(VDEZ)$. Then the point T is fixed by the rule $DE : DF = R : DT$ (so $R \times DF = DE \times DT$). Then he stated his theorem:

TF is tangent to the curve CIF.

Nor is it inappropriate to invoke the idea of motion, because elsewhere in his book Barrow often constructed tangents by motion arguments. By showing how a question about areas could be converted quite generally into one about tangents, Barrow was opening up a new line of attack on certain mathematical problems, as well as finding a result of independent value. What he did subsequently was to tackle some specific problems to do with finding tangents, then he looked in more detail at problems about areas, and finally he arrived at the converse of the result we have just studied: from a question about tangents he arrived at a question about areas.

The words with which the historian Dirk Struik concluded his notes on Barrow may serve equally well here, at the conclusion of our look at the situation immediately before Newton and Leibniz:[48]

> We end with a word of caution. Despite the fact that, in order to understand these seventeenth-century mathematicians, we are inclined to translate their reasoning into the notation and language with which we are familiar, we must constantly be aware that our point of view is not equivalent to theirs. They saw geometric theorems in the sense of Euclid, where we see operations and calculating processes. At the same time, just because these mathematicians applied their geometric notions in an attempt to transcend the static character of classical mathematics, their geometric thought has a richness that may easily escape observation in the modern transcription. If we were to rewrite Euclid in the notation of analytic geometry we would obtain a body of knowledge with a character different from that of Euclid and, despite all the advantages that the algebraic computations would bring, we would lose some of the more subtle and aesthetic qualities of Euclid.

We may also consider a more recent historian, Michael Mahoney, who, at the end of a long and careful study of Barrow's mathematical work, explicitly addressed the relationship between it and the calculus.[49] Mahoney argued that Barrow thought of tangents globally but not locally — he did not think of indefinitely small quantities as the fundamental terms of analysis — and he did not think of ratios as quantities, or ratios of indefinitely small quantities as possibly finite expressions. These steps were taken by Newton and Leibniz as part of their algebraic analysis of the geometrical problems tackled by Barrow. However,[50]

> Nothing in Barrow's Lectures or other works suggests ... that he shared the convictions about the analytical power of algebra that underlay it. For that reason, Lectures XI and XII remain historically an example of mid-seventeenth-century extensions of geometrical analysis by way of infinitesimal elements rather than the prototype of the calculus. Competent and well informed, but not particularly original, Barrow provided in his Lectures an inventory of the materials available to Newton and Leibniz. What it reveals about the origins of the calculus says more about the nature of their creativity than about his.

[48] See (Struik 1969, 263).

[49] See (Mahoney 1990).

[50] See (Mahoney 1990, 240).

2.4 Further reading

Feingold, M. (ed.) 1990. *Before Newton: The Life and Times of Isaac Barrow*, Cambridge University Press. This book contains much information on the intellectual and scientific scene in Britain before the time of Newton, and a lengthy essay by M. Mahoney, 'Barrow's mathematics: between ancients and moderns', that offers a very careful account of what he did and what he did not do.

Stedall, J. 2004. *The Arithmetic of Infinitesimals: John Wallis 1656*, Springer. A translation of the *Arithmetica Infinitorum* accompanied by a detailed yet highly readable account of this famous and controversial work that sheds new light on many aspects of British mathematical life in the 17th century.

Stedall, J. 2008. *Mathematics Emerging: A Sourcebook 1540–1900*, Oxford University Press. A treasure trove, and unique in providing generous reproductions of the original texts alongside their translations. It offers an extensive coverage of the calculus and quite an amount of material on many aspects of algebra.

Struik, D. 1969. *A Source Book in Mathematics: 1200–1800*, Harvard University Press. An invaluable collection of mathematical highlights from the fields of arithmetic, algebra, geometry and, most notably, analysis, by over sixty mathematicians, accompanied by pertinent comments from the doyen of historians of mathematics.

Yoder, J.G. 2004. *Unrolling Time: Christiaan Huygens and the Mathematization of Nature*, Cambridge University Press. A detailed account of Huygens' invention of a theoretically perfect pendulum clock and the reception of his masterpiece, the *Horologium Oscillatorium* of 1673, that also explores his relationship with the other scientists of his time.

3

Newton and Leibniz

Introduction

In this chapter we take our first look at Newton and Leibniz, and at the opening moves in their invention of two interestingly different versions of the calculus.

3.1 Newton

Newton was born on Christmas Day, 1642, in the manor house of Woolsthorpe, near Grantham in Lincolnshire. He was a sickly child, and his father had already died. When his mother remarried he was sent, at the age of three, to live with his grandmother whom he never came to like, much less love. At the age of ten he was briefly reunited with his mother before being sent away to live in Grantham, where he attended the Grammar School and developed a strong interest in mathematics. By 1661, when he entered Cambridge University, he was already confirmed in certain traits. He would immerse himself in thought to the point of neglecting his meals; he was estranged from others and kept his own company; his mind was never at rest. It was always to be so.

At Cambridge, the syllabus was undemanding and did not reflect the intellectual tumult of the age. But it left the young man, already eager to learn and explore, free to follow up what clues he could find. He kept a notebook, and in 1664 began to fill it with a rush of observations, entering both practical experiments and the fruits of his reading. Many notes are on Descartes: his theories of light, of planetary motion, and of the tides. Other passages mark the beginning of Newton's interest in mechanics, and it seems that he first read Keplerian astronomy at this time. Yet others reflect his absorption in the philosophical ideas of Henry More, the leading Cambridge Platonist of the day and an influential writer. More's central concern was to reconcile mechanical philosophy with Christian religion by resolving the conundrum: How can God act in a world which obeys physical laws? His analyses of this problem impressed themselves upon the devout Newton, and may well have helped him towards his own novel ideas of universal gravitation.

Figure 3.1. Isaac Newton (1642–1727)

Not least, Newton discovered advanced mathematics. Looking back, 35 years later, he recalled that:[1]

> By consulting an accompt of my expenses at Cambridge in the years 1663 and 1664 I find that in ye year 1664 a little before Christmas I being then senior Sophister, I bought Schooten's Miscellanies & Cartes's Geometry (having read this Geometry & Oughtred's Clavis above half a year before) & borrowed Wallis's works and by consequence made these Annotations out of Schooten & Wallis in winter between the years 1664 and 1665. At wch time I found the method of Infinite series. And in summer 1665 being forced from Cambridge by the Plague I computed ye area of ye Hyperbola at Boothby in Lincolnshire to two & fifty figures by the same method.

Newton's early self-taught struggles with Descartes's ideas were recorded at the end of his life by John Conduitt, who was born in 1688 and came to know Newton well when he married Newton's niece.[2]

> He bought Descartes's Geometry & read it by himself when he was got over 2 or 3 pages he could understand no farther than he began again & got 3 or 4 pages farther till he came to another difficult place, than he began again & advanced farther & continued so doing till he made himself Master of the whole without having the least light or instruction from any body.

---

[1]Westfall, *Never at Rest*, p. 98. *Miscellanies* is van Schooten's *Exercitiones Mathematicae* (1657), 'Cartes's Geometry' is van Schooten's edition of 1659–1661.

[2]Quoted in Westfall, *Never at Rest*, pp. 98–99.

For once, Cambridge may deserve some slight credit, for although Isaac Barrow, the first Lucasian Professor of Mathematics, was never Newton's tutor, it is hard to see who else could have stimulated Newton's interest in mathematics, or lent him a copy of Wallis's *Arithmetica Infinitorum.* Indeed, Barrow amassed a large collection of mathematical books, and Newton had free access to Barrow's library.[3] That said, this strange young man proceeded on his own, as was always his fashion, and in the eighteen months to Spring 1666 he thought intensively about mathematics. In less than two years he invented the calculus, and turned again to mechanics to grapple, unsuccessfully, with the motion of the planets. What caused these large and fast-moving objects to stay in orbit and not rush away from the Sun? The legend of the apple refers to this period. It was recorded by the physician William Stukeley, who wrote in his *Memoirs of Newton's Life*:[4]

> After dinner, the weather being warm, we went into the garden and drank tea, under the shade of some apple trees. He told me, he was just in the same situation, as when formerly, the notion of gravitation came into his mind. It was occasion'd by the fall of an apple, as he sat in contemplative mood. Why should that apple always descend perpendicularly to the ground, thought he to himself.

This famous story suggests that Newton, even then, sought to extend known, terrestrial mechanisms to celestial problems. Then came the dedicated explorations of optics and the nature of colour and light, which for many years were Newton's most public demonstrations of his genius.[5]

Figure 3.2. Newton's rooms in Trinity College, Cambridge, were to the right of the gatehouse

[3]A number of these books passed to Newton after Barrow's death in 1677. See (Feingold 1990, 336).

[4]The *Memoirs* were published in 1752, and are now available at www.royalsociety.org/library/moments/newton-apple/.

[5]Westfall, *Never at Rest*, p. 158, notes that it is difficult to discover when Newton took up the study of optics and the nature of colours, but it might have been by the summer of 1665, and he had an elaborate theory by 1670.

In 1669, when Barrow resigned his chair to pursue his chosen career as a divine at the court of Charles II, Newton was appointed his successor as Lucasian professor. Given the highly political atmosphere of Cambridge, Barrow presumably had a hand in it, although the oft-told story that he resigned in Newton's favour is improbable. In any event, Newton was now financially secure and free to devote himself to research. By the Statutes governing his chair he also had to lecture, on 'some part of Geometry, Astronomy, in Geography, Optics, Statics, or some other Mathematical discipline'.[6] This does not seem to have been a great success. It was recorded a few years later that:[7]

> So few went to hear Him, and fewer yt understood him, yt oftimes he did in a manner, for want of Hearers, read to ye Walls.

Nor did his individual teaching thrive any better. During his whole time at Cambridge Newton seems to have had only three serious students, although he was Lucasian Professor until 1696.

Before we follow him 'Voyaging', as Wordsworth was to put it 'through strange seas of Thought, alone' we turn to look at his complicated relationship to the work of René Descartes.[8]

Newton took his algebraic symbolism from Descartes. He wrote equations to describe curves, as Descartes had done. He cared, as Descartes had, for establishing the priority of the geometrical over the algebraic. And by his achievement he raised the creative tension we have traced in *La Géométrie* to a new level of fecundity.[9]

'At which time', he had written of the winter of 1664–1665, 'I found the method of *Infinite series*'. Harmless-sounding words, but they represented an enormous increase in the power of mathematics. Indeed, circumstances of publication (to which the solitary Newton was averse) meant that for many years Newton's reputation as a mathematician rested on his prowess with infinite series. Leibniz, for example, was to regard Newton as a first-rate mathematician on such grounds, without knowing that he had already invented the calculus. So what are infinite series, and what did Newton contribute?

All of Descartes's algebra involved finite expressions (polynomials). Every term and every equation in his work can be written down completely. An *infinite series* is typified by this one, which is one of the first that Newton was to write:

$$x - \frac{1}{3}.\frac{1}{2}x^3 - \frac{1}{5}.\frac{1}{8}x^5 - \frac{1}{7}.\frac{1}{16}x^7 - \frac{1}{9}.\frac{5}{128}x^9 + \ldots, etc.$$

It is something like a polynomial, then, but goes on for ever in a way that can only be described by a hand-waving gesture, or 'etc.', or dots (...). We receive a sense of the importance of infinite series from a fascinating document that we now briefly discuss: Newton's second letter to Leibniz or *Epistola Posterior*, of 24 October 1676.[10] (We consider the above series in more detail in Section 4.2, when we also see how Newton derived it.)

This is a famous source documenting some of Newton's earliest work, but it is not completely self-explanatory. Newton had been reading Wallis's *Arithmetica Infinitorum* where an ingenious *interpolation method*, as Wallis called it, was explained for

[6]See Westfall, *Never at Rest*, p. 208.

[7]See Westfall, *Never at Rest*, p. 209.

[8]Wordsworth, *Prelude*, Book III, line 63, in F&G 12.E3.

[9]Descartes's *La Géométrie* was discussed in *Volume* 1, Chapter 12.

[10]See (Turnbull 1960, 129–134) in F&G 12.C2.

finding the areas under certain curves (this was called finding the quadrature of the curve). If, for instance, the areas under the curves

$$y = 1, \ y = 1 - x^2, \ y = (1 - x^2)^2, \ldots$$

are known, as they were by this time, then the areas under curves with the more awkward equations

$$y = \sqrt{1 - x^2}\,, \ y = (1 - x^2)\sqrt{1 - x^2}, \text{ etc.}$$

might be calculated. This is done by noticing that all these equations form a patterned sequence, when expressed in exponential notation, in which known and unknown areas alternate: the right-hand sides are

$$(1 - x^2)^{0/2}, \ (1 - x^2)^{1/2}, \ (1 - x^2)^{2/2}, \ (1 - x^2)^{3/2}, \ (1 - x^2)^{4/2}, \ \ldots .$$

So the quadrature of $y = \sqrt{1 - x^2}$ — that is, of the circle $x^2 + y^2 = 1$ — must lie, by some principle of continuity, between those of $y = 1$ and $y = 1 - x^2$. In the first part of the *Epistola Posterior*, Newton described how he was following through Wallis's argument when he discovered that the quadratures of the awkward curves could be expressed as infinite series; he seems to have done this by a process of shrewd guesswork and pattern recognition. His significant leap of imagination was *to accept an infinite series as the answer*. This is going far beyond what Descartes would have allowed as acceptable.

A few weeks later, Newton looked again at his notes, and saw how to simplify his work, thereby eliminating the need for creative guesswork, and how to progress further. This he also described in the *Epistola Posterior*.

Newton now saw that his discovery essentially had nothing to do with finding quadratures (or areas), but was about expressing things with fractional exponents as infinite series. So he applied the same technique as before, to the algebraic expressions themselves, such as

$$(1 - x^2)^{1/2},$$

and came up in this case with the infinite series

$$1 - \frac{1}{2}x^2 - \frac{1}{8}x^4 - \frac{1}{16}x^6 + \ldots .$$

He had thus found a rule for 'the general reduction of radicals into infinite series'.

In order to check that this was correct, he squared the infinite series, multiplying it by itself term by term, and found that he did indeed get back to $1 - x^2$, and similarly for the others. Finally, Newton re-extracted the square root arithmetically, to show that the initial geometrical context was not essential.

Newton did not rest here. He now knew that he had a general rule for handling expressions like $(1 - x)^{m/n}$. They could be written as infinite series, and those series treated just like polynomials. This technique became even more powerful when coupled to the calculus, but it is quite striking on its own, for it brings transcendental curves into the orbit of algebraic analysis. A transcendental curve, such as the cycloid or the quadratrix, cannot be described by a polynomial equation but it can be handled by infinite series. Descartes's abhorrence of these curves was not shared by Newton, not least because they were, thanks to infinite series, amenable to algebraic analysis.

While still in the first phase of his voyage through the frontier of mathematics, Newton surpassed Descartes in another way. This was his spectacular enumeration of cubic curves, made during the late 1660s, and his first study of their geometrical properties.[11]

We see how completely Newton was indeed the 'master' of geometry and of *La Géométrie*, for this work is a Herculean feat of symbol manipulation. However much we may want to suppress the fact from history and mathematics courses alike, the truth is that success in mathematics is often a matter of calculating lots of interesting examples and being lucky. Newton would not have been Newton had he not been willing to calculate for weeks on end, but he also lived at a propitious time.

But there is another point. Where does geometry stop, and algebra take over? Later, in the 1680s and 1690s, Newton took to writing about this question. We have manuscripts of lectures, and a book, the *Arithmetica Universalis* (published in 1707, but written much earlier) in which he put down his thoughts on the matter.[12]

He began by reviewing Pappus's classification of curves (which he called 'lines'), and then went on as follows (the 'Moderns' he refers to are Descartes and his followers):

Newton on the nature of curves.

> But the Moderns advancing yet much farther, have received into Geometry all Lines that can be expressed by Equations, and have distinguished, according to the Dimensions of the Equations, those Lines into Kinds; and have made it a Law, that you are not to construct a Problem by a Line of a Superior Kind, that may be constructed by one of an inferior one.

Newton now turned critical:

> In the Contemplation of Lines, and finding out their Properties, I approve of their Distinction of them into Kinds, according to the Dimensions of the Equations by which they are defined. But it is not the Equation, but the Description that makes the Curve to be a Geometrical one. The Circle is a Geometrical Line, not because it may be expressed by an Equation, but because its Description is a Postulate. It is not the Simplicity of the Equation, but the Easiness of the Description, which is to determine the Choice of our Lines for the Construction of Problems. For the Equation that express a Parabola, is more simple than that that expresses a Circle, and yet the Circle, by reason of its more simple Construction, is admitted before it. The Circle and the Conick Sections, if you regard the Dimension of the Equations, are of the same Order, and yet the Circle is not numbered with them in the Construction of Problems, but, by reason of its simple Description, is depressed to a lower Order, viz. that of a right Line; so that it is not improper to construct that by a Circle that may be constructed by a right Line. But it is a Fault to construct that by the Conick Sections which may be constructed by a Circle. Either therefore you must fix the Law to be observed in a Circle from the Dimensions of Equations,

[11] Newton, *An Improved Enumeration ... of the General Cubic Curve*, 1695, in *MPIN* VII, 589, 635; here and elsewhere we abbreviate *The Mathematical Papers of Isaac Newton* to *MPIN*; in F&G 12.D2.

[12] See Whiteside, *MPIN* 7, and Newton, *Universal Arithmetic*, pp. 465–470, transl. J. Raphson, London, 1769, from which the extract below is taken, and F&G 12.D3. There are also several extracts from Newton's work in (Stedall 2008).

> and so take away as vitious the Distinction between Plane and Solid Problems; or else you must grant, that the Law is not so strictly to be observed in Lines of Superior Kinds, but that some by reason of their more simple Description, may be preferred to others of the same Order, and may be numbered with Lines of inferior Orders in the Construction of Problems.

We see that, in contrast to Descartes, Newton's criterion of simplicity was ease of generation, so the circle is, as the Greeks proposed, simpler than the parabola, and the cycloid is simpler than any algebraic curve of high degree.

Nor did Newton observe a distinction between algebraic and transcendental curves, and so he completely disagreed with Descartes's views on this aspect of the classification of curves.

> In Constructions that are equally Geometrical, the most simple are always to be preferred. This Law is beyond all Exception. But Algebraick Expressions add nothing to the Simplicity of the Construction. The bare Descriptions of the Lines only are here to be considered. These alone were considered by those Geometricians who joined a Circle with a right Line. And as these are easy or hard, the Construction becomes easy or hard. And therefore it is foreign to the Nature of the Thing, from anything else to establish Laws about Constructions. Either therefore let us, with the Antients, exclude all Lines besides a right Line, the Circle, and perhaps the Conick Sections, out of Geometry, or admit all, according to the Simplicity of the Description. If the Trochoid were admitted into Geometry, we might, by its Means, divide an Angle in any given Ratio. Would you therefore blame those who should make use of this Line to divide an Angle in the Ratio of one Number to another, and contend that this Line was not defined by an Equation, but that you must make use of such Lines as are defined by Equations? If therefore, when an Angle was to be divided, for Instance, into 10001 Parts, we should be obliged to bring a Curve defined by an Equation of above an hundred Dimensions to do the Business; which no Mortal could describe, much less understand; and should prefer this to the Trochoid, which is a Line well known, and described easily by the Motion of a Wheel or a Circle, who would not see the Absurdity? Either therefore the Trochoid is not to be admitted at all into Geometry, or else, in the Construction of Problems, it is to be preferred to all Lines of a more difficult Description.

However, Newton never arrived, as Descartes had done, at a consistent, general classification of curves.

We earlier suggested that Newton's ideas were incoherent. It is impossible to extract a coherent philosophy from their richness without injustice. On occasion, geometrical simplicity commended itself to Newton; at other times his working practice was an orgy of calculation. Newton was so impressed by his own power of thought (not unjustifiably) that he had great difficulty in acknowledging the contributions of others; in his account of infinite series, for example, he greatly played down the significant input of Wallis's ideas. Nor was Newton above administering pontifical rebukes, as this little comedy from the 1680s shows.[13]

Newton on curves.

> Descartes, in regard to his accomplishment of this problem, makes a great show as if he had achieved something so earnestly sought after by the Ancients and for whose sake he considers that Apollonius wrote his books on conics. With all respect to so great a man I should have believed that this topic remained not at all a mystery to the Ancients. For Pappus informs us of a method for drawing an ellipse through five given points and the reasoning is the same in the case of the other conics. And if the Ancients knew how to draw a conic through five given points, does any one not see that they found out the composition of the solid locus. To be sure, their method is more elegant by far than the Cartesian one. For he achieved the result by an algebraic calculus which, when transposed into words, following the practice of the Ancients in their writings, would prove to be so tedious and entangled as to provoke nausea, nor might it be understood. But they accomplished it by certain simple proportions, judging that nothing written in a different style was worthy to be read, and in consequence concealing the analysis by which they found their constructions. To reveal that this topic was no mystery to them, I shall attempt to restore their discovery by following in the steps of Pappus' problem. And to this end I propose these problems:
>
> 1. *To describe a conic through three given points* A, B, C *which shall have the given centre* O. ...
>
> 2. *To describe a conic through the five given points* A, B, C, D, E. ...

Newton's logic was polemical. On the basis of a speculation about what the ancients could do, Newton then turned to suggest that Descartes's solution, by contrast, was nauseating. He next proceeded to show how to solve these problems, before returning to his argument with Descartes.

> And this seems the most natural method of solving the problem, not merely because it is relatively simple but since the first part of the problem (in the form propounded by Descartes himself) is to find some point having the given condition, and thereafter, since there are an infinity of points of this sort, to determine the locus in which they are all found. What then is more natural than to reduce the difficulties of this latter part to those of the former by determining the locus from a few points after they are found? In consequence, since the Ancients did develop a procedure for constructing a conic through five given points, no one should have doubted that they composed solid loci by this means.

[13]See Whiteside, *MPIN* IV, 275–283, and F&G 12.D1.

Newton had just started to fill in the gaps in his education, having been largely self-taught, and he came late to the study of the Greeks. When he applied himself to the resolution of the Pappus problem of the locus to three or four lines he solved it as, he says, any (gifted) Greek would have done, and on those grounds impugned Descartes for making 'a great show as if he had achieved something' — not that Descartes yielded anything to Newton in his need to assert that he was mathematically beholden to no-one else.[14] Should we be grateful that Descartes missed this solution to the Pappus problem, or did he know that his algebraic methods were immensely more general than any classical method could be? Newton's riposte seems curiously inappropriate for the end of the 17th century.

If it is right to see a struggle in the mind of Newton to live with the discoveries of Descartes — and not all historians would pursue such a line of analysis — the central intellectual disagreement must be over the role and concept of motion, and this in two ways. First, for Newton curves were often generated by motion, in a more vivid, less abstract way than in Descartes's conception. Unlike Descartes, Newton allowed transcendental curves, and exploited the idea of motion for finding tangents to curves, where Descartes was more statically algebraic (as you will see in the next chapter). Second, Descartes had a theory of planetary motion that was widely accepted, and we shall see in Chapter 5 that Newton destroyed this theory in his *Principia*. It is worth noticing that this substantial disagreement with Cartesian cosmology occupied Newton in what was arguably the most important work of his life.

It is fitting to end this section by contemplating Newton's own assessment of his relationship to Descartes. He offered this comparison in a letter he wrote to Hooke on 15 February 1676:[15]

> 'If I have seen further, it is because I have stood on ye sholders of Giants'.

Should one insist on the 'If' with its suggestion that he has not? Should one suspect insincerity, on the grounds that there were no giants? And is it a compliment or an act of denigration to single out Descartes as the man you have surpassed? Newton knew as well as anyone that there was no one greater in his chosen fields — the comparison was as inevitable as it is instructive.

3.2 Newton and his calculus

We have already seen that Newton's first significant original discovery was his method of infinite series, which he tells us that he discovered in the Winter of 1664–1665 and applied the next Summer to evaluate the area under a hyperbola. It was this discovery that he was willing to confide to Leibniz; the discovery of the calculus he did not confide. It is now time to look at that discovery, to see how the method of infinite series fits in, and to see how Newton continued to amplify his methods and re-think them as the years went by.

At some point over the Summer of 1664, Newton began to immerse himself in mathematics, so much so that he seems to have thought of very little else. He made rapid progress through the second Latin edition of Descartes's *La Géométrie* with its many commentaries, and here he found Hudde's rule for finding repeated roots. By

[14] Descartes's solution of the Pappus problem is discussed in *Volume* 1, Section 13.4.

[15] See (Turnbull 1960), Vol. 1, p. 416.

Box 2. A note on centres of curvature.
In Figure 3.3, three circles C_1, C_2, and C_3 each touch the curve γ at the point P. But C_1 is more curved than γ, C_3 less curved, and C_2 (intermediate between the two) is as curved as γ. By making these ideas precise, Newton showed how to develop the theory of centres and radii of curvature algorithmically: the centre and radius of C_2 are said to be the *centre* and *radius of curvature* of γ at P.

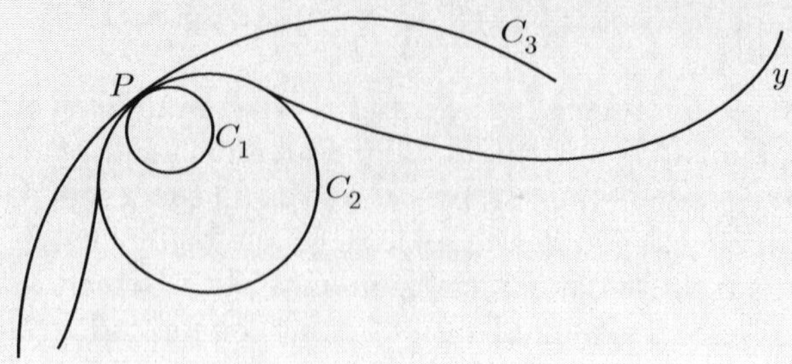

Figure 3.3. The curvature of a curve at a point

Figure 3.4 shows two of the points of greatest and least curvature on an ellipse; they are at the ends of the major and minor axes respectively.

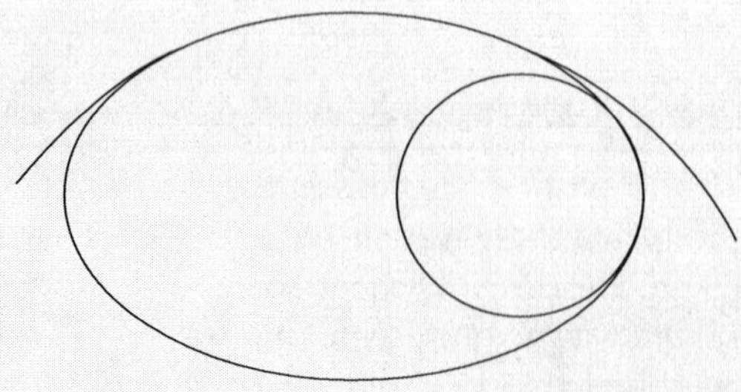

Figure 3.4. Extreme points of curvature on an ellipse

using it more flexibly than had its author, Newton obtained tangents to a great many curves, and found a pattern to his results in many cases. This led him to ask whether he could find not just one circle that touches a given curve at a given point (that is, meets it in two coincident points) but the 'best' one, the one that most closely approximates the curve at the given point (see Box 2). He found that he could, thus determining the centre of curvature of the curve at that point, and going well beyond anything he could have read. That Christmas he turned 22 years of age.

Now Newton investigated what he had learned from Wallis's method for finding areas, and in so doing discovered the method of infinite series. The *Epistola Posterior*, which we discussed in the previous section, is both reminiscent of Wallis and fresh with Newton. It was in the style of Wallis to investigate $y = (1 - x^2)^{1/2}$ by looking at other curves like $y = (1 - x^2)^{m/2}$ and guessing the answer by spotting a general pattern. So the idea of intercalation is something that Newton took over. But there is a difference. Wallis was after a definite area, a number, whereas Newton worked with polynomials

$$(1 - x^2)^{0/2},\ (1 - x^2)^{2/2},\ (1 - x^2)^{4/2},\ \text{etc.},$$

from which he obtained the polynomials:

$$x,\ x - \frac{1}{3}x^3,\ x - \frac{2}{3}x^3 + \frac{1}{5}x^5,\ \text{etc.}$$

His intercalations also gave him other algebraic expressions, in this case infinite series. For example, from the circle $y = (1 - x^2)^{1/2}$ he obtained

$$x - \frac{x^3}{2.3} - \frac{x^5}{5.8} - \frac{x^7}{7.16} - \frac{5x^9}{9.128} + \text{etc.}$$

(see the derivation in Section 4.2).

Newton's problem is more general because its answer is a variable expression, and in turn it made its contribution to revealing a greater pattern than any that Wallis had seen. Sometime in the Spring or Summer of 1665, Newton began to look closely at his results about areas as well as his results about tangents. Some things that were obvious to Newton when writing to Leibniz had already been so in 1665:

- the curve $y = (1 - x^2)^{0/2} = 1$ sits above an area of x
- the curve $y = (1 - x^2)^{2/2} = 1 - x^2$ sits above an area of $x - \frac{1}{3}x^3$
- the curve $y = (1 - x^2)^{4/2} = 1 - 2x^2 + x^4$ sits above an area of $x - \frac{2}{3}x^3 + \frac{1}{5}x^5$.

Newton spotted the pattern, as the following celebrated statement of his earliest findings in his *De Analysi* (On Analysis) shows.[16] This was written and shown to Barrow only a little while later, in Summer 1669. We shall concentrate on *De Analysi* because it was an almost polished version of his findings that Newton was willing to let some people read (see Figure 3.5).

Newton's rules for finding areas.

> The general method which I had devised some time ago for measuring the quantity of curves by an infinite series of terms you have, in the following, rather briefly explained than narrowly demonstrated. To the base AB of some curve AD let the ordinate BD be perpendicular and let AB be called x and BD y. Let again $a, b, c, \ldots$ be given quantities and m, n integers. Then
>
> *Rule 1*
>
> If $ax^{m/n} = y$, then will $(na/(m+n))x^{(m+n)/n}$ equal the area ABD. The matter will be evident by example.
>
> *Example 1:* If $x^2 (= 1 \times x^{2/1}) = y$, that is, if $a = n = 1$ and $m = 2$, then $(\frac{1}{3})x^3 = ABD$.
>
> *Example 2:* If $4\sqrt{x} (= 4x^{1/2}) = y$, then $\frac{8}{3}x^{3/2} (= \frac{8}{3}\sqrt{x^3}) = ABD$.

[16] Newton, *De Analysi per Aequationes Infinitas* was written in 1669 but first published in 1711; see *MPIN* II, 207–209, and in F&G 12.A2.

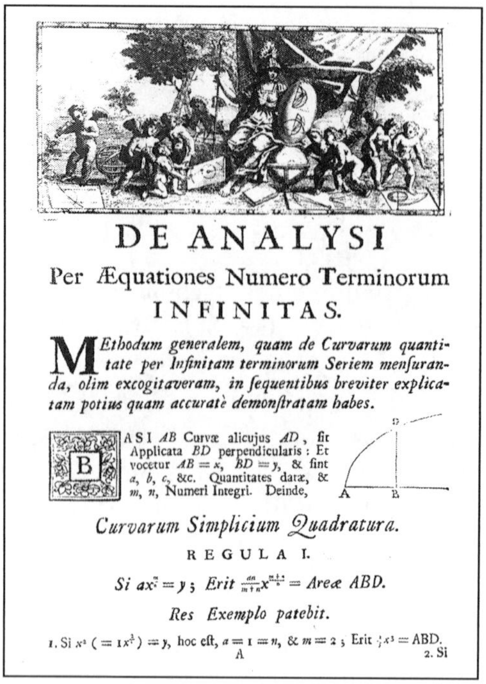

DE ANALYSI

Per Æquationes Numero Terminorum

INFINITAS.

Methodum generalem, quam de Curvarum quantitate per Infinitam terminorum Seriem mensurandam, olim excogitaveram, in sequentibus breviter explicatam potius quam accuratè demonstratam habes.

Basi *AB* Curvæ alicujus *AD*, sit Applicata *BD* perpendicularis: Et vocetur *AB* = *x*, *BD* = *y*, & sint *a*, *b*, *c*, &c. Quantitates datæ, & *m*, *n*, Numeri Integri. Deinde,

Curvarum Simplicium Quadratura.

REGULA I.

Si $ax^{\frac{m}{n}} = y$; *Erit* $\frac{an}{m+n}x^{\frac{m+n}{n}}$ = *Areæ ABD.*

Res Exemplo patebit.

1. Si $x^2 (= 1x^{\frac{2}{1}}) = y$, hoc est, $a = 1 = n$, & $m = 2$; Erit $\frac{1}{3}x^3$ = ABD.

A

2. Si

Figure 3.5. The opening page of Newton's *De Analysi* (1711)

Example 3: If $\sqrt[3]{x^5} = x^{5/3} = y$, then $\frac{3}{8}x^{8/3}(= \frac{3}{8}\sqrt[3]{x^8}) = ABD$.

Example 4: If $1/x^2 (= x^{-2}) = y$, that is, if $a = n = 1$ and $m = -2$, then $((1/-1)x^{-1}) = -x^{-1}(= -(1/x)) = \alpha BD$ infinitely extended in the direction of α: the computation sets its sign negative because it lies on the further side of the line BD.

Example 5: If $\frac{1}{\sqrt{x^3}}(= x^{-3/2}) = y$, then $(\frac{2}{-1})x^{-1/2} = -(2/\sqrt{x}) = BD\alpha$.

Example 6: If $(1/x)(= x^{-1}) = y$, then $(\frac{1}{0})x^{0/1} = (\frac{1}{0})x^0 = (\frac{1}{0}) \times 1 =$ infinity, just as the area of the hyperbola is on each side of the line BD.

Figure 3.6. Newton's 2nd rule for finding areas

Rule 2

If the value of y is compounded of several terms of that kind the area also will be compounded of the areas which arise separately from each of those terms (see Figure 3.6).

Let its first examples be these:

If $x^2+x^{3/2} = y$, then $\frac{1}{3}x^3+\frac{2}{5}x^{5/2} = ABD$. For if there be always $BF = x^2$ and $FD = x^{3/2}$, then by the preceding rule $\frac{1}{3}x^3$ = the surface AFB described by the line BF and $\frac{2}{5}x^{5/2} = AFD$ described by DF; and consequently $\frac{1}{3}x^3 + \frac{2}{5}x^{5/2}$ = the whole ABD.

Thus if $x^2 - x^{3/2} = y$, then $\frac{1}{3}x^3 - \frac{2}{5}x^{5/2} = ABD$:

and if $3x - 2x^2 + x^3 - 5x^4 = y$, then $\frac{3}{2}x^2 - \frac{2}{3}x^3 + \frac{1}{4}x^4 - x^5 = ABD$.

The content of Rule 1 is that the area under $y = ax^{m/n}$ is

$$\frac{na}{m+n}x^{(m+n)/n},$$

provided that m/n is not equal to -1, for, as Example 6 makes clear, the rule breaks down precisely in that case. Rule 2 is the simple extension of Rule 1 needed to deal with such examples as Wallis's problem produced. So it was easy for Newton to find certain areas — indeed he seems to have known Rule 1 since 1665 — and we have seen that Newton had a good grasp of how to find tangents. In that connection, we give a specific example below of what he could deal with quite generally by late 1665. What is exciting about it is that Newton used a new method, quite different from his earlier use of Hudde's rule. We shall discuss how Newton might have discovered his new method once we have come to grips with what it involves.

To find a tangent to $x^3 + y^3 - 3axy = 0$ (the folium of Descartes), Newton argued as follows (see Figure 3.7). (Here Newton used the symbol o to mean a very small quantity: it does not stand for zero.)

1. What is (x, y) at one instant will be $(x + po, y + qo)$ in the next, so

$$(x + po)^3 + (y + qo)^3 - 3a(x + po)(y + qo) = 0.$$

2. Expanding the brackets gives

$$x^3 + 3x^2po + 3x(po)^2 + (po)^3 + y^3 + 3y^2qo + 3y(qo)^2 + (qo)^3$$
$$-3axy - 3axqo - 3aypo - 3apoqo = 0.$$

3. The original equation is $x^3 + y^3 - 3axy = 0$, so we may deduce that the terms without o are together equal to 0. Cancel them and divide the rest by o, to obtain

$$3x^2p + 3xp^2o + p^3o^2 + 3y^2q + 3yq^2o + q^3o^2 - 3axq - 3ayp - 3apqo = 0.$$

This has the form

$$3x^2p + 3y^2q - 3axq - 3ayp + \text{terms in } o.$$

4. Drop the terms in o as being negligible, as indeed they are if o is very small, giving

$$3x^2p + 3y^2q - 3axq - 3ayp = 0.$$

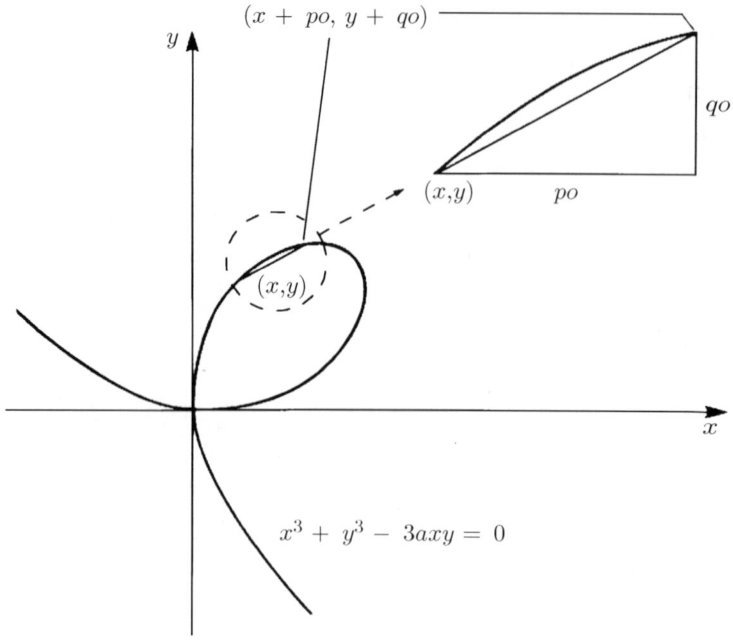

Figure 3.7. Finding a tangent to a folium: the chord closely approximates the tangent when o is very small

5. The slope of the tangent is always q/p, which in this case is

$$\frac{q}{p} = \frac{x^2 - ay}{ax - y^2}.$$

Why does this so-called o-method work? Because the slope of the chord joining (x, y) to $(x + po, y + qo)$ is q/p. When o is small enough to ignore, the chord is almost the same as the tangent. (We defer a more substantial criticism of Newton's approach to Section 4.2.)

In 1665 Newton wrote down only his rules and their results; proofs were generally lacking. In a further extract from *De Analysi*, Newton discussed how his Rule 1 might be proved.[17]

The example concerns a curve given by an equation (as yet unknown) between x and y (see Figure 3.8). The area underneath it is taken to be $z = \frac{2}{3}x^{3/2}$. After arguing much as before, Newton wrote:

> 'Conversely therefore if $x^{1/2} = y$, then will $\frac{2}{3}x^{3/2} = z$'.

Since this is the claim in Rule 1, it is worth sticking with it for a while. Newton argued that if the area z is given by $z = \frac{2}{3}x^{3/2}$ then the curve is $y = x^{1/2}$. The area determines the curve because the change in y as x changes is given by adding ys, and of course y determines z. So the converse must hold — that is, if $y = x^{1/2}$ then $z = \frac{2}{3}x^{3/2}$.

[17]See Whiteside, *MPIN* II, 243–245, and F&G 12.A4.

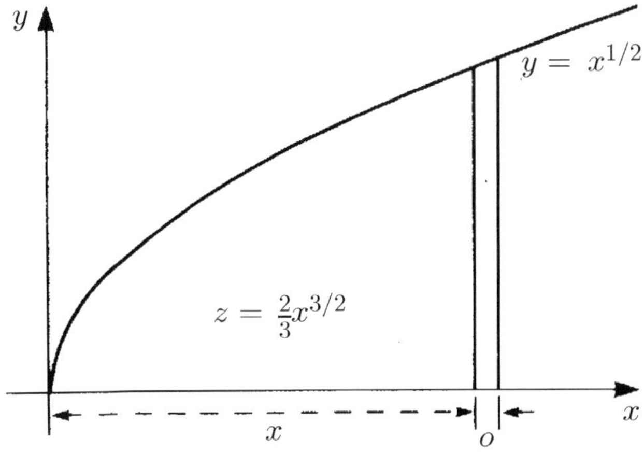

Figure 3.8. An infinitesimal increase in the area under a curve

Now this is an amazing result. To find an area, z, Newton argued that you applied the o-method to something else, the something else being a curve (an equation in y and x) whose *tangent* at x is y. He proved Rule 1 and argued that it is true in general — namely, to find an area is the opposite process or 'inverse' to finding a tangent.

Here, then, is Barrow's observation that there is a connection between area and tangency problems, but now raised to a vastly greater level of generality and clarity. As Newton had known for some years, but here remarked afresh,[18] whole lists of answers to area questions can now be written down. Every time you find a tangent, you also find an area under another curve by this same result — and given a curve, Newton's methods could often find the tangent at once. For example, he could find tangents to any curve given by a polynomial equation in x and y.

Newton went further, and based his very definition of quadrature on the idea that it is the inverse of finding tangents. In his view, to evaluate an area you would, eventually, look it up in a list of tangents derived from equations, and read off the area as the equation whose tangents you know (going from y to z, as it were). To this end he compiled long lists of this kind. So what for some people was (and would continue to be) a *theorem* that finding areas and tangents are inverse processes — the *Fundamental Theorem of the Calculus* — was taken by Newton as the very definition of quadrature. An example is given in Box 3.

Newton's o-method, and his way of regarding area questions as the opposite of tangency questions, are set down clearly in his *De Analysi*. He seems to have seen all along that finding tangents and finding areas are inverse processes. Perhaps he learned this from Barrow, perhaps he saw it just by looking at tables of results; the historical record is unclear. But we can say a little more about the origins of his o-method. By Summer 1665 his method for finding tangents to curves, originally based on his use of Hudde's rule, had become an algorithm resting on his observation of a crucial pattern in the results. Whatever doubts he might have had about the validity of his reasoning,

[18] See Whiteside, *MPIN* II, 245, and F&G 12.A4.

Box 3. Newton's use of the Fundamental Theorem of the Calculus. If we start with $y = x^{1/2}$, then we guess that the area under it is $z = \frac{2}{3}x^{3/2}$.

To check our guess, we take the curve $z = \frac{2}{3}x^{3/2}$ and find the slope of its tangent at x: the slope is $y = x^{1/2}$. This is the equation we started with, thus confirming our guess.

his algorithms dealt with the formal algebra of equations, and so were both easy and powerful to use.

In late 1665 he tackled the problem of finding tangents to mechanical curves. At that time this problem was generally approached by regarding the curves as curves of motion, but the results so obtained were not always correct. At first Newton made mistakes, too, but by thinking more carefully about the velocity of the generating point he found his way to a systematic method: his o-method. He found that this new method also enabled him to re-derive his old results about tangents to algebraic curves, in a conceptually clearer way than Hudde's rule, which he now dropped. He began, indeed, to think of all curves as generated by motion — preferring, if you wish, Roberval to Descartes. Newton seems to have come upon this idea on his own, although late in life he was prepared to speculate that he might have picked it up from Barrow's lectures.

To enable his algebraic algorithms to apply to mechanical curves he reintroduced his method of infinite series. In another passage in *De Analysi*[19] Newton started from the infinite series for $z = \arcsin x$, which he had found as a series in x, and now wrote down the series for x in powers of z (by what he called 'root extraction'). It was a great moment — the first time that something as useful as the series for the sine of an angle had been written down in the West.[20] Newton also wrote down the infinite series for cosine, and observed that its terms develop according to very simple rules.

From this and other remarks, as for instance on the cycloid,[21] we can see that Newton had obtained a uniform formal method for quadratures that was applicable to all known curves, whether algebraic or transcendental — or at least to those for which series expansions could be found.

His method was as general as this because it had one signal virtue over all previous methods. Suppose, for example, that it allowed Newton to find the tangent to a curve with equation $y = f(x)$ at the point (a, b), and that the slope of the tangent there is $f'(a)$, using more modern notation. Suppose that it also allowed him to find the tangent to a curve with equation $y = g(x)$ at the point (a, c), and that the slope of the tangent there is $g'(a)$. Then his method immediately implied that the tangent to a curve with equation $y = f(x) + g(x)$ at the point $(a, b + c)$ has slope $f'(a) + g'(a)$. Previous methods had no such simple way of yielding the answer, and so each problem about tangents had to be tackled afresh. Newton, one might say, had found the most

[19] See Whiteside, *MPIN* II, 237–239, and F&G 12.A3.

[20] More than two centuries earlier, the sine and cosine series had been discovered by the South Indian mathematician Mādhava, who seems to have been born around the middle of the 14th century. However, his work survives only in the form of a few verses recorded by some of his followers, and his discovery seems to have been made in a different way. See (Plofker 2009, Chapter 7).

[21] Newton, *De Analysi per Aequationes Infinitas*, 1669, in *MPIN* II, 237–239, F&G 12.A3.

efficacious way of defining where the tangent is at a point on a curve, one that allowed him to produce the sum rule: $(f+g)' = f' + g'$ (this is a particular case of his Rule 2).

Newton, however, did not publish his results. He finished an investigation into a method for finding tangents and curvatures on 13 November 1665, and then lost all interest in it.[22] Interest flickered in May and October of 1666, but then went out entirely for two years (during which time he was consumed with a passion for optics and the nature of light). Even his *De Analysi* was not published until late in his life (in 1711), although in 1669 he did 'leak' its contents a little. Others of his contemporaries had been loath to publish, notably Fermat and Roberval, but Newton carried this remoteness to extremes. This was ultimately to cause him much pain and difficulty when his claim to have discovered the calculus was disputed by the younger generation of Leibnizians.

It is time to leave Newton here and turn to the work of Leibniz. In the next chapter we shall investigate Newton's changing and increasingly sophisticated ideas about the foundations of the calculus, when a comparison with Leibniz is again instructive. But in 1666 there could be no comparison: in the entire world only Newton possessed a coherent package of new ways for dealing with area and tangency questions, a set of rules and methods that we can now call the *calculus*.

3.3 Leibniz and his calculus

Newton had only one equal in his lifetime, the remarkable Gottfried Wilhelm Leibniz. Although we have mentioned him briefly a few times already, now is the time to take a longer look. He was born in Leipzig, where his father was Professor of Moral Philosophy at the University, and attended the University when only 14 (which was not unusual then). He specialised in law, but earned his 'Habilitation' (the qualification that gave him the right to lecture, without pay, at a German university) with a lecture in the Philosophy Faculty on what he called 'The combinatorial art'. He applied for a tutorship in law to earn money, but was unsuccessful at Leipzig so he went instead to Altdorf, near Nuremberg, where he presented his finished doctoral thesis almost on arrival. He took his degree in February 1667, when only 20, and left to seek a career in the world.

Leibniz was amazingly quick — a polymath, and a generalist. He never held a position as a mathematician, and some aspects of his life can be explained by noticing that, once he failed in his attempts to get a research post at the Académie des Sciences in Paris, he had to make his living in the uncertain world of court and ducal politics. He was ambitious to learn everything, to catalogue everything that was known, to write or coordinate the production of a universal encyclopaedia, and to simplify discovery by creating an appropriately logical language. His thesis on the art of combinations was an exploration of the idea that all concepts are made up of combinations of simple concepts, together with some rules for enumerating the possible number of suitable combinations. He was a voracious correspondent, and over 15,000 of his letters still survive, which is just as well, for fate was to consign him from 1676 until 1714 to the rather minor position of Court Councillor in Hannover. This involved him first in engineering, then as a librarian, and finally as a diplomat. In none of these activities were the tasks in hand matched to his abilities, and his grand designs almost never

[22] See Whiteside, *MPIN*, I, 382.

Figure 3.9. Gottfried Wilhelm Leibniz (1646–1716)

succeeded. But in his spare time he became arguably the greatest theorist of logic and language since Aristotle, and a philosopher and mathematician of the first rank.

He hoped at one stage to obtain an academic post in Paris, where he lived between 1672 and 1676. Although he did not succeed in this, he promoted his ideas for a calculating machine and began to invent the calculus. He set out in earnest to learn mathematics in 1672, and managed to enlist Christiaan Huygens, then in Paris and at the height of his fame, to help him.

Huygens set him a most productive question: What is the sum to infinity of the following series?

$$\frac{1}{1}+\frac{1}{3}+\frac{1}{6}+\frac{1}{10}+\cdots .$$

Leibniz had already told Huygens that he could sometimes sum infinite sequences, by a neat method taking differences: To sum a series $b_1 + b_2 + b_3 + \cdots$, Leibniz tried to find a sequence $a_1, a_2, a_3, a_4, \ldots$, such that

$$b_1 = a_1 - a_2\ ,\ b_2 = a_2 - a_3\ ,\ \text{etc.}$$

He observed that

$$b_1 + b_2 + b_3 + \cdots + b_n = (a_1 - a_2) + (a_2 - a_3) + \cdots + (a_n - a_{n+1}) = a_1 - a_{n+1},$$

with most of the terms cancelling. If, for an infinite series, the term a_n tends to zero, then this sum is simply a_1, the first term. So the *sum* of the bs may be found by taking *differences* of two as.

This observation intrigued Huygens, whence his question (to which, of course, he knew the answer). Like a good supervisor, he was trying to lead Leibniz gradually into difficult, new mathematics. Leibniz's answer illuminated his method: Since

$$\frac{1}{1} = 2(1 - \tfrac{1}{2}), \quad \frac{1}{3} = 2(\tfrac{1}{2} - \tfrac{1}{3}), \quad \frac{1}{6} = 2(\tfrac{1}{3} - \tfrac{1}{4}), \quad \text{etc.,}$$

so

$$\frac{1}{1} + \frac{1}{3} + \frac{1}{6} + \frac{1}{10} + \cdots = 2\Big((1 - \tfrac{1}{2}) + (\tfrac{1}{2} - \tfrac{1}{3}) + (\tfrac{1}{3} - \tfrac{1}{4}) + \cdots\Big) = 2.$$

As Leibniz found out when he visited London in 1673, this result was not new, nor indeed was the technique — but Leibniz had discovered it himself. Moreover, in his mind it was but an illustration of a general principle: making sums and taking differences are inverse processes to one another. Put on his mettle by his rather embarrassing trip to London, which had shown him just how much he still had to learn, he pushed his insight far harder than had any of his contemporaries (except one).

He first applied it to geometry, by regarding an area as made up of many vertical lines (an idea that he had learned from Gregory of Saint-Vincent's *Opus Geometricum*). Here, the sum of the lines is the area, and the difference between successive lines (taken to be a tiny, fixed, distance apart) is a measure of the slope of the boundary curve (see Figure 3.10). This suggested to Leibniz that area and tangency questions might somehow be amenable to inverse processes.

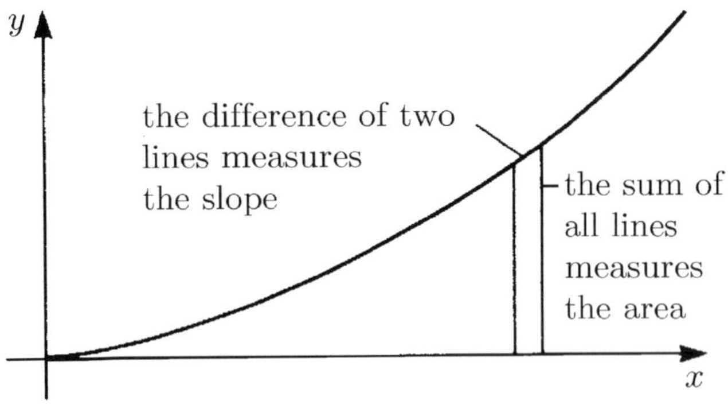

Figure 3.10. Leibniz on the connection between slopes and areas

Leibniz took Huygens' advice to read widely in the subject, and in the Spring of 1673 he came upon a passage in Pascal that he realised he understood far better than its author. Pascal, in his study of the circle, had noticed that the infinitesimal triangle BEF is similar to each of the triangles BAO and TBO (see Figure 3.11). Leibniz realised that this is true for a general curve and does not depend on the curve being a circle (see Figure 3.12).

This realisation was to lead him to many clever transformations between problems, as data from one triangle were transformed into data about another, culminating in his 'transmutation rule'. This was a rule for converting the quadrature of one curve to the quadrature of a second, which could be constructed from the first curve by means of its tangents.

In turn, the transmutation rule led to his famous infinite series for $\pi/4$, as follows. From the circle $y^2 = 2ax - x^2$ he obtained, by transmutation, the equation of a curve

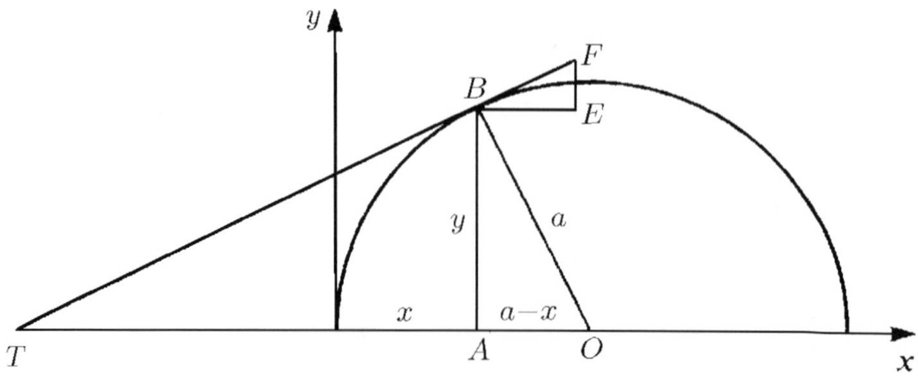

Figure 3.11. The characteristic triangle in Pascal's work

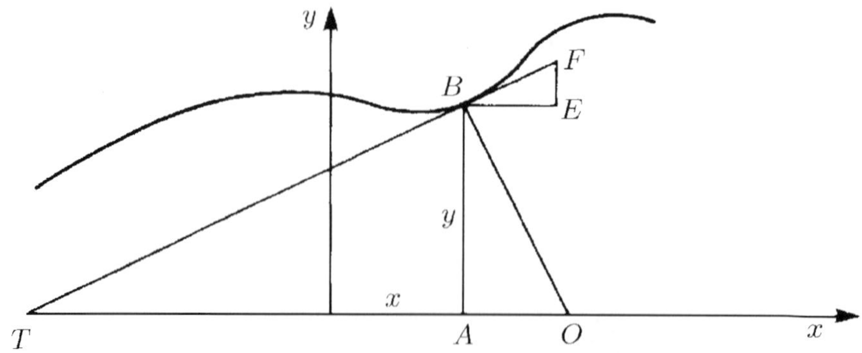

Figure 3.12. Leibniz's use of the infinitesimal triangle

called the *versiera*,

$$x = \frac{2az^2}{a^2 + z^2}.$$

His rule then gave him a connection between their areas, and he investigated this by expanding the expression in x as an infinite series, a technique that he had learned from Nicolaus Mercator's *Logarithmotechnia*.[23]

The *Logarithmotechnia*, which had appeared in 1668, was the first work to connected the hyperbola $y = \frac{1}{1+x}$ to the idea of logarithms. In it, Mercator began by dividing out the quotient by long division, which gave him

$$y = 1 - x + x^2 - x^3 + x^4 - \cdots, \text{etc.}$$

Then, by what has been called 'a very bald, loose use of indivisibles',[24] Mercator produced what is in effect a quadrature of that series term by term, to give

$$\log(1 + x) = x - \frac{1}{2}x^2 + \frac{1}{3}x^3 + \cdots, \text{ etc,}$$

[23]The Danish-born Nicolaus Mercator is not to be confused with the cartographer of the previous century, Gerardus Mercator.

[24]See (Whiteside 1961, 256).

This enlarged the concept of logarithm in two directions: logarithms could now be *defined* either as the area under a hyperbola or by Mercator's infinite series. However, only Newton saw the potential of that approach at first, and he was thoroughly alarmed by the publication of *Logarithmotechnia*; he had covered the same ground a couple of years previously but, of course, had not told anyone.

Leibniz found in this way that $\pi/4$ — the area of a quarter circle of unit radius — is

$$1 - \frac{1}{3} + \frac{1}{5} - \frac{1}{7} + \ldots,$$

a striking result.[25]

Again, although Leibniz did not know it, this result was not new in the West; James Gregory had found it in 1671. Nor was he the first to use transformations to solve area problems. But in Leibniz's mind it fitted into a general, abstract, rule-governed approach to all kinds of mathematical problems. When, in 1675, he coupled these ideas to his desire for a flexible and powerful symbolic language, indeed an *ars inveniendi* ('art of invention'), the result was to be the 'Leibnizian calculus'. We are in the fortunate position of almost being able to watch it happen, for Leibniz's working notes, made during the crucial period, 25 October to 11 November 1675, have survived.[26]

Here we consider a passage from the notes, dated 28 October, to which we have added a commentary. In his notes, Leibniz wrote omn.ℓ to mean all the lines of a figure, as Cavalieri had done before when calculating areas (see Section 2.2). The overline symbol that Leibniz used is to be read as brackets.

Leibniz's notation for the calculus.

To resume, $\frac{\ell}{a} = \frac{p}{\text{omn.}\ell} = y$, therefore $p = \frac{\overline{\text{omn.}\ell}}{\ell}$. Hence, $\text{omn.}y\frac{\ell}{a}$ does not mean the same thing as omn.y into omn.ℓ, nor yet y into omn.ℓ; for, since $p = \frac{y}{a}\ell$ or $\frac{\overline{\text{omn.}\ell}}{a}\ell$, it means the same thing as omn.ℓ multiplied by that one ℓ that corresponds with a certain p; hence, $\text{omn.}p = \text{omn.}\frac{\overline{\text{omn.}\ell}}{a}\ell$. Now I have otherwise proved $\text{omn.}p = \frac{y^2}{2}$, i.e., $= \frac{\overline{\text{omn.}\ell}^2}{2}$; therefore we have a theorem that to me seems admirable, and one that will be of great service to this new calculus, namely,

$$\frac{\overline{\text{omn.}\ell}^2}{2} = \text{omn.}\overline{\overline{\text{omn.}\ell}\frac{\ell}{a}},$$

whatever ℓ may be; that is, if all the ℓs are multiplied by their last, and so on as often as it can be done, the sum of all these products will be equal to half the sum of the squares, of which the sides are the sum of the ℓs or all the ℓs. This is a very fine theorem, and one that is not at all obvious.

Leibniz's equation above is possibly alarming, and is not immediately intelligible, but he proceeded to write it in a more lasting and recognisable form in a couple of paragraphs, as we shall see. Plainly we are catching Leibniz in the middle of working something out, and these pages were not intended for publication. We can see that

[25] Remarkably, this too had been discovered before, by the 14th-century Indian mathematician Mādhava, a fact that Leibniz could not possibly have known.

[26] See (Child 1920, 80–84), from which the next extract comes, and F&G 13.A1.

at this stage his notation for the area of a curve was Cavalieri's: he wrote omn.ℓ for omnes ℓ or 'all ℓ', for all the lines ℓ of the figure. After a tricky start, there is a theorem, which says $\frac{\text{omn.}\ell}{a} = \text{omn.omn.}\ell\frac{\ell}{a}$, whatever ℓ may be, and which he liked presumably because of its generality. Another such theorem is $\text{omn.}x\ell = x\text{omn.}\ell - \text{omn.omn}\ell$. Then a pleasant surprise: he writes $\int$ for omn. — the integral sign appears for the first time in the history of the calculus (see Figure 3.13).

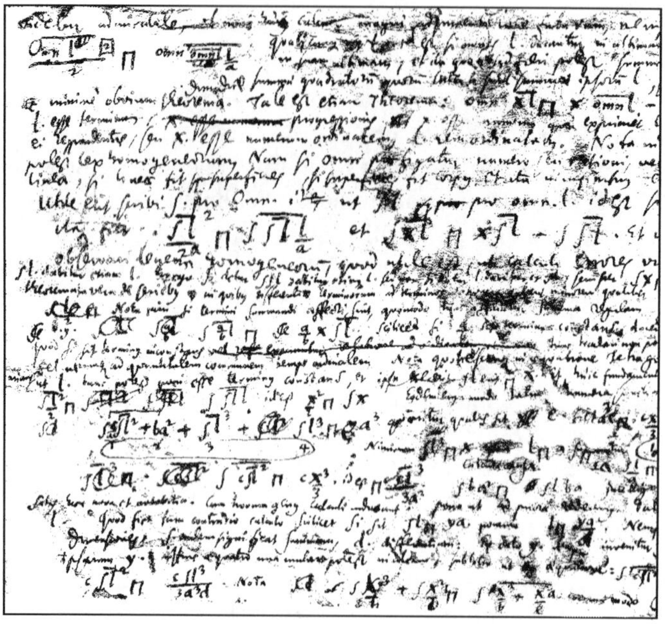

Figure 3.13. The first appearance of the integral sign

Another theorem of the same kind is: $\text{omn.}x\ell = x\text{omn.}\ell - \text{omn.omn.}\ell$, where ℓ is taken to be a term of a progression, and x is the number which expresses the position or order of the ℓ corresponding to it; or x is the ordinal number and ℓ is the ordered thing.

N.B. In these calculations a law governing things of the same kind can be noted; for, if omn. is prefixed to a number or ratio, or to something indefinitely small, then a line is produced, also if to a line, then a surface, or if to a surface, then a solid; and so on to infinity for higher dimensions.

It will be useful to write $\int$ for omn., so that $\int \ell = \text{omn.}\ell$, or the sum of the ℓs. Thus,

$$\frac{\int \overline{\ell}^2}{2} = \int \overline{\int \overline{\ell}\frac{\ell}{a}}$$

and $\int \overline{x\ell} = x\int \overline{\ell} - \int\int \ell$.

From this it will appear that a law of things of the same kind should always be noted, as it is useful in obviating errors of calculation.

N.B. If $\int \ell$ is given analytically, then ℓ is also given; therefore if $\int\int \ell$ is given, so also is ℓ; but if ℓ is given, $\int \ell$ is not given as well. In all cases $\int x = x^2/2$.

N.B. All these theorems are true for series in which the differences of the terms bear to the terms themselves a ratio that is less than any assignable quantity.

$$\int x^2 = \frac{x^3}{3}.$$

Now note that if the terms are affected, the sum is also affected in the same way, such being a general rule; for example, $\int \overline{\frac{a}{b}\ell} = \frac{a}{b} \times \int \ell$, that is to say, if $\frac{a}{b}$ is a constant term, it is to be multiplied by the maximum ordinal; but if it is not a constant term, then it is impossible to deal with it, unless it can be reduced to terms in ℓ, or whenever it can be reduced to a common quantity, such as an ordinal. ...

I propose to return to former considerations.

Given ℓ, and its relation to x, to find $\int \ell$. This is to be obtained from the contrary calculus, that is to say, suppose that $\int \ell = ya$. Let $\ell = ya/d$; then just as $\int$ will increase, so d will diminish the dimensions. But $\int$ means a sum, and d a difference. From the given y, we can always find y/d or ℓ, that is, the difference of the ys. Hence one equation may be transformed into the other; just as from the equation

$$\int \overline{c \int \overline{\ell 2}} = \frac{c \int \overline{\ell 3}}{3a^3}$$

we can obtain the equation

$$c \int \overline{\ell 2} = \frac{c \int \overline{\ell 3}}{3a^3 d}.$$

N.B.

$$\int \frac{x^3}{b} + \int \frac{x^2 a}{e} = \int \overline{\frac{x^3}{b} + \frac{x^2 a}{e}}.$$

And similarly,

$$\frac{x^3}{db} + \frac{x^2 a}{de} = \frac{\frac{x^3}{b} + \frac{x^2 a}{e}}{d}.$$

After he introduced the integral sign, Leibniz recognised that if you know $\int \ell$ then you know ℓ, but not conversely. Then it becomes obscure, until we reach 'Given ℓ, and its relation to x, to find $\int \ell$.' Leibniz was looking for a general approach so he passed from sums to differences, writing d for differencing on the bottom line, as in y/d, because it lowers dimensions. It is plainly his view that performing $1/d$ is the inverse of performing $\int$.

This is quite startling. The notation is streamlined in conformity with his liking for a logical symbolism. Moreover it is to obey certain rules, such as (on writing y for ℓ)

$$\int xy = x \int y - \int \int y,$$

which is how his transmutation rule looks in this example and in his new notation. We can see this as a special case of the formula for integrating by parts.

The idea of summation is allowed to dictate the choice of symbol, $\int$ is the long script *s* for summa ('sum'), and the inverse process of differencing is similarly denoted by *d*. At first he put the *d* below what it acted upon — looking as though it were dividing it — since this location seemed to suggest the lowering of the dimensions involved, from areas to lines. This contrasted nicely with the $\int$ sign which raises dimensions, going from lines to areas. But within two weeks (on 11 November) he dropped the dimensional overtones of the symbolism, writing *d* in front of what it acted upon: dx and dy entered the calculus. It is remarkable that the two principal pieces of notation that we still use in the modern calculus (*d* and $\int$) were introduced within three weeks of each other, in the latter part of 1675.

Leibniz's notation is astonishingly good at its designated task. With it, we can apply operations or processes to equations and variables, and find areas and tangents by mere calculation. For example, after a false start, Leibniz spotted how to differentiate a product: $d(uv) = du.v + u.dv$. (Notice that if we set $u = x$ and $v = \int y$ this rule says that $d(x \int y) = \int y + xy$, so if we integrate both sides we get the transmutation rule above. Leibniz did not notice results like this until later.)

Geometry could now be left behind and the algebraic methods that were making their way further and further into our story could begin to occupy centre stage. For the first time, we can speak of *integration* — to briefly and anachronistically use a term later introduced by Johann Bernoulli in 1691 — instead of finding areas, and *differentiation* instead of finding tangents. Each process is both geometrical and algebraic, with its own symbols and rules of operation, yoked together as inverse processes.

It was to be a long time before all these lessons were to be drawn, and we shall look at stages in this process later. It is easy to see why it took so long. For a start, Leibniz did not publish his discoveries until 1684. Moreover, the approach was still in some basic ways flawed. It is worth listing two of these difficulties.

- The *d*-operator took successive differences, but between what? It was not clear that one could speak of one ordinate and the next as if they came in sequence, nor was it clear what an infinitely small quantity, a *differential*, actually was.

- Consequently, many of the earlier logical difficulties that mathematicians had encountered passed over to the calculus. Can a curve be thought of as a polygon with infinitely many sides, for example, or does such talk make no sense?

These difficulties were not solved by the slick notation, but were merely postponed until the geometry raised them elsewhere. For example, just as one can take differences of differences, so one can take successive differentials, such as ddx. But while it was clear that these objects are even smaller than differentials, it was not clear that they were negligible. Experience was necessary before one could decide whether to include or exclude them in any given case.

That said, Leibniz's achievements were immense, as the ubiquity of his notation today attests. So it is appropriate to conclude this chapter by comparing it with Newton's (also in its unpublished form around 1675).[27]

---

[27] A fuller comparison between the calculi of Newton and Leibniz appears in Section 4.5.

Both versions of the calculus were algebraic; they applied processes to formulas in which variables are related by equations. In this they consummated and made explicit what was only covert, or at least not central, in the work of their predecessors.

Newton's approach was more geometrical, his fundamental quantity being that of a variable moving with the passage of time. He made frequent use of figures and appealed to intuition about motion; Leibniz tried to avoid that and placed more emphasis on formal rules.

Leibniz was more arithmetical, less involved on paper with intuition about the basic processes of integration and differentiation. It is no accident that Newton was to write his *Principia Mathematica*, the great bible of natural philosophy, and that Leibniz was to forge the language of the calculus.

Lastly, there is no independent theory of the integral in Newton's calculus; it is *defined* to be anti-differentiation. In Leibniz's calculus, integration is an autonomous process which is *shown* to be the opposite of differentiation.

However, it would be too simplistic, and wildly anachronistic, to see Leibniz as the pure and Newton the applied mathematician, for Newton was a great geometer and Leibniz was too universal for these restrictive classifications. But it proved impossible to do mathematics without working under their influence, and when we look at the subsequent developments we shall see that the Leibnizian style was to dominate the study of the calculus itself, while for the ensuing one hundred years the Newtonian style led the study of what else the calculus could do.

3.4 Further reading

Aiton, E.J. 1985. *Leibniz: A Biography*, Adam Hilger. A thorough study, covering much of Leibniz's work with considerable emphasis on his mathematics.

Cohen, I.B. and Smith, G.E. (eds.) 2002. *The Cambridge Companion to Newton*, Cambridge University Press. A variety of essays on a variety of aspects of Newton's life and work. More demanding than *Let Newton Be!* (see below), but going more deeply into the many issues it explores.

Dry, S. 2014. *The Newton Papers: The Strange and True Odyssey of Isaac Newton's Manuscripts*, Oxford University Press. A fascinating glimpse into the afterlife of documents and the implications for our study of the past.

Fara, P. 2002. *Newton, the Making of Genius*, Macmillan. Not a conventional history, but an exploration of how Newton has been presented down the centuries, and a stimulating illustration of the nature of history and how history gets written.

Fauvel, J., Flood, R., Shortland, M., and Wilson, R. (eds.) 1988. *Let Newton be!: A New Perspective on his Life and Works*, Oxford University Press. A delightful, well-illustrated, and informative exploration of many aspects of Newton, written at the level of this book.

Gleick, J. 2003. *Isaac Newton*, Fourth estate. A highly readable biography that looks at both the scientific discoveries and the life of this remarkable man.

Iliffe, R. 2007. *Newton: A Very Short Introduction*, Oxford University Press. A brief, but accurate, introduction to the life and work of Isaac Newton.

MacDonald Ross, G. 1984. *Leibniz*, Oxford University Press. This is still the best quick introduction to Leibniz, lucidly compressing a great deal into only 120 pages.

Westfall, R.S. 1983. *Never at Rest: A Biography of Isaac Newton*, Cambridge University Press. This book has been the definitive biography of Newton for a generation. It is remarkable not just for its discussion of many particular topics (the calculus, optics, dynamics, the Mint) but also for its dramatic flow.

Whiteside, D.T. (ed.) 1967–1981. *The Mathematical Papers of Isaac Newton* (abbreviated here to *MPIN*), Cambridge University Press. A mine of information on many more topics than Newton alone.

4

The Development of the Calculus

Introduction

In the 17th century, problems concerning tangents and areas were tackled by a variety of methods. Of these, by far the most powerful were the methods of the calculus discovered by Newton and Leibniz, as we described in Chapter 3. These methods generally made it easy to find the tangent to a given curve at a given point, and in this way the new calculus made some former problems straightforward to solve.

What then emerged to occupy the attention of mathematicians was the inverse problem: Given some property of the tangent at every point, find the curve. Such problems arose in many different variants, some of great utility, and so they were to be taken up enthusiastically throughout the early 18th century on both practical and theoretical grounds. At their simplest, inverse tangent problems were solved by an appeal to the Fundamental Theorem of the Calculus, so they are natural generalisations of questions about integration. But mathematician soon found inverse tangent problems that were not so simple, and in Section 4.1 we shall see how this happened and how it was dealt with.

Then in Section 4.2 we consider how Newton developed his method of infinite series and coupled it to his version of the Fundamental Theorem of the Calculus, thus obtaining, in his opinion, a perfectly general method for approaching any problem in the calculus. In Section 4.3 we look at his investigations of the fundamental principles of the calculus: his so-called o-method, his later ideas of prime and ultimate ratios, and his ideas about limits that he set out in his *Principia Mathematica*.

In Section 4.4 we turn to look at Leibniz's mature calculus, as he presented it in his hugely influential paper of 1684, that was to shape the Continental reception of the calculus for two generations. This chapter concludes with a comparison of the achievements and approaches of Newton and Leibniz that deepens the one with which we concluded the previous chapter.

4.1 Inverse tangent problems

The first person known to have formulated an inverse tangent problem was Florimond Debeaune. Debeaune was a wealthy member of the nobility in his home town of Blois, a hundred miles south-west of Paris, where he was born in 1601 and where he became a counsellor at the Court of Justice. He earned a reputation as a high-quality lens grinder, and in 1639 Descartes wrote to him to ask him to design a machine that would make hyperbolic lenses. The project failed, but the two men remained in touch, and Debeaune went on to write his *Notes Briéves* (Brief Notes), which were published in 1649 in the first Latin edition of Descartes's *La Géométrie*. In this work he showed that the equations $y^2 = xy+bx$, $y^2 = -dy+bx$, and $y^2 = bx-x^2$ represent a hyperbola, a parabola, and an ellipse, respectively. He died in 1652.

Debeaune was led to propose his inverse tangent problem in 1638 as a result of his study of Descartes's *La Géométrie*, which had been published the previous year — so soon did Descartes's ideas begin to transform geometry. Debeaune raised it as one of four problems that he presented to the mathematical community; it has come down to us in the form of a letter to Roberval. A little later, in March 1639, Debeaune explained to Mersenne why he was interested in the problem.[1]

> As for my curves, I don't pretend to prove by their means that a quadruple weight is necessary to raise the sound of a string by an octave, for I have a clear and simple proof of that, as of several other questions you have asked me. But I do need the curves, first of all to prove that a heavy body when suspended makes its oscillations through small arcs in the same time as through large ones, and likewise that the strings of a lute, or something similar, when taut make their oscillations in the same time when they are large and when they are small. For, while these are things about which one can experiment, and for the strings of the lute they are quite obvious because their sound has the same tone when their emotion is great and when it is small, nonetheless the thing being proved geometrically it would be of no little help with other speculations.

Debeaune was interested in explaining mathematically why a plucked string vibrates as it does and, specifically, in explaining why the frequency with which the string vibrates is independent of the force with which it is struck. This tells us that he identified frequency with the pitch of the note, and the size (or amplitude) of the vibration with its volume, connections that are familiar to every musician. Mersenne would have been a good person to consult, because he was interested in giving just such a physically based account of the nature of sound, and had conducted experiments on the subject, and also because he would have ensured that Debeaune's question was disseminated widely.

All this is very interesting, but it does not make Debeaune's problem any easier to solve. Indeed, it does not seem at all easy to connect his problem with the motion of a vibrating string. It was Debeaune's view, as he wrote to Mersenne, that 'It is not possible to acquire any solid knowledge of physical nature without geometry, and the best geometry consists of analysis, indeed without it, it is quite imperfect.' (By 'analysis' Debeaune meant 'taking apart', the then-customary use of the term.) But he did not go on to explain how his inverse tangent problem would help him in studying nature, and it is probable that he viewed it as a warm-up exercise, typical of the kind of problem that one might have to solve if one were to go on to give a mathematical analysis of motion.

[1] Debeaune to Roberval, in Mersenne, *Correspondance* VIII, 348, in F&G 11.B1(b).

Debeaune's problem. Our first task is to understand what Debeaune's problem says, and to see what makes it an inverse tangent problem. Debeaune stated it as follows in a letter to Roberval (see Figure 4.1).[2]

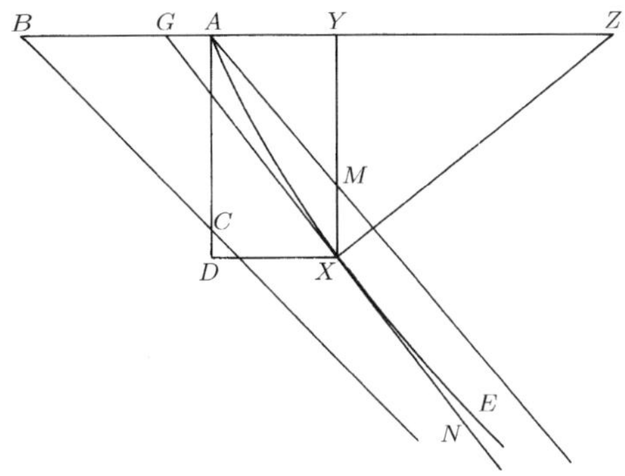

Figure 4.1. Debeaune's problem

> Let there be a curve AXE whose vertex is A, axis AYZ, and the property of this curve is that, having taken any point on it you wish, say X, from which the line XY is drawn as a perpendicular ordinate to the axis, and having taken the tangent GXN through the same point X, and extended the perpendicular XZ to it at X until it meets the axis, there will be the same ratio of ZY to YX as a given line, like AB, has to the line $YX - AY$.

We are given an axis, AYZ, and seek a curve, AXE, with some special property. So we draw the axis and the curve, and draw the tangent, GXN, at some point, X, on the curve. We erect the perpendicular, XY, as shown. To understand the problem, we locate the line segments ZY, YX, and AY. In addition, we need a line segment, AB, to play the role of a unit of length. Now we can follow Debeaune's statement of his problem. We are asked to find the curve, AXE, with the property that

$$\frac{ZY}{YX} = \frac{AB}{YX - AY}\,.$$

From this formulation, we can see that the problem is an inverse tangent one: we are required to find a curve given the above general property of the slope (ZY/YX) of its tangents (note that, as with Descartes's work, the slope of the tangent is captured by giving the ratio that measures the slope of the normal at X).

In what form might Debeaune have expected to get an answer? Probably not as an equation in some system of coordinates, as Descartes's ideas about curves were more complicated than that. It is more likely that Debeaune would have expected an answer in the form of a recipe for constructing points on the curve geometrically, but in the event he was to be unlucky, for no-one in his lifetime would be able to answer his question satisfactorily. Roberval was able to show that the solution curve has an asymptote (the line through B drawn at 45° to the axis), but that was all. In October 1638 Descartes gave his solution in the form of a mechanical description of how the curve might be drawn approximately, which was sufficient to confirm Roberval's result. Descartes, however, gave only an approximate construction because he was satisfied

[2] Debeaune to Mersenne, in Mersenne, *Correspondance* VIII, 142–143, in F&G 11.B1(a).

that the curve was what he called a 'mechanical curve'. We shall continue the story of attempts on this problem below, when we look at Leibniz's response to it.

So here is a problem that is easy to state but was very difficult to solve. The difficulty inherent in inverse tangent problems, which was often unexpected, was to become yet another factor that pushed people away from the geometrical type of reasoning (in which the answer is a curve given by a construction) and towards an algebraic style (in which the answer may be given as an equation). The contributions of the calculus were to be two-fold: it provided a formalism in which inverse tangent problems could be written down; and it provided a set of rules for manipulating the problem symbolically until the problem could (quite often) be solved, at least in the sense that the solution curve could be described via equations or formulas. This success was to become quite marked in the contemporary study of physical and astronomical problems where, as we shall see, such problems often arose naturally.

Other inverse tangent problems. Rather than make any attempt at completeness, we present a few physical problems that were to lend themselves to the inverse tangent formulation. Look them over and see whether anything strikes you about the circumstances in which they came to be solved.

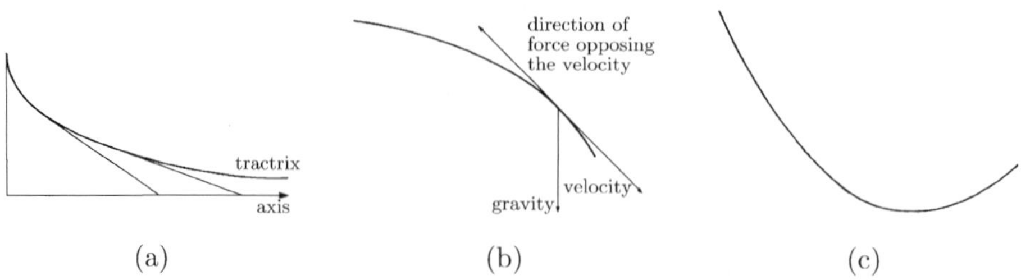

Figure 4.2. (a) A tractrix; (b) a trajectory; (c) a catenary

- In the 1670s Claude Perrault, who is generally remembered as the architect of the east wing of the Louvre Palace in Paris, asked this question: If you walk along a straight line dragging a heavy weight behind you (which is not itself on the line) along what path does the weight travel? The answer is a curve called the *tractrix* (see Figure 4.2(a)).[3] This geometrical curve was known to Newton and Leibniz in its own right (but was not published by them as a solution to this problem) and later to Huygens.

- In Newton's *Principia Mathematica* (1687) certain paths of particles were shown to arise from various types of force directed towards a central point. Subsequently, paths of particles moving under gravity and encountering various forms of air resistance were studied, by Newton and others. In such problems, the instantaneous direction in which the particle is accelerating is known, and the path is sought (see Figure 4.2(b)).

- What is the shape of a chain fastened at two points and hanging under its own weight? This was an old problem, and for a long time all that was known was

[3]From the Latin 'trahere', to drag, from which we get the word 'tractor'.

Huygens' discovery that the shape is not that of a parabola, as Galileo had claimed. The problem was solved by Leibniz, Jakob Bernoulli, Johann Bernoulli, and others in 1690. The solution curve is called a *catenary* — from the Latin for a chain (see Figure 4.2(c)). We shall return to this curve later.

The dates of these problems suggest that without the calculus, first published in the 1680s, they were found to be very difficult, but that when it became available some progress could be made on a broad front. This, in outline, was the case — but it was not quite as simple as that, as we shall see.

Leibniz's approach to Debeaune's problem. Let us return to Debeaune's problem, and see how Leibniz was to deal with it. We are fortunate to have Leibniz's original notes, from which we learn that he took up the problem in July 1676, when his calculus was not yet a year old.[4] They tell us about the personalities of Leibniz and Descartes, and give us a sense of the mathematics involved. It is not an easy passage, but it is a very instructive one.

Leibniz on Debeaune's problem.

> In the third volume of the *Correspondence* of Descartes, I see that he believed that Fermat's method of Maxima and Minima is not universal: for he thinks that it will not serve to find the tangent to a curve, of which the property is that the lines drawn from any point on it to four given points are together equal to a given straight line.
>
>> Descartes to Debeaune: I do not believe that it is in general possible to find the converse to my rule of tangents, nor of that which M. Fermat uses, although in many cases the application of his is more easy than mine; but one may deduce from it a posteriori theorems that apply to all curved lines that are expressed by an equation, in which one of the quantities, x or y, has no more than two dimensions, even if the other had a thousand. There is indeed another method that is more general and a priori — namely, by the intersection of two tangents, which should always intersect between the two points at which they touch the curve, as near one another as you can imagine; for in considering what the curve ought to be, in order that this intersection may occur between the two points, and not on this side or on that, the construction for it may be found. But there are so many different ways, and I have practised them so little, that I should not know how to give a fair account of them.
>
> Descartes speaks with a little too much presumption about posterity; he says that his rule for resolving in general all problems on solids has been without comparison the most difficult to find of all things which have been discovered in geometry up to the present, and one which will possibly remain so after centuries, 'unless I take upon myself the

[4] See (Child 1920, 116–122) and F&G 13.A2.

trouble of finding others' (as if several centuries would not be capable of producing a man able to do something that would be of greater moment).

Plainly, Leibniz thought that Descartes was rather arrogant. Leibniz said of him that he 'speaks with a little too much presumption', meaning that Descartes gave himself airs over the difficulty of the problems he had solved, and he claimed to have solved problems which, Leibniz implies, he had not. But notice that Leibniz also claimed to have solved problems that had baffled others — with what accuracy we shall see shortly.

We can reach quite a shrewd judgement as to whether Leibniz did indeed use the calculus to tackle Debeaune's problem — and whether he solved it — by reading not the mathematics but the commentary that Leibniz wrote for his own benefit. It is clear from the symbolism alone that he used the calculus. But did he solve the problem? You should read the following text over once, to get a sense of his argument, and then in more depth to understand the mathematics involved.

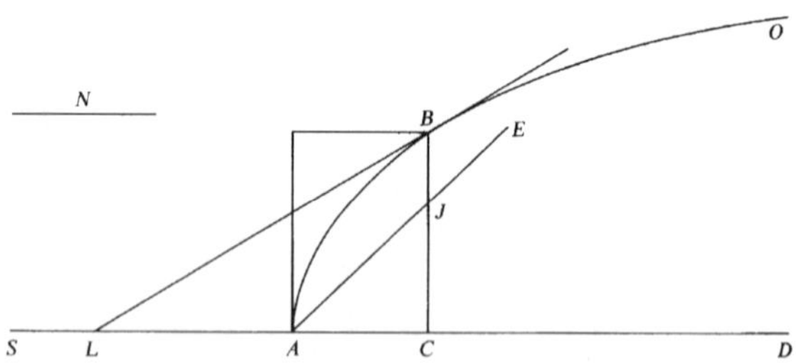

Figure 4.3. Leibniz's investigation of Debeaune's problem (1)

Leibniz shortly continued (a bar over an expression is Leibniz's notation for brackets):

The problem on the inverse method of tangents, which Descartes says he has solved: EAD is an angle of 45 degrees. ABO is a curve, BL a tangent to it; and BC, the ordinate, is to CL as N is to BJ. Then

$$CL = \frac{BC = ny}{BJ = y - x}, \quad CL = t,$$

hence

$$t = \frac{ny}{y-x}, \quad \frac{n}{t} = \frac{y-x}{y} = 1 - \frac{x}{y},$$

hence

$$\frac{x}{y} = \frac{t-n}{t}; \quad \text{but} \quad \frac{t}{y} = \frac{\overline{dx}}{dy};$$

therefore

$$\frac{\overline{dx}}{dy} = \frac{n}{y-x}, \quad \text{or } \overline{dx}y - x\overline{dx} = \overline{dy}n;$$

hence

$$\int \overline{dxy} - \int x\overline{dx} = -n \int dy.$$

> Now, $\int \overline{dy} = y$, and $\int \overline{\overline{xdx}} = x^2/2$, and $\int \overline{\overline{dxy}}$ is equal to the area $ACBA$, and the curve is sought in which the area $ACBA$ is equal to $(x^2/2)+ny = (AC^2/2) + nBC$.
>
> Let this $x^2/2$, i.e., the triangle ACJ be cut off from the area, then the remainder $AJBA$ should be equal to the rectangle ny.
>
> The line that Debeaune proposed to Descartes for investigation reduces to this, that if BC is an asymptote to the curve, BA the axis, A the vertex, AB, BC, fixed lines, for BAC is a right angle.

We can see that Leibniz has done three things. He has translated Debeaune's problem into the language of the calculus; gone a long way to solving it (although he has left the answer as an integral or an area); and given a geometrical interpretation of the final calculus expression. Now we shall see what he thought had still to be done (see Figure 4.4).

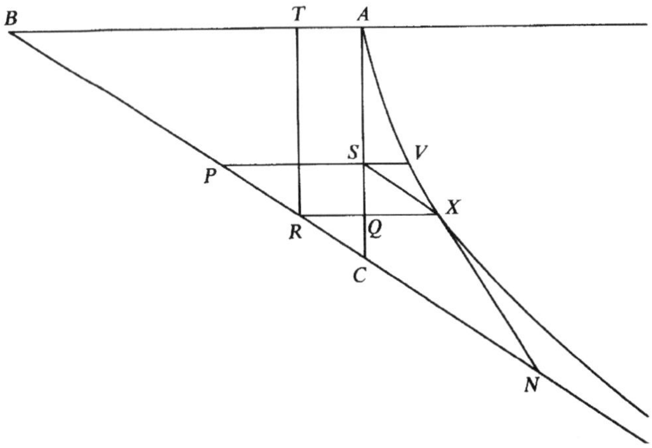

Figure 4.4. Leibniz's investigation of Debeaune's problem (2)

> Let RX be an ordinate, XN a tangent, then RN is always to be constant and equal to BC; required the nature of the curve.
>
> This is how I think it should be done. Let PV be another ordinate, differing from the other one RX by a straight line VS, found by drawing XS parallel to RN; then the triangles SVX, RXN are similar,
>
> $$RN = t = c, \text{ a constant, } RX = y, \ SV = dy,$$
>
> and therefore
>
> $$\frac{\overline{dy}}{dx} = \frac{y}{t = c};$$
>
> hence
>
> $$cy = \int \overline{\overline{ydx}} \text{ or } c\overline{dy} = \overline{ydx}.$$
>
> If AQ or $TR = z$, and $AC = f$, while $BC = a$; then,
>
> $$\frac{AC}{BC} = \frac{f}{a} = \frac{TR}{BR} = \frac{z}{x};$$
>
> and thus $x = az/f$.

If $\overline{dx}$ is constant, then $\overline{dz}$ is also constant. Hence

$$cdy = \frac{a}{f}y\overline{dz}, \text{ or } cy = \frac{a}{f}\int \overline{\overline{ydz}},$$

and $cy\overline{dy} = \frac{a}{f}y^2\overline{dz}$, therefore $c\frac{y^2}{2} = \frac{a}{f}\int y^2\overline{dz}$. Hence we have both the area of the figure and the moment to a certain extent (for something must be added on account of the obliquity); also $cz\overline{dy} = \frac{a}{f}yz\overline{dz}$, and therefore $c\int z\overline{dy} = \frac{a}{f}\int \overline{\overline{ydz}}$.

Also $\frac{\overline{cdy}}{y} = \frac{a}{f}dz$, and hence, $c\int \frac{\overline{dy}}{y} = \frac{a}{f}z$. Now, unless I am greatly mistaken, $\int \frac{\overline{dy}}{y}$ is in our power. The whole matter reduces to this, we must find the curve in which the ordinate is such that it is equal to the differences of the ordinates divided by the abscissae, and then find the quadrature of that figure. $\overline{d\sqrt{ay}} = \frac{1}{\sqrt{ay}}$.

Figures of this kind, in which the ordinates are $dy/y, dy/y^2, dy/y^3$, are to be sought in the same way as I have obtained those whose ordinates are $ydy, y^2\overline{dy}$, etc. Now $w/a = \overline{dy}/y$, and since $\overline{dy}$ may be taken to be constant and equal to β, therefore the curve, in which $w/a = \overline{dy/y}$, will give $wy = \alpha\beta$, which would be a hyperbola. Hence the figure, in which $dy/y = z$, is a hyperbola, no matter how you express y, and if y is expressed by ϕ^2 we have $dy = 2\phi$, and $\frac{2\phi}{\phi^2} = \frac{2}{\phi}$. Now, $c\int \frac{dy}{y} = \frac{a}{f}z$, and therefore $\frac{fc}{a}\int \frac{dy}{y} = z$, which thus appertains to a logarithm.

Thus we have solved all the problems on the inverse method of tangents, which occur in Volume 3 of the *Correspondence* of Descartes, of which he solved one himself; but the solution is not given; the other he tried to solve but could not, stating that it was an irregular line, which in any case was not in human power, nay not within the power of the angels unless the art of describing it is determined by some other means.

Even on our first reading, Leibniz's own comments tell us a great deal. It is almost as if he were talking to himself and we are privileged to be listening in. He set up some equations, presumably transcribing the problem into his own notation, and then he said what the problem reduces to, adding: 'This is how I think it should be done'. So he was pursuing an idea. More calculations follow, and then he wrote: 'Now, unless I am greatly mistaken, [something] is in our power'. He claimed to have reduced the problem to another one. But at the end he wrote only that the answer 'thus appertains to a logarithm'. 'Appertains' is a bit of a weasel word; Leibniz had not reached something that he regarded as an adequate solution to the problem.

So we see that Leibniz claimed to have solved both of the problems that had baffled Descartes. But as a 20th-century editor of Leibniz's notebooks, J. M. Child, remarked at this point: 'He has not solved either of them'.[5] Leibniz had not only made one of his many slips — he was a notoriously careless worker — but when he started again on the problem his description of the solution was certainly rather vague. However, he had come a long way towards finding the solution, as we shall now see.

Leibniz rewrote Debeaune's statement of the problem in a post-Cartesian notation. First, in the left-hand side of the equation that defines the problem (see Figure 4.1 which is clearer) let us denote by α the angle $\angle XGA$, where the tangent GXN meets the axis $GAYZ$. Since the triangles XGY and ZXY are similar, it follows that the angle $\angle ZXY = \alpha$. So the ratio $ZY : YX = \tan\alpha$. On the right-hand side, Leibniz wrote n for AB (the unit of length). He then took the point A as origin and, in a standard 17th-century way of choosing coordinates, he denoted AY by y and YX by x. So the problem becomes: find the curve for which

$$\tan\alpha = \frac{n}{x-y}.$$

Leibniz also re-lettered the diagram, so that CB became the old XY and his t/y became $\tan\alpha$. With his choice of x and y axes, $\tan\alpha$ is $dx : dy$, so he should have obtained the equation $dx : dy = n : (x-y)$. By mistake, however, he got the sign wrong and wrote $dx : dy = n : (y-x)$. He rewrote this as $ydx - xdx = ndy$ and hence, he said (now validly),

$$\int ydx - \int xdx = -\int ndy.$$

Note that he brought in another minus sign, whether by mistake or to correct his earlier error is not clear. The quadrature $\int ydx$ eluded him, although he could do the others, and he seems to have got into a muddle, from which he extricated himself by appealing to the fact that the curve has an asymptote. This assumption is rather hard to justify, and one wonders how Leibniz might have been able to prove it, other than by quoting Roberval. Be that as it may, Leibniz now started all over again. He used the asymptote BN as a new coordinate axis (let us call it the Y axis), rescaled the x-axis (we shall write $X = x/2$), and wrote down a new equation in the differentials dX and dY,

$$\frac{dX}{dY} = \frac{-\frac{1}{2}n}{Y} \quad \text{or} \quad \frac{-dX}{\frac{1}{2}n} = \frac{dY}{Y},$$

which he could integrate, in the sense that $\int dY : Y$ was, as he said, 'in his power'. This is the one that he said appertained to the logarithm, but notice that he did not give it explicitly. As it happens, he had written down $\int \frac{dy}{y} = \log y$ in November 1675. Moreover,

$$\int \frac{-dX}{\frac{1}{2}n} = \frac{-X}{\frac{1}{2}n}.$$

Leibniz's new coordinates are related to the old ones by:

$$X = x/2\,, \quad Y = y - x + n,$$

and so the equation that he wound up with is equivalent to the correct one that he should have obtained in the first place.

A month later Leibniz wrote to Henry Oldenburg, the Secretary of the Royal Society, to say that he believed himself to be the first to find the curve, having found it

[5]See (Child 1920, 122).

'on the day, indeed in the hour, when I first began to seek it', and that he had 'solved it at once by a sure analysis'.[6] More prudently, in 1684 Leibniz thought highly enough of this problem to put his treatment of it at the end of the famous paper in which he introduced his differential calculus — presumably he thought that the problem would show his new ideas to advantage.

We learn from all of this that Leibniz attached a great deal of importance to this problem and to his method of approaching it. We see that he was a quick, if careless, worker, and we get some idea of what he thought an answer to such a problem might be. This is worth spelling out. In the passage itself, Leibniz spoke of an integral as a 'quadrature' — that is, as the area of a certain figure, but he seems to have been unable to decide whether to present his answer as an expression in x or as a geometrical object involving an area.

This is typical of 17th- and early 18th-century writers. It is not a fudge. Rather, it is a wholly reasonable desire to get behind the formulas to an underlying geometrical reality, entirely of a piece with other instances in the no-man's-land between algebra and geometry that we have met so far. We shall meet more examples in this chapter, and each one reveals something of the author's point of view. In the present case, it is clear that Leibniz was moving towards a calculus as a set of formal algorithms, and it is just as clear that he had not yet arrived.

4.2 Newton's calculus and inverse tangent problems

The other obvious person for us to look to for a deepening sense of the importance of inverse tangent problems and the methods of the calculus is Isaac Newton. As we saw, he had put the essentials of his calculus together in the mid-1660s. Newton and Leibniz never met, although Leibniz visited London twice, in 1673 and 1676, but for a while they were in touch through Oldenburg. Oldenburg was a natural choice for both men to use, since he was based in London but was German by birth, and he used his position to disseminate scientific knowledge — much as Mersenne had done. When in 1676 Leibniz wrote to Oldenburg twice to report on his latest mathematical findings, in letters that were to be forwarded to Newton, Oldenburg pressed Newton to reply. Newton was reluctant to get involved, because he had been upset by comments being made in some circles about his theory of colours, and feared that his new mathematics would only elicit a similar response. But eventually he did compose two replies and sent them to Oldenburg to be forwarded to Leibniz.

In his first reply, the *Epistola Prior* (the *Earlier Letter*, to use the name that Newton gave it some forty years later when embroiled in a priority dispute over the discovery of the calculus), Newton chose to stress the power and breadth of his method of infinite series.[7]

[6]See (Scriba 1963, 122).

[7]See (Turnbull 1960, 2, 32–33); F&G 12.C1.

Newton's *Epistola Prior*.

> Though the modesty of Mr Leibniz, in the extracts from his letter which you have lately sent me, pays great tribute to our countrymen for a certain theory of infinite series, about which there now begins to be some talk, yet I have no doubt that he has discovered not only a method for reducing any quantities whatever to such series, as he asserts, but also various shortened forms, perhaps like our own, if not even better. Since, however, he very much wants to know what has been discovered in this subject by the English, and since I myself fell upon this theory some years ago, I have sent you some of those things which occurred to me in order to satisfy his wishes, at any rate in part.
>
> Fractions are reduced to infinite series by division; and quantities by extraction of the roots, by carrying out those operations in the symbols just as they are commonly carried out in decimal numbers. These are the foundations of these reductions: but extractions of roots are much shortened by this theorem,
>
> $$(P+PQ)^{m/n} = P^{m/n} + \frac{m}{n}AQ + \frac{m-n}{2n}BQ + \frac{m-2n}{3n}CQ + \frac{m-3n}{4n}DQ + etc.$$
>
> where $P+PQ$ signifies the quantity whose root or even any power, or the root of a power, is to be found; P signifies the first term of that quantity, Q the remaining terms divided by the first, and m/n the numerical index of the power of $P+PQ$, whether that power is integral or (so to speak) fractional, whether positive or negative. For as analysts, instead of aa, aaa, etc., are accustomed to write a^2, a^3, etc., so instead of $\sqrt{a}, \sqrt{a^3}, \sqrt{c : a^5}$, etc.[8] I write $a^{1/2}, a^{3/2}, a^{5/3}$, and instead of $1/a, 1/aa, 1/a^3$, I write a^{-1}, a^{-2}, a^{-3}. And so for $\frac{aa}{\sqrt{c:(a^3+bbx)}}$ I write $aa(a^3+bbx)^{-1/3}$ and for
>
> $$\frac{aab}{\sqrt{c : \{(a^3+bbx)(a^3+bbx)\}}}$$
>
> I write $aab(a^3+bbx)^{-2/3}$: in which last case, if $(a^3+bbx)^{-2/3}$ is supposed to be $(P+PQ)^{m/n}$ in the Rule, then P will be equal to a^3, Q to bbx/a^3, m to -2, and n to 3. Finally, for the terms found in the quotient in the course of the working I employ A, B, C, D, etc., namely, A for the first term, $P^{m/n}$; B for the second term, m/nAQ; and so on. For the rest, the use of the rule will appear from the examples.

This result is the binomial theorem for a fractional index, which Newton had found in 1664 but which he did not publish until 1704, as an Appendix in his *Opticks*.

> *Example 1*
>
> $$\sqrt{(c^2+x^2)} \text{ or } (c^2+x^2)^{1/2} = c + \frac{x^2}{2c} - \frac{x^4}{8c^3} + \frac{x^6}{16c^5} - \frac{5x^8}{128c^7} + \frac{7x^{10}}{256[c]^9} + \ etc.$$
>
> For in this case $P = c^2, Q = x^2/c^2, m = 1, n = 2,\ A(= P^{m/n} = (cc)^{1/2}) = c$, $B(= (m/n)AQ) = x^2/2c$, $C\left(= \frac{m-n}{2n}BQ\right) = -\frac{x^4}{8c^3}$; and so on.

[8] Note that Newton wrote $\sqrt{c : X}$ for the cube root of X.

Example 2

$$\sqrt[5]{(c^5 + c^4x - x^5)},$$

i.e.

$$(c^5 + c^4x - x^5)^{1/5} = c + \frac{c^4x - x^5}{5c^4}[+]\frac{-2c^8x^2 + 4c^4x^6 - 2x^{10}}{25c^9} + etc.$$

as will be evident on substituting 1 for m, 5 for n, c^5 for P and $(c^4x{-}x)^5/c^5$ for Q, in the rule quoted above. Also $-x^5$ can be substituted for P and $(c^4x + c^5)/(-x^5)$ for Q. The result will then be

$$\sqrt[5]{(c^5 + c^4x - x^5)} = -x + \frac{c^4x + c^5}{5x^4} + \frac{2c^8x^2 + 4c^9x + [2]c^{10}}{25x^9} + etc.$$

The first method is to be chosen if x is very small, the second if it is very large.

In his reply, Leibniz also stressed the importance of infinite series and displayed some of his knowledge of them, before raising some new questions. At this point, Leibniz's hope of obtaining a permanent position in Paris finally collapsed, and he had to return to Hannover. In October 1676 he made his way home via London, where he met John Collins (whom Barrow had called 'the English Mersenne'). Collins was so overwhelmed by Leibniz's intellect that he may have shown Newton's *De Analysi* to Leibniz by mistake; at all events he did not tell Newton of his indiscretion, and Newton found out about it only much later.[9] However, Leibniz took notes on only the half of it that concentrates exclusively on material about infinite series — none of the arguments about fluxions in *De Analysi* were transcribed. The inference is that at this stage Leibniz was interested only in what Newton had to say about infinite series.

On 24 October 1676 Newton sent his second letter (the *Epistola Posterior*) to Oldenburg for Leibniz, but Oldenburg delayed forwarding it until he knew that Leibniz was settled in Hannover, and Leibniz received it only in May 1677.

We now look in detail at this letter.[10] It opens with some polite remarks that also attest to Newton's high opinion of the method of infinite series.

Newton's *Epistola Posterior*.

I can hardly tell with what pleasure I have read the letters of those very distinguished men Leibniz and Tschirnhaus. Leibniz's method for obtaining convergent series is certainly very elegant, and it would have sufficiently revealed the genius of its author, even if he had written nothing else. But what he has scattered elsewhere throughout his letter is most worthy of his reputation — it leads us also to hope for very great things from him. The variety of ways by which the same goal is approached has given me the greater pleasure, because three methods of arriving at series of that kind had already become known to me, so that I could scarcely expect a new one to be communicated to us. One of mine I have described before; I now add another, namely, that by which I first chanced on these series — for I chanced on them before I knew the divisions and extractions of roots which I now use.

[9]Westfall, *Never at Rest*, p. 264.

[10]See Newton, *Correspondence*, Vol. 2, 129–134, and F&G 12.C2.

And an explanation of this will serve to lay bare, what Leibniz desires from me, the basis of the theorem set forth near the beginning of the former letter.

Newton then summarised his former route to his discovery of the method of infinite series. Even though what he wrote was doubtless carefully crafted with Leibniz in mind, it surely gives a reasonable impression of how Newton had come to his insights, and it dovetails well with the story that we told earlier, based on his notes made at the time. It is also an impressive account of what mathematicians often do: spot patterns that they then seek to explain. We will see that Newton was particularly concerned with understanding how the coefficients of powers of x are generated.

At the beginning of my mathematical studies, when I had met with the works of our celebrated Wallis, on considering the series by the intercalation of which he himself exhibits the area of the circle and the hyperbola, the fact that, in the series of curves whose common base or axis is x and the ordinates

$$(1-x^2)^{0/2}, (1-x^2)^{1/2}, (1-x^2)^{2/2}, (1-x^2)^{3/2}, (1-x^2)^{4/2}, (1-x^2)^{5/2}, etc,$$

if the areas of every other of them, namely

$$x^3, x-\frac{1}{3}x^3, x-\frac{2}{3}x^3+\frac{1}{5}x^5, x-\frac{3}{3}x^3+\frac{3}{5}x^5-\frac{1}{7}x^7, etc.$$

could be interpolated, we should have the areas of the intermediate ones, of which the first $(1-x^2)^{1/2}$ is the circle: in order to interpolate these series I noted that in all of them the first term was x and the second terms $\frac{0}{3}x^3$, $\frac{1}{3}x^3$, $\frac{2}{3}x^3$, $\frac{3}{3}x^3$, etc., were in arithmetical progression, and hence that the first two terms of the series to be intercalated ought to be $x-\frac{1}{3}(\frac{1}{2}x^3)$, $x-\frac{1}{3}(\frac{3}{2}x^3)$, $x-\frac{1}{3}(\frac{5}{2}x^3)$, etc. To intercalate the rest I began to reflect that the denominators $1, 3, 5, 7$, etc., were in arithmetical progression, so that the numerical coefficients of the numerators only were still in need of investigation. But in the alternately given areas these were the figures of powers of the number 11, namely of these, $11^0, 11^1, 11^2, 11^3, 11^4$, that is, first 1; then $1, 1$; thirdly, $1, 2, 1$; fourthly $1, 3, 3, 1$; fifthly $1, 4, 6, 4, 1$, etc. And so I began to inquire how the remaining figures in these series could be derived from the first two given figures, and I found that on putting m for the second figure, the rest would be produced by continual multiplication of the terms of this series,

$$\frac{m-0}{1}\times\frac{m-1}{2}\times\frac{m-2}{3}\times\frac{m-3}{4}\times\frac{m-4}{5}\times etc.$$

For example, let $m = 4$, and $4\times\frac{1}{2}(m-1)$, that is 6 will be the third term, and $6\times\frac{1}{3}(m-2)$, that is 4 the fourth, and $4\times\frac{1}{4}(m-3)$, that is 1 the fifth, and $1\times\frac{1}{5}(m-4)$, that is 0 the sixth, at which term in this case the series stops. Accordingly, I applied this rule for interposing series among series, and since, for the circle, the second term was $\frac{1}{3}(\frac{1}{2}x^3)$,

I put $m = \frac{1}{2}$, and the terms arising were

$$\frac{1}{2} \times \frac{\frac{1}{2}-1}{2} \text{ or } -\frac{1}{8}, \quad -\frac{1}{8} \times \frac{\frac{1}{2}-2}{3} \text{ or } +\frac{1}{16}, \quad \frac{1}{16} \times \frac{\frac{1}{2}-3}{4} \text{ or } -\frac{5}{128},$$

and so to infinity. Whence I came to understand that the area of the circular segment which I wanted was

$$x - \frac{\frac{1}{2}x^3}{3} - \frac{\frac{1}{8}x^5}{5} - \frac{\frac{1}{16}x^7}{7} - \frac{\frac{5}{128}x^9}{9} \textit{ etc.}$$

And by the same reasoning the areas of the remaining curves, which were to be inserted, were likewise obtained: as also the area of the hyperbola and of the other alternate curves in this series

$$(1-x^2)^{0/2},\ (1-x^2)^{1/2},\ (1-x^2)^{2/2},\ (1-x^2)^{3/2}, \text{ etc.}$$

And the same theory serves to intercalate other series, and that through intervals of two or more terms when they are absent at the same time. This was my first entry upon these studies, and it had certainly escaped my memory, had I not a few weeks ago cast my eye back on some notes.

But when I had learnt this, I immediately began to consider that the terms

$$(1-x^2)^{0/2},\ (1-x^2)^{2/2},\ (1-x^2)^{4/2},\ (1-x^2)^{6/2}, \textit{ etc.}$$

that is to say,

$$1,\ 1-x^2,\ 1-2x^2+x^4,\ 1-3x^2+3x^4-x^6, \textit{ etc.}$$

could be interpolated in the same way as the areas generated by them: and that nothing else was required for this purpose but to omit the denominators $1, 3, 5, 7$, etc., which are in the terms expressing the areas; this means that the coefficients of the terms of the quantity to be intercalated $(1-x^2)^{\frac{1}{2}}$, or $(1-x^2)^{\frac{3}{2}}$, or in general $(1-x^2)^m$, arise by the continued multiplication of the terms of this series

$$m \times \frac{m-1}{2} \times \frac{m-2}{3} \times \frac{m-3}{4} \textit{ etc.}$$

so that (for example)

$$(1-x^2)^{\frac{1}{2}} \text{ was the value of } 1 - \frac{1}{2}x^2 - \frac{1}{8}x^4 - \frac{1}{16}x^6 \textit{ etc.},$$

$$\text{and } (1-x^2)^{\frac{3}{2}} \text{ was the value of } 1 - \frac{3}{2}x^2 + \frac{3}{8}x^4 + \frac{1}{16}x^6 \textit{ etc.},$$

$$(1-x^2)^{\frac{1}{3}} \text{ was the value of } 1 - \frac{1}{3}x^2 - \frac{1}{9}x^4 - \frac{5}{81}x^6 \textit{ etc.}$$

So then the general reduction of radicals into infinite series by that rule, which I laid down at the beginning of my earlier letter became known to me, and that before I was acquainted with the extraction of roots.

By the end of this passage, Newton had turned to the relationship between integer and fractional powers. He knew perfectly well that for any expression A one has $(A^{1/2})^2 = A$, and that this applies to a binomial such as $A = 1 - x^2$. But he also had an infinite series expansion of $(1-x^2)^{1/2}$, so he proceeded to check that the square of this expression is indeed $1 - x^2$.

> But once this was known, that other could not long remain hidden from me. For in order to test these processes, I multiplied
>
> $$1 - \frac{1}{2}x^2 - \frac{1}{8}x^4 - \frac{1}{16}x^6, etc.$$
>
> into itself; and it became $1 - x^2$, the remaining terms vanishing by the continuation of the series to infinity. And even so $1 - \frac{1}{3}x^2 - \frac{1}{9}x^4 - \frac{5}{81}x^6$, etc. multiplied twice into itself also produced $1 - x^2$.

And then, to bring the matter to a conclusion, he checked that the arithmetical method for finding the square roots of numbers worked for binomial expressions as well.

> And as this was not only sure proof of these conclusions so too it guided me to try whether, conversely, these series, which it thus affirmed to be roots of the quantity $1 - x^2$, might not be extracted out of it in an arithmetical manner. And the matter turned out well. This was the form of the working in square roots.
>
> $$\begin{array}{lllll} 1 & -x^2(1 & -\frac{1}{2}x^2 & -\frac{1}{8}x^4 & -\frac{1}{16}x^6\text{, etc.} \\ 1 & & & & \\ \hline 0 & -x^2 & & & \\ & -x^2 & +\frac{1}{4}x^4 & & \\ \hline & & -\frac{1}{4}x^4 & & \\ & & -\frac{1}{4}x^4 & +\frac{1}{8}x^6 & +\frac{1}{64}x^8 \\ \hline & & 0 & -\frac{1}{8}x^6 & -\frac{1}{64}x^8 \end{array}$$
>
> After getting this clear I have quite given up the interpolation of series, and have made use of these operations only, as giving more natural foundations. Nor was there any secret about reduction by division, an easier affair in any case.

Newton's method of infinite series was therefore established on general grounds — what later became called the *binomial theorem* — and the guesswork part involving intercalation was dropped.

Newton next turned to indicate how infinite series could be used in the method of tangents. But now, perhaps because the problems were harder and his successes more recent, he became reticent, and although Leibniz could read that Newton had a general method for finding tangents, he would not learn on what it was based, for that was concealed by an anagram — anagrams were a common device at the time for establishing priority without revealing the discovery itself.

> But in that treatise infinite series played no great part. Not a few other things I brought together, among them the method of drawing tangents which the very skilful Sluse communicated to you two or three

years ago, about which you wrote back [to him] (on the suggestion of Collins) that the same method had been known to me also. We happened on it by different reasoning: for, as I work it, the matter needs no proof. Nobody, if he possessed my basis, could draw tangents any other way, unless he were deliberately wandering from the straight path. Indeed we do not here stick at equations in radicals involving one or each indefinite quantity, however complicated they may be; but without any reduction of such equations (which would generally render the work endless) the tangent is drawn directly. And the same is true in questions of maxima and minima, and in some others too, of which I am not now speaking. The foundation of these operations is evident enough, in fact; but because I cannot proceed with the explanation of it now, I have preferred to conceal it thus:[11]

$$6accdae13eff7i3l9n4o4qrr4s8t12vx.$$

On this foundation I have also tried to simplify the theories which concern the squaring of curves, and I have arrived at certain general Theorems. And, to be frank, here is the first Theorem.

For any curve let $dz^\theta \times (e + fz^\eta)^\lambda$ be the ordinate, standing normal at the end z of the abscissa or the base, where the letters d, e, f denote any given quantities, and θ, η, λ are the indices of the powers of the quantities to which they are attached. Put

$$\frac{\theta+1}{\eta} = r, \quad \lambda + r = s, \quad \frac{d}{\eta f} \times (e + fz^\eta)^{\lambda+1} = Q \quad \text{and} \quad r\eta - \eta = \pi,$$

then the area of the curve will be

$$Q \times \left(\frac{z^\pi}{s} - \frac{r-1}{s-1} \times \frac{eA}{fz^\eta} + \frac{r-2}{s-2} \times \frac{eB}{fz^\eta} - \frac{r-3}{s-3} \times \frac{eC}{fz^\eta} + \frac{r-4}{s-4} \times \frac{eD}{fz^\eta}, etc. \right)$$

the letters A, B, C, D, etc., denoting the terms immediately preceding; that is, A the term z^π/s, B the term $-\frac{r-1}{s-1} \times \frac{eA}{fz^\eta}$ etc. This series, when r is a fraction or a negative number, is continued to infinity; but when r is positive and integral it is continued only to as many terms as there are units in r itself; and so it exhibits the geometrical squaring of the curve. I illustrate the fact by examples.

Newton then moved on to claim that in principle the method of infinite series can solve every problem in mathematics, although he allowed that some problems will always be too complicated and burdensome to solve in practice. By this he meant that any problem involving curves, tangents, and areas, including inverse tangent problems, will have an answer that can be expressed as an infinite series.

When I said that almost all problems are soluble I wished to be understood to refer specially to those about which mathematicians have hitherto concerned themselves, or at least those in which mathematical arguments can gain some place. For of course one may imagine

[11] It conceals the statement "Data aequatione quotcunque fluentes quantitates involvente, fluxiones invenire; et vice versa", which translates as "Given an equation involving any fluent quantities whatever, to find the fluxions, and vice versa."

> others so involved in complicated conditions that we do not succeed in understanding them well enough, and much less in bearing the burden of such long calculations as they require. Nevertheless — lest I seem to have said too much — inverse problems of tangents are within our power, and others more difficult than those, and to solve them I have used a twofold method of which one part is neater, the other more general. At present I have thought fit to register them both by transposed letters, lest, through others obtaining the same result, I should be compelled to change the plan in some respects.
>
> $$5accdae10effh11i4l3m9n6oqqr8s11t9y3x : 11ab3cdd10eaeg10ill4m7n6o$$
> $$3p3q6r5s11t8vx, 3acae4egh5i4l4m5n8oq4r3s6t4v,$$
> $$aaddaececceiijmmnnooprrrsssssttuu.$$
>
> This inverse problem of tangents, when the tangent between the point of contact and the axis of the figure is of given length, does not demand these methods. Yet it is that mechanical curve the determination of which depends on the area of an hyperbola. The problem is also of the same kind, when the part of the axis between the tangent and the ordinate is given in length. But I should scarcely have reckoned these cases among the sports of nature. For when in the right-angled triangle, which is formed by that part of the axis, the tangent and the ordinate, the relation of any two sides is defined by any equation, the problem can be solved apart from my general method. But when a part of the axis ending at some point given in position enters the bracket, then the question is apt to work out differently.

The second anagram concealed the following message:

> Una Methodus consistit in extractione fluentis quantitatis ex aequatione simul involvente fluxionem ejus: altera tantum in assumptione Seriei pro quantitate qualibet incognita ex qua caetera commode derivari possunt, & in collatione terminorum homologorum aequationis resultantis, ad eruendos terminos assumptae seriei.
>
> (One method consists in extracting a fluent quantity from an equation at the same time involving its fluxion; but another by assuming a series for any unknown quantity whatever, from which the rest could conveniently be derived, and in collecting homologous terms of the resulting equation in order to elicit the terms of the assumed series.)

So it claims that the extraction of a fluent quantity from an expression involving its fluxions can always be carried out by substituting for the unknown an infinite series and then equating like terms. We shall examine these claims in more detail below.

Newton then ended with problems of a more general kind involving irrational powers of the unknown, which he proposed as a challenge to Leibniz. It is not clear, however, what Newton himself could have done with them.

> The communication of the solution of affected equations by the method of Leibniz will be very agreeable; so too an explanation how he comports himself when the indices of the powers are fractional, as in this

> equation $20+x^{3/7}-x^{6/5}y^{2/3}-y^{7/11}=0$, or surds, as in $(x^{\sqrt{2}}+x^{\sqrt{7}})^{\sqrt[3]{2/3}}=y$, where $\sqrt{2}$ and $\sqrt{7}$ do not mean coefficients of x, but indices of powers or dignities of it, and $\sqrt[3]{2/3}$ means the power of the binomial $x^{\sqrt{2}}+x^{\sqrt{7}}$. The point, I think, is clear by my method, otherwise I should have described it. But a term must at last be set to this wordy letter. The letter of the most excellent Leibniz fully deserved of course that I should give it this more extended reply. And this time I wanted to write in greater detail because I did not believe that your more engaging pursuits should often be interrupted by me with this rather austere kind of writing.

So Leibniz learned that Newton claimed to be able to solve inverse tangent problems by a two-fold method, one part of which he called 'neater' (the direct extraction of the fluent) and the other 'more general' (in which the answer would be given as an infinite series). However, Newton did not say what those methods were, again preferring the security of an indecipherable anagram. Since the fluxional material was thus concealed, Leibniz could at best only presume to know in general terms what these obscurities meant. If you look at the decoded message, you might well wonder if even that could have helped Leibniz much.

Leibniz's reply was more open, disclosing the essence of his differential calculus, but the correspondence that he so patently wanted with Newton never developed. Oldenburg died in September 1677, and in any case Newton did not want to continue. Richard Westfall, his 20th-century biographer, went so far as to speak of an 'unpleasant paranoia [that] pervaded the *Epistola Posterior*' and painted a convincing picture of Newton as someone secretive and impatient to consider other things.[12] Newton wrote a covering note to Oldenburg with the *Epistola Posterior* in which he said:[13]

> I hope this will so far satisfy M. Leibniz that it will not be necessary for me to write any more about this subject. For having other things in my head, it proves an unwelcome interruption to me to be at this time put upon considering these things.

So here, for a while, matters were to rest between them.

What was it that Newton was at such pains to conceal but of which he was so eager to establish ownership? Newton was claiming to have not one but two methods for tackling the inverse problem of tangents. But even decoded, the second anagram is not entirely clear, and because the first part of it involves Newton's terms, 'fluxions' and 'fluents', which are difficult to understand, we shall explain them in the next section, after we have dealt with the part about infinite series.

The Fundamental Theorem of the Calculus. Both Newton and Leibniz appreciated the importance of the Fundamental Theorem of the Calculus, the recognition that finding tangents and areas ('differentiation' and 'integration') are inverse processes. For example, to integrate $1/x$ is to find $\log x$, and conversely, differentiating $\log x$ yields $1/x$.

In the more geometrical language of the 17th century, we might interpret the first statement in this way: Find the curve whose tangent at the point (x, y) has a slope equal

[12]Westfall, *Never at Rest*, p. 266.

[13]See (Turnbull 1960, 110).

to the reciprocal of x. Newton would have written it as:

$$\text{slope of tangent} = \frac{\dot{y}}{\dot{x}} = \frac{1}{x},$$

so

$$\dot{y} = \frac{\dot{x}}{x}.$$

So anti-differentiation (finding the fluents from the fluxions) yields $y = \log x$.

Leibniz might have put it this way:

$$\text{slope of tangent} = \frac{dy}{dx} = \frac{1}{x},$$

so

$$dy = \frac{dx}{x}.$$

So, by the Fundamental Theorem, $y = \log x$.

This illustrates that the Fundamental Theorem of the Calculus solves some inverse tangent problems, and so inverse tangent problems can be viewed as a generalisation of that theorem. To see that inverse tangent problems can involve more than the Fundamental Theorem, let us return to Debeaune's problem. The solution to it that Leibniz so nearly found is:

$$x + n\log(y - x + n) = n\log n.$$

Differentiation yields

$$dx + \frac{nd(y - x + n)}{y - x + n} = 0,$$

which simplifies to

$$(y - x)dx + ndy = 0$$

or

$$\frac{dx}{dy} = \frac{n}{x - y},$$

the differential equation that Leibniz should have begun with.

So, differentiating the solution yields the original problem (a check on one's accuracy recommended by Newton) and therefore solving this inverse tangent problem is, in some sense, the opposite process to differentiation. Likewise, it is true that differentiation enables us to check on our solution to a problem in integration. But finding the solution to an inverse tangent problem generally involves us in more work than simply performing an integration. Whether we regard the above method as elegant presumably depends on how far practice with similar problems improves our grasp of it, but its answers are surely more elegant than those provided by the method of infinite series to which we now turn.

The method of infinite series. The method of infinite series assumes that the solution to a problem can be expressed in the form:

$$y = a_0 + a_1x + a_2x^2 + \cdots,$$

where $a_0, a_1, a_2, \ldots$ are unknown constants that have to be determined from the problem. The method consists of replacing everything that involves y in the statement of the problem with a (possibly infinite) series that involves only xs and the constants. We can then compare coefficients to find the constants and so obtain an expression for the solution.

In the case of Debeaune's problem, for example, the equation for y that has to be solved can be written as

$$\frac{dx}{dy} = \frac{n}{x-y}.$$

Clearing fractions allows us to write this as

$$n\frac{dy}{dx} = x - y.$$

When we substitute the above expression for y as an infinite series into this equation, we obtain

$$n(a_1 + 2a_2x + 3a_3x^2 + \cdots) = x - (a_0 + a_1x + a_2x^2 + \cdots),$$

$$\text{or}\quad n(a_1 + 2a_2x + 3a_3x^2 + \cdots) = -a_0 + (1 - a_1)x - a_2x^2 - \cdots).$$

We now equate corresponding powers of x and find this system of equations for the coefficients in the infinite series for y:

$$na_1 = -a_0, \quad n2a_2 = 1 - a_1, \quad n3a_3 = -a_2, \quad \ldots,$$

and for $j > 3$ we find $nja_j = -a_{j-1}$, or $a_j = -\frac{1}{nj}a_{j-1}$.

This system of equations can then be solved in terms of a_0:

$$a_1 = -\frac{1}{n}a_0, \quad a_2 = \frac{1}{n}.\frac{1-a_1}{2} = \frac{1}{n}.\frac{1-\frac{-1}{n}a_0}{2}, \quad \ldots.$$

So the solution is given by this infinite series:

$$y = a_0 - \frac{a_0x}{n} + \frac{(1+\frac{a_0}{n})x^2}{2n} + \cdots.$$

In the case at hand, the solution curve passes through the origin, so when $x = 0$ we have $y = 0$, and therefore $a_0 = 0$, so the solution becomes

$$y = \frac{x^2}{2n} - \frac{x^3}{6n^2} + \cdots.$$

Newton's preferred method would be something like what we have in Box 4 (where we have allowed ourselves the luxury of Leibniz's more familiar notation).

The greater elegance of the first method arises from the fact that it requires skill to manipulate, and therefore usually draws on one's understanding. On the other hand, as Newton said, the second method, the method of infinite series, is more general, and, in some sense, routine. If we could believe that every mathematical object can be given a power-series expansion, then it might even seem that every inverse tangent problem could be solved by the use of infinite series, which would account for Newton's early optimism.

But we can also judge these two methods by looking at the quality of the answers they tend to give. The first method generally gives much more comprehensible answers than the second method; for example, it can be difficult to plot points on the solution curve accurately when the second method is used, because it inevitably gives us an infinite addition sum. A curve described by an equation such as

$$x + n\log(y - x + n) = n\log n$$

may not be easy to draw, but to plot arbitrarily many points on it requires only that we have a table of logarithms available. Because of the form in which the solution is given, algebraic analysis lends itself to finding out further properties of the curve, such

Box 4. The solution of Debeaune's differential equation.
We write the equation to be solved in the form

$$\frac{dy}{dx} = \frac{x-y}{n},$$

and put $x - y = nu$, so $1 - \frac{dy}{dx} = n\frac{du}{dx}$.

This turns the equation into

$$1 - n\frac{du}{dx} = u,$$

from which we deduce that

$$\frac{du}{1-u} = \frac{dx}{n},$$

and so

$$-\log(1-u) = \frac{x}{n} + a,$$

where a is an arbitrary constant.

We now replace u by $\frac{x-y}{n}$, and after a little more work obtain this expression for the solution:

$$x + n\log(n - x + y) = n\log n - na.$$

If the solution is to satisfy $y = 0$ when $x = 0$ then $a = 0$, and the solution takes the simpler form

$$x + n\log(n - x + y) = n\log n,$$

as we claimed earlier.

as seeing whether it has an asymptote, and so on. However, if ingenuity fails us, we must fall back on the second method, that of infinite series. The price we pay is in the quality of the answers we get, but the generality of the latter method is considerable.

In what follows we shall not be much concerned with the detailed application of either of these methods, for their technicalities can obscure a more basic point: both left mathematicians with formal expressions that, to 17th-century taste, required a geometrical interpretation. There was then a tension between the kinds of answers that the new methods gave and those that were expected. How this tension was resolved is discussed below, after we have looked at how Newton and Leibniz developed their ideas about the calculus in general, and how its discovery was made public in the 1680s.

4.3 Newton's mature calculus

Newton (by 1666) and Leibniz (by 1676) both possessed routine methods based on formulas and general rules for dealing with questions about tangents and areas, and even with certain inverse tangent problems. We now look in more detail at these two versions of the calculus, concentrating on the justifications that their authors put forward for how and why their methods worked. It was clear that both did work, but the reasons for this were much less clear. What underpinned the formal systems? We start with the views of Newton.

The Newtonian calculus. In 1671 Newton wrote a manuscript, *The Methods of Series and Fluxions*, that gives us a good insight into his way of thinking about the calculus. All variables were conceived of as varying in time, and accordingly were called *fluents*, or flowing quantities. The rates at which the variables change were called their *fluxions*, and simple rules take us from fluents to their fluxions. Thus Newton was pursuing his o-method and his analysis of curves by motion, in a way that appears to rely on an underpinning conception of *time*. Naively, we might suppose that Newton's variables were all necessarily time-dependent, in the obvious sense of being measured by clocks (say), but Newton strove to avoid misunderstandings by spelling out just how he conceived of time as entering the theory of the calculus:[14]

Newton on the nature of time.

> We can, however, have no estimate of time except in so far as it is expounded and measured by an equable local motion, and furthermore quantities of the same kind alone, and so also their speeds of increase and decrease, may be compared one with another. For these reasons I shall, in what follows, have no regard to time, formally so considered, but from quantities propounded which are of the same kind shall suppose some one to increase with an equable flow: to this all the others may be referred as though it were time, and so by analogy the name of 'time' may not improperly be conferred upon it. And so whenever in the following you meet with the word 'time' (as I have, for clarity's and distinction's sake, on occasion woven it into my text), by that name should be understood not time formally considered but that other quantity through whose equable increase or flow time is expounded and measured.

To judge from this passage, how did Newton conceive of the role of time in his calculus? He did not base his fluxional methods on time 'formally', but one of the variables he was considering would be supposed to 'increase with an equable flow', and thus be analogous to time, even to the extent of being *called* 'time' now and again. It was this arbitrary variable that would be the basic independent variable upon which the others would depend, all other variables being referred to it 'as though it were time'. So his calculus did not rely on time as such (on a simple understanding of the concept), but rather assumed that an independent variable had the same flowing property as time.

This conception seems to have greatly aided Newton's intuition in developing his calculus. It also helps us to understand what Newton was doing when we see see the deep role that time-like motions had in his thoughts and in his way of expressing ideas. But the question would arise, sooner or later, as to whether this analogy with time was a sufficiently coherent conception on which to ground his calculus. Was it logically rigorous enough to give the right answers in new and untried situations?

Motion and the o-method. Let us first see how, in practice, the motion idea entered into Newton's detailed arguments. A few pages after the passage from *The Method*

[14]Newton, *De Methodis Serierum et Fluxionum* (The Methods of Series and Fluxions) in Whiteside, *MPIN* III, 32–353, quoted on p. 73.

of Series and Fluxions that we have just read, Newton justified applying his rules of the calculus to find the fluxions associated with the equation

$$x^3 - ax^2 + axy - y^3 = 0.$$

His argument went as follows, where, following Newton's modern editor, D.T. Whiteside, we have replaced Newton's letters *m* and *n* with $\dot{x}$ and $\dot{y}$, thereby providing the notation for relating fluents to their fluxions that Newton gave only twenty years later.

Each fluent (that is, x or y) increases during an infinitely small period of time by what he called its *moment*, which is proportional to its speed of flow, that is, its moment is $\dot{x}o$, say, where $\dot{x}$ is the speed at which the fluent x flows (and o is the infinitely small quantity we met in Section 3.2). Then, during the infinitely small period of time, the fluent x becomes $x + \dot{x}o$, and y has similarly become $y + \dot{y}o$.

If you think of the curve as being traced out by a moving point, then at one instant of time the point is at (x, y) and an infinitesimal amount o of time later it is at $(x + \dot{x}o,\ y + \dot{y}o)$, another point on the curve; the moments, $\dot{x}o$ and $\dot{y}o$, are thereby to be regarded as distances. So x and y in the original equation can be replaced by the expressions $x + \dot{x}o$ and $y + \dot{y}o$, and the new equation can then be manipulated algebraically to find the ratio $\dot{y}/\dot{x}$. Newton arrived at the equation

$$3x^2\dot{x} - 2ax\dot{x} + a\dot{x}y + ax\dot{y} - 3y^2\dot{y} = 0,$$

which can be rearranged[15] to give

$$\frac{\dot{y}}{\dot{x}} = \frac{3x^2 - 2ax + ay}{ax - 3y^2}$$

as the slope of the curve (that is, of its tangent) at the point (x, y).

What kind of justification did Newton provide for this? First of all, time (or the underlying variable analogous to it) makes no overt symbolic appearance: the flowing of x and y is in relation to time, but it does not formally appear in the mathematics. Newton's argument is entirely rigorous up to the point where he has a collection of terms in 'o' cluttering up the equation and which he would like to eliminate: 'I therefore cast them out', he said with panache, justifying this on the grounds that o is 'infinitely small'.

It is hard to find this entirely convincing. If o can now be discarded, how could it have functioned validly earlier in the argument? Nevertheless, it is plausible that if we understood better the properties of infinitely small things and processes, then Newton's reasonable-sounding intuition could be spelled out logically and rigorously. Or, to put the same point another way, if there is a problem in fully understanding what Newton was saying here, it would seem that he considered that the way forward would be to be clarify what was happening in infinitely small intervals of time.

There is, however, a significant problem with Newton's method. The motion of a point is described by the two fluents x and y. The equation relating x and y describes the curve of the point's motion. At any instant the point has speeds (fluxions) $\dot{x}$ and $\dot{y}$, so the slope of the tangent at that instant is $\dot{y}/\dot{x}$ (as a ratio of speeds) or indeed is $\dot{y}o/\dot{x}o$ as a ratio of moments (that is, of infinitely small distances), which comes to the same thing on cancelling the *o*s. As we can see from Figure 4.5, the point $(x + \dot{x}o, y + \dot{y}o)$ must appear higher up on the tangent line than is the point of tangency, (x, y).

---

[15]See Whiteside, *MPIN* III, 79–81, and F&G 12.A5.

But in his 1671 argument,[16] however, Newton insisted that the point with coordinates $(x + \dot{x}o, y + \dot{y}o)$ is a *point on the curve*, for the relation between the two new, increased, fluents is the same as before, namely, that described by the equation. Only on this assumption can $x + \dot{x}o$ and $y + \dot{y}o$ be substituted back into the equation in place of x and y.

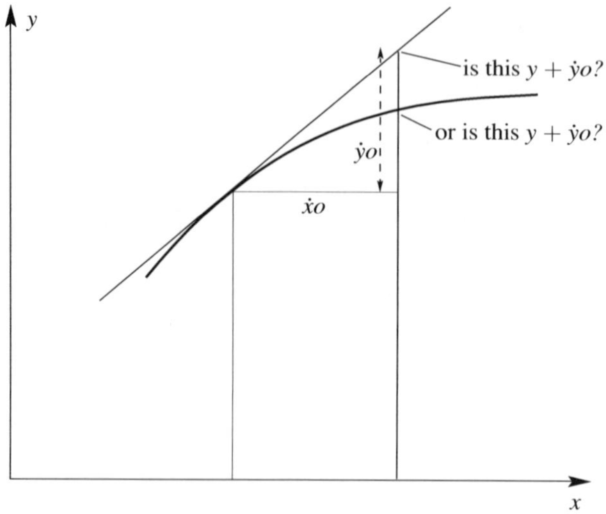

Figure 4.5. Newton's o-method

There appears, then, to be a fundamental contradiction: Is the point with coordinates $(x + \dot{x}o, y + \dot{y}o)$ on the curve or on the tangent line? Newton's escape in 1671 was his claim that when o is infinitely small, the difference between the two is negligible. This is surely true, but it diverts attention onto the shifty status of quantities mutating under our very gaze.

We can also ask whether Newton himself took steps to make his arguments convincing, as well as plausible? Presumably there was little need to do so, for as long as his calculus was known to (and used by) himself alone. But he seems to have felt some disquiet about his approach, because he gave a different account in 1687. In that year he published the major work on which his later fame mostly rests: his *Principia Mathematica*. This is *not* a book on the calculus. It is a book about the physical universe, written in the language of geometry. Insofar as it was about motion, Newton set down his thoughts as best he could in geometrical terms. But what he said related to his studies of variable quantities of any kind, and it became clear that even Newton needed to invoke the calculus on occasion, and so there are several passages in which he dropped hints about it. These passages were written in a style that might be called 'geometrical calculus' — essentially, geometrical reasoning about magnitudes and how they change in very small intervals of time — and were studied in depth for clues to Newton's ideas about his calculus. (More people knew that he had some general method than knew what it consisted of, and Newton was far from forthcoming in public or in private on the subject.) It is to one of these passages that we now turn, but we must remember

[16]See Whiteside, *MPIN* III, 79–81, and F&G 12.A5.

that throughout its length, *Principia Mathematica* establishes for the historian that the mind that wrote it could also invent the calculus: the calculus and *Principia Mathematica* are brother and sister, not parent and child.

'Prime' and 'ultimate' ratios. On pages 38–39 of *Principia Mathematica* Newton gave this account of the new foundations that he proposed for the study of curved lines and surfaces.[17]

Newton on limits in the *Principia*.

> What has been demonstrated concerning curved lines and the [plane] surfaces comprehended by them is easily applied to curved surfaces and their solid contents. In any case, I have presented these Lemmas before the propositions in order to avoid the tedium of working out 'lengthy' proofs by *reductio ad absurdum* in the manner of the ancient geometers. Indeed, proofs are rendered more concise by the method of indivisibles. But since the hypothesis of indivisibles is problematical and this method is therefore accounted less geometrical, I have preferred to make the proofs of what follows depend on the ultimate sums and ratios of vanishing and the first sums and ratios of nascent quantities, that is, on the limits of such sums and ratios, and therefore to present proofs of those limits beforehand as briefly as I could. For the same thing is obtained by these as by the method of indivisibles, and we shall be on safer ground using principles that have been proved. Accordingly, whenever in what follows I consider quantities as consisting of particles, or whenever I use curved line-elements [or minute curved lines] in place of straight lines, I wish it always to be understood that I have in mind not indivisibles, but evanescent divisibles, and not sums and ratios of determinate parts but the limits of such sums and ratios, and that the force of such proofs always rests on the method of the preceding lemmas.
>
> It may be objected, that there is no such thing as an ultimate proportion of vanishing quantities, inasmuch as before vanishing the proportion is not ultimate, and after vanishing, it does not exist at all. But by the same argument it could equally be contended that there is no ultimate velocity of a body reaching a certain place at which the motion ceases; for before the body arrives at this place the velocity is not the ultimate velocity; and when it arrives there, there is no velocity at all. But the answer is easy; to understand the ultimate velocity as that with which the body is moving neither before it arrives at its ultimate place and the motion ceases nor after it arrives there, but at the very instant when it arrives, that is, the very velocity with which the body arrives at its ultimate place, and with which the motion ceases. And similarly

[17]There were three editions of the *Principia* in Newton's lifetime, and there are translations into English of the third edition (1726). The first was done by Andrew Motte in 1729, and although revised and republished by Florian Cajori in 1930 it is generally agreed to be showing its age, so whenever possible we have used the translation by Bernard Cohen and Anne Whitman of 1999. When we have quoted material from earlier editions that was altered for later editions we have stated our sources explicitly.

> the ultimate ratio of vanishing quantities is to be understood not as the ratio of quantities before they vanish, or after they have vanished, but the ratio with which they vanish. Likewise also the first ratio of nascent quantities is the ratio with which they begin to exist [or come into being]. And the first and the ultimate sum is the sum with which they begin and cease to exist (or to be increased or decreased). There exists a limit which their velocity can attain at the end of the motion, but not exceed. This is their ultimate velocity. And it is the same for the limit of all quantities and proportions that come into being and cease existing. And since this limits is certain and definite, the determining of it is properly a geometrical problem. But everything that is geometrical is legitimately used in determining and demonstrating whatever else may be geometrical.
>
> It can also be contended that if the ultimate ratios of vanishing quantities are given, their ultimate magnitudes will also be given: and thus every quantity will consist of indivisibles, contrary to what Euclid has proved concerning incommensurables, in the tenth Book of his *Elements*. But this objection is based on a false supposition. Those ultimate ratios with which quantities vanish are not actually ratios of ultimate quantities, but limits which the ratios of quantities decreasing without limit are continually approaching; and which they can approach so closely that their difference is less than any given quantity, but which they can never exceed and can never reach before the quantities are decreased indefinitely. This matter will be understood more clearly in the case of quantities that are infinitely great. If two quantities whose difference is given are increased indefinitely, their ultimate ratio will be given, namely, the ratio of equality; and yet the ultimate or greatest quantities of which this is the ratio will not on this account be given. Therefore whenever to make things easier to comprehend I speak in what follows of quantities as minimally small or vanishing, or ultimate, take care not to understand quantities that are determinate in magnitude, but always think of quantities that are to be decreased without limit.[18]

Fluxions are not mentioned. Instead, Newton introduced a new method, based on 'evanescent divisible quantities', after implying that all could be proved rigorously by the ancient method of double reductio ad absurdum at too high a price in tedium, and after dismissing the speedier method of indivisibles as too crude. He also introduced the concept of the 'limits' of 'the first and last sums and ratios of nascent and evanescent quantities'. The meaning of this is not immediately clear. Evanescent quantities are vanishing quantities, sums and ratios of which can somehow be understood. One feels sympathy for early readers of *Principia* trying to reconstruct what Newton had in mind! How indeed can such quantities have a final ratio or 'ultimate proportion', as Newton called it, as their individual quantities become zero?

Newton addressed this difficulty in the second paragraph, and answered much as one might answer the paradox of Zeno that challenged the idea of a moving object

[18](Cohen and Whitman 1999, 441–443).

having a velocity at a point. In the third paragraph he tried to make the point more clearly, when he wrote: 'For those ultimate ratios with which quantities vanish are not truly the ratios of ultimate quantities, but limits towards which the ratios of quantities decreasing without limit do always converge'. Here he proposed that two quantities (a and b, say) each of which is diminishing to zero have a ratio a/b provided that b is not zero, and that as a and b decrease to zero, one is to study the limiting value of this ratio, and not to confuse it with the ratio $\frac{\text{limit of } a}{\text{limit of } b}$, which would be $0/0$.

The concept of a *last* or *ultimate ratio* is a sophisticated one. To see what Newton was trying to achieve, and the problem that this concept was designed to solve, we hone in on a problematic detail of fluxions that we indicated earlier.

In the 1680s, as his remarks in the *Principia* suggest, Newton refined his o-calculus. He continued to insist that $(x + \dot{x}o, y + \dot{y}o)$ is a point on the curve, but he now argued that $\dot{x}$ and $\dot{y}$ depend on o and so are no longer the instantaneous velocities (fluxions) at the point (x, y). Accordingly, $\dot{y}/\dot{x}$ is not the slope of the tangent at (x, y), but the slope of the chord joining the points (x, y) and $(x + \dot{x}o, y + \dot{y}o)$. This slope gets closer and closer to the slope of the tangent as o gets smaller and smaller, and in the limit the ratio $\dot{y}/\dot{x}$ is equal to the slope of the tangent. Newton called this limiting value the *last* (or *ultimate*) ratio (when he thought of o as decreasing to zero) or the *first* (or *prime*) ratio (when he thought of o as increasing from zero). So an approximation argument that asserts that we can ignore terms in o was replaced by a more plausible, if more difficult, argument to the effect that certain quantities are ultimately zero.

What Newton became concerned with was the ratio of 'augments' — that is, the little added distances that he used to call 'moments'. Typically, he considered the ratio of 'augment of ordinate' to 'augment of abscissa', of Δy to Δx (see Figure 4.6).

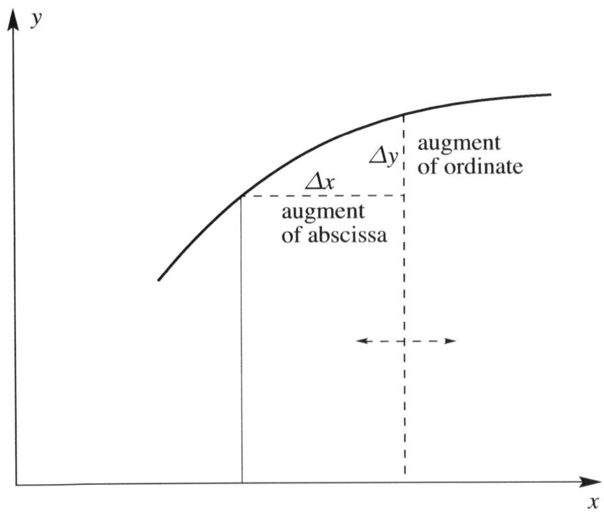

Figure 4.6. Newton's method for tangents

These augments get smaller and smaller, and become evanescent. But just before they vanish altogether, they have a last or *ultimate ratio* (Newton said), which is the ratio of the fluxions. Or, going the other way, the augments can be considered as growing from nothing, as becoming nascent. Then the first ratio they have will be the first or *prime ratio* of the nascent augments, which is again (he said) the same as the ratio of fluxions.

Newton's solution of grounding his fluxional calculus on the *ultimate ratios of evanescent augments* and the *prime ratios of nascent augments* did not prove adequate in the long run, for the question of whether these ratios actually exist was not cogently justified. (The Leibnizian calculus, it is fair to say, was hardly in better shape foundationally.) Newton's views became influential, however, because he finally published some account of his calculus in 1704, as an appendix to his *Opticks*. This treatise, *Tractatus de Quadratura Curvarum* (Treatise on the Quadrature of Curves) was a shortened version of a manuscript that Newton wrote in the early 1690s, and was nearly his last piece of creative mathematics, as well as virtually the first sample of his mathematics to be published, after what his biographer has described as 'more than thirty years of delay and evasion'.[19]

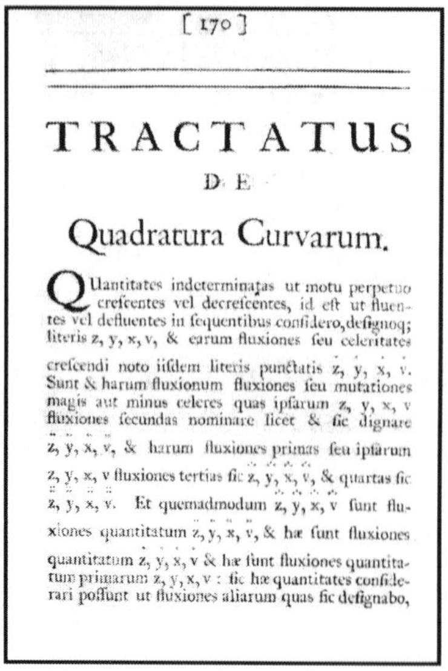

[170]

TRACTATUS

DE

Quadratura Curvarum.

Quantitates indeterminatas ut motu perpetuo crescentes vel decrescentes, id est ut fluentes vel defluentes in sequentibus considero, designoq; literis z, y, x, v, & earum fluxiones seu celeritates crescendi noto iisdem literis punctatis $\dot{z}$, $\dot{y}$, $\dot{x}$, $\dot{v}$. Sunt & harum fluxionum fluxiones seu mutationes magis aut minus celeres quas ipsarum z, y, x, v fluxiones secundas nominare licet & sic dignare $\ddot{z}$, $\ddot{y}$, $\ddot{x}$, $\ddot{v}$, & harum fluxiones primas seu ipsarum z, y, x, v fluxiones tertias sic $\dddot{z}$, $\dddot{y}$, $\dddot{x}$, $\dddot{v}$, & quartas sic $\ddddot{z}$, $\ddddot{y}$, $\ddddot{x}$, $\ddddot{v}$. Et quemadmodum $\dddot{z}$, $\dddot{y}$, $\dddot{x}$, $\dddot{v}$ sunt fluxiones quantitatum $\ddot{z}$, $\ddot{y}$, $\ddot{x}$, $\ddot{v}$, & hæ sunt fluxiones quantitatum $\dot{z}$, $\dot{y}$, $\dot{x}$, $\dot{v}$ & hæ sunt fluxiones quantitatum primarum z, y, x, v : sic hæ quantitates considerari possunt ut fluxiones aliarum quas sic designabo,

Figure 4.7. The opening page of Newton's 'Treatise on the Quadrature of Curves'

We conclude with Newton's account of prime and ultimate (or first and last) ratios from this treatise, in which the role of motion is also helpfully summarised.[20]

Newton on fluxions and fluents.

> I don't here consider Mathematical Quantities as composed of Parts *extreamly small*, but as *generated by a continual motion.* Lines are described, and by describing are generated, not by any apposition of Parts, but by a continual motion of Points. Surfaces are generated by the motion of Lines, Solids by the motion of Surfaces, Angles by the

[19] Westfall, *Never at Rest*, p. 639.

[20] See (Whiteside 1964, 141).

> Rotation of their Legs, Time by a continual flux, and so in the rest. These Geneses are founded upon Nature, and are every Day seen in the motion of Bodies ...
>
> Therefore considering that Quantities, encreasing in equal times, and generated by this encreasing, are greater or less, according as their Velocity by which they encrease, and are generated, is greater or less; I endeavoured after a Method of determining the Quantities from the Velocities of their Motions or Increments, by which they are generated; and by calling the Velocities of the Motions, or of the Augments, by the Name of *Fluxions*, and the generated Quantities *Fluents*, I (in the Years 1665 and 1666) did, by degrees, light upon the Method of Fluxions, which I here make use of in the *Quadrature of Curves*.
>
> Fluxions are very nearly as the Augments of the Fluents, generated in equal, but infinitely small parts of Time; and to speak exactly, are in the *Prime Ratio* of the nascent Augments: but they may be expounded by any Lines that are proportional to 'em.

4.4 Leibniz's mature calculus

We now turn to one of the most celebrated and important papers ever published in mathematics: Leibniz's account of his differential calculus, published in the newly founded journal, the *Acta Eruditorum Lipsiensium* (Acts of the Scholars of Leipzig).[21]

First a few words about it. Leibniz's account is long, and it is certainly not an easy piece (Leibniz's knack of making trivial mistakes did not help). Those who learned the calculus from it were very good mathematicians in their own right. But a published work has more influence than a mere manuscript, and this paper is a convenient peg on which to hang a date for the public birth of the calculus.

But it is more than that. Leibniz's calculus was a vigorous creature by the time it came of age, not only capable of disparaging Newton's (valid) claims to priority but, much more importantly, of asserting its own powerful methods of argument. A movement got under way with this paper and its companion on the integral calculus, published in 1686: a movement full of influences, debates, discussions, controversies, and even feuds, for it was Leibniz's calculus, not Newton's, that came to dominate the next half-century or more. That is the first significance of this short paper, which we now consider.[22]

Leibniz's differential calculus.

> *A new method for maxima and minima as well as tangents, which is neither impeded by fractional nor irrational quantities, and a remarkable type of calculus for them.*
>
> Let an axis AX [Figure 4.9] and several curves such as VV, WW, YY, ZZ be given, of which the ordinates VX, WX, YX, ZX, perpendicular to the

[21] The journal was founded in 1682 by Leibniz and Otto Mencke. It carried articles in Latin, many on mathematical and scientific subjects, and was published monthly. Over the years many European scholars published in its pages.

[22] See Struik, *A Source Book*, 272–280, F&G 13.A3, and (Stedall 2008, 120–133) where a copy of the Latin original is also given.

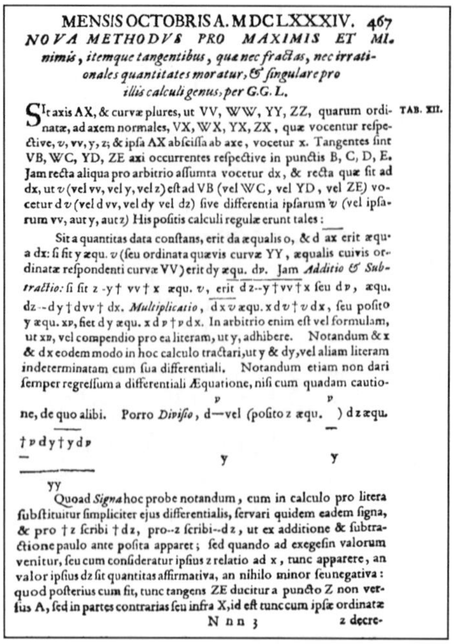

MENSIS OCTOBRIS A. MDCLXXXIV. 467

NOVA METHODUS PRO MAXIMIS ET MI-nimis, itemque tangentibus, quæ nec fractas, nec irrationales quantitates moratur, & ſingulare pro illis calculi genus, per G.G. L.

SIt axis AX, & curvæ plures, ut VV, WW, YY, ZZ, quarum ordinatæ, ad axem normales, VX, WX, YX, ZX, quæ vocentur reſpective, *v*, vv, y, z; & ipſa AX abſciſſa ab axe, vocetur x. Tangentes ſint VB, WC, YD, ZE axi occurrentes reſpective in punctis B, C, D, E. Jam recta aliqua pro arbitrio aſſumta vocetur dx, & recta quæ ſit ad dx, ut *v* (vel vv, vel y, vel z) eſt ad VB (vel WC, vel YD, vel ZE) vocetur d*v* (vel d vv, vel dy vel dz) ſive differentia ipſarum *v* (vel ipſarum vv, aut y, aut z) His poſitis calculi regulæ erunt tales: TAB. XII.

Sit a quantitas data conſtans, erit da æqualis 0, & d ax erit æqu. a dx: ſi ſit y æqu. *v* (ſeu ordinata quævis curvæ YY, æqualis cuivis ordinatæ reſpondenti curvæ VV) erit dy æqu. d*v*. Jam *Additio & Subtractio:* ſi ſit z -y † vv † x æqu. *v*, erit d z--y † vv † x ſeu d*v*, æqu. dz --d y † d vv † dx. *Multiplicatio*, d x *v* æqu. x d *v* † *v* dx, ſeu poſito y æqu. x*v*, fiet d y æqu. x d *v* † *v* d x. In arbitrio enim eſt vel formulam, ut x*v*, vel compendio pro ea literam, ut y, adhibere. Notandum & x & d x eodem modo in hoc calculo tractari, ut y & dy, vel aliam literam indeterminatam cum ſua differentiali. Notandum etiam non dari ſemper regreſſum a differentiali Æquatione, niſi cum quadam cautione, de quo alibi. Porro *Diviſio*, d $\frac{v}{y}$ vel (poſito z æqu. $\frac{v}{y}$) d z æqu. $\frac{\dagger v\,dy \dagger y\,dv}{yy}$

Quoad *Signa* hoc probe notandum, cum in calculo pro litera ſubſtituitur ſimpliciter ejus differentialis, ſervari quidem eadem ſigna, & pro † z ſcribi † d z, pro--z ſcribi--d z, ut ex additione & ſubtractione paulo ante poſita apparet; ſed quando ad exegeſin valorum venitur, ſeu cum conſideratur ipſius z relatio ad x, tunc apparere, an valor ipſius dz ſit quantitas affirmativa, an nihilo minor ſeu negativa: quod poſterius cum fit, tunc tangens ZE ducitur a puncto Z non verſus A, ſed in partes contrarias ſeu infra X, id eſt tunc cum ipſæ ordinatæ

N n n 3 z decre-

Figure 4.8. The opening page of Leibniz's first article on the calculus

axis, are called v, w, y, z respectively. The segment AX, cut off from the axis is called x. Let the tangents be VB, WC, YD, ZE, intersecting the axis respectively at B, C, D, E. Now some straight line selected arbitrarily is called dx, and the line which is to d as v (or w, or y, or z) is to XB (or XC, or XD, or XE) is called dv (or dw, or dy, or dz), or the difference of these v (or w, or y, or z). Under these assumptions we have the following rules of the calculus.

If a is a given constant, then $da = 0$, and $d(ax) = adx$. If $y = v$ (that is, if the ordinate of any curve YY is equal to any corresponding ordinate of the curve VV), then $dy = dv$.

Now *addition* and *subtraction*: if $z-y+w+x = v$, then $d(z-y+w+x) = dv = dz - dy + dw + dx$.

Multiplication: $d(xv) = xdv + vdx$, or, setting $y = xv$, $dy = xdv + vdx$. It is indifferent whether we take a formula such as xv or its replacing letter such as y. It is to be noted that x and dx are treated in this calculus in the same way as y and dy, or any other indeterminate letter with its difference. It is also to be noted that we cannot always move backward from a differential equation without some caution, something which we shall discuss elsewhere.

Now *division*:

$$d\frac{v}{y} \quad \text{or} \quad \left(\text{if } z = \frac{v}{y}\right) dz = \frac{\pm v dy \mp y dv}{yy}.$$

Here Leibniz has set out the rules for differentiating algebraic expressions, without indicating how they might be proved. Next he discussed the relationship between

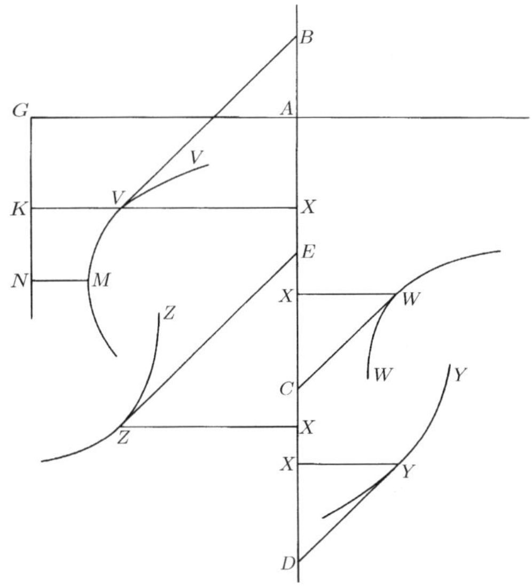

Figure 4.9. Leibniz on differentiation

a quantity and its differential (regarded now as a small increment in the variable) and explained that when dx is positive the value of x may be increasing, or decreasing, or stationary, in which case the variable has a local maximum or minimum, and the tangent is parallel to the x-axis. He then discussed the case where the curve traced by the variable v has an inflection point, and showed how second differentials (differentials of differentials) are involved.

> The following should be kept well in mind about the *signs*. When in the calculus for a letter simply its differential is substituted, then the signs are preserved; for z we write dz, for $-z$ we write $-dz$, as appears from the previously given rule for addition and subtraction. However, when it comes to an explanation of the values, that is, when the relation of z to x is considered, then we can decide whether dz is a positive quantity or less than zero (or negative). When the latter occurs, then the tangent ZE is not directed toward A, but in the opposite direction, down from X. This happens when the ordinates z decrease with increasing x. And since the ordinates v sometimes increase and sometimes decrease, dv will sometimes be positive and sometimes be negative; in the first case the tangent VB is directed toward A, in the latter it is directed in the opposite sense. None of these cases happens in the intermediate position at M, at the moment when v neither increases nor decreases, but is stationary. Then $dv = 0$, and it does not matter whether the quantity is positive or negative, since $+0 = -0$. At this place v, that is, the ordinate LM, is *maximum* (or, when the convexity is turned to the axis, *minimum*), and the tangent to the curve at M is directed neither in the direction from X up to A, to approach the axis, nor down to the other side, but is parallel to the axis. When dv is infinite with respect to dx, then the tangent is perpendicular to the

axis, that is, it is the ordinate itself. When $dv = dx$, then the tangent makes half a right angle with the axis. When with increasing ordinates t; its increments or differences dv also increase (that is, when dv is positive, $d\,dv$, the difference of the differences, is also positive, and when dv is negative, $d\,dv$ is also negative), then the curve turns toward the axis its concavity, in the other case its convexity. Where the increment is maximum or minimum, or where the increments from decreasing turn into increasing, or the opposite, there is a point of inflection. Here *concavity* and *convexity* are interchanged, provided the ordinates too do not turn from increasing into decreasing or the opposite, because then the concavity or convexity would remain. However, it is impossible that the increments continue to increase or decrease, but the ordinates turn from increasing into decreasing, or the opposite. Hence a *point of inflection* occurs when $d\,dv = 0$ while neither v nor $dv = 0$. The problem of finding inflection therefore has not, like that of finding a maximum, two equal roots, but three. This all depends on the correct use of the signs.

Sometimes it is better to use *ambiguous signs*, as we have done with the division, before it is determined what the precise sign is. When, with increasing x, v/y increases (or decreases), then the ambiguous signs in

$$d\frac{v}{y} = \frac{\pm v dy \mp y dv}{yy}$$

must be determined in such a way that this fraction is a positive (or negative) quantity. But $\mp$ means the opposite of $\pm$, so that when one is $+$ the other is $-$ or vice versa. There also may be several ambiguities in the same computation, which I distinguish by parentheses. For example, let

$$\frac{v}{y} + \frac{y}{z} + \frac{x}{v} = w;$$

then we must write

$$\frac{\pm v dy \mp y dv}{yy} + \frac{(\pm) y dz (\mp) z dy}{zz} + \frac{((\pm)) x dv ((\mp)) v dx}{vv} = d[w],$$

so that the ambiguities in the different terms may not be confused. We must take notice that an ambiguous sign with itself gives $+$, with its opposite gives $-$, while with another ambiguous sign it forms a new ambiguity depending on both.

Leibniz then indicated how his four rules apply to powers and roots of a variable, and observed that his conclusions, although not new results, were sufficiently useful to merit being stated explicitly. He could have said the same earlier about subtraction and division, and observed explicitly that $x^{-1} = 1/x$, but he did not.

Powers. $dx^a = ax^{a-1}dx$; for example, $dx^3 = 3x^2dx$. $d\frac{1}{x^a} = -\frac{adx}{x^{a[+]1}}$; for example, if $w = \frac{1}{x^3}$, then $dw = -\frac{3dx}{x^4}$.

Roots. $d\sqrt[b]{x^a} = \frac{a}{b}dx\sqrt[b]{x^{a-b}}$ (hence $d\sqrt[2]{y} = \frac{dy}{2\sqrt[2]{y}}$, for in this case $a = 1$, $b = 2$), therefore $\frac{a}{b}dx\sqrt[b]{x^{a-b}} = \frac{1}{2}\sqrt[2]{y^{-1}}$, but y^{-1} is the same as $\frac{1}{y}$; from

> the nature of the exponents in a geometric progression, and
>
> $$\sqrt[2]{\frac{1}{y}} = \frac{1}{\sqrt[2]{y}}, \qquad d\frac{1}{\sqrt[b]{x^a}} = \frac{-adx}{b\sqrt[2]{x^{a+b}}}.$$
>
> The law for integral powers would have been sufficient to cover the case of fractions as well as roots, for a power becomes a fraction when the exponent is negative, and changes into a root when the exponent is fractional. However, I prefer to draw these conclusions myself rather than relegate their deduction to others, since they are quite general and occur often. In a matter that is already complicated in itself it is preferable to facilitate the operations.

Only now did Leibniz call his method the *differential calculus* and connect it to the study of differential equations, which are, literally, equations between differentials. He claimed that his calculus was entirely general, and that it applied to both transcendental as well as algebraic curves, and that it enabled one to find tangents at any point on any curve.

> Knowing thus the *Algorithm* (as I may say) of this calculus, which I call *differential calculus*, all other differential equations can be solved by a common method. We can find maxima and minima as well as tangents without the necessity of removing fractions, irrationals, and other restrictions, as had to be done according to the methods that have been published hitherto. The demonstration of all this will be easy to one who is experienced in these matters and who considers the fact, until now not sufficiently explored, that dx, dy, dv, dw, dz can be taken proportional to the momentary differences, that is, increments or decrements, of the corresponding x, y, v, w, z. To any given equation we can thus write its differential equation. This can be done by simply substituting for each term (that is, any part which through addition or subtraction contributes to the equation) its differential quantity. For any other quantity (not itself a term, but contributing to the formation of the term) we use its differential quantity, to form the differential quantity of the term itself, not by simple substitution, but according to the prescribed Algorithm. The methods published before have no such transition. They mostly use a line such as DX or of similar kind, but not the line dy which is the fourth proportional to DX, DY, dx — something quite confusing. From there they go on removing fractions and irrationals (in which undetermined quantities occur). It is clear that our method also covers transcendental curves — those that cannot be reduced by algebraic computation, or have no particular degree — and thus holds in a most general way without any particular and not always satisfied assumptions.
>
> We have only to keep in mind that to find a *tangent* means to draw a line that connects two points of the curve at an infinitely small distance, or the continued side of a polygon with an infinite number of angles, which for us takes the place of the *curve*. This infinitely small distance can always be expressed by a known differential like dv, or by

a relation to it, that is, by some known tangent. In particular, if y were a transcendental quantity, for instance the ordinate of a cycloid, and it entered into a computation in which z, the ordinate of another curve, were determined, and if we desired to know dz or by means of dz the tangent of this latter curve, then we should by all means determine dz by means of dy, since we have the tangent of the cycloid. The tangent to the cycloid itself, if we assume that we do not yet have it, could be found in a similar way from the given property of the tangent to the circle.

In the next few paragraphs Leibniz took an artificial and deliberately complicated example to display the power of his rules and to indicate how they can be used. Then, as he put it, 'it remains to show their use in cases easier to grasp', and he gave the problem of refraction that had earlier been studied by Descartes, Snell, and Fermat. The problem after that is a more contrived example, but is typical of the geometrical problems of Leibniz's time.

Now I shall propose an example of the calculus, in which I shall indicate division by $x : y$, which means the same as x divided by y, or x/y. Let the *first* or given equation be[23]

$$x : y + (a+bx)(c-xx) : (ex+fxx)^2 + ax\sqrt{gg+yy} + yy : \sqrt{hh+lx+mxx} = 0.$$

It expresses the relation between x and y or between AX and XY, where a, b, c, e, f, g, h are given. We wish to draw from a point Y the line YD tangent to the curve, or to find the ratio of the line DX to the given line XY. We shall write for short $n = a+bx$, $p = c-xx$, $q = ex+fxx$, $r = gg+yy$, and $s = hh+lx+mxx$. We obtain

$$x : y + np : qq + ax\sqrt{r} + yy : \sqrt{s} = 0,$$

which we call the *second* equation. From our calculus it follows that

$$d(x : y) = (\pm xdy \mp ydx) : yy,$$

and equally that

$$d(np : qq) = [(\pm)2npdq(\mp)q(ndp+pdn)] : q^3,$$
$$d(ax : \sqrt{r}) = +axdr : 2\sqrt{r} + adx\sqrt{r},$$
$$d(yy : s) = ((\pm))yyds((\mp))4ysdy : 2s\sqrt{s}.$$

All these differential quantities from $d(x : y)$ to $d(yy : \sqrt{s})$ added together give 0, and thus produce a *third* equation, obtained from the terms of the second equation by substituting their differential quantities. Now $dn = bdx$ and $dp = -2xdx$, $dq = edx + 2fxdx$, $dr = 2ydy$, and $ds = ldx + 2mxdx$. When we substitute these values into the third equation we obtain a *fourth* equation, in which the only remaining differential quantities, namely dx, dy, are all outside of the denominators and without restrictions. Each term is multiplied either by dx or by dy, so that the law of homogeneity always holds with respect to these two quantities, however complicated the computation may be. From this

[23]Note that the left-hand side of the equation is made up of three ratios added together.

we can always obtain the value of $dx : dy$, the ratio of dx to dy, or the ratio of the required DX to the given XY. In our case this ratio will be (if the fourth equation is changed into a proportionality):

$$\mp x : yy - axy : \sqrt{r}(\mp)2y : \sqrt{s}$$

divided by

$$\mp 1 : y(\pm)(2npe + 2fx) : q^3(\mp)(-2nx + pb) : qq + a\sqrt{r}((\pm))yy(l + 2mx) : 2s\sqrt{s}.$$

Now x and y are given since point Y is given. Also given are the values of n, p, q, r, s expressed in x and y, which we wrote down above. Hence we have obtained what we required. Although this example is rather complicated we have presented it to show how the above-mentioned rules can be used even in a more difficult computation. Now it remains to show their use in cases easier to grasp.

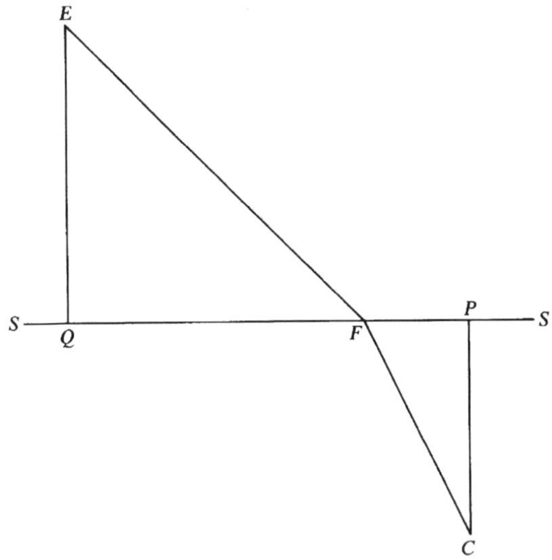

Figure 4.10. Leibniz on refraction

Let two points C and E [Figure 4.10] be given and a line SS in the same plane. It is required to find a point F on SS such that when E and C are connected with F the sum of the rectangle of CF and a given line h and the rectangle of FE and a given line r are as small as possible. In other words, if SS is a line separating two media, and h represents the density of the medium on the side of C (say water), r that of the medium on the side of E (say air), then we ask for the point F such that the path from C to E via F is the shortest possible. Let us assume that all such possible sums of rectangles, or all possible paths, are represented by the ordinates KV of curve VV perpendicular to the line GK [Figure 4.9]. We shall call these ordinates w. Then it is required to find their minimum NM. Since C and E [Figure 4.10] are given, their perpendiculars to SS are also given, namely CP (which we call c) and EQ (which we call e); moreover PQ (which we call p) is given. We denote $QF = GN$ (or AX) by x, CF by f, and EF by g. Then $FP = p - x$, $f = \sqrt{cc + pp - 2px + xx}$ or $= \sqrt{l}$ for short; $g = \sqrt{ee + xx}$ or $= \sqrt{m}$ for short. Hence

$$w = h\sqrt{l} + r\sqrt{m}.$$

The differential equation (since $dw = 0$ in the case of a minimum) is, according to our calculus,

$$0 = +hdl : 2\sqrt{l} + rdm : 2\sqrt{m}.$$

But $dl = -2(p - x)dx$, $dm = 2xdx$; hence

$$h(p - x) : f = rx : g.$$

When we now apply this to dioptrics, and take f and g, that is, CF and EF, equal to each other (since the refraction at the point F is the same no matter how long the line CF may be), then $h(p - x) = rx$ or $h : r = x : (p - x)$, or $h : r = QF : FP$; hence the sines of the angles of incidence and of refraction, FP and QF, are in inverse ratio to r and h, the densities of the media in which the incidence and the refraction take place. However, this density is not to be understood with respect to us, but to the resistance which the light rays meet. Thus we have a demonstration of the computation exhibited elsewhere in these *Acta*, where we presented a general foundation of optics, catoptrics, and dioptrics. Other very learned men have sought in many devious ways what someone versed in this calculus can accomplish in these lines as by magic.

This I shall explain by still another example. Let 13 [Figure 4.11] be a curve of such a nature that, if we draw from one of its points, such as 3, six lines 34, 35, 36, 37, 38, 39 to six fixed points 4, 5, 6, 7, 8, 9 on the axis, then their sum is equal to a given line.

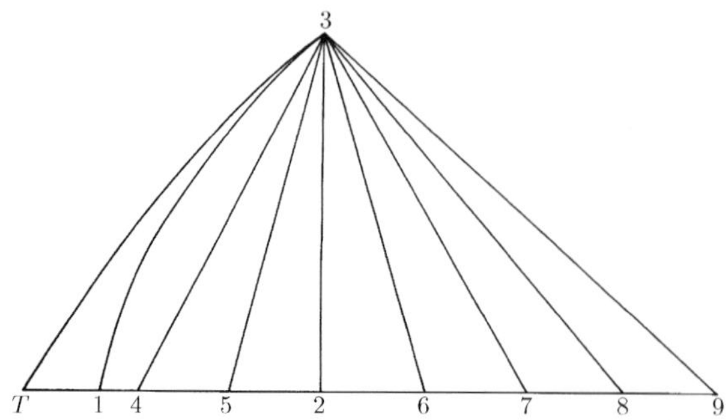

Figure 4.11. Leibniz on differentiation, concluded

> Let $T14526789$ be the axis, 12 the abscissa, 23 the ordinate, and let the tangent $3T$ be required.[24] Then I claim that $T2$ is to 23 as
>
> $$\frac{23}{34}+\frac{23}{35}+\frac{23}{36}+\frac{23}{37}+\frac{23}{38}+\frac{23}{39} \quad \text{is to} \quad -\frac{24}{34}-\frac{25}{35}+\frac{26}{36}+\frac{27}{37}+\frac{28}{38}+\frac{29}{39}.$$
>
> The same rule will hold if we increase the number of terms, taking not six but ten or more fixed points. If we wanted to solve this problem by the existing tangent methods, removing irrationals, then it would be a most tedious and sometimes insuperable task; in this case we would have to set up the condition that the rectangular planes and solids which can be constructed by means of all possible combinations of two or three of these lines are equal to a given quantity. In all these cases and even in more complicated ones our methods are of astonishing and unequaled facility.

Leibniz ended his presentation with a reference to the problem that had started the research into inverse tangent problems (see Figure 4.9).

> And this is only the beginning of much more sublime Geometry, pertaining to even the most difficult and most beautiful problems of applied mathematics, which without our differential calculus or something similar no one could attack with any such ease. We shall add as appendix the solution of the problem which Debeaune proposed to Descartes and which he tried to solve in Volume 3 of the *Letters*, but without success. It is required to find a curve WW such that, its tangent WC being drawn to the axis, XC is always equal to a given constant line a. Then XW or w is to XC or a as dw is to dx. If dx (which can be chosen arbitrarily) is taken constant, hence always equal to, say, b, that is, x or AX increases uniformly, then $w = \frac{a}{b}dw$. Those ordinates w are therefore proportional to their dw, their increments or differences, and this means that if the x form an arithmetic progression, then the w form a geometric progression. In other words, if the w are numbers, the x will be logarithms, so that the curve WW is logarithmic.

Leibniz seems to have hoped for great things with this paper. In the title alone, he claimed that the method was new; that it was of great generality for solving the geometrical problems of maxima, minima, and tangency; that it dealt with both fractions and irrationals; and that it was a calculus — a set of rules. In fact the Latin proudly claims that the method was not obstructed by fractions or irrationals which had been obstacles in the path of other approaches.

Did he keep these promises? The first paragraph is utterly obscure at the crucial point, the introduction of dx. It is apparently finite! Since this agrees neither with Leibniz's earlier ideas nor with the methods of the integral calculus, one must suppose

[24] In the equation that follows, Leibniz gave some segments a minus sign because all his segments are positive. The tangent is found by replacing each distance with an expression in x and y and differentiating, and does not require that the equation of the curve 13 be found explicitly.

Box 5. Using Leibniz's rules for differentiation.
Let $v = x^3 - ax^2 + axy - y^3 = 0$. Then

$$dv = d(x^3 - ax^2 + axy - y^3) = d0 = 0$$

(by rule (1) because 0 is a constant)

$$= d(x^3) - d(ax^2) + d(axy) - d(y^3) \quad \text{(by rule (2))}$$

$$= x^2dx + xd(x^2) - ad(x^2) + a(ydx + xdy) - y^2dy - yd(y^2)$$

(by rule (3) considering x^3 as x^2x, etc.).

But, by rule (3),

$$d(x^2) = d(xx) = xdx + xdx = 2xdx \quad \text{and} \quad d(y^2) = 2ydy, \text{ so}$$

$$dv = x^2dx + 2x^2dx - 2axdx + aydx + axdy - y^2dy - 2y^2dy$$
$$= (3x^2 - 2ax + ay)dx + (ax - 3y^2)dy = 0.$$

So

$$\frac{dy}{dx} = \frac{3x^2 - 2ax + ay}{3y^2 - ax}.$$

that Leibniz was trying to head off some logical difficulties at the start of the paper — at the cost of having to eat his words later. Things improve in the second paragraph where he presented his 'rules of the calculus'.

1. If a is constant, then $da = 0$, and $d(ax) = adx$
2. $d(x + y) = dx + dy$
3. $d(xv) = vdx + xdv$
4. $d\left(\frac{v}{y}\right) = \frac{ydv - vdy}{y^2}$.

This is dramatic stuff, ruthlessly putting the algebraic rules before the geometrical content. It has no precedent in print — this is symbol crunching for which no prior understanding of geometry was required. As such, it has something of the shocking quality of any technological solution to a supposedly philosophical problem. Leibniz's calculus, even more obviously than Newton's, is a matter of carrying out routine tasks (although the conceptual difficulties have only been covered up, not resolved). As a consequence, the method is relatively easy to employ. You can well imagine carrying it out even if you do not understand it, just as one might use a computer today without knowing how it works. An example of it in use is given in Box 5.

To find tangents, maxima, and minima using his method, Leibniz explained, not too lucidly, that given an equation in x and y that describes a curve the ratio $dy : dx$ gives the slope of the tangent to the curve at a point. When $dy = 0$, the slope is horizontal, and so the curve may have a local maximum or minimum. You can glean this from the first half of the third paragraph, but it is certainly a help if you know what he is trying to say. A little lower down, there are some more rules.

5. $dx^a = ax^{a-1}dx$; for example, $dx^3 = 3x^2dx$

6. $d\sqrt[b]{x^a} = \frac{a}{b}\sqrt[b]{x^{a-b}}$. Using Newton's system of fractional indices, this is more recognisable as $dx^{a/b} = \frac{a}{b}x^{(a-b)/b}dx$. (Fractional and irrational powers were a problem for earlier mathematicians, but not for Leibniz.)

But then come two more paragraphs that raise more problems than they solve. The first paragraph advertises his method, but his claim that 'It is clear, that our method also covers transcendental curves ...' is simply not clear. Nor did Leibniz go on to make it plausible, because his six rules seem very general, and the claim is neither proved (how could it be?) nor illustrated by a sufficiently profound example. In the second of these paragraphs, even if the idea of a curve as an infinite-sided polygon does not stop you in your tracks, the 'infinitely small distance ... dv' might. Differentials were introduced as finite things. It is rare that an exercise in papering over the cracks fails quite so quickly, although Leibniz did go on to explain how matters could be put right: the finite differentials vary in proportion to the infinitely small differences. By reverting to differentials of an infinitesimal size Leibniz was doing nothing that would have upset his skilled contemporary readers: as we shall see in the next section, little fuss was made about such things. But by changing his mind in mid-article, Leibniz did nothing to help his reader to understand what he might have understood by infinitesimals either.[25]

Leibniz on the integral calculus. Although we shall not discuss it in depth, we should mention the article in which, among other things, Leibniz sketched his approach to the integral calculus. It was in this later paper that Leibniz considered the inverse relationship between differentiation and integration, d and $\int$, and showed how his method dealt with transcendental curves, treating the cycloid explicitly.[26]

Leibniz on the integral calculus.

> For transcendental problems, wherever dimensions and tangents occur that have to be found by computation, there can hardly be found a calculus more useful, shorter, and more universal than my *differential calculus, or analysis of indivisibles and infinites,* of which only a small sample or corollary is contained in my Method of Tangents published in the *Acta* of October 1684. It has been much praised by Dr. Craig, who has also suspected that there is more to it, and on p. 29 of his little book[27] has made an attempt to prove Barrow's theorem (that the sum of the intervals between the ordinates and perpendiculars to a curve taken on the axis and measured in it is equal to half of the square of the final ordinate). In trying this he deviates a bit from his goal, which does not surprise me in the new method: so that I believe that I may oblige him and others by publishing here an addition to a subject that seems to have so wide a use. From it flow all the admirable theorems and problems of this kind with such ease that there is no more need

[25] Most likely he thought of them as a convenient way to talk about limits, and not as really existing (mathematical or physical) objects, see (Arthur 2013).

[26] See Leibniz, *Mathematische Schriften*, V, 230–231. This translation is taken from Struik, *A Source Book*, 281–282.

[27] See (Craig 1687), which refers to Leibniz's differential calculus.

to teach and retain them than for him who knows our present algebra to memorize many theorems of ordinary geometry.

I proceed to this subject in the following way. Let the ordinate be x, the abscissa y, and the interval between perpendicular and ordinate, described before, p. Then according to my method it follows immediately that $pdy = xdx$, as Dr. Craig has also found. When we now subject this differential equation to summation we obtain $\int pdy = \int xdx$ (like powers and roots in ordinary calculations, so here sum and difference, or $\int$ and d, are each other's converse). Hence we have $\int pdy = \int xdx$, which was to be demonstrated. Now I prefer to use dx and similar symbols rather than special letters, since this dx is a certain modification of the x and by virtue of this it happens that — when necessary — only the letter x with its powers and differentials enters into the calculus, and transcendental relations are expressed between x and some other quantity. Transcendental curves can therefore also be expressed by an equation, for example, if a is an arc, and the versed sine[28] x, then $a = \int dx : \sqrt{2x - x^2}$ and if the ordinate of a cycloid is y, then $y = \sqrt{2x - xx} + \int dx : \sqrt{2x - xx}$, which equation perfectly expresses the relation between the ordinate y and the abscissa x. From it all properties of the cycloid can be demonstrated. The analytic calculus is thus extended to those curves that hitherto have been excluded for no better reason than that they were thought to be unsuited to it. Wallis's interpolations and innumerable other questions can be derived from this.

We give one further example of Leibniz's approach, his proof of the Fundamental Theorem of the Calculus that he published in the *Acta Eruditorum* in 1693.[29]

Leibniz on the Fundamental Theorem of the Calculus.

I shall now show *that the general problem of quadratures can be reduced to the finding of a line that has a given law of tangency (declivitas),* that is, for which the sides of the characteristic triangle have a given mutual relation. Then I shall show how this line can be described by a motion that I have invented. For this purpose [Figure 4.12] I assume for every curve $C(C')$ a *double characteristic triangle,* one, TEG, that is assignable, and one, GLC, that is inassignable, and these two are similar. The inassignable triangle consists of the parts GL, LC, with the elements of the coordinates CF, CB as sides, and GC, the element of arc, as the base or hypotenuse. But the assignable triangle TBC consists of the axis, the ordinate, and the tangent, and therefore contains the curve at the given point C. Now let $F(H)$, the region of which the area has to be squared, be enclosed between the curve $H(H)$, the parallel lines FH and $(F)(H)$, and the axis $F(F)$; on that axis let A be a fixed point, and let a line AB, the conjugate axis, be drawn through A

[28]The versed sine of a is $1 - \cos a$. So $dx = \sin a$ and $2x - x^2 = 1 - \cos^2 a = \sin^2 a$.

[29]In *Acta Eruditorum*, 1693, 385–392, reprinted in Leibniz, *Mathematische Schriften*, V, 294–301. This extract is on pp. 298–299, and is taken from Struik, *A Source Book*, 282–284.

perpendicular to AF. We assume that point C lies on HF (continued if necessary); this gives a new curve $C(C')$ with the property that, if from point C to the conjugate axis AB (continued if necessary) both its ordinate CB (equal to AF) and tangent CT are drawn, the part TB of the axis between them is to BC as HF to a constant [segment] a, or a times BT is equal to the rectangle AFH (circumscribed about the trilinear figure $AFHA$).

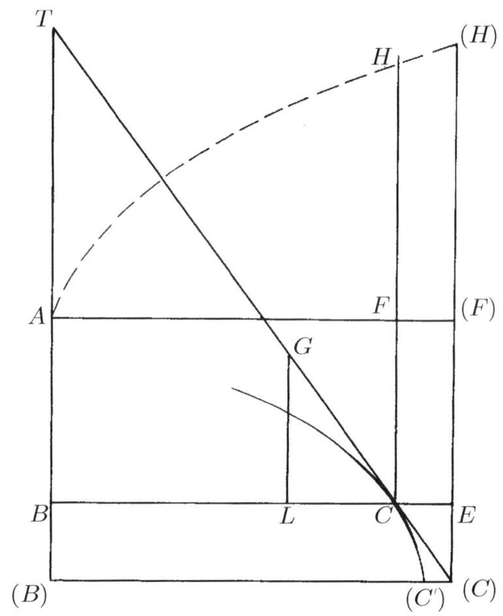

Figure 4.12. Leibniz on the Fundamental Theorem of the Calculus

This being established, I claim that the rectangle on a and $E(C)$ (we must discriminate between the ordinates FC and $(F)(C)$ of the curve) is equal to the region $F(H)$. When therefore I continue line $H(H)$ to A, the trilinear figure $AFHA$ of the figure to be squared is equal to the rectangle with the constant a and the ordinate FC of the squaring curve as sides. This follows immediately from our calculus. Let $AF = y$, $FH = z$, $BT = t$, and $FC = x$; then $t = zy : a$, according to our assumption; on the other hand, $t = ydx : dy$ because of the property of the tangents expressed in our calculus. Hence $adx = zdy$ and therefore $ax = \int zdy = AFHA$. Hence the curve $C(C')$ is the quadratrix with respect to the curve $H(H)$, while the ordinate FC of $C(C')$, multiplied by the constant a, makes the rectangle equal to the area, or the sum of the ordinates $H(H)$ corresponding to the corresponding abscissas AF. Therefore, since $BT : AF = FH : a$ (by assumption), and the relation of this FH to AF (which expresses the nature of the figure to be squared) is given, the relation of BT to FH or to BC, as well as that of BT to TC, will be given, that is, the relation between the sides of triangle

TBC. Hence, all that is needed to be able to perform the quadratures and measurements is to be able to describe the curve $C(C')$ (which, as we have shown, is the quadratrix), when the relation between the sides of the assignable characteristic triangle TBC (that is, the law of inclination of the curve) is given.

Perhaps the most important thing to note is the fine quality of Leibniz's intuition. He had the insight to do the do-able, and avoid the unsolvable. His rules really do simplify the problem; they are easy to use and of immense generality, and consequently his calculus could be taken up by others. All in all, it was a spectacular case of selecting the right problem to solve.

If it seemed impossible to say with complete rigour why the rules worked, Leibniz did not, however, settle for a comforting, if superficial, sense of plausibility, that our account may have suggested. Instead, throughout his life he outlined and defended a theory of infinitesimals that is outlined in Box 6. Not all of his correspondents and successors were persuaded — the calculus was to defy those who sought to rigorise it for at least another century.

4.5 A comparison

We conclude our account of the mature calculi of Newton and Leibniz with two comparisons drawn from recent historical studies. In the first, the Italian historian Domenico Bertoloni Meli offers a vindication of the position that Newton and Leibniz did indeed invent different things.[30]

Newton

- Curves have continuous curvature. If they are treated as polygons, this procedure must be seen either as an approximation, or as a preliminary step in the calculations; ultimately the sides of the polygon become vanishingly small.
- Fluxions express the speed of change of a variable and are finite. They result from variables flowing continuously, almost always with respect to time. Hence kinematics is part of the foundations of the Newtonian calculus.
- Fluxions and fluents are attributes of a variable. They leave the order of infinity unchanged; that is, they remain finite. Dimensions however, vary according to the variable with respect to which they are calculated. The fluxion of a velocity with respect to time is an acceleration; the fluent of an area with respect to a length is a volume.

Leibniz

- Curves are conceived as infinitangular polygons consisting of incomparably many rectilinear segments. Tangents are the prolongations of such segments.
- Variables range over a discrete sequence of incomparably near values, and differentials are the differences between contiguous pairs of such variables, such as ordinates, abscissae, or arc lengths. Differentials are indeterminate because the sequences of the relevant variable or the associated polygon can be chosen in infinitely many ways. Moreover, differentials can be given arbitrarily small values in the calculations so that by neglecting them in the appropriate circumstances the error in the result is less than any given quantity.

[30]See (Bertoloni Meli 1993, 68 and 72–73).

Box 6. Leibniz on infinitesimals

Leibniz often discussed mathematical problems using the language of infinitesimals. Even so, it was his view throughout his mature working life that infinite and infinitesimal quantities do not exist. How can this apparent contradiction be explained?

His argument that infinite quantities do not exist ran as follows: It was true of any quantity that the whole quantity is greater than any proper part of it, but this is not true of infinite quantities (as Galileo's Paradox (see Section 17.2) of the natural numbers and their squares makes clear). Therefore infinite things are not mathematical quantities. He likewise believed that talk of infinitesimal quantities led to a contradiction in which a whole was equal to a part, and so infinitesimals are not mathematical quantities.

However, that did not mean that they could not be used in mathematics. They could be used if it could be shown that each time they occurred there was an equivalent argument involving only finite quantities.

His various defences of this position cannot be fully analysed here, but they rest on what he called the 'Law of Continuity'. This is the idea that if two quantities differ by an arbitrarily small amount then they are equal. Thus, *if* the area under a curve was shown to differ from a sum of some finite areas (say, rectangles underneath the curve) by an arbitrarily small amount as the number of rectangles was increased, *then* the area under the curve was equal to that sum.

On the basis of this insight he deduced two conclusions. One was that treating infinitesimals in this way would lead to a suitable reductio proof of the conclusion. The other was that infinitesimals could be understood as arbitrarily small finite magnitudes and treated directly using the Law of Continuity.

Unfortunately, Leibniz never published a unified defence of his views, which are scattered through sundry publications, letters, and unpublished drafts written throughout his life. As a result, literalist interpretation of his words have been published by several authors (see (Blåsjö 2017) and (Jesseph 2015)). It seems, however, that Leibniz's position was that infinitesimals are useful fictions, as he called them, very useful and reliable but not quantities, and as such not mathematical objects with a real existence. The view taken here follows a paper by David Rabouin and Richard Arthur.[a]

[a] See (Rabouin and Arthur, 2020)

- Differentiation and integration are operations on variables and change the order of infinity, not the dimension of a variable. The differential of a length is an incomparably small length, whereas the integral of an incomparably small velocity is a finite velocity.

Bertoloni Meli concludes:

> These brief and schematic observations show that although the Newtonian and Leibnizian formulations of the calculus could perform analogous operations, their equivalence presents considerable difficulties and cannot be accepted without major qualifications. The notion of equivalence becomes misleading unless similarities and differences with regard to their conceptual basis and notation are spelt out.

Bertoloni Meli's Italian compatriot, Niccolò Guicciardini, has offered a somewhat different view, but one that tends to the same conclusion.[31] He notes that both Newton and Leibniz stated that

- actual infinitesimals do not exist, they are useful fictions employed to abbreviate proofs;
- infinitesimals should rather be defined as varying quantities in a state of approaching zero;
- infinitesimals can be completely avoided in favour of limit-based proofs, which constitute the rigorous formulation of calculus.

As for those slippery differentials, he commented that one can infer from both Leibniz's early manuscripts and his mature works, that for Leibniz actual differentials were just 'fictions', symbols without referential content, and he quoted from a letter that Leibniz wrote in 1706.[32]

> 'Philosophically speaking, I no more believe in infinitely small quantities than in infinitely great ones, ..., I consider both as fictions of the mind for succinct ways of speaking, appropriate to the calculus, as also are the imaginary roots in algebra.'

Guicciardini then moved to this rich conclusion, which we break into sections.[33] First, the similarities:

> The similarities between Newton's and Leibniz's approach to the foundation of the calculus are striking. Newton's approach to the question of the existence of infinitesimals is similar to Leibniz's. For Newton too, infinitesimals ('moments' or 'indefinitely little quantities') can be used as a shorthand for longer and more rigorous proofs given in terms of limits. Both Newton and Leibniz speak of infinitesimals as 'vanishing' quantities in such a way that these quantities seem to be defined as something in between zero and finite, as quantities in the state of disappearing, or coming into existence, in a fuzzy realm between nothing and finite. More often they make it clear that infinitesimals can be replaced in terms of limits.

Next, he remarked on a difference in the mathematical practices of the two men:

> I do not find a strong conceptual opposition between Leibniz and Newton, but rather different 'policies'. Both agreed that limits provide a rigorous foundation for the calculus. However, for Leibniz this was more a rhetorical move to defend the legitimacy of the differential algorithm, while for Newton this was a programme that should be implemented. Newton developed explicitly a theory of limits, publishing it in analytical and synthetic forms. Leibniz simply alluded to the possibility of building the calculus on the basis of such a theory .... The mature Leibniz felt the need to elaborate publicly the theory of limits mainly as a retrospective justification: typically when occupied in defending the calculus from critics. His idea was that such 'metaphysical' questions should not interfere with the successful development of the algorithm of differentials.

[31] See (Guicciardini 1999, 159).
[32] See (Guicciardini 1999, 159).
[33] See (Guicciardini 1999, 164).

Finally, he gave this conclusion:

> We thus begin to grasp why it is fruitful to conceive the Newtonian and the Leibnizian calculi as 'not equivalent in practice'. Notwithstanding the similarities regarding the justification of the algorithm, in practice the approach of the two men was different. While Newton spent much effort in developing limits as his 'rigorous' language, Leibniz preferred to promote talk about infinitesimals as a means of discovery of, and a way of talking about, new truths.

So we see that the calculi as they grew in the hands of Newton and Leibniz were two rather different things. As we shall see in the next chapter, one significant influence on its development was Newton's remarkable account of motion in his *Principia Mathematica*, and that in grappling with that work Leibniz was persuaded to rethink some of his ideas about the foundations and practice of the calculus.

4.6 Further reading

Bertoloni Meli, D. 1993. *Equivalence and Priority; Newton versus Leibniz*, Oxford University Press. This book, and Guicciardini's listed at the end of the next chapter, give two fascinating views of the struggles over the calculus and the mathematical analysis of nature waged between Newton and Leibniz. They may not be easy reading, but they are well written and state of the art.

Boyer, C.B. 1959. *The History of the Calculus and its Conceptual Development*, Dover. A thorough study and an informative book, especially on the work of people other than Newton (about whom it has been superseded by the work of Whiteside) and Leibniz.

Edwards, C.H., Jr. 1979. *The Historical Development of the Calculus*, Springer. More up-to-date than Boyer, it similarly goes from ancient to modern times. Helpful for its commentaries, mostly on the mathematical aspects which are discussed in some detail.

Grattan-Guinness, I. (ed.) 1980. *From the Calculus to Set Theory, 1630–1910*, Duckworth. This valuable collection of essays includes: K. M. Pedersen, 'Techniques of the calculus, 1630–1660'; H. J. M. Bos, 'Newton, Leibniz and the Leibnizian tradition'; I. Grattan-Guinness, 'The emergence of mathematical analysis and its foundational progress, 1780–1880'; T. Hawkins, 'The origins of modern theories of integration'; J. W. Dauben, 'The development of Cantorian set theory'; and R. Bunn, 'Developments in the foundations of mathematics, 1870–1910'.

5

Newton's *Principia Mathematica*

Introduction

In this chapter we look at the creation of one of the most influential and important works of mathematical physics ever written, Newton's *Philosophiae Naturalis Principia Mathematica* (*Mathematical Principles of Natural Philosophy*). This book, written in Latin, was published in three volumes in 1687, and two revised editions were published in Newton's lifetime.

Prior to Copernicus in the 16th century, there was a perfectly reasonable theory of why things fall: they are attracted by their weight to the centre of the universe, which was situated at the centre of the Earth. There was another law entirely for the region beyond the Moon, which was obviously very different. So a major consequence of any heliocentric theory was the need for a theory of weight that could explain why objects fall as they do when, plainly, they were not falling towards the Sun. It is this conundrum that Newton solved (and is the point behind the famous story of the apple).

In Section 5.1 we look at some attempts to explain the motion of the planets that were current in Newton's youth, and then investigate how he set about collecting the astronomical data and organising his own ideas so as to present an alternative theory. We shall also see that the astronomer Edmond Halley was crucial in seeing Newton's manuscripts into print over a period of some years.

In Section 5.2 we consider the content of the *Principia*. On the one hand it is based around the idea of gravity as a force obeying an inverse-square law. On the other hand, and this is the business of Book III of the *Principia*, it is a matter of showing in great detail how the motion of planets and satellites is consistent with this inverse-square law and can therefore be said to be explained by it. This involved Newton in a number of technical feats, and we draw attention to one of them: his proof that a solid spherical mass behaves as though it is a point mass concentrated at the centre of the sphere. We also remark that, contrary to some later stories, Newton did not write the *Principia* in the language of the calculus and then conceal it behind a display of geometry.

In Section 5.3 we look at the initial reception of the *Principia*: this was very positive in Britain, but much more disputed on the Continent, where Descartes's ideas had deeper roots. We conclude with an account of the final years of Newton's life, by which time he had become an establishment figure.

5.1 The creation of Newton's *Principia*

To understand the *Principia Mathematica* and its reception, it is helpful to begin by looking at the context in which Newton proceeded, for his book was the culmination of twenty years' work during which he struggled to master both mechanics and astronomy.

Newton had four major 17th-century predecessors in his studies: Galileo, Kepler, Descartes, and Huygens. From Galileo's *Dialogue Concerning the Two Chief World Systems*, translated into English in 1661, Newton seems to have learned the laws governing the descent of bodies under Earthly gravity and also about inertia, a topic we discuss below. But historians agree that Newton never read the *Discourse Concerning the Two New Sciences*, Galileo's last work, in which his ideas about motion were given their most thorough mathematical exposition.[1]

From Kepler's work, Newton could have learned the most careful and profound description of the motion of the planets. But he seems never to have read any of Kepler's own accounts, and professional astronomers were divided in their assessment of Kepler's planetary laws. They agreed in the main with his first law (that planets travel in ellipses with the Sun at one focus) and with the third law (which relates the period of the orbit to its mean radius). But astronomers had little use for the second law (that the line joining any planet to the Sun sweeps out equal areas in equal times), because they found it impossible to calculate with. As a result, the books that Newton read tended to state this law — if they mentioned it at all — only to reject it in favour of one that was better adapted to calculation. Moreover, Kepler's physical explanation of the motion of the planets, in which he supposed a kind of magnetic force that emanated from the Sun and somehow pushed the planets round, never won any support. Indeed, the works of Kepler and Galileo were curiously unconnected, and it was to be others who tried to produce a theory of planetary motion that could explain on more convincing mechanical grounds why the planets moved as Kepler claimed they did.[2]

The most successful proponent of such a theory was Descartes, who argued for it in his *Principia Philosophiae* (Principles of Philosophy, 1644), and it seems that Newton derived his first law of motion from reading this book, as we discuss below. But otherwise, the physics of Descartes and Newton were very different. Descartes argued that the Universe was full of invisible little particles that swept round the Sun like a cosmic whirlpool or vortex. The action of this vortex was to drive the planets round, and the result was that all the planets should lie in the same plane and orbit the Sun in the same direction. This theory had three things to recommend it: it explained why the orbits of the planets all lie more or less in the same plane; it explained why they all go round the Sun in the same direction; and it explained this by invoking an intuitively

[1] See Westfall, *Never at Rest*, p. 89, following Whiteside, *MPIN* 1974, VI, 3.

[2] See the discussion in *Volume* 1, Chapter 10.

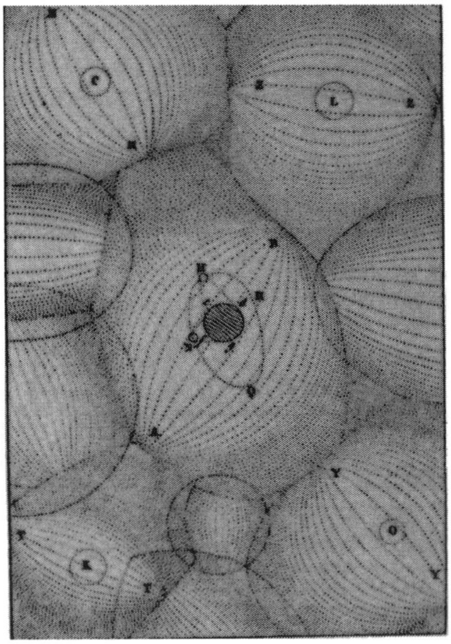

Figure 5.1. Descartes's vortices, from his *Principia Philosophiae* (1644)

simple mechanism for celestial motion: collision. For these reasons, Descartes's 'vortex theory' was widely accepted, especially on the Continent of Europe.

Christiaan Huygens was later to say that when he first read Descartes's book — he was then 15 or 16 — he thought everything was 'splendid' but that later he 'recovered a good deal from the infatuation I had for it'.[3]

Indeed, as a theory it left much to be desired. In particular, it did not account for the motion of the planets with the precision required by Kepler's laws. Nor did Descartes's detailed account of collisions stand up to much scrutiny, and his book is remarkable in its disdain for experiment and observation. It is, rather, an a priori physics based on some ideas that were clear and immediate to the mind of its author. For this reason, Huygens, like Pascal, called Descartes's book a 'romance', full of conjectures and fictions that came to be accepted as truths because of their intrinsic charm. As we shall see, this book provided Newton with his first instruction in the theory of motion, but was later to be decisively attacked by him in the *Principia*.

Huygens' own ideas were more precise. For example, he gave a careful account of circular motion, based on the idea that there is what he called a *centrifugal force* (see Box 7). He imagined a body being whirled rapidly around a fixed central point like a stone in a sling, and found a quantitative relationship between the radius of the circle, the size of the centrifugal force, and the velocity of the body. But Huygens was unhappy with the idea of a force as a fundamental principle; throughout his life he remained an enthusiast for the Cartesian idea of collisions, as the kind of mechanistic concept that ultimately made it possible to explain phenomena.[4]

---

[3]Quoted in (Hall 1983, 295).

[4]The historian George Smith (2002) has pointed out that Huygens' discussion of centrifugal force is the only place that his axioms referred to the concept of force.

Box 7. Centripetal and centrifugal forces

According to Newton, all bodies attract one another by a force, which he called gravity. The effect of a force acting on a body is to cause it to accelerate in the direction of the force — so Newton's force has both a size and a direction.

As a simple example of when two bodies attract one another, consider the case in which one body is so large that it remains at rest; this is a reasonable first approximation to the situation where the bodies are the Sun and a planet. Then the smaller body is attracted to the larger one, and this force therefore always pulls it towards the same point, the common centre of gravity of the two bodies. This is an example of what Newton called a *centripetal force*: a force directed towards a fixed point, which, in some intuitive sense, is the centre of the orbit.

This differs from Huygens' idea of a force that pushes a body away from the centre, which Huygens had called a *centrifugal force*; such a force causes the body to flee from the centre. This is the force that you experience if you are swung round, say at the end of a rope, or in a rotating drum. The term is used today in connection with such objects as a spin dryer, where the water is driven out by the centrifugal force generated by the rotating drum.

It is clear from this brief outline that the task facing Newton was immense. A causal theory of motion capable of yielding accurate quantitative descriptions was still lacking, even for terrestrial physics, and such a theory covering both small objects near the Earth and the motion of the planets must have seemed even further out of reach. Kepler's idea of a solar force was rejected, and Descartes's more plausible ideas had not been made sufficiently precise. Let us see how Newton came to create just such a unified theory.

Newton's route to the *Principia*. In England the theory of celestial mechanics was much debated in the late 1670s by Fellows of the Royal Society, such as Robert Hooke, Halley, and Wren. Their debate was quickened by the magnificent comet of 1680, which could be seen in daytime for weeks, and at its largest stretched over nearly one-third of the visible sky, making it at least 60 million miles long (see Box 8).

But the final impulse that set the *Principia* in motion was a visit that Edmond Halley paid to Newton in August 1684. By January of that year, Hooke, Wren, and Halley had all concluded that Kepler's third law and Huygens' formula for centrifugal force together imply that the Sun attracts each planet according to an inverse-square law — that is, by a force whose size varies inversely with the square of the distance of the planet from the Sun.

However, this was more of an insight than a rigorous mathematical theory, and none of them could derive Kepler's laws from dynamical principles. Halley therefore decided to travel to Cambridge and ask for Newton's help. Much later, in 1722, Abraham De Moivre wrote down his account of the visit:[5]

[5]Quoted from Westfall, *Never at Rest*, p. 403.

Box 8. Comets

The significance of comets in the 17th century arises from the fact that they were beginning to be regarded as true celestial objects. Galileo, for example, was somewhat behind the times in holding to the old, Aristotelian belief that comets were phenomena of the upper atmosphere. Once they were recognised as being really up in the heavens — beyond the Moon — they raised some interesting questions. Along what paths, for example, did they travel? Most people thought, like Kepler, that they travelled along straight paths, either into or out of the sun, although not at a uniform speed.

In 1680, Flamsteed discussed the so-called 'great comet' in correspondence with Newton, and raised the idea that it orbited the sun and changed its direction as it made its closest approach. But Newton did not accept Flamsteed's ideas until 1682, when he made some observations of another comet — the one that we now call Halley's comet — which tells us that in 1680 Newton was some way from regarding gravitation as universal. But when writing *De Motu Corporum*, he observed excitedly that the inverse-square law 'allowed [one] to define the orbit of comets and thereby their periods of revolution'.[a]

Newton was, in fact, the first to compute the orbit of a comet (he took the great comet of 1680) and his treatment of this problem in the *Principia* is one of the great successes of the work (see Figure 5.3). Moreover, once the orbit of comets had become better understood, it emerged that some (including the great comet) move around the sun in the direction opposite to those of the planets. This is difficult to reconcile with Descartes's vortex theory.

And Halley? A keen astronomer and a skilful mathematician, it was his achievement to read the existing literature in the light of the new, orbital theory of comets, and in 1705 to pronounce some of them regular visitors to our skies. In particular, he estimated that the comet of 1682 had a period of about 75 years. Accordingly he predicted its return in 1758, hoping that, if he were proved correct, 'candid posterity will not refuse to acknowledge that this was first discovered by an Englishman'.

[a]See Whiteside, *MPIN* VI, 57.

> Dr [Halley] asked him what he thought the Curve would be that would be described by the Planets supposing the force of attraction towards the Sun to be reciprocal to the square of their distance from it. Sr Isaac replied immediately that it would be an Ellipsis, the Doctor struck with joy & amazement asked him how he knew it, why saith he I have calculated it, whereupon Dr Halley asked him for his calculation without any farther delay, Sr Isaac looked among his papers but could not find it, but he promised him to renew it, & then to send it him.

Newton set himself to re-derive the solution, and in November he sent the Royal Society a nine-page tract called *De Motu Corporum* (On the Motion of Bodies). In it he showed how an elliptical orbit and a centripetal force together imply an inverse-square law for the force. But at this stage he did not even sketch the converse result, which would have answered Halley's question, and he offered no argument that an inverse-square law for a centripetal force implies that the orbit must be a conic section. Newton's tract caused a great stir, and was eagerly read by members of the Royal Society. Halley immediately went back to Cambridge to urge Newton perhaps to publish *De Motu Corporum*, perhaps to amplify it. But Newton had already allowed himself to become totally immersed in the problems of physics and celestial mechanics. Until the spring of 1686 he seems to have thought of nothing else.

In order to obtain the most accurate data, Newton wrote repeatedly to John Flamsteed at the Royal Observatory in Greenwich. Flamsteed was the first Astronomer Royal, and he had been put in post by Charles II in March 1675 when the Observatory was founded, with the mission of improving navigation at sea and in particular solving the problem of determining longitude, a vital topic that we shall discuss in Section 6.3. This was one of the family of challenges that had led to the creation of the Royal Society. Flamsteed was an observational astronomer of unrivalled precision, but his later life was marred by disputes with Newton and Halley over the publication of his results. Newton also continued to lecture at Cambridge, but otherwise he disappeared entirely from intellectual life for two years. In so doing he returned to the habits of his first adult years, and also to his original questions — but this time he was to publish his results and give the world evidence of his brilliance.

Figure 5.2. The Royal Observatory, Greenwich

The quantity of motion. Newton's first task was to produce a good theory of dynamics. In *De Motu Corporum* he had derived elliptical orbits on dynamically incorrect grounds. It took him six months to elucidate the correct notion of force to put into the

Figure 5.3. The great comet of 1680, from Newton's *Principia*, Book III

Principia, and to isolate the concept of inertia with which to express a body's tendency to keep on going with a uniform velocity if not acted on by an external force.[6]

Another problem that Newton had to solve was this: if a theory is to be quantitatively accurate, the basic terms must be quantifiable — but how was the *quantity of motion* to be measured? This question had stumped Newton's contemporaries. Newton proposed *momentum*, the product of mass and velocity. In his second law he expressed the way in which force acts so as to cause a change in momentum, and in this way he established the dynamical principles that he needed.

In the *Principia* Newton stated three laws of motion.[7]

Newton's laws of motion.

Law 1

Every body perseveres in its state of being at rest or of moving uniformly straight forward except insofar as it is compelled to change its state by forces impressed.

Projectiles persevere in their motions, except insofar as they are retarded by the resistance of the air and are impelled downward by the force of gravity. A spinning hoop, which has parts that by their cohesion continually draw one another back from rectilinear motions, does not cease to rotate, except insofar as it is retarded by the air.

[6] A theory of motion is called *dynamical* if it invokes a concept of force. A theory that merely considers speeds is called *kinematical*.

[7] See *Principia*, Axioms, or the Laws of Motion, pp. xvii–xix in the Motte–Cajori translation and F&G 12.B2.

> And larger bodies — planets and comets — preserve for a longer time both their progressive and their circular motions, which take place in spaces having less resistance.
>
> Law 2
>
> *A change in motion is proportional to the motive force impressed and takes place along the straight line in which that force is impressed.*
>
> If some force generates any motion, twice the force will generate twice the motion, and three times the force will generate three times the motion, whether the force is impressed all at once or successively by degrees. And if the body was previously moving, the new motion (since motion is always in the same direction as the generative force) is added to the original motion if that motion was in the same direction or is subtracted from the original motion if it was in the opposite direction or, if it was in an oblique direction, is combined obliquely and compounded with it according to the directions of both motions.
>
> Law 3
>
> *To any action there is always an opposite and equal reaction; in other words, the actions of two bodies upon each other are always equal and always opposite in direction.*
>
> Whatever presses or draws something else is pressed or drawn just as much by it. If anyone presses a stone with a finger, the finger is also pressed by the stone. If a horse draws a stone tied to a rope, the horse will (so to speak) also be drawn back equally toward the stone, for the rope, stretched out at both ends, will urge the horse toward the stone and the stone toward the horse by one and the same endeavor to go slack and will impede the forward motion of the one as much as it promotes the forward motion of the other. If some body impinging upon another body changes the motion of that body in any way by its own force, then, by the force of the other body (because of the equality of their mutual pressure), it also will in turn undergo the same change in its own motion in the opposite direction. By means of these actions, equal changes occur in the motions, not in the velocities — that is, of course, if the bodies are not impeded by anything else. For the changes in velocities that likewise occur in opposite directions are inversely proportional to the bodies because the motions are changed equally. This law is valid also for attractions, as will be proved in the next scholium.

In this celebrated passage, placed towards the beginning of the *Principia*, Newton stated his axioms or laws of motion — the laws are not proved, as indeed the word 'axioms' implies. The text following the statement of each law is further explanation and elucidation of it, not justification. These laws, the foundations on which subsequent deductions are to be built, are about the behaviour of moving bodies and bodies with forces acting on them. It is interesting that Newton's laws are not visibly mathematical, or at any rate algebraic: they do not have an appearance of equations involving symbols and their manipulation. If you have studied modern applied mathematics you may be surprised at how far we are from modern conceptual formulations such as 'force = rate of change of momentum' for Law 2.

Newton also had to decide what were the crucial astronomical ideas that he needed to explain on the basis of his theory of dynamics. He chose to go for all three of Kepler's laws. In Section 5.2 you will see how Newton's vast generalisation of Kepler's second (equi-area) law, prominently placed near the front of the book, was to play a vital role in the theory he presented. To state and prove this law in such generality may indeed be Newton's greatest contribution to dynamics. Certainly everyone before him had thought that Kepler's laws applied only to the planets and were hard to justify further — Newton showed that their explanation rested on quite general dynamical grounds.

Of course, as Kepler had known and as Newton more-or-less proved, any real use of the equi-area law to describe or calculate an orbit must itself be approximate. Newton called the curve that measures the area of a sector of an ellipse 'geometrically irrational' — his phrase for what Descartes had called mechanical curves. In the context of practical astronomy, Newton's remarks confirmed that astronomers would have to continue to settle for only approximate answers. While this may partially vindicate those who replaced Kepler's law with their own, it does nothing to diminish Newton's achievement. The Newton scholar I. B. Cohen has observed that:[8]

> It was an unusual and a very daring step to erect an astronomical system encompassing Kepler's three laws, as Newton did. Following the imaginative leap forward that Newton made, in showing the physical meaning and conditions of mathematical generality or applicability of each of Kepler's laws, this whole set of three laws gained a real status in exact science.

It is also worth noticing that the *Principia* takes it for granted that the solar system is heliocentric, because the first drafts did not. Newton had written lengthy justifications for heliocentrism, but then suppressed them.

The birth of the *Principia*. The circumstances in which the *Principia* came into the world were dramatic. In April 1686 the manuscript of Book I was presented to the Royal Society. Halley continued his role as the work's midwife, and successfully pursued his goal of ensuring that Newton's master work was printed and published in its entirety, against obstacles that would have exhausted a less forceful person.

The Royal Society, whose finances were somewhat delicately poised throughout these years, had nearly bankrupted itself through the publication of a very handsome, expensive, and unsellable *History of Fishes*; there was certainly no money spare for printing Mr Newton's thoughts upon the Cosmos. Indeed, the Royal Society was reduced to recompensing Halley — an employee of the Society — with fifty copies of the *History of Fishes*. In the end Halley had to pay for the publication of the *Principia* himself — fortunately the work was to make a profit, unlike the *History of Fishes*, so he did not lose by his generous gesture.

But all along he had a greater problem to contend with: Would the author allow his work to be published at all? The ever-touchy Newton had been roused to thunderous rage by learning that Robert Hooke was claiming not only priority in discovering the inverse-square law, but also that Newton had taken it from him, and more besides. As the year went on, Newton announced to Halley that he would withdraw Book III altogether, if the reward of his labours was to be this irritation of false claims. At length

[8]See (Cohen 1980, 229).

Halley, a model of saintly tact and diplomacy, soothed Newton, and the printing went on its way.

The moral of this tiresome squabble points up nicely the significance of the *Principia*. Hooke's claim was actually true, in the limited sense that he had considered an inverse-square law earlier than Newton, as others had done, too, and he anticipated Newton also in the realisation that a force must be acting on a body in orbit to deflect it constantly out of a straight line. Hooke was an ingenious man with bright and appropriate ideas, but he was quite unable to distinguish between having an isolated, well-informed idea and the massive systematic endeavour of deducing a whole 'System of the World' on dynamical principles.

At last, on 5 July 1687, Halley's task was done: the *Principia* was a printed book. With its publication, Newton's status as a mathematician, and the nature of mathematical physics, were changed utterly. Perhaps it was Newton, agreeably taking himself less seriously than usual, who spread the story that he passed a student in the Cambridge street who said: 'there goes the man that writt a book that neither he nor any body else understands'.[9]

5.2 The content of the *Principia*

Imagine that Newton's *Principia* has just come into your hands, one day in late 1687. You first turn over the 547 pages, written in scholarly Latin, to get an impression of what it contains. This book had begun to create a stir even before it was published. What is it actually about? The notes that follow provide an overview, a guide of sorts, with which to begin to answer that question. We then consider various passages of the *Principia* in more detail in order to get a more precise grasp of certain topics from this remarkable work.

Book I is preceded by definitions and axioms: definitions of quantity of matter; quantity of force; forces of various kinds; a distinction between relative and absolute motion; relative and absolute time. Then there are the three axioms or laws of motion we looked at above, and some of their elementary consequences. It is worth stressing that Newton's concept of force is novel, vastly more general than anything discussed by Galileo or Huygens, and fundamental to the *Principia*; the work is about forces at least as much as it is about the motion of physical objects.[10]

Book I: The Motion of Bodies. This book opens with a long, careful, cumulative discussion of 'the method of first and last ratios of quantities'. This is a geometrical study of curves and their tangents in the spirit in which Newton conducted his investigations of the calculus. There follows a lavish study of the motion of a point under a centripetal force. Newton established that the line joining a fixed point to a moving one sweeps out equal areas in equal times if and only if the force on the moving point is directed towards the fixed point. This is a very general result: the size of the force can depend in any way on the length of the radius line; the orbit can be any shape determined by the law, not just a circle or an ellipse. Numerous special cases are then worked out, including this one: If the moving body traverses a conic section under a centripetal force directed towards one focus, then the magnitude of the force is inversely proportional to the square of the distance.

---

[9]Quoted in Westfall, *Never at Rest*, p. 468.

[10]See (Smith 2002) for an extended discussion of this point.

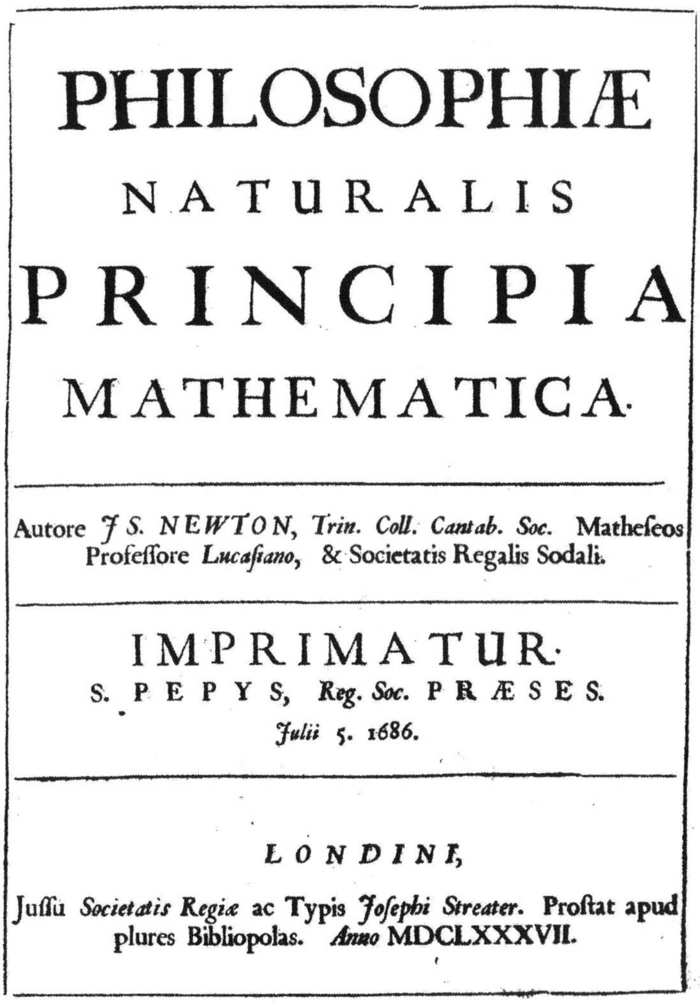

PHILOSOPHIÆ
NATURALIS
PRINCIPIA
MATHEMATICA.

Autore *J S.* NEWTON, *Trin. Coll. Cantab. Soc.* Mathefeos
Profeffore *Lucafiano,* & Societatis Regalis Sodali.

IMPRIMATUR.
S. PEPYS, *Reg. Soc.* PRÆSES.
Julii 5. 1686.

LONDINI,

Juffu *Societatis Regiæ* ac Typis *Jofephi Streater.* Proftat apud
plures Bibliopolas. *Anno* MDCLXXXVII.

Figure 5.4. The title page of Newton's *Principia*

This, as Newton well knew, is the relevant case in astronomy. In the first edition of the *Principia* he also stated the converse (it is Corollary 1 to Proposition 13): under an inverse-square law, bodies move in curves that are conic sections with the centre of force as one focus (see Box 9). He enriched the statement with a skeleton proof in the second edition (1713).

Now an astronomer needs to know not only what orbit a planet has as a whole, but where along that orbit it can be found at any particular time. Newton tackled this problem next, and showed in principle how to solve it using Kepler's equi-area law. He also showed that if planets traverse ellipses under the action of a force that obeys an inverse-square law, then they necessarily obey Kepler's third law, the '3/2-power law' (which says that the time taken to complete an orbit is proportional to the average radius of the orbit raised to the power 3/2). Newton even showed how in principle to determine the orbit given any centripetal law of force.

Box 9. Newton's inverse-square law.
Newton considered the special case of two bodies mutually attracting one another, and investigated various ways in which the size of this force can depend on the distance between them. If the size of the force is inversely proportional to the square of the distance between the bodies, then the force is said to obey an inverse-square law. This means that if the distance between the two bodies is tripled, then the size of the force is reduced to a ninth of what it was, and so on; so nearby bodies exert more of an influence than distant ones.

In the special case raised in Box 7 of a large, stationary body and a small, moving one, then, because the force on the moving body is always directed towards the fixed one, it is natural to wonder why they do not eventually collide, according to this theory of motion. In Hooke's view, this was somehow because the smaller body has a velocity directed away from the larger one (if it had not, then they would certainly collide). One of Newton's achievements was to show on mathematical grounds that Hooke's insight was correct.

The inverse-square law formula asserts that the force exerted on a body of mass m_1 by another body of mass m_2 at a distance r apart is

$$\frac{Gm_1m_2}{r^2},$$

where G is a constant that does not depend on the bodies concerned (known as the gravitational constant).

We pass over a number of propositions, including several that concern conic sections, and come to Newton's discussion of the attraction between solid bodies under an inverse-square law. Here he established that a spherical shell exerts no force on a point inside it and attracts a point outside it in the same way as a point mass concentrated at the centre — see Box 10. This result, which surprised him as much as his contemporaries, enabled him to reduce the study of large spherical objects, such as planets, to the study of points and centripetal forces, which he had already described.

Book II: The Motion of Bodies in Resisting Media. This book consists of a discussion of several topics: the motion of bodies subject to various forms of air resistance; hydrostatics and the density and pressure of liquids; pendulums oscillating in resisting media; liquids emptying through holes and flowing through canals; and the transmission of motion through a fluid. These are important subjects, but all we need to note here is that at the end of Book II Newton rejected Descartes's theory of motion in vortices:

> Hence it is clear that the planets are not carried along by corporeal vortices.

Box 10. Spherical bodies and point masses.
The gravitational pull of a spherical shell For much work in astronomy, the assumption that planets and the Sun are spherical in shape is entirely reasonable. Newton obtained the remarkable result that a thin spherical shell of matter of uniform density attracts bodies outside it exactly as would a point mass, situated at the centre of the shell and having the same mass as the shell. This means that a solid sphere, thought of as a nest of such shells, also attracts bodies outside it exactly as does a point mass. So in Newton's theory of gravity, large solid spheres may be replaced by points (of the same mass) — a considerable simplification of the theory.

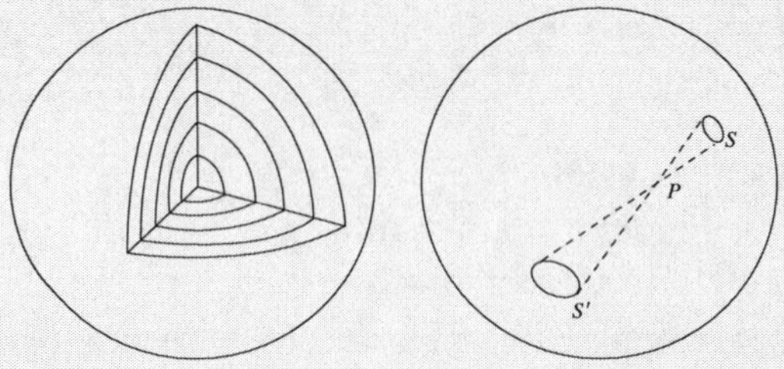

Figure 5.5. Nested spherical shells and the attraction of a shell on an interior point

There is no gravitational pull inside a spherical shell
To understand this result, consider a point, P, inside a spherical shell (not necessarily at the centre, or there is nothing to prove!). Consider the attraction on it of a small piece of the surface, S, and the piece of the surface, S', opposite to S. Although one piece, say S, is nearer to P, it is, by the same token, smaller than S'. Because S is nearer than S', it exerts a stronger pull on P; but because it is smaller, it exerts a weaker pull.

The question is to decide how these two aspects balance, and Newton showed that, on the assumption that the force of attraction obeys an inverse-square law, they exactly cancel out. So P is pulled towards neither S nor S'. By regarding the sphere as made up of infinitely many of these little double cones, we can see that P is pulled in no direction at all: there is no net gravitational attraction inside a spherical shell.

Book II ends by dismissing Descartes's ideas in these words:[11]

> the hypothesis of vortices can in no way be reconciled with astronomical phenomena, and serves less to clarify the celestial motions than to obscure them. But how these motions are performed in free spaces without vortices can be understood from book 1; and will now be shown more fully in book 3 on the system of the world.

[11] See Newton, *Principia*, Book II, 789, 790. For the Motte–Cajori version, see F&G 12.B9.

This is a central accomplishment of the *Principia.* Newton showed that motion of the planets can be explained in terms of an intuitively implausible hypothesis (attraction at a distance by a force) which is mathematically derived from, and consistent with, observation. At the same time, an intuitively plausible physical hypothesis (the vortex theory of planetary motion) is shown — with what degree of rigour we discuss below — to be false on purely mathematical grounds. This is a sizable conjunction of intellectual events, and was to generate a fierce controversy. But, first, how did Newton 'more fully treat of it'?

Book III The System of the World. Book III contains Newton's mathematical demonstration that the theory of an inverse-square law of gravity acting as a force between bodies can explain the motion of the planets and of their satellites, the motion of comets, and can begin to explain the motion of the Moon and to determine the shape of the Earth.

In the opening pages, he argued for four rules. The first three are:

Newton's rules in natural science.

> Rule 1: No more causes of natural things should be admitted than are both true and sufficient to explain their phenomena.
>
> Rule 2: Therefore, the causes assigned to natural effects of the same kind must be, so far as possible, the same.
>
> Rule 3: Those qualities of bodies that cannot be intended and remitted [that is, qualities that cannot be increased and diminished] and that belong to all bodies on which experiments can be made should be taken as qualities of all bodies universally.

He concluded his discussion of Rule 3 with these remarks:

> Finally, if it is universally established by experiments and astronomical observations that all bodies on or near the earth gravitate ... toward the earth, and do so in proportion to the quantity of matter in each body, and that the moon gravitates ... toward the earth in proportion to the quantity of its matter, and that our sea in turn gravitates ... toward the moon, and that all planets gravitate ... toward one another, and that there is a similar gravity [heaviness] of comets toward the sun, it will have to be concluded by this third rule that all bodies gravitate toward one another. Indeed, the argument from phenomena will be even stronger for universal gravity than for the impenetrability of bodies, for which, of course, we have not a single experiment, and not even an observation, in the case of the heavenly bodies. Yet I am by no means affirming that gravity is essential to bodies. By inherent force I mean only the force of inertia. This is immutable. Gravity is diminished as bodies recede from the earth.
>
> Rule 4: In experimental philosophy, propositions gathered from phenomena by induction, should be considered either exactly or very nearly true notwithstanding any contrary hypotheses, until yet other phenomena make such propositions either more exact or liable to exceptions. This rule should be followed so that arguments based on induction may not be nullified by hypotheses.

He then reached his remarkable conclusions in just a few pages. First, he listed various phenomena:

1. The satellites of Jupiter sweep out equal areas in equal times and obey the 3/2-power law (so they obey Kepler's 2nd and 3rd laws).

2. Likewise, the satellites of Saturn sweep out equal areas in equal times and obey the 3/2-power law (so they obey Kepler's 2nd and 3rd laws).

3. The planets Mercury, Venus, Mars, Jupiter, and Saturn encircle the Sun.

4. All the planets obey Kepler's 3rd law (as does the Earth when considered as orbiting the Sun).

5. And all obey Kepler's 2nd law.

He remarked about the motion of the Moon only that it obeys the equi-area law.

Next, Newton drew on Book I to prove some propositions that establish that the above motions imply an inverse-square law of gravity. Only the motion of the Moon caused him any difficulties, but Newton was able to use the known distance of the Moon from the Earth (60 Earth-radii) and the time it takes to complete an orbit (27 days, 7 hours, 43 minutes) to calculate the distance that the Moon would fall towards the Earth in one minute 'if deprived of all its motion'. (His argument involves two 60s: the number of Earth-radii that gives the distance of the Moon from the Earth and the number of seconds in a minute; we have kept track of which is used in what follows.)

He found that the distance the Moon would fall is 15 feet 1 inch, and 1 and 4/9 of a twelfth of an inch. It follows (on the inverse-square law) that the force at the surface of the Earth is 60^2 times as great (60 Earth-radii), and that the distance a body would fall in a minute if released near the surface of the earth should also be multiplied by 60^2. Therefore in a second (60 seconds in a minute) such a body should fall 15 feet 1 inch, and 1 and 4/9 of a twelfth of an inch — which it does (within the limits of error built in to this calculation). Newton clinched the argument by assessing the force near the surface of the Earth in terms of measurements on pendulums.

He concluded:[12]

Newton on the motion of the planets.

> **Proposition 5 Theorem 5** *The circumjovial planets [or satellites of Jupiter] gravitate toward Jupiter, the circumsaturnian planets [or satellites of Saturn] gravitate toward Saturn, and the circumsolar [or primary] planets gravitate toward the Sun, and by the force of their gravity they are always drawn back from rectilinear orbits and kept in curvilinear orbits.*
>
> For the revolutions of the circumjovial planets about Jupiter, of the circumsaturnian planets about Saturn, and of Mercury and Venus and the other circumsolar planets about the Sun are phenomena of the same kind as the revolution of the Moon about the Earth, and therefore (by rule 2) depend on causes of the same kind, especially since it has been

[12]See (Cohen and Whitman 1999, 805–806). Note that Newton's propositions, numbered consecutively, consist of 'Theorems' or 'Problems', numbered independently.

> proved that the forces on which those revolutions depend are directed toward the centers of Jupiter, Saturn, and the Sun, and decrease according to the same ratio and law (in receding from Jupiter, Saturn, and the Sun) as the force of gravity (in receding from the Earth).
>
> COROLLARY 1. Therefore, there is gravity toward all planets universally. For no one doubts that Venus, Mercury, and the rest [of the planets, primary and secondary,] are bodies of the same kind as Jupiter and Saturn. And since, by the third law of motion, every attraction is mutual, Jupiter will gravitate toward all its satellites, Saturn toward its satellites, and the Earth will gravitate toward the Moon, and the Sun toward all the primary planets.
>
> COROLLARY 2. The gravity that is directed toward every planet is inversely as the square of the distance of places from the center of the planet.
>
> COROLLARY 3. All the planets are heavy toward one another by corols. 1 and 2. And hence Jupiter and Saturn near conjunction, by attracting each other, sensibly perturb each other's motions, the Sun perturbs the lunar motions, and the Sun and Moon perturb our sea, as will be explained in what follows.
>
> **Scholium** Hitherto we have called 'centripetal' that force by which celestial bodies are kept in their orbits. It is now established that this force is gravity, and therefore we shall call it gravity from now on. For the cause of the centripetal force by which the Moon is kept in its orbit ought to be extended to all the planets, by rules 1, 2, and 4.
>
> **Proposition 6 Theorem 6** All bodies gravitate toward each of the planets, and at any given distance from the center of any one planet the weight of any body whatever toward that planet is proportional to the quantity of matter which the body contains.

How neatly it all fits together. The mathematical centripetal force in Book I is identified with gravity in Book III and is shown to be the cause of the weight of all bodies, terrestrial or astronomical. Everything in the universe is bound together by the universal force of gravity.

Newton was also able to extend his theory to provide a theory of the tides and the precession of the equinoxes.

His work left some questions obscure: Why, for instance, do the planets all lie in roughly the same plane, and orbit the Sun in the same direction? However, because of the retrograde motion of certain comets, this uniformity was no longer complete, and so failure to account for it was not a major criticism of his theory. Indeed, major changes in scientific conceptual models often involve ignoring questions that were previously thought important and relevant. Thus, just as an observation that Kepler thought very significant — why there are only six planets — played no part in Descartes's vortex theory, so some Cartesian questions were of no particular concern (and certainly were unanswerable) in the Newtonian system.

This brief summary of the *Principia* is intended to be helpful, in collecting in one place what you need to know at this stage. It is also meant to overwhelm you, much as the *Principia* probably overwhelmed its first readers. Much has been omitted: the mathematical technicalities, all the physics (including Newton's discussions of the transmission of sound) and all the comparisons between theory and observation that Newton conducted so thoroughly. It would be hard to overestimate the difficulty and originality of the work. It was not just another contribution to the debate about natural philosophy. It was intended to change the grounds of that debate — and so, eventually, it did.

The content of Books I and III. In due course, we shall discuss the reception of *Principia*. But the rest of this section stays with the work itself, investigating it in greater detail. Book II has been rather brusquely summarised, for reasons that have to do with its subject matter, while Books I and III form an oddly contrasting pair. Book III is the one for the astronomer: not only does it account for planetary orbits, but it presents a theory of the motion of the Moon and of comets; these theories are tied exceptionally closely to observations. On the other hand, the mathematician would like Book I. The very general account of motion, as well as the specific analyses of motion under various kinds of centripetal force, and the many results about conic sections, are all impressive, as is the dramatic reduction of solid bodies to points for the purposes of calculation. The inverse-square law of gravity (which is what makes possible this dramatic reduction) is not made as an assumption, but is here derived from Kepler's laws as a theorem.

We shall see later that it is the existence of two such approaches in one work that accounts in large part for the varied ways in which the *Principia* was received. But let us now look at some passages in more detail, to see how these different approaches actually appear.

After the introductory definitions, axioms (the laws of motion), and corollaries (further implications of the axioms), Book I leads off with a series of lemmas about the first and last ratios of quantities.[13] We can recognise issues of concern to Newton here. These lemmas express the ideas of the geometrical calculus that Newton first published in the *Principia*. They express, in a subtle and plausible form, the intuitive idea that we get very close to a curvilinear area by covering it with narrow rectangles and shrinking their widths more and more. In fact, Newton went further than this, and claimed that under these conditions the two areas concerned (the curvilinear area and its covering by narrow rectangles) become 'ultimately equal'. Lemma 1 defines this important notion as what happens when quantities or their ratios 'approach nearer to each other than by any given difference', thereby providing a perfectly workable concept of limit upon which geometrical arguments can be based, without recourse to infinitesimal considerations.

However, the opening eleven lemmas of the *Principia* have generated a large amount of commentary over the centuries, much of it obscure and inconclusive. Newton claimed that they are fundamental, being an alternative to both the tedium of proofs by contradiction and the use of indivisibles (which he found harsh and insufficiently geometrical). They are based, he noted, on a defence of the method of limits. But Whiteside has argued that these lemmas were not central to the arguments in the *Principia*, and were 'undeniably a retrospective gloss'.[14]

---

[13]See Newton, *Principia*, Book I, 433–444. For the first three of these lemmas in the Motte–Cajori version, see F&G 12.B3.

[14]See Whiteside, *MPIN*, Vol. 6, p. 108.

However, in a later provocative article, Bruce Pourciau has argued that 'each lemma *is itself* (a natural replacement for) an elementary and basic definition, property, or theorem of the calculus'.[15] [Pourciau's italics].

To form an opinion, we must look at some of the lemmas in detail. Pourciau translated the first one as follows:

> For those ultimate ratios ... are limits towards which the ratios ... approach nearer than by any given difference.

This, as he points out, is very close to Cauchy's definition of limit over a century later.[16]

Pourciau continued in this fashion through the lemmas. Lemma II, for example, may seem complicated:

> LEMMA II: If in any figure *AacE*, terminated by the right lines *Aa*, *AE*, and the curve *acE*, there be inscribed any number of parallelograms *Ab*, *Be*, *Cd*, &c comprehended under equal bases *AB*, *BC*, *CD*, &c and the sides, *Bb*, *Cc*, *Dd*, &c parallel to one side *Aa* of the figure; and the parallelograms *aKbl*, *bLcm*, *cMdn*, &c are completed: then if the breadth of those parallelograms be supposed to be diminished, and their number to be augmented in infinitum, I say, that the ultimate ratios which the inscribed figure *AKbLcMdD*, the circumscribed figure *AalbmcndoE*, and the curvilinear figure *AabcdE*, will have to one another, are the ratios of equality.

But with the accompanying figure (Figure 5.6) it can be understood as a geometrical form of a working definition of the integral. For it says that the area under a curve is approximated indefinitely well by increasing sequences of inscribed and circumscribed parallelograms.

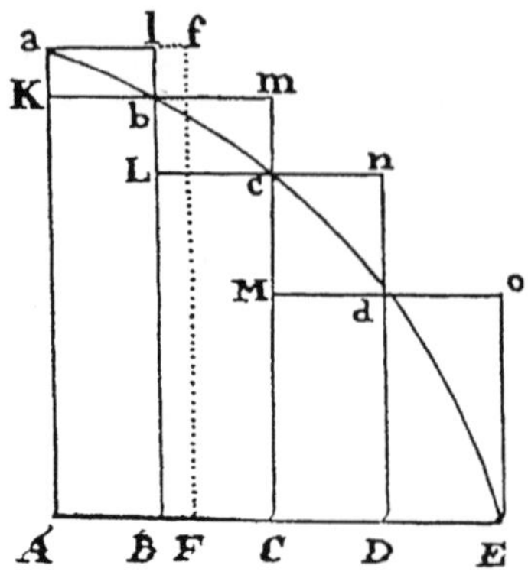

Figure 5.6. Newton: parallelograms approximating the area under a curve

With the same understanding of limits, Pourciau read Lemma VI as providing a definition of derivative (see Figure 5.7).

> LEMMA VI: If any arc *ACB*, given in position, is subtended by its chord *AB*, and in any point *A*, in the middle of the continued curvature, is touched by a right line *AD*, produced both ways; then if the points *A* and *B* approach one another and meet, I say, the angle *BAD*, contained between the chord and the tangent, will be diminished in infinitum, and ultimately will vanish.

[15]See (Pourciau 1998, 281).

[16]We discuss Cauchy's ideas below, in Section 16.1.

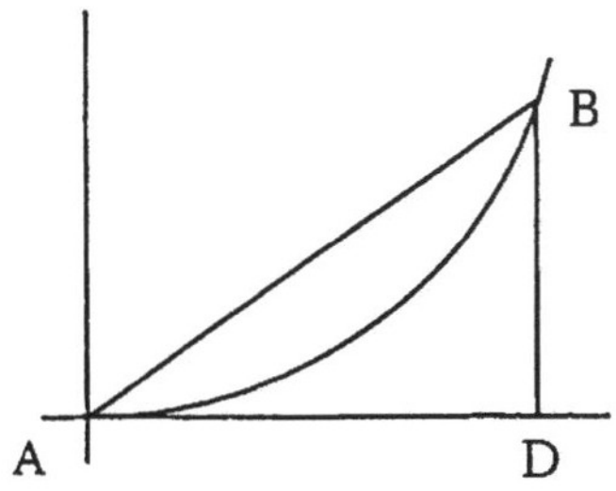

Figure 5.7. A chord approximating a tangent

Pourciau continued in this fashion until he had interpreted the final lemma as a geometrical version of the Fundamental Theorem of the Calculus.

Pourciau was not claiming that Newton's statements were the exact equivalent of the modern definitions and theorems. Indeed, partly because they lack modern algebraic and arithmetical interpretations they remain a little vague. He was certainly not claiming that they were understood by Newton's readers in the fashion he suggested, only that they could serve — for Newton — as functioning equivalents for specific results in the modern calculus.

He therefore walked a tight line — on the one hand, respecting the failure of Newton's contemporaries and many later commentators to understand them (Leibniz, he tells us, even regarded the last three lemmas as false), and on the other hand, rejecting the idea that the lemmas were fundamentally obscure and even incoherent. On this view, we are confronted with a series of lemmas that make good sense, but which only Newton could grasp.

Pourciau concluded his paper with some tentative explanations of how Newton came to these lemmas. He offered two suggestions: either Newton was so immersed in this work that he saw these geometrical expressions much in the way later writers would see the calculus (but not as a translation of pre-existing calculus notions into geometry) or he was simply inspired and came to them in an epiphany. We might at least agree that one should never underestimate Newton.

Newton's equi-area law. The strength and utility of the concept of ultimate equality was soon made manifest in the very first theorem of the *Principia*. This is the important result that generalises Kepler's second (equi-area) law. Its proof is so remarkably simple, using the notion of ultimate equality, and the result is such an important and elegant one, that it can stand as a paradigm example of a Newtonian theorem.[17]

Newton's equi-area law.

> **Proposition 1 Theorem 1** *The areas which bodies made to move in orbits describe by radii drawn to an unmoving center of force lie in unmoving planes and are proportional to the times.*
>
> Let the time be divided into equal parts, and in the first part of the time let a body by its inherent force describe the straight line AB. In the second part of the time, if nothing hindered it, this body would (by law 1)

[17]See (Cohen and Whitman 444-445). For the Motte–Cajori version, see F&G 12.B5.

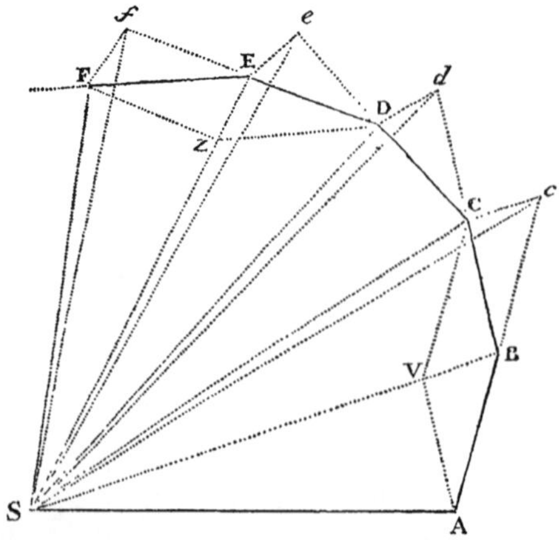

Figure 5.8. A particle obeying a central force sweeps out equal areas in equal times, from Newton's *Principia*, Book I

> go straight on to c, describing line Bc equal to AB, so that — when radii AS, BS, and cS were drawn to the center — the equal areas ASB and BSc would be described. But when the body comes to B, let a centripetal force act with a single but great impulse and make the body deviate from the straight line Bc and proceed in the straight line BC. Let cC be drawn parallel to BS and meet BC at C; then, when the second part of the time has been completed, the body (by corol. 1 of the laws) will be found at C in the same plane as triangle ASB. Join SC; and because SB and Cc are parallel, triangle SBC will be equal to triangle SBc and thus also to triangle SAB. By a similar argument, if the centripetal force acts successively at $C, D, E, \ldots$, making the body in each of the individual particles of time describe the individual straight lines CD, DE, EF, $\ldots$, all these lines will lie in the same plane; and triangle SCD will be equal to triangle SBC, SDE to SCD, and SEF to SDE. Therefore, in equal times equal areas are described in an unmoving plane; and by composition [or componendo], any sums $SADS$ and $SAFS$ of the areas are to each other as the times of description. Now let the number of triangles be increased and their width decreased indefinitely, and their ultimate perimeter ADF will (by lem. 3, corol. 4) be a curved line; and thus the centripetal force by which the body is continually drawn back from the tangent of this curve will act uninterruptedly, while any areas described, $SADS$ and $SAFS$, which are always proportional to the times of description, will be proportional to those times in this case. Q.E.D.

It is worth spending some time getting used to Figure 5.8 first. We are to imagine that the Sun is at the point S in the lower left-hand corner, and that it exerts a force on the moving body A, which starts off in the lower-right-hand corner. At that moment, the body is moving with a certain velocity along the line AB. We imagine first that the effect of the Sun is felt at discrete intervals of time (say, every second). The first time that this happens is when the body has reached B. It now receives an impulse pushing it towards the Sun along the line BVS. This knocks it off the line AB, and it no longer heads towards the point *c* but towards the point *C*. At the point *C* it receives another blow, and travels for a further second to *D*, where it is again given an impulse, and so on. So the path of the body is the polygonal path *ABCDEF*....

Newton argued as follows. He noted that all this happens in the plane defined by the line *AB* and the point *S*, and he checked that the impulse delivered each second always acts in this plane; this reduces the problem to one in plane geometry. Next, to determine the precise position of the point *C*, he argued that it must be the fourth vertex of the parallelogram *VBcC*; this is because he thought of the impulse along *BS* as giving a stationary body at *B* a certain velocity, to which must be added the velocity that the body already has at *B*.

Now for areas. The area swept out by the radius in the first second is *ASB*. In the next second, because of the impulse at *B*, the area is *BSC*. Had it not been for the impulse, the area swept out would have been *BSc*. But look at the triangles *BSC* and *BSc*: they are on the same base, *BS*, and lie between the same parallels, *BS* and *Cc*, and so they are equal in area. Moreover, the triangles *BSc* and *ASB* have the same height and equal bases, because $AB = Bc$. So these triangles are also equal in area. Newton therefore deduced that the triangles *ASB* and *BSC* are equal in area. The same is true at every stage, no matter what the law of force is, so the triangles *ASB*, *BSC*, *CSD*, ... are all equal in area.

Finally, he supposed that the impulses happened ever more frequently, thus increasing the number of triangles and diminishing their breadths, until in the limit the body *A* traces out a curved path *ABCDEF*.... It is now the case that the body sweeps out equal areas in equal (infinitesimal) amounts of time, and so (by an argument that Newton omitted) it sweeps out equal areas in equal times, whatever the law of force may be, provided that it is directed towards the centre.

Newtonian gravity. We now turn to the relationship between Books I and III of the *Principia*. Newton addressed this point in his preface to the whole work. He argued that Book I (and also Book II) describe how to pass from visible phenomena to invisible forces by means of mathematics. Book III is 'an example of this', in that the mathematical propositions of the first Books are there made to 'derive ... the forces of gravity', from which forces are deduced 'the motions of the planets, the comets, the Moon and the sea'. It is clear from this preface why Book III is so important. It is only there that the force of gravity is shown to give a really accurate account of the motion of the planets.

What might a contemporary reader have found most difficult to understand in Book III? One non-trivial difficulty would be the technical business of dealing with astronomical data. But suppose we consider a competent practitioner; then the real difficulty would be the kind of explanation that Newton proposed. *A physically real attractive force* reaching out over astronomical distances was a colossal, even shocking, novelty in 1687, and is still a bizarre concept when you think about it. Leibniz and Huygens could never accept it: Huygens wrote to Leibniz (on 18 November 1690), 'I am by no means satisfied [by] ... his Principle of Attraction, which to me seems absurd'.[18]

Huygens wished to retain Cartesian vortices, not because he could use them to describe orbits to a high degree of quantitative accuracy (he never could, and Newton had shown it to be impossible), but because he preferred the conception of dynamics

[18]Quoted in (Cohen 1980, 80).

that this theory offered. For him, the *Principia* was elegant geometry, but it was not physics. It is this puzzle — by which Newton's most gifted immediate readers were not persuaded, but their successors were — that we now discuss.

In his book, *The Newtonian Revolution*, Cohen argued that the answer to this puzzle lies in the bold extension of scientific reasoning that the *Principia* contains. The whole structure of the work was designed to persuade its readers of the effects of this mysterious force of gravity, not in virtue of its essence (whatever that might be) but by a logical mathematical argument. Books I and II are hypothetical: *Suppose*, said Newton, that there is such a centripetal force; *then* such-and-such conclusions follow. So far, so mathematical — the supposition is a mathematical idea being entertained for its own sake. But why suppose such a mysterious force in Nature?

It might help to think about how one might come to believe in the existence of anything that one cannot experience directly. One way is the discovery that the idea (that this thing exists) makes sense of a number of disparate phenomena while not simultaneously making nonsense of something else. In the present case, taking as the disparate phenomena the sundry elliptical orbits of planets and satellites as described by Kepler's laws, and as the idea that there may be some kind of centripetal force at work, the shapes of the orbits, and the speeds with which they are traversed, turn out to be consequences of the idea under investigation. In fact, all the orbits turn out to be consistent with the more precise idea that the force obeys an inverse-square law. So this idea brings together the astronomical data encapsulated in Kepler's laws by showing that his laws are *mathematical* consequences of supposing there to be an inverse-square law of force in operation.

The judgement that this hypothesis does not meanwhile make nonsense of something else is rather more subjective. On the one hand, the only place where Newton was forced into difficulties in merging theory and observation was the motion of the Moon, but there are good reasons (which we discuss below) for why this does not imperil his overall argument. On the other hand, a reader might feel that the very idea of a force acting at a distance was absurd and could offer no explanation of anything — such an objection might be categorised as 'metaphysical', in the best sense of the word.

On this analysis, the *Principia* aimed to convince its readers of the truth of a physical hypothesis by arguing that it helped to make mathematical sense of certain natural phenomena. Without appreciating this fact, one was naturally liable to read the book as an exercise in mathematics, rather than in physics — and readers looking for a physical explanation (that is, one that invokes an intuitively plausible physical or mechanical process) would not find one. So the differing receptions of the *Principia* might derive from the differing expectations of its readers about what it had to offer.

However, a reader who could follow Newton's line of thought through all three books would see presented a marvellously well-developed dynamics — truly a System of the World, and one that a vortex theory could never be, as Newton showed. Cohen called the Newtonian style 'a special blend of imaginative reasoning plus the use of mathematical techniques applied to empirical data',[19] and he documented the steady rise in complexity of the objects that the *Principia* discusses: a point and a central force; two points mutually attracting each other; two solid bodies; three solid bodies; and the Earth, Moon, and Sun. By the end, it is the real world that is being described, and not a simplified model of it — or so Cohen argued Newton's view to have been.

[19]See (Cohen 1980, 62).

Cohen's point about increasing complexity is well taken. A wealth of natural observations were shown to fit a simple mathematical law to a surprising degree of accuracy. But here it is the role of mathematics that deserves our special attention. Three aspects are noteworthy:

1. technical dexterity (formidable)
2. mathematical novelty (or, how the calculus nearly appears)
3. hypothetical reasoning (the crucial point).

'Hypotheses non Fingo', 'I do not feign hypotheses', has become one of Newton's best known sayings. But nothing could be more open to misunderstanding, for the *Principia* is *full* of hypotheses. Here is what Newton wrote.[20]

Newton on the nature of gravity.

> Thus far I have explained the phenomena of the heavens and of our sea by the force of gravity, but I have not yet assigned a cause to gravity. Indeed, this force arises from some cause that penetrates as far as the centers of the sun and planets without any diminution of its power to act, and that acts not in proportion to the quantity of the surfaces of the particles on which it acts (as mechanical causes are wont to do) but in proportion to the quantity of solid matter, and whose action is extended everywhere to immense distances, always decreasing as the squares of the distances. Gravity toward the sun is compounded of the gravities toward the individual particles of the sun, and at increasing distances from the sun decreases exactly as the squares of the distances as far out as the orbit of Saturn, as is manifest from the fact that the aphelia of the planets are at rest, and even as far as the farthest aphelia of the comets, provided that those aphelia are at rest. I have not as yet been able to deduce from phenomena the reason for these properties of gravity, and I do not feign hypotheses. For whatever is not deduced from the phenomena must be called a hypothesis; and hypotheses, whether metaphysical or physical, or based on occult qualities, or mechanical, have no place in experimental philosophy. In this experimental philosophy, propositions are deduced from the phenomena and are made general by induction. The impenetrability, mobility, and impetus of bodies, and the laws of motion and the law of gravity have been found by this method. And it is enough that gravity really exists and acts according to the laws that we have set forth and is sufficient to explain all the motions of the heavenly bodies and of our sea.

All Newton meant here was that he framed ('feigned' or made) no hypotheses about how gravity had its effects. He did not like uncontrolled speculation about mechanisms intended to cause changes, such as those he found in Descartes's physics. But he did believe strongly in testing speculations by drawing out their consequences, and the *Principia* is full of arguments of the form 'If X, then Y', where X is a speculation

[20] For the Motte–Cajori version, see F&G 12.B13.

(a law, say), Y is a conclusion (such as an orbit), and the deduction is firmly mathematical. To us, but not to Newton, such reasoning is hypothetical. To Newton it was framing hypotheses to slip in a causal explanation without testing it or to do so in an ad hoc way so that, as he put it in Rule 4, arguments 'be nullified by hypotheses'. With this distinction in mind, Newton's *Principia* can be called a showcase of mathematical hypothesis testing.

As such it was novel, and accordingly difficult to read. Books I and II (400 of the 547 pages) are mathematical. They show how almost any plausible theory of motion can be tested mathematically by being made to predict orbits, via his theorem on equal areas in equal times. What emerges is that no simple laws of motion, other than motion under an inverse-square law of attraction, has any chance of corresponding to Kepler's laws. Having thus driven rival theories from the field, it is only in Book III that Newton turned to physics, and showed that inferences on the assumption of gravity do indeed correspond exceptionally well to empirical evidence. Newton was clear that one had to accept gravity *because* there were no other theories that were mathematically sound, and gravity yielded descriptions well in accordance with observations, *even though* there was no physical explanation available for how gravity operated.

This is quite different from how the vortex theory operated. There, a simple mechanical explanatory model (collision) was intended to give a rough fit with observational data, and the model retained its appeal even though no-one could make it yield a good fit. The idea that a range of mathematical theories could be put on offer, and the best fit selected, is a novelty that one sees first with Newton. If it resembles anything, it resembles Kepler's choice of the ellipse as a curve for fitting planetary orbits, but raised to a higher level of sophistication.

Newton distinguished carefully between mathematical principles and their physical application. What is described in Books I and II, including an attractive force obeying an inverse-square law, is mathematical and need not relate to the real world. His attitude to the nature of gravity was to be phrased very carefully — not in the first edition, it is true, but in the second and third editions (of 1713 and 1726), which were the ones to be read most diligently by subsequent generations. We may presume that the lack of much comment by him on the subject initially was noticed by his first readers, and that what he was later to say about gravity was the product of much attentive thought.

What did Newton believe about gravity? Did he consider that it really exists, or that it is but a useful descriptive device? Newton argued that gravity explains the motion of the planets, although he had been unable to discover how it acts ('the cause of this power'), nor would he speculate. But he was adamant that gravity really exists and obeys the laws he had set down, and was not just a mathematical artifice.

The *Principia* and the Newtonian calculus. Before considering how the *Principia* was received, we turn to look at the mysterious case of the non-appearing calculus. It was Newton himself (writing anonymously) who started the story that the contents of *Principia* were discovered by using the calculus, and then rewritten in more traditional geometrical language:[21]

> By the help of the new Analysis Mr Newton found out most of the propositions in his *Principia Philosophiae*: but because the Ancients for making things certain admitted

[21]See Westfall, *Never at Rest*, pp. 423–424.

> nothing into Geometry before it was demonstrated synthetically, he demonstrated the Propositions synthetically, that the Systeme of the Heavens might be founded upon good Geometry. And this makes it now difficult for unskilful men to see the Analysis by which those Propositions were found out.

This is an interesting example of how primary historical testimony can be misleading if taken at face value. It appears that this version of history is what Newton wished to be believed, for reasons associated with the sad dispute of his later years concerning priority in inventing the calculus. But, as Whiteside wrote:[22]

> Nowhere, let me repeat, are there to be found extant autograph manuscripts of Newton's, preceding the *Principia* in time, which could conceivably buttress the conjecture that he first worked the proofs in that book by fluxions before remoulding them in traditional geometric form.

Nor, we should note, was Newton a particularly well-read geometer; in this, as in so much else, he was very much his own man.

What then of Pourciau's claim, raised earlier, that fundamental aspects of the calculus appear in geometrical dress in the opening lemmas of the *Principia*? Pourciau also dismissed the claim that there was an explicit translation of the calculus into the language of the *Principia*; there is simply no documentary evidence for this. He inclined somewhat to the view that the same gifted mind that could produce a theory of fluxions produced the *Principia* — and in the same way, by dealing geometrically with finite quantities and approximation arguments. Whether the lemmas were inserted as a pedagogic strategy (and an unsuccessful one at that) or emerged as one coherent whole from the depths of Newton's thought, Pourciau did not decide.

It is also true that Newton could derive some propositions in the *Principia* more easily with his calculus than without it. In a famous scholium (an explanatory or historical remark — Newton inserted several of these into the *Principia*) he indicated that since at least 1671 he had possessed a general method for finding tangents to curved lines and for 'the resolving other abstruser kinds of problems about the crookedness, areas, lengths, centres of gravity of curves &c'.[23]

Most importantly, on several occasions Newton reduced a question to a quadrature, and then gave the result of performing that quadrature, and he also relied on his method of infinite series in several places.[24] So his method of fluxions was used in the *Principia*, if under the table, so to speak. There was never a text written in the method of fluxions that Newton later rewrote as the *Principia* — but the famous book has more than a hint of that method, nonetheless.

Although the *Principia* is not a calculus book, the questions that it discusses are appropriate for the calculus. Newton's successors, lacking his geometrical brilliance, put the calculus-type arguments that he had eschewed into their understanding of his work. Mathematical physics came to be written in terms of the calculus not because of Newton's example, but because without it no-one else could easily follow him.

---

[22]See (Whiteside 1970, 125–126).

[23]Lemma II of Book II, to which this note was attached, amounts to differentiating $x^{m/n}$ to get $\frac{m}{n}x^{(m-n)/n}$, and is indicative of the high level of generality to which Newton was working.

[24]See, for example, Newton, *Principia*, Book I, Proposition 45, 539–545, where Newton wrestled with the motion of the Moon. For the Motte–Cajori version, see the extract in F&G 12.B8.

5.3 Responses to the *Principia*

We conclude this chapter with a brief look at the earliest British and Continental responses to the *Principia*, postponing until the next chapter the later, more critical, and ultimately more profound, responses on the continent of Europe.

At home, the appearance of the *Principia* caused a sensation. Westfall has noted some contemporary reactions. David Gregory wrote from Edinburgh that Newton had

> been at the pains to teach the world that which I never expected any man should have knowne.[25]

Edmond Halley said that

> so many and so Valuable Philosophical Truths, as are herein discovered and put past Dispute, were never yet owing to the Capacity and Industry of any one Man.[26]

Halley also wrote a Latin poem that graces the opening pages of the *Principia*. It begins[27]

> Behold the pattern of the heavens, and the balances of the divine structure;

It then proceeds to draw attention to the major achievements of the work, and concludes by urging the reader to

> Join me in singing the praises of Newton, who reveals all this,
> Who opens the treasure chest of hidden truth,
> Newton, dear to the Muses,
> The one in whose pure heart Phoebus Apollo dwells, and whose mind he has filled with all his divine power;
> No closer to the gods can any mortal rise.

John Aubrey, in vain pursuit of the priority claim on behalf of his friend Robert Hooke, spoke of

> the greatest Discovery in Nature that ever was since the World's Creation. It never was so much as hinted by any man before.[28]

The *Principia* also had a dramatic impact on the philosopher John Locke, then a political exile in Europe. Locke found the mathematics beyond him, but was assured by Huygens that he could take it on trust, so he went directly to the physics, which greatly impressed him. Quite generally, the scope of the work, coupled with its obvious difficulty, gave it an impressive reputation far outside the limited circle of mathematicians competent to read it.

The first Continental responses were partly coloured by the adherence to Cartesian ideas. As we saw, Descartes's vortex theory sought to describe how the planets are carried round in an invisible medium, without which they would travel in straight lines, and Newton analysed this theory in Book II of the *Principia* in order to refute it. More mundanely, objects can move through the air and ships through the sea, and this suggests the problem of analysing motion in a resisting medium.

---

[25] D. Gregory to Isaac Newton, 2 September 1687, in *The Correspondence of Isaac Newton*, H. W. Turnbull (ed.), Vol. II, p. 484. Gregory was sent a copy of the *Principia*, and as soon as he received it he began writing a commentary.

[26] Westfall, *Never at Rest*, p. 470.

[27] See (Cohen and Whitman, 1999, 379–380).

[28] See (Dick 1962, 245).

One of Leibniz's first observations on seeing a review of Newton's *Principia* in 1688 was that his (Leibnizian) differential calculus would be invaluable in building a theory of motion in a resisting medium. He immediately published a paper along those lines, in which announced some results although he suppressed his methods. The topic was a difficult one, and it continued to interest him; in particular, he studied the special case in which the medium exerts a resisting force proportional to the square of the velocity. This is the case that Newton had considered in the *Principia*, and it is the most interesting one on physical grounds. A solid body moving through a swarm of particles (perhaps a planetary vortex) experiences a resistance that is proportional to the relative velocities of the body and particle on impact, and also to the number of collisions in each unit of time. Since this number is proportional to the distance travelled, the total resistance is proportional to the square of the velocity.

Leibniz discussed the problem with Huygens, and soon after Huygens' death in 1695, Pierre Varignon, another of Leibniz's correspondents, took up the theme. Varignon was an engaging man, who later strove energetically to bring peace to the controversy over the invention of the calculus. He stood close enough to Newton to be able to commission Sir Godfrey Kneller to paint Newton's portrait for him in 1720, but he was also close enough to Leibniz and Johann Bernoulli to be able to correspond steadily with them.

Leibniz and Varignon had been drawn together in 1700 when the Académie des Sciences in Paris felt called upon to defend the calculus against the attacks of one Michel Rolle, who like Varignon was a salaried mathematician at the Académie itself. Rolle was an algebraist, well-trained in Cartesian techniques, who believed that the new Leibnizian calculus was not properly grounded and could lead to the wrong results. This was embarrassing for the Académie because Leibniz had become a foreign member earlier that year. As the Marquis de l'Hôpital, the leading Parisian supporter of the calculus, was away, the Académie looked to Varignon; when Varignon published his defence of the Leibnizian calculus Leibniz wrote to say how satisfied he was with it. Rolle conceded the argument some years later.

Figure 5.9. Pierre Varignon (1654–1722)

By this stage, Varignon had become quite expert at applying the differential calculus to problems of motion, and had shown how several of Newton's theorems about motion under a centripetal force could be derived in this way. He could show, for example, how to determine the law of force given the orbit, at least in simple cases. Leibniz wrote to encourage him in his endeavours, and to suggest that he try the more interesting converse problem: given several bodies attracting one another, find their orbits. This problem proved intractable, because determining the orbit given the law of force requires at least a grasp of the integral calculus — a subject not then known to Varignon. Moreover, his request to Bernoulli for instruction was turned down as a consequence of the contract that Bernoulli had signed with the Marquis de l'Hôpital (as we describe in Section 6.2). Nor could he have learned to integrate from Leibniz's paper on motion in a resisting medium, for Leibniz had not revealed his methods there. However, over the next few years Varignon patiently caught up, and did much to show how the calculus could solve many problems to do with motion. Specifically, he showed how the formalism of differential equations could be used to advantage in the study of dynamics, presenting a general technique in place of Newton's ingenious but ad hoc geometrical arguments.

Another success for the Leibnizian calculus came when Johann Bernoulli and Jakob Hermann independently tackled a problem that Newton had skated over in his *Principia*. This, perhaps surprisingly, was the determination of the orbit of a planet under the assumption that it is subject to an attractive force obeying an inverse-square law. Newton had shown how to find the orbit, so when Bernoulli and Hermann took up the problem it was not, strictly speaking, unsolved, but Newton's solution was written in the geometrical style of the *Principia*, and Bernoulli seems to have felt that it lacked something in generality and directness.

Then, in 1709, Johann Bernoulli discovered a mistake in one of Newton's theorems concerning motion through a medium that resists in proportion to the square of the velocity. Being unable to see where Newton had gone wrong, he produced a better proof of his own. But he also convinced himself (wrongly) that the mistake was evidence that Newton did not understand the subtleties of the calculus — with, if true, an obvious bearing on the dispute then raging over whether Newton or Leibniz had priority in the discovery of the calculus (which we discuss in Section 6.2). So he decided to raise the stakes and to put forward his alternative to Newton's treatment of the inverse-square law.[29] This is firmly Leibnizian, but his letter to de l'Hôpital shows signs of another agenda, because he was obviously pleased to surpass Newton, as he saw it.

> You will see, Monsieur, that I have reached at a stroke a differential equation of the first degree, in which there is no mixing up of indeterminates; and so the geometric construction can be easily deduced from it, the quadratures of the curved spaces being given, and even more conveniently than Mr Newton found it in his *Principia*.
>
> Moreover, my equation displays whether the sought-for trajectory is algebraic or not, depending on what hypothesis is given for the force.

At this time Newton was engaged with Roger Cotes in producing the second edition of the *Principia*. This edition was delayed, and when Nicolaus Bernoulli, Johann's nephew, came to London in October 1712, he told Newton that his uncle had found a mistake in his treatment of motion in a resisting medium. Newton accordingly corrected the mistake in the second edition, and believing Johann Bernoulli to have acted out of generosity, proposed him as a foreign member of the Royal Society. Johann

[29]See (Bernoulli, J. 1742b, 474), in F&G 13.B3.

Bernoulli, however, was trying to arrange for his two papers to be published just after the second edition was to appear, thereby correcting these mistakes publicly, but his plan misfired. When Newton later discovered the trap that had been prepared for him, he was less than pleased.

Nonetheless, Bernoulli's achievement was a real one, for he showed that calculus did indeed simplify arguments that could be carried out in other ways only with difficulty, and his suspicion that the calculus would be essential to making progress was to be amply confirmed.[30]

As these examples show, in the early decades of the 18th century many mathematicians worked hard to understand the *Principia*, and in their attempts to clarify it were led to increasing use of the calculus. The culmination of this line of approach was Euler's early book on mechanics (*Mechanica*, 1736), in which he redefined the agenda for mechanics for the day and began the work of rewriting the calculus to meet the task.[31]

5.4 Newton's final years

Newton's *Principia* was first published in 1687. The next year was to prove equally important to Newton, for it was the year of the 'Glorious Revolution', in which James II was overthrown in favour of William III and his wife Mary, who came over from the Netherlands, and the prospect of a Roman Catholic returning to the throne of England was ended. Newton's own strongly held religious views were risky for him: he adhered to the Arian heresy that placed the Son of God below God himself, a position that contradicted Anglican teaching on the Trinity. However, Newton had to do little to avoid being discovered, except attend a few Anglican services, and was only to reveal his true views on his deathbed. Accordingly, nothing prevented him from changing his career and moving into more public walks of life with the attendant oaths that would have to be sworn, and this was what he did.[32]

Newton was elected as a Member of Parliament for the University of Cambridge in 1689, and in that capacity he attended the proclamation of William and Mary as monarchs on 13 February 1689. If that did not excite him, life in London did. He met Christiaan Huygens at meetings of the Royal Society, and in the same year, 1689, he met the philosopher John Locke, who was an early enthusiast for the *Principia*, and they began a correspondence on many topics, including religion. One of Newton's abiding interests was in Biblical chronology, a subject upon which he wrote extensively. He was expert in the many ancient sources that had somehow to be squared with the right interpretation of the Bible and whatever ancient astronomical observations could be used.[33]

Newton, now he was a Member of Parliament, also began to be talked about as a possible member of the government. He was already able to help place his friends and supporters in important positions, and in 1691 he had secured the Savilian Professorship of Astronomy at Oxford for David Gregory over Halley; Halley's atheist views

[30] See (Bernoulli, J. 1742b).

[31] We discuss the work of Euler and his successors in Chapter 7 and subsequent chapters.

[32] Later, in 1714, Newton's friend and successor as Lucasian Professor, William Whiston, lost his positions in Cambridge because he publicly promoted his Arian views; Newton kept quiet.

[33] See (Buchwald and Feingold 2012).

may have swayed Newton, although it has been suggested that Gregory's were scarcely less strong, and later (in 1703) he placed Halley in the Savilian Chair of Geometry, following the death of John Wallis.

Newton's enthusiasm for mathematics waxed and waned in these years. He put his treatise on optics into the form that he would publish a decade later, and began to think of publishing his earlier works that had circulated only in manuscript. This was partly because he was drawn into the poisonous priority dispute, in which Newtonians and Leibnizians disputed claims for their man to be the true discoverer of the calculus, while seeking to place him above such a sordid matter. Newton also resumed his work on alchemy, and in the autumn of 1693 he fell into a depression that lasted for 18 months. Those who knew about it (John Locke and Samuel Pepys among them) took care not to spread the news, so our sources for it are few and conflicting explanations for it are too easy to give. Westfall, in his biography of Newton, notes the theory that mercury poisoning could have caused it (mercury vapour often arose in Newton's alchemical experiments), but inclines to the view that sheer intellectual exhaustion overwhelmed Newton.[34]

By 1696, when he was fully recovered, Newton realised that Cambridge life had little further to offer him. Scholarship was not necessary for success there, or even very common. Instead, on 19 March 1696, Newton became Warden of the Royal Mint. As so often is the case, the Treasury was coping with a national financial crisis. This had been provoked by a war with France that the accession of William and Mary had accelerated, and which threatened England with bankruptcy. The Mint, a relatively small branch of the Treasury, was struggling to cope with an epidemic of forgery.

Forgers had long been able to clip small pieces from the edges of coins and use the pieces to make new coins. The remedy had been introduced: machine-milled coins with a ribbed edge would replace the old, hand-hammered coins. But no measure had been passed to withdraw the old coins from circulation, and the result was a boon for coin clippers — so much so that the value of English coin began to decline and people no longer accepted coins at face value. Newton, who had been consulted, was among those who advocated the twin remedies of devaluation of the currency and the withdrawal of hand-hammered coins from circulation, which had to be surrendered but not at face value. The rich could and did benefit from the system, while the many poor lost.

This process was well under way before Newton became Warden of the Mint, but he carried it out with notable zeal, and saw that counterfeiters were prosecuted and some executed — he became a much loathed figure among them as a result. Newton, a born administrator, was regarded by all sides as an exceptional Warden of the Mint, and in December 1699 he was promoted to Master of the Mint.

He was now drawn further into government, as a man with a reputation for getting things done, and in April 1705 he was knighted by Queen Anne on her trip to Cambridge 'not for his contributions to science, not for his service at the Mint, but for the far greater glory of party politics in the election of 1705.'[35] Newton had already been elected President of the Royal Society in November 1703, possibly after Wren had declined to stand and suggested Newton, but apparently with no great enthusiasm from

[34]See Westfall, *Never at Rest*, pp. 531–545.
[35]See Westfall, *Never at Rest*, p. 625.

the other Fellows.[36] He proved to be as energetic as he had been at the Mint, behaviour that was quite the opposite of his immediate predecessors in office. He restored the Society's finances and re-dedicated the Society to the pursuit of science. In 1704 he revised and finally published his *Opticks*, partly as a way of reminding the Society what his idea of science was. He also had to manage relations with Gresham College, where the Royal Society had mainly been housed since its beginning. But Gresham College now wanted the Society to move out so that they could seek permission from Parliament to demolish and replace the College buildings. Eventually the Royal Society moved to new premises in Crane Court in 1710.[37]

Towards the end of his reign as President of the Royal Society, Newton became embroiled in a controversy with Flamsteed, the Astronomer Royal. Flamsteed wanted his life's observations published by the Royal Society in a catalogue that would extend published locations of the stars from about 1000 to approximately 3000, but Halley, supported by Newton, sought to publish only an abridged version that omitted many of Flamsteed's observations and replaced them with some of Halley's that Newton had found useful — Newton was preparing the second edition of his *Principia* at the time. Newton wrote a preface that falsely implied that Flamsteed had not wanted to see his observations made public and did so only on orders from Prince George.[38] Only politics saved the day for Flamsteed. A change of government occasioned by the death of Queen Anne in 1714 brought his friends to power and marginalised the now-elderly Newton. Although Flamsteed died in 1719 before the job was done, his former assistants saw to it that the *Historia Coelestis Britannica*, as the catalogue was called, appeared, in three volumes and essentially as Flamsteed had wanted it.

Newton's long life ended on 20 March 1727. His body lay in state in the Jerusalem Chamber in Westminster Abbey and was then taken and interred in the nave. The Lord Chancellor, and the Dukes and Earls who were Fellows of the Royal Society, attended the funeral; later Voltaire, with perhaps a pardonable degree of exaggeration, wrote of Newton that[39]

> 'He lived honoured by his compatriots and was buried like a king who had done well by his subjects'.

More extravagantly, the inscription on the monument to Newton that was erected in the Abbey in 1731, concludes:

> 'Let Mortals rejoice That there has existed such and so great an Ornament to the Human Race'.

But, as Westfall wrote at the end of his biography of Newton, 'only hyperbole can hope to express the reality of the man'.[40]

5.5 Further reading

Guicciardini, N. 1999. *Reading the* Principia. *The Debate on Newton's Mathematical Methods for Natural Philosophy from 1687 to 1736*, Cambridge University Press. A

[36] See Westfall, *Never at Rest*, p. 629.

[37] It moved to its present home in Carlton House Terrace, London, only in 1967, after several further moves.

[38] See Westfall, *Never at Rest*, p. 692.

[39] See (Voltaire 1734), English transl. (1984, 68–71), and F&G 12.F2.

[40] See Westfall, *Never at Rest*, p. 874.

rigorous and stimulating examination of how Newton's ideas were received by his contemporaries.

Guicciardini, N. 2009. *Isaac Newton on Mathematical Certainty and Method*, MIT Press. The author ranges across all of Newton's work to see how he responded to, and separated himself from, the work of Descartes and Leibniz.

The Newton Project is an astonishing collection of Newton documents that put together a rich, almost overwhelming picture of him. It is dedicated to publishing in full an online edition of all Newton's writings, whether they were printed or not. All the texts we discuss, and many more, are here.

`http://www.newtonproject.ox.ac.uk/`

6

The Spread of the Calculus

Introduction

In this chapter we look at how the generation of mathematicians after Newton and Leibniz responded to the invention of the calculus. This is largely a Continental story, one prominently associated indeed with a single family, the Bernoullis, and even more with one man, Leonhard Euler. We shall see how their work and that of their immediate successors help us to understand the reception of Newton's *Principia* in Continental Europe. Its mechanical principles conflicted with the widely accepted Cartesian theory of planetary motion, which Newton had been at pains to demolish, and his ideas initially met with some bafflement and then growing resistance. We consider how these ideas were made mathematically acceptable through the application of new ideas in the calculus, and how they were confirmed by two decisive sets of observations based on predictions drawn from Newton's theory: the observed flattening of the Earth at the poles, and a much improved analysis of the motion of the Moon.

Two of the mathematicians most associated with this successful fusion of the calculus with mechanics were Alexis-Claude Clairaut and Leonhard Euler. The third important mathematician in the middle of the 18th century was Jean le Rond D'Alembert, and we conclude this chapter with one of his most important discoveries, a mathematical description of the vibrating string.

6.1 The next generation

It is time to look at the people who learned the calculus, rather than those who invented it, and at the state of the practitioners rather than the state of the art.

The most prominent figures of the next generation were the Swiss mathematicians Jakob and Johann Bernoulli. Jakob was born in 1654 in Basel, where he became the professor of mathematics in 1687. Johann was born twelve years later, and learned advanced mathematics in Groningen in the Netherlands, and returning to Basel only on the death of his brother, whom he succeeded in the Chair. They learned the calculus in correspondence with Leibniz but, perhaps surprisingly, never met him.

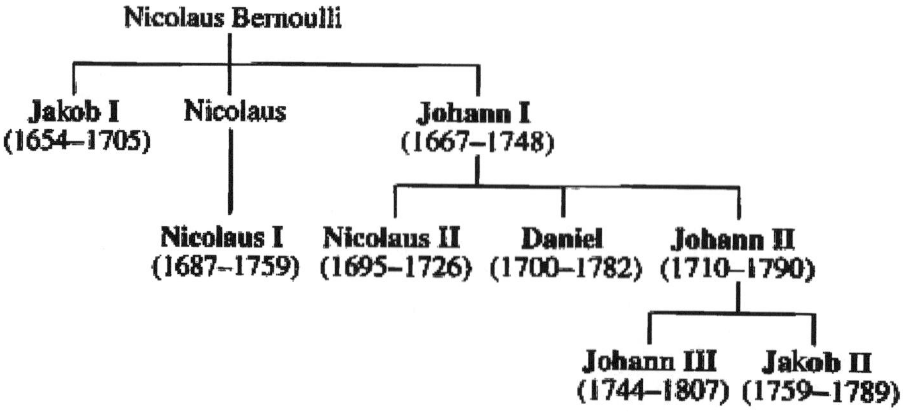

Figure 6.1. A genealogical tree of the Bernoulli family

The relationship between the two brothers soon became competitive when it was not hostile, and the rivalry extended to other members of the family, notably including Nicolaus I (the nephew of Jakob and Johann) and Johann's son Daniel. The extended family lasted for several generations, which is why the Bernoullis are sometimes given numbers, like royalty and the members of some American families. There were also other mathematicians of note who came from Basel, such as Jakob Hermann, who was eleven years older than Johann Bernoulli and rose to be the leading applied mathematician of his day. They all typically published their discoveries in the journal that Leibniz had founded in Leipzig, the *Acta Eruditorum*.

Figure 6.2. Jakob Bernoulli (1654–1705)

Figure 6.3. Johann Bernoulli (1667–1748)

In England there gradually emerged a group of capable Newtonians who were loosely organised around the Royal Society. Two of them are worth mentioning: Roger Cotes, who helped Newton to give the *Principia* the nearly definitive form it took in its second edition, but who died in 1716 at the age of 33; and Brook Taylor, who had perhaps the deepest grasp of the Newtonian calculus. But unlike Leibniz, Newton was unwilling to share his ideas openly, and few of Newton's immediate followers in Britain were very good mathematicians compared with those on the Continent: his best British descendants appeared only later.

In Paris, there was a distinguished group of mathematicians. Although its leading light was Pierre Varignon, the one who chiefly interests us now is another member of the landed aristocracy, Pierre Rémond de Montmort. For it was Montmort who most actively bridged the gap between the two increasingly hostile camps of Newtonians and Leibnizians, and who, through his correspondence, did most to give historians a sense of what it might have felt like at the time. However, the French lacked a forceful central figure, and for a time languished under the baleful effects of Descartes's ideas about physics, which kept them from fully appreciating the ideas of Newton.

Brook Taylor, who was born into an affluent land-owning family near Canterbury, graduated from Cambridge with a degree in Civil Law. He then moved to London, where in 1712 he was elected a Fellow of the Royal Society. In 1714 he became the Secretary of the Society, and in 1715 he travelled to Paris to meet the scientists there. One of those he met was the 37-year old Montmort, who later wrote to Nicolaus I Bernoulli that:[1]

> This Mr. Taylor is a very polite young man who will splendidly sustain the reputation the English have acquired through their extensive knowledge of the profound geometry. Like all other Englishmen he is terribly biased in favour of Newton's philosophy.

[1] See (Feigenbaum 1992, 387).

Figure 6.4. Brook Taylor (1685–1731)

Figure 6.5. Marquis de l'Hôpital (1661–1704)

Montmort himself was a competent mathematician, but he felt acutely the lack of real ability, and he confided to Taylor: 'I'm very upset that my ignorance reduces me to the simple rank of spectator'. But, as he wrote to Taylor on another occasion, he consoled himself with the thought that people like Bernoulli, being professionals:[2]

> always have their heads full of figures and theorems, whereas a man of the world like you is distracted by business, pleasures, the obligations of the Royal Society, from which I conclude that there has to be a lot more talent and genius in a person like you than in a professor in order to make great progress in the sciences.

Be that as it may, the future was to lie with the mathematical professionals on both sides of the Channel.

6.2 The calculus, 1690–1730

The priority dispute. In the early 1700s a fierce and sordid dispute developed between the English and the Continental mathematicians over the invention of the calculus.[3] Newton, who knew he had discovered the calculus first, was unable to believe that Leibniz had independently come to many of the same ideas. The followers of Leibniz, on the other hand, knew that he had published first and became increasingly tired of British harping on a priority that Newton endlessly failed to establish, and were more and more inclined to doubt it. Both sides became increasingly virulent and less honest, with Newton and Leibniz occasionally directing their lieutenants, who mainly fought their battles on their own initiative. Although the details of the dispute need not detain us, the style of argument is worth noting. For, insofar as rival claims to have invented the calculus were tested by proposing problems for the other side to solve, the idea was that failure to solve a problem would indicate a lack of insight into the calculus, and might accordingly be evidence that one's opponents had not invented it.

[2] See (Feigenbaum 1992, 388).

[3] On the priority dispute in detail, see (Hall 1980).

Box 11. Taylor Series.
It is helpful to begin with an example, such as the infinite series for the sine function:

$$\sin x = x - \frac{x^3}{3.2} + \frac{x^5}{5.4.3.2} - etc.$$

This could be called 'the Taylor series for the sine function expanded about the point $x = 0$'.

A topic that much engaged the attention of 18th-century mathematicians was finding similar infinite series expansions for functions they knew much less about than the sine function. What Brook Taylor established was that any function to which the calculus applies can be written in the form of an infinite series, and that the successive coefficients of the series are closely related to the successive fluxions (or derivatives) of the function.

At the time that Brook Taylor proposed his series, it was widely believed that every function of a variable was so obliging as to let itself be differentiated infinitely often, and so could be written as an infinite power series in x. So his discovery seemed to establish that the formalism of infinite series would be sufficient for all mathematical investigations. As we shall see, this was too good to be true, but the restrictions that had to be imposed on the use of Taylor series were also to mark the limits which the formal calculus of the 18th century could reach.

Taylor was by no means the only person to have obtained such a result; Bernoulli, for instance, had a related one which he used to evaluate integrals. Recent scholarship has disclosed that Newton was in possession of Taylor's result as early as the 1690s, although Whiteside is certain that Newton never let the relevant manuscript out of his hands. But we should note that the historian Lenore Feigenbaum, in her study of Taylor, writes: 'I have not come across any evidence that Newton actually used [Taylor's theorem] as [a formula] for generating power series or that he applied them as Taylor did in solving fluxional equations.'[a]

[a] See (Feigenbaum 1985, 77).

During this priority dispute over the calculus, Montmort corresponded with members of each side and conveyed information — not all of it true — that the protagonists wanted the others to hear. In particular, we learn from him that Newton turned against his erstwhile supporter and was instrumental in the cabal that finally decided Taylor to resign as Secretary of the Royal Society in 1718; it seems that Newton wanted the more pliant John Machin in the post. Yet Brook Taylor was a good mathematician. He is still remembered for the Taylor series (see Box 11); he wrote two original books on the mathematical theory of perspective; he was the first person to get anywhere near answering Mersenne's old call for a mathematical theory of the vibrating string; and he was the chief defender of English mathematical prowess when the Leibnizians challenged the English to match them at developing the calculus.

The point of the challenges was, as Leibniz once put it to Johann Bernoulli, 'to test the pulse of the English'.[4] So hot ran the passions that, should they fail, their calculus, and even their claim to priority of invention, might seem to be inadequate.

One of these problems will give their flavour. Asked by Leibniz to propose a problem that required a general, theoretical breakthrough for its solution, Bernoulli challenged the English to solve this problem:

> Given: a family of curves through a point defined by a certain condition (which Bernoulli stated explicitly).
>
> To find: a family of curves each meeting every member of the given family at right angles. These curves are called the orthogonal trajectories to the given family.

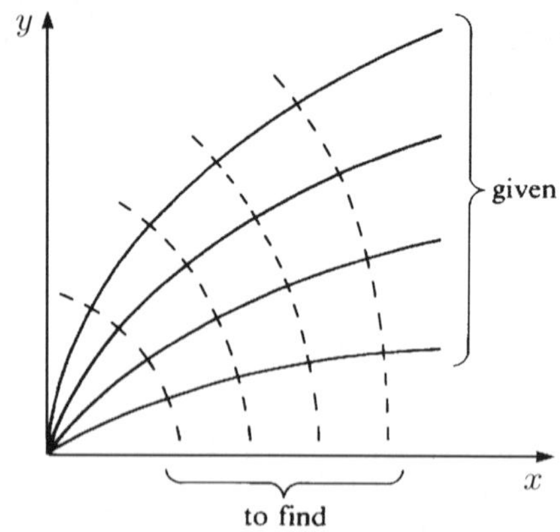

Figure 6.6. Orthogonal trajectories (shown dashed) to a family of curves

Unfortunately for Bernoulli, this was not a suitable question for its purpose. Although deriving the curves in the family from their defining property required considerable skill at formulating and solving differential equations, finding their orthogonal trajectories in the particular case proposed by Bernoulli did not require a theoretical breakthrough. The curves of the given family were too special, and several mathematicians answered the challenge: Hermann, two Bernoullis (Nicolaus I and II), and Taylor.

We can see from this example that the solving of differential equations was becoming a well-developed skill. What we cannot see from this is that by now Johann Bernoulli and Leibniz had a new theoretical tool: they were beginning to see how to extend the calculus to deal with expressions or functions involving more than one independent variable. This was to be a vital development, which we shall look at in more detail in Section 10.1.

Interest in the priority dispute lapsed with the deaths of Leibniz in 1716 and Newton in 1727, when it began to become apparent that by far the greater number of good mathematicians were the Leibnizians in Continental Europe. Indeed, it was eventually to be the case that many of the English cut themselves off from Continental mathematicians. But the matter is complicated by the uneven development of the two calculi,

[4]Quoted in Whiteside, *MPIN* 7, 154.

for any deficiencies that the Newtonian calculus may have had on the theoretical side were more than compensated for by the great successes of Newton's *Principia* and, as we shall see in Chapter 9, progress in that area depended on Euler's rewriting of that book in his simplified Leibnizian calculus.

In the 1720s and 1730s the methods of the calculus began to supplant, and not merely to augment, the traditional techniques of geometry. Accordingly, the works of such as Johann Bernoulli exhibit a curiously transitional character, partly geometrical, partly formal. By following the journey from inverse tangent problems to differential equations, and from the secrecy of Newton to the openness of Euler, we shall pursue as clear and direct a route as there is through the history of the calculus in this period.

Johann Bernoulli. Johann I Bernoulli's career provides the best illustration of how things developed. He has left us a vivid impression of himself and of a second important figure, the Marquis de l'Hôpital.[5] The meeting that Bernoulli described took place in 1691, when the Marquis was 30.

Johann Bernoulli meets the Marquis de l'Hôpital.

> Immediately on arriving in Paris in late Autumn 1691, Bernoulli visited Father Malebranche who once a week played host to the best known scholars of the city. Rather as an admission card, so to speak, he showed the famous philosophers, who were also good mathematicians, his construction of the catenary — on a single piece of paper — that had just been published a little earlier in the June volume of the *Acta Eruditorum* and which was his first important achievement. Because of this he was invited by Malebranche to take part regularly at his meetings. So the 24-year-old student appeared in the illustrious circle on the next occasion and was immediately introduced to the Marquis de l'Hôpital, to whom Malebranche had showed his paper a few days before.
>
> 'From the conversation' [Bernoulli wrote to his friend Montmort in 1718] 'which I had with M. le Marquis I knew right away that he was a good geometer for what was already known, but that he knew nothing at all of the differential calculus, of which he scarcely knew the name, and still less had he heard talk of the integral calculus which was only just being born, the little that there was of this calculus in the *Acta* of Leipzig having not yet reached him because of the war.' The Marquis, who found it difficult to see the conqueror of the catenary in the young man, examined him backwards and forwards 'but he saw soon enough that I was neither an adventurer nor the pretender that he believed I wanted to play at being. The conversation finally fell to the developed curve [evolute] or osculating circle, for the study of which he prided himself on an entirely particular rule drawn from M. Fermat's method of max. and min. To test him, I proposed an example of an algebraic curve (for this supposedly general rule only worked for algebraic curves and only gave the radius at the maximum).'

[5]See (Spiess 1955, 136–137) and F&G 13.B4.

> 'M. l'Hôpital took paper and ink, and began to calculate. After he had used up nearly an hour in scribbling over several pieces of paper he finally found the correct value of the radius at the maximum of the curve.'
>
> Bernoulli then said to him that there was a formula that would find the radius of curvature of any curve at any point in a few minutes, and put forward a curve for which he could find the sought-for value at once 'which struck him so much with surprise that from that moment he became charmed with the new analysis of the infinitely small and excited with the desire to learn it from me'. Bernoulli visited the Marquis the very next day, who asked him to visit four times a week 'to explain to him on each occasion and then to deliver a lesson based on the paper which I had written at home the evening before'. And, what is of particular importance now 'one of my friends from Basel who was lodging with me had the kindness to copy each of the papers I was to take to M. le Marquis, so I have preserved them all'. ...
>
> The lessons in Paris went on from the end of 1691 to the end of July 1692, so for over half a year; then l'Hôpital took his young instructor to his estate in Oucques where Bernoulli presented his lectures in daily contact with the Marquis and his spirited wife. 'I didn't hesitate' he wrote in the letter to Montmort 'to give to M. l'Hôpital new memoirs always written in my own hand whenever I found appropriate material, and he furnished me himself with the occasion for all sorts of questions.'

It is an attractive picture at first sight: the Marquis scribbling away, making error after error with what was obviously a pretty creaky method; the young Swiss, perhaps a little arrogantly, concealing his superior one for a while. Johann liked contests, and he liked to win. Nonetheless, it was an effective strategy — the Marquis was hooked.

As for the problem (to find centres of curvature) it is best explained by referring back to Descartes's method of normals. In that approach, the normal to a given curve at a specified point was found by looking at circles passing through the specified point. Such a circle with its centre on the normal would necessarily touch the curve at the given point. In some intuitive sense, one of these circles is a better approximation to the curve than any other — its centre is said to be the centre of curvature of the given curve at the specified point. Without the calculus it is a chore to find it, as l'Hôpital's experience makes plain. With it, it is much easier, as Bernoulli showed so convincingly.

The Marquis de l'Hôpital was an important catch, for Johann Bernoulli was soon contracted to teach him the Leibnizian calculus, and the result was the first text-book ever written on the calculus, the Marquis de l'Hôpital's *Analyse des Infiniment Petits, pour l'Intelligence des Lignes Courbes* (*The Analysis of the Infinitely Small, for the Understanding of Curved Lines*), published in Paris in 1696.[6] The terms of the contract are amusing to modern eyes, and must be viewed in the spirit of business ethics (they include a form of non-disclosure agreement). On 17 March 1694 l'Hôpital wrote:[7]

---

[6] Nicolaus I Bernoulli's copy of these lectures also survives.

[7] See (Spiess 1955), Vol. 1, Letter No. 20.

> I will be happy to give you a retainer of 300 livres .... I promise shortly to increase this ... which I know is very modest, as soon as my affairs are somewhat straightened out ... I am not so unreasonable as to demand in return all of your time, but I will ask you to give me at intervals some hours of your time to work on what I request and also to communicate to me your discoveries, at the same time asking you not to disclose any of them to others. I ask you even not to send here to M. Varignon or to others any copies of the writings you have left with me; if they are published I will not be at all pleased.

In due course the arrangement broke up, and in 1695 Bernoulli became Professor of Mathematics at Groningen, while waiting for something better to turn up. In due course it did, and on the death of his brother Jakob in 1705 he became Professor of Mathematics at Basel.

When the Marquis published his *Analyse* he acknowledged all the instruction that he had received from Johann Bernoulli, but in such a way that Bernoulli took offence.[8] L'Hôpital first praised Leibniz for his breakthrough into the differential calculus, and then went on:

> The Messieurs Bernoulli were the first who perceived the Beauty of the Method; and have carried it to such a length, as by its means to surmount difficulties that were before thought insuperable.

Next, l'Hôpital observed that this calculus was only half the story, for there was also an inverse method, called the 'calculus integralis', but he had abstained from writing about it on hearing from Leibniz that he was at work on the subject.

Then he went on:

> I must own myself very much obliged to the labours of the Messieurs Bernoulli, but particularly to those of the present Professor at Groeningen, as having made free with their Discoveries as well as those of Mr. Leibnitz: So that whatever they please to claim as their own I frankly return them.

To Bernoulli, the acknowledgement seemed rather slight. He seems to have felt that whatever generosity is contained in the phrase 'very much obliged' was diminished by the 'frank return' of 'whatever they please to claim as their own', which might suggest that very little was original with them. But what is most striking is that the considerable contributions of Johann are not separated from the much smaller ones of Jakob. Given that Johann did not particularly like his brother — indeed, he came to detest him — the Marquis's remarks did not please him, and he found them rather off-hand. We are very much in the world of the aristocracy, where the landed rich would buy the skills of gentlemen, draughtsmen, painters, even mathematicians, and appropriate them.[9]

The *Analyse des Infiniment Petits*. Let us now look in more detail at the Marquis's important book. The subject is Leibniz's differential calculus, but not his integral calculus. One may well suspect that the Marquis knew nothing of Newton's calculus, except what he could glean from the *Principia*. As for his opinion of the *Principia*, when the Marquis was shown the book:[10]

---

[8] See (Stone 1730), Preface and pp. vii, x, in F&G 13.B5, where a fuller extract from Edmund Stone's translation of the *Analyse* can be found.

[9] As an indication of what this might have meant to Johann Bernoulli, what we now call l'Hôpital's rule in the calculus was discovered by Bernoulli.

[10] Westfall, *Never at Rest*, p. 473.

> he cried out with admiration Good God what a fund of knowledge there is in that book? he then asked … every particular about Sr I. even to the colour of his hair, said does he eat & drink & sleep. is he like other men?

This is fine, but it does not suggest that he had mastered the work.

L'Hôpital's book opens with an indication of the methods employed and the problems addressed.[11] There is then a potted history, in which Descartes has a prominent role, and then there are ten chapters that cover such topics as the finding of tangents, maxima and minima, centres of curvature, evolutes, and caustics. In the final chapter Descartes's and Hudde's methods are re-derived by means of the calculus. In short, the calculus is applied to some traditional and some more recent problems in geometry, and is shown to be broader in scope and more powerful than any other methods available.

Let us look at how l'Hôpital (or rather, Johann Bernoulli) got down to details:[12]

L'Hôpital on the foundation of the calculus.

> 1. *Definition I. Variable quantities are those that continually increase or decrease; and constant or standing quantities, are those that continue the same while others vary.* As the ordinates and abscisses of a parabola are variable quantities, but the parameter is a constant or standing quantity.
>
> *Definition II. The infinitely small part whereby a variable quantity is continually increased or decreased, is called the differential of that quantity.*
>
> For example: let there be any curve line AMB [Figure 6.7] whose axis or diameter is the line AC, and let the right line PM be an ordinate, and the right line pm another infinitely near to the former.

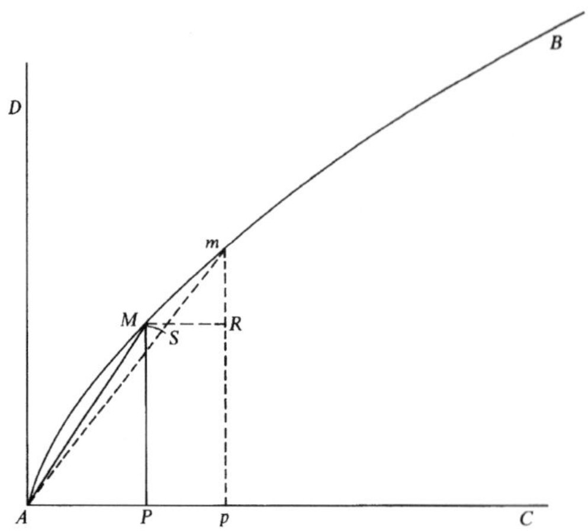

Figure 6.7. The curve AMB and an associated differential

[11] See (l'Hôpital 1696) and F&G 13.B5.

[12] See (Stone 1730) and Struik, *A Source Book*, pp. 313–315.

Now if you draw the right line MR parallel to AC, and the chords AM, Am; and about the centre A with the distance AM, you describe the small circular arch MS: then shall Pp be the differential of PA; Rm the differential of PM; Sm the differential of AM; and Mm the differential of the arch AM. In like manner, the little triangle MAm, whose base is the arch Mm, shall be the differential of the segment AM; and the small space $MPpm$ will be the differential of the space contained under the right lines AP, PM, and the arch AM.

Corollary It is manifest, that the differential of a constant quantity (which is always one of the initial letters a, b, c, etc. of the alphabet) is 0: or (which is all one) that constant quantities have no differentials.

Scholium The differential of a variable quantity is expressed by the note or characteristic d, and to avoid confusion this note d will have no other use in the sequence of this calculus. And if you call the variable quantities $AP, x; PM, y; AM, z$; the arch AM, u; the mixtlined space APM, s; and the segment AM, t: then will dx express the value of Pp, dy the value of Rm, dz the value of Sm, du the value of the small arch Mm, ds the value of the little space $MPpm$, and dt the value of the small mixtlined triangle MAm.

2. *Postulate I.* Grant that two quantities, whose difference is an infinitely small quantity, may be taken (or used) indifferently for each other: or (which is the same thing) that a quantity, which is increased or decreased only by an infinitely small quantity, may be considered as remaining the same.

For example: grant that Ap may be taken for AP; pm for PM; the space Apm for APM; the small space $MPpm$ for the small rectangle $MPpR$; the small sector AMS for the small triangle AMm; the angle pAm for the angle PAM, etc.

3. *Postulate II.* Grant that a curve line may be considered as the assemblage of an infinite number of infinitely small right lines: or (which is the same thing) as a polygon of an infinite number of sides, each of an infinitely small length, which determine the curvature of the line by the angles they make with each other [Figure 6.8].

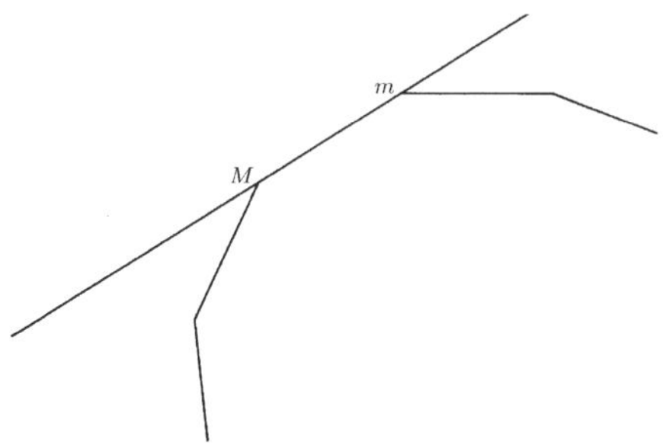

Figure 6.8. A tangent to a curve regarded as a polygon with infinitely many sides

For example: grant that the part Mm of the curve, and the circular arch MS, may be considered as straight lines, on account of their being infinitely small, so that the little triangle mSM may be looked upon as a right-lined triangle.

4. *Proposition I. To find the differentials of simple quantities connected together with the signs + and −.*

It is required to find the differentials of $a + x + y - z$. If you suppose x to increase by an infinitely small part, viz. till it becomes $x + dx$; then will y become $y + dy$; and z, $z + dz$: and the constant quantity a will still be the same a. So that the given quantity $a + x + y - z$ will become $a + x + dx + y + dy - z - dz$; and the differential of it (which will be had in taking it from this last expression) will be $dx + dy - dz$; and so of others. From whence we have the following.

Rule I. For finding the differentials of simple quantities connected together with the signs + and −.

Find the differential of each term of the quantity proposed; which connected together by the same respective signs will give another quantity, which will be the differential of that given.

5. *Proposition II. To find the differentials of the product of several quantities multiplied, or drawn into each other.*

The differential of xy is $ydx+xdy$: for y becomes $y+dy$, when x becomes $x + dx$; and therefore xy then becomes $xy + ydx + xdy + dxdy$. Which is the product of $x + dx$ into $y + dy$, and the differential thereof will be $ydx+xdy+dxdy$, that is, $ydx+xdy$: because $dxdy$ is a quantity infinitely small, in respect of the other terms ydx and xdy: For if, for example, you divide ydx and $dxdy$ by dx, we shall have the quotients y and dx, the latter of which is infinitely less than the former.

Whence it follows, that the differential of the product of two quantities, is equal to the product of the differential of the first of those quantities into the second plus the product of the differential of the second into the first.

We see that the *Analyse* opens, as any good textbook should, with a description of the foundations of the subject. The two main ideas are variable quantities and their infinitely small parts, called differentials, which are taken to be real, existing objects. So they are not quite Leibniz's differentials, which are (at least intuitively) infinitely small differences between successive values of a variable, but in some intuitive way they are the increments by which the variables vary. These two kinds of quantity are to obey two postulates. The first is some kind of approximation procedure that allows us to neglect infinitely small quantities on occasion (and is very vague). The second is one that we have met before, and presents a curve Leibniz-style as an infinite-sided polygon. Finally, rules are given for finding differentials: the differential of a sum or difference; and the differential of a product. These are Leibniz's algorithms.

You may well feel that this presentation is easier than the one first given by Leibniz himself in 1684, and it seems that the book did well. It is clear and interesting, and was to help considerably in disseminating the differential calculus because it showed how useful the calculus can be in solving problems about extreme values, locating centres of curvature, and explaining Hudde's rule for finding limits. In 1730 Stone thought it worthwhile to translate it into English.

Edmund Stone is an interesting figure. He was the son of a gardener on the estate of the second Duke of Argyll at Inveraray in Scotland. Apparently the Duke found Stone one day reading a copy of Newton's *Principia*, and learned that Stone had taught himself Latin and French in order to study mathematical works.[13] The Duke then allowed Stone to pursue his studies and sponsored his scientific activities, which generally took the form of translating French works into English. Stone's edition of l'Hôpital's *Analyse* went further, because it also converted Leibniz's mathematical terms and symbols into those of Newton's fluxional method, and so produced a dictionary between the two systems of the calculus. Stone then went on to write an introduction to the integral calculus and, although it was soon superseded by better treatises published in the 1730s and 1740s, it was translated into French in 1735; this is indicative of a growing interest in Newton's work in France during the 1730s. In 1735 Stone was elected a Fellow of the Royal Society, but he resigned in 1743 after the death of the Duke of Argyll, which seems to have put an end to his scientific career.

Johann Bernoulli was the author of the most thorough treatment of the integral calculus up to that time. As with the differential calculus, the subject matter of his integral calculus coincided with the topics that he had taught the Marquis de l'Hôpital. But this time the material, composed for the most part by 1700, was not published until Johann published his collected works in 1742. It is arguable that his contract with the Marquis might have prevented him from publishing this material earlier — but the Marquis had died in 1704, so perhaps there is some other explanation. Johann Bernoulli was a secretive man when it came to revealing his methods, although not his results.

From inverse tangent problems to differential equations. Bernoulli's definition of integration is interesting; he defined it, as Newton had done, as the inverse of differentiation and not as an infinite sum, as Leibniz had. He gave several techniques for evaluating integrals, some of which we look at below, and explained that the main use of this calculus was to find areas. Then he turned to its use in solving inverse tangent problems, and showed how to solve a variety of examples. Finally, he showed how to translate problems in geometry or mechanics into the new language of the calculus. This last point is perhaps the most important, because puzzling out how to formulate a given problem in the language of the calculus can be the hardest part of a mathematician's work.

In rough outline, this translation procedure goes as follows.

1. Suppose that the solution to the problem can be expressed as a curve; formulate the problem in terms of equations involving the coordinate variables, x and y.
2. Make the problem yield a statement about how neighbouring points on the curve are related; usually this takes the form of an equation involving differentials.
3. Finally, endeavour to pass from the differentials up to the finite quantities (x and y) and so determine the precise form of the equation that describes the solution curve.

[13]See Guicciardini's article on Stone in the *Oxford Dictionary of National Biography*.

We have seen an example of this process already, when we looked at how Leibniz tackled Debeaune's problem. We saw that, however clumsily, Leibniz first managed to re-state the problem first as a statement about x, y, and t: he wrote, incorrectly as we saw,

$$\frac{t}{y} = \frac{n}{y - x}$$

— this is stage (1) above. Then he introduced the differentials (dx and dy) — stage (2) above; and from that he deduced a statement about finite quantities — stage (3).

Bernoulli raised this insight to the status of a *method*, a systematic way of tackling many such problems (as for example in his solution of the catenary problem, see Box 12), and this moment of transition is marked by a change of name. Henceforth, inverse tangent problems became called *differential equations*, because they are translated literally into equations involving differentials. This change of name shows how closely the new methods of the calculus became associated with that particular sort of problem. For the same reason, British mathematicians came to speak more and more of 'fluxional equations'.

Although being able to express a problem as a differential equation was often remarkably useful, it was obviously even better if you could also solve it. Indeed, without the new solution techniques that Bernoulli introduced, any translation of a problem into the language of the calculus might have been done in vain. But this raises the vexed question: What is meant by a solution?

To Bernoulli and his contemporaries, a solution ideally meant a geometrical description of the required curve, but the calculus did not always lend itself to providing one. It is easy to see why this should be. To pass from the sort of solutions that the calculus provided, which were equations involving variables, to a geometrical description of the solution is the opposite of stage (1) above. But, as we saw with inverse tangent problems, such problems are much easier to state than to solve; describing the solution is much harder than describing the problem. So reversing stage (1) is much harder than traversing it, and since that was often hard enough, going backwards proved on occasion to be too difficult. The upshot of this was that mathematicians were under pressure to accept equations as solutions and not to look beyond them — and gradually they were to succumb. Mathematics became more formal and algebraic, and less geometrical in nature.

It remains for us to comment briefly on the techniques that Bernoulli introduced. He noticed, as Newton had much earlier, that the simplest kind of differential equation that we could hope to get was of the form

$$(\text{something in } x) \times dx = (\text{something in } y) \times dy,$$

because we could then hope to integrate both sides. Arranging for this to happen he called the *method of separation of variables*. However, it might happen that with one set of variables the problem could not be solved in such a way. Arguing that there is nothing in a name, Bernoulli then suggested looking for new variables with respect to which the differential equation does separate. Such a change of variable is the kind of thing that a mathematician can learn to spot with practice, and Bernoulli's package of techniques proved to be very powerful. Indeed, if it does not sound too banal, we might insist that one skill that a mathematician possesses is the ability to go by the book. In this case, this meant setting aside all qualms about the nature of differentials, and manipulating them formally, precisely as one does finite quantities in elementary

Box 12. Johann Bernoulli's solution of the catenary problem.
The catenary problem, which asks for the shape of a heavy chain hanging between two points, was raised by Jakob Bernoulli in the *Acta Eruditorum* in 1690 as a challenge to the mathematical community.

Johann Bernoulli gave the solution in his paper of 1691 but withheld his method for finding it. He later hinted at his method in his lectures on integration, and our account follows the reconstruction given by the historian Henk Bos.[a]

Bernoulli first expressed the force on an infinitesimal piece, AB, of the chain in terms of the weight of the chain hanging below A. He then calculated the force at the neighbouring point B, and then argued that since AB does not move, the effects of the forces at A and B must cancel out. So the raw ingredients are: the length, s, of the chain from its far end to the point A — which is proportional to its weight — and the differentials dx and dy that enter because the chain is curved, and so the forces at A and B do not point in quite the same direction. This concludes stage (1).

The differential equation that Bernoulli obtained is

$$\frac{dy}{dx} = \frac{s}{a},$$

where a is a constant. Stage (2) was completed when the variable s, which depends on the values of x and y, was eliminated in favour of an explicit expression involving x and y. We omit the details of how this was done, and pass straight to the resulting equation:

$$dy = \frac{adx}{\sqrt{x^2 + 2ax}}.$$

Stage (3), the way the solution of this equation is found, is discussed below.

[a]See (Bos 1980, 80–82).

algebra. Bernoulli's techniques are simple algebraic devices — his insight was to see that such methods still apply in their new setting. In so doing he was following the lead of Newton and Leibniz.

These ideas are well illustrated by Bernoulli's discussion of Debeaune's problem, to which we now turn.[14] Our commentary on Bernoulli's remarks follow the extract.

Bernoulli on Debeaune's problem.

> Another such example is the problem set to M. Descartes by M. Debeaune, the solution to which is not in his works but can be found in his *Letters* (vol. III, No. 71). The solution of it does not appear to be very easy according to our method, indeed at first sight the problem appears impossible by this method. But we shall see that by a change of variables it becomes easy to separate them, and that this problem can

[14]See Bernoulli, J., *Opera Omnia* 1, 1742, 62–64, and F&G 13.B1. We discussed Debeaune's problem in Section 4.1.

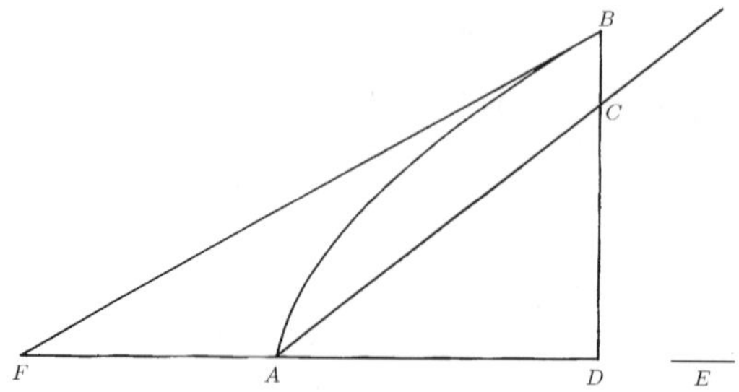

Figure 6.9. Bernoulli's construction of the curve that solves Debeaune's problem (1)

be solved completely once the quadrature of the hyperbola is given, for the curve is mechanical.

The problem goes like this [see Fig 6.9]: a line AC makes an angle of half a right angle with the axis AD, and E is a given constant line segment; what is the nature of the curve AB in which the ordinates BD are to the subtangents FD as the given E is to BC?

Solution. Let $AD = x, DB = y, E = a$, suppose by hypothesis that $dy : dx = a : (y - x)$, then $adx = ydy - xdy$. From this equation the nature of the curve is to be found, either by integration or by rewriting y with dy on one side and x with dx on the other, for then two areas can be found and by comparing them the nature of the curve can be found. But the equation just found cannot be integrated, nor can x and dx be separated from y and dy; however, it can be changed into another by substituting the value of another variable. Therefore let $y - x = z$, $y = x + z$ and $dy = dz + dx$. The equation just found transforms into this:

$$adx = zdz + zdx \quad \text{or} \quad adx - zdx = zdz$$

and so $dx = zdz : (a - z)$. Therefore these two variables separate, and we are led to the curve on multiplying by a,

$$adx = azdz : (a - z).$$

We would now expect Bernoulli to solve the differential equation by integrating both sides, but he did not. Instead, he embarked on a long, and to us somewhat obscure, account, because the answer he was looking for is a curve, not an equation; therefore he gave a construction for drawing it. This shows how strong the influence of Descartes still was at the time.

And dropping normals GT and NH, $GN = GH = a$ [see Figure 6.10], and drawing through the points H and N HV and MR parallel to GT, $NR = NG$; erecting a perpendicular RS and asymptotes RM, RS and drawing a hyperbola LKG through G: then $GO = z$ & $GQ = x$, and $KO = az : (a - z)$, and because QI always equals a, the hyperbolic space KGO will equal the rectangle HQ & producing the lines IQ, KO, the point P where they meet draws out the curve GPW which satisfies the equation just found $adx = azdz : (a - z)$. Having constructed the curve AB there is then no more work; for, QP being produced to Z [see Figure 6.11], as PZ shall equal the abscissa GQ the point Z will lie on the curve AB. Since $PZ = GQ = x - AD$ and $QP = z$, $QP + PZ$ will $= z + x = y = DB$. Q.E.I.[15]

Corollary I. NR [see Figure 6.10] is asymptotic to GPW & $QP = BC$ [see Figure 6.9], This curve AB has its asymptote parallel to AC.

Corollary II. The space $ADB = xy + ax - \frac{1}{2}yy$.

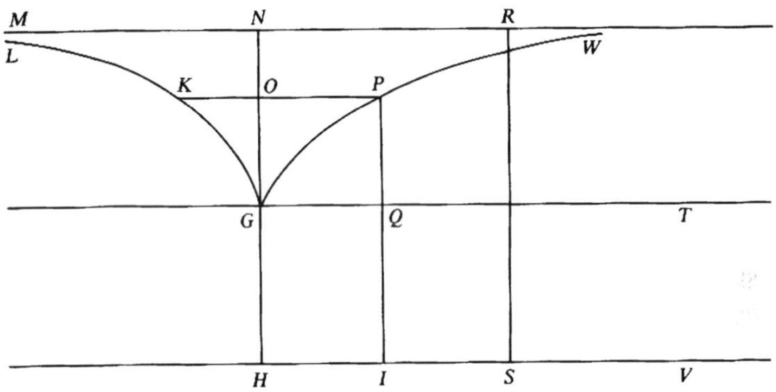

Figure 6.10. Bernoulli's construction of the curve that solves Debeaune's problem (2)

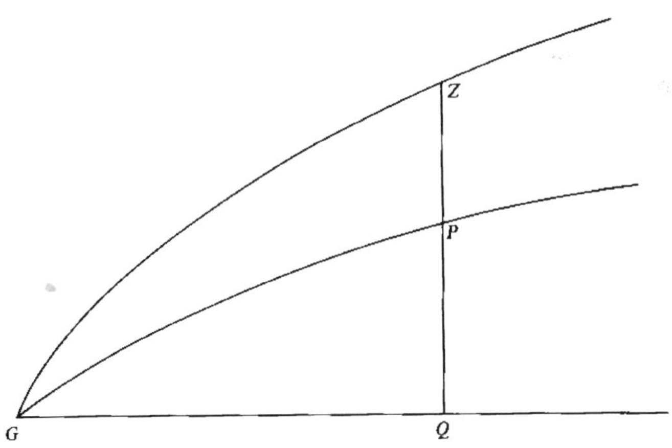

Figure 6.11. Bernoulli's construction of the curve that solves Debeaune's problem (3)

First Bernoulli stated the problem, and to solve it he introduced coordinates (stage (1)), and then differentials (stage (2)). Next he noted, as he had already warned in the first paragraph, that the method of separation of variables does not seem to apply to the differential equation $adx = ydy - xdy$. So he looked for a change of variable, and

[15]Q.E.I. stands for quod erat inveniendum, that which was to be found.

saw that $y - x = z$ (and so $dy - dx = dz$) will work. This yields an equation in the variables x and z in which the variables *do* separate,

$$dx = \frac{z}{a - z}\, dz.$$

Next, in stage (3), he found that he could give a geometrical description of the solution. So, true to the tradition in which he had grown up, he gave that answer, and not a formal algebraic one, which might have taken the form of an expression involving a logarithm.

We can see the growth of formalism by looking at some ideas that Johann Bernoulli wrote down only a few years later, in 1702.[16] He was by now quite clear that expressions like $\frac{dx}{x + c}$, in which c is a constant, are differentials of logarithms, and hence that

$$\int \frac{dx}{x + c}$$

is a logarithm. He showed by a variety of changes of variables that certain integrals can be expressed in terms of logarithms or circular arcs. Included in his list are the integrals that we have met before: the integral

$$\int \frac{dx}{\sqrt{x^2 + 2ax}},$$

which arose in the catenary problem, and the logarithm lurking in Debeaune's problem. Bernoulli seems to have regarded his changes of variables (which involved him in introducing complex numbers) as enabling him to pass from circular arcs to arcs of hyperbolas and back, thereby helping to make his geometrical interpretations of analytic formulas more flexible.

6.3 The Continental reception of the *Principia*

Although, as we saw in Chapter 5.3, the *Principia* was received with admiration in Britain, its reception on the Continent of Europe was to be another matter. In fact, Continental readings of the *Principia* are not easy to characterise briefly. Mathematically, the *Principia* is fiercely difficult, and readers lacking Newton's brilliance at geometry had to resort to (and thus to learn) the calculus in order to master the work. In this way, it marks a turning point in the development of mathematics.

Moreover, to practitioners of science the book was a conundrum, the problem being the nature of the *attractive force of gravity*. We have already noted Huygens' refusal to accept such a force. Although in 1688 he wrote 'Vortices destroyed by Newton', one could still pick holes in Newton's refutation of Cartesian vortices, and it was perhaps possible that one could try to replace them with other vortices of a different, non-Cartesian, kind. This Huygens attempted to do, but understandably his thoughts about novel vortices remained vague. Meanwhile Leibniz, basing himself on the *Principia* (and not, as he disingenuously suggested, only on a review of it published in *Acta Eruditorum* in 1688), attempted a more detailed defence of vortices but failed, as he himself admitted. Even when Huygens tried again, he had to point out that his revised vortex theory could not accommodate Kepler's third law (the 3/2-power law). What these attempts by the two leading Continental mathematicians tells us is how firmly

[16]See (Bernoulli 1702) and F&G 13.B2.

they resisted the fundamental concept of an attractive force of gravity. To them it was an 'absurd' idea (Huygens' word), incapable of explaining anything.

Twenty-five years later, when bad feelings between Newton and Leibniz erupted into an open feud, they fought over the nature of gravity as well as the invention of the calculus. Leibniz continued to argue that it was not enough to say, as Newton had done, that the cause of the properties of gravity had not been discovered. In his view, this was an evasion designed to avoid confronting the implausible mechanism by which gravity operates. Newton thought that this was mere playing with words, and when Leibniz criticised the *Principia* for not explaining gravity, a force that Leibniz found implausible and obscure ('occult' was his word), Newton drafted a reply in 1715 complaining that:[17]

> He denys conclusions without telling me the fault of the premisses ... His arguments against me are founded upon metaphysical & precarious hypotheses & therefore do not affect me: for I meddle only with experimental Philosophy ... He changes the signification of the words Miracles and Occult qualities that he may use them in railing at universal gravity.

The issue was never to be completely resolved. Action at a distance is still thought to require an explanation (Einstein's general theory of relativity being the most widely accepted candidate), but collision mechanisms rather fell out of favour. To understand why, it is not enough to look at their poor showing, even in the hands of Leibniz and Huygens. We must also look at the French school of mathematicians during this time.

The French were neutral in the Anglo–Hannoverian squabble over who invented the calculus, inclining to one side or the other as they saw fit. On matters of physics, however, they were firm Cartesians, so it was their change of mind that is the most interesting. In 1688, Pierre-Sylvain Regis's review of the *Principia* (in the *Journal des Sçavans*) set a rather hostile tone:[18]

> one cannot regard these demonstrations otherwise than as only mechanical; indeed, the author recognises himself that he has not considered their Principles as a Physicist, but as a mere Mathematician.

That this is a misreading should be clear. Regis correctly saw that Books I and II are hypothetical (in our sense) and mathematical, but he failed to see how their mathematical results are applied to the natural world in Book III.

Malebranche. How did Newton's use of mathematics come to be better appreciated? The central figure is the Oratorian Father Nicolas Malebranche, who came to mathematics in his mid-20s via his encounter with Descartes's philosophy in 1664. A well-read mathematician, he considered mathematics to be 'the foremost and fundamental discipline of all the human sciences',[19] and he gathered a notable group of people around him: Jean Prestet, Charles-Rene Reyneau, Pierre Rémond de Montmort, the Marquis de l'Hôpital, and, peripherally, Pierre Varignon. Of these, Varignon was to be the one who dealt most ably with the motion of a body under various forms of central force.

It was Malebranche himself who made the first break with Cartesian ideas. After writing an *Elémens des Mathématiques* with Jean Prestet in 1675, he subsequently disavowed its denial, which had followed Descartes, that mathematics could deal with

---

[17]Quoted in (Cohen 1980, 61–62).

[18]Quoted in (Cohen 1980, 16–17).

[19]See (Guerlac 1979, 81)

the infinite. Malebranche thereby opened up a willingness to employ calculus-type reasoning in many problems. His emphasis on mathematics as 'the most exact and unimpeachable form of knowledge' seems to have inclined Malebranche and his school to appreciate Newton's abandonment of hypothetical mechanisms in physics in favour of mathematical deduction.[20] They quickly realised that Newton's views had to be taken seriously, and were in the forefront of attempts to elucidate them by means of the Leibnizian calculus.

Under Malebranche's influence Varignon moved towards considering forces in a Newtonian fashion, while never quite freeing himself of Cartesian vortices. His position involved a species of agnosticism about underlying physical causes. Slowly but steadily this school came to appreciate the logical structure of the *Principia*, while attempting to retain something of Descartes's vortex theory.

This process was continued by the next generation. Jean-Jacques Dortous de Mairan and Joseph Privat de Molières, both disciples of Malebranche, made ingenious attempts at reconciling the incompatible theories of Descartes and Newton. From this emerged a lively sense of the power of mathematics, and an incipient realisation of the validity of the Newtonian style of reasoning. This produced an eclectic and unstable liaison of Cartesian, Malebranchist, and Newtonian ideas.

The generation of the 1730s was to sweep Cartesian physics away, as it had been effectively undermined. As Emilie du Châtelet wrote of Cartesianism in 1738: 'It is a house collapsing into ruins, propped up on every side ... I think it would be prudent to leave'.[21]

The most important figure in bringing the Parisian Académie des Sciences round to Newtonianism was Pierre-Louis Moreau de Maupertuis. He was well placed to do so, being independently wealthy and with good connections at Court. He was also personally charming, a capable mathematician, and very ambitious. As the French debated Cartesianism and Newtonianism he proved highly capable of keeping the sympathy of both sides, and when, as often, the mathematics of Newton's *Principia* proved too difficult for him, he was able to turn to no less a figure than Johann Bernoulli for help. He had visited Basel from September 1729 to July 1730, and Bernoulli was flattered by the younger man and was happy to advise him thereafter, seeing, correctly, that this would assist in the promotion of the Leibnizian calculus. All these carefully cultivated pieces of good fortune were to help Maupertuis make his name and his career when the question of the shape of the Earth became important, as we shall see.

Although the greatest success of Newton's *Principia* was its discussion of planetary motion, it was flawed by an imperfect account of the motion of the Moon. The treatment of several other topics was also somewhat contrived, although an improvement on contemporary practice. For example, the calculation of the shape of the Earth (which Newton took to be treatable as a fluid) depended on implausible and ad hoc assumptions about the nature of fluid pressure. Rather more surprisingly, Newton's laws of motion did not immediately present themselves in a form appropriate for mathematics, a point we look at in more detail in the last section of this chapter.

The creation of the Newtonian paradigm. Two specific problems illustrate how the *Principia* was received early in the 18th century: the shape of the Earth and the

[20] For the quotation from Malebranche, see (Guerlac 1981, 59).

[21] See (Besterman 1958, 261).

motion of the Moon. Each of these, for different reasons, was a problem for Newton's successors to tackle. Together they enable us to decide how Newtonian mechanics became the dominant scientific paradigm of the 18th century.

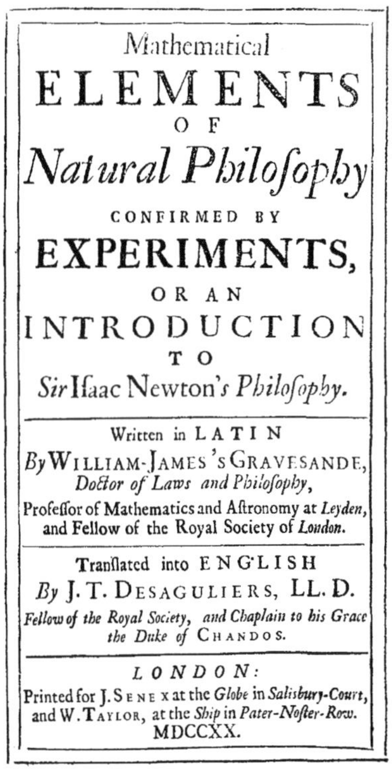

Mathematical
ELEMENTS
OF
Natural Philosophy
CONFIRMED BY
EXPERIMENTS,
OR AN
INTRODUCTION
TO
Sir Isaac Newton's *Philosophy*.

Written in LATIN
By WILLIAM-JAMES 's GRAVESANDE,
Doctor of Laws and Philosophy,
Professor of Mathematics and Astronomy at *Leyden*,
and Fellow of the Royal Society of *London*.

Translated into ENGLISH
By J. T. DESAGULIERS, LL. D.
Fellow of the Royal Society, and Chaplain to his Grace the Duke of CHANDOS.

LONDON:
Printed for J. SENEX at the *Globe* in *Salisbury-Court*,
and W. TAYLOR, at the *Ship* in *Pater-Noster-Row*.
MDCCXX.

Figure 6.12. The title page of Willem James 'sGravesande's book

The British took to Newtonian mechanics very quickly. So did the Dutch school of mathematicians — Willem 'sGravesande wrote a successful book on Newton's ideas in 1720 — and so did the Italians. But these groups were smaller and less important than the French one, based in Paris, and the essentially Swiss mathematical community centred around the widespread Bernoulli family. Here again we concentrate chiefly on the French, because it was their conversion to Newtonianism that was to prove decisive.

Newtonian ideas ultimately became widely accepted. Nature in all its guises, as it was presented in books in the later 18th century, such as Algarotti's *Newtonianism for Ladies* (see Figure 6.13), became increasingly full of attractive or repulsive forces centred on numerous hard bodies — for all the world like miniature solar systems — and by 1800 the greatest exponent of this point of view was the French mathematician Pierre-Simon Laplace, as we shall see.

An interesting and vivid light was thrown on Anglo–French relations in an early essay by the French essayist and polemicist François-Marie Arouet (who is much better known by his nom-de-plume of Voltaire). In it he described what he saw on his visit to England in 1727, the year of Newton's death.[22] Voltaire was struck by the extent to

[22]See (Voltaire 1734), English transl., pp. 68–71, and F&G 12.F2.

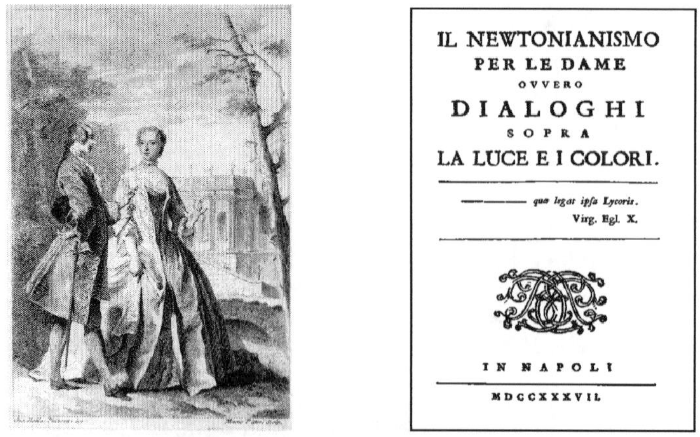

IL NEWTONIANISMO
PER LE DAME
OVVERO
DIALOGHI
SOPRA
LA LUCE E I COLORI.

quae legat ipsa Lycoris.
Virg. Egl. X.

IN NAPOLI
MDCCXXXVII.

Figure 6.13. Francesco Algarotti's *Newtonianism for Ladies* (1737), frontispiece (left) and title page (right)

which Newtonians and Cartesians disagreed. While the national chauvinism amused him considerably, he took Newton's side against Descartes. He praised Descartes for being the first to tackle certain problems, but he also said that Descartes had been wrong about dynamics and the nature of matter. The implication is that Newton's views were an improvement.

The question of the shape of the Earth, which was raised by Voltaire, is far from an idle one, and it was taken up energetically by Maupertuis. He argued that it was vital to know the shape of the Earth, and helpful to understand the motion of the Moon.[23] His concern, in an age of competing imperialisms, was navigation. A prerequisite for accurate maps is knowing the shape of the Earth; without that knowledge, exploration could be perilous. Sailors also require to know their position, and Maupertuis noted that the determination of longitude seemed to require knowing how the Moon moves. So both problems had an important practical side.

The shape of the Earth. Because the question of the shape of the Earth is so critical, it had been investigated for some time. Voltaire gave a characteristically polemical account of the story.[24] He rightly suggested that because a pendulum beats faster in the latitudes of France than it does near the equator, it follows — for a Newtonian — that the Earth is broader at the equator. This is because the greater the force of gravity upon a pendulum the faster it beats, and this force increases as one moves closer to the centre of the Earth. The most plausible hypothesis would be that the Earth would continue to flatten as one travelled North, so that the poles would be even nearer the centre than France is; consequently a Newtonian would predict that a pendulum would beat even faster at the poles.

Other theories of matter and weight led to different predictions. Huygens' theory did so, for example, predicting a flattening, but less than Newton had predicted. Alternative Cartesian theories, and the observations of Jacques Cassini in France, suggested that the Earth is narrower at the poles. So it was possible for observation to discriminate between them, but delicate. Moreover, one could look at other planets for clues:

[23] See (Maupertuis 1738, 3, 9, 11, 14, 224–225), and F&G 14.B1.

[24] See (Voltaire 1738, 166–167), and F&G 12.F3.

Newton observed that Jupiter is flatter at the poles.[25] However, later measurements of latitude made in France again suggested the opposite result — namely, that the polar circumference of the Earth is greater than that of the equator. So, was the Earth shaped like an onion, as Newton predicted, or like a lemon, as Descartes had suggested?

Finally, Maupertuis proposed to resolve this question, and he and Clairaut led an expedition to Lapland to measure the size of a degree of latitude in those inhospitable northern conditions, while others performed the same task in Peru, on the equator. Maupertuis was keen to lead the expedition because he knew that a successful outcome would make his name in France. He took Clairaut with him because he recognised that Clairaut's great mathematical gifts would be essential, and he took Anders Celsius with him as an accomplished observer, together with two other astronomers and a surveyor.[26]

The survey itself is worth a detailed look, because it contributed so decisively to the turn towards Newtonianism in France. In theory, the task was to measure the extent of land that measures one degree when seen from the centre of the Earth. For simplicity, both in theory and practice, this was best done by going North–South along a degree of latitude. Because one cannot stand at the centre of the Earth, measurements were taken that determined the vertical at the two ends of a straight path of a known length.

What would these measurements reveal? If the Earth were a perfect uniform sphere, then heading due South by a distance of $\frac{1}{360}$ of the Earth's circumference (about 70 miles) would result in a change of one degree in the direction of the vertical. If the Earth were very nearly flat, where you are, then the difference in the direction of the vertical at the two ends would be less than one degree, but if the Earth were strongly curved, where you are, then that difference would be greater. So a Newtonian would expect the difference in the vertical in Lapland (for a given length of a line) to be less than it would be in France, and they would expect a pendulum to beat faster in Lapland than it would in France.

The practical difficulties involved, however, were considerable. One can assume that over a large span of ordinary ground the curvature of the Earth will dominate local variations (hills and valleys), so one would need to make measurements over as long a line as possible. This is done by triangulation. First, one would establish a net of triangles whose vertices are prominent landmarks (hilltops, church spires, and the like). The angles between the vertices would then be carefully measured, and the lengths between some of them would be measured — the remaining lengths could then be calculated by trigonometry. In this way a path would be obtained separating the most distant vertices, which could well be out of sight of one another. One would then locate stars that are as near as possible to being directly overhead at one end of the path and measure their departure from the vertical at the other end of the path.

In practice, numerous difficulties could arise. There may be no convenient net of landmarks that could assist in a triangulation, and, if there is, then the landmarks themselves might be obscured by intervening trees. Every measurement taken contains an error, and these will spread through the trigonometrical calculations. A known length has to be placed on the ground, but every measuring rod expands and contracts with changes in temperature. The vertical has to be found and departures from it measured accurately. Given that the supposed flattening of the poles was likely to be small,

[25]See Newton, *Principia*, III, Proposition 18, 821; for the Motte–Cajori version, see F&G 12.B11.

[26]The Celsius (centigrade) scale of temperature is named after Anders Celsius.

there was a real risk that it could not be measured, because of the inevitable experimental errors involved.

All these difficulties would attend any surveying expedition. Going to Lapland raised even more. When the expedition left France in May 1736 the hope had been to use a chain of islands in the Gulf of Bothnia as landmarks, but they proved to be below each other's horizons. The expedition had to move, but then they got lucky and found a long valley running more or less due North from the top of the Gulf of Bothnia, near Tornea.[27] However, they now had to cope with heat and swarms of mosquitoes in the Summer, raging torrents on the river, and snow and ice in the Winter. To cope with this, they had the support of the King of Sweden, and therefore of the local militia who cut down trees and carried loads, including the bulky nine-foot-high sector used for making the stellar observation (a state-of-the-art instrument that had been specially commissioned and constructed in England). Eventually the scientists were able to lay down a known length of five miles and complete a good triangulation.

They also conducted elaborate and careful measurements of the oscillations of pendulums over a period of 24 hours, using the fixed stars as a clock. They took particular care to adjust the lengths of the pendulums to allow for the effect of temperature, because they knew that the longer a pendulum is, the slower it beats. When Winter set in they retired to Tornea, where they were well housed by the locals and became very popular. There they did their calculations, and in the Spring they repeated their measurements, checked their calculations, and returned to Paris in 1737.

The Peruvian expedition took a much longer time to report, but in 1737 Maupertuis was able to show that Newton was right: the Earth is flatter at the poles. In particular, their table of results showed that 'Gravitation increases from the Equator to the Pole, very nearly in the Ratio of the Squares of the Sines of the Latitude', as Newton had predicted.[28]

Controversy ensued as Cassini sought to defend his earlier, contrary findings, and Cartesians disputed the theory underlying Maupertuis's work, but Maupertuis and his team had done their work well and he presented it skilfully in his book, *La Figure de la Terre* (1738).[29] As a result, not only were their conclusions widely accepted — well beyond the bounds of those capable of discussing it with any competence — but Maupertuis became a celebrity. In his portrait he is shown wearing a fur hat to signify the trip to Lapland, with his left hand pressing down firmly on a globe to remind everyone of what he had established, and he became known thereafter as 'The Great Flattener'. Clairaut also benefitted from the fame of the expedition, and enjoyed a successful life as a mathematician (and as a gourmand who further enjoyed a contemporary fashion for geometry and geometers among high-society ladies) before he died of a fever in 1765 at the age of 52. But he detested D'Alembert — the feeling was mutual — apparently because of D'Alembert's greater success in salon society.[30]

It was in this charged atmosphere that Voltaire entered the story. A personal friend of Maupertuis, he was well placed to ride the tide of Newtonianism in France with his increasingly partisan expositions of Newton's ideas.

---

[27]Tornea, now called Tornio, is at the top of the Gulf of Bothnia, which separates Sweden from Finland.

[28]Maupertuis, *The Figure of the Earth*, p. 228.

[29]It is a measure of its importance that the book was translated into English, as *The Figure of the Earth* (1738), and published in the same year that it was published in French.

[30]See his *Éloge* by Diderot and Grimm in (Diderot 1981), but Diderot was closer to D'Alembert because of their work on the *Encyclopédie*, as we discuss in Section 7.2.

Figure 6.14. Pierre-Louis Moreau de Maupertuis (1698–1759)

Historians have assessed the significance of Maupertuis's work in different ways. The English Newton scholar Rupert Hall regarded it as having 'removed the last serious factual obstruction to the universal acceptance of Newtonian mechanics', although he immediately went on to say 'Acceptance, that is, as a basis for further investigations'.[31] However, the historian Clifford Truesdell thought that matters were not so simple. Writing of Maupertuis's result, he said:[32]

> While in the popular view the Newtonian system was by this one blow proved correct, the measured eccentricity did not agree with Newton's value, and the geometers were moved to read more critically the passage in which Newton derived his result. They found his argument insecure.

It emerged that although his prediction was qualitatively correct, Newton's ad hoc assumptions about fluid pressure were wrong, and gradually attempts were made to tackle this problem too.

However, as so often in life, the honeymoon of Newtonian theory in France did not last. By the end of the 1740s, Newton's theory of gravity was under sustained attack from the three people who best understood it: D'Alembert, Clairaut, and Euler. The problem they raised is our second one, concerning the motion of the Moon, but before turning to it we look at the contributions of the Marquise du Châtelet.

[31] See (Hall 1983, 352).

[32] See (Truesdell 1960, 19).

Figure 6.15. Voltaire (1694–1778)

Figure 6.16. Émilie de Breteuil, Marquise du Châtelet (1706–1749)

Marquise du Châtelet. Gabrielle-Emilie le Tonnelier de Breteuil, Marquise du Châtelet, was a well-respected intellectual. By 1741, the year she turned 35, she was known for her accounts of Leibniz's metaphysics and the first popular French account of Newton's work on optics and gravitation. Algarotti had visited her when he was finishing his *Newtonianismo per le Dame*, and the frontispiece of that book (see Figure 6.13) shows them in conversation. As a woman du Châtelet could attend only the public meetings of the Parisian Académie des Sciences, but, like her male colleagues, she could learn the new calculus only from its original publications and by talking to other mathematicians — in her case, Maupertuis in 1733 and 1734 and Clairaut a decade later.

Her connection to Maupertuis may be how she came into contact with Voltaire, with whom she was to enjoy a relationship for many years. Voltaire was advocating, at some risk to himself, that France reform its political system along English Parliamentary lines and tolerate religious diversity, and was using Newton's ideas about gravity in preference to Descartes's as a scourge with which to reprimand the French élite. This drew him to Maupertuis, and he may have arranged for Maupertuis to give lessons to du Châtelet, but if Maupertuis expected another society lady dabbling in ideas (which was all that patriarchal society expected or encouraged) he was soon surprised. Although Voltaire also tried to master Newton's ideas, all agreed that the Marquise was by far the better student.

In 1738 Voltaire published his *Elémens de la Philosophie de Neuton* in Holland and in England, an intermittently erroneous book that nonetheless did much to advance Newtonian ideas (it was soon translated into English as well, as *The Elements of Sir Isaac Newton's philosophy*). Voltaire's name alone appeared on the cover, and he took on the personal risk associated with the book, but he regarded du Châtelet as the co-author, as the preface makes clear. What was controversial was not so much the book's

Figure 6.17. Frontispiece of Voltaire's *Elemens de la Philosophie de Neuton*

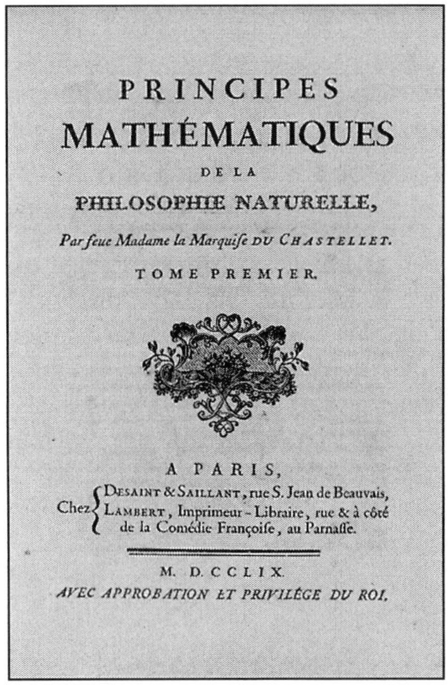

PRINCIPES
MATHÉMATIQUES
DE LA
PHILOSOPHIE NATURELLE,
Par feue Madame la Marquise DU CHASTELLET.
TOME PREMIER.

A PARIS,
Chez { DESAINT & SAILLANT, rue S. Jean de Beauvais, LAMBERT, Imprimeur-Libraire, rue & à côté de la Comédie Françoise, au Parnasse.

M. D. CCLIX.
AVEC APPROBATION ET PRIVILÉGE DU ROI.

Figure 6.18. Du Châtelet's French translation of Newton's *Principia*

attack on Cartesian physics as its philosophical stance, which seemed to advocate materialism (a position that Newton did not hold) and so to challenge the theology of the Catholic church. A revised edition of the book became a success, and once again Newton's ideas were promoted in France in a reforming and anti-establishment context.

The frontispiece to Voltaire's book (see Figure 6.17) shows Voltaire at his table busily writing, guided by the light of Newton that du Châtelet reflects down upon him with her mirror.

The Marquise du Châtelet seems to have decided to translate Newton's *Principia* into French in 1744 or 1745, using the first two (Latin) editions. She worked on her translation over the next four years, whenever she could take time away from incessant lobbying for her son to be given a regiment in the French army (she eventually succeeded). Much of the writing was done in the early hours of the morning, and to judge by surviving drafts she was exceptionally thorough.[33] In due course she sent her translations to Clairaut, along with her own original commentary.

Her work was published in two volumes, and her commentary occupies almost two-thirds of the second volume. It begins with a long overview of the Newtonian theory of the heavens, based on Newton's account of the system of the world. She began with the general theory, and then dealt with its applications to planetary motion, the shape of the Earth, the tides, and estimations of the mass of the Earth and the Moon; other aspects of the *Principia* were, for the moment, dismissed with the remark that 'M.

[33] See (Zinsser 2001), who argues that the extent of the material is evidence for her exclusive authorship of the final work.

Newton composed this book in order to destroy Descartes's vortices'.[34] She explained each mathematical term as it was introduced, a practice she continued into the second part of the commentary where she discussed the motion of three bodies under their mutual gravity, and the shape of the Earth, and here she made much more use of the calculus than Newton had done — in her case, the Leibnizian calculus — drawing on earlier work of Clairaut. This surely helped to make the theory more accessible to a contemporary audience.

The work was finished in 1749, but not published. In September of that year she went into childbirth. The baby was born on the 4th of the month, and at first all seemed well and the mother resumed work on the proofs of the book, but on the 10th she suddenly fell ill and died during the night. Voltaire, whose long affair with the Marquise had ended some years earlier, was devastated, and left France for the Court of Frederick the Great in Prussia. The prospects for the publication of the translation seemed bleak.

The translation was eventually published ten years later, in 1759, edited by Clairaut; an incomplete edition had appeared in 1756.[35] A stimulus for the edition is likely to have been the reappearance, almost as predicted, of Halley's comet at the end of 1758. The Marquise's death, and probably her sex, prevented the book from having the impact that it might otherwise have done, but the *Encyclopédie* entry on Newtonianism in 1779 paid her the compliment of listing her as one of seven authors 'who tried to make Newtonian philosophy easier to understand'.[36] Even then, as the entry makes clear, Newton's ideas were still securely established only in Great Britain.

The motion of the Moon. The motion of the Moon is difficult to understand because the Moon is part of a system of three bodies: the Earth, the Moon, and the Sun. While Newton dealt well with two bodies acting on each other by gravity, the three-body problem, as it became known, is unsolved to this day:

> Given three arbitrary bodies acting on each other by gravity and released initially with given velocities, what will be their orbits?

We do not know, for example, whether the Moon will always orbit the Earth, or move away from it, or one day collide with it.[37] This mathematical problem is too difficult to solve exactly. On the other hand, we can reach some conclusions if some simplifying assumptions are made. If we assume, for example, that the only effect of the Sun is to perturb slightly an otherwise elliptical orbit of the Moon around the Earth, we can try to calculate that perturbation exactly. This is what Newton did, and it is what many people have done since, but in his case success was less than complete. The Moon is an easy object to observe, and the mismatch between his mathematical predictions and physical reality was apparent.

The problem was a technical one. If there were just the Moon and the Earth (a two-body problem), then the Moon's orbit would be an ellipse. The effect of a third body (the Sun) is that the whole elliptical orbit rotates slowly around the centre of gravity of the Earth and the Moon (see Figure 6.19). The question was: How slowly? Newton's calculations in the *Principia* (Book I, 45) were based on some simplifying assumptions,

[34] Modified from (Zinsser 2001, 232).

[35] Du Chatelet's is still the only complete translation of the *Principia* into French.

[36] Quoted in (Zinsser 2001, 238).

[37] What we now know of the early history of the solar system reminds us that, for long-term predictions, the Moon is part of a system that involves all the other planets, not just the Sun and the Earth.

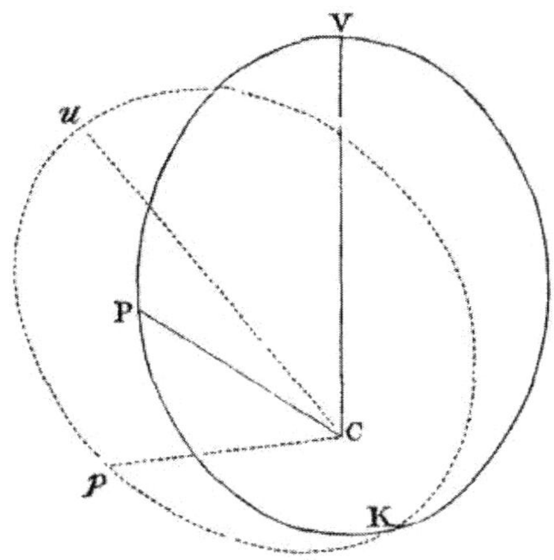

Figure 6.19. Two positions of an elliptical orbit, adapted from a figure in Newton's *Principia*

in order to make the problem tractable at all, and they showed the orbit returning to its original position every eighteen years.[38] But Nature, it appeared, prefers nine. Newton conceded this in later editions of the *Principia*, in a single crisp sentence: 'The apse of the Moon is about twice as swift'.[39] Newton tried, and failed, to find a more complicated and more accurate theory; we do not know whether he felt pleased that his approximate theory was out by a factor of only 2. Later, in papers that he never published, Newton showed how to get a much better approximation to the Moon's motion, but difficulties remained.

The problem is not merely technical; its implications were profound. This failure of Newton's, it was thought, might be the loose thread unravelling his whole theory. We have seen how unpopular his idea of universal gravitation was initially in some quarters. Newtonian gravity (or, more precisely, the inverse-square law) now came to stand or fall by its ability to describe sufficiently accurately the motion of the Moon.

There was an eminently practical reason for caring about the Moon: the determination of longitude on ships at sea. Latitude is relatively easy to determine from the Sun or the stars, but longitude is not. In the worst of many naval accidents, Admiral Sir Cloudesley Shovell, returning from Gibraltar in 1707 with his fleet of five ships, became lost in bad weather with heavily overcast skies for twelve days. The navigators on board largely agreed that the fleet was to the West of the Brittany peninsula, but on the night of 22 October in heavy fog they ran aground on the Isles of Scilly, and two thousand men drowned, the Admiral among them. This catastrophe, at a time when the British were at war with the French, precipitated a move to solve the problem of longitude once and for all. In 1714 the British Board of Longitude was established,

---

[38]See Newton, *Principia* I, Section 9, 534–545, and for an extract from the Motte–Cajori version see F&G 12.B8.

[39]The apse is a point on its orbit at which the Moon is its greatest distance from the Earth, the extremity of a major axis, such as V or u in Figure 6.19.

and was authorised to offer a large cash prize of £20,000 for the solution of the problem of determining longitude at sea, which is a measure of how seriously the British government took the issue.[40]

The problem was acute for any seafaring nation, since the methods then in use produced very large errors (of 10% or so). Newton's opinion was sought, and he pronounced that a solution would be difficult to find and that the methods he knew about faced severe obstacles. Could the heavens be used as a celestial clock? If so, and they could be read accurately, then the problem of longitude would be solved, whence the demand for an accurate theory of the motion of the Moon. Or perhaps the moons of Jupiter could function as a clock, as the French expert Domenico Cassini had hoped. Newton was doubtful about this, too.

Significant progress on the longitude problem had to wait until 1747, when Clairaut proposed to modify the inverse-square law by adding on a small extra term. That such a proposal could be made shows the uncertain hold that the inverse-square law still had on some working mathematicians. Indeed, Clairaut now held, as he put it to Euler, that it was 'a proven fact that Newtonian gravitation is inadequate to account for the [lunar] phenomena'.[41] Their rival D'Alembert came independently to the conclusion that Newton's theory was incorrect, and so did Euler in 1748, after he had studied a different three-body system (the Sun, Jupiter, and Saturn). Each had a different remedy in mind, and for a while Newton's theory of gravity seemed about to fall.

Figure 6.20. Alexis-Claude Clairaut (1713–1765)

The Euler–Clairaut correspondence documents this exciting and dramatic debate quite closely. It began when Euler had just submitted his essay on the motion of Saturn to the Académie des Sciences in Paris for consideration in their prize competition. Clairaut was one of the judges, and he read the essay in September 1747. He had

[40]It is always difficult to estimate the value of sums of money over a gap of centuries, but all the results of the Economic History calculator turn this into a sum in the millions, and even the tens of millions, of pounds. This sounds large — and it is — but one might also note that none of these estimates reaches even half the cost of a single modern passenger aircraft.

[41]Quoted in (Kopolevich 1966, 655).

recognised Euler's handwriting, and he wrote to Euler on 11 September to say that he was delighted that Euler had thought about Newtonian attraction. In his essay Euler expressed his doubts about the inverse-square law, citing Newton's failure with the motion of the apse of the Moon, and indicated that he wished to reintroduce vortices. Clairaut also had his doubts, and went on:[42]

> It is true that on adding some other term one feels that the theory will better accord with the phenomena. But it seems to me that this term must be such that at the distances of Mercury, Venus, the Earth and Mars it must be almost insensible, in view of the extreme smallness of the motion of the apsides. And if, as it seems initially from your work, the law of squares is palpably in error at the distance of Saturn and Jupiter it would still be necessary to add terms which were significant only at that distance. I confess that the whole of gravitation seems to me to be only a speculative hypothesis.

Clairaut then remarked that:

> It seems to me, and I am not a candidate for the prize, much more important to know if Newtonian attraction holds or not than to treat simply of Saturn. And in seeing if the square law of attraction must suffer some correction which can only be for small distances it seems to me to be necessary to begin by finishing the theory of the moon.

When he addressed the Paris Académie des Sciences in November 1747 Clairaut suggested replacing the inverse-square law with something like r^{-2} plus a term negligible when r is large, such as r^{-4}. But he soon withdrew the suggestion that the modifying term should be an inverse-fourth power, because it predicted that objects near the surface of the Earth should be heavier than they are.

By then Euler had written to Clairaut suggesting that the English astronomer James Bradley's many observations of the Moon's motions implied that vortices 'or some other material cause' was responsible for the motion. Clairaut replied on 7 December that he had no confidence in Euler's vortices, which he thought Euler himself had shown to be no help at all.[43]

Elsewhere in 1747, in a passage that gives a clear picture of a first-rate scientific mind at work, Clairaut set down what he thought made the *Principia* difficult to understand.[44] He first praised it, which was still a controversial thing to do in France, by saying:

> The famous book *The Mathematical Principles of Natural Philosophy* has been the occasion of a great revolution in Physics. The method which Mr Newton, its illustrious author, has followed to derive facts from their causes, has shed the light of mathematics on a science which up till then had been in the shadows of conjectures and hypotheses.

Then he turned to say what had to be done next. No matter, said Clairaut, that Newton concealed his fluxional calculus, for he pointed out that the calculus is now so familiar that it was easy to repair that omission. However

> is it not right to reproach him for another wrong which without doubt has struck all those who have studied his book with a true desire to understand it? Namely, that in most of the difficult places he employed too few words to explain his principles ....

One gets a vivid picture of even Clairaut struggling to work his way through the *Principia* in order to understand it, rather than merely to admire it. But for Clairaut it was

[42]For this exchange of letters, see Euler, *Opera Omnia* (4A) 5, 173–175, and F&G 14.B2 and 14.B4.

[43]See Euler, *Opera Omnia* (4) 5, letter 421, and F&G 14.B2(b).

[44]See (Clairaut 1749), and the extract in F&G 14.B3.

a matter of regret that Newton had not explained his principles, especially when one of his predictions had turned out to be obviously wrong. However, Clairaut reflected, so much else was right: 'Kepler's laws ..., the movement of the nodes of the moon ..., the tides, ... and finally several other questions equally favourable to attraction it appeared to me as difficult to reject it as to accept it.'

But then came the surprise. On 17 May 1749 he simply announced that, by taking a new point of view, he had found that the problem disappeared; the inverse-square law could give the correct prediction for the apse line of the Moon.

Euler was not immediately persuaded. He had moved to Berlin eight years before (as we describe in Section 7.1), and the St Petersburg Academy was in the doldrums. Its observatory had burned down in 1747, and the publication of its journal was in disarray. In 1749 Euler suggested that they have a prize competition and proposed several propitious topics, all of an astronomical nature. They chose:[45]

> To demonstrate whether all the inequalities observed in lunar motion are in accordance with Newtonian theory — and if they are not, to demonstrate the true theory behind all these inequalities, such that the exact position of the Moon at any time can be computed by means of it.

Euler became one of the judges, indeed the decisive one.

By the time that the Academy announced its competition, Clairaut had publicly announced that he now believed Newtonian gravitation to be adequate, and Euler wrote to the Russian academicians to tell them so, noting that he was still not persuaded. 'Hence', he went on, 'you may judge for yourself that this question is not only the most profound of all those the Academy had to select from, but that the most complete answer may be expected from it as well'.

Meanwhile, Clairaut hesitated over whether to enter the competition. He published his paper, which gave D'Alembert a chance to find fault with it, and he submitted his entry in December 1750. By then the Academy had decided to extend the competition to 1 June 1751, but when they sent Clairaut's entry to Euler he replied that it 'is superb, and it is hardly likely that anything better will be received prior to June 1'. It is striking that Euler was so quickly persuaded of a view that he had openly dismissed not long before. He repeated his endorsement, and his own change of opinion, in the official statement he wrote on 5 June, and the result was announced on 6 September 1751. Clairaut then published his own theory of the motion of the Moon in 1753.

In his paper of 1749 Clairaut had completely retracted his idea of modifying the inverse-square law. Now, and again in 1753, he proposed instead that the error lay in the poor way in which exact, unsolvable equations for the motion of the Moon had been reduced to inexact, approximate, but solvable equations. D'Alembert, who had waited for Clairaut to declare his hand, thereupon retracted too.

What, then, was the new viewpoint that was so powerful that both D'Alembert and Euler were so quickly converted? It was an improved mathematical analysis of the problem — specifically a new approach to the differential equations that were taken to describe the motion of the Moon — so it was another victory for the calculus.

We can follow Clairaut's analysis to its surprising conclusion without getting too deeply into the mathematics in Boxes 13 and 14. Put simply, Clairaut attempted to determine the motion of the Moon by the method of successive approximations, and

---

[45]See (Kopelevich 1966, 655).

Box 13. Clairaut's formulation of the motion of the Moon.
Clairaut formulated the problem of the motion of the Moon in terms of differential equations, and after integrating twice found this expression for the solution:

$$\frac{f^2}{Mr} = 1 - g \sin v - q \cos v + \sin v \int \Omega \cos v dv - \cos v \int \Omega \sin v \, dv,$$

where f, g, and q are constants of integration, M is the sum of the masses of the Earth and the Moon, v is an astronomical quantity called the true anomaly and may be taken to represent the velocity of the Moon and, most importantly, Ω is an unknown function of r (the radial distance of the Moon), and is the perturbative force of the Sun. Clairaut's problem was how to express Ω.

He decided on a process of successive approximations. It was already known that the apse of the Moon moves rather as though the Moon precesses on an ellipse. So Clairaut first wrote the equation of an ellipse in polar coordinates,

$$\frac{k}{r} = 1 - e \cos mv.$$

In this expression k, e, and m are constants that are either to be determined from the constants f, g, and q, or otherwise found from observation. In particular, e was already known empirically to be about 0.05. Because $\cos mv$ varies between -1 and $+1$, this means that

$$0.95 \le \frac{k}{r} \le 1.05,$$

so

$$\frac{k}{1.05} \le r \le \frac{k}{0.95}.$$

Thus far, Clairaut was following the path first taken by Newton.

was misled by the fact that the terms entering the first approximation were too small to capture any precession. Only on passing to the second approximation did he discover his error.

The practical consequences were immediate. In 1755 Tobias Mayer, an astronomer at Göttingen, calculated a set of lunar tables based on Euler's theory, and eventually, in 1765, the British Board of Longitude rewarded his widow with £3,000, and paid Euler £500 for his contributions. At the same time, they paid the English inventor John Harrison £20,000 for his usable solution to the longitude problem: an accurate, spring-driven, pocket watch, just over five inches in diameter, that is now on display in the Royal Observatory at Greenwich.

Even Euler found the details of Clairaut's method difficult.[46] But in the end Newtonianism became accepted very much as Newton had presented it, in that

- the theory of the solar system was highly mathematical
- its predictions rested on a highly theoretical analysis

[46]See Euler, *Opera Omnia* (4) 5, 195–196, and F&G 14.B4.

Box 14. Clairaut's defence of the inverse-square law.
Clairaut now took the expression $k/(1-e\cos mv)$ as the first approximation to r, substituted it into his original equation, and obtained this better approximation to r:

$$\frac{k}{r} = 1 - e\cos mv + \beta\cos\frac{2v}{n} + \gamma\cos\left(\frac{2}{n} - m\right)v + \delta\left(\frac{2}{n} + m\right)v.$$

Here n is another quantity determined by observation (and therefore known) and β, γ, and δ are constants that are determined from the earlier constants. The point to notice is that the new terms describe the way that the ellipse slowly changes.

Now for the trap into which Clairaut had briefly fallen. He evaluated β, γ, and δ and found that, to nine decimal places,

$$\beta = 0.007090988,\ \gamma = -0.00949705,\ \text{and}\ \delta = 0.00018361.$$

These numbers are much smaller than e, and although the plan of successive approximations called for them to be employed in an expression and used to calculate the next, and better, approximation to k/r, Clairaut felt that they were already too small to allow his method to double the value of m. Consequently, he believed that the error must lie in the inverse-square law, which therefore needed tweaking.

But during the Spring of 1749 he calculated the next approximation, and found that he had been wrong. The contributions coming from the γ term were quite large, and were proportional to the transverse perturbing force, whereas the initial contribution to m was related to the radial perturbing force. Only by going to the second approximation could Clairaut pick up the effect that was making the Moon's ellipse precess. Now, on calculating the numbers, Clairaut found that the monthly apsidal motion was $3^\circ\ 2'\ 6''$, which was just $2'$ less than the empirical value that he accepted.[a]

[a]See (Wilson 2002, 215).

- its theoretical presuppositions seemed inevitable if its conclusions were accepted — in particular, if the mysterious force of gravity was accepted as really existing.

The role of mathematics in science was now much greater than it had been, for it became the glue that held the vast new edifice together. For historians of mathematics, this is perhaps the most important consequence of the growing acceptance of Newtonianism in the period 1687–1760. We therefore look at this role in more detail in the next chapter, concentrating on the significant breakthrough that extended the capacity of mathematicians to describe problems involving more than one independent variable, such as time and one or more coordinates of space. This enabled them to deal for the first time with many of the most interesting aspects of the physical world.

6.4 Further reading

Arianrhod, R. 2012. *Seduced by Logic: Émilie du Châtelet, Mary Somerville and the Newtonian Revolution*, Oxford University Press. This is a pacy and informative account of the work of two of the author's 'heroines' and is also good on the context of their lives.

Brown, Lloyd A. 1956. The Longitude, in *The World of Mathematics*, Vol. 2, J.R. Newman (ed.), Allen and Unwin, 780–819. This is a rich account, covering the methods used and the social and political background, in a four-volume set of essays that is an unrivalled classic in the mathematical literature.

Greenberg, J.L. 1995. *The Problem of the Earth's Shape from Newton to Clairaut; the Rise of Mathematical Science in Eighteenth-Century Paris and the Fall of 'Normal' Science*, Cambridge University Press. A rich, stimulating, and demanding book, it is recommended not just for the exceptionally thorough treatment of the topic in its main title but for the argument about the nature of history of science referred to in its subtitle.

Terrall, M. 2002. *The Man Who Flattened the Earth*, Chicago University Press. This is a very readable account of the life, work, and significance of Maupertuis.

Shank, J.B. 2008. *The Newton Wars and the Beginning of the French Enlightenment*, Chicago University Press. This book argues that it was the polemics around Newton's work, rather than the work itself, that led to Newton's influence on the Enlightenment.

Sobel, D. 1995. *Longitude*, Fourth Estate. A highly readable account of the topic, it is particularly good on Harrison.

The Papers of the Board of Longitude,
`cudl.lib.cam.ac.uk/collections/rgo14`.
As the website says: 'The sixty-eight volumes of the papers of the Board of Longitude document in rich detail many of this public institution's remarkably wide-ranging activities over more than a century (1714–1828).'

7

The 18th century

Introduction

The next five chapters concentrate on developments in the 18th century, the period after the discoveries of Newton and Leibniz. The name of Descartes could also be added to the list of essential precursors, because his introduction of algebraic methods into geometry influenced the work not only of Newton and Leibniz but of everyone who came after them. Indeed, one can profitably think of the century as the working-out of many of the implications of the advent of Cartesian algebra and the calculus.

The century was dominated by the work of Leonhard Euler, a Swiss mathematician whose output not only dwarfs that of his contemporaries quantitatively (his collected works run to more than 80 volumes) but more than matches it qualitatively. The only comparable figures were his French rival, Jean le Rond D'Alembert, and his Italian/French successor, Joseph-Louis Lagrange, and the rest of this chapter sets the scene for what follows by sketching their lives. We shall see that scientific activities were dominated by the scientific academies that were established by the courts of France, Prussia, and Russia, where leading mathematicians could work with a considerable degree of freedom with a view to harnessing the power of science. The chapter ends by considering the work of Lagrange and others on the solution of equations and the so-called Fundamental Theorem of Algebra.

In Chapter 8 we look in more detail at Euler's revival of number theory, which Fermat had tried unsuccessfully to promote and which led in Lagrange's hands to the beginnings of a theory of what are called quadratic forms. This can be thought of partly as a succession of generalisations of the simple question: Which numbers can be written as a sum of two squares?, and more importantly as the creation of a mathematical theory governing such problems. Then we look at some of the remarkable results that Euler obtained involving infinite series, including

$$1 + \frac{1}{2^2} + \frac{1}{3^2} + \frac{1}{4^2} + \cdots = \frac{\pi^2}{6}.$$

This result, which is striking in its own right and answered a long-standing problem, also has a bearing on the properties of prime numbers. Then we turn to Euler's contributions to geometry, the only subject that he did not more or less rewrite.

We return to the calculus in Chapter 9. It was felt early on that, for all its power and its successes, the calculus, as Newton and Leibniz had left it, lacked sufficiently good grounds for acceptance. It might be easy to use, but it was much harder to understand *why* it worked, and neither Euler nor Lagrange was able to give it lasting foundations.

But even so, they both extended the calculus remarkably. One of Euler's insights was to replace the geometrical language that Newton and Leibniz had used (involving variables and curves) with a language of functions. He also made great use of the idea of expressing functions as infinite series, and in this way he advanced the theory of the exponential, logarithmic, and trigonometric functions. He also clarified and extended the theory of differential equations, but one distinction belongs to D'Alembert, who was the first to apply the calculus to the motion of a vibrating string. As Euler realised, this opened the way to studying functions of more than one variable, and thereby to the application of the calculus to many problems in the natural sciences.

The theme of applications is taken up in Chapter 10. We look in detail at D'Alembert's work on the vibrating string, and then at Euler's lifelong programme to extend the calculus and make it yield results about point masses, solid masses, elastic bodies, and the motion of fluids. Then we pass to perhaps the most important mathematical topic of the period, the study of celestial mechanics. First Euler, then Lagrange, and then Laplace at the start of the 19th century, took up the challenge of investigating whether Newton's mysterious force of gravity could explain the subtle motions of the solar system.

7.1 Euler

Basel. Leonhard Euler was born in Basel on 15 April 1707. His father Paul (or Paulus) was a vicar who had studied theology at Basel University. His mother Margaretha was a vicar's daughter with distinguished classicists among her ancestors. Leonhard was the first of their four children, all of whom survived past the age of 30.

When Euler was one-and-a-half years old, the family moved to Riehen, about an hour's walk to the north of Basel. His father gave him some mathematical instruction at home, using a rather difficult book, Stifel's second edition of Rudolff's *Coss* (1553), and sent him to the local school when he was 7.[1] At the age of 13, he went to Basel University; this was quite usual at that time — the University provided the function of what later became the higher reaches of school. In 1723 Euler graduated as Magister from the Philosophy Faculty, and on that occasion he gave a public speech in Latin, comparing the Cartesian and Newtonian systems of natural philosophy. This was a topic of some interest because Continental Europeans were then struggling to rescue Descartes's vortex theory from the stringent criticisms of Newton.

Euler then enrolled in the Theology Faculty, as his father wished, but gravitated at once to mathematics and the lectures of Johann Bernoulli. Euler was a friend of Johann's son, Johann II (who had graduated at the same time as Euler) and so was admitted to the private extra sessions that Johann I put on. Bernoulli recognised his

[1] Stifel's work, and the tradition to which it belongs, were described briefly in *Volume* 1, Chapter 8.

Figure 7.1. Leonhard Euler (1707–1783)

talents at once, and went on over the years to praise him far more strongly and unreservedly than he ever praised anyone in his own highly talented mathematical family until, in 1745, Bernoulli called Euler 'incomparable, the prince of mathematicians'.[2]

At the age of 20 Euler entered one of the prize competitions organised by the learned academies of the day, the Académie des Sciences in Paris. The topic was the optimal way to place masts on a ship to ensure both speed and stability, and Euler submitted a highly theoretical essay based, he said, on the 'surest principles of mechanics'. He came second, which was quite an achievement for he had seen only barges, ferry boats, and canoes on the Rhine — but never a sailing ship! The work is listed as the 4th in a catalogue, compiled by the 19th-century historian of mathematics Gustav Eneström, of over 860 works, from short papers to long books, that Euler was to write.[3]

St Petersburg. Euler's career took off in 1727 with a call to the new Academy being set up in St Petersburg — indeed, Russia was to be his home for two long spells lasting several years. In the years 1700 to 1721 Tsar Peter the Great had fought the Swedish empire for access to the Baltic Sea, and so to trade routes with Europe and beyond. This secured, he built a new city on the swamps that formed the mouth of the Neva river,

---

[2]See (Calinger 2016, 241).

[3]Eneström's catalogue is an invaluable guide to Euler's works, and is accessible online through the Euler Archive. This is a growing and well-arranged collection of reproductions of Euler's original publications, sometimes with commentaries or translations into English and other information. We refer to Euler's publications by their E (for Eneström) numbers.

a prodigious and brutal enterprise. St Petersburg was created as the face of modern Russia, in many ways the opposite of Moscow, and in 1724 Peter the Great resolved to found an Academy of Science there. Its purpose was to bring the benefits of modern science to Russia, and accordingly to attract foreign talent.

Peter never lived to see his Academy. He died in 1725 and his widow Catherine I took it over. Johann Bernoulli declined an invitation to move there, but his sons Nicolaus II and Daniel I did go. Euler's invitation came later, as a result of lobbying by the young Bernoullis, Johann I, and Christian Goldbach, the first permanent secretary of the Academy. Unfortunately, Catherine I died on 17 May 1727 and Euler arrived a week later to an Academy with an uncertain future as dynastic battles for the throne of Russia broke out. Matters did not improve until the reign of the Empress Anna Ivanova, from 1730 to 1740.

The mathematicians among the Academicians were Daniel I Bernoulli (but not Nicolaus, who died from an intestinal ulcer after only nine months in Russia) and Jakob Hermann, who was the leading applied mathematician of the day, but who returned to Basel in 1730. Other scientists, and there were many, included Joseph-Nicolas Delisle, an astronomer and geographer, and Goldbach (although Court politics forced him to be in Moscow from 1728 to 1732). From Delisle, Euler learned rigorous spherical trigonometry, state-of-the-art celestial mechanics, and the mathematical principles of cartography, all topics that he later took up. Goldbach was a worldly diplomat, fluent in several languages, and much more interested in mathematics than the term 'Goldbach's conjecture' would suggest.[4] He corresponded extensively with Euler.

Hermann's departure led to Daniel Bernoulli's promotion, and a year later Euler became the Professor of Physics when that post fell vacant. In 1732 Daniel also returned to Basel, and Euler took over as Professor of Mathematics, ceding the physics chair to his friend Georg Krafft. With his increased salary, Euler could now afford to marry, and in January 1734 he did so. His wife was Katharina Gsell, who came from a family of artists in Basel. Their first child, Johann Albrecht, was born in November 1734; Goldbach was one of the godfathers. The next year, however, was perilous for Euler. He caught a life-threatening illness, and although he recovered, the illness was in some way connected with a problem that flared up again in 1738 and cost Euler the sight of his right eye. Stories of Euler having lost his sight through overwork are not to be believed.

The year 1739, when Euler was 32 and his life resumed an even keel, is a good point in his career for us to take stock of his progress. Making every allowance for his distressing illnesses, it is clear that Euler was not a prodigy. In that year he wrote three works, bringing his total to 35 or so. Of these 35, only three mark his entry to the first rank of mathematicians: his remarkable solution in 1734 of the so-called 'Basel problem' (which we discuss below in Section 8.2), his proof that the sum of the reciprocals of the prime numbers diverges (which implies that there are infinitely many primes), and the two-volume *Mechanica* of 1736. The rest are diverse but it can be argued that if illness had killed Euler in 1738 he would be known only to historians of mathematics, and none of us would have any idea what we would have lost. Most of his influential material was to come much later.

[4] Goldbach's conjecture says that every even number greater than 2 is the sum of two prime numbers. Substantial results are known in this direction, but at the time of writing it is still unproved in general; see Section 18.2.

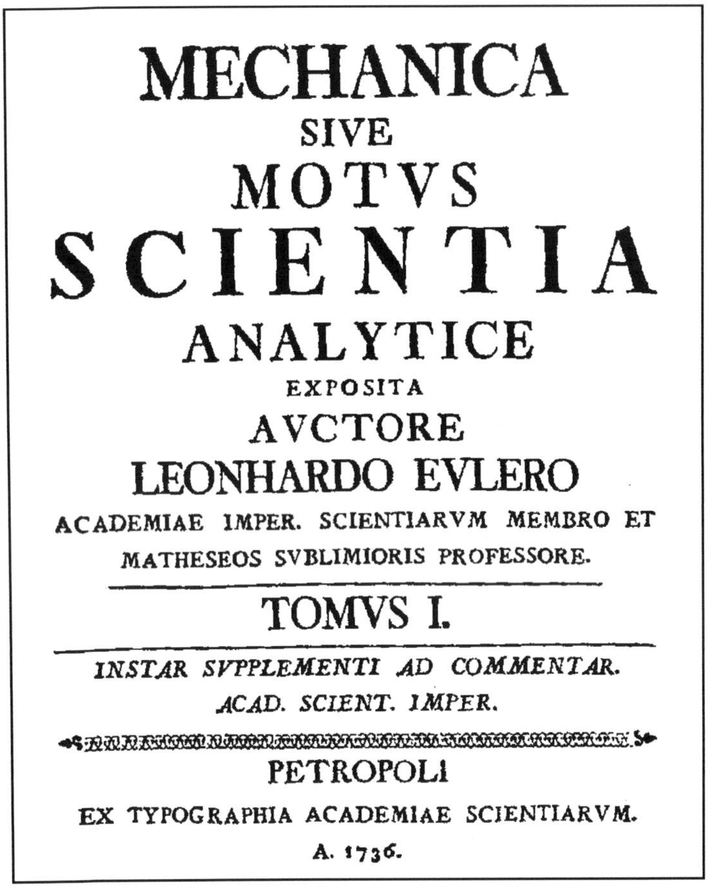

MECHANICA
SIVE
MOTVS
SCIENTIA
ANALYTICE
EXPOSITA
AVCTORE
LEONHARDO EVLERO
ACADEMIAE IMPER. SCIENTIARVM MEMBRO ET
MATHESEOS SVBLIMIORIS PROFESSORE.

TOMVS I.

INSTAR SVPPLEMENTI AD COMMENTAR.
ACAD. SCIENT. IMPER.

PETROPOLI
EX TYPOGRAPHIA ACADEMIAE SCIENTIARVM.
A. 1736.

Figure 7.2. Euler's *Mechanica* (1736)

The *Mechanica* (E15, E16), however, is a substantial work (see Figure 7.2). It indicates that Euler had already formed a project of studying mechanics at six levels: point masses, solid bodies, elastic and flexible bodies, bodies of non-constant volume, continua, and fluids. At this stage Euler could deal only with point masses and their motion under forces, possibly in resisting media (Volume 1), and their motion on surfaces (Volume 2). The book is written throughout in the language of the Leibnizian calculus, and not in the essentially geometrical style of Newton, Johann Bernoulli, and Hermann. The importance of this move, which was continued in many other works, cannot be over-estimated, and we shall return to it.

Euler continued to be moderately productive, writing extensively on musical harmony as part of his interest in the nature of sound. But in 1740 the Empress Anna Ivanova died, and renewed dynastic turmoil exacerbated a situation that many foreigners found worrying. Euler, his wife, and his family always disliked the cold winters, but they were even more worried about the ever-present risk of fire. Fires regularly swept through the wooden houses of the new town, so much so that people were advised to keep trunks of belongings ready packed for speedy evacuation, and the Eulers' spacious house on the banks of the Neva was one of these at risk. So in 1741 Euler accepted an invitation to move to Berlin and the Academy of Sciences that was being established there by Frederick the Great of Prussia, and after his friends had lobbied energetically to persuade the St Petersburg authorities to let Euler go, as his contract permitted, he made the move to Berlin.

Berlin. Euler was appointed to manage the new Academy, which was being created from the near ruin of the Academy established in 1700 in Berlin and headed for a time by Leibniz. But it had been deliberately run down by Frederick Wilhelm I, the father of Frederick the Great, who despised Leibniz and all intellectuals — and indeed anything except the military. The Berlin Literary Society, founded in 1743, was added to the mix, apparently at Euler's urging, and after a further change of name the Académie Royale des Sciences et Belles-Lettres was established in Berlin with Maupertuis as its first President; the use of French was entirely deliberate as we shall see. By the mid-18th century Academies of Science were widely seen as a force for modernisation.

Euler and his family settled into a house near the Académie and the Royal Palace, numerous financial irregularities of the Académie were sorted out over the next few years, and Euler's publications began to pour out.

This is perhaps the time to describe just how much Euler wrote. The academic publishing world worked much more slowly even than it does today. The journals were those of the learned academies, and they made little distinction between the various branches of mathematics and science, although the humanities subjects were often published separately. Euler left behind in St Petersburg several works that came out only after his departure, and immediately on his arrival he started to fill up the newly refounded Berlin journal, the *Mémoires de l'Académie Royale des Sciences et des Belles-Lettres de Berlin*; the first issue carrying his name has seven articles by him. But since books and articles take time to write, it is difficult to say when Euler was writing, even if we know the date when the article or book was submitted and the much later date when it was published. So it is not easy to say what Euler was working on in the 1740s, but by the end of 1749 the Eneström index of works published had reached number 120, up from 32 in 1739.

Among the 88 items published in that decade we find numerous works on topics raised in the *Mechanica*; works on astronomy doubtless inspired by the presence of Delisle in the St Petersburg Academy but continuing in Berlin with a focus on comets; works on integration and the summation of series; the Königsberg bridges problem (submitted in 1736 but not published until 1741) which we discuss in Section 8.3; a work on the tides that won him the prize of the Académie des Sciences in Paris in 1741; a book on the calculus of variations in 1744 (E65) entitled *Methodus Inveniendi Lineas Curvas, etc.* (Method for Finding Curved Lines, etc.); his greatly enlarged and enriched 'translation' (E77) of a major English work on gunnery that became the definitive treatment of the subject and was at once translated back into English and put into French; work on optics, light, and colour; a defence of religious revelation in 1747 that was greatly out of tune with Emperor Frederick's freethinking ideas; some further work on number theory; the two-volume *Introductio in Analysin Infinitorum* (*Introduction to the Analysis of the Infinite*) of 1748 that over the years was translated into French, German, and English (E101, E102); the two-volume *Scientia Navalis* (Naval Science) of 1749 (E110, E111); and the study of the motion of Saturn and Jupiter (a three-body problem) that won him the prize of the Académie des Sciences in Paris in 1749 (E120).

Euler's publication figures now ran steadily at about 120 items a decade, or one a month, for the rest of his life, so that Eneström's total by 1783, the year of Euler's death, stood at 563. The posthumously published ones then brought the number up to 866 in 1911. It is hard to believe that Euler ever felt the need to revise; more likely, by the time that Euler sat down to write, it was all clear in his mind and it was just a matter of describing carefully what he saw.

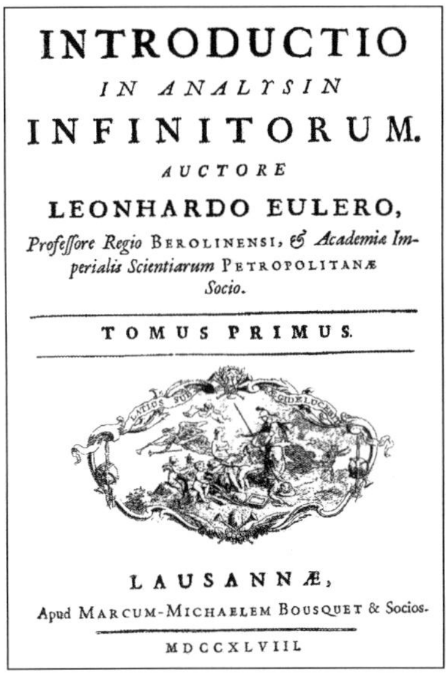

INTRODUCTIO

IN ANALYSIN

INFINITORUM.

AUCTORE

LEONHARDO EULERO,

Professore Regio BEROLINENSI, & *Academiæ Imperialis Scientiarum* PETROPOLITANÆ *Socio.*

TOMUS PRIMUS.

LAUSANNÆ,

Apud MARCUM-MICHAELEM BOUSQUET & Socios.

MDCCXLVIII.

Figure 7.3. Euler's *Introductio in Analysin Infinitorum* (1748)

Two relatively early books from the Berlin years stand out: the *Introductio in Analysin Infinitorum*, and the *Scientia Navalis*. Despite its title, the *Introductio* is not what we would think of as an introductory calculus textbook. What it introduces is the method of infinite series, which Euler applies in the first volume to functions and to the theory of partitions, and in the second volume to the study of curves (and, in an appendix, to surfaces). Throughout this work, the functions that Euler treated are algebraic functions, trigonometric and exponential functions, and their expressions as infinite series.

It is here that Euler presented the power series expansions of the trigonometric functions and connected them to the exponential function through the use of complex numbers, obtaining

$$e^{i\theta} = \cos\theta + i\sin\theta,$$

even if, it seems, he never wrote the simple but profound consequence, $e^{i\pi} = -1$. (We look at this discovery in more detail in Section 9.2.) He also obtained infinite-product expansions for the sine function, and his famous formulas for the sums of the reciprocals of all even powers of the integers (see Section 8.2). Euler was also willing to accept arbitrary infinite series, even if they came without a guarantee of convergence.

The *Scientia Navalis* is another remarkable book. It was completed apparently some ten years before it was published, so it must be seen as a contribution to Russia's urgent desire to become a naval power. Volume 1 has an account of the equilibrium of ships, goes on to the new topic of stability and small vibrations, and gives the definition of an ideal fluid. Volume 2 deals specifically with ships, including the working of sails and rudders. Euler even proposed driving a ship by a propeller, although there were no engines sufficiently powerful to do so at the time, and he designed a turbine that was shown in 1944 to work at 71% efficiency; modern turbines run at just over 80%.

The 1750s continued in this vein. Among other highlights — all of which we look at later — are his resolution of the controversy about the definition of the logarithm of a negative quantity, which we consider in Section 9.2; progress on the dynamics of rigid bodies in Section 10.2; a book (E187) on the motion of the Moon aimed at solving the problem of determining longitude, which we have already noted in Section 6.3 and will again in Chapter 11; work on the principle of least action, which we consider in what follows; the *Institutiones Calculi Differentialis* (E212) of 1755; and hydrodynamics, including what are now called Euler's equations for a perfect fluid (E225), which we look at in Section 10.2.

The least action controversy. The principle of least action generated one of the largest, and nastiest, scientific controversies of the century. Maupertuis, the head of the Académie Royale in Berlin, was a vain man, famous for leading the French expedition to Lapland that verified the flattening of the Earth at its poles and helped to swing Continental Europe over to the Newtonian cause, as we described in Chapter 6. In 1744 he presented a paper to the Académie Royale (it was published the next year) on what became known as the 'principle of least action': the idea that a mechanical system evolves in a way that continually minimises a quantity called its action. He noted that Euler had made a statement of the same kind in his *Methodus Inveniendi*, but Maupertuis proposed to take it much further philosophically, and make it the animating principle of all of nature and the basis of a proof of the existence of God. Maupertuis gave Euler's argument a profoundly teleological spin — that is, he suggested that the mechanical system evolved to meet a preassigned goal.

However, when Samuel König, another Swiss Academician in Berlin, published a paper in 1751 that criticised the idea and suggested that it could be traced back to Leibniz's use of a principle of Aristotle's, Maupertuis went on the attack. He accused König of plagiarism, and with Euler's support he tried to have König driven out of the Academy. The official charge was that there was no letter from Leibniz that made such a claim, and that König had produced, or perhaps only relied on, a forgery. After an extensive search no letter could be found stating the principle of least action, but König produced fragments of what he claimed was a copy of one, and when the charge could not be proved, Maupertuis used all his influence to get the result he wanted. Eventually, König was subjected to an official censure of the Academy in 1752, but was not expelled.

Maupertuis's attack on König enraged Voltaire, who published a pamphlet entitled *Diatribe du Docteur Akakia Medicin du Pape*. The background here is that when Voltaire, Maupertuis, and König were friends in the late 1730s, Voltaire had persuaded König to teach algebra to his then-lover, Émilie du Châtelet. Emperor Frederick had meanwhile publicly supported Maupertuis, and Voltaire's pamphlet so enraged him that he had it burned by the hangman in public places in Berlin. The affair was the end of the courtship between Voltaire and the Emperor.

Euler's role in this is agreed by many historians to be one of the few stains on his otherwise attractive character, although Calinger points out that Maupertuis's position had the support of D'Alembert and others at the Académie in Paris.[5] It seems as though Euler was humouring Maupertuis, and thereby getting his way on numerous issues, but felt that he could not avoid giving something back on this occasion. Maupertuis's vanity did not make it easy for him to take advice from the much more gifted Euler, whom he once described (before the affair broke) as 'an exceedingly peculiar personality, a relentless pest, who likes to meddle in all affairs'. Euler, for his part, was by now chafing under the continued esteem that Frederick had for the French, which, he wrote to Goldbach, was 'in no way commensurate with the intellectual weight that was due to them'.[6]

Even before the least action affair, Euler's relationship with Frederick was poor. Euler lacked the wit, or the French style, that delighted the Emperor, who wrote to his brother that Euler's 'epigrams consist of calculations of new curves, some sorts of conic sections, or of astronomical calculations'. Such people, he said, are 'used as Dorian columns in architecture. They belong to the subfloor'.[7]

All that Frederick wanted from Euler were solutions to the most practical of problems, such as the design of canals and the design of wind and water mills. When Maupertuis's health declined in the aftermath of the least action affair, and he eventually took leave as President of the Academy, Euler seems to have wanted the position. He took over many of the duties on an acting basis, but Frederick would have none of it, and made it clear that he wanted Euler's only rival in the mathematical world, Jean le Rond D'Alembert.

The relationship between Euler and D'Alembert was not good, but when D'Alembert agreed to come and talk with the Emperor Frederick he insisted that he did not want the position and that the only suitable person for it was Euler. Understandably, this turned the relationship between the two mathematicians to one of friendship, but D'Alembert's arguments did not convince Frederick and the whole matter dragged on from Maupertuis's death in 1759 until 1763, by which time Euler realised that there was no future for him in Berlin. The free-thinking and otherwise Catholic religious nature of Frederick's entourage was also difficult for the devoutly Protestant Euler, and he set about looking for somewhere else to go. Back in 1748 he had briefly considered England, writing to a friend then in London that 'there is no other country in which I would rather like to settle than in England',[8] but by the mid-1760s there was only one place to go: back to St Petersburg under the enlightened leadership of the Empress Catherine II (now known as Catherine the Great). He left Berlin on 1 June 1766, and arrived in St Petersburg on 28 July to a triumphant reception.

St Petersburg revisited. The early 1760s had found Euler at work on many topics: number theory (see Section 8.1); elliptic integrals; hydrodynamics; ordinary differential equations (see Section 9.3); his definitive book on the motion of a solid body (E289) (see Section 10.2); and the propagation of sound. Publications that date from his return to St Petersburg include the three-volume *Institutiones Calculi Integralis* (E342, E366, E385) and his ever-popular *Letters to a German Princess* (E343, E344) — she was Frederick's niece, the Princess of Anhalt-Dessau.

[5] See (Calinger 2015), Chapters 10 and 11, notably pages 372–373.
[6] For both quotes, see (Fellmann 2007, 90).
[7] See (Fellmann 2007, 92).
[8] See (Fellmann 2007, 100).

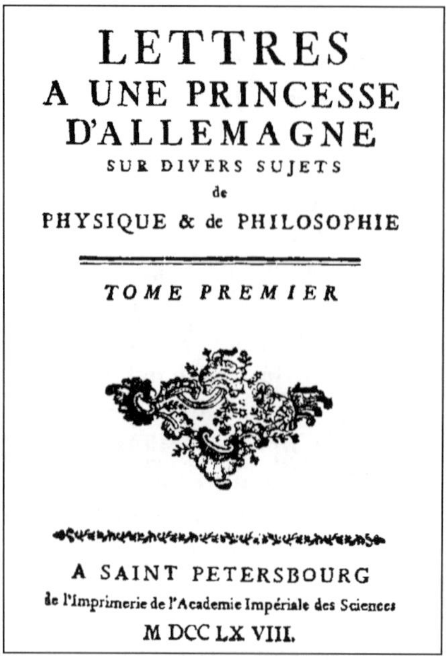

LETTRES
A UNE PRINCESSE
D'ALLEMAGNE
SUR DIVERS SUJETS
de
PHYSIQUE & de PHILOSOPHIE

TOME PREMIER

A SAINT PETERSBOURG
de l'Imprimerie de l'Academie Impériale des Sciences
M DCC LX VIII.

Figure 7.4. Euler's *Letters to a German Princess* (1768)

The *Institutiones Calculi Integralis* is the greatest, and inevitably the most difficult, of his works on the theory of the calculus. The emphasis, as ever, is on what one can do with it, and the approach is step by step. Euler was not good at foundational issues to do with the calculus, but once that could be left behind, what came out is a lucid account of how to integrate expressions that can be integrated, how to solve ordinary differential equations of many different kinds, linear and non-linear, and how to tackle partial differential equations. An appendix is devoted to the calculus of variations, with much praise for Joseph-Louis Lagrange, a young man who was to become Euler's true successor.[9]

Euler's initial happiness on his return to Russia was to turn to pain. He was granted a fine house to live in, but an operation to remove a cataract from his left eye went badly wrong, and he became almost completely blind in 1771. In the same year Euler's house caught fire in a blaze that engulfed over five hundred houses in the city, and Euler's life was saved thanks only to the bravery of one Peter Grimm, a craftsman. Then, in November 1773, Euler's beloved wife Katharina died at the age of 66.

Before Euler lost most of his sight he dictated his wonderful two-volume *Vollständige Einleitung zur Algebra* (*Elements of Algebra*) (E387, E388) to his former tailor, who reported in later life that he now understood the subject very well. It was published in 1770, although, as with all of Euler's books, much of it had been written and published earlier in various papers. It became a bestseller, and was translated immediately into several languages, including English, French, Latin, and Dutch. It was chosen as the

[9]The calculus of variations concerns the search for functions that maximise or minimise a given integral. It is used to verify that the shortest curve joining two points in the plane is the straight line segment between them, to investigate problems involving the principle of least action, and, to this day, to provide theoretical formulations for mechanics and physics.

first of his works to appear in the edition of Euler's *Opera Omnia*, and it does so with a work that Lagrange had written as its companion piece for the French edition, his 300-page 'Additions'.

After the failed cataract operation, Euler's sight permitted him to write normal-sized letters only on a blackboard, whence they could be transcribed, and in this fashion he continued to work with uninterrupted productivity. Indeed, there are signs that, with approaching old age, he was rushing to get things published while there was still time.

In 1776 Euler also found time to get married again, to the great consternation of his children who doubtless feared that the family money would be lost, and to the mirth of others who think they know what is going on when an old, blind man remarries. Euler had settled on the half-sister of his first wife, Salome Abigail Gsell, who was sixteen years his junior, and it seems that all went smoothly.

Euler's death is among the best known in the history of mathematics, and has been described by two people who were present. In September 1783 he was subject to attacks of dizziness, but he was still able to work on the mathematics of balloon flight in the aftermath of the Montgolfiers' sensational ascent in June of that year. On the morning of 18 September he gave a lesson to one of his grandchildren, and over lunch discussed with his assistants Anders Lexell and Nicholas Fuss the orbit of the planet Uranus, discovered by William Herschel just two years earlier. According to Fuss's account, around five o'clock he was sitting on the couch playing with his grandchild and smoking his pipe when the pipe slipped from his hand. He cried out 'My pipe!', and bent forward to pick it up, but could not. He grabbed his forehead with both hands, cried out 'I am dying' and fell unconscious. He did not recover from this massive stroke and, in the words of Nicolas de Condorcet, at around 11 o'clock that night 'The great Euler ceased to calculate and to breathe'.[10]

Conclusion. This account of Euler's career omits a great deal. Euler's work on applied mathematics is far from the dry collection of applicable topics that the subject was to dwell on in the 19th and 20th centuries. Even a list of the topics is stimulating: the nature and propagation of sound, including the theory of musical harmony; the nature of light and optical technology; the motion of gases and liquids (see Chapter 10.2), with an account of the shape and flow of rivers; ship design; cartography; planetary astronomy and celestial mechanics (see Chapter 11). Add the seemingly duller topic of the motion of solid bodies (also in Section 10.2), and we begin to realise that, for many of these topics, Euler's was the first proper mathematical account, and certainly the first to exploit the resources of the calculus systematically. That also makes something else suddenly clear: by and large, Euler had also to invent the relevant mathematics. It was Euler who showed how to apply the calculus, and for that he had to develop it extensively.

In this connection, the twelve prizes that he won from various Academies are interesting. In addition to the two mentioned already, there were essays on the nature of fire, on anchors for ships, on terrestrial magnetism, and on how to make the best use of wind when sailing; there were also two further attempts at solving the three-body problem.

---

[10]Nicolas de Caritat, Marquis de Condorcet, wrote a celebrated eulogy of Euler, in *Letters to a German Princess*, 1795, xxv–lxiii, and F&G 14.C4. See also (Fuss 1783).

Euler was not the only one to win prizes, but he had only one equal for most of his life: D'Alembert. For all Euler's work on extending the calculus from one to two or more independent variables — and he was a major figure there too — it was D'Alembert who applied the calculus to the vibrating string.

A similar list of topics in 'pure mathematics' could be given that were barely alluded to above. Foremost among these was number theory: Fermat's last theorem for the case $n = 3$, as well as essentially initiating the general study of quadratic forms (see Section 8.1); the study of prime numbers; special functions; the beta and gamma functions, and the functional equation for the zeta function; work on elliptic integrals; summation formulas; a systematic theory of ordinary differential equations (see Section 9.3); the differential geometry of surfaces; topology (the Königsberg bridges and the Euler formula for polyhedra, see Section 8.3); the calculus of variations; and Latin and magic squares.

Number theory alone reminds us how much could be said about the relationship between Euler and Lagrange. Euler's extraordinary inventiveness led him to formulate discoveries well in advance of their acquiring a proof. One of Lagrange's talents, particularly in the calculus of variations, mechanics, and number theory, was to turn Euler's deep insights into proper theorems. Lagrange was Euler's immediate successor at the Académie Royale in Berlin, before he was brought to Paris in 1786, but his considerable shyness prevented him from ever meeting Euler in person.

It is of enormous importance for the history of mathematics that Euler published copiously — and at every level from the most advanced to the most elementary. He was not among the elitists, such as Descartes or Newton, who wrote for only a few and who are valued for that. He was among those who chose to speak to any audience, and did so very well. But more than that, he rode the 18th-century wave of publishing research and contributed greatly to its success.

Euler had the gift of making his work look easy, even when the result is spectacular. His characteristic way of working was to start with a simple example, build up slowly and steadily from there, one step at a time, and keep going. This was surely an important part of what Euler wished to put across: doing mathematics using the systematic tools he presented may not be child's play, but brilliance is not necessarily required. It seems that he always did this — it was not something he slipped into when writing expository works, it was his habit in any substantial piece of research. His enormous productivity — Euler is by far the most prolific mathematician of all time — and the lucidity of his writing are two reasons why he had the influence that he did, as was his remarkable depth of insight.

Even today, publishing is not the usual way to proceed outside academia: commercial companies' publishing practices in many areas (such as medicine) may be secretive, selective, and at times barely honest. Publishing was far from usual in the 17th and early 18th centuries; Newton was secretive, and Johann Bernoulli's contract to work with the Marquis de l'Hôpital made it almost impossible for him to publish anything he had discovered, as it was the property of his employer. Only at the end of the 17th century did Leibniz start to agitate for a culture in which researchers published routinely, while submitting to some sort of a review process. It was a major step for a field to force its practitioners to share their hard-won knowledge. One mission of the Academies that sprang up was to foster this open publication — the prize competitions can be seen as a way of sustaining individual effort, and individual egos, in the new, less personal world. The sheer profusion of Euler's work, its evident quality, and the way that it became essential to future work by other people, all moved the practice of mathematics and science firmly into the world of publishing.

How may we characterise the genius of Euler? Certainly, he had an exceptional technical facility, an ability to work to a high level of abstraction, and a phenomenal memory, which he retained right into old age. He must have been able to keep a vast amount of information in his head, and to present it lucidly on demand. More remarkably, he could thoroughly reformulate ideas that others found difficult to understand at all.

In Chapter 5 we saw how thoroughly synthetic and geometrical Newton's *Principia* is. Beneath the Cartesian symbolism and the infinite series the mode of thought is geometrical. Geometry and the calculus were not tidily separated when Euler was young. Leibniz's calculus, which Euler learned from Johann Bernoulli, was based on manipulating indivisibles. In its geometrical formulation, much of mathematics is about is two variable quantities connected in some way that defines a curve. The ways that this connection was specified were many and various, and were by no means tied to Cartesian coordinates. The rule defined the curve, which it gave *globally*. Typically, but not by any means always, if you expressed the rule in coordinates and obtained an equation for the curve, then it would be written as an implicit equation, such as the one for a circle ($x^2 + y^2 = r^2$), and seldom as an explicit equation for y as a function of x.

By contrast, the calculus works *locally*. Differentiation starts at a point and allows us to look a small distance on either side. It works only if there is a unique value of y for each value of x in the small region that we are looking at.

What Euler saw clearly, and others only obscurely, was the centrality of the function concept. But this was not the input–output definition that we teach today; Euler's function concept was still tied to the ways in which functions presented themselves to him, but he saw that the calculus applied to functions rather than curves. It followed that to apply the calculus in any given setting, one should go function hunting, and express one's problem in function language. After that, there was an abundance of things that Euler could do, because the calculus was already quite sophisticated by the 1730s, and it made sense for mathematicians to add to the resources of the calculus.

An 18th-century view of mathematics was that it was the science of quantity — more precisely, of number (discrete quantities) and measurement (continuous quantities). Euler added to that a vision of mathematics as the science of functions. As two illustrations of this, we shall consider Euler's account of the logarithms of negative quantities (in Section 9.2) and his reformulation of Newton's laws of motion (in Section 10.2).

Euler's account of the foundations of the calculus is inadequate, but his uses of the calculus generated several different branches of mathematics. These may be said to be based around the extension of algebra to deal with infinite series, and so to connect the function concept on which Euler relied with one of Newton's great insights. The combination is in many ways more algebraic than geometrical, and Lagrange's approach to mathematics was even more markedly algebraic, so it is appropriate to start our more detailed look at mathematics in the 18th century by looking at algebra. But before we do, we introduce Lagrange himself, and his mentor in Paris, Jean le Rond D'Alembert.

Figure 7.5. Jean le Rond D'Alembert (1717–1783)

Figure 7.6. Joseph-Louis Lagrange (1736–1813)

7.2 D'Alembert and Lagrange

As we have seen on several occasions, Euler's work often led to competition or rivalry with his French contemporary, Jean le Rond D'Alembert. This was particularly true when it concerned mechanics, D'Alembert's consuming passion.

D'Alembert had been abandoned as an infant in 1717 on the steps of the church of St Jean le Rond, near Notre Dame in Paris — whence his name — and he was brought up by a glazier's wife, to whom he remained devoted all his life, with money supplied by his father. He was apparently a brilliant conversationalist, gifted with a superb memory, and was much sought after in the world of Parisian salons. He first learned mathematics from the intellectual descendants of Malebranche, and by the 1750s he had become a leading member of the group of philosophers who produced that bible of the Enlightenment, the *Encyclopédie* or *Dictionnaire Raisonné des Sciences, des Arts, et des Metiers* (Encyclopaedia, or a Systematic Dictionary of the Sciences, Arts, and Crafts), 1751–1772 .[11]

The original leader of the team was one Jean-Paul de Gua de Malves, but the true mentor of the group was Denis Diderot, who was a few years older than D'Alembert. Their philosophy of rationalism, promulgated through the *Encyclopédie*, brought the group into heated conflict with the forces of Catholic conservatism, especially the Jesuits. But they had energetic supporters — notably Voltaire, and later Condorcet — and were ultimately able to take control of the Académie Royale des Sciences in Paris.

Unhappily for us, D'Alembert's quickness of mind seems to have prevented him from adequately finishing a work, so his papers were often rushed and are consequently difficult to read. He would embark on subjects that were topical and publish quickly in order to establish priority. Moreover, he was quick to quarrel (with Clairaut, Euler, and Daniel Bernoulli, among others) and his papers have a disagreeable air of claiming too much. In 1841, the major German mathematician Carl Jacobi remarked that: 'it is impossible to choke down a single line of D'Alembert's mathematics, while most of Euler's work can be read with delight', and most of his contemporaries would probably have agreed.[12] Nonetheless, some of his papers are of exceptional distinction, none

[11]The Enlightenment is often taken as the high point of European rationalism and is regarded as a crucial precursor of the French revolution.

[12]Quoted in (Hankins 1970, 63).

more so than his first paper on the vibrating string, the work with which he made his name.

Euler's relationship with Lagrange was free of rivalry, if only because Lagrange was a shy man who avoided meeting him, despite corresponding with him over the calculus of variations, mechanics, and number theory.

Joseph-Louis Lagrange was born in Turin in 1736. At the time Turin belonged to the Duchy of Savoy, and he grew up speaking an Italian dialect with strong affinities to French. This can be seen even in his name: his father's surname was Lagrangia, but the young man chose the French spellings LaGrange or Lagrange, emphasising the French origin of the male side of his family.

Lagrange's father had originally intended his son to study the law, but courses in physics made Lagrange aware of his talent for mathematics, and he devoted himself to the exact sciences. His first publications appeared when he was 18, by which time he had come across a copy of Euler's *Methodus Inveniendi* (1744), which was to be his introduction to the new subject of the calculus of variations. In August 1755 Lagrange wrote to Euler setting out a new analytical method for formulating and solving problems in the subject, and Euler replied in September that he was very interested in the technique. By then, Lagrange's fame had begun to spread, and he was appointed a professor at the Royal Artillery School in Turin by royal decree. Euler then shared more of Lagrange's ideas with Maupertuis who was sufficiently impressed to arrange for Lagrange to be offered a Chair in Mathematics in Prussia. This would have been a better position than the one in Turin, but Lagrange turned it down, thus forfeiting his first opportunity to meet Euler.

Lagrange's earliest works, including some on the calculus of variations, were published in the first four volumes of a Turin journal called the *Miscellanea Taurinensis ou Mélanges de Turin* (Turin Miscellany), covering the 1760s. Even the earlier volumes reveal how very well read Lagrange had become, and how varied his interests were. They now embraced the theory of differential equations, the theory of sound and other topics in fluid mechanics, and various problems in celestial mechanics: the motion of Jupiter and Saturn, and the motion of the Moon.

In 1763 Lagrange left his native Turin for the first time and travelled to Paris, where D'Alembert already knew of his work and had corresponded with him. The trip was unsuccessful, because Lagrange fell ill, but D'Alembert was moved to set about trying to find a better position for him than his poorly paid professorship in Turin. In 1765, when Lagrange won a prize of the Académie Royale in Berlin on the motion of the satellites of Jupiter, D'Alembert wrote to Frederick the Great, who valued his advice highly, to suggest that he offer Lagrange a position at the Académie Royale. Lagrange again turned the opportunity down, writing that 'It seems to me that Berlin would not be at all suitable for me while Mr. Euler is there'.[13]

The next year, D'Alembert informed Lagrange that Euler was leaving Berlin to return to St Petersburg and asked him whether he would agree to be Euler's successor. The letter was followed by a generous offer from Frederick himself, and Lagrange now agreed to move. He travelled via Paris, where he met D'Alembert, and then London, where his long-time supporter, the former Ambassador from the Court of Naples to Turin, now resided, and arrived in Berlin at the end of October 1766 after a journey of over two months.

---

[13]See Itard, 'Lagrange', *Biographical Dictionary of Mathematicians* 3, p. 1304.

In Berlin Lagrange became friends with Johann Lambert (whom we shall meet in Section 14.1). He also met Johann III Bernoulli (a grandson of Johann Bernoulli) who was in charge of reorganising the astronomical observatory of the Académie Royale in Berlin. Lagrange's duties were to present a memoir once a month — over the years more than sixty of these were published — and to supervise the mathematical work of the Academy. He often entered the prize competitions of the Académie des Sciences in Paris, especially those on celestial mechanics, but soon after his arrival in Berlin he took up another topic entirely: the theory of numbers.

This was inspired by his close reading of the work of Euler. One of Lagrange's first discoveries, in 1768, was a proof that the equation

$$x^2 - ay^2 = 1$$

always has integer solutions, whenever a is not a perfect square. This is an old problem of Brahmagupta, Bhaskara II, and Fermat, that had been investigated by Wallis and Brouncker.[14]

In 1770 Lagrange was the first to prove that every positive integer is the sum of at most four squares; for example,

$$19 = 16 + 1 + 1 + 1 = 9 + 9 + 1; \quad 23 = 9 + 9 + 4 + 1.$$

Number theory was to become a lifelong interest, and he often contributed proofs and systematic expositions of matters that Euler had illuminated with examples and conjectures.

1770 is also the year in which Lagrange published a long memoir on the solution of polynomial equations by radicals. This was devoted to analysing why there was no known formula for the solution of a general polynomial equation of degree five or more.

In the 1770s Lagrange began work on the theory of partial differential equations, which had hitherto been studied only in particular cases — Euler too was at work on a systematic account. Lagrange also started to write up his ideas about mechanics. We know from a letter he wrote to Pierre-Simon Laplace[15] that a book-length account was nearly finished in 1782. It was sent to Paris, where the young Adrien-Marie Legendre was brought in to check the proofs, and the book, the *Traité de Mécanique Analytique* (Treatise on Analytical Mechanics), was finally published in 1788.

By then Lagrange's situation had changed dramatically. His wife, whom he had married within a year of arriving in Berlin, died in 1783 after a long illness, and in August 1786 the Emperor Frederick died. This robbed Lagrange of his most powerful supporter at Court, and the leaders of other national Academies began to intrigue to hire him. Among these was the Comte de Mirabeau, a popular French politician who was also influential at the French Court, and he succeeded in bringing Lagrange to Paris in 1787. Lagrange now became a senior member of the Académie des Sciences in Paris, having been a corresponding (or foreign) member since 1772. Perhaps surprisingly, he found himself temporarily unable to work creatively, although he pleased

[14]Lord Brouncker was the first President of the Royal Society of London. The equation $x^2 - ay^2 = 1$ is usually known as Pell's equation, due to a misattribution by Euler; John Pell was an English mathematician of the 17th century who corresponded with Leibniz. Brahmagupta and Bhaskara II were Indian mathematicians and astronomers of the 7th and 12th centuries who dealt with this problem; see *Volume* 1, Chapter 6.

[15]Laplace and his work are discussed in Chapter 11.

his new colleagues by his wide range of knowledge of many topics, including history, religion, botany, and medicine.

In 1790, as the French Revolution began, Lagrange joined the Commission that advised on a new system of weights and measures, from which the metric system emerged, and eventually he became Chair of the Commission. When the revolution turned violent and the French fought wars against their Austrian and Italian neighbours, he was fortunate to gain immunity from a law that allowed for the imprisonment of all foreigners born within enemy borders and the confiscation of their property. In 1795, after the Terror was over, he was appointed to the Bureau des Longitudes.

He also taught at the École Normale de l'An III (year three of the revolutionary calendar).[16] This enterprise lasted three months and eleven days, but out of its ashes grew the École Normale Supérieure, which has lasted to the present day. Lagrange also lectured at the newly founded École Polytechnique, which had been created in 1794. When Napoléon came to power Lagrange's career prospered still further. Lagrange died in 1813 and he was buried in the Panthéon in Paris. His funeral oration was given by Laplace, and the event was marked by other ceremonies in universities in Italy — but not in Berlin, the capital of Prussia, one of the countries arrayed against the French.

It is possibly in the field of mechanics that Lagrange made his greatest impact. His *Traité* attempted, with some success, to give all of mechanics a rigorous foundation using the methods of the calculus, and it contained a better defence of D'Alembert's method of virtual work than D'Alembert had ever been able to give. But during his time in Paris Lagrange also set out his bold attempt to base the calculus entirely on algebraic principles. This was his *Théorie des Fonctions Analytiques* of 1797 (second edition 1813), which we discuss in Section 9.4. As we shall see, the attempt failed, and was replaced by Cauchy's very different, and much more arithmetical, account (which we discuss in Section 16.1), but it is testimony not only to the profoundly algebraic tenor of Lagrange's work but also to the growing perception that the calculus had finally to be put on secure foundations.

7.3 Algebra

We conclude this chapter with a look at two topics: Lagrange's work on the solution of equations, and the Fundamental Theorem of Algebra.

The solution of equations. In 1770, Lagrange took up the question of whether polynomial equations always have solutions, and whether there is an algebraic formula for finding them. As he put it:[17]

> I propose in this memoir to examine the different methods which have been found up till now for the algebraic solution of equations, to reduce them to general principles, and to examine a priori why these methods succeed with the third and fourth degrees but fail for higher degrees.
>
> This examination will have a two-fold advantage: on the one hand it will serve to shed greater light on the known solutions to the third and fourth degrees; on the other hand it will be useful to those who wish to occupy themselves with the solution of higher degrees, in providing them with different views of this object and above all in saving them a great number of steps and useless attempts.

---

[16]This calendar replaced the usual one and the months had even been renamed.

[17]See (Lagrange 1770, 206–207), and F&G 14.D4.

His hope was that a proper understanding of why the methods for solving cubic and quartic equations work might suggest ways of tackling quintic equations (equations of degree 5), for which no formula for the solution was known. Since such a formula was to comprise nothing more than addition, subtraction, multiplication, division, and extraction of arbitrary roots, it would, if it could be found, show that the quintic equation was what is called *solvable by radicals* (from the Latin 'radix' for a root).

Lagrange, of course, knew that the quadratic equation

$$ax^2 + bx + c = 0,$$

with coefficients a, b, and c, has the well-known solutions

$$x' = \frac{-b + \sqrt{D}}{2a} \quad \text{and} \quad x'' = \frac{-b - \sqrt{D}}{2a},$$

where $D = b^2 - 4ac$. Moreover, the two values of $\sqrt{D}/a$ are $x' - x''$ and $x'' - x'$.

He also knew that a cubic equation can always be transformed into one without a term in x^2, and written in the form

$$x^3 + cx + d = 0.$$

The solutions of this equation are of the form

$$\sqrt[3]{-\frac{d}{2} + \sqrt{D'}} + \sqrt[3]{-\frac{d}{2} - \sqrt{D'}},$$

where

$$D' = (d/2)^2 + (c/3)^3.$$

It turns out that $\sqrt{D'}$ can be expressed in terms of the solutions x', x'', and x''' of the cubic equation:

$$\sqrt{D'} = (x' - x'')(x'' - x''')(x''' - x').$$

What about quartic equations? Lagrange knew that a quartic equation has a similar, if even more forbidding, set of formulas for its solutions. They involve a square root, which forms part of an expression inside a cube root, that in turn is part of an expression involving two successive square roots.

Lagrange's insight began with the well-known remark that any k quantities are the roots of an equation of degree k. Let us illustrate this when $k = 3$ with the three quantities x', x'', and x'''. Each of them satisfies the cubic equation

$$\Delta^3 - (x' + x'' + x''')\Delta^2 + (x'x'' + x''x''' + x'''x')\Delta - (x'x''x''') = 0.$$

Now, suppose that one were given an expression Δ in three quantities y', y'', and y'''. Suppose also that Δ takes only 3 different values when y', y'', and y''' are permuted, say the values x', x'', and x'''. Then Δ would satisfy a cubic equation — in fact the above equation — but notice that the coefficients of this equation do not change as y', y'', and y''' are permuted. This is interesting, because one would expect any expression in 3 quantities to take $3! = 6$ values when those three quantities are permuted.

Let us look at this in a little more detail in the case of the cubic equation with solutions x', x'', x'''. There are expressions taking various values as the solutions are permuted. For example:

- the expression $x' - x''$ takes 6 values when x', x'', and x''' are permuted:
$$x' - x'', x'' - x', x' - x''', x''' - x', x'' - x''', x''' - x''$$
- the expression $x' + x''$ takes 3 values: $x' + x''$, $x'' + x'''$, $x''' + x'$
- the expression $(x' - x'')(x'' - x''')(x''' - x')$ takes 2 values, of which one is the negative of the other, unless its value is zero
- expressions such as $x' + x'' + x'''$, $x'x'' + x''x''' + x'''x'$, and $x'x''x'''$ each take precisely 1 value.

Lagrange took the view that the reason that the solution formulas for polynomial equations of degrees two, three, or four involve square and cube roots is that there are expressions in two, three, or four quantities that take fewer values than one might expect, and so could be expressed as solutions of polynomial equations of lower degree.

In the case of a cubic equation, it is because there is an expression $\sqrt{D'}$ in the solutions that takes only two values when the roots are permuted, and another that takes only three values, that square roots and cubic roots can be used to express the solution to the cubic. A similar story would hold, he believed, for a quartic equation.

Lagrange had to demonstrate this in detail, which he did. His question then was: Are there similar expressions in five quantities that take only two, three, or four values? He would also allow the extraction of fifth roots, but not quantities that merely satisfy an equation of degree five, because that would be no advance at all.

We must, however, take note of a crucial distinction in the way that Lagrange analysed these polynomial equations. Sometimes he was writing about expressions in the coefficients, and sometimes he was considering the same expression as a function of the solutions. Lagrange was interested in solving polynomial equations, so the expressions that he wanted had to be written in terms of the coefficients of the equation. They could not be written in terms of the solutions, because the solutions are unknown until the equation is solved. On the other hand, as he saw it, the fact that there are expressions in the solutions taking fewer values than one might expect caused Lagrange to find them interesting. In the case of the cubic equation, it is essential that D' is expressible in terms of the coefficients.

Expressions in the solutions that take fewer values than expected he called 'resolvents' because they helped him to resolve the equation into simpler ones. To be useful, a resolvent must have an expression in terms of the coefficients of the equation. But his insight was to see that the study of the quintic equation might depend on the existence or non-existence of expressions in five symbols that took five or fewer values — and that this might be an easier problem.

We sum up our account so far.

In the case of the cubic equation, Lagrange connected the expression in the solutions taking only two values with the appearance of square roots in a formula for the solution, and he called this expression a resolvent.

In the case of the quartic equation with solutions x', x'', x''', x'''', Lagrange identified the expression $D'' = (x' - x'')(x' - x''')\dots(x''' - x'''')$, and also the expression or resolvent

$$(x' + x'')(x''' + x'''')$$

which takes only three values under the 24 permutations of the solutions. It therefore satisfies a cubic equation, and it can also be expressed in terms of the coefficients of the quartic equation. Accordingly, Lagrange connected the existence of this second resolvent with the fact that the quartic equation can be solved by first solving a quadratic equation and then a cubic equation — after which two more square roots have to be taken, a fact that Lagrange explained with yet more resolvents.

Lagrange summed up his lengthy investigation of the quartic equation in these words:[18]

Lagrange on the solution of quartic equations.

> We conclude our analysis of the methods which concern the solution of equations of the fourth degree here. Not only have we related these methods to one another and show their interconnections and their mutual dependence, but we have also, and this is the principal point, given the *a priori* reason why they lead, some to resolvents of the third degree, others to resolvents of the sixth, but which can be reduced to the third. One has seen how this derives in general from the fact that the roots of these resolvents are functions of quantities x', x'', x''', x'''', which, on making all the possible permutations of these four quantities, only receive three different values, like the function $x'x'' + x'''x''''$, or six values of which two are equal and of opposite sign, like the function $x' - x'' - x''' - x''''$, or even six values which, on dividing them into three pairs and taking the sum or the product of the values of each pair, the three sums or the three products are always the same, whatever permutation one makes of the quantities x', x'', x''', x'''', ... It is precisely the existence of such functions on which the solution of equations of the fourth degree depends.

Accordingly, to solve a quintic equation, Lagrange therefore looked for a resolvent, an expression in five quantities $x', x'', x''', x'''', x'''''$, that takes either three or four values. Failing to find one — the best he could do was an expression that takes only six values — he became doubtful that the quintic equation is solvable by radicals. The difficulty here was the increasing complexity and number of the expressions that he had to consider.

A conclusive negative answer would show that the quintic equation is not solvable by radicals, but the complexities of the expressions he obtained defeated even Lagrange:[19]

[18] See (Lagrange 1770, 307), and F&G 14.D4.

[19] See (Lagrange 1770, 307), and F&G 14.D4. This passage immediately follows the passage just quoted.

Lagrange on the solution of quintic equations.

> It follows from these reflections that it is very doubtful if the methods of which we have been speaking can give a complete solution of equations of the fifth degree, and still more so to those of higher degrees. And this uncertainty, coupled with the length of the calculations which these methods involve, must repel in advance all those who would seek to use them to solve one of the most famous and important problems in algebra. Also we observe that the authors of these methods have themselves been content to apply them to the third and fourth degrees and that no one has yet undertaken to push their work further.
>
> It would therefore be very desirable if one could judge *a priori* the success that one can expect in applying these methods to degrees higher than the fourth. We are going to try and give the means for this by an analysis similar to that which has served us up till now in respect of the known methods for the solutions of equations of the third and fourth degree.

After much more work he concluded that it was unlikely that the general quintic equation is solvable by radicals, but he could not be sure.

Lagrange's whole memoir makes it clear that he thought he had found the reason why something works, and that this had to do with resolvents that take only a certain number of values as the variables are permuted. But it became only too clear that the study of these resolvents was liable to become very involved, so perhaps it is not surprising that a programme of such complexity was not readily taken up. If Lagrange, who might be supposed to be committed to the approach, and who was the leading mathematician of his generation, sounded so doubtful, why should anyone else choose to get involved?

In the event, despite the independent publication in 1771 of another thoughtful memoir along these lines by one Alexandre-Théophile Vandermonde, the question was not pursued until almost the end of the century. Its eventual resolution was to be one of the most surprising and fertile episodes in the rich mathematics of the 19th century, as we shall describe in Chapter 19.

The Fundamental Theorem of Algebra. In the early 18th century, with the establishment of the calculus, interest attached to such integrals as

$$\int \frac{dx}{x+a}, \int \frac{dx}{x^2+a^2}, \int \frac{dx}{x^2-a^2}, \text{ and } \int \frac{(x+a)\,dx}{x^2+bx+c}.$$

This interest led naturally to the question of evaluating integrals of the form

$$\int \frac{P(x)}{Q(x)}\,dx,$$

where $P(x)$ and $Q(x)$ are polynomials. If the denominator Q could be factorised into linear and quadratic factors (say $L_1, L_2, \ldots$, and $Q_1, Q_2, \ldots$, respectively), then the question would reduce to one already answered. This is because P/Q can then be written

Box 15. Complex conjugation.
The *complex conjugate* of the complex number $\alpha = a + ib$ is $\bar{\alpha} = a - ib$.

A number is real (sometimes we say 'purely real') if and only if it equals its complex conjugate.

Notice that the sum and product of a number and its complex conjugate are always real: $\alpha + \bar{\alpha} = 2a$ and $\alpha\bar{\alpha} = a^2 + b^2$. It follows that α and $\bar{\alpha}$ are the roots of a polynomial equation with real coefficients: $x^2 - 2ax + a^2 + b^2 = 0$.

by an algebraic process (akin to long division) as a sum of the form

$$\frac{p_1}{L_1} + \frac{p_2}{L_2} + \cdots + \frac{P_1}{Q_1} + \frac{P_2}{Q_2} + \cdots + R,$$

where $p_1, p_2, \ldots$ are constants, $P_1, P_2, \ldots$ are linear terms in x, and R is a polynomial in x.

So people became interested in whether every polynomial with real coefficients can be factorised into linear and quadratic factors. If the answer was yes, then this would be expressed as a theorem:

> *The Fundamental Theorem of Algebra* (version 1): Every polynomial of degree n with real coefficients can be factorised into a number of linear and quadratic factors, possibly with repetitions, whose coefficients are real numbers.

Later on, as mathematicians became more comfortable with complex numbers, it became natural to factor the quadratic terms into two linear factors with complex entries. For example:

$$x^2 + 3 = (x + i\sqrt{3})(x - i\sqrt{3}),$$

where $i^2 = -1$ and the right-hand side is the product of a complex number and its complex conjugate.

Once this step was taken mathematicians began to pose the original question as one about polynomials with complex coefficients, and this led to another form of the Fundamental Theorem:

> *The Fundamental Theorem of Algebra* (version 2): Every polynomial of degree n with real or complex coefficients can be factorised into n linear factors, possibly with repetitions, whose coefficients are complex numbers.

Throughout the 18th century mathematicians of the period were unsure of what to make of complex numbers, and always sought real answers. For them the question addressed by the Fundamental Theorem of Algebra was whether every polynomial can be factorised into linear and quadratic terms with real coefficients.[20]

How did they attempt to understand the Fundamental Theorem of Algebra? Indeed, did they even believe it? As we have already remarked, they knew that quadratic equations can be solved, and that Italian mathematicians of the 16th century had solved

[20] In 1629 Albert Girard had claimed, without proof, that every equation of degree n has n roots, see (Stedall 2011, 44–45).

the general cubic and quartic equations. They also knew that no-one had found a general method for solving equations of degree 5 or more.

Euler wrote to his mentor Johann Bernoulli on 15 September 1739 to say that he believed that every polynomial equation could be resolved into either the right number of linear factors, or at least into the appropriate number of quadratic factors. This seems to be the first time that the Fundamental Theorem of Algebra was affirmed; previously Leibniz had invoked the claim, but only in order to deny it. On 1 September 1742, Euler wrote to Johann's nephew, Nicolaus Bernoulli, repeating the claim, but Nicolaus was not convinced, and offered as a counter-example

$$x^4 - 4x^3 + 2x^2 + 4x + 4 = 0,$$

whose solutions are

$$1 + \sqrt{2 + \sqrt{-3}},\ 1 - \sqrt{2 + \sqrt{-3}},\ 1 + \sqrt{2 - \sqrt{-3}},\ \text{and}\ 1 - \sqrt{2 - \sqrt{-3}}.$$

Nicolaus had found these solutions, but denied that they could be written in the form $a + b\sqrt{-1}$, so he denied the truth of the Fundamental Theorem of Algebra.

Euler replied on 10 November. He exhibited the required quadratic factors from which the complex solutions can be found, and he claimed to have a general argument that proved the Fundamental Theorem of Algebra for all polynomials of degree at most 4. While he waited for Bernoulli to reply he discussed the matter with Christian Goldbach, who was also doubtful and queried the theorem for polynomials of the form $x^4 + px + q$, but Euler corrected him. When Nicolaus Bernoulli replied, it was to withdraw his claim about the quartic polynomial and to agree with Euler that the Fundamental Theorem of Algebra was true. He even claimed that he could possibly prove it, provided that it was true that every imaginary quantity is an elementary function of quantities of the form $a + b\sqrt{-1}$, which, he added, nobody denies.

In his *Introductio in Analysin Infinitorum* (Volume 1, Chapter II), Euler asserted the Fundamental Theorem of Algebra in the following terms. Polynomial functions (with real coefficients tacitly understood) are said to have either linear or complex factors, the number of complex factors is always even, and the complex factors can always be paired in such a way that the product of these pairs is real. A polynomial of odd degree always has a real factor and, in fact, an odd number of real factors, and a polynomial of even degree has an even number of real factors (possibly none). In Chapter IX Euler returned to this theme and offered partial proofs of these claims — but he never attempted to justify them fully.

For this, he was criticised by his great rival, D'Alembert, and Euler explained to him that the book had been written three years earlier, in 1745, and that he had made great progress since, 'although I freely confess that I do not yet have a solid demonstration that every algebraic expression can be resolved into real trinomial factors'.[21] By then, D'Alembert had published his own proof of the Fundamental Theorem of Algebra, which was submitted to the Académie Royale in Berlin in 1746 and published in 1748. It grew out of an earlier account (from 1745) that had not been published. In the 1745 version D'Alembert argued that any algebraic function of a complex number takes values that are complex numbers, and so, on the grounds that a solution of a polynomial equation is such a function, a polynomial equation has complex solutions. Because the

[21] Quoted in (Gilain 1991, 111); trinomial (three-term) factors are quadratics.

Box 16. Complex solutions of real polynomials.
Let $p(x)$ be the polynomial

$$a_n x^n + a_{n-1}x^{n-1} + \cdots + a_1 x + a_0.$$

Let $\bar{p}(x)$ be the polynomial obtained from $p(x)$ by replacing each coefficient with its complex conjugate:

$$\bar{p}(x) = \bar{a}_n x^n + \bar{a}_{n-1}x^{n-1} + \cdots + \bar{a}_1 x + \bar{a}_0.$$

Whatever value x has, $p(x)$ is a real or complex number, and we denote its complex conjugate by $\overline{p(x)}$, so

$$\overline{p(x)} = \bar{p}(\bar{x}).$$

Now let $p(x)$ have real coefficients and let r be a solution of the equation $p(x) = 0$, so $p(r) = 0$. Then $\bar{r}$, the complex conjugate of r, is also a solution of $p(x) = 0$, because

$$p(r) = 0 \Rightarrow \overline{p(r)} = 0 \Rightarrow \bar{p}(\bar{r}) = 0 \Rightarrow p(\bar{r}) = 0;$$

the final deduction follows because $p(x)$ has real coefficients. So if a solution of a polynomial equation with real coefficients is complex, then its complex conjugate is another solution.

equation has real coefficients, it follows that if $a+b\sqrt{-1}$ is one solution then $a-b\sqrt{-1}$ is another (see Box 16). In 1746, presumably finding this account altogether too naive, D'Alembert tried to prove the existence of complex solutions directly.

D'Alembert considered a polynomial $p(x)$ with constant term a, so

$$p(x) = a + q(x).$$

To solve the equation he replaced a by the variable y to obtain the curve with equation

$$y + q(x) = 0.$$

He then applied the methods of the calculus and infinite series to write x as a power series in y, from which he deduced that, when y is small, x is necessarily complex. He deduced that x must be complex for any value of y, and so found a complex value for x when $y = a$.

Historians of mathematics generally consider this to be the first serious attempt at proving the Fundamental Theorem of Algebra. It certainly has serious weaknesses: there is no attempt to prove the convergence of the series for x, nor to see whether the claim is valid for all values of y or just some of them.

In 1746 Euler wrote his great memoir on the subject, which was eventually published in 1751 (E170). He now aimed to show directly that every solution of a polynomial equation is of the form $a + b\sqrt{-1}$, where a and b are real.

In the first part of his memoir, Euler argued that numbers of this form are closed under addition, subtraction, multiplication, division, and the extraction of roots. Moreover, he claimed that these are the only operations required to solve polynomial equations. He then attempted to show that every polynomial $p(x)$ can be factorised completely into linear terms.

More precisely, Euler first argued that every polynomial equation of odd degree has at least one real solution, because when x and $-x$ are very large the value of the polynomial is large and negative at one of them, and large and positive at the other. This being the case, the graph of the polynomial must cross the x-axis somewhere, and that value of x is a solution of the equation.

Euler also argued similarly that when the polynomial is of even degree with a negative constant term it has at least two real solutions, because it is positive when x is large, and negative when x is zero.

He next aimed to show that a polynomial of degree $2^k m$, where m is odd, can be reduced to one of degree $2^{k-1}m'$, where m' is odd. He directed attention to polynomials whose degree is exactly a power of 2, because other cases can be reduced to this one by simply multiplying by a suitable power of x. So he aimed to show that a polynomial of degree 2^k can be factorised into two polynomials of degree 2^{k-1}.

He first dealt with the cases where the polynomial is of degree 4. To solve such a polynomial equation, Euler first reduced it to the form

$$x^4 + bx^2 + cx + d = 0.$$

If this can be factorised, then it can be written as

$$(x^2 + ux + \lambda)(x^2 - ux + \mu) = 0.$$

Euler expanded this product and compared it with the given equation. This gave him four equations, one for each power of x, and by eliminating λ and μ from them he deduced that u must satisfy the equation

$$u^6 + 2bu^4 + (b^2 - 4d)u^2 - c^2 = 0.$$

This equation is a cubic in u^2 and so can be solved, but the method hints at a pattern upon which Euler was to rely when he got to the general case. Euler now observed that this equation for u can be solved according to the second of his basic assumptions (that the constant term, $-c^2$, is negative), so he could say that u was known. He concluded by showing that λ and μ can be found as rational functions of b, c, d, and u.

Euler also argued that u must be the sum of two of the solutions because, from the formal identity

$$x^4 + bx^2 + cx + d = (x^2 + ux + \lambda)(x^2 - ux + \mu) = (x - \alpha_1)(x - \alpha_2)(x - \alpha_3)(x - \alpha_4),$$

some expression of the form $u = \alpha_1 + \alpha_2$ must be true. So u takes on 6 values as the solutions are permuted, and therefore satisfies an equation of degree 6. In fact, the possibilities are

$$\begin{aligned} u_1 &= \alpha_1 + \alpha_2 = p = -u_4 = -(\alpha_3 + \alpha_4), \\ u_2 &= \alpha_1 + \alpha_3 = q = -u_5 = -(\alpha_2 + \alpha_4), \\ u_3 &= \alpha_1 + \alpha_4 = r = -u_6 = -(\alpha_2 + \alpha_3), \end{aligned}$$

and so the equation for u is

$$(u^2 - p^2)(u^2 - q^2)(u^2 - r^2) = 0.$$

This is indeed an equation of even degree with a non-positive constant term $-p^2q^2r^2$, and Euler checked that pqr is real. He showed in fact that

$$pqr = (\alpha_1 + \alpha_2)(\alpha_1 + \alpha_3)(\alpha_1 + \alpha_4),$$

which is invariant under all permutations of the αs, and can be expressed in terms of the coefficients of the original equation, and hence is real.

Euler then carried out a similar analysis of polynomial equations of degrees 8 and 16. The proof in the case of a polynomial equation of degree 2^k was only sketched by Euler. Whatever its defects, it carried some force because he had first done the equations of degrees 4, 8, and 16 explicitly. The passage from the computational argument to the conceptual one is much more plausible when it is seen to be true in the first few cases. His argument was a serious attempt to show *why* the computations work in these cases, and why they can therefore be trusted to work in general. The whole idea of explaining why a computation works, although necessary if the general computation cannot be carried out, is nonetheless an attractive one, and one that was to prove powerful.

Gauss's criticisms. There were several occasions in the 18th century when an approach opened up by Euler was made more rigorous by Lagrange, and the Fundamental Theorem of Algebra is one of these. Here, Lagrange set himself the task of completing Euler's proof of the Fundamental Theorem of Algebra.[22] The resulting argument is reasonably convincing, but it was left to the 20-year-old Carl Friedrich Gauss in his doctoral dissertation of 1797 to make the decisive criticism.[23] Gauss addressed it chiefly to D'Alembert's attempts, but, as he indicated elsewhere, it also applied to all the others.

The criticism that Gauss regarded as decisive was that D'Alembert assumed the existence of the solutions and showed only that they then had to be complex; he should first have proved that the solutions actually exist. To put the point another way, D'Alembert assumed that there were solutions, possibly hyper-complex ones (meaning that they could be manipulated like numbers but were not necessarily complex numbers), and showed that if the roots were 'hyper-complex' then they were necessarily complex. He did not consider other possibilities: that the solutions exist but could not be manipulated like numbers, or that the solutions did not even exist. That said, Gauss agreed that such problems could be overcome.

Gauss also criticised D'Alembert's use of infinite series in his proof, and showed by means of an example that it was unsound, again politely admitting that it was perhaps capable of being re-cast in a more reliable form. This criticism notwithstanding, Gauss admitted that 'the true strength of the proof sees to me not to have been weakened at all by all the objections', and he even stated that one could build a rigorous proof on that foundation.[24]

Gauss then turned to the second of Euler's arguments, which reduced the problem to the factorisation of polynomials whose degree is a power of 2. If factorisation can be assured, then the Fundamental Theorem of Algebra follows immediately by induction on the degree of the polynomial. Gauss first observed that Euler's approach tacitly assumes that polynomial equations do have solutions.

Gauss on the Fundamental Theorem of Algebra.

> Euler tacitly supposes that the equation $X = 0$ has $2m$ roots, of which he determines the sum to be $= 0$ because the second term in X is

[22]See (Lagrange 1772b).

[23]See (Gauss 1799) and the extracts in F&G 15.A2.

[24]See Gauss, *Werke* III, 11. For an evaluation of D'Alembert's proof and Gauss's criticism of it, see Gilain's introduction in (D'Alembert 1746, lxxvi–xcii), and for a re-evaluation of Lagrange's proof, see (Suzuki 2006).

> missing. What I think of this licence I have already declared in art. 3. The proposition that the sum of all the roots of an equation is equal to the first coefficient with the sign changed, does not seem applicable to other equations unless they have roots; now although it ought to be proved by this same demonstration that the equation $X = 0$ really does have roots, it does not seem permissible to suppose the existence of these.

He then predicted, and refuted, a likely reply:

> No doubt those people who have not yet penetrated the fallacy of this expression will reply, 'Here it has not been demonstrated that the equation $X = 0$ can be satisfied (for this expression means that the equation has roots) but it has only been demonstrated that the equation can be satisfied by values of x of the form $a + b\sqrt{-1}$; and indeed that is taken as axiomatic'. But although types of quantities other than real and imaginary $a+b\sqrt{-1}$ cannot be conceived of, it does not seem sufficiently clear how the proposition awaiting demonstration differs from that supposed as axiomatic. ... Therefore that axiom can have no other meaning than this: Any equation can be satisfied either by the real value of an unknown, or by an imaginary value expressed in the form $a+b\sqrt{-1}$, or perhaps by a value in some other form which we do not know, or by a value which is not totally contained in any form. But how such quantities which are shadowy and inconceivable can be added or multiplied is certainly not understood with the clarity which is required in mathematics.

In other words, it must be shown that a polynomial equation has either real or complex solutions, and no solutions of any other kind. This seems to be the first time that the possibility of numbers of a new kind was raised, if only to be refuted.

Gauss then remarked that Lagrange had thoroughly resolved some of the objections to Euler's argument, but that gaps in the proof remained — notably the assumptions that polynomial equations have solutions and that the only problem is to show that they are complex numbers. He wrote:

> Finally, Lagrange has dealt with our theorem in the commentary *Sur la Forme des Racines Imaginaires des Equations*, 1772. This great geometer handed his work to the printers when he was worn out with completing Euler's first demonstration ... However, he does not touch upon the third objection at all, for all his investigation is built upon the supposition that an equation of the mth degree does in fact have roots.

We return to the history of the Fundamental Theorem of Algebra in Chapter 18.

7.4 Further reading

Calinger, R. 2016. *Leonhard Euler: Mathematical Genius in the Enlightenment*, Princeton University Press. Likely to become the standard biography of Euler, with full discussions of many aspects of his work.

Dunham, W. 1999. *Euler: The Master of Us All*, Dolciani Mathematical Expositions, 22, Mathematical Association of America. A good short introduction to Euler and many of the topics that we also touch on, that assumes a certain amount of mathematical knowledge.

Euler, L. 1988, 1990. *Introduction to the Analysis of the Infinite*, transl. J. D. Blanton, two vols., Springer. The first volume is on the calculus, the second on geometry, and in these good translations they show just what a lucid, and patient, communicator Euler was.

Fellmann, E.A. 2007. *Leonhard Euler*, Birkhäuser. This book covers a great deal of the life and remarkable work of Euler in less than 200 pages, an achievement that only someone who had worked for years in the Euler Archives in Basel could have made.

Stedall, J. 2011. *From Cardano's* Great Art *to Lagrange's* Reflections*: Filling a Gap in the History of Algebra*, European Mathematical Society. This book looks at many advances between the work of Cardano and Lagrange that are generally neglected in the wake of Lagrange's somewhat dismissive remarks.

The Euler tercentenary in 2007 produced a number of books about him, his work, and its implications to the present day. We note here the five-volume set produced by the Mathematical Association of America, which is particularly accessible, and contains a wide variety of material.

Vol. 1. Sandifer, C.E. (ed.). *The Early Mathematics of Euler.*

Vol. 2. Dunham, W. (ed.). *The Genius of Euler: Reflections on His Life and Work.*

Vol. 3. Sandifer, C.E. (ed.). *How Euler did it.*

Vol. 4. Bogolyubov, N.N., Mikhailov, G.K., and Yushkevich, A.P. (eds.). *Euler and Modern Science.*

Vol. 5. Bradley, R.E., D'Antonio, L.A., and Sandifer, C.E. (eds.). *Euler at 300: An Appreciation.*

A further volume appeared in 2015, with short articles on Euler's work in a wide variety of areas:

Sandifer, C.E. (ed.). *How Euler did Even More*, Mathematical Association of America, Washington.

8

18th-century Number Theory and Geometry

Introduction

It is possible to regard the 18th century as the century of algebra, not just because of the successes that mathematicians achieved in that field but also because it entered more and more into the foundations of the subject. Here we look at its uses in number theory, a subject that Euler and Lagrange did much to revive, and in geometry, where algebra grew from a useful method to become the dominant partner.

8.1 Number theory

It was Euler, and Lagrange after him, who brought number theory into the mainstream of mathematics, where it has remained ever since. Euler's involvement drew on ancient sources, as well as the writings of Fermat who had often written about problems involving numbers but had largely failed to interest his contemporaries.[1]

One claim by Fermat, inspired by noticing that the first five numbers in the sequence

$$2^1 + 1 = 3,\ 2^2 + 1 = 5,\ 2^4 + 1 = 17,\ 2^8 + 1 = 257,\ 2^{16} + 1 = 65,537,\ \ldots,$$

are prime numbers, was that all such numbers, where the exponents are themselves powers of 2, are prime. In December 1729, Euler received a letter from Christian Goldbach, who challenged him to find out whether Fermat's claim that all such numbers of the form $2^{2^n} + 1$ are prime. Euler replied in January to say that he had nothing to add to Fermat's discovery, but some time later he found an ingenious argument to prove that the next Fermat number

$$2^{32} + 1 = 4{,}294{,}967{,}297$$

[1] Fermat's work in number theory is discussed in *Volume* 1, Chapter 11.

is not prime, but is divisible by 641. Since then, no other 'Fermat number' has been shown to be prime, so Fermat's conjecture was particularly unfortunate. Euler presented his discovery to the St Petersburg Academy in 1732, without indicating how he found it. This paper (E26) was published only in 1738, and Euler published a proof only in 1747/48 (E134), which shows how slowly publications were being handled at the time.

Fermat had also claimed that if a prime number p does not divide an integer a, then p must divide $a^{p-1} - 1$; for example, the prime number $p = 5$ does not divide the integer $a = 6$, but 5 does divide $6^{5-1} - 1 = 6^4 - 1 = 1295$. But he never published a proof of the result, which is often called Fermat's 'little theorem'. Euler listed it in E26 in a series of results that he could not prove, and then gave a proof in 1736 that relies on little more than the binomial expansion of $(1 + a)^p$ and induction on a, which he later replaced with a simpler proof.[2]

Perfect numbers. Certain numbers, such as 6 and 28, have the property that the sum of their factors (excluding the number itself), equals the original number. For example, the factors of 6 are 1, 2, and 3, and $1 + 2 + 3 = 6$, and the factors of 28 are 1, 2, 4, 7 and 14, and $1 + 2 + 4 + 7 + 14 = 28$. Numbers with this property were held to have special religious or mystic significance, and were considered 'perfect': thus, a number is *perfect* if it is the sum of its factors (other than the number itself). After 6 and 28, the next perfect numbers are 496 and 8128, and then there are no more until 33,550,336.

The study of perfect numbers can be traced back to Greek times, and in Book IX, Proposition 36, of the *Elements*, Euclid had proved that whenever $2^n - 1$ is a prime number, the number $2^{n-1}(2^n - 1)$ must be perfect;[3] the five perfect numbers above correspond to $n = 2, 3, 5, 7, 13$. If $2^n - 1$ is not prime, then $2^{n-1}(2^n - 1)$ is not perfect; for example, $2^4 - 1 = 15$ is not prime, and $2^3(2^4 - 1) = 8.15 = 120$ is not perfect.[4]

The search for perfect numbers led Mersenne to investigate when numbers of the form $2^n - 1$ are prime.[5] These numbers, when prime, have become known as *Mersenne primes*.[6]

When studying perfect numbers in the 1740s, Euler observed that all known perfect numbers are even, and he wondered whether this was the case for all perfect numbers. He was unable to prove this, observing in a paper written in 1747, but published only posthumously in 1849, that 'whether ... there are any odd perfect numbers is a most difficult question', but he was at least able to prove that all *even* perfect numbers must have the form $2^{n-1}(2^n - 1)$, where $2^n - 1$ is prime.[7]

[2]The earlier paper, E54, was published only in 1741. In 1758 Euler (E271) gave his simpler proof and generalised the result by showing that, for any integers n and a, the number $a^{\varphi(n)} - 1$ is always divisible by n, where $\varphi(n)$ is the number of integers less than n that have no factors in common with n.

[3]See *Volume* 1, Chapter 5.

[4]We note that if n is composite then $2^n - 1$ is composite, but if n is prime it does not follow that $2^n - 1$ is prime. For example, $2^{11} - 1 = 2047 = 23 \times 89$.

[5]See *Volume* 1, Chapter 11.

[6]Currently, the largest currently known (October 2019) is $2^{82,589,933} - 1$; it is not known whether there are infinitely many Mersenne primes. For an extract from Mersenne on these primes, see (Stedall 2008, 158–159).

[7]See (Euler 1849, 88), E798. We still do not know whether there exist any odd perfect numbers, but it is known that if they do they must be huge, indeed greater than 10^{1500}, which is a number greater than the number of atoms in the visible universe.

Fermat's last theorem. Fermat had managed to prove, using his 'method of infinite descent', that there are no positive integers a, b, c that satisfy $a^4 + b^4 = c^4$, and rashly wrote in the margin of his copy of Bachet's edition of Diophantus's *Arithmetica* that he had a marvellous proof that for no integer $n > 2$ are there positive integers a, b, c that satisfy $a^n + b^n = c^n$, but the margin was too small to contain it. It is unlikely that he had a proof, but the claim eventually grew into Fermat's famous 'last theorem'.[8]

A charming indication of how Euler viewed his work, his aims, and his partial achievements, can be gleaned from a letter describing his success with $x^3 + y^3 = z^3$ that he wrote to Goldbach on 4 August 1753.[9]

Euler on a case of Fermat's last theorem.

> There's another very lovely theorem in Fermat whose proof he says he has found. Namely, on being prompted by the problem in Diophantus, find two squares whose sum is a square, he says that it is impossible to find two cubes whose sum is a cube, and two fourth powers whose sum is a fourth power, and more generally that this formula $a^n+b^n = c^n$ is impossible when $n > 2$. Now I have found valid proofs that $a^3 + b^3 \neq c^3$ and $a^4 + b^4 \neq c^4$, where $\neq$ denotes cannot equal. But the proofs in the two cases are so different from one another that I do not see any possibility of deriving a general proof from them that $a^n + b^n \neq c^n$ if $n > 2$. Yet one sees quite clearly as if through a veil that the larger n is, the more impossible the formula must be. Meanwhile I haven't yet been able to prove that the sum of two fifth powers cannot be a fifth power. To all appearances the proof just depends on a brainwave, and until one has it all one's thinking might as well be in vain. But since the equation $aa + bb = cc$ is possible, and so also is this possible, $a^3 + b^3 + c^3 = d^3$, it seems to follow that this, $a^4 + b^4 + c^4 + d^4 = e^4$, is possible, but up till now I have been able to find no case of it. But there can be five specified fourth powers whose sum is a fourth power.

We give two examples to illustrate Euler's observations:[10]

$$3^3 + 4^3 + 5^3 = 6^3 \text{ and } 30^4 + 120^4 + 272^4 + 315^4 = 353^4.$$

It is clear from Euler's comments that he was dissatisfied. Although he had found a proof, indeed one by infinite descent, it seemed to him that he would not be able to solve the general case ($x^n + y^n = z^n$, for any $n > 2$) in this way, so we see that he had both a specific and a general aim in mind.

Euler's claim, that he had proved Fermat's Last Theorem in the case $n = 3$, rested on an argument that he did not publish until his *Algebra* appeared in 1770.[11] A brief look at this ultimately inconclusive argument is sketched in Box 17 and brings out some important features of Euler's number theory.

Perhaps the most interesting thing about Euler's argument — which is more important than the 'proof' of Fermat's Last Theorem for $n = 3$ — is that Euler extended

---

[8]See *Volume* 1, Chapter 2, Section 2. The theorem was eventually proved, see (Wiles 1995) and (Wiles and Taylor 1995).

[9]This is letter 169 in the Euler–Goldbach correspondence; see (Euler 2015) for a different translation.

[10]The sum of the fourth powers was found only in 1911.

[11]See Euler, *Algebra*, Part II, Chapter XV, §243.

Box 17. Euler on $x^3 + y^3 = z^3$.
Euler first observed that if $x^3 + y^3 = z^3$, then the integers x, y, and z cannot all be even, for then they would have a common factor, and we can assume without loss of generality that they do not. Nor can they all be odd, because the sum of two odd numbers is even. So exactly one of x, y, z is even, and Euler assumed that z is even. There is no loss of generality here, provided that x, y, and z are allowed to be negative.

So Euler assumed that $x + y = 2m$, say, and $x - y = 2r$, say. It follows that $x^3 + y^3 = 2m(m^2 + 3r^2)$ is a cube. Therefore, said Euler, either $2m$ and $m^2 + 3r^2$ are both cubes, or they are not, and in the latter case m must be a multiple of 3, because this is the only way that $2m$ and $m^2 + 3r^2$ are not relatively prime.

Euler was able to show that this last possibility cannot happen, and he next considered how a number of the form $m^2 + 3r^2$ can be a cube. Inspired by the identity

$$(a^2 + 3b^2)^3 = (a^3 - 9ab^2)^2 + 3(3a^2b - 3b^3)^2,$$

which says that a cube of a number of the form $m^2 + 3r^2$ is itself of that form, Euler claimed that if a number of the form m^2+3r^2 is a cube, then it is a cube of a number of that form — that is, there are integers a and b such that $(a^2+3b^2)^3 = m^2 + 3r^2$.

Euler now appealed to Fermat's method of infinite descent. The numbers $u = a - 3b$, $v = a + 3b$, and $w = 2a$ are cubes with the appropriate sum, $u + v = w$. They can be shown to be relatively prime (for if they were not, then m and n would not be) and $w = 2a < z^3$. So the descent can begin, and the contradiction is immediate.

Euler believed that this showed that Fermat's last theorem follows in the case $n = 3$, and indeed it would do, if one could be certain that there are integers a and b for which $(a^2 + 3b^2)^3 = m^2 + 3r^2$. Euler argued that one can factorise $m^2 + 3r^2$ as

$$m^2 + 3r^2 = (m + r\sqrt{-3})(m - r\sqrt{-3}),$$

and claimed that if $m \pm r\sqrt{-3}$ is a cube then it is cube of a number of the same form $a \pm b\sqrt{-3}$. This is true, as it turns out, but not for the reasons that Euler gave. In his *Algebra* (1770, Part II, §182) he claimed that $x^2 + cy^2$ could always be regarded as a square, and this is false when $c = 5$ and in many other cases. So there is a difference between $x^2 + 3y^2$ and $x^2 + 5y^2$, which Euler was unable to explain.

his reasoning about integer numbers to new numbers of the form $x + y\sqrt{-c}$. Euler boldly proposed to discuss prime, relatively prime, square, and cube numbers of this kind, treating then as if they were integers, and presuming that concepts such as 'prime' would similarly apply. He did so when imaginary quantities were still a source of controversy in mathematics.

Sums of two squares. Yet another subject that Fermat had broached and that Euler took up to lasting effect is called the 'theory of quadratic forms'. It began with the observation, known, it is said, to Plato, that all the integer solutions of $x^2+y^2=z^2$ are of the form $x = p^2 - q^2, y = 2pq, z = p^2 + q^2$ (or multiples of these). This had led mathematicians to ask: Which non-zero numbers are the sum of two squares? The smallest answers are $1, 2, 4, 5, 8, 9, 10, 13, \ldots$ — note that we allow one of x or y to be zero. It was soon seen that these numbers are either 1, 2, squares, primes of the form $4n + 1$, or products of numbers of these kinds.

It is also clear that no number of the form $4n + 3$ can be the sum of two squares. For, each of the two squares is the square of either an even number or an odd number. An even number has, by definition, the form $2k$ so its square is of the form $4k^2$, and an odd number has, by definition, the form $2k+1$ so its square is of the form $4k^2+4k+1$. Sums of two such numbers are therefore of the form $4n$, $4n + 1$, or $4n + 2$, but never $4n + 3$.

Fermat was the first to claim that every prime number of the form $4n+1$ is uniquely a sum of two squares; for example, $41 = 4^2+5^2$. He mentioned it in a letter to Mersenne in 1640, where he said that his proof used his method of infinite descent; but no proof of his survives.

Euler took up this problem a century later, in 1747, when he wrote to Goldbach about it.[12] The identity

$$(x^2 + y^2)(u^2 + v^2) = (xu + yv)^2 + (xv - yu)^2$$

shows that a product of two numbers, each of which is a sum of two squares, is itself a sum of two squares. So Euler looked at the *divisors* of numbers that are a sum of two squares, and attempted to show that these divisors are also of the same form. In his letter, he presented a proof that every prime divisor of a sum of two squares is itself a sum of two squares. His argument relied on the method of infinite descent.

It remained to show that if a prime p is of the form $4n + 1$ then there are integers x and y such that $x^2 + y^2 = p$, but it was only in 1749 that Euler could conclude the argument, writing to Goldbach on 12 April 'Now I have finally found a valid proof'.[13]

We can see this using modern notation.[14] By Fermat's little theorem, any two integers x and y satisfy

$$x^{p-1} \equiv y^{p-1} \pmod{p},$$

and so, in the present case,

$$x^{4n} \equiv y^{4n} \pmod{p}.$$

But

$$x^{4n} - y^{4n} \equiv (x^{2n} + y^{2n})(x^{2n} - y^{2n}) \equiv 0 \pmod{p},$$

so if there are values of x and y for which

$$x^{2n} - y^{2n} \not\equiv 0 \pmod{p}$$

then they must satisfy

$$x^{2n} + y^{2n} \equiv 0 \pmod{p},$$

and this shows that p divides a sum of two squares, since $x^{2n} + y^{2n} = (x^n)^2 + (y^n)^2$. This shows that p is a sum of two squares.

---

[12] See (Euler 2015, nr. 115); the letter was written on 6 May 1747.

[13] See (Euler 2015, nr. 138); the letter was also published in (Fuss 1843, 493).

[14] See Box 55 in Chapter 18.

Today the existence of the above x and y is immediate, because mathematicians after Euler developed a theory of polynomial equations modulo a prime number. But for Euler it required a novel argument involving finite differences, which we omit.

Fermat had also claimed — once again, without any proofs surviving — that he could tackle new problems that were to enlarge the field of enquiry considerably. He had claimed, for example, that prime divisors of numbers of the form $x^2 + 2y^2$ are also of the form $x^2 + 2y^2$, and that prime divisors of numbers of the form $x^2 + 3y^2$ are also of the form $x^2 + 3y^2$. It took Euler about 30 years to establish these claims, and to characterise the primes of these forms. He found that an odd prime is of the form $x^2 + 2y^2$ if and only if it is of the form $8n + 1$ or $8n + 3$; and a prime is of the form $x^2 + 3y^2$ if and only if it is of the form $6n + 1$.

But Fermat had also come up against a problem that he could not solve, and this was to prove much more interesting. He saw that with $x^2 + 5y^2$, something unexpected happens. The identity

$$(x^2 + 5y^2)(u^2 + 5v^2) = (xu + 5yv)^2 + 5(xv - yu)^2$$

shows that the product of two numbers of the form $x^2 + 5y^2$ is again of this form. But it is not true that every divisor of a number of this form must also be of this form. For example,

$$21 = 1^2 + 5.2^2 = 3.7, \quad \text{and} \quad 161 = 6^2 + 5.5^2 = 7.23,$$

but 3, 7, and 23 are not of the required form. Fermat conjectured, but admitted that he could not prove, that primes that are congruent to 3 or 7 (modulo 20) do have products that are of the form $x^2 + 5y^2$. Primes of the form $x^2 + 5y^2$ other than 5 are all of the form $20n + 1$ or $20n + 9$, but Fermat seems not to have noticed this.[15]

This puzzle was to be elucidated by Lagrange, as we now discuss.

Lagrange's work on quadratic forms. If Euler's work often had the character of a deep but informal exploration and opening-up of a subject, then Lagrange's represents the next stage: the systematic study and rigorous development of the main ideas.[16] Such was to be the case with his study of quadratic forms. In Lagrange's theory, a quadratic form is an expression of the form $ax^2 + 2bxy + cy^2$, where a, b, c are integers and the *discriminant* $\Delta = b^2 - ac$ is not a square (because otherwise the form would factorise as the product of two linear terms).

Lagrange addressed Fermat's problem with numbers of the form $x^2 + 5y^2$, which has discriminant $\Delta = 0^2 - 1.5 = -5$, by introducing a second quadratic form $2x^2 + 2xy + 3y^2$ which has the same discriminant, $\Delta = 1^2 - 2.3 = -5$. He then showed that these are essentially the only two quadratic forms with this discriminant, using a definition of when two forms represent the same collection of numbers, that we explain in Box 18.

More precisely, Lagrange showed that any quadratic form with discriminant $\Delta = -5$ is equivalent to one of the two inequivalent forms $x^2 + 5y^2$ and $2x^2 + 2xy + 3y^2$. Each of these forms represents integers that the other one does not. For example, $2x^2 + 2xy + 3y^2$ represents 3 when $x = 0, y = 1$, and 7 when $x = 1, y = 1$. So the anomalous behaviour of primes of the form $x^2 + 5y^2$ is explained by observing that it is not the

[15] See Fermat, *Oeuvres*, Vol. 2, p. 432.

[16] See (Lagrange 1773/1775).

Box 18. Equivalence of forms
Lagrange took a quadratic form $ax^2 + 2bxy + cy^2$ and considered changes of variable, such as

$$x' = \alpha x + \beta y, \quad y' = \gamma x + \delta y,$$

where $\alpha, \beta, \gamma, \delta$ are integers, and $\alpha\delta - \beta\gamma = 1$.

Suppose this transforms the quadratic form into a new quadratic form $a'x'^2 + 2b'x'y' + c'y'^2$. The transformation is invertible, because

$$x = \delta x' - \beta y', \quad y = -\gamma x' + \alpha y'.$$

Therefore any integer represented by the form $ax^2 + 2bxy + cy^2$ is also represented by the corresponding form in x' and y', and vice versa. For, if x_0 and y_0 are such that

$$ax_0^2 + 2bx_0y_0 + cy_0^2 = m,$$

then $x' = \alpha x_0 + \beta y_0$ and $y' = \gamma x_0 + \delta y_0$ satisfy

$$a'x'^2 + 2b'x'y' + c'y'^2 = m.$$

Lagrange decided that these two forms should be regarded as equivalent.

only quadratic form of its kind. In fact, as you can check, setting $b^2 - ac = \Delta$,

$$(ax^2 + 2bxy + cy^2)(au^2 + 2buv + cv^2) = (axu + xbv + byu + cyv)^2 - \Delta(xv - yu)^2,$$

which goes some way to explaining Fermat's observations when $\Delta = -5$.

Lagrange's observation opened the way to the study of all quadratic forms of the general form $ax^2 + 2bxy + cy^2$. In this broader setting he showed that, for an odd number m to be represented by a form $ax^2+2bxy+cy^2$ with a given value of $\Delta = b^2-ac$, it is necessary that Δ is a square (modulo m).

This result shows that progress with the problem of deciding which numbers m are represented by a quadratic form $ax^2+2bxy+cy^2$ with a given discriminant $\Delta = b^2 - ac$ depends on being able to decide which numbers are squares (modulo m). This was a difficult question that neither Lagrange nor Euler was able to solve, although Euler had written to Goldbach as early as 1742 with an impressive array of evidence that pointed to the correct theorem.[17] This and other open questions raised by the work of Euler and Lagrange were to be a powerful stimulus to the work of Gauss in the next generation.

8.2 Infinite series

Euler's work in number theory led him into related areas that Fermat had not looked at, such as the summation of infinite series.

---

[17]See (Fuss 1843), Vol. 1, pp. 144–153, and (Edwards 1983).

Box 19. The harmonic series has no finite sum.

$$1 + 1/2 + (1/3 + 1/4) + (1/5 + 1/6 + 1/7 + 1/8) + (1/9 + 1/10 + \cdots + 1/16) + \cdots$$
$$> 1 + 1/2 + 1/2 + 1/2 + 1/2 + \cdots$$

because each of the bracketed terms is greater than $1/2$. So the successive partial sums eventually exceed any finite amount, and therefore the harmonic series does not converge.

By Euler's time it was well known that some infinite series have finite sums. For example, the infinite series

$$1 + 1/2 + 1/4 + 1/8 + 1/16 + \ldots$$

converges to 2, in the sense that we can get as close to 2 as we wish by adding together enough terms of the above series. In the same way, the sum of

$$1 + 1/3 + 1/9 + 1/27 + \ldots,$$

where each term is one-third of the previous one, is $1\frac{1}{2}$. In fact, for any number $n > 1$, the sum of

$$1 + 1/n + 1/n^2 + 1/n^3 + \ldots$$

is $n/(n-1)$; we shall need this result later.

It was also known that some series cannot be summed. It had been realised in the 14th century that the *harmonic* series

$$1 + 1/2 + 1/3 + 1/4 + 1/5 + \cdots$$

has no finite sum, because an arbitrarily large number can be exceeded (such as one million) by adding together enough terms of the series (see Box 19).

Euler was particularly fascinated by the harmonic series and series similar to it, such as the sum of the reciprocals of the squares. The problem of finding the sum of the reciprocals of the perfect squares,

$$1 + (1/2)^2 + (1/3)^2 + (1/4)^2 + (1/5)^2 + \cdots = 1 + 1/4 + 1/9 + 1/16 + 1/25 + \cdots,$$

exercised many minds around the mid-1730s. It was already an old problem; it seems to have been posed first by the Italian mathematician and clergyman Pietro Mengoli in 1644. When it defeated Jakob Bernoulli in Basel in 1689 he communicated it to the mathematical community as an important challenge, after which the problem became known as the *Basel problem*.

Simple calculations like the one used to handle the harmonic series show that the sum of the reciprocals of the squares lies between 1 and 2, and by summing the first few terms, we can begin to suspect that this series converges to a number that is a little over 1.64. Indeed, in one of his earliest papers, Euler showed that the sum is approximately 1.644934.[18]

[18]See (Euler 1731/1738, 33, E20). Later Euler calculated this sum to 20 decimal places.

Box 20. Solving the Basel problem.
Euler used the usual power series expression for $\sin x$, noting that the coefficient of x^3 is $-1/6$:

$$\sin x = x - x^3/3! + x^5/5! - x^7/7! + \cdots.$$

Now a polynomial equation such as $1 - x^2/4 = 0$, with solutions $x = 2$ and $x = -2$, can be factorised as

$$1 - x^2/4 = (1 - x/2)(1 + x/2).$$

Similarly (Euler claimed), since $\sin x = 0$ when

$$x = 0,\ \pi \text{ or } -\pi,\ 2\pi \text{ or } -2\pi,\ 3\pi \text{ or } -3\pi,\ \ldots,$$

that is, when

$$x \text{ or } (1 - x/\pi) \text{ or } (1 + x/\pi) \text{ or } (1 - x/2\pi) \text{ or } (1 + x/2\pi) \text{ or } \ldots = 0,$$

we can then 'factorise' $\sin x$ as follows:

$$\sin x = x(1 - x/\pi)(1 + x/\pi)(1 - x/2\pi)(1 + x/2\pi)(1 - x/3\pi)(1 + x/3\pi)\cdots.$$

Combining terms in pairs gives

$$\sin x = x(1 - x^2/\pi^2)(1 - x^2/4\pi^2)(1 - x^2/9\pi^2)\cdots.$$

We now find the coefficient of x^3 in this expression. This is formed from the x term, together with each of the single terms $-x^2/k^2\pi^2$ in turn when $k = 1, 2, 3, \ldots$, the remaining terms (from the other brackets) all being 1. This gives

$$-1/\pi^2 - 1/(4\pi^2) - 1/(9\pi^2) - 1/(16\pi^2) - \cdots.$$

But the coefficient also equals $-1/6$ (from the series for $\sin x$ given earlier).

Equating these gives

$$-1/6 = -1/\pi^2(1 + 1/4 + 1/9 + 1/16 + 1/25 + \cdots),$$

from which it follows that

$$1 + 1/4 + 1/9 + 1/16 + 1/25 + \cdots = \frac{\pi^2}{6},$$

as required.

Can the exact sum of this series be found? One of Euler's earliest achievements, in 1734, was to show (initially by somewhat dubious means) that this sum is $\pi^2/6$, and this brought him international fame; his original method is outlined in Box 20.[19] The method consists of 'factorising' $\sin x$ and looking at the coefficient of x^3.

By similarly considering the coefficients of x^5 and x^7, Euler extended his calculations to find the sum of the reciprocals of the 4th powers and the 6th powers, obtaining the results

$$1 + (1/2)^4 + (1/3)^4 + (1/4)^4 + \cdots = 1 + 1/16 + 1/81 + 1/256 + \cdots = \frac{\pi^4}{90},$$

[19] See (Euler 1735/1740b, E41) for the original; there is an English translation in the Euler Archive. For instructive and more detailed accounts than we can provide here, see (Dunham 1999) and (Sandifer 2007, 157–165).

and

$$1 + (1/2)^6 + (1/3)^6 + (1/4)^6 + \cdots = 1 + 1/64 + 1/729 + 1/4096 + \cdots = \frac{\pi^6}{945}.$$

He then continued in this way, concluding with the sum of the reciprocals of the 12th powers, and later went on to the 26th powers, obtaining the answer:

$$1 + (1/2)^{26} + (1/3)^{26} + (1/4)^{26} + \cdots = \frac{691}{6825 \times 93555}\pi^{26},$$

which is indeed correct!

Euler's way of treating the infinite series for sine as though it were a finite polynomial expression requires more justification than he gave it. It seems to have worried him too, for, as the historian Ed Sandifer has pointed out, Euler later gave a much more rigorous account (E63) that was published in 1743 in an obscure French journal. His method involves little more than a clever use of the formula for the arc length of the circle and term-by-term integration of an infinite series.[20]

Another important result of Euler's, published in his paper (1734/1740a, E43), is that the sum of the first n terms of the harmonic series

$$1 + 1/2 + 1/3 + 1/4 + 1/5 + \cdots + 1/n$$

is exceedingly close to the natural logarithm of n. In fact, as n becomes large, their difference

$$(1 + 1/2 + 1/3 + 1/4 + 1/5 + \cdots + 1/n) - \log n$$

approaches a fixed number, which is about 0.5772; this number is now known as 'Euler's constant', or the 'Euler–Mascheroni constant'. Little is known about it — we do not even know for certain whether it is an irrational number, although this seems likely.

Euler continued his investigations, and in (1737/1744, E72) and again in his *Introductio in Analysin Infinitorum* (1748) he observed that there was what has become known as a product formula. The later account is clearer. Euler began by observing that if one expands an infinite product of the form

$$(1 + \alpha x)(1 + \beta x)(1 + \gamma x)\cdots$$

as the infinite series

$$1 + Ax + Bx^2 + Cx^3 + \cdots$$

then $A = \alpha + \beta + \gamma + \cdots$ is the sum of the individual numbers in the product, B is the sum of pairs $\alpha\beta + \alpha\gamma + \beta\gamma + \cdots$, C is the sum of triples, and so on.

The quotient

$$\frac{1}{(1 - \alpha x)(1 - \beta x)(1 - \gamma x)\cdots} = 1 + Ax + Bx^2 + Cx^3 + \cdots$$

is particularly interesting. When each term $1/(1 - kx)$ is expanded by the binomial theorem and the results are multiplied together, then once again A is the sum of the numbers $\alpha, \beta, \gamma, \ldots$ taken singly, B is now the sum of the products taken two at a time (squares included), C is the sum of the products taken three at a time (repeats included), and so on.

[20]We shall not describe this proof of Euler's here — see (Sandifer 2007, 209–211) — because it does not seem to have had much impact in its day.

Euler now took $x = 1$ and $\alpha = 1/2, \beta = 1/3, \gamma = 1/5, \ldots$ — the reciprocals of the primes — and found that

$$\frac{1}{(1-1/2)(1-1/3)(1-1/5)\cdots} = 1 + 1/2 + 1/3 + 1/4 + \cdots,$$

which diverges. He next remarked that the same expansion works when any power of a prime is used, and wrote

$$\frac{1}{(1-(1/2)^n)(1-(1/3)^n)(1-(1/5)^n)\cdots} = 1 + 1/2^n + 1/3^n + 1/4^n + \cdots,$$

'where all natural numbers occur without exception'.[21]

In his 1737 paper he proved the amazing result that, even if we throw away most of the terms of the harmonic series and keep only those that correspond to prime numbers,

$$1 + 1/2 + 1/3 + 1/5 + 1/7 + 1/11 + \ldots,$$

then there is still no finite sum: adding the reciprocals of the primes gives a divergent series. Although Euler did not say so explicitly, this result implies that there are infinitely many primes, and this theorem is sometimes referred to as Euler's proof of the infinitude of primes.

In these remarkable papers Euler brought together for the first time two seemingly unrelated topics: infinite series and prime numbers. This is the basis for the subject now known as analytic number theory.

8.3 Euler and geometry

Among the geometrical topics that Euler investigated were polyhedra. In a letter to Christian Goldbach, written partly in German and partly in Latin and dated 14 November 1750 (see Figure 8.1), he observed that the numbers of vertices S (angulae solidae, or solid angles, in Latin), H faces (hedrae), and A edges (acies) seem to be related by the simple formula:[22]

$$H + S = A + 2.$$

For example,

- a cube has 6 faces, 8 vertices, and 12 edges, and $6 + 8 = 12 + 2$
- a triangular prism has 5 faces, 6 vertices, and 9 edges, and $5 + 6 = 9 + 2$.

At first, Euler was unable to obtain a proof of his formula, but a year later he produced a dissection argument that involved slicing out tetrahedral pieces. Unfortunately, his proof was deficient, and a complete proof was not given until more than forty years later, in 1794, by the analyst and number-theorist Adrien-Marie Legendre.

[21] This result is known today as Euler's product formula. It was generalised by Riemann to the case where the positive integer n is replaced by a complex number s. The resulting function of s is called Riemann's zeta function, and it is the gateway to deep properties of prime numbers, as we briefly discuss in Section 18.2. As Sandifer remarked (2007, 256), Euler, unlike Riemann, never thought of the expressions as functions of the exponent.

[22] For a reprint of this letter, see the Euler–Goldbach correspondence (Euler 2015, nr. 149), and for a translation and commentary, (Biggs, Lloyd, and Wilson 1998, 74–89). This formula is sometimes incorrectly credited to Descartes, who did not have the terminology or motivation to derive it: indeed, it was Euler who first introduced the concept of an edge.

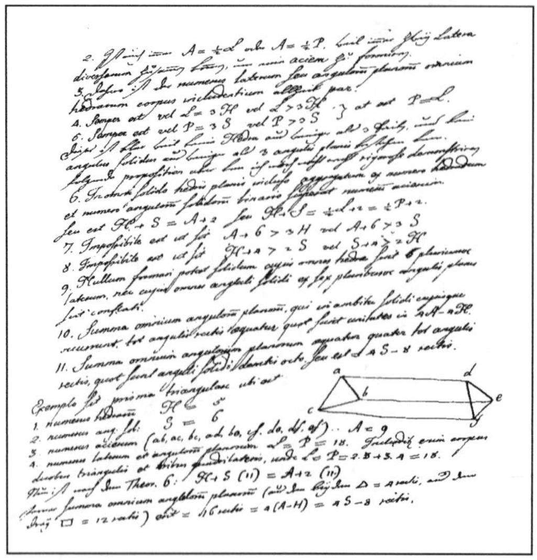

Figure 8.1. Euler's letter to Goldbach, 14 November 1750

The geometry of position. In September 1679, Leibniz had written to Christiaan Huygens, remarking:[23]

> I am still not satisfied with algebra, because it does not give the shortest proofs or the most beautiful constructions in geometry. That is why I believe that, so far as geometry is concerned, we need still another analysis which is distinctly geometrical or linear and which will express *situation* [situs] directly, as algebra expresses *magnitude* directly.

It is not clear what Leibniz had in mind, but it was interpreted by some, including Euler, as referring to a type of geometry in which metrical ideas, such as distance, length, and angle, do not arise. This subject is now known as topology.

On 26 August 1735 Euler presented to his colleagues at the Academy of Sciences at St Petersburg his solution of 'a problem relating to the geometry of position'. It concerned the medieval Prussian city of Königsberg which was divided into four areas by the river Pregel. Figure 8.2, taken from a 17th-century map of the city, shows the four areas and the seven bridges connecting them; Euler's own diagram (see Figure 8.3) is clearer because it ignores the strictly geometrical features of length and angle.[24] The problem was: To go for a connected walk around the city, crossing each of the seven bridges only once. Euler proved that the task is impossible, and showed how his arguments can be extended to any problem of a similar type.

Euler seems to have been intrigued by the fact that the Königsberg problem apparently belonged to the geometry of position (as he interpreted it), and on 13 March 1736 he wrote to Giovanni Marinoni, Court Astronomer in the court of Kaiser Leopold I in Vienna, in the following words:[25]

> This question is so banal, but seemed to me worthy of attention in that geometry, nor algebra, nor even the art of counting was sufficient to solve it. In view of this, it occurred to me to wonder whether it belonged to the geometry of position, which Leibniz had

[23] See (Leibniz 1969, 248–249).

[24] See (Euler 1736/41, 128, E53).

[25] Euler, *Opera Omnia* (4), Vol. 1, nr. 1468.

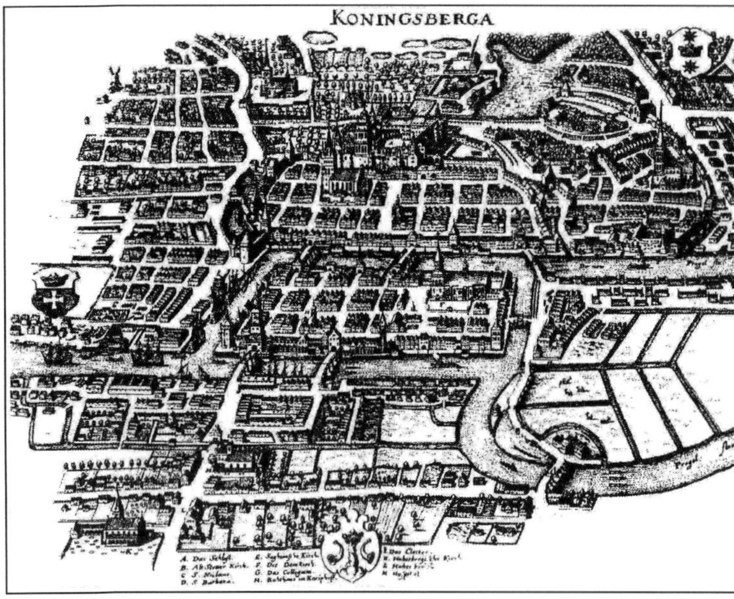

Figure 8.2. A map of the city of Königsberg.

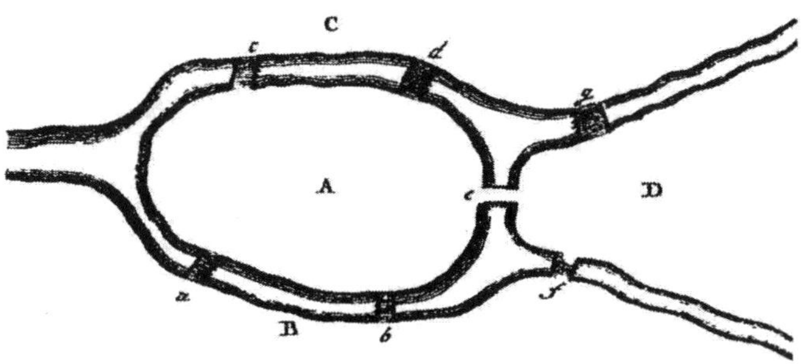

Figure 8.3. Euler's map of the seven bridges.

> once so much longed for. And so, after some deliberation, I obtained a simple, yet completely established, rule with whose help one can immediately decide for all examples of this kind, with any number of bridges in any arrangement, whether such a round trip is possible, or not.

In another letter, sent to Karl Ehler, an amateur mathematician and the Mayor of Danzig,[26] dated 3 April 1736, Euler observed:[27]

> Thus you see, most noble Sir, how this type of solution bears little relationship to mathematics, and I do not understand why you expect a mathematician to produce it, rather than anyone else, for the solution is based on reason alone, and its discovery does not depend on any mathematical principle ... In the meantime, most noble Sir, you have assigned this question to the geometry of position, but I am ignorant as to what this new

[26]Danzig, now Gdansk in Poland, is some 80 miles west of Königsberg.

[27]Euler, *Pis'ma k ucenym*, Izd. Academii Nauk SSSR, Moscow–Leningrad (1963). See (Hopkins and Wilson 2007).

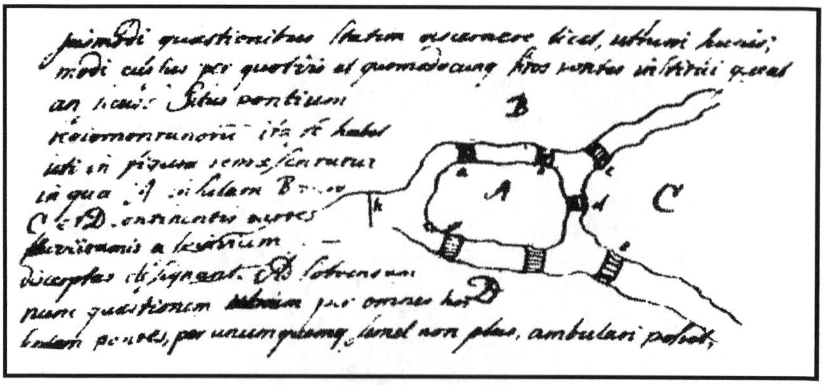

Figure 8.4. Euler's letter to Marinoni

discipline involves, and as to which types of problem Leibniz and Wolff expected to see expressed in this way.

Euler wrote up his solution in a paper that has become celebrated.[28] It is in this paper that Euler first referred to the geometry of position (geometria situs) as the geometrical analysis mentioned by Leibniz.

Although several writers have claimed that Euler solved the problem by drawing a network or graph with four vertices (representing the four areas of the city) and seven edges (representing the bridges), he did not do so. His approach to the problem was to count the number of bridges emerging from each area and to note how many of these numbers are odd. His conclusions were as follows:

- If there are more than two areas to which an odd number of bridges lead, then such a journey is impossible.
- If, however, the number of bridges is odd for exactly two areas, then the journey is possible if it starts in either of these two areas.
- If, finally, there are no areas to which an odd number of bridges lead, then the required journey can be accomplished starting from any area.

For the Königsberg bridges problem, the numbers of bridges are 3, 3, 3, and 5; since this list contains more than two odd numbers the problem has no solution. However, for the system of bridges in Figure 8.5, Euler observed that the numbers of bridges that emerge from the six areas A–F are 8, 4, 4, 3, 5, 6, respectively. Since just two of these numbers (corresponding to areas D and E) are odd, a walk is possible as long as it begins in one of these areas and ends in the other.

8.4 The study of curves

We shall pursue two themes: first, the study of where two curves intersect, which has its origins in the 17th–century study of the solution of equations; second, the study of curves in their own right. We shall see that what ensued took the form of almost reversing the roles of algebra and geometry, until algebraic reasoning replaced geometrical reasoning as the source of validity in mathematics.

[28] See (Euler 1736/1741, E53), and (Biggs, Lloyd, and Wilson 1998) for a translation.

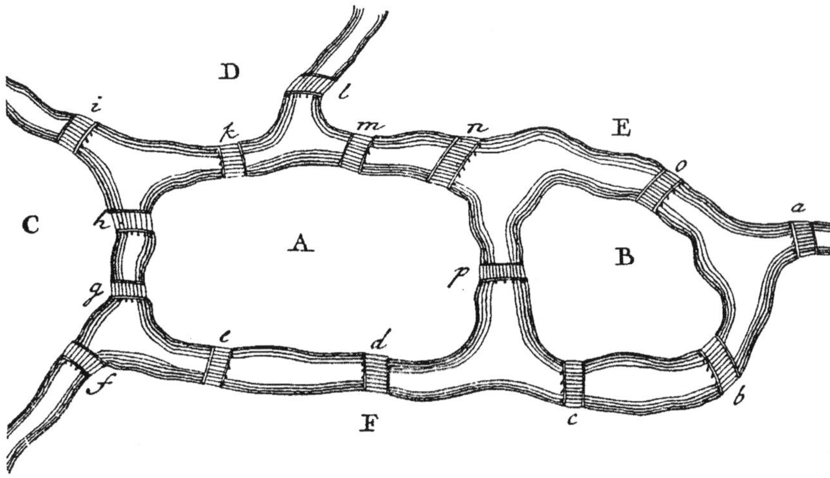

Figure 8.5. A new system of bridges

In the 17th century mathematicians generally solved equations in the manner of Descartes, namely, via a construction that exhibited their roots as the coordinates of points of intersection of curves.[29] By the middle of the 18th century this approach had begun to seem more trouble than it was worth. However, there was still interest in the question of saying something about the points in which two curves meet one another, because this question has both a geometrical and a more practical significance. To see why, we start as they did by asking this question: Given two curves, how many meeting points are there?

A case of this that we have already met is when one curve is given by a polynomial equation

$$y = a_n x^n + a_{n-1}x^{n-1} + \cdots + a_0$$

and the other curve is the straight line $y = 0$. Eliminating y between these equations gives

$$a_n x^n + a_{n-1}x^{n-1} + \cdots + a_0 = 0$$

as an equation specifying the x-coordinates of the points of intersection, and we have seen that the answer was widely believed to be that there are n solutions to this equation and so n points of intersection of the two curves. This case is already not easy to analyse and other cases are much harder, so the central topic in geometry became that of finding where one curve meets another (see Figure 8.6). More precisely, it became the algebraic version of this: Find the points that simultaneously satisfy two equations in x and y.

The method of attack was always the same. There are two equations, one for each curve, and we first eliminate y, obtaining an equation in x alone (known as the 'resultant' of the original equation). Then we solve this resultant equation in x. The branch of algebra involved in the first stage was called elimination theory, and it entailed dealing with various other problems, such as: Can we eliminate y? If we could rewrite any equation in x and y so that y occurs only once (and the equation has been written in the form $y = f(x)$) then the problem would be easy to solve — but this cannot always

[29]See *Volume* 1, Section 11.2.

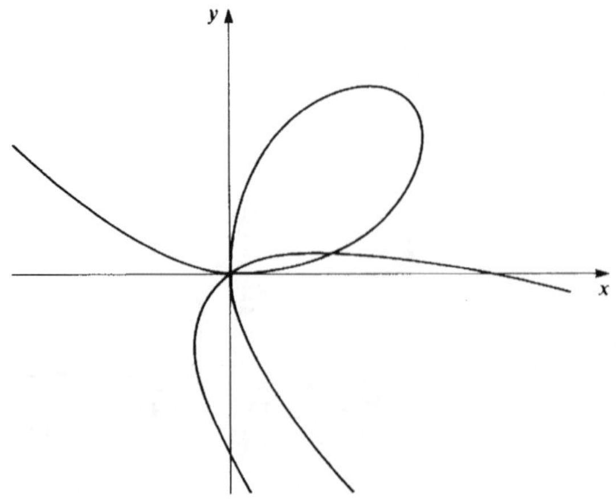

Figure 8.6. Two curves crossing — but in how many points?

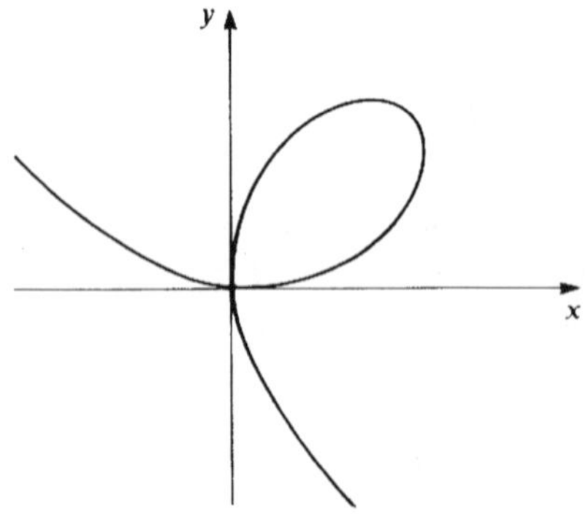

Figure 8.7. The folium of Descartes

be done, as is shown by the folium of Descartes (see Figure 8.7), which is given by the equation

$$x^3 + y^3 = 3xy.$$

We can get some idea of the importance of elimination theory from this call for a good attack on the problem in 1770 by the Académie des Marines, the scientific school attached to the French Navy:[30]

> The elimination of unknowns is one of the most important parts [of mathematics] to perfect, both because the extreme length of ordinary methods makes it so repugnant and because the general resolution of equations depends on it.

The first thing that mathematicians sought to establish was how many common points two curves should have. This obviously depends on the curves, but there was a widespread 17th-century belief that if one curve is of degree m and the other is of degree

[30]Quoted in (Rider 1981, 168).

n then they have mn points of intersection. In the 18th century this 'truth' gradually changed its status from a useful heuristic principle to something requiring a proof.

But before this principle could be proved true, there were some difficulties to be overcome. For a start, it was evidently false in general: curves of degree m and n do not always appear to intersect in precisely mn points. Two circles (curves of degree 2) may meet in 0 or 2 points (and perhaps in 1 point on occasion, depending on how you count points of tangency), but not visibly in the $2 \times 2 = 4$ points that the formula implies. So the principle seems to have collapsed even before it gets going. On the other hand, it is certainly true in many cases. Two conics meet in at most four points, and that they do sometimes meet in precisely four is evident from Figure 8.8.

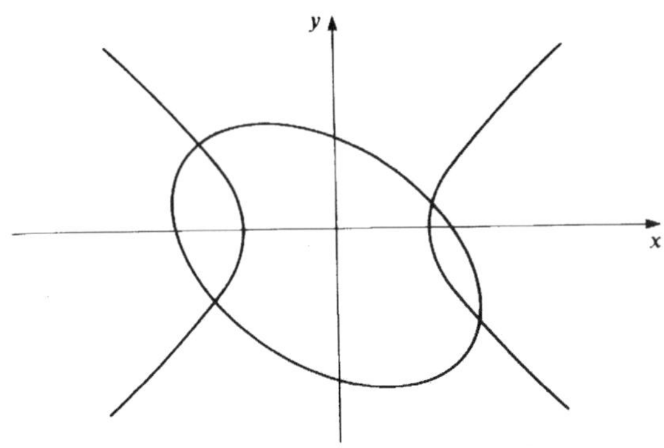

Figure 8.8. Two conics that meet in four points

In fact, the principle that curves of degrees m and n intersect in precisely mn points is a rare and interesting example of something more common in science than in mathematics: a true result that seems superficially to be false. On such occasions it often transpires that one is not looking at the situation in quite the right way: with a slight redefinition of terms, the apparent exceptions can be seen to have been correct all along. Here, the key issue is: How do you count a point of intersection? If one is sufficiently convinced that the principle must be true, then it may not be beyond human wit to count intersections in such a way as to ensure this.

Consider each of the figures in Figure 8.9, showing various intersections of a circle (degree 2) and a line (degree 1). By the principle there should be $2 \times 1 = 2$ points of intersection.

In figure (a) there are clearly two points of intersection. In figure (b) we can persuade ourselves that the point of tangency is a 'double point' and count it as two points — wobble the line a bit as in figure (c) if you are still in doubt.[31] But what of figure (d)? How can the most vivid imagination conjure two points of intersection out of curves that patently do not intersect at all?

However, 18th-century mathematicians were gradually becoming accustomed to using complex numbers. Here the answer seemed perfectly straightforward: the fourth figure is a picture of a line meeting a circle in two complex points. This may be hard to see if you have a purely geometrical perception, but to mathematicians of a firmly

[31] Descartes and Fermat's ideas about tangents were discussed in *Volume* 1, Section 9.1.

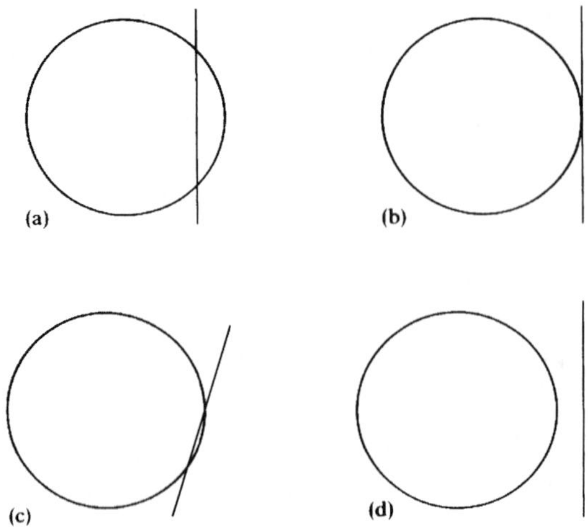

Figure 8.9. Four possibilities for a circle meeting a line

algebraic turn of mind it made excellent sense. For, in just the same way as a quadratic equation with no real roots (such as $x^2+2x+5=0$) turns out to have two complex roots — thus preserving the principle that every polynomial equation has as many roots as its degree — so here complex points can be imagined to preserve the *mn* principle. We see vividly here the motivation for people to shift geometrical questions into an algebraic form, and how algebra and complex numbers came to be preferred to geometry and only real intersection points.

This is illustrated very elegantly for us by a short paper (E148) that Euler published in 1748, where he pointed out that for the theorem to be true one must be able to say a number of rather implausible things. The problems that Euler saw were that:

- even parallel lines meet in a point — this requires the invention of points at infinity
- the parabola in Figure 8.10 meets the line ℓ_1 in two points, both imaginary, and meets the line ℓ_2 in two points, one of which is at infinity.

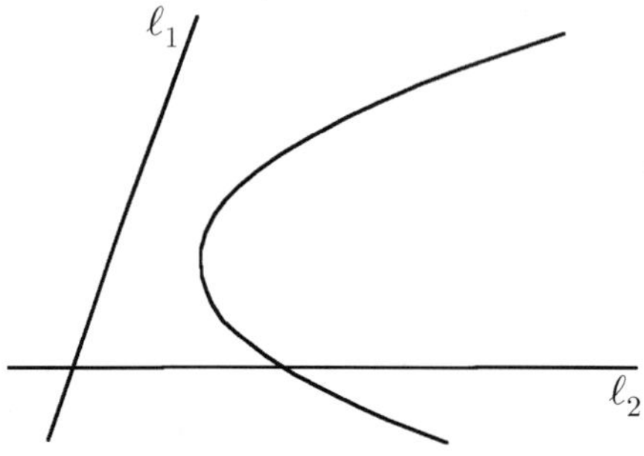

Figure 8.10. A parabola and two lines

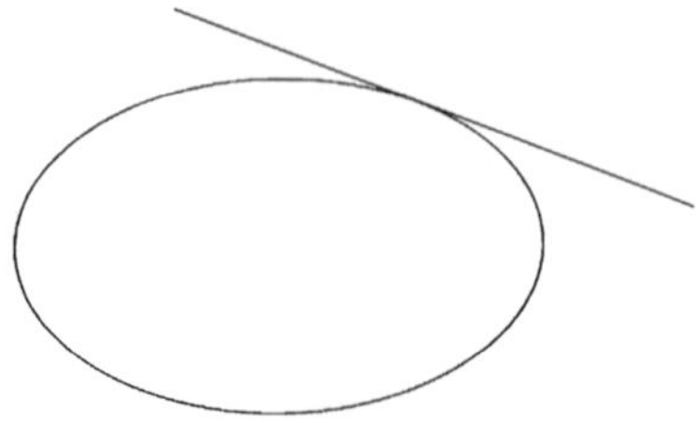

Figure 8.11. An ellipse and a tangent

A case that Euler did not mention, that of tangency, is exemplified in Figure 8.11. For the ellipse to meet the line in two points, common points to the two curves must be counted the 'correct' number of times — in this case, two, not one. (The term for this is 'multiplicity'.) In other words, we must alter the way we count, in order to make the theorem correct.

Combinations of these difficulties are also possible. As we have already remarked, two circles should meet in $2 \times 2 = 4$ points, but it is clear they meet in at most 2. Where are the others? In this case the resolution is that the other points are both complex and at infinity! A simple algebraic conjecture is turning into something rather difficult. The hardest thing to define was multiplicity, and Euler was never to find a suitable definition of it: the problem was too complicated. In November 1751 Euler wrote to the mathematician Gabriel Cramer (who had also considered the problem) that it led to 'such tangled formulas that one completely loses patience in pursuing the calculations'.[32]

The Académie des Marines disparaged the attempts by Euler and Étienne Bézout, another person who had tackled the problem, because they had not shown that the resultant had the right degree (mn). But in 1779 Bézout announced a solution that satisfied his contemporaries, and the principle has been called Bézout's theorem ever since. A remarkable feature is that it does not determine the solutions of the resultant equation and so locate the common points. It is an example of a non-constructive existence proof: it tells you that certain points necessarily exist, but it does not tell you where they are or how to find them. All that it does is to establish that the resultant has degree mn, and so the two curves meet in mn points.

Bézout's proof of his theorem was ingenious but artificial, and it was soon replaced by one due to a leading French mathematician of the early 19th century, Siméon-Denis Poisson. But his proof also fails to clinch the result by modern standards, for the same reason that attempts on the Fundamental Theorem of Algebra were flawed: a rigorous theory of complex numbers is required in order to deal with these existence questions.

Note how algebraic the original problem about curves has become. A complicated algebraic exercise was required to show in how many points two curves meet, by exhibiting a polynomial of the right degree. The answer to a geometrical question is now the degree of a polynomial! A hundred years earlier, the answer to a question about a polynomial was sought in the behaviour of suitably chosen curves. So over the course of a century the roles of geometry and algebra became completely reversed.

But if questions about plane curves were no longer to be raised in order to solve equations, what about those questions about curves that arose naturally? The Académie des Marines was in no doubt about their importance, and the geometry of curves exerted its own fascination, independently of any applications. The two most influential textbook presentations were due to Euler and Cramer; Euler's was his *Introductio in Analysin Infinitorum*. Euler had originally asked Cramer to oversee the printing of the *Introductio*, but because of a conflict of interests that job eventually went to someone else, for Cramer was by then at work on his own *Introduction à l'Analyse des Lignes Courbes Algébriques* (Introduction to the Analysis of Algebraic Curved Lines), which

[32]The Euler–Cramer correspondence is published in Euler, *Opera Omnia* (IVA) Vol. 7, number O0474, quoted in (Rider 1981, 168).

appeared in 1750. Cramer's was, if anything, the more influential treatment of geometry, but we start with that of Euler.

The first significant difference between the 17th and 18th centuries in this respect is the much greater readiness of later mathematicians to define curves by means of equations, and not merely to introduce the algebra as a technical convenience. This is made quite clear in Euler's *Introductio*, Book II. Here he began by describing how any function y of a variable quantity x can be exhibited as a curve, by imagining x to vary along a straight line ℓ, and plotting the y-values corresponding to each x along a line at right angles to ℓ. Conversely, each curve defines a (possibly many-valued) function when we reverse the process. So it is Euler who standardised what are now inappropriately called Cartesian coordinates (though others before him had used right-angled coordinate systems when it suited them). In so doing, Euler put into a textbook a way of working that he had already found in dealing with mechanics, as we shall see in Section 10.2.

A good example of the benefits of his new approach can be seen in Euler's discussion of conic sections in the *Introductio*.[33] The balance of algebra and geometry in Euler's text is most interesting. Euler started from 'a general equation for lines of the second order' and reached the division of conics into the three basic kinds: hyperbola, ellipse, and parabola.

There is one bit of algebraic technique that Euler slipped through in a line, but is rather important so we dwell on it longer. At the beginning of the second paragraph, Euler spoke of moving from his starting point of the 'general equation for lines of the second order', by which he meant an equation of the form

$$axx + bxy + cyy + dx + ey + f = 0,$$

to the equation that he used for distinguishing the different conics, which was

$$yy = \alpha + \beta x + \gamma xx.$$

This latter equation was just as general for his purposes — 'all the lines of the second order may be contained in this equation' — and once he had reached it his results on conic sections fell out very rapidly, by considering the different values γ could have, as the next extract shows.

But first, let us see how he got from one equation to the other. This is where Euler's new algebraic perspective on coordinate geometry came into play: what he had done was to transform the coordinates. Euler recognised that it made life much simpler to set up coordinate axes beforehand and at right angles — this is the advance he made on Descartes — and also that a useful technique could be to transform the coordinate axes in mid-problem. This can be thought of geometrically, as shifting the figure about in relation to its background axes, or algebraically, as changing the labelling of points in relation to the axes. In either way one can reach a simpler equation that retains all the crucial properties.[34]

[33] See Vol. 2, Chapter 5.

[34] See Euler, *Introductio in Analysin Infinitorum*, II, and F&G 14.A4.

Euler on conic sections.

> However the greatest difference in the curved lines which are included in the equation $yy = \alpha+\beta x+\gamma xx$ is produced by the character of the coefficient γ, depending on whether it has a positive or a negative value. For if γ has a positive value, assume that the abscissa x is infinite, in which case the term γxx turns out to be infinitely greater than the remainder $\alpha+\beta x$ and for that reason the expression $\alpha+\beta x+\gamma xx$ acquires a positive value. The ordinate y will likewise acquire two infinitely large values, one positive, one negative, because the same thing happens, if $x = -\infty$. In this case, however, the expression $\alpha + \beta x + \gamma xx$ assumes an infinitely great positive value. On account of this, if y becomes a positive quantity, the curve will have four branches stretching out to infinity, two corresponding to the abscissa $x = +\infty$ and two corresponding to the abscissa $x = -\infty$. Therefore these curves which have four branches stretching out to infinity, are thought to constitute one type of lines of the second order, and are called 'hyperbolas'.
>
> If, however, the coefficient γ has a negative value, then if $x = +\infty$ or $x = -\infty$, the expression $\alpha + \beta x + \gamma xx$ will have a negative value and therefore the ordinate y becomes imaginary. Therefore nowhere in these curves will an abscissa or an ordinate be able to be infinite. For that reason no part of the curve will be able to extend to infinity, but the whole curve will be contained in a finite and limited space. So this type of lines of the second order acquires the name of 'ellipses', on account of the fact that their character is contained in this equation $yy = \alpha + \beta x + \gamma xx$, if γ is a negative quantity.
>
> Therefore if the value of γ produces such a different character of lines of the second order, depending on whether it is positive or negative, that on this account two different sorts are rightly created: if $\gamma = 0$, a value which is midway between affirmative and negative numbers, the curve resulting from this also constitutes a certain type midway between hyperbolas and ellipses, called the 'parabola', which therefore expresses its nature by the equation $yy = \alpha + \beta x$.

Notice the dramatic change in perception consolidated by Euler's work. Even those quintessentially geometrical objects, the conic sections, were barely geometrical any longer, but were defined by equations: in their definition, treatment, and the style of argument in deducing properties, the conic sections had become algebraic objects. There is still a faintly geometrical haze, it is true, in that the algebra refers to geometrical figures and their properties. Also, Euler's argument for distinguishing the conic sections according to the value of γ amounts to what we would think of as 'curve sketching' — that is, making general inferences about the form of a curve from the form of its equation. But we are now close to a situation where the geometry of conic sections could drop out of sight altogether, for all the difference that would make to the richness and power of the algebraic analysis. Indeed, Euler went on to derive almost all the elementary properties of conics algebraically — properties such as the existence and location of their centres, foci, axes, diameters, and asymptotes — and for better or worse succeeding generations could, if they chose, discard the geometrical methods in geometry altogether.

Newton would surely not have approved of the way that Euler began by deriving equations for the curves and then deducing their properties from these equations. He might, however, have approved of Chapter IX of the *Introductio*, in which Euler treated cubic curves in the same way, because even Newton had needed to rely on algebra then — plainly, Newton knew when not to take his own advice!

Newton had worked out a classification of cubic curves by type in 1667, but he published it only in 1704, as an Appendix to his *Opticks*. His account was daunting and sometimes obscure, but at one point he remarked that it followed from his classification that there are five basic cubic curves; any other cubic can be obtained as the shadow of one of these five under projection from a point source of light.[35]

Here is a fascinating blend of geometry and algebra. A cubic curve is a curve, and so a geometrical object, and the adjective 'cubic' refers to geometrical ideas — it meant, according to Newton, that a straight line meets the curve in 3 points. But the analysis of cubics starts from their equations and proceeds by symbol manipulation until each equation is reduced to its simplest form — it is algebra all the way. Their forms were then drawn and described geometrically, and above all the argument about shadows, being about shape, is geometrical. So, when we think of the way that algebra was to grow until geometry was swamped, it is interesting to see how Newton's successors took their stands on the question of geometry versus algebra. What do we find?

Euler's approach to the study of algebraic curves was geometrical in its aims, but algebraic in its methods. It was geometrical, in that he investigated what might happen in general to a curve as x and y become very large — or go to infinity, as he put it — and he looked in particular for curves of lower degree that closely approximate the curve near infinity. It was algebraic, in that his investigation was firmly rooted in coordinate methods.

When Euler applied these techniques to cubic curves he confirmed Newton's results, as did Cramer in his *Introduction*. He noted that the subdivision into species was rather arbitrary, and gave canonical equations for each of the five main families. He then turned to properties of cubic curves. Here, several results were already known, of which the most striking may be this (although neither Euler nor Cramer mentioned it):

> any line through two inflection points on a cubic curve meets the curve again in a third inflection point.

This result had been proved first by de Gua (1740), and again by the Scottish mathematician Colin MacLaurin in an Appendix to his *Treatise on Algebra* of 1748.[36] But Euler did not take up the idea of shadows, nor did Cramer, and nor did anybody else apply it to the theory of curves for over half a century afterwards. It is tempting to believe that its neglect by Euler and Cramer led to this idea disappearing for such a long time. Instead, and in line with his treatment of geometrical approach to curves in terms of their behaviour for large values of the x and y coordinates, Euler classified cubics in 16 species, and noted which of Newton's curves belong to which species.

---

[35] See Whiteside, *MPIN* VII, 589, 635, and F&G 12.D2.

[36] We discuss MacLaurin's work in greater detail in Chapter 9.

Euler also attempted a classification of curves of degree 4, but found that there would be 146 different types, not all of which he could analyse, and accordingly he did not progress to curves of degree 5 or more.

Gabriel Cramer was Swiss, like Euler, and seems to have been on good terms with him. At the age of 20, with another scholar, Giovanni Calandrini, he applied for the professorship in philosophy at the Académie de Calvin in Geneva in 1724. The Académie was so impressed with both of them that a new Chair in mathematics was created, and both men were appointed and instructed to share the work. They were both privately wealthy, and they took it in turns to travel and study while the other stayed in Geneva and taught (and drew the salary). Eventually, in 1734, Calandrini was appointed Professor of Philosophy and Cramer became Professor of Mathematics, a post he retained until his unexpected death at the age of 48.

On his travels Cramer spent five months in Basel, where he got to know Johann Bernoulli, Daniel Bernoulli, and Euler, before going on to London and Paris. He was asked by Johann Bernoulli to be the sole editor of his works, which he published in four volumes in 1742, and also of his brother Jakob's works, of which he published two volumes in 1744.

Cramer's best work was his *Introduction à l'Analyse des Lignes Courbes Algébriques*, which he published in 1750. Although he had had a draft copy of Euler's *Introductio* for a month in 1744, scholars incline to the view that his own work was largely independent, and in any case, Cramer's became the better-known treatment of geometry. The French mathematician Michel Chasles, in his history of geometry of 1837, called it 'the most complete and even today the most highly regarded treatise of this vast and important branch of geometry".[37]

The success of Cramer's *Introduction* derived partly from being written in French (Euler wrote his in Latin), partly from being very thorough (it is twice the size of Euler's), and partly from being lavishly and beautifully illustrated (over 300 curves were drawn, see Figure 8.12). It necessarily overlaps considerably with Euler's treatise. Cramer defined curves algebraically and studied them algebraically, as had Euler — as Cramer put it, 'his object being almost the same as mine, it is not surprising if we are often together in our conclusions'. Nonetheless, Cramer emphasised different topics. He was more interested in points where the curve intersects itself, and in how it looks 'near infinity' (when x or y is very large). But he was not interested in the way in which those two topics interrelate, a question on which another mathematical writer, de Gua, was particularly insightful. Still, Cramer's questions seem to be a geometer's questions, because they are about shapes. The direction of his thought was: Given this equation, what does it mean and what does it describe?

We can gain a vivid sense of how geometry and algebra stood if we look at the prefaces to the books by de Gua and Cramer.

De Gua had more than his share of personal misfortune — his parents plunged from riches to bankruptcy when he was a child. At first he contemplated a religious life before becoming a mathematician and scientist. In 1742 he was appointed Professor of Greek and Latin Philosophy at the Collège Royal in Paris (later the Collège de France) — apparently on the strength of his mathematical work of 1740 — and lectured on mathematics and Newtonianism until he resigned in 1748. During this period he worked with D'Alembert and Diderot in setting up the *Encyclopédie*. Then he turned

[37]See (Chasles 1837, 152).

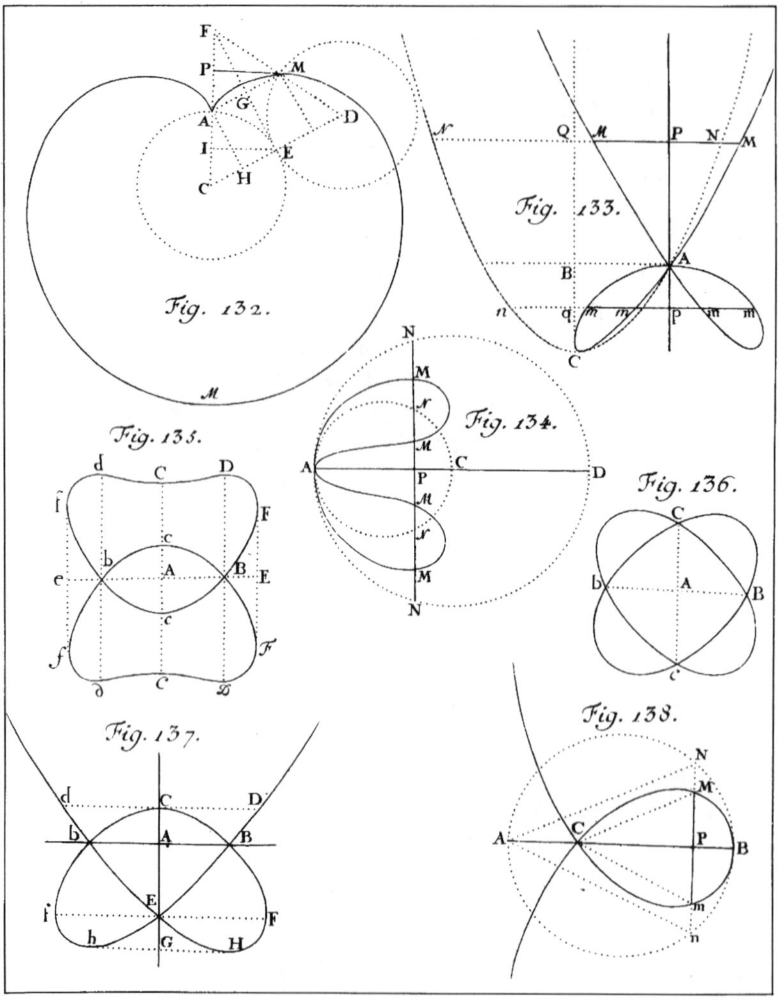

Figure 8.12. Some algebraic curves, from Cramer's *Introduction*

to economic theory, became a gold prospector in the Languedoc, which nearly ruined him, and as an old man wrote voluminously on mineralogy and conchology.

De Gua's study of curves has a title that is almost a story in itself: *Usages de l'Analyse de Descartes pour Découvrir, Sans le Secours du Calcul Differentiel, Les Propriétés, Ou Affections Principales des Lignes Géometriques de Tous les Ordres.* (Uses of Cartesian Analysis for Discovering, without the Help of Differential Calculus, the Properties of Geometrical Lines of all Orders) (1740). The avoidance of the calculus is indicative of a distinctive geometrical preference.

The book was influential. De Gua had absorbed not only the principles of Descartes's algebraic analysis, but also Newton's approach to cubics, the hints of projective shadow geometry. Putting these together he was able to reach some novel perceptions about aspects of curves and ways of approaching them. In particular, he saw the virtues of transforming coordinates so as to produce different equations relating to a curve, equations that would highlight some aspect of the curve's properties — and in making the coordinate transformations he was guided by the Newtonian vision of lights and shadows, so as to make what we might call 'projective transformations'; an example is given in Box 21.

It is clear from the preface of de Gua's book that although (in his opinion) the calculus is more general — it can deal with mechanical or transcendental curves as well as algebraic ones — algebraic analysis is much better when it comes to dealing with geometrical curves, and geometrical properties, such as singular points and infinite branches.[38] In this spirit, he set himself the task of understanding many of the things that Newton had said about cubic curves. His methods were algebraic: he used coordinate transformations to transform the equations until the properties of the curves became more apparent: this led him to recognise an unexpected use for the differential calculus. But when he came to consider his main result, the connection between singular points and infinite branches, he found that 'the source of the analogy [lay] in the theory of shadows'. Thus the techniques were algebraic, but he considered the underlying explanation to be geometrical.

By contrast, Cramer's preface balances the merits of algebra and geometry differently.[39] Whereas de Gua had listed the possible kinds of singular points and infinite branches of curves of the first five degrees, Cramer attempted to go further and sort them into types, but he stopped with quartics because they proved too complicated. In this sense, de Gua gave more detail about the 'pieces' a curve comes in, but Cramer did more to fit the pieces together to provide a global picture. Only de Gua connected singular points with infinite branches, and invoked the idea of projection, but Cramer studied many more curves. Cramer seems to have been nearer to Euler's position, namely that geometrical properties should be explained by algebraic arguments.

We note in passing how full and warm were the tributes that both de Gua and Cramer paid to their predecessors; Cramer, especially, comes over as a very courteous and friendly person, although even he let slip a barbed comment about Newton preferring 'the pleasure of being admired to that of instruction'.

We conclude, however, on a downbeat note. Under the influence of Euler and Cramer, the study of curves became more and more algebraic. Other authors did not disdain the subject either, although none wrote books with such authority as these two. But the study of geometry proceeded at a much slower pace than that of other subjects, so that it is legitimate to speak of a real relative decline. Geometry was no longer the source of rigour — algebra supplied that — nor of the most powerful techniques — the calculus supplied those. Geometry survived because of its own interest and charms, but in the course of the 18th century the new algebraic analysis became the most vital branch of mathematics.

Indeed, if there was anything that Euler did not do, or on which his influence was not wholly generous and beneficial it was the geometry of curves. His study of algebraic curves was considered very complicated by Julius Plücker,[40] the geometer who took up the subject most energetically in the early 19th century, and arguably it contributed to a feeling that algebraic geometry was just too hard beyond degree 4. In particular, it contributed to the growing neglect of projective methods in geometry, after

[38] See (de Gua 1740), and F&G 14.D1.

[39] See (Cramer 1750) Preface, and F&G 14.D2.

[40] For details of Plücker's work in geometry, see Section 15.2.

Box 21. De Gua's projective transformations.
We consider the curve defined by the equation $y^2 = x^3$, depicted below.

How can we begin to understand the properties of this curve? In particular, what happens as it goes off to infinity? De Gua found that a suitable projective transformation would give him the curve $y = x^3$ (also below) — and by this process the curve was transformed into a new curve with an inflectional tangent.

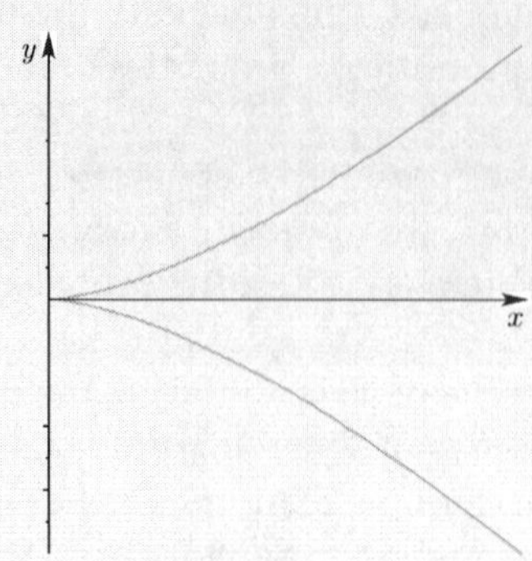

Figure 8.13. Graph of $y^2 = x^3$

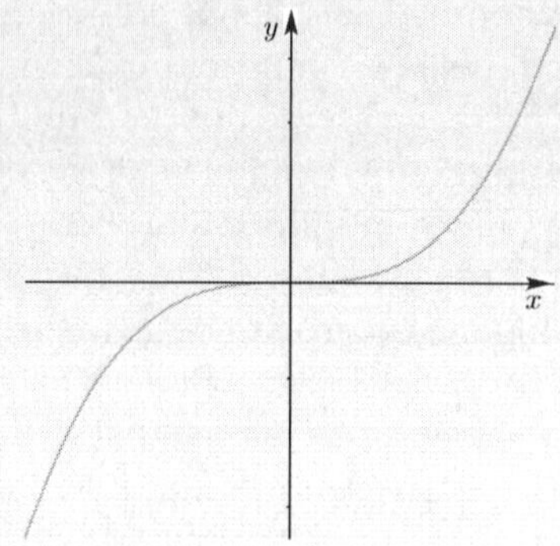

Figure 8.14. Graph of $y = x^3$

To picture this remarkable finding, think of a torch as casting the shadow of the x-axis to infinity. Then all of the curve has a shadow image, except for the cusp point itself (because it lies on the x-axis). To understand which points are transformed we argue as follows. On the original curve we can travel continuously along, albeit with a somewhat abrupt reversal, or turn, at the cusp point; in the transformed curve we travel steadily in an increasingly north-easterly direction when suddenly we find ourselves coming in from the south-west. This is also disconcerting in its way, but we have learned more about the original curve: it is a curve with an inflectional tangent in projective disguise.

Observations of this kind were the basis of de Gua's analogy, through projective transformation, between singular points and infinite branches of curves.

the start given to them by Desargues, Blaise Pascal, de la Hire, and Newton. The revival of projective methods, and the creation of projective geometry, were to be among the achievements of the early 19th century, as we shall see in Chapter 15.

8.5 Further reading

Bradley, R.E. and Sandifer, C.E. 2007. *Leonhard Euler: Life, Work, and Legacy*, Studies in the History and Philosophy of Mathematics, Vol. 5, Elsevier. This is a varied collection of useful articles, including Hopkins and Wilson's 'The Truth about Königsberg'.

Richeson, D.S. 2008. *Euler's Gem: The Polyhedron Formula and the Birth of Topology*, Princeton University Press. The author looks at the study of polyhedra from the ancient Greeks to the modern theory of surfaces.

Takase, M. 2007. Euler's theory of numbers, in *Euler Reconsidered*, R. Baker (ed.), Kendrick Press, 377–421. This is not an easy read for a beginner, but is one of a number of useful essays in a rich volume.

Watkins, J.J. 2013. *Number Theory: A Historical Approach*, Princeton University Press. This book combines an approach to number theory through its problems with an account of its history.

Weil, A. 1984. *Number Theory: An Approach through History from Hammurapi to Legendre*, Birkhäuser. André Weil was one of the leading mathematicians of the 20th century, and his surprisingly readable and stimulating account repays careful study.

9

Euler, Lagrange, and 18th-century Calculus

Introduction

One of the great puzzles for mathematicians in the 18th century was to explain why the powerful methods of the calculus worked. In this chapter we look at how they tackled this problem, starting with the explanations offered by Newton and Leibniz. Newton's calculus in particular was criticised by the Irish theologian Bishop Berkeley, and was then defended by the Scottish mathematician Colin MacLaurin, but not in a way that settled the disquiet felt by many. Leibniz's calculus was no more secure, and in Section 9.2 we examine how Euler rewrote it in the language of functions. This did not resolve the issue but it was a productive move, and we take the opportunity to present some of Euler's contributions to the calculus — notably, his theory of the trigonometric, exponential, and logarithmic functions. In Section 9.3 we see how these ideas led to breakthroughs in the theory and use of differential equations, and then review some of the ways in which Euler advanced this branch of mathematics. In Section 9.4 we return to the troubled issue of the foundations of the calculus and briefly examine the ideas put forward by D'Alembert and Lagrange. As we shall see, many interesting insights were generated down the century, but the problem remained for later generations to tackle.

9.1 Early critiques of the calculus

Difficulties with the foundations of the calculus have caused problems for mathematicians since its invention. Indeed, problems with the nature of motion may be said to have been around since the time of Zeno in the 5th century BC. We can usefully reformulate one of his famous paradoxes as follows.

> A hare sets off to catch a tortoise, which is 100 yards ahead. The hare runs at 100 yards a minute, the tortoise at one-tenth of that speed. After a minute,

the hare has reached where the tortoise began, but the tortoise is now 10 yards away. After a further one-tenth of a minute, the hare is again where the tortoise was, and the tortoise is now only 1 yard away. After a further one-hundredth of a minute, the hare is closer still, but has not caught up. How can the hare catch up with the tortoise if the tortoise is ahead at every stage of the race?

The obvious, and surely compelling answer is that the hare catches up after

$$1 + 0.1 + 0.01 + \cdots = 1.111\ldots \text{ minutes,}$$

by which time it has run

$$100 + 10 + 1 + 0.1 + 0.01 + \cdots = 111.111\ldots \text{ yards,}$$

that is, after $1\frac{1}{9}$ minutes and in $111\frac{1}{9}$ yards. So our answer involves us in being able to form the sum of an infinite series of numbers, and if we can make sense of this infinite sum, then we can resolve the paradox.

The difficulty is not with the answer, which involves rational numbers, but with the method by which it is arrived at, which involves an infinite summation; the answer is obtained as a limit of a sequence of ever close approximations. This is a problem for the calculus, as a brief look back at our account of the early Newtonian calculus reveals. We saw in Section 4.2 that, in his first letter to Leibniz, Newton placed great emphasis on the utility of infinite series, to the point of believing that his method of solving inverse tangent problems was entirely general once infinite series were admitted. Leibniz agreed so completely that when he studied Newton's unpublished tract, *De Analysi*, he took notes only on the material about infinite series. Infinite series involving a variable can be turned into infinite series of numbers merely by replacing the variable with a number, so understanding infinite sums would seem to be fundamental to the early calculus.

Other fundamental ideas in the early calculus were equally suspect, such as the infinitely small. If we look back at how Newton described his early work in his 'Method of Series and Fluxions' of 1671 (see Section 4.3) and the *Principia*, and at what Leibniz wrote in his 'New Method, etc.' (see Section 4.4), what do we find? We encounter references to indefinitely small parts, infinitely small periods of time, infinitely small additions, and something equivalent to nothing. We also meet evanescent divisible quantities, ultimate proportions of evanescent quantities, ultimate velocities, the ratios with which quantities vanish, limits towards which the ratios of quantities decreasing without limit converge, momentary differences, differential quantities, infinitely small distances, polygons with infinitely many angles, and differential quantities which when added together give 0.

The same passages make it clear that these ideas worried Newton and Leibniz, even as they wrote them down. Newton plainly expected objections and offered refutations in advance; significantly, these are not without problems of their own. Both suggested that it is necessary to evaluate expressions that, if not of the form 0/0, were distinguishable from such expressions only in an obscure way. In the Leibniz extract (see Section 4.4) the mathematical expression dx is introduced as an arbitrary finite segment, but later it becomes an infinitesimal. It cannot be both, so what is it?

Newton initially based his calculus on the idea of fluxions. In his manuscript *The Method of Series and Fluxions* (1671), Newton regarded all variables as varying in time (which is why he called them *fluents*) and called the rates at which these variables change their *fluxions*. In working with them, Newton argued that a fluent increases during an infinitely small period of time by an amount proportional to its speed of flow $\dot{x}$, so x increases to $x + \dot{x}o$, where o is an infinitesimal amount of time. His arguments are entirely rigorous until, as he put it, he 'casts out' unwanted os, on the grounds that o is 'infinitely small' — but what is an infinitely small quantity? If you reply, 'one that is arbitrary small', then Newton was assuming that in such intervals the velocity is constant, which rules out any kind of acceleration.

By the time that Newton came to write his *Principia* he had replaced infinitesimal arguments with a more sophisticated idea of limits, such as the limiting value of an infinite sum: he now spoke of 'first and last ratios' of quantities. In Book I of the *Principia* he introduced the intuitive idea that a curvilinear area can be evaluated by covering it with narrower and narrower rectangles, and asserted that under these conditions the two areas concerned (the curvilinear area, and its covering by narrow rectangles) *become ultimately equal*. According to Lemma 1 — the first lemma in the *Principia* — two quantities are ultimately equal when their ratios 'approach nearer to each other than by any given difference'. This is a workable idea, and we shall see that something like it was to prove satisfactory to later generations of mathematicians, but it is not entirely adequate. To raise just one objection, what might it mean when the quantities themselves become infinite, — for example, n and $n + 1$, as n increases (a case that Newton directly addressed)?

Leibniz's approach spoke more openly of infinitesimal quantities, in which he seems to have placed more trust than Newton did, if only as a manner of speaking. As such, it appealed to those who found such talk intelligible (a large group of mathematicians including many geometers) and it left cold those seeking a less intuitive but more precise approach. But in Leibniz's hands, even more than in Newton's, the calculus displayed the great virtue that it solved problems in a routine way. So the paradox arose that the most powerful advance in mathematics since the Greeks, with methods that were quite easy to learn and apply, made insufficient sense.

One particular aspect of the problem is worth noting now, so that the magnitude of a later step can be appreciated. Both Newton and Leibniz dealt with variables and limiting values of variables. This is understandable, because the concept of an arbitrary variable quantity was one that had come to prominence only in the 17th century, but it concealed a latent appeal to quantities varying in time that was to prove part of the problem and eventually, and profitably, to be eliminated from the discussion.

Mathematicians were uncomfortably aware of the problem, but the most famous attack on the difficulty came from outside their circle. In the early years of the 18th century the mathematical world had been drawn into an ugly debate about the invention of the calculus (which we discussed in Chapter 6.2). Had it been discovered first by Newton or Leibniz? We now know that priority belongs with Newton, but that it was established only in unpublished or privately printed documents. Leibniz was the first to publish articles on the calculus, and his supporters argued that he had therefore discovered it first. The minutiae of the debate need not concern us, but as the battle raged and national pride was laid on the line, other issues were drawn in. Of these, the most momentous in their day were theological. Newton's universe was permeated with gravity, a mysterious force that acted, in some unknown way, over great distances. Newton was content to imagine that God could be active in the universe, not least as the agent of gravity. Leibniz preferred to imagine that God, in His perfection, had created a universe that did not need his continual intervention. To the Newtonians, and indeed to some later philosophers, the Leibnizian universe was one in which God could not intervene, and was therefore almost a Godless place.

Once theological issues were at stake in a debate about the nature and scope of mathematics, theologians could naturally join in. The best-known and most important of these was George Berkeley, Bishop of Cloyne in Ireland. As a young man of 24 he had already attacked Newton's cosmology in his *Essay Towards a New Theory of Vision* (1709), and in 1734, in his tract *The Analyst*, he launched a powerful attack on the calculus for its lack of adequate foundations. It is clear from its subtitle (see Figure 9.1) that his prime motive in writing this tract was the vindication of theology. The 'infidel mathematician' to whom the *Discourse* is addressed is usually taken to have been Newton's supporter, Edmond Halley, although several other English mathematicians of the period had religious views to which Berkeley took exception. The suggestion of Halley's name acquires a certain potency when one learns that Halley had argued that certain Christian dogmas were inconceivable and could not be justified by reason, and this had so persuaded a friend of Berkeley's that on his deathbed he refused the Bishop's administration of the last rites.

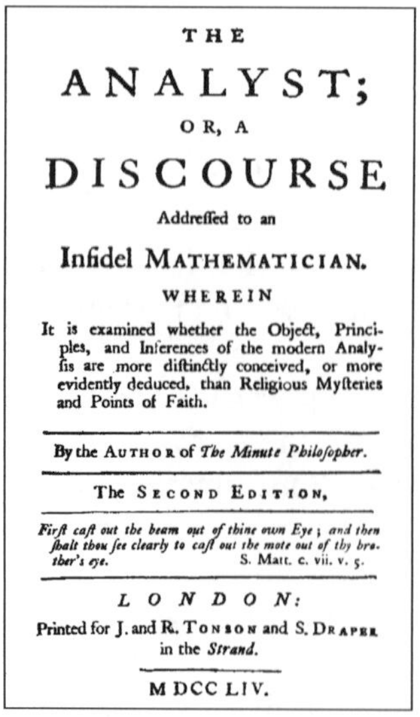

THE

ANALYST;

OR, A

DISCOURSE

Addreffed to an

Infidel MATHEMATICIAN.

WHEREIN

It is examined whether the Object, Principles, and Inferences of the modern Analyfis are more diftinctly conceived, or more evidently deduced, than Religious Myfteries and Points of Faith.

By the AUTHOR of *The Minute Philofopher.*

The SECOND EDITION,

Firft caft out the beam out of thine own Eye; and then fhalt thou fee clearly to caft out the mote out of thy brother's eye. S. Matt. c. vii. v. 5.

LONDON:

Printed for J. and R. TONSON and S. DRAPER in the *Strand.*

MDCCLIV.

Figure 9.1. Title page to Bishop Berkeley's *The Analyst*, 2nd edn. (1754)

In *The Analyst* Berkeley did not attempt to deny the usefulness of the calculus, but was keen to point out that its foundations were no less intuitive than those of theology. The following extracts from Berkeley's argument first show what particular aspect of the calculus was under attack and the nature of the Bishop's argument. They make it clear that Berkeley was attacking the rule for finding the fluxion of x^n in Newton's *De Quadratura*. His argument is that increments are initially assumed to be something (non-zero) and then in the same argument they are assumed to be nothing (zero). These assumptions are inconsistent, yet consequences of both are used simultaneously. Berkeley also pointed out a similar situation in the Leibnizian calculus: infinitely small quantities are both assumed and rejected; in particular, products of differences are rejected merely on the assumption that such a rejection will not introduce errors.[1]

Berkeley's attack on the calculus.

> §XIII Now the other method of obtaining a rule to find the fluxion of any power is as follows. Let the quantity x flow uniformly, and be it proposed to find the fluxion of x^n. In the same time that x flowing becomes $x+o$, the power x^n becomes $\overline{x+o}|^n$, i.e., by the method of infinite series
> $$x^n + nox^{n-1} + \frac{nn-n}{2}oox^{n-2}\ \&c.,$$
> and the increments
> $$0 \quad \text{and} \quad nox^{n-1} + \frac{nn-n}{2}oox^{n-2} + \&c.$$
> are one to another as
> $$1 \quad \text{to} \quad nx^{n-1} + \frac{nn-n}{2}ox^{n-2}\&c.$$
> Let now the increments vanish, and their last proportion will be 1 to nx^{n-1}. But it should seem that this reasoning is not fair or conclusive. For when it is said, let the increments vanish, i.e. let the increments be nothing, or let there be no increments, the former supposition that the increments were something, or that there were increments, is destroyed, and yet a consequence of that supposition, i.e. an expression got by virtue thereof, is retained. Which, by the foregoing lemma, is a false way of reasoning. Certainly when we suppose the increments to vanish, we must suppose their proportions, their expressions, and every thing else derived from the supposition of their existence to vanish with them.

In other words, Berkeley was claiming that mathematicians first divide by o (which is assumed to be non-zero) and then let o vanish (that is, become zero).

Nor did Leibniz's way of proceeding escape Berkeley's scorn.

> §XVIII ...The notion or idea of an *infinitesimal quantity*, as it is an object simply apprehended by the mind, hath been already considered. I shall now only observe as to the method of getting rid of such quantities, that it is done without the least ceremony. As in fluxions the point of first importance, and which paves the way to the rest, is to find the fluxion of a product of two indeterminate quantities, so in the *calculus differentialis* (which method is supposed to have been borrowed from the former with some small alterations) the main point is to obtain the difference of such product. Now the rule for this is got by rejecting the product or rectangle of the differences. And in general it is supposed that no quantity is bigger or lesser for the addition or subduction of its infinitesimal: and that consequently no error can arise from such rejection of infinitesimals.

[1]See (Berkeley 1734, 69–71, 72, 75–76, 100–102), and F&G 18.A1.

Berkeley's comments call for a reply. There is indeed some inexactness in the way that the arguments he was criticising had been expressed. His criticisms rose to the following celebrated passage about the difficulties in explaining the Newtonian calculus:

> §XXXV The great Author of the Method of Fluxions felt this Difficulty, and therefore he gave in to those nice Abstractions and Geometrical Metaphysics, without which he saw nothing could be done on the received Principles; and what in the way of Demonstration he hath done with them the Reader will judge. It must, indeed, be acknowledged, that he used Fluxions, like the Scaffold of a building, as things to be laid aside or got rid of, as soon as finite Lines were found proportional to them. But then these finite Exponents are found by the help of Fluxions. Whatever therefore is got by such Exponents and Proportions is to be ascribed to Fluxions: which must therefore be previously understood. And what are these Fluxions? The Velocities of evanescent Increments? And what are these same evanescent Increments? They are neither finite Quantities nor Quantities infinitely small, nor yet nothing. May we not call them the Ghosts of departed Quantities?

But it soon becomes apparent that his agenda has nothing to do with improving mathematical understanding, as the next passage gradually reveals.

> §L Qu. 59 If certain philosophical virtuosi of the present age have no religion, whether it can be said to be for want of faith?
>
> Qu. 62 Whether mysteries may not with better right be allowed of in Divine Faith than in Human Science?
>
> Qu. 63 Whether such mathematicians as cry out against mysteries have ever examined their own principles?
>
> Qu. 66 Whether the modern analytics do not furnish a strong *argumentum ad hominem* against the philomathematical infidels of these times?
>
> Qu. 67 Whether it follows from the above-mentioned remarks, that accurate and just reasoning is the peculiar character of the present age? And whether the modern growth of infidelity can be ascribed to a distinction so truly valuable?

Here, Berkeley compared arguments about faith with those in mathematics and science, and suggested that it was illegitimate to accept obscurity in one but not the other, ending with a clear call to defend the faith.

The publication of *The Analyst* provoked a number of pamphlets supporting Newton's calculus. Berkeley validly dismissed some of these as defences of arguments that their authors did not understand — for example, James Jurin's *Geometry no Friend to Infidelity: Or a Defence of Sir Isaac Newton and the British mathematicians* (1734). The feebleness of these arguments leant strength to Berkeley's reiterated criticisms of the Newtonian calculus. Benjamin Robins's reply, expressed in a book and a number of tracts, went deeper, because Robins, unlike Jurin, appreciated that quantities need not be considered as actually reaching a limit — a limit was something to be regarded as only potentially reached. Nonetheless, Berkeley's serious criticisms of the calculus

Figure 9.2. Colin MacLaurin (1698–1746)

called for better replies if a proper presentation of the calculus were ever to be given. The first mathematician to go some way to meeting this need was the Scotsman Colin MacLaurin, in his *Treatise of Fluxions* of 1742.

After Newton, MacLaurin was the best-known British mathematician of the 18th century. He was born in Glasgow and became a child prodigy, entering Glasgow University at the age of 11 to study divinity, but Robert Simson, the Professor of Mathematics there and an enthusiast for Euclidean geometry, persuaded him to study mathematics. MacLaurin went on to become a Professor of Mathematics at Marischal College, Aberdeen, at the age of 19. He was elected a Fellow of the Royal Society at 21, and a year later he published his first major work, his *Geometria Organica*, with Newton's approval; it dealt with the properties of algebraic curves in the plane. He then left Scotland to become a tutor in France for some years, and while he was there in 1724 he won a Prize from the Académie des Sciences in Paris for a paper entitled 'On the percussion of bodies', which is a study of collisions and the nature of force. He returned to Scotland in 1725 to become Professor of Mathematics at Edinburgh, where his candidacy was greatly helped by a strong recommendation from Newton, and he remained there until his death. He took an active part in defending the city from the forces of the Jacobite rebellion of 1745, planning and supervising the erection of fortifications, but still the city fell, and MacLaurin had to flee to York. The severe exhaustion that he incurred as a result permanently weakened his health, and he died in 1746 not long after he had returned to Edinburgh.

MacLaurin's fundamental idea in his *Treatise of Fluxions* (1742) was to go back to the methods of Archimedes, who had defended his insights by the use of double *reductio ad absurdum*, and then to defend the results of the calculus in similar terms. The following extract shows how MacLaurin found the fluxion of A^2. This method yields valid proofs, but it is necessary to discover the correct result in advance, and

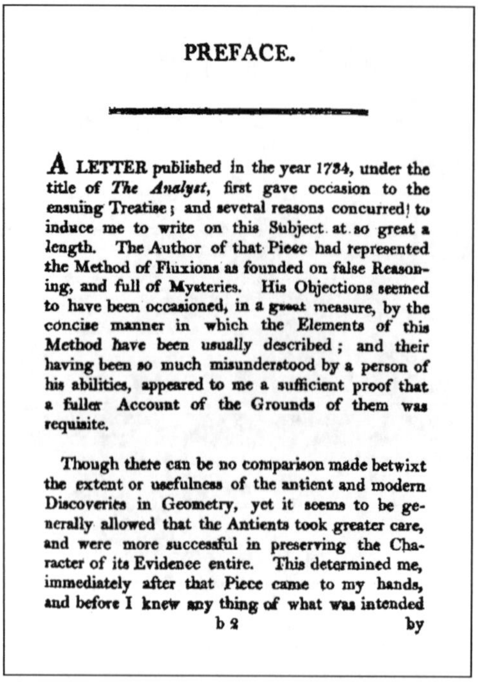

PREFACE.

A LETTER published in the year 1734, under the title of *The Analyst*, first gave occasion to the ensuing Treatise; and several reasons concurred to induce me to write on this Subject at so great a length. The Author of that Piece had represented the Method of Fluxions as founded on false Reasoning, and full of Mysteries. His Objections seemed to have been occasioned, in a great measure, by the concise manner in which the Elements of this Method have been usually described; and their having been so much misunderstood by a person of his abilities, appeared to me a sufficient proof that a fuller Account of the Grounds of them was requisite.

Though there can be no comparison made betwixt the extent or usefulness of the antient and modern Discoveries in Geometry, yet it seems to be generally allowed that the Antients took greater care, and were more successful in preserving the Character of its Evidence entire. This determined me, immediately after that Piece came to my hands, and before I knew any thing of what was intended

b 2 by

Figure 9.3. MacLaurin's explanation of why he wrote his *Treatise of Fluxions*

therefore by some other method.[2] Accordingly, while MacLaurin may have gone a long way to defend the calculus from the criticisms of Berkeley, he did not provide it with a defence that was adapted to the research methods of his time.

MacLaurin defends the Newtonian calculus.

> *The fluxion of the root A being supposed equal to a, the fluxion of the square AA will be equal to $2A \times a$.*
>
> Let the successive values of the root be $A - u$, A, $A + u$, and the corresponding values of the square be $AA - 2Au + uu$, AA, $AA + 2Au + uu$, which increase by the differences $2Au - uu$, $2Au + uu$, etc. and because those differences increase it follows from art. 704 that if the fluxion of A be represented by u, the fluxion of AA cannot be represented by a quantity that is greater than $2Au + uu$, or less than $2Au - uu$. This being premised, suppose, as in the proposition, that the fluxion A is equal to a; and if the fluxion of AA be not equal to $2Aa$, let it first be greater than $2Aa$ in any ratio, as that of $2A + o$ to $2A$, and consequently equal to $2Aa + oa$. Suppose now that u is any increment of A less than o; and because a is to u as $2Aa + oa$ to $2Au + ou$, it follows that if the fluxion of A should be represented by u, the fluxion of AA would be represented by $2Au + ou$, which is greater than $2Au + uu$. But it was shown from art. 704 that if the fluxion of A be represented by u the fluxion of AA

[2]See (MacLaurin 1742), II, 581–582, and F&G 18.A2.

cannot be represented by a quantity greater than $2Au + uu$. And these being contradictory, it follows that the fluxion of A being equal to a, the fluxion of AA cannot be greater than $2Aa$. If it can be less than $2Aa$, where the fluxion of A is supposed equal to a, let it be less in any ratio of $2A - o$ to $2A$, and therefore equal to $2Aa - oa$. Then because a is to u as $2Aa-oa$ is to $2Au-ou$, which is less than $2Au-uu$ (u being supposed less than o, as before) it follows that if the fluxion of A was represented by u, the fluxion of AA would be represented by a quantity less than $2Au - uu$, against what has been shown from art. 704. Therefore the fluxion of A being supposed equal to a, the fluxion of AA must be equal to $2Aa$.

9.2 Euler's calculus

In contrast to MacLaurin, Euler, the fastest-moving research mathematician of the day, did not seem unduly bothered with the niceties of foundational arguments. He casually used expressions in the form of quotients with both numerator and denominator ultimately having the value zero, apparently without any qualms about the lack of rigorous definition of the quantities concerned. In his *Institutiones Calculi Differentialis* (Foundations of the Differential Calculus) of 1755, one of his major treatises that defined the calculus for the later 18th century, he was quite prepared to commit himself in print as follows:[3]

Euler on the foundations of the calculus.

There is no doubt that any quantity can be diminished until it vanishes and is transformed into nothing. But an infinitely small quantity is nothing else but a vanishing quantity and, therefore, actually will be $= 0 \ldots$ If a quantity were so small that it is smaller than any given one, then it certainly could not be anything else but zero; for if it were not $= 0$, then a quantity equal to it could be shown, which is against the hypothesis. To those who ask what the infinitely small quantity in mathematics is, we answer that it is actually $= 0$.

...why do we not always characterise the infinitely small quantities by the same sign 0 instead of using particular symbols to designate them? ...It is true that any two zeros are equal in such a way that their difference is zero, yet, since there are two methods of comparison, one arithmetic, the other geometric, we see this difference between them (depending on the origin of the quantities to be compared) the arithmetic ratio of two arbitrary zeros is equality, but not the geometric one. This can best be understood from the geometric proportion $2 : 1 = 0 : 0$... It is in the nature of a proportion that when the first term is twice the second, then the third term must also be twice the fourth.

This, however, is also clear in ordinary arithmetic. It is known that a zero multiplied by an arbitrary number gives zero and that $n.0 = 0$ as well as $n : 1 = 0 : 0$. From this it seems possible that two quantities,

[3]See pp. 69–72, given in Struik, *A Source Book*, 384–385.

whatever their geometric ratio may be, will always be equal if we look at them from the arithmetic point of view. Hence if two zeros can have an arbitrary ratio, then I judge that different signs should be applied, especially when we have to consider a geometric ratio of different zeros. The calculus of the infinitely small is therefore nothing but the investigation of the geometric ratio of different infinitely small quantities. This enterprise will be thrown into the greatest confusion unless we use different signs to indicate these infinitely small quantities ...

Hence, if we introduce into the infinitesimal calculus a symbolism in which we denote dx an infinitely small quantity, then $dx = 0$ as well as $adx = 0$ (a an arbitrary finite quantity). Notwithstanding this, the geometric ratio $adx : dx$ will be finite, namely $a : 1$, and this is the reason that these two infinitely small quantities dx and adx (though both $= 0$) cannot be confused with each other when their ratio is investigated. Similarly, when different infinitely small quantities dx and dy occur, their ratio is not fixed though each of them $= 0$.

It would seem from this extract from Euler's work that, in arguments about the value of 0/0, quantities appear to be simultaneously zero and non-zero, and that more confidence is being expressed in the methods of the calculus than clarity about its foundations.

The abstract, formal treatment in the algebraic language of the calculus came to be called 'analysis'. In his three great textbooks, published from the mid-century onwards, Euler determined the scope and style of analysis for at least the next fifty years and left an irrevocable imprint on the subject. We shall look at the first of these influential textbooks, the *Introductio in Analysin Infinitorum*. Here Euler showed how to treat in an algebraic way all manner of things that had previously had stronger geometrical aspects; he virtually garbed mathematics in algebraic dress. In doing so he consolidated and gave definitive form to the trend that had developed over the previous hundred years, in which the formal techniques of the calculus increasingly came to dominate the subject, until the subject itself became perceived differently. What started as an approach to geometrical problems became an autonomous subject defined on its own terms, and Euler set the seal on this development. He went on to follow it up with two more great texts on the differential and integral calculus, his *Institutiones Calculi Differentialis* (1755), and his *Institutiones Calculi Integralis* (Foundations of the Integral Calculus) in three volumes, (1768–1770).

One aspect of Euler's contribution was simply to provide a new statement of what analysis is about. He stated explicitly that analysis is the subject concerned with analytical expressions, and in particular with *functions*. He defined the concept of a function as follows:[4]

> A *function* of a variable is any analytical expression whatever of that variable quantity and numerical or constant quantities.

Here

> a *variable quantity* is an indeterminate or universal quantity which is determined by all the values that it contains.

---

[4]See Euler, *Introductio in Analysin Infinitorum*, Chapter 1. For an extract from it and one from Euler's *Institutiones Calculi Differentialis*, see (Stedall 2008, 233–243).

This may seem somewhat unilluminating! We make two comments before we take it further. First, the concept of function was not original with Euler — nor indeed was this particular definition, which was due to Johann Bernoulli — but Euler perceived how crucial the concept was: he grasped its importance and brought it to the centre of the mathematical stage. Second, if the words used in the definitions seem to mean very little, that is perhaps in the nature of definitions! The crucial test is what counts as being instances of the definition, so looking at some examples will help us to see what Euler meant.

Variables were denoted by letters, such as x, y, and z, so a function of a variable quantity z was something like these *analytical expressions*: $a + 3z$, $az + 4zz$, $az + \sqrt{bb - zz}$, and c^z. (Here a, b, c are unspecified but fixed numbers.) One immediate corollary of the above definitions, Euler noted, was that a function of a variable quantity is itself a variable quantity, so a 'function of a function' is a legitimate object of study and use in analysis.

Euler categorised functions as Leibniz had categorised curves, as being either *algebraic* or *transcendental.* Functions were *algebraic* if they could be expressed by using only algebraic operations on the variable: addition, subtraction, multiplication, division, and related operations such as raising to a fixed (integral or rational) power and extracting roots. Functions that are not algebraic are *transcendental.* Examples are the exponential function e^z, the cosine ($\cos z$) and the logarithm ($\log z$); these are not expressible in terms of finite algebraic operations on the variable z. But although they are much harder to handle, Euler showed how transcendental functions are also amenable to algebra.

There is much that can be done to a function to transform it in various ways: one can manipulate it algebraically, possibly factorise it, change the variable, expand it as a power series, differentiate it, and so on. Euler was brilliant at such manipulations. To see what was involved in this, and begin to gauge the importance of his mathematical style, let us look at one of Euler's greatest achievements, the redefining of elementary transcendental functions as power series. The formulas we are about to meet were not new — they were known to Newton, for instance — but it was their derivation, and their use in *defining* functions, that formed Euler's keen-sighted contribution.

In his *Introductio*, Chapter VII, Euler obtained a power series expansion for the *exponential function* a^z; that is, some fixed number a (greater than 1) raised to the power (or *exponent*) z, where z is a variable. So what Euler was trying to achieve looks like

$$a^z = a_0 + a_1 z + a_2 z^2 + \cdots + a_n z^n + \cdots.$$

What are the coefficients a_k? We outline his answer in Box 22.

Euler's procedure was surprising and noteworthy in a number of ways. First, he used the concepts of infinitely small quantities (infinitesimals) and infinitely large numbers. Second, his handling of these and the whole chain of inferences involving power series seems natural, but also quite magical. Third, note the throwaway abandon with which Euler produced a result to 23 decimal places — there was no mathematical cause for such accuracy, simply a joyful exultation in his abilities.

We should spend a little longer considering the significance of Euler's use of infinitesimals. Recall that Leibniz expressed his calculus in the language of infinitesimals, although he was aware that this raised deep philosophical problems. There

Box 22. Euler's definition of the exponential function.
Euler let a be a number greater than 1, and ω be what he called 'an infinitely small number, or a fraction so small that, although not equal to zero, still $a^\omega = 1 + \psi$, where ψ is also an infinitely small number'. He let $\psi = k\omega$, so $a^\omega = 1 + k\omega$, where k depends on a. He then observed that 'We have $a^{j\omega} = (1 + k\omega)^j$, whatever value we assign to j'.

By the binomial theorem,

$$a^{j\omega} = 1 + \frac{j}{1}k\omega + \frac{j(j-1)}{1.2}k^2\omega^2 + \frac{j(j-1)(j-2)}{1.2.3}k^3\omega^3 + \cdots.$$

He then let $j = z/\omega$ 'where z denotes any finite number', which makes j infinitely large. He substituted z/j for ω, and in the resulting equation

$$a^z = (1 + kz/j)^j = 1 + \frac{1}{1}kz + \frac{1(j-1)}{1.2j}k^2z^2 + \frac{1(j-1)(j-2)}{1.2j.3j}k^3z^3 + \cdots$$

he observed that, because j is infinitely large, all the ratios of the form $(j-n)/j$ equal 1, and so

$$a^z = 1 + \frac{kz}{1} + \frac{k^2z^2}{1.2} + \frac{k^3z^3}{1.2.3} + \cdots.$$

He now chose a so that the corresponding value of k is 1. So, when $z = 1$,

$$a = 1 + \frac{1}{1} + \frac{1}{1.2} + \frac{1}{1.2.3} + \cdots.$$

In this case, $a_k = 1/k!$.

Euler observed that this value of a is approximately 2.71828182845904523536028... (he would have summed the first 24 terms of the series) and stated that the symbol e will be used for this number for the sake of brevity. The symbol e used for this purpose first appeared in Euler's *Mechanica*, Vol. I (1736).[a]

[a] It seems that Leibniz was the first to calculate the value of e, which he did in an unpublished letter to one Rudolph Christian von Bodenhausen on the study of the catenary in 1691, where he gave a value correct to 8 significant figures. See (Raugh and Probst 2019). In 1714 Cotes improved on this when he published a value correct to 12 significant figures in his *Harmonia Mensuarum* (1722), p.7.

was still discussion about infinitesimals among mathematicians of the next generation: Bernard de Fontenelle, for example, showed that they were useful in analysing curves at places where the x- or y- coordinate is very large, if one replaces x by $1/x$ and y by $1/y$. He then argued that if x is infinite then $1/x$ is infinitesimal. Euler's attitude towards infinitesimals is unclear. He seems to have regarded them as really existing, rather than as standing as a shorthand expression for 'very small but finite', or 'arbitrarily small' quantities, but he had no great liking for philosophical questions and never set out a significant foundational position. What mattered to him was that infinitesimals could be handled algebraically, and that they worked.

Now for the pay-off. To derive a power series expansion for a^z, or e^z, may not seem in itself very deep, or even interesting, but when he had treated a few more transcendental functions in like vein, some most extraordinary and exciting things began to emerge. Unexpected connections came to light between different functions, and in Euler's hands the functions themselves began to change their nature. Thus, Euler re-wrought the concept of sine until it scarcely belonged with circles or triangles any more: it was now almost completely algebraic or analytic[5] – all because of the remarkable introduction of $\sqrt{-1}$.

Euler on the trigonometric functions.

> §126. After logarithms and exponential quantities we shall investigate circular arcs and their sines and cosines, not only because they constitute another type of transcendental quantity, but also because they can be obtained from these very logarithms and exponentials when imaginary quantities are involved. Let us therefore take the radius of the circle, or its sinus totus, $= 1$. Then it is obvious that the circumference of this circle cannot be exactly expressed in rational numbers; but it has been found that the semicircumference is by approximation
>
> $$= 3.141592653589793\ldots$$
>
> [127 decimal places are given] for which number I would write for short π, so that π is the semicircumference of the circle of which the radius $= 1$, or π is the length of the arc of $180°$ degrees.[6]
>
> §127. If we denote by z an arbitrary arc of this circle, of which I always assume the radius $= 1$, then we usually consider of this arc mainly the sine and cosine. I shall denote the sine of the arc z in the future in this way $\sin A.z$, or only $\sin z$; and the cosine accordingly $\cos A.z$, or only $\cos z$. Hence we shall have, since π is the arc of $180°$, $\sin 0 = 0$, $\cos 0 = 1$ and $\sin \frac{1}{2}\pi = 1$, $\cos \frac{1}{2}\pi = 0$.

After a whole set of trigonometric formulas and identities, Euler continued as follows:

> §132. Since $(\sin z)^2 + (\cos z)^2 = 1$, we shall have by factorisation
>
> $$(\cos z + i \sin z) \times (\cos z - i \sin z) = 1,$$
>
> which factors, although imaginary, still are of great use in combining and multiplying sines and cosines.
>
> Consider the following product: $(\cos z + i \sin z)(\cos y + i \sin y)$, which results in
>
> $$\cos y \cos z - \sin y \sin z + (\cos y \sin z + \sin y \cos z)i.$$
>
> Since
>
> $$\cos y \cos z - \sin y \sin z = \cos(y + z)$$
>
> and
>
> $$\cos y \sin z + \sin y \cos z = \sin(y + z)$$
>
> we can express this product as
>
> $$(\cos y + i \sin y)(\cos z + i \sin z) = \cos(y + z) + i \sin(y + z)$$

[5]See Euler, *Introductio in Analysin Infinitorum*, Chapter 8, and F&G 14.A2.

[6]The symbol π was introduced by William Jones in 1706. Jones became a Fellow of the Royal Society in 1711 and created a very substantial mathematical library.

and likewise

$$(\cos y - i \sin y)(\cos z - i \sin z) = \cos(y+z) - i \sin(y+z)$$

also

$$(\cos x \pm i \sin x)(\cos y \pm i \sin y)(\cos z \pm i \sin z) = \cos(x+y+z) \pm i \sin(x+y+z).$$

§133. It now follows that

$$(\cos z \pm i \sin z)^2 = \cos 2z \pm i \sin 2z$$

and

$$(\cos z \pm i \sin z)^3 = \cos 3z \pm i \sin 3z.$$

Generally we have

$$(\cos z \pm i \sin z)^n = \cos nz \pm i \sin nz.$$

This last formula is today known as De Moivre's formula.[7] As Euler immediately explained, it is an excellent way to find the relationships between $\cos nz$ or $\sin nz$ and $\cos z$ or $\sin z$. The *Introductio* is not the first place where the connections between the exponential function and the functions sine and cosine were written down — they had been known to the experts for some time — but it is the place where it was put at the centre of the theory. §133 then continues:

It follows that

$$\cos nz = \frac{(\cos z + i \sin z)^n + (\cos z - i \sin z)^n}{2}$$

and

$$\sin nz = \frac{(\cos z + i \sin z)^n - (\cos z - i \sin z)^n}{2}$$

When we develop these binomials in a series we shall get

$$\cos nz = (\cos z)^n - \frac{n(n-1)}{1.2}(\cos z)^{n-2}(\sin z)^2 + \cdots \text{ etc.}$$

and

$$\sin nz = \frac{n}{1}(\cos z)^{n-1} \sin z - \frac{n(n-1)(n-2)}{1.2.3}(\cos z)^{n-3}(\sin z)^3 + \cdots \text{ etc.}$$

§134. Let the arc z be infinitely small; then we get $\sin z = z$ and $\cos z = 1$; let now n be an infinitely large number, while the arc nz is of finite magnitude.

Take $nz = v$; then since $\sin z = z = v/n$ we shall have

$$\cos v = 1 - \frac{v^2}{1.2} + \frac{v^4}{1.2.3.4} - \cdots \text{etc.}$$

and

$$\sin v = v - \frac{v^3}{1.2.3} + \frac{v^5}{1.2.3.4.5} - \cdots \text{etc.}$$

[7] Abraham De Moivre was a French mathematician who came to London with the Huguenots to escape religious persecution. He is best known for his work on probability, life expectancy, and the binomial distribution. The geometrical version of the formula that bears his name appears in a paper of his from 1722; he had published something similar in 1707. Euler seems not to have known of these works.

Euler next showed how to use these expressions to calculate values of the sine and cosine functions, and therefore of the other trigonometric functions.

> §138. Let us now take in the formulas of §133 the arc z infinitely small and let n be an infinitely great number ε such that εz will take the finite value v. We thus have $\varepsilon z = v$ and $z = v/\varepsilon$, hence $\sin z = v/\varepsilon$ and $\cos z = 1$. After substituting these values we find
>
> $$\cos v = \frac{\left(1 + \frac{vi}{\varepsilon}\right)^{\varepsilon} + \left(1 - \frac{vi}{\varepsilon}\right)^{\varepsilon}}{2},$$
>
> $$\sin v = \frac{\left(1 + \frac{vi}{\varepsilon}\right)^{\varepsilon} - \left(1 - \frac{vi}{\varepsilon}\right)^{\varepsilon}}{2i}.$$
>
> In the previous chapter we have seen that
>
> $$\left(1 + \frac{z}{\varepsilon}\right)^{\varepsilon} = e^{z},$$
>
> where by e we denote the base of the hyperbolic logarithms; if we therefore write for z first iv, then $-iv$, we shall have
>
> $$\cos v = \frac{e^{iv} + e^{-iv}}{2} \text{ and } \sin v = \frac{e^{iv} - e^{-iv}}{2i}.$$
>
> From these formulas we can see how the imaginary exponential quantities can be reduced to the sine and cosine of real arcs. Indeed, we have
>
> $$e^{iv} = \cos v + i \sin v,$$
> $$e^{-iv} = \cos v - i \sin v.$$

It is interesting to note that putting $v = \pi$, which Euler did not, gives the famous formula $e^{i\pi} = -1$.

Here we see that Euler connected the power series for the sine and cosine functions that he had deduced in §134 to the exponential function. Thus the sine and the cosine functions, the exponential function, and an imaginary number, are all closely related, a fact only obscurely realised by Johann Bernoulli. Nor was this all. Euler proceeded to bring into the picture another great transcendental function, the logarithm function. Looking at Euler's remarks on this will form the culmination of this section, as they show very well the profundity of his views.

Euler on logarithms. To get to Euler's conception of logarithm, we make a detour and go back a hundred years. Logarithms had been a helpful calculating device since the early 17th century, admired and used by anyone who had complicated calculations to do. But in mid-century a curious connection came to light between the idea of a logarithm and something in quite another part of mathematics, the area under a hyperbola.[8]

In 1647 Gregory Saint-Vincent, a Belgian Jesuit priest and mathematics teacher, published a work on the quadrature of the circle and other conic sections. Entitled *Opus Geometricum Quadraturae Circuli et Sectionum Coni* (A Geometrical Work on the Quadrature of the Circle and Conic Sections), it was a weighty tome of over 1250

[8]See our discussion of Nicolaus Mercator's contribution in Section 3.3.

folio pages, its style of analysis severely geometrical, using both classical exhaustion procedures and infinitesimal techniques similar to those of Cavalieri. One of its readers, a student called A.A. Sarassa, noticed something that had escaped Saint-Vincent, that his exhaustion proof of the properties of areas under a hyperbola implied a relationship between them and the properties of logarithms. What this was may be seen in Figure 9.4.

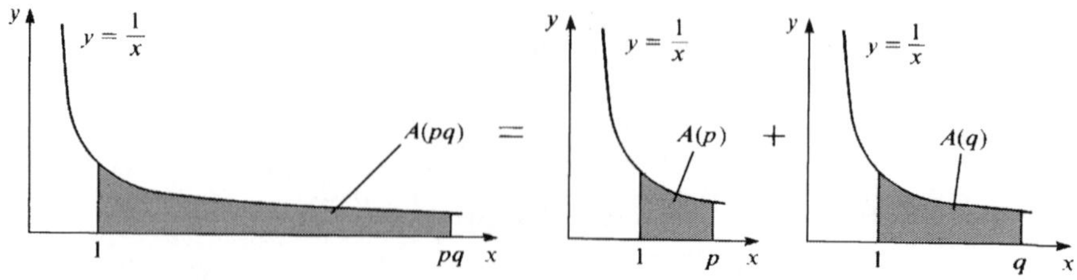

Figure 9.4. The relationship between the area under a hyperbola and the logarithm function: $A(pq) = A(p) + A(q)$

If we denote by $A(t)$ the area under the hyperbola between the values $x = 1$ and $x = t$, then the property emerging from Saint-Vincent's work is that

$$A(pq) = A(p) + A(q),$$

for any two values of p and q greater than 1. In other words, these areas behave precisely as logarithms do:

$$\log(pq) = \log p + \log q.$$

During the 1650s this insight was built upon, as mathematicians began to realise that the very tedious business of computing logarithms might be transferred into the computation of areas under a hyperbola. This was easier to think of than to carry out, however. It would be necessary to find out just what the area $A(t)$ under a hyperbola is, in numerical terms, for a whole range of values of t. It was by no means clear that this was an easier problem than that of computing logarithm values in the first place. Several mathematicians worked on this problem, including Isaac Newton (who had calculated the area under a hyperbola to 55 decimal places), Lord Brouncker, and John Wallis.

As with so much of mathematics, the study of logarithms acquired an extra twist when the calculus was introduced. A logarithm might now be *the answer* to certain integration problems (because of its definition as an area), and this led to another strange connection becoming known. In 1702 Johann Bernoulli observed that the integral of $1/(b^2 + x^2)$ could be calculated in two ways: one way gave a result to do with trigonometry (the inverse tangent function), while the other arrived at the area under a hyperbola. So the answer was something involving logarithms, and in this particular case the logarithm of an imaginary number.[9]

In fact,

$$\int_0^t \frac{dx}{b^2 + x^2} = b^{-1} \tan^{-1}(t/b),$$

[9]See (Bernoulli 1702, 400), and F&G 13.B2.

where $\tan^{-1}$ is the inverse tangent function, as can be seen from the substitution $x = b \tan y$. And as a simple piece of algebra

$$\frac{1}{b^2 + x^2} = \frac{1}{(b + ix)(b - ix)} = \frac{1}{2b}\left(\frac{1}{b + ix} + \frac{1}{b - ix}\right),$$

and so (after a little calculation)

$$\int_0^t \frac{1}{2b}\left(\frac{1}{b + ix} + \frac{1}{b - ix}\right) dx = \frac{i}{2b}(-\log(1 + (i/b)t) + \log(1 - (i/b)t))$$

$$= \frac{i}{2b}\frac{\log(b - it)}{\log(b + it)}.$$

But why is it true that

$$b^{-1}\tan^{-1}(t/b) = \frac{i}{2b}(-\log(1 + (i/b)t) + \log(1 - (i/b)t))\ ?$$

Bernoulli's answer was purely formal, but his discovery was perplexing, because his mathematical procedure had given rise to results whose meaning was unavailable within the initial context of the problem. Furthermore, it hinted at a connection between the trigonometric functions and the logarithm function, which must in turn be defined for imaginary values of the variable.

In the next decade these considerations led to a protracted debate between Bernoulli and Leibniz over the logarithms of negative and imaginary numbers. What could they be? Bernoulli maintained that the logarithm of $-x$ is the same as that of x, which Leibniz denied. Bernoulli's view implied that $\log(i) = 0$, for then necessarily $\log(-1)^{1/2} = \frac{1}{2}\log(-1)$, which equals $\frac{1}{2}\log(1) = 0$ if positive and negative numbers have the same logarithm.

An edition of the correspondence between Leibniz and Bernoulli was published by Gabriel Cramer in 1745. Euler read it and at once took up the question. His first paper on logarithms (E168), which he published in 1751, with its summary and critique of each protagonist's views, reads as though Euler read the correspondence pen in hand and wrote down his immediate response. Euler had been clear by 1744, when he began to write his *Introductio*, of a major fact about the logarithms of positive numbers: the logarithm and exponential functions are *inverse functions*:

$$\text{if } \log a = b \text{ then } a = e^b, \quad \text{and} \quad \text{if } e^b = a \text{ then } b = \log a.$$

Thus, the logarithm function could be defined as shown in Box 23.

Euler saw, too, an implication of Bernoulli's argument, that the key to solving the problem of defining $\log(-1)$ lies in generalising it to the complex case and solving that. What, then, is $\log(\alpha + i\beta)$?

Euler's solution is remarkable in both the simplicity of its reasoning and the startling nature of its conclusion: the logarithm function cannot be a single-valued function. As he wrote to Cramer in 1746,[10]

> I have finally discovered the true solution: in the same way that to one sine there correspond an infinite number of different angles I have found that it is the same with logarithms, and each number has an infinity of different logarithms, all of them imaginary unless the number is real and positive; there is only one logarithm which is real, and we regard it as its unique logarithm.

[10]Quoted in (Speziali 1983, 428).

Box 23. Euler's definition of logarithms, 1744

In Chapter 6 of his *Introductio*, Euler began by arguing that just as we know that $a^2 = a.a$ and $a^3 = a.a.a$, we also know that $a^{1/2} = \sqrt{a}$, and $a^{p/q} = \sqrt[q]{a^p}$, and in this way we understand what is meant by a^x for any x. We also know that $a^{x_1}.a^{x_2} = a^{x_1+x_2}$.

He then defined the logarithm of y to the base a as x, where $y = a^x$, and wrote $\log_a y = x$.

We now check that this yields the characteristic property of logarithms: the log of a product is the sum of the logs. If

$$y_1 = a^{x_1} \text{ and } y_2 = a^{x_2}$$

then

$$x_1 = \log_a y_1 \text{ and } x_2 = \log_a y_2.$$

Now,

$$y_1.y_2 = a^{x_1+x_2}, \text{so } x_1 + x_2 = \log_a(y_1.y_2),$$

and so

$$\log_a(y_1.y_2) = \log_a y_1 + \log_a y_2.$$

So Euler's logarithm has the characteristic property of logarithms.

To find the logarithm of $\alpha + i\beta$ Euler argued that

$$\alpha + i\beta = \gamma(\cos x + i \sin x),$$

where $\gamma = \sqrt{\alpha^2 + \beta^2}$ and x is the angle whose cosine is α/γ and whose sine is β/γ. So taking the logarithm of both sides gives

$$\log(\alpha + i\beta) = \log(\gamma(\cos x + i \sin x)) = \log \gamma + \log(\cos x + i \sin x).$$

But Euler already knew that

$$\cos x + i \sin x = e^{ix},$$

so

$$\log(\cos x + i \sin x) = ix,$$

because log and exp are inverse functions. Therefore

$$\log(\alpha + i\beta) = \log \gamma + ix,$$

where γ and x are as defined above. Euler pointed out that this means that x can be replaced by any angle of the form $x + 2k\pi$, where k is an integer, because these angles have the same values for their cosines and for their sines, and so the logarithm function takes infinitely many values, all differing by multiples of 2π.

According to Euler, the logarithm of a positive real number is found as follows. A real number α is the same as the complex number $\alpha + i\beta$, where $\beta = 0$; so what Euler had shown in this case is that α has infinitely many logarithms:

$$\log \alpha, \quad \log \alpha + 2\pi i, \quad \log \alpha + 4\pi i, \quad \ldots$$

The first of these, the usual logarithm, is real — the rest are complex.

The logarithm of a negative number likewise has infinitely many values, and, as Euler observed, $\log(-1)$ is found from the above formulae by setting

$$x = -\pi + 2k\pi, \text{ for which } \cos x = -1 \text{ and } \sin x = 0.$$ [11]

So $\log(-1) = i\pi + 2ki\pi$, where k is an integer.

Euler's 1746 clarification of logarithms did not appear in the *Introductio*, for that book was written and in the publisher's hands by 1745 (even though not it was published until 1748). Euler seems to have needed the stimulus of Cramer's publication to link up the two ideas that he already knew, the inverse nature of exp and log and his discovery that $e^{ix} = \cos x + i \sin x$. So the *Introductio* discussed the exponential function for complex values of the variable, but logarithms only in the case of a positive real variable. But by the late 1740s Euler had shown that the logarithm for other values is a many-valued function, and that this conception resolves the paradoxes that had plagued Leibniz and Bernoulli. This did not quite settle the issue to everyone's satisfaction, however. D'Alembert, for one, came up with a criticism reminiscent of Bernoulli's views. To the argument that $\log(-a) = \log(a)$, because $(-a)^2 = a^2$, Euler had claimed that it is always possible to choose pairs of values among the infinitely many values of $\log(a)$ and $\log(-a)$ for which the equation $2\log(-a) = 2\log(a)$ is true.

D'Alembert was scathing in his attack on what he saw as sophistry:[12]

> This formula means, or all analytical pretensions must be renounced, that twice the logarithm of $+a$ equals twice the logarithm of $-a$, and not that the sum of two different logarithms of $+a$ equals the sum of two different logarithms of $-a$ .... There is no argument or calculation, however subtle, which is capable of refuting such a simple proposition.

D'Alembert preferred, as Bernoulli had done, to take the logarithm as having the same value for positive and negative numbers.[13] This ensured that the logarithm was an unambiguous expression (where it was defined) but it prevented him from being able to define it for complex numbers. Euler's position was preferable in the latter respect, while being harder to understand, and indeed beyond the comprehension of many of his contemporaries.

What Euler had achieved by the mid-18th century is remarkable. He had produced a unified theory of the exponential, logarithmic, and trigonometric concepts, and had done this around the idea of a function and the algebra of infinite series. The means that he used for doing this included anything that his intuition suggested, including infinitesimals, regardless of any wider philosophical implications. He thereby transformed a collection of disparate bits of mathematical knowledge with geometrical overtones into the powerfully clear and coherent algebraic subject of analysis.

9.3 Differential equations

What actually *is* the calculus? If we take it in its Leibnizian incarnation, it is a set of algorithms for handling problems about curves, but these algorithms also apply to

---

[11]See (Euler 1751b, 162).

[12]See (D'Alembert 1761, 198).

[13]See (D'Alembert 1796, 180–209).

formal expressions involving variables. There are two basic operations, differentiation and integration, d and $\int$, which obey such rules as these:

$$d(uv) = udv + vdu \quad \text{and} \quad d \int u = u.$$

These basic operations can be given geometrical interpretations — for example, d has to do with finding tangents, and $\int$ with finding areas — and mathematicians appealed to these meanings when studying curves.

Euler's methodological insight can be expressed as saying that the calculus is not about curves: it is about *formal expressions and functions*. It is about expressions that can be differentiated and integrated. The calculus is only about curves insofar as they can be described by formal expressions. This description of a curve permits us to use differentiation to find tangents and so forth, but the concept of a formal expression is, if not more basic, then at least more general.

Euler's treatment of trigonometry and of logarithms is very much in this spirit, although the rules of the calculus are not employed. The analysis of the sine, cosine, and exponential functions is quite formal: they are expressed as power series and treated algebraically, and there is very little geometry. Above all, the analysis of logarithms and their connection with exponentials depends crucially on the concept of a function, for the logarithm is treated as the inverse function to the exponential. It is the idea of a function that unifies the whole package.

Euler's reformulation acquired significance because he made it do new work, accompanied as it was by his amazing technical versatility. This point was stressed above and we shall not labour it again here, but it should always be borne in mind. Euler both proposed a new way of formulating mathematical concepts, and provided it with a way of making them work. The point to see here is how his emphasis on mathematics as a science of formal expressions restructured mathematical theory.

It may be helpful to think of a body of mathematical work as having three aspects: problems, methods, and results. This trichotomy may not get one very far, because one person's method can be somebody else's problem, and so forth, but it is useful in helping us to focus attention on particular aspects of the story. When applied to Euler's work, the trichotomy is particularly useful in elucidating what he regarded as an answer or a result.

Debeaune's inverse tangent problem provides a good example. Whereas Bernoulli conducted a formal analysis, but couched his answer in the form of a construction, Euler went from the differential equation

$$dx = \frac{zdz}{a - z}$$

to an answer in this form:

$$x + z + a\log(a - z) = \text{ constant}.$$

To Euler, but not to his predecessors, such expressions are answers: there is no need for a geometrical interpretation.

So Euler's mathematics is full of investigations of objects defined by differential equations or integrals, and the problems are *solved* by finding power series expansions or other algebraic reformulations. Indeed, in his definitive account of the integral calculus he did not even mention Debeaune's equation by name when he gave a complete account of how to solve all differential equations of the form

$$(\alpha + \beta x + \gamma z)dx = (\delta + \varepsilon x + \mu z)dz$$

(this equation reduces to Debeaune's when we set $\alpha = a$, $\beta = 0$, $\gamma = -1$, $\delta = \epsilon = 0$, and $\mu = 1$).

Even from this more formal point of view there are good answers and not-so-good answers. An interesting example is given by the three-way correspondence between Euler, Johann Bernoulli, and Daniel Bernoulli. The problem they discussed concerned the vibrations of a rod clamped at one end. Daniel raised the problem in a letter to Euler of 18 December 1734, and he wrote to Euler again on 4 May 1735 to say that he had found a differential equation that describes its shape, but that the only solutions he could find to the equation (sines and exponentials) seemed inappropriate. At that stage, Euler could only reply with a solution in the form of a power series, and in October 1735 he presented a paper to this effect to the St Petersburg Academy of Sciences. This is not a good answer. The power series solution is unilluminating, and Euler failed to see that the rod can actually vibrate in several distinct ways.

There the matter rested, and the three of them busied themselves with other problems, until Euler spotted a marvellous simplification. He described it in a letter he wrote to Johann Bernoulli on 15 September 1739 and at much greater length in a paper published in 1743.[14]

As we can see below, he claimed in his letter that he had a simple, general method for all differential equations of the form he described. The method reduces such problems to the solution of a polynomial equation, and the answer to the differential equation was always given as a sum of exponentials, sines, and cosines.

We make a few preliminary comments on Euler's argument. We do not know how Euler hit upon his idea, but it could have been by trying to see whether the differential equation is solved by functions of the form $y = e^{-px}$. Note that if $y = e^{-px}$ then

$$\frac{dy}{dx} = -py, \quad \frac{d^2y}{dx^2} = p^2y, \quad \frac{d^3y}{dx^3} = -p^3y, \quad \text{and so on,}$$

so when these expressions are substituted into the equation, the result, as we shall see, is a polynomial equation multiplied by an exponential term e^{-px}. This term is never zero, so we can divide by it, and therefore, as Euler claimed, $y = e^{-px}$ is a solution of the differential equation if p is a solution of the polynomial equation.

Euler on linear differential equations.

> I have recently found a remarkable way of integrating differential equations of higher degrees in one step, as soon as a finite [algebraic] equation has been obtained. Moreover this method extends to all equations which, on setting dx constant, are contained in this general form:
>
> $$y + \frac{ady}{dx} + \frac{bddy}{dx^2} + \frac{cd^3y}{dx^3} + \frac{dd^4y}{dx^4} + \frac{ed^5y}{dx^5} + \text{etc.} = 0.$$
>
> To find the integral of this equation I consider this equation or algebraic expression:
>
> $$1 - ap + bp^2 - cp^3 + dp^4 - ep^5 + \text{etc.} = 0.$$

[14]See (Euler 1998, nr. 213), in F&G 14.A1(a).

If possible this expression is resolved into simple real factors of the form $1 - \alpha p$: if, however, this cannot be done resolve it into factors of two dimensions of this form $1 - \alpha p + \beta pp$, which resolution can always be done in reals, for whatever form the equation may have it can always be put in the form of a product of factors either simple, $1 - \alpha p$, or of two dimensions $1 - \alpha p + \beta pp$, all real. This resolution being done, I say that the value of y is a finite expression in x and constants, obtained from all the members which have been factors of the algebraic expressions, and singular members supply singular terms of the integral.

The question then becomes: How do we find the values of p? Euler claimed that the polynomial equation can always be factorised into linear terms of the form $1 - \alpha p$ and quadratic terms of the form $1-\alpha p+\beta pp$, by the Fundamental Theorem of Algebra, (which we discussed in Section 7.3, and which was widely believed at the time but not proved). Euler went on to relate the roots of the equations $1-\alpha p = 0$ and $1-\alpha p+\beta pp = 0$ to solutions of the differential equation via the equation

$$p = \pm\frac{1}{y}\frac{dy}{dx}.$$

The first equation is straightforward: if

$$p = -\frac{1}{y}\frac{dy}{dx} = \frac{1}{\alpha},$$

then

$$\frac{dy}{dx} = -\frac{1}{\alpha}y,$$

which has the general solution $y = Ce^{-x/\alpha}$. But Euler merely stated what happens in the second case, as he could because he was writing to Bernoulli. We have supplied the details in Box 24.

Certainly the simple factor $1-\alpha p$ gives a member of the integral $Ce^{-x/a}$, and a composite factor $1 - \alpha p + \beta pp$ gives this member of the integral

$$e^{-\alpha x/2\beta}\left(C\sin A.\frac{x\sqrt{4\beta-\alpha\alpha}}{2\beta} + D\cos A.\frac{x\sqrt{4\beta-\alpha\alpha}}{2\beta}\right)$$

where for me $\sin A.$ and $\cos A.$ denote the sine and the cosine of arcs in a circle of radius $= 1$: however it is to be noticed that if the expression $1-\alpha p+\beta pp$ cannot be resolved into simple real factors, when $4\beta > \alpha\alpha$, still the integrals are real.

Euler then gave an example of what his approach establishes.

Let the following be taken as a suitable example

$$ydx^4 = k^4 d^4 y, \quad \text{or} \quad y = k^4\frac{d^4 y}{dx^4} = 0;$$

this gives rise to the algebraic expression $1 - k^4 p^4$, whose real factors are these three $1-kp$, $1+kp$, $1+k^2p^2$; and from these spring the integrals of the equation

$$y = Ce^{-x/k} + De^{x/k} + E\sin A.\frac{x}{k} + F\cos A.\frac{x}{k};$$

Box 24. Quadratic factors.
The equation $1 - \alpha p + \beta pp = 0$ has the roots

$$p = \frac{\alpha \pm \sqrt{\alpha^2 - 4\beta}}{2\beta},$$

and because Euler knew that this quadratic has complex conjugate roots he wrote them as

$$p = \frac{\alpha \pm i\sqrt{4\beta - \alpha^2}}{2\beta}.$$

For simplicity, we write the two roots as

$$u = \frac{\alpha + i\sqrt{4\beta - \alpha^2}}{2\beta} \quad \text{and} \quad v = \frac{\alpha - i\sqrt{4\beta - \alpha^2}}{2\beta}.$$

As we have seen, the corresponding differential equation is

$$p = -\frac{1}{y}\frac{dy}{dx},$$

which has the solution $y = Ce^{-px}$, and this suggested to Euler that the solutions to the original differential equation will be of the form

$$Ae^{ux} + Be^{vx}.$$

This expression written out in full is

$$Ae^{\alpha x/2\beta}e^{i(\sqrt{4\beta-\alpha^2}/2\beta)x} + Be^{\alpha x/2\beta}e^{-i(\sqrt{4\beta-\alpha^2}/2\beta)x},$$

which reduces to Euler's expression on setting

$$A = \tfrac{1}{2}(D - Ci), \; B = \tfrac{1}{2}(D + Ci),$$

using the result

$$e^{\alpha+i\beta} = e^{\alpha}(\cos\beta + i\sin\beta).$$

in which expression, because a four-fold integration has been done in one operation, there are four new constants as the nature of the integration demands. If it would please you, most excellent sir, I shall write down the method of proof on another occasion.

Euler's new solution to the vibrating rod equation is much better than his former one, which was given as a power series. The answer is now given as a simple combination of well-known functions (exponentials, sines, and cosines). Daniel Bernoulli had noticed that such functions were among the solutions, but only Euler saw that every solution could be written in terms of them. You can see what some of the solutions actually look like at any moment in time (see Figure 9.5), and this is very instructive. For example, each of these is separately a solution.

1. $y = e^{-x/k}$,
2. $y = e^{x/k}$,
3. $y = \sin(x/k)$,
4. $y = \cos(x/k)$.

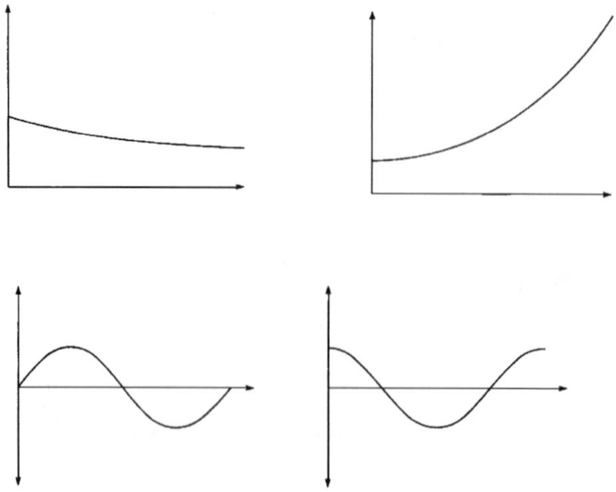

Figure 9.5. Euler's four basic solutions

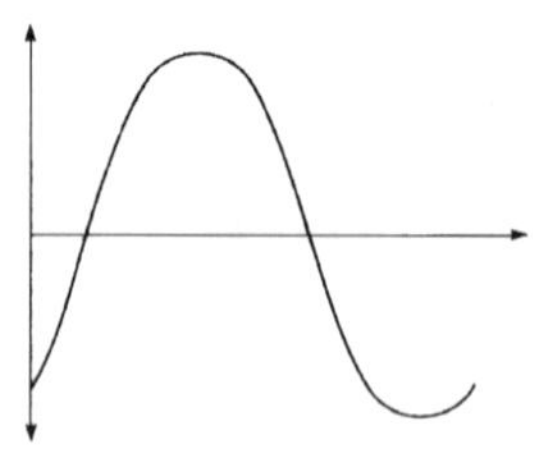

Figure 9.6. Another solution

Any sum of these basic solutions, such as $y = 2\sin(x/k) - \cos(x/k)$, which looks like the graph in Figure 9.6, is also a solution.

Each of these ways in which the rod can vibrate is different, and the new approach brings them to light in a way that the power series analysis does not. They are called its *basic modes of vibration*, and at any instant the shape of the rod is a certain sum of these basic modes.

If you look at these pictures you may wonder how the rod could perform all these oscillations while still fastened to the wall. The answer involves what are called the 'initial conditions': Euler and the Bernoullis assumed that the rod is always horizontal at the wall (that is, the mortar is secure and immovable), so any solution is subject to the two initial conditions that when $x = 0$, necessarily $y = 0$ and $\frac{dy}{dx} = 0$. The allowed solutions are consequently of the form

$$y = \alpha e^{x/k} + \beta e^{-x/k} - (\alpha + \beta)\cos(x/k) - (\alpha - \beta)\sin(x/k).$$

Euler did not send Johann Bernoulli a proof of his claims about this class of differential equations, but this did not unduly inconvenience his 72-year old former professor, who replied with a proof in early December 1739. Bernoulli began by showing how to pass from the differential equation to the polynomial equation, much as we did above.[15] Interestingly, he adhered to the geometrical language that Euler would

[15] See (Euler 1998), in F&G 14.A1(b)).

gradually drive out, and spoke of the solution curves, which he wrote as $y = n^{x/p}$, as 'logarithmic curves whose subtangent is to be found'. When he came to the equation $1 - k^4p^4 = 0$, he immediately said that its solution was $p = 1/k$. Bernoulli noticed that whereas he had one solution, Euler had exhibited several, and commented that, for this to be the case,

> my logarithms will be impossible or imaginary, but it is also the same in your solution, allowed to be more general, for you must let k be impossible or non-real.

These remarks should be interpreted as indicating that complex numbers were puzzling when they occurred in problems involving real quantities, but were nonetheless accepted, perhaps as something that could be better understood.

Euler's solution also explained something that Daniel Bernoulli had noted experimentally: a thin rod — Bernoulli used a needle — clamped to a wall can be made to emit several different sounds at once when it is struck. This corresponds to the fact that if $y = f(x)$ and $y = g(x)$ are solutions of Euler's differential equation, then so too are sums of the form $y = af(x) + bg(x)$, for arbitrary constants a and b. So at any given moment of time it may be any shape given as a sum of the basic modes. (We discuss the variation of the shapes with time in Section 10.1).

This analysis of Euler's approach to the study of differential equations clarifies one crucial way in which he differed from Johann Bernoulli over what he considered an answer. The methods that they used to solve differential equations were not very different — changes of variable, cunning substitutions, and so on — but where Bernoulli would then try to provide a geometrical interpretation, Euler did not. As we have seen, Euler also developed new methods, but the real departure was at the level of what is an acceptable answer — and it seems that he had some success in convincing Bernoulli of the value of this way of thinking.

The result was the virtual disappearance of geometrical analysis. In its place came a *theory* of differential equations. The emergence of this subject as an important part of mathematics in its own right can be explained in two ways. On the one hand, differential equations were of increasing utility in all parts of applied mathematics. We saw an example of this with the vibrating rod, but Taylor's discussion of the vibrating string also invoked a differential equation, and so did the analysis put forward by D'Alembert in the 1740s, as will be seen in Section 10.1. On the other hand, differential equations form an attractive subject in their own right, being simple to state but difficult to solve. Indeed, differential equations are still a challenge to mathematicians today.

We can get a glimpse of what a *theory* of such things might be, and how it was developed, by looking at the last of Euler's three great textbooks on the calculus, the *Institutiones Calculi Integralis*. The sheer size of the work suggests that Euler was presenting either a coherent theory or a complete zoo. In fact it was a theory, couched in terms of key ideas such as: differentials of the second or higher degree; homogeneous equations; the dimension of a variable; solution by introducing multipliers; solution by the method of infinite series; construction by quadrature; integration by approximations; and reduction and transformations, showing how heavily the formal side of the calculus is being deployed. Examples occur for the first time only on page 355 — this is usually a good sign that we are in the presence of a theory rich enough to keep mere examples at bay.

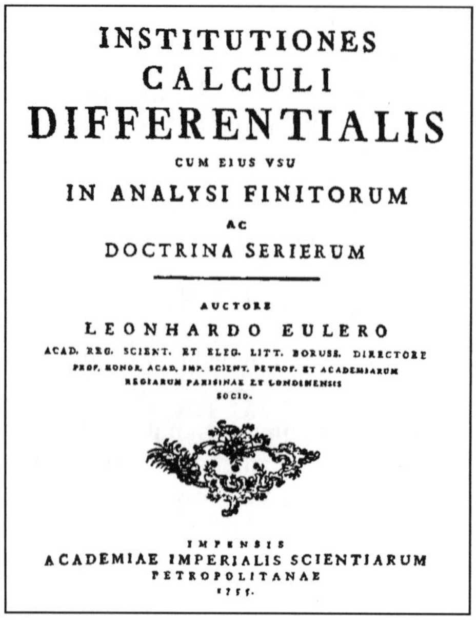

INSTITUTIONES
CALCULI
DIFFERENTIALIS
CUM EIUS VSU
IN ANALYSI FINITORUM
AC
DOCTRINA SERIERUM

AUCTORE
LEONHARDO EULERO
ACAD. REG. SCIENT. ET ELEG. LITT. BORUSS. DIRECTORE
PROF. HONOR. ACAD. IMP. SCIENT. PETROP. ET ACADEMIARUM
REGIARUM PARISINAE ET LONDINENSIS
SOCIO.

IMPENSIS
ACADEMIAE IMPERIALIS SCIENTIARUM
PETROPOLITANAE
1755.

Figure 9.7. Euler's *Institutiones Calculi Differentialis* (1755)

Euler did not disdain geometrical constructions, but they plainly comprised only one of his techniques. It was less than a hundred years since Leibniz had struggled to master a single inverse tangent problem, and already Euler felt able to present a theory of many different kinds of differential equation.

Even if one stands back and looks at Euler's work on the calculus as a whole, two other points are so obvious that they are easy to miss. The *Introductio in Analysin Infinitorum* had introduced a theory of formal expressions and functions, and the *Institutiones Calculi Differentialis* had explained what differentiation is and how to do it. However, the *Institutiones Calculi Integralis* presented the integral calculus as being about many kinds of function, including some defined by integrals, and in large part as being about differential equations. The essential subject matter is not, for example, curves, nor were foundational questions central to Euler's concerns. He was mainly interested in presenting a sophisticated and general set of methods in the style of algebraic analysis: the central results that one would learn from this book were *methods*.

We also note that Euler *published* his results and also his methods — not just in his books but also in a stream of papers. The contrast with the defensive Roberval, the secretive Newton, and even the combative Johann Bernoulli, could not be more marked. Euler went to enormous pains to make himself understood. He took great care to present his new ideas carefully, with well-chosen examples, and the theoretical developments gradually following one after another, as Condorcet rightly eulogised.[16] By the mid-18th century, the time when mathematicians could still advance their careers by concealing their methods, or leaking them under restricted circumstances to a few friends, was drawing to a close. It became increasingly necessary for them to publish their arguments, and to submit them to the judgement of their peers, before their results could be accepted. This aspect of mathematical life, which is sometimes taken

[16]See the extract from Condorcet, *Elogium of Euler* (1802, xxv–lxiii), in F&G 14.C4.

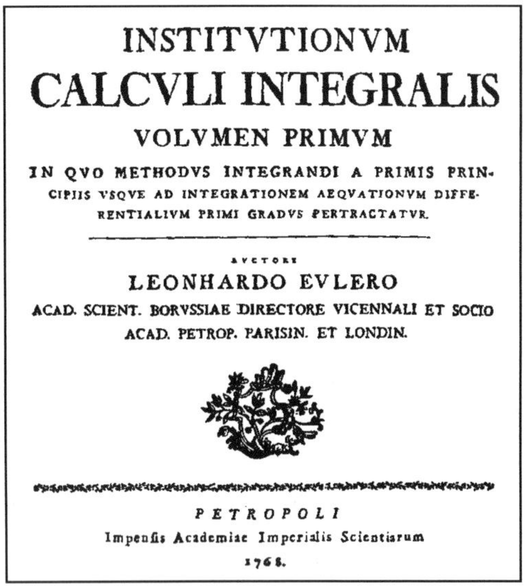

INSTITVTIONVM
CALCVLI INTEGRALIS
VOLVMEN PRIMVM
IN QVO METHODVS INTEGRANDI A PRIMIS PRINCIPIIS VSQVE AD INTEGRATIONEM AEQVATIONVM DIFFERENTIALIVM PRIMI GRADVS PERTRACTATVR.

AVCTORE
LEONHARDO EVLERO
ACAD. SCIENT. BORVSSIAE DIRECTORE VICENNALI ET SOCIO
ACAD. PETROP. PARISIN. ET LONDIN.

PETROPOLI
Impensis Academiae Imperialis Scientiarum
1768.

Figure 9.8. Euler's *Institutiones Calculi Integralis*, Vol. 1 (1768)

to be crucial to the practice of any kind of science, is a creation of the 18th century. We have seen it exemplified by Euler's work, but it was eventually to become the common practice.

The reasons for this change, as with many a social development, are hard to present convincingly. Undoubtedly, one was the sheer growth of the profession. There were now mathematicians in most of the countries of Europe, and even some in America. It was impossible to sway them by staging a mathematical contest — there were too many individuals, and to reach fellow-mathematicians through one of the many scientific journals set up in the 18th century required that one supported one's claims with proofs.

Competitions *per se* did not go away, but they changed their form. Now a scientific society would announce a topic that they thought worthy of serious attention — the motion of the Moon was one such topic (see Chapter 10) — and invite anonymous entries by a date perhaps two years hence. This focused the community's attention on the chosen topic. It was a good way to generate progress in the science, and for the candidate it was a good way to make one's name — not least because competition could be tough: Condorcet tells us that Daniel Bernoulli won the Paris Prize no fewer than thirteen times (and Euler won it twelve)! But if one dared to enter, then one's arguments would be scrutinised by the judges, and winning entries would be published and not secreted away. Finally, we might speculate that once a community of scholars grows beyond a certain size it cannot communicate its results alone, but would have to reveal its methods also; otherwise, it would have become increasingly difficult to acquire a reputation that put one's answers beyond reproach.

There was also the question of producing the next generation of mathematicians. At that time, the universities did not provide much of an education, at least in advanced mathematics. Many 18th-century mathematicians displayed early promise, based on their study of the available literature, and then, like Euler, attached themselves to a tutor from the older generation. The more that new techniques were widely available in the literature, the more good new mathematicians could be produced in this way. Whether or not the trend towards publication was deliberately pursued with this aim in mind, it certainly became easier for young mathematicians to catch up with their elders in the field, and the profession grew accordingly.

The profession was also international and non-aristocratic. Books and journals were sent to libraries across Europe, even when the national Academies of Science might have preferred to keep their discoveries to themselves, and although a private income helped, it became increasingly possible to earn one's living as a full-time mathematician. In all of these ways, Euler appears as the quintessential figure of his time. Only in the production of students did Euler achieve relatively little, but as Laplace (the leading mathematician in Napoleonic Europe) was later to say of him: 'Read Euler, Read Euler, he is our master in everything.'[17]

9.4 The foundations of the calculus

We have seen that MacLaurin defended the calculus against Berkeley's criticisms by lengthy arguments involving double reductio ad absurdum, and that Euler advocated an inconsistent collection of intuitive responses. A more constructive way forward was proposed by D'Alembert, who went directly for an account of the foundations of the calculus that was based on the use of limiting arguments, such as Newton had preferred. In his article 'Limite' in the *Encyclopédie Méthodique*, D'Alembert claimed that 'the theory of limits is the basis of the true metaphysics of the calculus', and he interpreted Newton's prime and ultimate (or first and last) ratios as limiting values, rather than as quantities springing into being or vanishing. He also argued, as Leibniz had done on occasion, that the Leibnizian derivative was likewise a limit of ratios of increments.

D'Alembert addressed directly the question of how to defend a limiting argument that attempts to make sense of an expression of the form 0/0.[18] Unfortunately his answer is not clear. He got off on the wrong foot by suggesting that 0/0 can take any value, including the correct one, and then shifted to arguing that the limit is not the ratio 0/0. Moreover, his limit concept was imprecise, and rested on geometrical ideas in an unclear way. He defined it in this way:

D'Alembert on limits.

> One magnitude is said to be the *limit* of another magnitude when the second may approach the first within any given magnitude, however small, though the first may never exceed the magnitude it approaches; so that the difference of such a quantity to its limit is absolutely inassignable.

The image, one might say, is of a fish rising to the surface, except that it never touches the air — it gets as close as you like, and may touch the surface, but it can never break

[17]This comment is much repeated, because it is so apt, but it often travels without a source being given. The original quote is from Giuglielmo Libri, who wrote in the *Journal des Savants*, January 1846, p. 51: 'These memorable words which we heard from his own lips: 'Read Euler, Read Euler, he is our master in everything''.

[18]See D'Alembert, 'Différentiels', *Encyclopédie* 4, 1754, in Struik, *A Source Book*, 342–345, and F&G 18.A3.

it. On such a definition, the number 2 is the limit of the sequence

$$1.9,\ 1.99,\ 1.999,\ \ldots,$$

but it is not the limit of the sequence

$$1.9,\ 2.01,\ 1.9999,\ 2.000001,\ \ldots,$$

because there are values in the sequence that exceed 2. D'Alembert's examples do not allow us to decide whether 2 is the limit of the sequence:

$$1.9,\ 1.999,\ 2.0000,\ 2.0000,\ \ldots,$$

in which 2 is actually reached; in all of his examples the limit is greater than every member of the sequence. It is not clear what mathematical point D'Alembert was seeking to make by insisting that 'the first may never exceed the magnitude it approaches', but it makes it clear to the historian of mathematics that his mind was working along geometrical or kinematic lines.

In 1784 Lagrange, who was then at the Académie Royale in Berlin, proposed that the Academy award a prize for a successful attempt to put the calculus on rigorous foundations. In the judges' words, the prize was to be for 'a clear and precise theory of what is called Infinity in mathematics', because[19]

> It is well known that higher mathematics continually uses infinitely large and infinitely small quantities. Nevertheless, geometers, and even the ancient analysts, have carefully avoided everything which approaches the infinite.

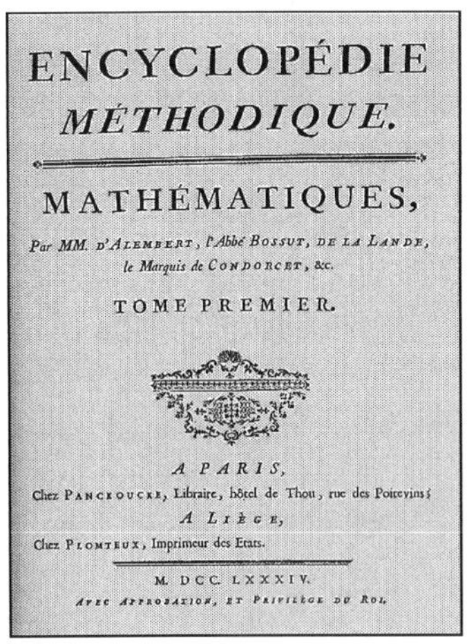

ENCYCLOPÉDIE
MÉTHODIQUE.

MATHÉMATIQUES,

Par MM. D'ALEMBERT, *l'Abbé* BOSSUT, DE LA LANDE, *le Marquis de* CONDORCET, &c.

TOME PREMIER.

A PARIS,
Chez PANCKOUCKE, Libraire, hôtel de Thou, rue des Poitevins;
A LIÈGE,
Chez PLOMTEUX, Imprimeur des Etats.

M. DCC. LXXXIV.
AVEC APPROBATION, ET PRIVILÈGE DU ROI.

Figure 9.9. The *Encyclopédie Méthodique* (1784)

The specification of the prize required that the matter be 'treated with all possible rigour, clarity, and simplicity'. It is clear from the terms of the competition that the judges were not asking for a few well-phrased definitions, but recognised that something much more substantial was required to vindicate the calculus. And although the prize was awarded to a certain Simon-Antoine-Jean l'Huilier for a work later published as *Exposition Élémentaire* (1786), none of the submissions really satisfied Lagrange and his fellow judges. In their report the judges complained of a lack of rigour and deplored

[19]Quoted in (Grabiner 1981, 41).

the candidates' failures to deal with the problem of deduction from contradictory assumptions (one of Berkeley's main criticisms); the candidates had not explained 'how so many true theorems have been deduced', and they had not even seen that the principle desired had to be 'extended to Algebra, and Geometry treated in the manner of the Ancients'.

Lagrange had a special interest in the foundations of the calculus. When he found himself teaching analysis at the recently founded École Polytechnique in 1797 he published his ideas as his *Théorie des Fonctions Analytiques* (Theory of Analytical Functions). True to his algebraic principles and preferences, he described his book as

> containing the principles of the differential calculus, relieved of all consideration of the infinitely small or of evanescent quantities, of limits or of fluxions, and reduced to the algebraic analysis of finite quantities.

His book began with a critique of earlier attempts to provide foundations for the calculus. Infinitesimals, he said, rather unfairly, were not at all rigorous, but earlier mathematicians had been concerned only with obtaining solutions to problems, and had therefore neglected the basic principles of the calculus. The criticism was unfair, not only because his preferred alternative was no more rigorous, but also because the method of infinitesimals contained more of the kernel of the eventual solution than did Lagrange's own methods.

Nonetheless, the kernel of his criticism was valid. He echoed Berkeley in deploring the fact that results were obtained by 'the compensation of errors'. He rejected fluxions because of their dependence upon motion (a 'foreign idea' belonging to physics, not mathematics) and because there was no adequate definition of 'instantaneous variable velocity'. He regarded the limit concept as too vaguely geometrical, and geometry, like motion, was 'foreign' to the very spirit of analysis. Having rejected all previous attempts, he set as his main objective the goal of placing all previous results in the calculus within a rigorous framework — he was not extending the subject but consolidating it.

In a lengthy study of Lagrange's approach, the historians Giovanni Ferraro and Marco Panza have observed that:[20]

> Throughout his theory, Lagrange certainly pursued an ideal of conceptual clarity involving the elimination of any sort of infinitesimalist insight. This has been often noticed, and was emphasised by Lagrange himself.

As such, Lagrange was not the first.

> This was a sweeping project rooted in a mathematical program ... whose *manifesto* was the first volume of Euler's *Introductio in Analysin Infinitorum*. Its main purpose was the development of a fairly general and formal theory of abstract quantities: quantities conceived merely as elements of a net of relations, expressed by formulas belonging to an appropriate language and subject to appropriate transformation rules.

The great success of the calculus was the biggest challenge to this project. Infinitesimals, and issues such as when to notice them and when to ignore them, seemed impossible to explain in formal terms. Ferraro and Panza quote Lagrange to this effect, from his short report:[21]

---

[20] See (Ferraro and Panza 2012, 96).

[21] See (Lagrange 1799).

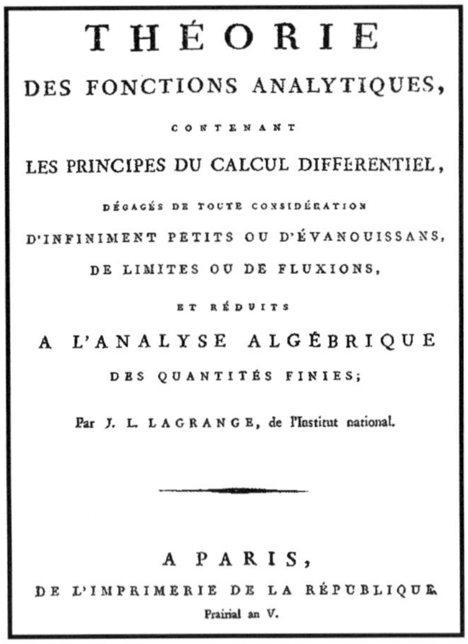

THÉORIE

DES FONCTIONS ANALYTIQUES,

CONTENANT

LES PRINCIPES DU CALCUL DIFFERENTIEL,

DÉGAGÉS DE TOUTE CONSIDÉRATION

D'INFINIMENT PETITS OU D'ÉVANOUISSANS,

DE LIMITES OU DE FLUXIONS,

ET RÉDUITS

A L'ANALYSE ALGÉBRIQUE

DES QUANTITÉS FINIES;

Par J. L. LAGRANGE, de l'Institut national.

A PARIS,

DE L'IMPRIMERIE DE LA RÉPUBLIQUE.

Prairial an V.

Figure 9.10. The title page of Lagrange's *Théorie des Fonctions Analytiques* (1797)

> I do not deny that one could rigorously prove the principles of the differential calculus through the consideration of limits envisaged in a particular way, as *MacLaurin*, *D'Alembert* and several others after them did. But the kind of metaphysics that has to be applied for this purpose is, if not contrary, at least foreign to the spirit of analysis, which should have no metaphysics but that which consists in the first principles and in the fundamental operations of calculation.

Still following the work of these historians, we can capture Lagrange's dilemma in this way. He had to show that derived functions can replace differential quotients. Specifically, given a function $y = f(x)$, he had to find an infinite sequence of other functions $f'(x), f''(x), \ldots$ that provide, apart from numerical factors, the coefficients of the power series expansion of $f(x + \xi)$. Formally, these must coincide with the differential quotients $\dfrac{d^k y}{dx^k}$ — but the notion of differential quotient has no place in Lagrange's theory.

Indeed, Euler's *Introductio* is not entirely formal: it involves the used of infinitely large and infinitely small numbers. At the very least, Lagrange needed a way to regard those aspects of Euler's work as shorthand for some purely algebraic procedure.

We can now look, albeit briefly, at what Lagrange did.[22]

Lagrange on the foundations of the calculus.

> Now let us consider a function $f(x)$ of a variable x. If we replace x by $x + i$, i being any arbitrary quantity, it will become $f(x + i)$ and, by the theory of series, we can expand it in a series of the form
>
> $$f(x) + pi + qi^2 + ri^3 + \cdots ,$$

[22]See Lagrange, *Théorie des Fonctions Analytiques*, 202ff., and extracts in Struik, *A Source Book*, 389–391, and F&G 18.A4.

> in which the quantities $p, q, r, \ldots$, the coefficients of the powers of i, will be new functions of x, which are derived from the primitive functions of x, and are independent of the quantity i.
>
> But, in order to prove what we claim, we shall examine the actual form of the series representing the expansion of a function $f(x)$ when we substitute $x+i$ for x, which involves only positive integral powers of i.
>
> This assumption is indeed fulfilled in the cases of various known functions; but nobody, to my knowledge, has tried to prove it *a priori* — which seems to me to be all the more necessary since there are particular cases in which it is not satisfied. On the other hand, the differential calculus makes definite use of this assumption, and the exceptional cases are precisely those in which objections have been made to the calculus.

Here, Lagrange has set himself the task of proving a priori that every function can almost always be expanded as a power series, partly in order to isolate and explain the cases where it was already known that it cannot be done — for example, the function $f(x) = x^{1/2}$ does not have a power series expansion round $x = 0$.

He began by proving that in the series arising from the expansion of the function $f(x + i)$, no fractional power of i can occur, except for particular values of x. Having accomplished this, he continued as follows:

> We have seen that the expansion of $f(x + i)$ generates various other functions ..., all of them derived from the original function $f(x)$, and we have given the method for finding these functions in particular cases. But in order to establish a theory concerning these kinds of functions we must look for the general law of their derivation.

What he then wrote is a slick piece of formal algebra:

> For this purpose, let us take once more the general formula
>
> $$f(x+i) = f(x) + pi + qi^2 + ri^3 + \cdots,$$
>
> and let us suppose that the undetermined quantity x is replaced by $x+o$, o being any arbitrary quantity independent of i. Then $f(x+i)$ will become $f(x+i+o)$, and it is clear that we shall obtain the same result by simply substituting $i + o$ for i in $f(x + i)$. The result must also be the same whether we replace the quantity i by $i+o$ or x by $x+o$ in the expansion $f(x)$.
>
> The first substitution yields
>
> $$f(x) + p(i+o) + q(i+o)^2 + r(i+o)^3 + \cdots,$$
>
> or, expanding the powers of $i + o$ and writing out for the sake of simplicity no more than the first two terms of each power (since the comparison of these terms will be sufficient for our purpose):
>
> $$f(x) + pi + qi^2 + ri^3 + si^4 + \cdots + po + 2qio + 3ri^2o + 4si^3o + \cdots.$$
>
> In order to carry out the other substitution, we note that we obtain
>
> $$f(x) + f'(x)o + \cdots, \; p + p'o + \cdots, \; q + q'o + \cdots, \; r + r'o + \cdots$$

> when we replace x by $x+o$ in the functions $f(x), p, q, r, \ldots$, respectively; here we retain in the expansion only the terms that include the first power of o. It is clear that the same expression will become
>
> $$f(x) + pi + qi^2 + ri^3 + si^4 + \cdots + f'(x)o + p'io + q'i^2o + r'i^3o + \cdots.$$
>
> Since these two results must be identical whatever the values of i and o may be, comparison of the terms involving $o, io, i^2o, \ldots$, will give:
>
> $$p = f'(x),\ 2q = p',\ 3r = q',\ 4s = r',\ \ldots.$$

Lagrange now drew a potent conclusion:

> Now it is clear that in the same way that $f'(x)$ is the first derived function of $f(x)$, p' is the first derived function of p, q' the first derived function of q, r' the first derived function of r, and so on. Therefore, if, for the sake of greater simplicity and uniformity, we denote by $f'(x)$ the first derived function of $f(x)$, by $f''(x)$ the first derived function of $f'(x)$, by $f'''(x)$ the first derived function of $f''(x)$, and so on, we have $p = f'(x)$, and hence $p' = f''(x)$; consequently $q = \frac{p'}{2} = \frac{f''(x)}{2}$, hence $q' = \frac{f''(x)}{2}$; consequently $r = \frac{q'}{3} = \frac{f'''(x)}{2.3}$, hence $r = \frac{f'''(x)}{2.3}$, consequently $s = \frac{r'}{4} = \frac{f^{iv}(x)}{2.3.4}$, hence $r = \frac{f^{iv}(x)}{2.3.4}$, and so on.
>
> Then by substituting these values in the expansion of the function $f(x+i)$, we obtain
>
> $$f(x+i) = f(x) + f'(x)i + \frac{f''(x)}{2}i^2 + \frac{f'''(x)}{2.3}i^3 + \frac{f^{iv}(x)}{2.3.4}i^4 + \cdots.$$
>
> This new expression has the advantage of showing how the terms of the series depend on each other and above all how we can form all the derived functions involved in the series provided that we know how to form the first derived function of any primitive function.

At this stage in his analysis, Lagrange had convinced himself that all the problems he faced reduce to obtaining the first derived function of a given function. He continued:

> Thus, provided that we have a method of computing the first [derived] function of any primitive function, we can obtain, by merely repeating the same operation, all the derived functions, and consequently all the terms of the series that result from expanding the primitive function.
>
> Finally, only a little knowledge of the differential calculus is necessary to recognise that the derived functions $y', y'', y''', \ldots$ of x coincide with the expressions
>
> $$\frac{dy}{dx}, \frac{d^2y}{dx^2}, \frac{d^3y}{dx^3}, \ldots$$
>
> respectively.

We see that Lagrange had shown, to his satisfaction, that every function f can almost always be expanded as a power series of the form

$$f(x+i) = f(x) + f'(x)i + \frac{f''(x)}{2}i^2 + \frac{f'''(x)}{2.3}i^3 + \cdots,$$

where i stands for an arbitrary increment, provided that one could somehow derive the coefficient of the first power of i from f itself. He then observed that the coefficients (f', f'', and so on — this notation was introduced by Lagrange) agreed with the calculus expressions $\frac{df}{dx}$ and $\frac{d^2f}{dx^2}$, respectively. He deduced that it would be possible to define $\frac{df}{dx}$ as f', and in this way base the calculus on the algebra of power series without appealing to geometrical concepts.

His line of reasoning was thoroughly algebraic, and his attempt to prove that it can always be done was new. He made a strong case that the meaning of $\frac{df}{dx}$ does not have to depend upon limits, infinitesimals, or first and last ratios: $f'(x)$ is simply the coefficient of i in the series expansion of $f(x+i)$. Such a claim, if valid, would go a long way towards making the calculus rigorous. However, Lagrange's ideas were not completely convincing and they were soon replaced, as you will see in Chapter 16. But the force of his example, and his eminence as a mathematician, generated much debate over the foundations of the calculus: those mathematicians who were not satisfied with merely obtaining results would now find it necessary to justify their methods much more rigorously.

Most mathematicians and historians who have written about Lagrange's work have tended to observe that his whole approach was flawed because not every function has a power series expansion. Ferraro and Panza, however, make the important observation that the historian's task is to see whether it stood on its own terms, with a late 18th-century understanding of the concepts involved. They distinguish between the modern view — which, as we shall see, came in with Bolzano and Cauchy — and an 18th-century algebraic view. In the modern view, the calculus deals with real numbers, perhaps regarded as geometrical or mechanical magnitudes. The alternative view is harder to articulate. Its notion of quantity is intended to be more abstract, if not indeed entirely so (real numbers and geometrical magnitudes being only special cases). Put crudely, mathematics deals with any kind of quantity with which one can do algebra (add, subtract, multiply, etc.). Functions are what express relations between these quantities. In the view of Ferraro and Panza:[23]

> Lagrange's theory is not algebraic insofar as it carries out a reduction to algebra understood as a separate elementary field, considered more primitive than this theory. Rather, it is algebraic insofar as it deals with algebraic quantities, that is, insofar as it is not concerned with particular quantities, but with quantities in general. And it is formal, insofar as these quantities are identified through the relations they have with each other, which are in turn displayed by appropriate formulas.

So the questions become: Can such a broad concept of quantity be made to work? Can the whole of mathematics be based on it? The crucial requirement is that the general theory Lagrange proposed makes sense when it is interpreted in the language of numbers and magnitudes. Ferraro and Panza conclude that:[24]

> Our basic point in explaining Lagrange's failure is precisely that his notion of algebraic quantity does not guarantee that algebraic quantities meet this crucial requirement, so much so that he cannot but surreptitiously suppose that they do meet it.

[23]See (Ferraro and Panza 2012, 100).

[24]See (Ferraro and Panza 2012, 100).

The conclusion that Ferraro and Panza draw from the failure of Lagrange's approach on its own terms is dramatic. They discuss the attempt to reduce all mathematics to algebraic rules for algebraic quantities and to give foundations to the calculus in such terms, and comment:[25]

> Lagrange's failure marked, at least on the Continent, the end of the program of eighteenth-century algebraic analysis begun by Euler. By taking the ideal of purity pervading this program to its extreme consequences, Lagrange's theory made it clear that, if conceived as Euler and Lagrange suggested, purity was incompatible with reductionism and foundationalism: if algebraic analysis had to be pure in this sense, the goal of recovering the whole edifice of mathematics within its limits and of grounding this edifice on it could never be reached. This makes the historical interest of this failure and of its explanation clear. Such an interest does not merely rest on the pre-eminent role Lagrange had in his time and continues to have in the history of mathematics. It is also related to the fact that this failure brings with it the end of a way of doing and conceiving mathematics that characterised a long season of its history. Reacting to Lagrange's foundational perspective and to his ideal of purity was then also a way of promoting a new idea of what mathematics should be.

9.5 Further reading

Euler Reconsidered, 2007. R. Baker (ed.), Kendrick Press. The book contains some valuable and relevant essays. The following two essays show how much can be said on the basis of a fresh examination of Euler's approach to the calculus:

Ferraro, G. Euler's treatises on infinitesimal analysis, 39–101.

Panza, M. Euler's *Introductio in Analysin Infinitorum*, 119–166.

Jahnke, N.H. 2002. Algebraic analysis in the 18th century, in *A History of Analysis* 105–136, American and London Mathematical Societies. This is a valuable survey of the calculus in the 18th century that covers much more than its title might suggest.

Wilson, R. *Euler's Pioneering Equation*, Oxford University Press, 2018. This useful short book describes the history and significance of Euler's equation $e^x = \cos x + i \sin x$ in more detail than we could include here, and looks at the contributions of Cotes and Johann Bernoulli among others.

[25] See (Ferraro and Panza 2012, 101).

10

18th-century Applied Mathematics

Introduction

In this chapter and the next we look at a variety of topics in which mathematics was used to study the natural world. The first is the motion of a vibrating string, with its implications for the production of musical sound. This leads us to a fascinating mid-century debate about the nature of mathematical functions. Then we look at Euler's work on the motion of rigid bodies and of fluids. In the next chapter we look at the contributions of Lagrange and Laplace to celestial mechanics.

It is reasonable to wonder why all this mathematics was done. One answer is its utility. Astronomy, especially the study of the motion of the Moon, had implications for navigation. Euler's *Scientia Navalis* (Naval Science), first published in 1749, is a fundamental work on hydrostatics and ship design. A later treatment of the same subject, published in 1773, was translated into English in 1776, which suggests that it was regarded as important by those who ran the world's foremost naval power.[1] Gunnery was another stimulus. In 1742, Frederick the Great asked Euler to look at Benjamin Robins' practical tract, *New Principles of Gunnery*.[2] Euler responded in 1745 with a much extended version of the tract in German, which in its turn was translated back into English in 1777; Condorcet in his *Éloge* of Euler found, however, that this work advanced nothing except the science of calculation.

As Condorcet's remark might suggest, not all of this applied work was actually useful. To give two more examples, in 1754 Euler gave a design for a turbine that was never built, and his important work on hydrodynamics was quite independent of the design of canals on an industrial scale — Euler went no further in that direction than to design an aqueduct to amuse Frederick the Great. There are also examples of topics

[1] See (Euler 1773).

[2] Robins was an English engineer who worked for the East India Company.

whose practical importance was neglected: until the 1790s the study of magnetism and electricity was regarded as an experimental science and not as a mathematical one.

There are very few examples of other eminent mathematicians busying themselves more than Euler did with questions of applied science, but how important were his contributions? A typically vigorous description of the situation was given by the historian Clifford Truesdell. In a celebrated article, written in 1960 as part of a campaign to generate historical studies of the exact sciences, he said:[3] '[anyone] who is trained in physics today will ask, what were the fundamental experiments upon which 'classical' mechanics was founded?' His answer, founded on his extensive reading of the 18th-century literature, was stark and simple: 'I have never been able to find any'. In what will be the theme for this section, Truesdell went on:

> What was, then, the method? Rational mechanics was a science of experience, but no more than geometry was it experimental ... While some great mechanical experiments were done in the Age of Reason ... [and there] were also large, cooperative projects ... the effect of all this expense on what we now consider the achievement of the period was nil. The method used in the great researches was entirely mathematical, but the result was not what would now be called pure mathematics. Experience was the guide; experience, physical experience and the experience of accumulated previous theory. If we were to seek a word for what was done, it would not be physics and it would not be pure mathematics; least of all would it be applied mathematics: it would be rational mechanics.

There are several things to tease out of this paragraph — for example, the distinction between experience and experiment. But, perhaps surprisingly, the arbiter in scientific debate throughout the 18th century was generally the theorist, and seldom the experimenter. The balance shifted only in the 19th century, and the evidence that the theorists would adduce was experience, not experiment.

Why might experience predominate over experiment in the investigation of the matters raised by Newton? One answer is that celestial mechanics does not lend itself to experimental study, only to passive observation. The motion of the Moon and the shape of the Earth are not to be altered in the laboratory, but can be dealt with by a combination of measurement and mathematics.

Truesdell's rejection of the term 'applied mathematics' derived partly from a wish to avoid introducing a term that was not in popular use until the 19th century, but also because applied mathematics is a subject in which mathematics plays a subservient role. In it, mathematics is applied, usually as a technique to harmonise discoveries already made. It might make the theory more elegant, but it is not creative. The term that was in frequent use, 'mixed mathematics', was a catch-all for the various hybrid disciplines that combined mathematics and engineering, such as architecture, ballistics, navigation, dynamics, etc., and as such did not fit Truesdell's conception either. The term 'rational mechanics' is meant to direct our attention to the subject matter (mechanics) and to the manner in which it was discussed: rationally, theoretically, and, in a significant way, mathematically.

Truesdell's distinction between physics and pure mathematics was meant to separate out an approach to the study of nature that is driven by experiment from a formal, logical system of deductive reasoning. This distinction was widely observed in the 18th

[3]See (Truesdell 1960, 35–36).

century as lying between the classical theoretical sciences and the new experimental ones.

The size of the rational mechanical enterprise is itself interesting. The historian John Heilbron quotes one observer as reckoning in 1762 that there were twenty mathematicians for every physicist, and records the Swiss polymath Johann Heinrich Lambert as saying in 1770 that 'For many years young people have emerged from universities knowing scarcely anything more than pure mathematics'.[4] With such a training one could take up rational mechanics far more easily than experimental physics. As we have seen, rational mechanics engaged the attention of every important mathematician of the century, including Clairaut, D'Alembert, and, above all, Euler.

It would seem, then, that mathematics was at the centre of 18th-century science, or at least of the dominant theoretical part of that science, rational mechanics. D'Alembert even referred to his century as 'the age of mechanics' — which might tell us something about his own interests and priorities. His view was that the centrality of mathematics derived from its place in the acquisition of knowledge.

In his *Traité de Dynamique* (Treatise on Dynamics) of 1743, and again more thoroughly in his *Discours Préliminaire* (Preliminary Discourse) to the *Encyclopédie* in 1751, he argued that algebra, geometry, and mechanics are related via a chain of ideas.[5] From the senses one learns of extension or space filled by bodies (impenetrable shapes). Drop the idea of impenetrability and one has the idea of pure extended magnitudes, to which geometry is appropriate. By abstracting the physical, and looking only at the rules for manipulating numbers (which might be the measurements of physical things), we arrive at algebra, the science of magnitudes in general. Or we can move in the other direction and pass from geometry to mechanics by enriching the conceptual mix.

So for D'Alembert mathematics played a central role in science because certainty, he argued, is obtained by reasoning based on true and self-evident principles: specifically, algebra is the most certain. Because geometry (for him) was a special case of algebra, algebraic certainty infuses into geometry up to the point where the idea of extension is opaque. Similarly, mechanics is a special case of geometry, made a little more problematic by the obscurities inherent in the concept of impenetrability. So mathematics, and algebra in particular, is central because the other activities are special (if not entirely clear) cases of it, and the role of mathematics is to bring certainty to scientific thought.

There is a markedly Cartesian flavour to all this. But D'Alembert was willing to drop the dubious ideas of Descartes (such as vortices) and was a Cartesian only in clarity of spirit. Indeed, he carried this scepticism further than most, and regarded the concept of a force as hopelessly unclear. He always refused to ascribe physical reality to the force of gravity: for him, gravity was an effect, whose cause was both unknown and unnecessary for the scientist to seek. When he wrote that 'the nature of movement is an enigma for the philosophers', he was quite prepared for it to remain their problem.

As we have seen, the most original exponent of Newtonianism was Clairaut. When philosophical rigour demanded it, he would, like Newton himself, claim that the concept of gravity did not explain anything because it was not understood. But when writing to Euler he could be less guarded. The evidence suggests that he regarded gravitational attraction as real, questioning only the precise law by which it was to be

[4]See (Heilbron 1982, 9).

[5]See (D'Alembert 1751, 19–22, 26–27), and F&G 14.D3.

described.[6] Euler, too, was always willing to talk about forces, but less certain about the precise nature of gravity.

As this view spread, a certain philosophical laziness crept in: people began to feel that the effectiveness of Newtonian physics legitimised the theoretical constructs that it contained. This differs from Newton's own mature views in that what Newton had treated mathematically, but regarded as unexplained, was now regarded as explained by the mathematics. Thus the simplest way to summarise the situation during the 18th century would be to say that mathematics was generally taken to be the way that nature was to be analysed and made to yield up its secrets. On this sophisticated and essentially Newtonian view, all scholars were agreed.

10.1 The vibrating string

The successful analysis of the problem of the vibrating string provides an excellent illustration of how differential equations (of a novel kind) enabled mathematicians to understand a phenomenon previously understood only empirically. But, rather surprisingly, it was also to demonstrate some of the weaknesses of the algebraic style of mathematics. We are interested in it because it shows both progress made and progress needed.

The problem of the vibrating string derives from the study of musical sounds. Why is the sound of a violin string so predictable? Why does tightening a string raise the pitch (which is what makes tuning possible)? Why does shortening the string also raise the pitch? In 1638 Marin Mersenne had stated the following law for determining the frequency of vibration of a string:

$$\nu = \frac{\sigma}{\ell}\sqrt{T},$$

where ν denotes the frequency of the note, ℓ the length of the string, T the tension in it, and σ is a constant, determined by the material of which the string is made. But when Mersenne raised the question of why this rule should be true he found that no one could explain it. Although Debeaune, for example, was interested in his inverse tangent problem because he thought that it would help him to explain the motion of the string, he got nowhere in that direction — and neither did Christiaan Huygens, some years later. There the matter rested until 1713, when it was taken up by Brook Taylor.

Taylor's contribution. Brook Taylor came from a musical family — whence, presumably, his interest in the topic. His father regularly played host to leading musicians of the day, and Taylor himself was a harpsichordist. He even backed up his theoretical conclusions with ingenious experiments designed to measure the rates at which harpsichord strings vibrate, because they do so too fast for anyone to count, but he carried out this work only after he had presented his theoretical analysis to the Royal Society in September 1712. And although his subsequent paper was rightly to earn him the reputation of being the first person to derive Mersenne's law mathematically, it was written in his usual cryptic style, and was hard to understand.[7]

[6] See, for example, Euler, *Opera Omnia* (4) 5, extracts in F&G 14.B2 and 14.B3.

[7] See (Taylor 1714).

Nonetheless, his main points were clear. He began by making two simplifying assumptions.

- the amplitude of oscillation of a string is independent of its frequency — that is (as every musician knows), volume is independent of pitch.
- the string vibrates in such a way that all of the string crosses the axis simultaneously (see Figure 10.1).

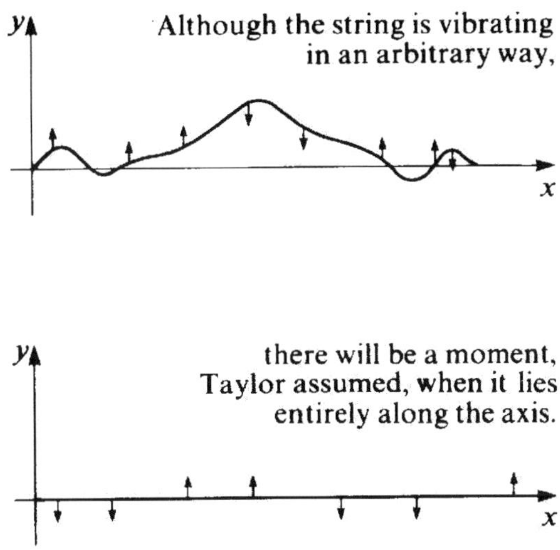

Figure 10.1. Two positions of a vibrating string, according to Taylor

From these assumptions, Taylor argued that each point of the string *therefore* behaves like a simple pendulum, and that each piece moves up and down with the *same* period. This left him with the question of what determines this period. By considering infinitesimal pieces of the string he convinced himself that:

- the force on each piece is determined by the curvature of the string[8]
- this force is equal to the force which, if acting on a simple pendulum, would cause it to oscillate with the same period as the string.

Finally, he determined the shape of the string and its frequency of vibration, and so derived Mersenne's law.

For historians, it is interesting that Taylor's second assumption is wrong — strings do not necessarily behave like this. Moreover, in deducing that each point of the string behaves like a simple pendulum he made a further error. One may think that this does not leave much, but his analysis commanded general respect until it was superseded by a new approach in the late 1740s, so we have the interesting task of working out

[8]The curvature of a curve at a point is the reciprocal of the radius of the best-fitting circle at that point; the concept had been introduced by Newton in the early 1660s (see Box 2). He used the concept to describe the acceleration of a point that traces out a curve; here it is being used, conversely, to describe a force.

why an incorrect argument, especially about a mechanical problem, might nonetheless command respect.

Taylor's argument contains more than a grain of truth, however, even though it fails to clinch the question, so various possibilities arise. Such an argument might seem a step forward, especially when the problem had proved so intractable; this could be the case if, for example, it yielded some correct, interesting, or useful results. But even if that were not the case, the argument might turn out upon closer examination to contain the germ of a correct idea; if no-one else could do any better, then the argument might be regarded as wrong but still useful. The nature of the perceived source of the error can also matter. If a premise in the argument is wrong, it might be hard to demonstrate this; it might strike one as plausible but in need of a proof, or implausible and likely to be refutable. It might seem reasonable to believe that it is often true, or at least that it is true in interesting cases. If it was a deduction that was invalid, it might still be possible to salvage the argument as a whole.

To investigate further, we consider a vibrating string, and look at a point, M, on it. Suppose that when the string is at rest the point is at P, x units along the axis, as shown in Figure 10.2.

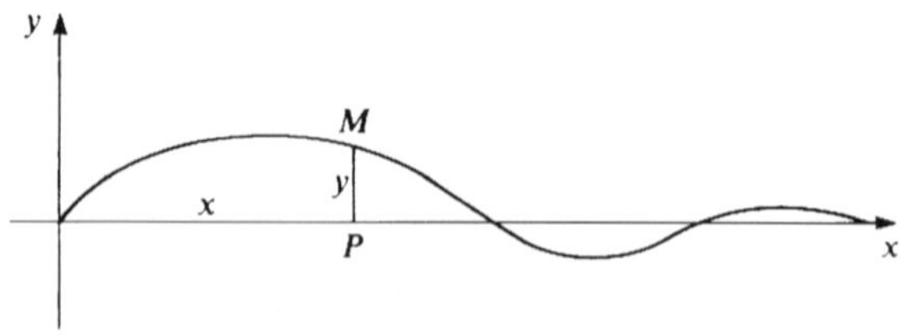

Figure 10.2. The shape of a vibrating string at a moment of time t

At any instant when the string is in motion, the point M is displaced by an amount y from P. This displacement y certainly depends on x, but the string is vibrating, so y also depends on the time t. So the shape of the string depends on two variables, x and t.

The real difficulty confronting Taylor was that the problem of the vibrating string called for a type of calculus that could deal with two independent variables, but no such theory was available in 1713, even to a mathematician of Taylor's competence. So he needed to make some simplifications to get started, and the purpose of his second assumption (that the entire string crosses the axis simultaneously) was to enable him to reduce the problem to one that essentially depends on only a single variable, time. This assumption is not obviously wrong: we cannot see what happens, because the string moves too fast. But whether one regarded it as plausible or implausible was plainly a matter of taste, some such assumption seemed essential, and his argument was clearly a major improvement on what had gone before. The invalid deduction that his argument contained needs to be set against the honour that he obtained: the first good mathematical evidence for the validity of Mersenne's law. Taylor was surely on the right track.

What this story tells us is that the Newtonian calculus was a considerable advance on earlier methods for analysing the problem of the vibrating string, but that it was still inadequate to the task as it stood. This was not the first time that the calculus had been found wanting. In Section 6.3 we saw how Leibniz and Johann Bernoulli had begun to find ways of dealing with problems in two variables, but that their methods were rather ad hoc. Accordingly, anyone who could see a way to enrich the calculus so that it could solve such problems would be taking a big step forward.

D'Alembert's breakthrough. The beginnings of a general theory of functions of two variables had been worked out by Daniel Bernoulli and Euler in the 1740s (see Box 25). The mathematician who took the decisive step to apply them was D'Alembert.

In a paper written in 1747 and published in 1749, D'Alembert wrote, after some preliminary defining of terms:[9]

> Let t be the time elapsed from the moment when the string started to vibrate: it is certain that the ordinate PM can only be expressed by a function of the time t and of the abscissa of the corresponding arc s or AP. Let, therefore, $PM = \phi(t, s)$, that is, let it be equal to an unknown function of t and s.

Note that D'Alembert assumed that the vibrations of the string are very small, so that the length s of the string from one point to another is measured by the difference in the x-coordinates of the two points. He therefore used x and s to mean the same thing (and we shall use x in our account).[10] It is conventional to draw figures in which the vibrations of the string are drawn much larger, so as to be visible.

So the height $PM = \phi(t, x)$ is a function of *two* variables, t and x, and D'Alembert was able to show how the techniques of differential equations could be made to apply to such things. He first considered the situation at a fixed moment in time as x varies — that is, he looked at the instantaneous shape of the string. Then he looked at a fixed point M on the string as the time t varies — that is, he looked at the oscillations of each point on the string.

Like Taylor, he now argued that the tension in the string caused each piece of it to be accelerated towards the axis by an amount proportional to how curved the string was at that point. In scarcely a page of work he deduced an equality between two second partial derivatives that described the motion of the string. Curiously, he never wrote this equation out explicitly, but proceeded at once to solve it. Today, we would write this differential equation as

$$\frac{\partial^2 \phi}{\partial t^2} = c^2 \frac{\partial^2 \phi}{\partial x^2},$$

and it has since come to be called the *wave equation*. Here c is a constant determined by the string.

This was a dramatic moment. It was the first time that the most powerful branch of the calculus, that of *differential equations*, was shown to extend to significant problems with more than one independent variable. Nature abounds in such problems, so to solve the problem of the vibrating string was to suggest that many other problems might similarly be solved.

Just as we saw with differential equations involving only one independent variable, the equation provides a convenient summary of the problem. If a problem is derived from the natural world, then it can be difficult even to obtain the equation, so the formalism of differential equations offers the researcher a convenient half-way house. But

[9]See (D'Alembert 1749); the extract is taken from Struik, *A Source Book*, 353.

[10]This has the effect of restricting the validity of D'Alembert's analysis to strings that move only infinitesimally from their rest positions — not a condition to which the players of stringed instruments could submit.

Box 25. Partial differentiation.
Let $f(t,x)$ be a function that is defined for all values of the two variables t and x and that can be differentiated as a function of t for any fixed value of x, and as a function of x for any fixed value of t. In the 18th and 19th centuries mathematicians tried many notations for this before settling on the one in use today and which we have imposed on the texts we consider.

The derivative of f with respect to x for a fixed value of t is denoted $\frac{\partial f}{\partial x}$, and the derivative of f with respect to t for a fixed value of x is denoted $\frac{\partial f}{\partial t}$.

For example, let $f(t,x) = x^2 \sin t$. Then

$$\frac{\partial f}{\partial x} = 2x \sin t \quad \text{and} \quad \frac{\partial f}{\partial t} = x^2 \cos t.$$

If the function can be differentiated twice, then

- The second derivative of f with respect to x is written $\frac{\partial^2 f}{\partial x^2}$.
- The derivative of f with respect to x and then t is written $\frac{\partial^2 f}{\partial t \partial x}$.
- The derivative of f with respect to t and then x is written $\frac{\partial^2 f}{\partial x \partial t}$.
- The second derivative of f with respect to t is written $\frac{\partial^2 f}{\partial t^2}$.

So when $f(x,t) = x^2 \sin t$, we have

$$\frac{\partial^2 f}{\partial x^2} = 2 \sin t, \quad \frac{\partial^2 f}{\partial t \partial x} = 2x \cos t, \quad \frac{\partial^2 f}{\partial x \partial t} = 2x \cos t, \text{ and } \frac{\partial^2 f}{\partial t^2} = -x^2 \sin t.$$

Euler was the first to show that, generally,

$$\frac{\partial^2 f}{\partial t \partial x} = \frac{\partial^2 f}{\partial x \partial t}.$$

it is obviously best if one can go on to find solutions to the equation, and in this case D'Alembert succeeded. He found that the solutions take the form (see Box 25):

$$\phi(t,x) = f(ct + x) + g(ct - x),$$

where f and g are *arbitrary* functions (provided that they can be differentiated twice) and that, for each value of t, the graph of ϕ depicts a string fastened at each end, as the original problem requires. This solution may seem very general, even vague, but on reflection we can see why it *should* be very general, for the string can be released from any initial shape and with any initial velocity. As D'Alembert noted, 'this equation includes an infinity of curves'.[11]

[11] Quoted in Struik, *A Source Book*, p. 355.

Box 26. A solution of the wave equation.
If we substitute $\phi(t,x) = F(t) \times G(x)$ into the equation

$$\frac{\partial^2 \phi}{\partial t^2} = c^2 \frac{\partial^2 \phi}{\partial x^2},$$

we obtain

$$\frac{\partial^2 F(t)}{\partial t^2} G(x) = c^2 \frac{\partial^2 G(x)}{\partial x^2} F(t),$$

which implies that

$$\frac{1}{F(t)} \frac{\partial^2 F(t)}{\partial t^2} = c^2 \frac{1}{G(x)} \frac{\partial^2 G(x)}{\partial x^2}.$$

But a function of t (on the left-hand side) can be equal to a function of x (on the right-hand side) only if they are both constant, which we shall call k^2c^2, where k^2 is some constant, as yet undetermined, related to the nature of the string. This yields the ordinary differential equations

$$\frac{d^2F}{dt^2} = k^2c^2F(t) \quad \text{and} \quad \frac{d^2G}{dx^2} = k^2G(x).$$

The solutions of these equations are of the form:

$$F(t) = \cos kct \text{ or } \sin kct, \quad G(x) = \cos kx \text{ or } \sin kx,$$

and so the solution of the wave equation is a product of these.

In a second memoir on the subject, published in 1752, D'Alembert had the idea of assuming that some solutions might be of the form

$$\phi(t,x) = F(t) \times G(x)$$

— that is, a function depending only on time multiplied by a function depending only on distance. This reduced his differential equation in two independent variables to two separate differential equations each in a single variable, as we explain in Box 26.

For the first time, recognisable solutions had appeared: functions of the form $\cos kct \cos kx$ or $\cos kct \sin kx$ are solutions of the wave equation, although they are by no means the most general ones.

We now look at how Euler also found that the wave equation has very general solutions. His solution is also important because it is one of the first occasions where the equality of mixed partial derivatives was used and understood.

Euler first rederived the equation of the vibrating string in a way that he felt led more simply to the solution.[12] He wrote the wave equation in the form

$$\left(\frac{\partial^2}{\partial t^2} - c^2 \frac{\partial^2}{\partial x^2}\right) y = 0$$

and then factorised it as

$$\left(\frac{\partial}{\partial t} + c\frac{\partial}{\partial x}\right)\left(\frac{\partial}{\partial t} - c\frac{\partial}{\partial x}\right) y = 0.$$

[12]See (Euler 1755a) E213.

This holds because the mixed partial derivatives $\frac{\partial^2 f}{\partial x \partial t}$ and $\frac{\partial^2 f}{\partial t \partial x}$ are equal. He argued that the equation of motion of the string can therefore be regarded as a pair of first-order differential equations of the form

$$\frac{\partial y}{\partial t} = c\frac{\partial y}{\partial x} \quad \text{and} \quad \frac{\partial y}{\partial t} = -c\frac{\partial y}{\partial x}.$$

He then showed that solutions of these two equations are functions of the form $g(ct+x)$ and $f(ct-x)$ respectively, and that the solutions of the wave equation are therefore also of the form D'Alembert had given.

Mersenne's law and the phenomenon of modes. D'Alembert's new ideas led to some successes: the first satisfactory deduction of Mersenne's law and the explanation of another property of vibrating strings known as the existence of modes. These are the simplest shapes in which a string can vibrate. The explanation of modes also showed precisely what was wrong with Brook Taylor's assumption that the entire vibrating string always crosses the axis simultaneously.

Mersenne's law claims that a given string vibrates with a specific frequency. On D'Alembert's analysis, if we take the solution

$$\phi(t,x) = \cos Nct \sin Nx$$

and look at a particular point x on the string, then as time varies this point moves according to the equation

$$\phi(t) = a \cos Nt,$$

where $a = \sin Nx$ is a constant. This means that it behaves as though it were the bob of a simple pendulum. Moreover, it oscillates with a frequency that is the same, whatever point of the string is taken. So the frequency of the whole string is determined — indeed, it is determined by c and so by the string itself.

Mersenne's law also claims that the frequency is inversely proportional to ℓ, the length of the string. This follows from the fact that both ends of the string are fixed, so $\phi = 0$ when $x = \ell$. This means that $\sin N\ell = 0$, so $N\ell$ must be a multiple of π, say $N\ell = k\pi$. So $N = k\pi/\ell$, and the frequency is inversely proportional to the length of the string, as Mersenne had claimed. If we then check through the technical details relating c to the length, tension, and shape of the string, as D'Alembert did, we derive the rest of Mersenne's law.

The phenomenon of modes had been demonstrated visually by John Wallis at Oxford in the 1660s. He festooned a string with paper rings and noticed that when it was struck in a way that made it emit a higher note, the rings bunched together at the middle of the string. Musicians recognise that halving a string, which doubles the frequency, results in a note that is an octave above the original note. The explanation of this phenomenon helped to confirm the mathematicians of the 18th century in their belief that they had finally understood the physical basis of pitch and harmony, and this impelled them to look more carefully at their ideas about functions.

To explain modes in the manner of D'Alembert, consider the two solutions[13]

$$\phi(t,x) = \cos(\pi ct/\ell)\sin(\pi x/\ell)$$

corresponding to $N = \pi/\ell$, and

$$\phi(t,x) = \cos(2\pi ct/\ell)\sin(2\pi x/\ell)$$

[13] A similar approach was taken by Euler (1755b, §41).

corresponding to $N = 2\pi/\ell$. In the second case the string AB, which now has twice the frequency, always has the shape shown in Figure 10.3: it behaves as though it were two strings, each of half the original length, and joined at a fixed point C in the middle. This explains how the same string can be made to play certain different notes without being tightened or changed in length, and indeed that it will naturally vibrate in a variety of ways — but not in any arbitrary way. The only tones it can emit (its *harmonics*) are those notes whose frequencies are multiples of a basic frequency.

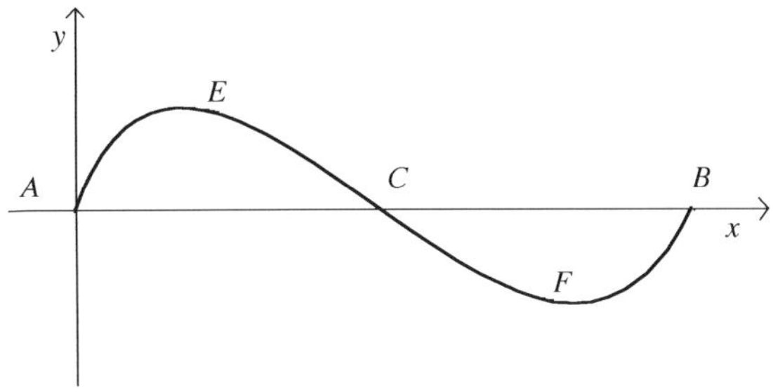

Figure 10.3. A string vibrating as though it were two strings each of half the original length

Finally, we can similarly explain how a string can emit several notes at once, thus resolving the question that had most disturbed Mersenne. For it is easy to show that if $\phi(t,x)$ and $\psi(t,x)$ are solutions of the wave equation, then so is any combination of the form $a\phi + b\psi$, where a and b are arbitrary constants. So a string may vibrate in two or more ways simultaneously, emitting two or more different notes as it does so. This phenomenon of simultaneous motion in several modes had been detected in the mid-1730s by Daniel Bernoulli and Euler in their analysis of a vibrating clamped rod. As in the case of the vibrating string, the prediction was made on mathematical rather than physical grounds. It was the existence of this phenomenon that made it easy for Euler and Bernoulli to explain why Taylor's assumption that the whole string crosses the axis simultaneously must be wrong.

What is the solution to the wave equation? How general can such a solution curve be? This was the major question opened up by D'Alembert's work on this topic, in what became one of the most famous mathematical controversies of the century.

The problem for mathematicians became how to reconcile the great generality of solutions that he had shown to exist with the particular cases of products of sines and cosines. This was to involve D'Alembert in a dispute with Euler that also embroiled Daniel Bernoulli.

D'Alembert must have sent a copy of his first paper on the vibrating string to Euler in 1748, because Euler replied to him before the paper was published with a paper of his own (E140), in which he praised D'Alembert for giving a 'very beautiful solution'. But in later papers he went on to disagree with him about the generality of the solutions, as we shall see.

D'Alembert's original claim was that any function of the form

$$\phi(t,x) = f(ct + x) + g(ct - x)$$

is a solution if it describes a string fixed at each end of the string (so $\phi(t, 0) = 0 = \phi(t, \ell)$ for a string of length ℓ). One may suppose that the functions f and g are required to be twice-differentiable, and then it is a simple matter of differentiating with respect to x and t to verify D'Alembert's claim.

It will help to focus on three aspects of the problem:

1. D'Alembert had claimed (see Box 26) that a solution can be found that is of the form $\cos kct \times \cos kx$, where k is an integer. This, coupled with the observation that the sum of any two solutions is also a solution, implies that there are solutions that are finite sums of the form

$$h(t, x) = a_1 \cos ct \cos x + a_2 \cos 2ct \cos 2x + a_3 \cos 3ct \cos 3x + \cdots,$$

   and even that there might be solutions of this last form which are infinite sums. But can every function of the form $f(ct + x) + g(ct - x)$ be described by a sum of this type — even an infinite sum, whatever that might mean?

2. A string can be put in many different shapes before it is released, so a natural question to ask is: Is every possible shape described by an expression of the above form, or are there shapes that cannot be described in this way? If so, what happens when a string is released from such an initial shape?

3. More generally, what is the relationship between a curve and a function? Is every curve the graph of a function, or must the curve be described by some formula or expression before it can be associated with a function. (We suppose that when it is drawn with respect to t, x-axes there is exactly one point on the curve for each value of x — at least in some given interval.)

In the absence of answers to questions of this kind, and in particular of any clear understanding of what was meant by the phrases '*any* function of the form $\phi(t, x) = f(ct + x) + g(ct - x)$' and '*any* curve', mathematicians had a lot to try to clarify, and a lot to disagree about.

D'Alembert believed that any shape taken by a string could be expressed as a (possibly infinite) sum of sines or cosines, and Daniel Bernoulli agreed in his (1753), but Euler did not. Rather, in 1755 (E213), he drew the more radical conclusion that any curve that is the graph of a piece of a function defined on an interval could be made to yield a solution to the wave equation.

We saw above how Euler solved the wave equation in this paper. He concluded:

> §28 Therefore taking ϕ and ψ as arbitrary functions, either of $y = \phi(ct + x)$ and $y = \psi(ct - x)$ satisfies the equation which gives the motion of the string.

Having found the general form of the solution to the equation, Euler next showed how to regard it as a sum of other solutions. First he explained how once the graphs of the functions $y = \phi(x)$ and $y = \psi(x)$ are known, the graphs of the functions $y = \phi(x + ct)$ and $y = \psi(x - ct)$ can be drawn, as well as the graphs of such functions as $y = \phi(x + ct) + \psi(x - ct)$ and $y = a\phi(x + ct)$ for any multiple a. Euler seems to have had in mind that t was fixed, and his method was nothing more than to observe that the values of the functions $y = \phi(ct + x)$ and $y = \psi(ct - x)$ are then known and can be added together and multiplied by constants.

Euler then explained what he meant by solution curves being arbitrary: they need not be restricted to those defined by explicit expressions, even infinite sums of sines and cosines. Indeed, he showed how two entirely arbitrary curves can be used to generate solutions to the wave equation

> §30 ... whether they are expressed by some equation or whether they are traced in any fashion, in such a way as not to be subject to any equation. The reader is asked to reflect carefully on this circumstance, which is the basis of the universality of my solution contested by M. D'Alembert.

The problem is the unclear relation between arbitrary curves and explicit expressions. Euler considered possible extensions of the curve which gives the initial shape of the string, and concluded that he needed to find ϕ and ψ for large values of t. He then went on:

> §37. The different parts of this curve are thus not joined to each other through any law of continuity, and it is only by the description that they are joined together. For this reason it is impossible that all this curve should be comprised in any equation, unless perchance the [initial] figure be such that its natural continuation entails all these repeated parts; and this is the case when the figure is Taylor's sine curve or a mixture of such curves according to Mr Bernoulli. This is also, according to all appearances, the reason that Messrs Bernoulli and D'Alembert have believed the problem soluble in these cases only. But the manner in which I have just carried out the solution shows that it is not necessary for the directing curve to be expressed by any equation, and the shape of the curve is itself enough to let us infer the motion of the string, without subjecting it to calculation. I will make it plain also that the motion is not the less regular than if the initial shape were a sine curve, and thus the regularity of the motion cannot be alleged in favour of the sine curves to the exclusion of all others, as Mr Bernoulli seems to claim.

This obscure paragraph addresses an apparent paradox: on the one hand, given a piece of a curve defined on an interval there is no unique way to extend the curve beyond the interval, but on the other hand, if the curve is defined by some equation then that equation specifies a unique extension of the curve. Furthermore, the paradox cannot be avoided by restricting the curves to regular curves, such as the sums of sine curves that Bernoulli had considered.

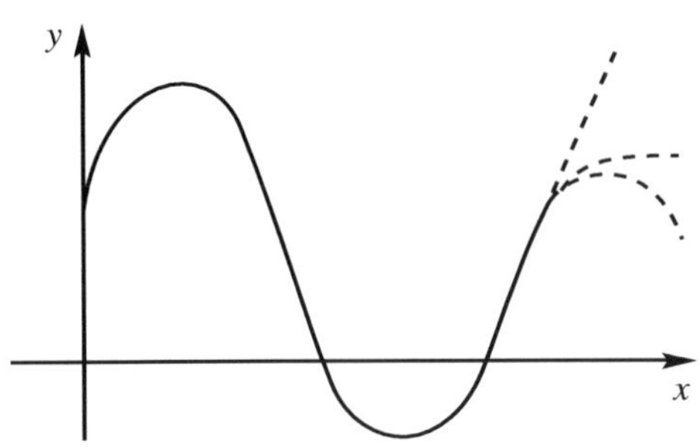

(a)

Figure 10.4. The graph of a function can be continued in many different ways

To take the first point first, suppose that we are given the piece of a curve for which x lies between 0 and 1. Can we draw the rest of it? Clearly, no: see Figure 10.4. Evidently the curve can be continued in any way we like: the idea that all of it can be determined from just one piece is absurd. On the other hand, if we believe that every curve is described by an analytic expression (such as a Taylor series might provide) then the answer would seem to be 'yes'. From the curve we obtain the analytic expression that defines it, and this is valid not just for the values of x between 0 and 1, but for all x, so this expression enables us to draw the rest of the curve.

The only way out, as Euler realised, was to admit that the techniques employed in the proof of Taylor's theorem (see Box 11) restrict its validity to a limited class of functions — those to which the processes of the calculus apply unreservedly. Consequently, there are other functions to which the calculus does not apply, or applies only with reservations, and among these will be some of those defined by graphs drawn with a free motion of the hand.[14]

D'Alembert did not want to accept this conclusion, because in his view all of mathematics rested on the universal validity of reasoning algebraically about expressions. There was one way out for him, which was to deny that the wave equation admitted solutions of such generality. In the position he came to hold, solutions to the vibrating string problem were necessarily given by equations, and what we might call analytical expressions.

It is interesting to see how the German mathematician Bernhard Riemann summarised the situation a hundred years later, in the course of a major study of his own. He wrote:[15]

Riemann on trigonometric series.

> The opinions of the prominent mathematicians of this time were, and remained, divided on the matter; for in later work everyone essentially retained his own point of view.
>
> In order to finally arrange his views on the problem of arbitrary functions and their representation by trigonometric series, Euler first introduced these functions into analysis, and supported by geometrical considerations, applied infinitesimal analysis to them. Lagrange considered Euler's results (his geometrical construction for the course of the vibration) to be correct, but he was not satisfied with Euler's geometrical treatment of the functions. D'Alembert, on the other hand, acceded to Euler's way of obtaining the differential equation and restricted himself to disputing the validity of his result, since one could not know for an arbitrary function whether its derivatives were continuous. Concerning Bernoulli's solution, all three agreed not to consider it as general. While D'Alembert, in order to explain Bernoulli's solution as less general than his own, had to assert that an analytically given periodic function cannot always be represented by a trigonometric series, Lagrange believed it possible to prove this.

---

[14] In (Euler 1770a, III, §301) (E385).

[15] See (Riemann 1867, 263). This translation is taken from Riemann, *Collected Papers*, p. 223. For another translation, see (Birkhoff and Merzbach 1973, 19).

Issues concerning the generality of the solutions, and the ways in which they can be represented, are deep. Truesdell discussed them in the following terms. His argument rests on an interpretation of what he called Leibniz's law, which he explained as follows:[16]

> To clarify D'Alembert's viewpoint it thus remains only to explain why he requires $z = \Phi(u)$ to be an 'equation'. He himself, while never giving any reason, shows by his obstinate repetitions from now on until the end of his life that he regards it as entirely obvious that 'mechanical' functions are to be exiled from mathematics, or at least from mathematical physics. This is a consequence of Leibniz's law of continuity as it was widely interpreted in the eighteenth century: Only 'continuous' functions occur in the solution of physical problems. While nowadays this seems a merely arbitrary prejudice, we must bear in mind that the majority of the geometers and more particularly the physicists of the day shared it. E.g. John [Johann] Bernoulli and D'Alembert invoked Leibniz's law in order to justify the application of the laws of physics to infinitesimal elements. Less obvious, perhaps, is the advantage of the resultant uniqueness theorem, indeed not proved but nevertheless correctly believed at the time, by which each soluble physical problem has but a single solution, determinate in principle up to a singularity resulting from its very nature, and indeed such a metaphysics would furnish a basis for regarding differential equations as a correct means of formulating natural laws.

Truesdell spoke of 'continuous' functions, in inverted commas, because the term was not precisely defined in the 18th century, and its informal meaning is not the one used in modern work. We may take it to mean that the graph of the function has no breaks, and also that the function is given by a single law — say by an expression to which the calculus applies unreservedly (for which Truesdell used the modern word 'analytic'). His interpretation was that if the calculus applies unreservedly then the functions that arise must be infinitely differentiable. But this is not necessarily true of functions whose graphs are 'drawn by a free motion of the hand'.

Truesdell continued:[17]

> While the difference between Euler's view and D'Alembert's might seem a matter of pure mathematics, in fact it is the very opposite. Today it is plain that *the phenomenon of wave motion contradicts Leibniz's law*. This was surely not obvious to Newton despite his enormous physical insight, nor to any other early physicist; rather, it is a discovery of Euler, by purely mathematical means. The differential equation
>
> $$\frac{ddy}{dt^2} = cc\frac{ddy}{dx^2}$$
>
> has solutions that are not analytic; D'Alembert's formula
>
> $$y = \phi(x + ct) + \psi(x - ct),$$
>
> as Euler interprets it, gives them at will. If [the differential equation] is the entire statement of the physical principle governing the motion of the vibrating string, then it follows that non-analytic functions occur in the solutions of physical problems. Since to this everyone today agrees without question, it is now hard to understand that Euler's refutation of Leibniz's law was *the greatest advance in scientific methodology in the entire century*. Both Euler and D'Alembert realized immediately what was at issue in the otherwise rather tedious problem of the vibrating string. This is the only scientific reason for the sharpness of the controversy that Euler and D'Alembert were to carry on until their deaths at the end of the century.

---

[16]See (Truesdell 1955, 244, 247–248) in F&G 14.C3(b).

[17]The emphases are due to Truesdell.

In Truesdell's view, the point at issue between Euler and D'Alembert was whether every function without breaks is necessarily given by a single law, and is therefore continuous in this sense. This was the prevailing, and almost unquestioned, belief, and was the 18th-century mathematician's interpretation of the influential Leibnizian philosophical tradition that Nature, and therefore the mathematics describing it, could have no gaps, jumps, or discontinuities. Accordingly, the chief significance of the debate between D'Alembert and Euler about the nature of solutions to the wave equation was that it resulted in the refutation of a crucial part of the Leibnizian philosophical tradition — an event that Truesdell called *'the greatest advance in scientific methodology in the entire century'* (his italics). It was not just that non-analytic functions must be admitted into mathematics; they must also be admitted into natural science. In some sense these functions occur in nature and cannot be kept out by some a priori statute of limitations. This was contrary to a long tradition of optimistic assumptions that the mathematical analysis of nature proceeds according to some pre-assigned harmony between the natural world and the mental world of mathematical discourse.

Although we have had to omit many developments, we can come away with an impression of the topic of the vibrating string as another one in which the formal calculus, based more and more on the apparent validity of algebraic reasoning, achieved striking successes beyond the reach of the geometrical analysis of earlier times. But, as its study also shows, formal mathematics was beginning to raise more questions than it could answer. In Chapter 9 we looked at some of the 18th-century debates about why the calculus worked, and how much reliance could be placed on algebraic analysis; we shall soon see that mathematicians of the 19th century gradually chose to stand on other ground. But first we turn to consider a branch of mathematics that prospered as the 18th century proceeded: the theory of mechanics.

10.2 Euler's vision of mechanics

Mechanics is the name given to the study of objects acted upon by forces; it is conventionally divided into statics and dynamics, according as the objects are at rest (in equilibrium) or in motion. We noted earlier that Euler had sketched in his *Mechanica* (1736) a plan that would cover point masses, solid rigid masses, elastic bodies, fluids, and gases. Much of this was unknown territory at the time, and Euler's contributions to various parts of this programme significantly enlarged the reach of mathematics. In this section, we look at Euler's account of rigid bodies and outline his work on the motion of fluids.

Rigid bodies. We can obtain a vivid picture of the 18th-century understanding of dynamics by once again comparing the contributions of D'Alembert and Euler.

D'Alembert's most important theoretical work on dynamics was his *Traité de Dynamique* (1743). It is not an easy book to read, partly because D'Alembert was frequently a very poor expositor of his own ideas, and partly because it was rushed out to secure priority when its author was worried that Clairaut was about to publish some of the same conclusions — a mistaken belief, as it transpired.

D'Alembert sought to avoid the concept of force, which he felt was unclear and explained nothing. Whenever possible, he preferred to reduce everything to the study of impact, which shows that the Cartesian tradition in mechanics still had some life in it. Part One of his *Traité*, on the general laws of motion, opens with three laws. The first, taken from Newton, says that a body remains in a state of rest if it is not acted upon by an external cause, because a body cannot put itself into motion. The second law, also taken from Newton, asserts that a body in motion remains moving uniformly in a straight line, at least until a new cause, different from the one that set it in motion, acts upon it; the resulting motion is then found by adding the velocities obtained from each cause separately by the parallelogram law. The ways in which these causes are described strongly suggest that he thought of force as really a succession of impacts.

A discussion of the measurement of motion and of time followed, culminating in an explanation of the formalism of the calculus for expressing velocity and acceleration. Here D'Alembert explained that he regarded the rate of change of velocity as all that was meant by an accelerating force.

D'Alembert's third law of motion can be paraphrased as asserting that momentum is conserved in an impact. Because momentum is defined as mass times velocity, this law amounts to a concealed definition of mass.

In the much longer second part of his *Traité*, D'Alembert introduced what was to become known as 'D'Alembert's principle'. This was obscurely expressed, it became contentious, and it owes its survival to the variational form in which Lagrange later wrote it. Yet, to quote the historian Pierre Crépel, 'it is this principle that posterity has universally accepted as one of D'Alembert's main contributions to science'.[18]

D'Alembert's principle.

> Bodies act as one another in only three different ways that are known to us: by immediate impulse, as in the case of an ordinary impact; by the interposition between them of some body to which they are attached; by virtue of mutual attraction, as in the Newtonian system of the Sun and the Planets.
>
> ...
>
> Just as the motion of a body which changes direction can be regarded as composed of the motion it had originally and a new motion that it has acquired, so the motion that the body had originally can be regarded as composed of a new motion that it has acquired, and another that it has lost. It follows from this that the laws of a motion changed by obstacles depend only on the laws of the motion destroyed by these obstacles. For it is obviously sufficient to decompose the motion of the body before meeting an obstacle into two other motions, one of which is unaffected by the obstacle while the other is annihilated.

We see that after a preamble, D'Alembert concentrated on the first two forms of action, before stating his principle.

Crépel's helpful interpretation of this passage is contained in his account of why the principle is useful:

> D'Alembert deduces from it that the determination of all motions reduces to applying the principle of equilibrium and that of composite motion. That is why it is often said that D'Alembert's principle reduces dynamics to statics. The simplest example is that of a body without elasticity obliquely striking a fixed impenetrable wall: the only component of motion preserved after the impact is that parallel to the wall, the component

[18] See (Crépel 2005, 163).

> perpendicular to the wall being destroyed (Part I, Chapter III). A typical theorem from Chapter II is as follows: 'The state of motion or rest of the centre of gravity of many bodies does not change under the mutual action of these bodies provided that the system is entirely free, that is, it is not subject to motion around a fixed point".

Whatever we make of D'Alembert's *Traité de Dynamique*, it seems clear that its greatest influence was on his protégé Lagrange, whose *Méchanique Analitique* (1788) is pervaded by D'Alembert's ideas, recast in a clearer and more rigorous form. But even Lagrange, after praising the generality of the principle, observed that:[19]

> the difficulty of determining the forces which must be destroyed, as well as the laws of equilibrium among these forces, often renders the application clumsy and troublesome.

Truesdell, although a relentless critic of D'Alembert, gave one of the best summaries of D'Alembert's principle when he noted that it involves two independent ideas:[20]

- the product of the mass times the acceleration of a body, if reversed in sign, may be regarded as a force on a par with the applied forces
- the forces exerted by the constraints need not be considered except insofar as they restrict the actual accelerations.

'Its merit', Truesdell continued, 'is the perception that *those ideas are general and may be used to obtain differential equations of motion for a large class of dynamical systems*'.

Euler's equations of motion. In 1752 Euler published a decisive reformulation of the theory of mechanics that brought it into line with the practice of the calculus as he understood it, and which has survived with little change in elementary accounts to this day.[21] Truesdell, called this paper 'a great masterpiece', and observed that 'it has dominated the mechanics of extended bodies ever since'.[22] He went on:

> This paper contains the first proposal of the so-called Newton's equations, $\mathbf{f} = m\mathbf{a}$ in rectangular Cartesian coordinates, a 'new principle of mechanics', the common origin of all the several other principles then in use.

It is worth observing that Euler connected this study to the precession of the equinoxes, which is a difficult problem in celestial mechanics concerning a slight wobble in the Earth's rotation about its axis that is caused by the Sun.

Euler began by setting out the plan of his paper. He explained that a solid body is one whose parts do not move with respect to each other, and promised to show that at any instant the motion of a solid body can be understood as the motion of its centre of gravity and the rotation of the solid around an axis through the centre of gravity. His first task, therefore, was to study how the forces acting on the solid affect the motion of its centre of gravity, and for this existing principles would suffice. His second task was then to understand the rotation about an axis that itself was varying, and for this new principles would be needed. He observed that one could begin by studying rotations around a fixed axis, but that it would be necessary to consider axes of rotation that do not pass through the center of gravity of the body.

---

[19] See Euler, *Opera Omnia* (2), 11.2, 188.

[20] See Euler, *Opera Omnia* (2), 11.2, 190, 191.

[21] See (Euler 1752), E177.

[22] See (Truesdell 1984, 317).

He then turned to the new principle upon which he proposed to base all of mechanics. It should, he said, be derived[23]

> from first principles, or rather axioms, on which all the theory of motion is based. The axioms relate to infinitely small bodies that can only have a progressive motion; and all other principles of motion must be deduced from these, those which serve to determine the motion of solids as well as of fluids; all other principles will be nothing but the application of these axioms in various ways.

There are several such principles in use, he went on, but he proposed to derive them all from a *single* principle, which he now put forward.

Consider, he said, an infinitely small body of mass M that is moving under the action of some forces. This motion can be understood by choosing a fixed but arbitrary plane, and considering the height x of the point mass above this plane. One then decomposes the forces acting on the point mass in directions parallel to the plane and perpendicular to it. Let P be the force perpendicular to the plane. After a time dt the point mass will be at a distance $x + dx$ from the plane,[24]

> and taking the element of time dt as constant, it will be the case that $2Mddx = \pm\, dt^2$, according as the force P tends to move the body away from or towards the plane. It is this single formula that contains all the principles of mechanics.

Euler proceeded to explain his formula in these terms. The quantity M is measured in units that mean that the point mass has a weight of M near the surface of the Earth, and the force P is then the weight of the body. The speed with which the body moves away from the plane is then dx/dt, and if this is the speed that it would acquire by falling through a height of h, then

$$\left(\frac{dx}{dt}\right)^2 = h, \quad \text{and so} \quad dt = \frac{dx}{\sqrt{h}}.$$

To explain the general motion of the point mass, Euler next supposed that it was measured with respect to three mutually perpendicular planes, and supposed that the forces acting were P, Q, and R. He then wrote down these equations of motion:[25]

$$2Mddx = Pdt^2; \quad 2Mddy = Qdt^2; \quad 2Mddz = Rdt^2.$$

This is the first time that Newton's equations of motion were expressed in the formalism of the calculus. We can see that they have been expressed with respect to three mutually perpendicular, but otherwise arbitrary, axes. We should also note an important difference between Euler's formulation and Newton's: Newton spoke of bodies, Euler of infinitesimal elements out of which bodies are formed.

Next Euler noted that if no forces are acting then $P = 0$, $Q = 0$, and $R = 0$, and so the above equations can be integrated and the point mass shown to move in a straight line. This establishes that a body at rest remains at rest, and that one in motion continues to move uniformly in the same direction unless it is acted upon by a force. (The separation of rest from uniform motion seems not to have been a pedagogical position of Euler's, but rather to reflect a naive belief that rest and motion of any kind are somehow different.)

---

[23]See (Euler 1752, 194).

[24]See (Euler 1752, 195).

[25]Note that Euler's conventions about units produce a factor of 2 in formulas where our conventions do not.

Motion with a fixed centre of gravity: the Euler angles. Euler next considered the motion of a body whose centre of gravity is fixed. A more complicated argument of the same kind as before, comparing the position of the body at times t and $t + dt$, then allowed Euler to deduce that at any instant the body is rotating about an axis through the centre of gravity.

He then set about describing the motion. He did so by supposing that there are three mutually perpendicular axes in the body that meet at the centre of gravitythese axes are moving with respect to the earlier choice of coordinates, which were taken with respect to three fixed planes. The difficulty is that the axis about which the body is rotating itself changes with time. Euler showed that it is enough to know how the three axes change with time; this depends on the shape of the body and the distribution of mass within it. Euler was able to find the differential equations of motion, but even he found them 'too long', and he concluded with a discussion of some special cases.

Some years later, however, Euler was able to return to this question and show in his book, *Theoria Motus Corporum Solidorum seu Rigidorum* (Theory of the Motion of Solid or Rigid Bodies) (E289) of 1765, that every rigid body has a set of axes with respect to which its behaviour is particularly simple.[26]

His first task was to describe a rotation about a fixed but arbitrary axis through the centre of gravity of the body. His description of how to describe the rotation that takes a sphere from one position to another was new, and is remembered to this day when we refer to the 'Euler angles' of a rotation. It is inevitably complicated, and is best followed pen in hand, drawing as we go, and consulting the translation in the Euler Archive.

To define the position of the axis, Euler considered a sphere of radius s with its centre I at the centre of gravity of the body. Euler then chose an arbitrary set of three mutually perpendicular axes IA, IB, IC in the body. He supposed that the points A, B, C lie on the sphere and define a triangle on it with three right angles (see Figure 10.5). We know where the body is when we know where the sphere is, and we know where the sphere is when we know where the three axes are.

The axis of rotation, denoted by IO, meets the sphere at two points, one of which, O, with coordinates (x, y, z) with respect to the coordinate axes in the body. This gave him three arcs on the sphere,

$$AO = \alpha, \; BO = \beta, \; CO = \gamma,$$

that specify the position of the axis of rotation with respect to the axes in the body. These arcs are the *Euler angles*: they specify the position of the axis of a rotating body with respect to angles in the body itself. In this way, we know where the axis of rotation is with respect to the three axes. It remained for Euler to describe how points on the body move as they rotate about the axis IO.

Euler next considered the position of an arbitrary point of the body, which he denoted by Z (see Figure 10.6). The line IZ meets the sphere in two points, and he considered the arcs AZ, BZ, and CZ, for which

$$\cos AZ = \frac{x}{s}, \quad \cos BZ = \frac{y}{s}, \quad \cos CZ = \frac{z}{s}.$$

He could now discuss the motion of the point Z with respect to the axes in the body. He decomposed the velocity along the three coordinate axes, and considered the change

[26]Our account concentrates on Vol. 1, Ch. 5. There is an English translation of much of this book by Ian Bruce in the Euler Archive.

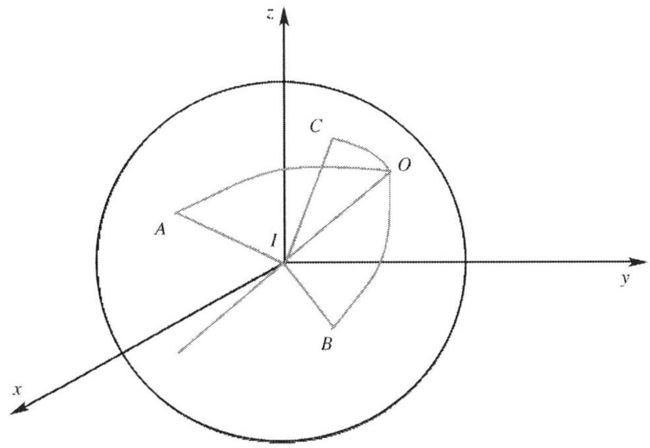

Figure 10.5. The Euler angles of a body rotating about an axis through its centre of gravity, I

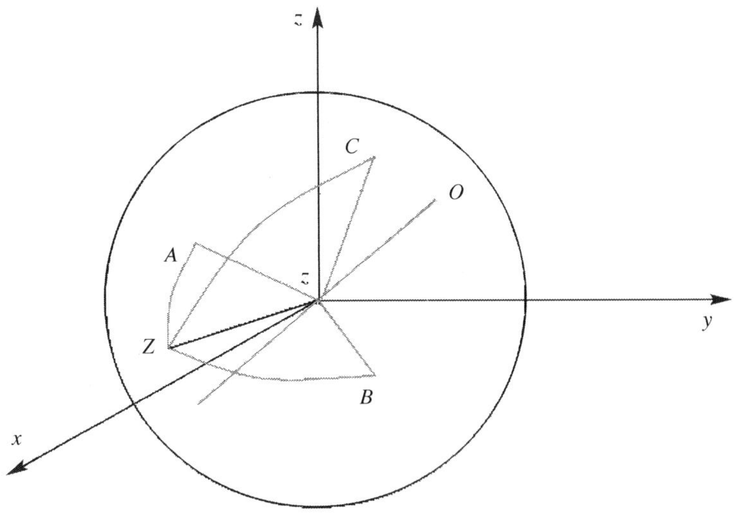

Figure 10.6. A point Z on a body rotating about an axis through its centre of gravity

in position of Z between the times t and $t + dt$. The mathematics was complicated rather than difficult. The important thing to note is that once Euler had the velocity of Z as a function of time, and therefore also the acceleration, his fundamental principle immediately allowed him to connect the motion with the forces acting on the body and therefore to do dynamics.

This gave Euler the coordinates of an arbitrary moving point on the sphere.

Quantifying the moment of inertia. Euler then addressed the question of how to describe and quantify the motion of a rotating body. We first illustrate the problem with an example of a wheel rotating about an axis. We shall think of an idealised wheel in which a very thin rim of some mass is attached to the centre by weightless spokes.

We surely want to say that the faster the wheel goes, the greater is the energy it contains, and the greater is the impact it would make on anything (such as a brake) coming into contact with the rim. We also want to say that if two wheels of the same mass but different sizes rotate at the same angular velocity then there is more energy in the larger wheel, and that if two wheels of the same size rotate at the same angular velocity but with different masses then the more massive wheel has the greater energy.

If we now consider a point on the rim of a rotating wheel, then we know that its linear velocity v at any instant is the product of its angular velocity ω and its distance r from the centre:

$$v = r\omega.$$

If the point has mass m then the momentum of the point is $mv = mr\omega$ and its kinetic energy is $\frac{1}{2}mv^2 = \frac{1}{2}mr^2\omega^2$. These formulas agree with what we know about bodies moving in straight lines.

Euler's task was to put these ideas into a systematic theory of bodies rotating about axes (whether in the body or not).[27] In Chapter 1 he defined the *centre of inertia* of a body as

> §285. a point in any body, around which the mass or inertia is equally distributed in some manner according to the equality of the moments.

He commented that it is the same point as the one commonly called the *centre of gravity*.

In Chapter 3 he set himself this task:

> §361. If a rigid body at rest and mobile about a fixed axis is acted upon by some forces, to find the motion arising in the first instant of time.

He addressed the problem in this way:

> The moments of all the forces are gathered together with respect to the axis of gyration, with attention paid to whatever sense they turn, and let the sum of all the moments be equal to Vf, and from the sense of this motion the first direction to be impressed is known. Then let $d\omega$ be the angle, through which the body is urged forwards [accelerated] about the axis in the element of time dt, and the individual elements of the body dM are multiplied by the square of their distances from the axis rr and from the calculation there is gathered the integral $\int rrdM$.
>
> With which put in place it is necessary that
>
> $$\frac{dd\omega}{2gdt^2}\int rrdM = Vf.$$
>
> Thus now in turn the angle $d\omega$ is elicited, through which the body turns in the element of time dt from the moment of the forces Vf.

This is not perhaps obvious, but it is a statement of the form

$$\text{acceleration} \times \text{mass} = \text{force}$$

with which we are familiar from Newton's work:

(the acceleration in the angle) × (something to do with the mass and its distribution in the body) equals the applied force.

Euler summed up his finding as follows, introducing for the first time the concept of *moment of inertia*:

---

[27]Throughout this section, all translations are by Ian Bruce. Our additions are in square brackets.

Euler on moment of inertia.

> §362. Hence the angle completed $d\omega$ in the element of time dt varies directly as the moment of the forces Vf and inversely as $\int rrdM$, which is the sum of all the elements of the body dM multiplied by the square of their distances from the axis of gyration.
>
> §363. This formula is similar to that, by which the generation of progressive [linear] motion is expressed, while here in place of the forces, the moment of the forces and in place of the mass of the body M the value of the integral $\int rrdM$ is taken, which value henceforth we will call the moment of inertia.
>
> ...
>
> §422. The moment of inertia of a body with respect to some axis is the sum of all the products which arise, if the individual elements of the body are multiplied by the square of their distances from the axis.

There are two things to note here. First, the moment of inertia, being an integral of terms dM and r^2 that are always positive, is also necessarily positive. Second, this calculation can be done with respect to any axis, and is not restricted to any axis about which one might suppose that the body is 'really' rotating.

Euler's next innovation was to introduce the concept of the principal axes of rotation of the body. This was the breakthrough that extended Newtonian mechanics from the study of point masses to arbitrary bodies — everything from cars and orbiting satellites to the bones in our bodies.

The principal axes of rotation. Euler now embarked on an examination of what happens when a different axis is taken. First he considered a new axis parallel to the first one, and found that the further the axis is from the centre of inertia, the greater is the corresponding moment of inertia. Therefore, said Euler, one should take the axis through the centre of inertia because it yields a minimum for the moment of inertia among all axes in that direction.

Next, Euler considered axes through the centre of inertia I, but pointing in different directions. He took a set of three axes IA, IB, and IC that are at right angles to each other. He then took a small piece of the body dM with coordinates x, y, z with respect to these axes, and he introduced a new axis through I, which he called IG.

He then rotated the coordinate axes so that one of them coincided with IG. He derived the formulas for the position of dM with respect to the new axes and deduced the expression for the moment of inertia of the body around the new axis. The formulas are somewhat complicated but, as Euler had already observed, the moment of inertia is always positive, and so it makes sense to look for an axis IG for which the moment of inertia is a minimum or a maximum. This is a calculus problem involving the angles between the axes IA, IB, IC, and the axis IG, which Euler proceeded to solve.

The result is a cubic equation, which must have either one real root or three. Euler was unable to deduce from the equation itself that there are always three real roots, and instead gave an obscure argument to support this. The result, as Euler proclaimed, was that every rigid body has three axes mutually at right angles with respect to which the moments of inertia are either a maximum or a minimum. These he called the *principal axes*:

> §446. The principal axes of any body are these three axes passing through the centre of inertia of this body, with respect to which the moments of inertia are either a maximum or a minimum.

Euler then spelled out how to analyse the motion of a rotating solid body with respect to any axis, in terms of its motion with respect to the principal axes. He also explained how the action of forces affects the motion, and how to solve many problems in the dynamics of rigid bodies.

Fluid mechanics. The theoretical and mathematical study of fluid flow may be said to have begun with Galileo and, more profoundly, with Newton's attempt to explain the tides and the shape of the Earth. It was extended by Daniel Bernoulli in his *Hydrodynamica* (Hydrodynamics) of 1738, who obtained a formula connecting the pressure on the walls of a container of a volume of fluid with the velocity of the fluid. This provoked his father, Johann Bernoulli, to publish his *Hydraulica* (Hydraulics) in 1742, replete with the false claim that he had written it in 1732. A year later, Clairaut discussed the shape of the Earth in his *Théorie de la Figure de la Terre* (Theory of the Shape of the Earth), and in 1747 D'Alembert outlined a new theory of the tides in his *Réflexions sur la Cause Générale des Vents* (Reflections on the General Cause of Winds).

But it was Euler's work on fluids that showed how the calculus could be most successfully applied to bodies that are not solid and inflexible. In particular, Euler gave the equations of motion for what is called a perfect fluid — one that is incompressible and inviscid (without viscosity or 'stickiness').[28] Although he did provide a discussion of equilibrium figures of a fluid that suggests that some of them might approximate the shape of a planet, he was particularly pleased to derive a theory of fluids based on the idea that a fluid is composed of infinitesimal solid bodies, because this extended his version of Newton's mechanics to fluids.

He began his first paper (E225) by offering what he regarded as the first clear distinction between solids and liquids: a solid can be held in equilibrium by two equal and opposite forces, but a fluid is in equilibrium only if it is held in place by an equal force at each point of its surface which acts perpendicular to the surface. He deduced that in an incompressible fluid in equilibrium the pressure in a body of fluid at a point depends only on the depth of the point, and noted the changes that have to be made if the fluid is elastic or compressible. But Euler found further analysis very difficult, and he contented himself with a study of particular cases, such as the theory of the barometer.

He made more progress in his next paper (E226), where he considered the motion of an infinitesimal cube of an incompressible fluid during an infinitesimal moment of time. He argued that what drives the motion is the differences in pressure between each pair of opposite faces of the cube. If the pressures are not equal, then the pressure difference manifests itself as a force that causes the cube to move. Euler showed that this analysis leads to three equations, one in each of the x-, y-, and z-directions, that describe a pressure difference as an acceleration in that direction, which causes an acceleration at each point and also stretches and squashes the coordinate axes.

[28] Euler published three memoirs (E225, E226, E227) on this in 1757 and a further paper (E258) in 1761.

The result is the three equations:[29]

$$\begin{aligned}
\frac{\partial u}{\partial t} + u\frac{\partial u}{\partial x} + v\frac{\partial u}{\partial y} + w\frac{\partial u}{\partial z} &= -\frac{1}{\rho}\frac{\partial p}{\partial x},\\
\frac{\partial v}{\partial t} + u\frac{\partial v}{\partial x} + v\frac{\partial v}{\partial y} + w\frac{\partial v}{\partial z} &= -\frac{1}{\rho}\frac{\partial p}{\partial y},\\
\frac{\partial w}{\partial t} + u\frac{\partial w}{\partial x} + v\frac{\partial w}{\partial y} + w\frac{\partial w}{\partial z} &= -\frac{1}{\rho}\frac{\partial p}{\partial z} - g.
\end{aligned}$$

In these equations, (u, v, w) denotes the velocity at a point in the fluid with coordinates (x, y, z), and each of u, v, and w is a function of x, y, z and the time t.[30]

There is also an equation stating that the volume of every part of the fluid remains unchanged. Euler explained this better, and expressed the conclusion more elegantly, in his *Principia Motus Fluidorum* (Principles of the motion of fluids) (E258), which he published in 1761. This equation is

$$\frac{\partial u}{\partial x} + \frac{\partial v}{\partial y} + \frac{\partial w}{\partial z} = 0.$$

This is known as the *continuity equation* for the flow of an incompressible liquid in space, and it expresses the idea that as the liquid flows its volume does not change.

In the three papers we have looked at, Euler established necessary conditions that any flow of an incompressible liquid must obey. These equations must describe a great variety of possible motions, however, and Euler was able to deal only with special cases, although he did become the first person to describe vortex motion in mathematical terms.

In fact, Euler's equations of motion for a perfect fluid are still far from being adequately understood, and the problems that are raised by the equations for a general fluid (the so-called Navier–Stokes equations) are among the 'millennium problems' whose solutions could earn a mathematician a million dollar prize from the Clay Mathematics Institute.[31]

10.3 Further reading

Darrigol, O. 2005. *Worlds of Flow*, Oxford University Press. The mathematics in this book is not easy, but the phenomena that it describes are rich, diverse, and well explained in their historical settings.

Hankins, T.L. 1985. *Science and the Enlightenment*, Cambridge University Press. This is a valuable overview, and particularly useful on D'Alembert.

Romero, A. 2007. 'Physics and analysis: Euler and the search for fundamental principles of mechanics', in *Euler Reconsidered*, R. Baker (ed.), Kendrick Press.

Truesdell, C.A. 1984. *An Idiot's Fugitive Essays on Science*, Springer. Truesdell was for many years the doyen of Euler scholars, and he combined deep learning with strong views on many aspects of the history of science.

[29] See (E226, p. 286).

[30] We have modernised his notation: Euler at this point gave a careful explanation of how his expressions are to be interpreted as partial derivatives, but he lacked a specific notation to express them.

[31] See (Carlson, Jaffe, and Wiles 2006).

11

18th-century Celestial Mechanics

Introduction

In Chapter 5 we saw how Newton provided an explanation for Kepler's laws in the *Principia* and thereby transformed the subject of celestial mechanics. In this chapter we look at how Newton's theory was used to investigate some of the principal problems in celestial mechanics during the second half of the 18th century.

11.1 Testing the *Principia*

Whereas Newton's work was essentially geometrical in character, the work of his successors was based on the new language of the calculus, as Lagrange pointed out in 1768:[1]

> A century will soon have elapsed since [the *Principia*] saw the light of day, and a great number of authors have worked to clarify and improve it; but it does not seem that the parts that in fact stand in need of being improved have been perfected in a proper way to form a true commentary. These are above all those that treat of the movement of fluids, and of the effect of the mutual attraction of the planets, that is, a part of the second book and nearly all of the third, where one no longer finds the rigor and precision that characterised the rest of the work. The problems that Newton was unable to resolve with the aid supplied by his century and his genius, have subsequently been resolved in a large part by the geometers [mathematicians] of this century; but their solutions, based on different principles and on analyses more or less long and complicated, are scarcely proper to form a sequel to a work which sparkles everywhere with the elegance and simplicity of its demonstrations. It would thus be very interesting to undertake to translate, so to speak, these solutions into the language of *Principia Mathematica*, to add the others that are still lacking, and to give thus to the greatest production of the human mind the perfection of which it is capable.

[1] Quoted in (Wilson 1985, 25–26). We add that in 18th-century French the word 'géomètre' or geometer applied to a gentleman, and 'mathématicien' or mathematician to a tradesman.

While it is not clear that the 'translation' suggested by Lagrange would be possible in practice, the fact that Lagrange considered such a translation desirable is interesting. Even though he saw the calculus as the tool for achieving results — as it was for him and his contemporaries in celestial mechanics — he still saw geometry as the purest form of mathematics, which it was impossible to surpass.

11.2 Academy prizes

The 18th century was a period in which research in celestial mechanics was strongly encouraged by the Paris Académie des Sciences. From 1720 the Académie, prompted in part by a need to find a solution to the longitude problem, had begun to award prizes for work in astronomy and in navigation. Many of the contests, which were open exclusively to non-members of the Académie, featured planetary problems, with Euler and Lagrange usually dividing the spoils.

But as the century wore on Laplace emerged as a new force in celestial mechanics, and from the 1770s onwards, progress in the subject was no longer dominated by Euler and Lagrange, but by Lagrange and Laplace, as the historian Curtis Wilson has observed:[2]

> Neither Lagrange nor Laplace can be looked to today for the kind of mathematical rigour that would be developed after them, by such mathematicians as A.-L. Cauchy and K.T.W. Weierstrass. Lagrange's concern for symmetric form may be appear to us to be at times almost a fetish; Laplace's sharp eye for results may seem opportunistic. But in the interactive relation between the two men, opposing qualities complemented each other, and accounted for the main advances of celestial mechanics during the last forty years of the eighteenth century.

Prizes in celestial mechanics awarded by the Paris Académie des Sciences, 1748–1780

Year	*Topic*	*Winner*
1748	Inequalities of Saturn and Jupiter	L. Euler
1750	Inequalities of Saturn and Jupiter	–
1752	Inequalities of Saturn and Jupiter	L. Euler
1764	Libration of the Moon	J.L. Lagrange
1766	Inequalities of Jupiter's four satellites	J.L. Lagrange
1768	Lunar theory	–
1770	Lunar theory	L. Euler and J.-A. Euler
1772	Lunar theory	L. Euler and J.L. Lagrange
1774	Secular equation of the Moon	J.L. Lagrange
1776	Theory of perturbations of comets	J.L. Lagrange
1778	Theory of perturbations of comets	N. Fuss (student of L. Euler)
1780	Theory of perturbations of comets	J.L. Lagrange

[2]See (Wilson 1995, 130).

For much of this time Lagrange was in Berlin, having succeeded Euler at the Académie des Sciences there in 1766. In 1787 he moved to Paris where Laplace, the younger of the two by some thirteen years, had been active since the early 1770s. Laplace was much influenced by Lagrange and did not fight shy of telling him so, writing to him in November 1778:[3]

> No one reads you with more pleasure than I, because no geometer [mathematician] appears to me to have carried to as high a point as you all the parts that go to make a great analyst. Permit me this avowal of my gratitude and respect, since it is principally by an assiduous reading of your excellent works that I have formed myself.

11.3 Laplace

Figure 11.1. Pierre-Simon Laplace (1749–1827)

Pierre-Simon Laplace began his career as a professor of mathematics at the École Militaire in Paris, where he taught from 1769 to 1776, and where, while acting as an examiner in 1785, he first encountered the young Napoléon Bonaparte. Their paths would cross again later, and when Napoléon seized power in 1799 he appointed Laplace as Minister of the Interior. The appointment was not a success, and after only six weeks Laplace was replaced by Napoléon's brother. Laplace was later appointed to the Senate and in 1803 became Chancellor of the Senate, a position of little power but carrying a substantial salary. Napoléon, having recognised Laplace's shortcomings, later said of him that 'he sought everywhere for subtleties, had only problematic ideas, and in short carried the spirit of the infinitesimal into administration'.[4]

---

[3]Quoted in (Wilson 1995, 109).

[4]Quoted in (Gillispie 1997, 176).

Laplace was also the unofficial leader of the Bureau de Longitudes, the national organisation founded in 1795 in the aftermath of the Revolution to help with practical astronomy and navigation. But it was not as a politician or as an administrator that Laplace made his mark upon the world, but as a mathematician and scientist, recognition that began with his election to the Académie des Sciences in 1773.

It was during the 1770s that Laplace established his mathematical reputation, embarking on a programme of research in celestial mechanics (a term that he coined) and in probability. In 1796 he published a two-volume work entitled *Exposition du Système du Monde* (Account of the System of the World), which was a resounding success. It was a semi-popular treatment of his work in celestial mechanics, and a model of French prose that had its origins in a course of lectures that he had been due to deliver at the École Normale, but which was never given, because of the closure of the École. But his major work in celestial mechanics was a monumental five-volume treatise, his *Traité de Mécanique Céleste* (Treatise on Celestial Mechanics) which took him twenty-six years to write, and in which he aimed to set down all there was to know about the subject, providing a review of all the significant achievements since the time of Newton. As he stated in the Introduction, he wanted to establish the validity of Newton's laws as far as astronomy could reach:[5]

> Astronomy, considered in the most general manner, is a great problem of mechanics, in which the elements of the motions are the arbitrary constant quantities. The solution of this problem depends, at the same time, upon the accuracy of the observations, and upon the perfection of the analysis. It is very important to reject every empirical process, and to complete the analysis, so that it shall not be necessary to derive from observations any but indispensable data.

His general approach was first to develop the theory and then to combine observation and calculation to demonstrate its usefulness. Unfortunately, he totally neglected to mention the works on which his own work depended, which makes it extremely difficult for us to sort out the relationships between his own contributions and those of Euler, Clairaut, D'Alembert, and Lagrange. Although the *Mécanique Céleste* included techniques and results that people wanted to use, as indeed Laplace expected them to, the mathematics was extremely difficult and the text contained very few diagrams to help. An idea of its difficulty can be gleaned from the fact that when the American mathematician Nathaniel Bowditch provided a masterful translation of the first four volumes into English, he included an extensive commentary of similar length to the original treatise.[6] As Bowditch himself famously remarked:[7]

> Whenever I meet in La Place with the words 'Thus it plainly appears', I am sure that hours, and perhaps days of hard study will alone enable me to discover *how* it plainly appears.

So high was the standard of Bowditch's commentary, and such was the demand for it, that even French mathematicians wanted it, and they wanted it to be translated into French.[8] Meanwhile Laplace's *Exposition du Système du Monde*, although a work in its own right was also an outline or prospectus for the *Mécanique Céleste*, and Laplace

[5] P.-S. Laplace, *Traité de Mécanique Céleste* (transl. N. Bowditch), Vol. 1, xxiii.

[6] N. Bowditch, *Mécanique Céleste by the Marquis de Laplace, translated with commentary*, 4 vols., Boston (1829–39).

[7] Bowditch, Vol. 4, p. 62.

[8] Bowditch, Vol. 4, p. 64.

arranged for a second edition of it to be published specifically to accompany the launch of the treatise.

For those in England who were unable to cope with unexpurgated Laplace, even with the help of Bowditch's translation and commentary, there was a more digestible version entitled the *The Mechanism of the Heavens* (1831) by Mary Somerville, one of a rare band of 19th-century women who publicly engaged in mathematical or scientific pursuits. Somerville was the first woman to have experimental results published by the Royal Society, and her rendering of Laplace was much admired, not least by Laplace himself, who considered her an enlightened judge of his work[9] and is reported to have said that she was the only woman who understood it.[10]

Figure 11.2. Mary Somerville (1780–1872)

The other field in which Laplace made an important contribution is the theory of probability. We cannot discuss this here, except to draw attention to its connection with his work on celestial mechanics. This is made explicit in his popular work *Essai Philosophique sur les Probabilités* (*A Philosophical Essay on Probabilities*) (1814) in which he describes how 'the probablility calculus' led him to many of his important results in planetary theory. In the same work, Laplace gave full expression to his philosophy of determinism, describing an 'intelligence':[11]

---

[9]See (Hahn 2013, Vol. 2, 1250), and (Stenhouse 2020, 20–21).
[10]See (Somerville 1873, 156).
[11]See (Laplace 1814/1995, 2).

Laplace's demon.

> We ought then to consider the present state of the universe as the effect of its previous state and as the cause of that which is to follow. An intelligence that, at a given instant, could comprehend all the forces by which nature is animated and the respective situation of the beings that make it up, if moreover it were vast enough to submit these data to analysis, would encompass in the same formula the movements of the greatest bodies of the universe and those of the lightest atoms. For such an intelligence nothing would be uncertain, and the future, like the past, would be open to its eyes.

This passage reproduces almost exactly one that Laplace had published almost forty years previously, in a 1776 essay on the integration of differential equations and its application to the theory of chance, indicating its significance for him. It is strikingly reminiscent of a passage by Leibniz, which Laplace may well have known:[12]

> From this one sees then that everything proceeds mathematically — that is, infallibly — in the whole wide world, so that if someone could have a sufficient insight into the inner parts of things, and in addition had remembrance and intelligence enough to consider all the circumstances and to take them into account, he would be a prophet and would see the future in the present as in a mirror.

Laplace's all-comprehending intelligence is now famously known as 'Laplace's Demon'. It is called a Demon because it is supposed to be a secular entity and not a divine intelligence. Although it is not known who first used and popularised the term, the notion became well known in the 19th century, feeding into the belief that ultimately no motion could defy prediction. As we shall see in Chapter 21, by the end of the 19th century this belief was to be completely shattered.

11.4 The stability of the solar system

One of the fundamental questions of celestial mechanics, and one that particularly occupied Laplace, concerns the stability of the solar system. Will the planets in the distant future continue to move in the way that they do now, or will something catastrophic, such as a collision or an escape, eventually occur? Or, to put it another way, if we know the present positions and velocities of the planets, can we predict their motions for all future time (and also deduce them for all past time)? In 1718 Newton himself had voiced doubts about this stability:[13]

> For while Comets move in very excentrick Orbs in all manner of Positions, blind Fate could never make all the Planets move one and the same way in Orbs concentrick, some inconsiderable Irregularities excepted, which may have risen from the mutual Actions of Comets and Planets upon one another and which will be apt to increase, till this System wants a Reformation.

Under Newton's law of gravitation, each planet moves (to a first approximation) in an elliptical orbit around the Sun, the Sun being at one focus of the ellipse. This description is a first approximation because it allows only for the interaction between the

[12] See (Leibniz 1906, 129–134).

[13] See (Newton 1718, 378).

Sun and the particular planet whose motion is being described, and does not take into account the forces between the individual planets. Although these planetary forces are weak because the masses of the planets are very small relative to the mass of the Sun — the ratio of the mass of Jupiter (the largest planet) to the mass of the Sun is approximately 1 : 1000 — they cause perturbations to the original elliptical orbit so that it changes very slowly. It is conceivable that these slow changes could, after a very long period of time, alter the present orbits in such a way that a planet could be thrown out of the system or a collision could occur.

Although such a scenario does not agree with observations made over the last millennium, it is much harder to prove mathematically that this could never happen, and it is the search for such a mathematical proof that provides the connection with the three-body problem. If the solar system is considered as a system of bodies, and if only gravitational forces are taken into account (all other forces such as solar winds or relativistic effects being ignored), then it can be modelled as an n-body problem, for some whole number n. If we want to solve the n-body problem, a reasonable strategy is to start with a low value for n, solve the problem, and then see whether the solution can be generalised.

By the middle of the 18th century the two-body problem had been completely solved. As Kepler had famously said of Mars, a planet will move round the Sun in an ellipse. Newton had given a geometrical solution in his *Principia* and Johann Bernoulli had provided the first analytical solution in 1710. However, it was clear from early on that the three-body problem was not going to give up its secrets so easily. Newton, who had first formulated the problem in the *Principia*, was only too well aware of the difficulties that it entailed, and later speculated:[14]

> By reason of the deviation of the Sun from the centre of gravity, the centripetal force does not always tend to that immobile centre, and hence the planets neither move exactly in ellipses nor revolve twice in the same orbit. There are as many orbits of a planet as it has revolutions, as in the motion of the Moon ...but to consider simultaneously all these causes of motion and to define these motions by exact laws admitting of easy calculation exceeds, if I am not mistaken, the force of any human mind.

The basic definition of the problem is as follows:

> Three particles move under their mutual gravitational attraction: Given their initial positions and velocities, determine their subsequent motion.

In three-dimensional space the position and velocity of each body are defined by three position coordinates and three velocity components, which makes a total of eighteen independent variables.

Using Newton's laws, we can describe the problem using nine second-order differential equations (see Box 27). This means that a complete solution would involve 18 arbitrary constants (known as 'integrals'). However, some of these constants are determined by fundamental mechanical properties that a three-body system shares with a two-body one: the system's centre of mass moves in a straight line with constant velocity (this provides six integrals), its angular momentum is conserved (giving three more integrals) and the energy in the system is conserved (see Box 28). This provides ten integrals in total, but finding further integrals proved difficult, and mathematicians turned to other methods.

[14] Quoted in (Smith 2002, 153).

Box 27. The equations of motion of the three-body problem

In the three-body problem are three particles P_1, P_2, P_3 with masses m_1, m_2, m_3, respectively. The ith particle P_i has coordinates (x_{i1}, x_{i2}, x_{i3}). The Pythagorean theorem gives us the distance r_{ij} between the ith and jth particles.

Newton's laws of motion tell us that the force that particle P_j exerts on particle P_i is given by the gravitational constant (in units in which this is 1) multiplied by the product of the masses m_j and m_i and divided by the square of r_{ij}. When we resolve this into its components in the directions of the coordinate axes, this is written as

$$m_i m_j \left(\frac{x_{j1} - x_{i1}}{r_{ij}^3}, \frac{x_{j2} - x_{i2}}{r_{ij}^3}, \frac{x_{j3} - x_{i3}}{r_{ij}^3} \right).$$

The appearance of the *cube* of the distance may be unexpected, but notice that each fraction has the dimension of the square of the reciprocal of the distance. Notice also that this formula is obviously correct when the positions of the particles differ in only one coordinate.

The acceleration of particle P_i is

$$(\ddot{x}_{i1}, \ddot{x}_{i2}, \ddot{x}_{i3}).$$

Newton's law, that force equals mass times acceleration, tells us that to understand the motion of particle 1, say, we look at the force on it exerted by particles 2 and 3 in each of the component directions, and we equate this force to the corresponding component of the above product (in which we put $i = 1$ and take $j = 2$ and then $j = 3$). When we do this, the coefficient m_1 cancels, and we obtain, for the first component,

$$\ddot{x}_{11} = m_2 \frac{x_{21} - x_{11}}{r_{12}^3} + m_3 \frac{x_{31} - x_{11}}{r_{13}^3}.$$

The equations for the other components come out the same way, and are

$$\ddot{x}_{12} = m_2 \frac{x_{22} - x_{12}}{r_{12}^3} + m_3 \frac{x_{32} - x_{12}}{r_{13}^3},$$

$$\ddot{x}_{13} = m_2 \frac{x_{23} - x_{13}}{r_{12}^3} + m_3 \frac{x_{33} - x_{13}}{r_{13}^3}.$$

The same argument, but with an arbitrary i and with i, j, k all different, gives this equation for the nth component of particle i, for i and $n = 1, 2, 3$:

$$\ddot{x}_{in} = m_j \frac{x_{jn} - x_{in}}{r_{ij}^3} + m_k \frac{x_{kn} - x_{in}}{r_{ik}^3}.$$

This is a set of nine second-order differential equations. These equations are harder to solve than they look, because the positions are hidden in the distances r_{ij} and the variables cannot be separated.

In 1747 Euler was developing his ideas concerning the Sun–Earth–Moon system — the most famous example of a three-body problem. He was one of the first to formulate the general equations of motion of the problem (it was done simultaneously

Box 28. Consequences of the equations of motion.
We now use the symmetries in the equations in Box 27. The equations

$$m_1\ddot{x}_{1n} + m_2\ddot{x}_{2n} + m_3\ddot{x}_{3n} = 0,$$

for $n = 1, 2$, or 3, are obtained by multiplying the equation for $\ddot{x}_{1n}$ by m_1, the equation for $\ddot{x}_{2n}$ by m_2, and the equation for $\ddot{x}_{3n}$ by m_3, and adding. These three equations can then be integrated to give

$$m_1x_{1n} + m_2x_{2n} + m_3x_{3n} = A_nt + B_n,$$

where A_n and B_n are constants of integration. These equations show that the centre of mass of the three particles remains at rest or moves uniformly in space in a straight line.

Similarly, we find that

$$m_1x_{11}\ddot{x}_{12} + m_2x_{21}\ddot{x}_{22} + m_3x_{31}\ddot{x}_{32}$$
$$- (m_1x_{12}\ddot{x}_{11} + m_2x_{22}\ddot{x}_{21} + m_3x_{32}\ddot{x}_{31}) = 0.$$

Two similar equations are found by cycling the numbers 1, 2, 3.

These equations can then be integrated to give

$$\sum_i m_i\,(x_{i2}\dot{x}_{i3} - x_{i3}\dot{x}_{i2}) = C_i,$$

where C_i is a constant, and two similar equations that are found as before.

Together, these equations tell us that the angular momentum of the system about each of the coordinate axes is constant throughout the motion.

and independently by Clairaut and D'Alembert), which he reformulated by introducing trigonometric series and using the method of undetermined coefficients. In this method, which goes back to Newton, a power series $\sum_n a_nx^n$ is substituted into the differential equation and a recurrence relation obtained that leads to expressions for the coefficients in terms of some arbitrary constants.

Although Clairaut did not publish his lunar theory until 1753, his procedures first appeared in a memoir on the perturbations of Saturn and Jupiter, which won a prize of the Académie des Sciences in 1748 and which we discuss below.

The first use of Euler's theories was in his construction of lunar tables published between 1744 and 1750. In 1754 these tables were superseded by the more accurate tables of the German astronomer Tobias Mayer (as we saw in Section 6.3). Mayer had used Euler's theory to construct his tables, but he had also developed additional techniques to bring his tables into better agreement with observations.

Mayer's tables won him a longitude prize of £3000 from the British government, but Euler's contribution was also recognised, as is evident from a letter of 13 June 1765 from the Longitude Commissioners to the Navy Board:[15]

[15]National Maritime Museum, London, ADM/A/2572; see *Papers of the Board of Longitude*, `https://cudl.lib.cam.ac.uk/collections/rgo14/1`.

> ...great Progress has been made towards discovering the Longitude at Sea by a set of Lunar Tables constructed by Tobias Mayer deceased late Professor at Goetingen in Germany upon the principles of Gravitation laid down by Sir Isaac Newton, in the Construction of which Tables he was considerably assisted from Theorems furnished by Professor Euler of the University of Berlin ... And that the said Professor Euler is also deserving of an honorary and pecuniary Acknowledgement for his useful and ingenious Labours towards the discovery of the Longitude ... it is therefore enacted that a Reward or Sum of Money, not exceeding Three hundred pounds in the whole, shall be paid to the said Professor Euler ...

Euler continued to work on the lunar theory and in the following decade was responsible for an important simplification of the three-body problem. In a paper presented to the St Petersburg Academy in 1762 (published in 1766) he gave the first formulation of what came to be known as the 'restricted three-body problem'.[16] It appears in the fourth section of the introduction to the paper, the scene having been set with an explanation of why celestial bodies appear to move in Keplerian ellipses (where one gravitational interaction dominates all the rest), and in particular why mathematicians have been able to achieve certain success in their determination of the motion of the Moon (it is sufficiently close to the Earth and its orbit is not too eccentric):[17]

> 4. An investigation of this sort, which almost seems to transcend the forces of human ingenuity, should certainly not be undertaken suddenly, but rather it will be proper that our efforts be directed step by step through it. Therefore, the general problem of three bodies mutually attracting each other will thus most conveniently be restricted: so that the mass of one might, as it were, vanish before the remaining two. In this way, we will understand that two bodies, the larger two of course, should move following the *Keplerian* laws; and all perturbation in the motion of the third should be consumed; if the position and motion of the third body were set from the beginning as if it should be attracted to each of the larger two by an equal force, then in this way we will have a case whose investigation requires a completely new method. This problem is far from ready to be approached, so that rather, I might be forced to admit to having exerted myself in vain in pursuing it. However, truly I have observed a completely singular case, with memorable simplicity, where the motion of the Moon could be treated in such a way that it would perpetually appear either attached to or opposite the Sun; consideration of such a case, which is not lacking in use for this most difficult business, should, it seems, not cause any displeasure.

So Euler's idea was to start with a two-body system whose motion is known, and then add to it a third body of negligible mass moving in the same plane as the first two bodies. Since the third body is attracted by the other two, but does not influence their motion, the problem is reduced to that of determining the motion of the third body from the equations that Euler derived (see Box 29).

In Figure 11.3 the two finite bodies, P_1 of mass m_1 and P_2 of mass m_2, are rotating in a common plane about their centre of mass which is located at O, the origin of the coordinate system. The position of the third body P in the same plane is given by the distances r_1 and r_2 of the body from P_1 and P_2, respectively.

This may appear to be rather an artificial formulation, but, as Euler recognised, it is rather a good approximation to certain real problems — in particular, the Sun–Earth–Moon problem, if we neglect the eccentricity of the Earth's orbit and the inclination of the Moon's orbit on the ecliptic (the projection of the Earth's orbit on the celestial

[16] See (Euler 1764/1766, 544–558) (E304).

[17] See (Euler 1764/1766, 546–547), transl. R. Cretney.

Box 29. The equations of motion for the restricted three-body problem. Let $m_1 = 1 - \mu$ and $m_2 = \mu$ be the masses of the two finite bodies P_1 and P_2, choose the unit of distance so that the constant difference between the two finite bodies is 1, and choose the unit of time so that the gravitational constant k^2 is also 1. If the coordinates of P_1, P_2 and the third (massless) body P are $(\xi_1, \eta_1), (\xi_2, \eta_2)$, and (ξ, η), respectively, and if

$$r_1 = \sqrt{(\xi - \xi_1)^2 + (\eta - \eta_1)^2},$$
$$r_2 = \sqrt{(\xi - \xi_2)^2 + (\eta - \eta_2)^2},$$

then the equations of motion of the third body are

$$\frac{d^2\xi}{dt^2} = -(1-\mu)\frac{(\xi - \xi_1)}{r_1^3} - \mu\frac{(\xi - \xi_2)}{r_2^3},$$
$$\frac{d^2\eta}{dt^2} = -(1-\mu)\frac{(\eta - \eta_1)}{r_1^3} - \mu\frac{(\eta - \eta_2)}{r_2^3}.$$

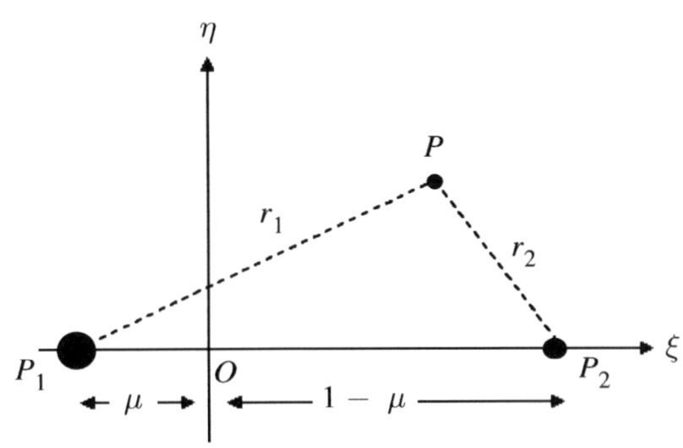

Figure 11.3. The restricted three-body problem

sphere). It can also be used for the Sun–Jupiter–small planet problem, if we neglect the eccentricity of Jupiter and the inclination of the orbits.

Euler's paper contained other important results. In particular, he showed that certain initial conditions give rise to a special class of solutions, now known as 'particular solutions'. These are periodic solutions — solutions in which the motion repeats itself in equal intervals of time — with the added property that the geometrical configuration of the bodies does not change over time.

A particular solution can occur in two ways. Either the configuration simply rotates in its own plane around the centre of mass, or an expansion or contraction takes place in which the ratio of the distance between the first and second bodies to the distance between the second and third bodies remains constant. The particular solutions found by Euler are collinear (see Figure 11.4). In these solutions, all the bodies are set in motion from positions on a straight line and, given the appropriate initial conditions, they will stay on that line while the line itself rotates in a plane about the centre of mass of the bodies. Thus, the bodies move periodically in ellipses, but maintain a collinear configuration.[18]

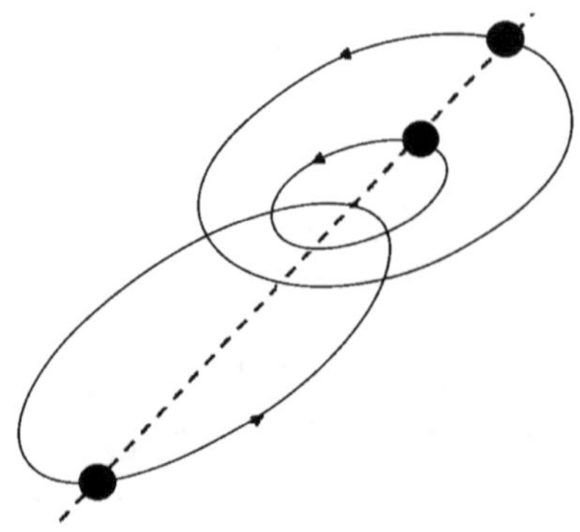

Figure 11.4. A collinear solution to the restricted three-body problem

Euler continued to work on the restricted problem, and in the early 1770s he had the clever idea of formulating it using a rotating coordinate frame. Contrary to what one might expect, this greatly simplifies the problem, although making the necessary coordinate transformation requires some additional work at the beginning (see Box 30). This is because, when one axis lies on the line through the centre of gravity of the two large bodies, these bodies appear stationary — it is as though the two large bodies are fixed to the x-axis. Moreover, the transformation results in autonomous equations of motion — that is, equations that do not explicitly depend on the independent variable. This enables one to gain a greater insight into the dynamics of the system than if a stationary coordinate frame were used. Euler did not solve the restricted problem, but, as we shall see in Chapter 21, he laid an important foundation for others to work on.

Lagrange's most celebrated work on the three-body problem appeared in a memoir that he wrote for the Paris Académie prize contest of 1772. The contest asked for a contribution to the lunar theory, and was won jointly by him and Euler. In his memoir, in which he first developed the theory and then applied it to the motion of the Moon, Lagrange tackled the three-body problem in an entirely new way, by trying to determine the motion of the bodies relative to each other, solely in terms of the distances between them. As he explained:[19]

> These researches contain a method for resolving the three-body problem which is different from all those presented up to now. It consists of employing in the determination of the orbit of each body elements none other than the distances between the three bodies — that is to say, the triangle formed by these bodies at each instant. For this, it is necessary first to find the equations which determine these distances by the time; next, supposing the distances known, it is necessary to deduce the relative motion of the bodies with respect to an arbitrary fixed plane. It will be seen, in the first Chapter, how I am able to fulfil these two objectives, although the second in particular demands a delicate and very complicated analysis.

---

[18] In the following year, 1765, Euler showed that collinear solutions also exist for the general three-body problem (E400).

[19] See (Lagrange 1772a, 229).

Box 30. The equations of motion of the restricted three-body problem in rotating coordinates.

If the equations of motion are now referred to a new system of axes x and y, with the same origin as the old axes, but rotating in the $\xi\eta$-plane in the direction in which the finite bodies move with uniform angular velocity, then the coordinates (ξ, η) of the third body in the new system are given by

$$\xi = x\cos t - y\sin t, \qquad \eta = x\sin t + y\cos t,$$

with similar sets of equations for (ξ_1, η_1) and (ξ_2, η_2).

Differentiating these equations twice and using some algebra gives

$$\frac{d^2x}{dt^2} - 2\frac{dy}{dt} = x - (1-\mu)\frac{(x-x_1)}{r_1^3} - \mu\frac{(x-x_2)}{r_2^3},$$

$$\frac{d^2y}{dt^2} + 2\frac{dx}{dt} = x - (1-\mu)\frac{(y-y_1)}{r_1^3} - \mu\frac{(y-y_2)}{r_2^3}.$$

Because we can always choose the direction of the x-axis so that the two finite bodies lie on it, we have $y_1 = y_2 = 0$ and the equations become

$$\frac{d^2x}{dt^2} - 2\frac{dy}{dt} = x - (1-\mu)\frac{(x-x_1)}{r_1^3} - \mu\frac{(x-x_2)}{r_2^3},$$

$$\frac{d^2y}{dt^2} + 2\frac{dx}{dt} = x - (1-\mu)\frac{y}{r_1^3} - \mu\frac{y}{r_2^3}.$$

These are the equations of the motion of the third body with respect to the rotating coordinates. Because they are a pair of second-order differential equations, they represent a system of *order* 4 (or four *degrees of freedom*, because the solution depends on four arbitrary constants that are determined by the initial conditions). However, as was shown later, they admit an integral that reduces the system to one of order 3.

Although Lagrange's concern with applying his method to the motion of the Moon meant that he did not exploit the symmetry of the equations to its full potential — he defined the three relative velocities as those of bodies B and C with respect to body A, and that of body B with respect to C — he nonetheless achieved considerable success. He reduced the general problem to one of order 7, giving eleven integrals of the motion: the ten mentioned above together with the integral that comes from eliminating time from the equations. He also rediscovered Euler's collinear solutions and found a new set of particular solutions in which the bodies are always at the vertices of a moving equilateral triangle (see Figure 11.5).

Lagrange believed that his particular solutions, although interesting mathematically, were only a curiosity, and unlikely to occur in the real world. However, it has since been discovered that at least one type, and possibly both types, of particular solutions do exist in the solar system; the discovery of these 'real' solutions is related to equilibrium points associated with the particular solutions. From a physical point of

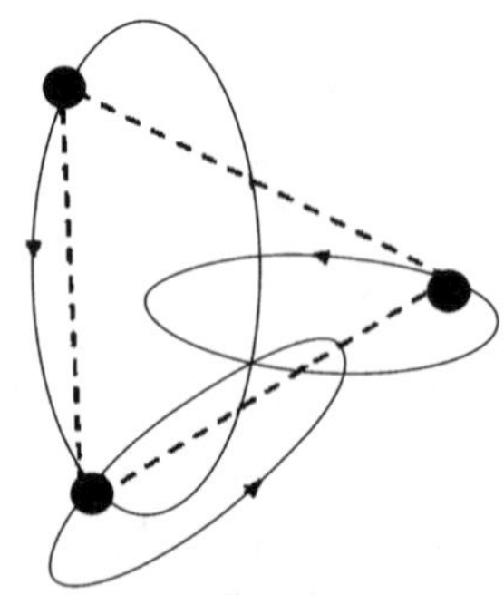

Figure 11.5. An equilateral triangle solution of the restricted three-body problem

view, the five equilibrium points, now known as 'Lagrangian points', are the points where the forces acting on the third body in a rotating system are balanced, and so there is no motion relative to the rotating system.

Lagrange proved the existence of triangular Lagrangian points in the Sun–Jupiter system and thus effectively predicted the presence of the 'Trojan asteroids' — two groups of asteroids that travel in the same orbit around the Sun as Jupiter, clustered around the Lagrangian points 60° ahead and 60° behind the planet, and named after the heroes in the Greek tales of the Trojan War — although they were not actually observed until 1906. Later, in the early 1980s, the Voyager missions to Saturn led to the discovery of other equilateral triangle solutions.[20] With respect to the collinear solution, it is believed that the *Gegenschein* (counter-glow) — the faint light in the sky sometimes observed after sunset in the plane of the ecliptic, diametrically opposite the Sun — is due to the Sun's illumination of a build-up of meteor particles at a Lagrangian point.

As far as finding further integrals of the three-body problem was concerned, the situation remained where Lagrange had left it until 1843 when Carl Gustav Jacob Jacobi, a professor of mathematics at Königsberg, found a twelfth integral by using a procedure known as the 'elimination of the nodes' (see Box 31). Later in the 19th century it would be proved that no other integral of the problem exists, and consequently that a solution, should it exist, would have to be in the form of an infinite series. (We continue this part of the story in Chapter 21.)

11.5 Jupiter and Saturn

One of the main challenges for Lagrange and his contemporaries was to explain the departures from the actual orbits of the planets from the elliptical orbit they would be in if they were the sole planet orbiting the sun. These differences are collectively known as 'inequalities' and astronomers distinguished between those that can be shown to be periodic and the others, called 'secular', that are not known to be periodic. The periodic

[20]Satellites at the Lagrangian points for the Earth–Sun system are used in astronomical research.

Box 31. Jacobi's integral.
Let

$$U = \frac{n^2}{2}(x^2 + y^2) + \frac{(1-\mu)}{r_1} + \frac{\mu}{r_2}.$$

Then the equations of motion of the planar circular restricted three-body problem in rotating coordinates can be written as

$$\ddot{x} - 2\dot{y} = \frac{\partial U}{\partial x},$$

$$\ddot{y} + 2\dot{x} = \frac{\partial U}{\partial y}.$$

On multiplying the first equation by $\dot{x}$ and the second equation by $\dot{y}$ and adding, we obtain an equation which can be integrated to give

$$\dot{x}^2 + \dot{y}^2 = 2U - C,$$

where C is a constant of integration. This constant, which can be written as

$$C = n^2(x^2 + y^2) + 2\left(\frac{(1-\mu)}{r_1} + \frac{\mu}{r_2}\right) - (\dot{x}^2 + \dot{y}^2),$$

was first derived by Jacobi in 1836, and is known as 'Jacobi's constant' or 'Jacobi's integral'.

inequalities depend on the planets' positions relative to one another, and they compensate for each other within a relatively short period of time. The secular inequalities are the most troubling because, in principle, they might exert an ever-growing influence on the motion of the planet that could lead to a collision or an escape. However, these changes happen extremely slowly, possibly taking centuries, their effects being imperceptible during the course of a single revolution.

The inequalities in the mean motions of Jupiter and Saturn, the two largest planets, had first been recognised by Kepler. Newton had derived values for their planetary masses relative to the mass of the Sun (something that he had not been able to do for the other planets, apart from the Earth) and had computed the force that Jupiter exerts on Saturn, for any particular positions of the two planets about the Sun. But he was unable to compute the cumulative effect of the perturbing force as they continually changed their positions relative to each other and to the Sun. The inequalities were readily detectable — the perturbations were the largest known in the solar system — and the failure to account for them (that is, the failure to match Newtonian theory with observation) became of increasing concern.

In 1748, and again in 1750 and 1752, the Paris Académie made the inequalities of Jupiter and Saturn the subject of a prize contest. Once again, Euler was in the frame, winning the first and the third contests, with no prize being awarded in 1750. At the beginning of his first memoir on the topic, Euler outlined the problem:[21]

Euler on the problem of Jupiter and Saturn.

[21] Quoted in (Golland and Golland 1993, 56).

> The Royal Academy of Sciences of Paris, proposed as a subject for the prize of the year 1748, a theory of Saturn and Jupiter, by which one could explain the inequalities of the two planets which is provided by their mutual cause, principally about their conjunction.
>
> We know, first of all, that there is no doubt that the Royal Academy is of the view that the theory of Newton, founded on universal gravitation, which is found to be quite admirably well in accord with all of the celestial motions, that those which are the inequalities which are discovered in the motion of the planets, one is boldly able to maintain, that the mutual attraction of the planets is the cause. Therefore as the Astronomers had perceived the various inequalities in the motion of Saturn, one concludes, very likely, that they are caused by the force with which this planet is attracted to Jupiter which not only is closest to Saturn but also exceeds it in mass, and by consequence in attractive force all of the other planets together, such that their effects are indefinitely small compared to that of Jupiter. For the same reason, the force of Saturn on Jupiter so exceeds that of all of the other planets, that to determine the disturbances to which the motion of Saturn and Jupiter are subjected, one can without error, neglect the forces of the other planets.
>
> Now following this theory, the cause of the inequalities which the Astronomers have observed in the motions of Saturn and Jupiter, is made known, and in order to answer the proposed question, one will have only to determine the motion of three bodies which are mutually attracted in ratios composed of their masses, and by the inverse-square ratio of their distances, and then put in place of one of the three bodies the Sun, and the bodies Saturn and Jupiter in lieu of the other two. By this, one sees the question proposed is reduced to the solution of a problem purely of mechanics: but it is necessary to admit that this problem is one of the most difficult ones of mechanics and hence one must not seek a perfect solution, until much more progress is made in analysis.

If Euler considered this problem to be 'one of the most difficult', then certainly it was difficult. So rather than seeking a 'perfect solution', which, as he pointed out, could not be done without new developments in analysis, he looked for an approximation.

First, he determined a set of four differential equations that describe the motion of the two planets. He then reduced the problem by introducing a number of simplifying assumptions — for example, assuming that the motion of the two planets takes place in the same plane and that the orbit of Jupiter is a circle.

Eventually he was left with a pair of differential equations that he was unable to solve directly. The difficulty was a troubling term of the form $(1 - g\cos\omega)^{-3/2}$, where g, which is a constant defined in terms of the mean distances from the Sun of the two planets, has a value of 0.8404, and ω is the angular distance between Saturn and Jupiter as viewed from the Earth. Recognising that a Taylor expansion would be unsatisfactory — the resulting Taylor series converges too slowly for practical purposes and the powers in $\cos\omega$ would have to be transformed before the series could be integrated — Euler took the original step of expanding expressions of the above form as a trigonometric series:

$$(1 - g\cos\omega)^{-\mu} = A + B\cos\omega + C\cos 2\omega + D\cos 3\omega + \cdots,$$

where the coefficients $A, B, \ldots$, are themselves infinite series in ω. Here, as he showed, each successive coefficient can be calculated from the two preceding ones, and so he had only to find an approximation for A and B, which he was able to do. To find the approximate solution, he then performed the integration term by term, using the method of undetermined coefficients.

Unfortunately, Euler's calculations were dogged by errors. For that reason, and because he failed to take into account higher-order perturbations, neither of Euler's prize-winning memoirs provided a satisfactory solution to the problem. Nevertheless, despite its shortcomings, the memoir of 1748 was of great importance because of its introduction of trigonometric series. The expressions given by these series for the perturbations proved to be very effective at predicting planetary positions. In 1756 Euler used the same series to attack the more general problem of determining the motion of any two planets, winning yet another Paris Académie prize. Throughout the rest of the 18th century, Clairaut, D'Alembert, Lagrange, and Laplace all used these series to deal with planetary perturbations, and put much effort into finding quicker ways of calculating the coefficients.

For Euler, it was enough that these series provided good results up to the accuracy of the observations, and he was not concerned about their convergence. However, although it is possible to order the terms of the series so that, on average, the coefficients of successive terms decrease in size, there may be a distant term in the series that turns out to be surprisingly large.[22] Therefore, given any finite number of terms in the series, we cannot be sure how closely their sum will approximate the correct solution. How good the approximation actually is has to be determined empirically and also depends on how well the differential equations describe the motion concerned.

In 1766 Lagrange won the Paris Académie prize for a memoir on the moons of Jupiter. Originally it was thought that the four moons move with uniform circular motion around the planet, but in the 1740s the Swedish astronomer Pehr Wargentin, having observed a long sequence of the eclipses of the moons, found a periodic inequality of around 437.6 days for each of the three inner moons.

Having set up the equations of motion, Lagrange developed expressions for the perturbing forces due to the action of the Sun and the mutual actions of the moons which he approximated as trigonometric series. The resulting expressions were then inserted into the differential equations. Eventually, he arrived at differential equations of the form

$$\frac{d^2u}{dt^2} + M^2u + T = 0,$$

where M is a constant and T is a function composed of sines and cosines of multiples of the time t. These equations were sufficient to predict the anomalies detected by Wargentin, because, for each moon, one term in the solution dominates all the rest. In the case of Jupiter's first moon, Io, it is the term with argument $2(\mu_2-\mu_1)t$, where μ_1 and μ_2 are the mean motions of the first two moons, Io and Europa, and M in this instance is identical to μ_1. The coefficient of this term has $4(\mu_2-\mu_1)^2-\mu_1^2$ in the denominator, and

[22]The series are not what are today known as *uniformly convergent*, a notion that was not recognised or understood until the beginning of the 19th century.

since the mean motion of Io is almost exactly twice that of Europa ($\mu_1 : \mu_2 = 2.0075 : 1$), this denominator is extremely small, and therefore the coefficient is extremely large. By considering this term alone, Lagrange was able to account for the anomalies in the motion of Io, and he applied a similar analysis with equal success to the other moons.

In 1774 Lagrange submitted a memoir to the Paris Académie that contained a new method for dealing with secular inequalities. Laplace, who had been elected to the Académie the year before, read the memoir as soon as it arrived and immediately realised that the same method could be employed to determine other secular inequalities. He registered his idea with the Académie's secretary and his work was published in the following year — three years before that of Lagrange. However, just before it was printed, he received a letter from Lagrange that provided details of exactly the same idea, and so he added the letter as an addendum to his memoir. Having received Laplace's manuscripts, Lagrange initially decided to hand over the subject to Laplace, writing to him:[23]

> Long ago I proposed to myself to take up once more my old work on the theory of Jupiter and Saturn, to push it further and to apply it to other planets ... But as I see that you yourself have undertaken this research, I willingly renounce it, and I am very happy that you have relieved me of the necessity of undertaking this work, for I am persuaded that the sciences can only gain much thereby.

But a few weeks later Lagrange changed his mind and returned to work on the topic, suggesting to Laplace that they engage in a friendly rivalry, exchanging only published work. Thus began their joint dominance of the field.

Meanwhile, concerned by the uncertainties surrounding the masses of the planets, Laplace turned to other gravitational topics — the theory of the tides, the determination of cometary orbits, and the attraction of spheroids — and it was a decade before he returned to the mutual perturbation problem. But when he did, he returned in spectacular fashion. His 'Mémoire sur les inégalités séculaires des planètes et des satellites' (Memoir on the secular inequalities of planets and satellites), which was presented to the Académie in 1785 and published two years later, is now acknowledged to be 'one of the signal memoirs in his oeuvre'.[24] This memoir is largely concerned with the motions of Jupiter and Saturn, and an important motivation for Laplace in this particular work was that the anomalies in these motions provided the greatest impediment to a proof of the stability of the solar system.

In his memoir, Laplace announced previously unknown inequalities in the motions of Jupiter and Saturn, provided an explanation for the pattern of the dance of Jupiter's satellites, and gave an a priori proof of the stability of the solar system. It was the first of four memoirs, written between 1785 and 1788, in which he conveyed his most important discoveries in planetary theory. The second and third memoirs were devoted to the improvement of his theory of Jupiter and Saturn, which compared theory with observation and contained detailed calculations, while the fourth was on the secular equation of the moon. We shall concentrate on the first memoir, since it contains his main findings.

---

[23] Quoted in (Wilson 1995, 122).

[24] See (Gillispie 1997, 124).

Laplace opened the paper, a relatively brief one by his standards, with a short history of the state of knowledge with respect to the differences between the theoretical and observed positions of celestial bodies. In the case of Jupiter and Saturn, he wrote:[25]

> It is, however, impossible not to recognise very appreciable variations in the orbits of Jupiter and Saturn. If we compare the observations of these two planets made since the renewal of astronomy [that is, since Tycho Brahe], we find that the motion of Jupiter is constantly faster and that of Saturn constantly slower than [the motions calculated] from a comparison between modern and ancient observations.[26]

Having eliminated the action of comets as the cause for these effects (because their masses are too slight), Laplace established the interdependence between the motions of the two planets by using general properties of the mutual action of the planets, taking account only of quantities with very long periods. He then declared:[27]

> It is thus very probable that the observed variations in the motions of Jupiter and Saturn are an effect of their mutual action, and since it is established that this action can produce no inequality which is either constantly increasing or is periodic with a very long period and independent of the position of the planets, and since it can only cause inequalities dependent on their mutual configurations, it is natural to think that there exists within their theory a large inequality of this type, whose period is very long and from which these variations result.
>
> In examining the circumstances of the motion of Jupiter and Saturn, we easily perceive that their mean motions are very nearly commensurable, and that five times the mean motion of Saturn is almost equal to twice the mean motion of Jupiter; from which I have concluded that the terms, which, in the differential equations of motion of the planets, have for argument five times the mean longitude of Saturn, less twice that of Jupiter, could become sizeable as a result of the integrations, although they are multiplied by the cubes and products of three dimensions of the eccentricities and the inclinations of the orbits.

The fact that the mean motions are, as Laplace observed, approximately in the ratio 5 : 2, was well known. This near-commensurability means that if λ is the mean longitude of Jupiter and λ' is the mean longitude of Saturn, then the quantity $5\lambda' - 2\lambda$, which appears in the denominator of the integral of the perturbation function, is very small. This is another example of the problem of *small divisors*, so called because it leads to the appearance of large terms in the perturbations. The period of such terms is $\frac{2\pi}{5\lambda' - 2\lambda}$, hence the term *long period inequality*. Laplace continued:[28]

> Consequently, I have regarded these inequalities as the very probable cause of the variations observed in the motions of Jupiter and Saturn. The probability of this cause and the importance of the subject made me determined to undertake the long and laborious calculation necessary to assure myself of it. The result of this calculation has fully confirmed my conjecture in making me see:
> (1) that there exists in the theory of Saturn an equation of about 47′, whose period is almost 877 years, and which depends on five times the mean motion of Saturn less twice that of Jupiter;
> (2) that in the theory of Jupiter there exists an equation of opposite sign, of about 20′, and whose period is the same.

[25]See (Laplace 1784/1787, 2). When Jupiter and Saturn are at their closest, the gravitational pull of Jupiter on Saturn is about 1/200th of the pull of the Sun on Saturn, a not insignificant amount.

[26]We discussed the work of Tycho Brahe in *Volume* 1, Chapter 12.

[27]See (Laplace 1784/1787, 4–5).

[28]See (Laplace 1784/1787, 5–6).

Thus he had calculated the long period inequality to have a period of approximately 877 years.[29]

By computing higher-order perturbations in the eccentricities and inclinations, Laplace had reconciled Newton's theory with observations. As Laplace himself would later write:[30]

> these inequalities had seemed inexplicable on the basis of universal gravitation; they are now one of its most striking proofs.

Earlier, both Euler and Lagrange had been put off from attempting to compute the higher-order perturbations by the huge amount of work involved. As the historian Curtis Wilson has described, Laplace's achievement was in finding a way to make these calculations manageable:[31]

> Laplace went round the wall rather than attempting to go over it: he showed how to make a reasonable guess as to which of the higher-order perturbations might prove sizeable, and then devised procedures whereby such perturbations could be extracted from the differential equations, one at a time.

Next, Laplace turned his attention to the satellites of Jupiter, showing that the two results that had earlier been detected by observation — the relation between the mean motions of the first three satellites, and another found by Lagrange — were both rigorously exact. Furthermore, he calculated that, even if the satellites had not originally started out in positions that (according to Kepler's laws) would result in their mean motions satisfying the former of the two relations, their mutual attractions would be sufficient to bring them into the stated relationship, provided that they started out close enough.

Having shown that the mean motions of the satellites were subject only to periodic inequalities, Laplace felt secure in extrapolating his conclusions to the planetary system as a whole, providing arguments to show that neither the mean distances nor the orbital eccentricities and inclinations were subject to secular change. As a result, he was able to make his first claim for the stability of the solar system:[32]

> From this we can in general conclude that the expressions for the eccentricities and the inclinations of the orbits of the planets contain neither arcs of a circle [angles increasing proportionally with time] nor exponentials, and thus that the system of the planets is confined within invariant limits, at least with respect to their mutual action.

Three years later, he would state his claim even more concretely:[33]

> Thus the system of the world only oscillates round a mean state from which it never departs except by a very small quantity. By virtue of its constitution and the law of gravity, it enjoys a stability that can be destroyed only by foreign causes, and we are certain that this action is detectable from the time of the most ancient observations until our own day. This stability in the system of the world, which assures its duration, is one of the most notable among all phenomena, in that it exhibits in the heavens the same intention to maintain order in the universe that nature has so admirably observed on earth for the sake of preserving individuals and perpetuating species.

---

[29] In his subsequent memoirs on Jupiter and Saturn, Laplace recalculated the period to be 929 years.

[30] Laplace, *Mécanique Céleste*, Vol. V, 362.

[31] See (Wilson 1985, 286–287).

[32] See (Laplace 1784/1787, 50).

[33] Quoted in (Gillispie 1997, 145).

That Laplace had 'proved' the stability of the solar system seemed clear to Mary Somerville. She reported on it in her account of Laplace's work on the great inequalities of Jupiter and Saturn, where she made some informative remarks about the historical observations of these planets:[34]

> The formulae of the motions of Jupiter and Saturn determined by La Place, agree with their oppositions, the error not amounting to 12″.96, when it is to be recollected that only twenty years ago the errors in the best tables exceeded 1296″. These formulae also represent with great precision the observations of Flamstead [sic], of the Arabian astronomers, and of Ptolemy, leaving no grounds to doubt that La Place has succeeded in solving this difficulty, by assigning the true cause of these inequalities, which had for so many ages baffled the acuteness of astronomers; so that anomalies which seemed at variance with the law of gravitation, do in fact furnish the strongest corroboration of the universal influence it exerts throughout the solar system. Such, says La Place, has been the fate of that brilliant discovery of Newton, that every difficulty which has been raised against it, has formed a new subject of triumph, the sure characteristic of a law of nature.
>
> The precision with which these two greatest planets of our system have obeyed the laws of mutual gravitation from the earliest periods at which we have records of their motions, proves the stability of the system, since Saturn has experienced no sensible action of foreign bodies from the time of Hipparchus, although the sun's attraction on Saturn is about a hundred times less than that exerted on the earth.

Both Lagrange and Laplace wrote many more memoirs on celestial mechanics, often providing refinements to one another's work, with Laplace bringing all the results together in his *Mécanique Céleste.* However, as mentioned earlier, Laplace did not name his sources, and the extent to which he relied on Lagrange's results when producing those of his own is not evident, either there or in his earlier publications. Curtis Wilson, who made a deep study of the work of both Lagrange and Laplace on celestial mechanics, believed that Lagrange's influence was much greater than might be thought at first sight, particularly in the context of Laplace's work on Jupiter and Saturn.[35] Furthermore, as we have already noted, the results obtained by Lagrange and Laplace were only approximations — good approximations, but approximations, nonetheless. However, the success of Laplace's results gave him great faith in the theory, and he did not hold back from articulating as much, claiming in the *Exposition du Système du Monde* that:

> We shall see that this great law of nature [universal gravitation] represents all the celestial phenomena, down to their smallest details; there is not a single one of their inequalities that does not follow from it with admirable precision ...

11.6 Further reading

Gillispie, C.G. 1997. *Pierre-Simon Laplace 1749–1827. A Life in Exact Science,* Princeton University Press. This is a fine book by a major historian on a major figure in the histories of science and mathematics.

Linton, C.M. 2007. *From Eudoxus to Einstein: A History of Mathematical Astronomy,* Cambridge University Press. This is an accessible book for everyone from amateur astronomers to professional historians.

[34] See (Somerville 1831, 325).
[35] See (Wilson 1985, 289–290).

Taton, R. and Wilson, C. (eds.) 1995. *Planetary Astronomy from the Renaissance to the Rise of Astrophysics, Part B: The Eighteenth and Nineteenth Centuries*, Cambridge University Press. This set of essays is selective and technical, but is the best account of the topics that it covers.

Part II

The 19th Century

12

Introduction: The 19th Century

We now turn to the history of mathematics in the 19th century. Such is the scale of what was done, and such is its difficulty, that we have had to be highly selective. After an introductory chapter in which we sketch the social context, we devote two chapters to each of four themes: geometry, analysis, algebra, and applied mathematics.

Geometry revived in the early 19th century, and in unexpected ways. Two mathematicians on the periphery of the mathematical community — Nicolai Ivanovich Lobachevskii in Russia and János Bolyai in Romania/Hungary — found a new geometry, later called non-Euclidean or hyperbolic geometry. This is an example of a metrical geometry — it is about lengths, angles, shapes, and sizes — that had every reason to be considered as true as Euclidean geometry, but with very different theorems.

As we discuss in Chapter 14, this geometry was not accepted at first. When it was, it provoked a wholesale revision of mathematicians' ideas about the nature of geometry. This then merged with the investigations of a different type of geometry that we discuss in Chapter 15, where we see how a largely French, and then German, and Italian enterprise explored a non-metrical geometry that came to be called projective geometry, and found good reasons to consider it as fundamental. We follow this story to around 1900, when the outlines of a new consensus begin to emerge in a resurgence of the axiomatic approach to geometry reminiscent of Euclid's *Elements*.

As we saw in Chapter 9, the many successes of the calculus in the 18th century rested uneasily on unsatisfactory foundations. We briefly review some more of these successes in Chapter 16 before turning to the work of Augustin-Louis Cauchy, who found the way to rigorise the calculus that has been accepted ever since. This was a remarkable achievement, even if he did not complete the process; nor was everything he did correct or entirely clear, and we look at one of his deepest mistakes as an indication of how difficult this subject was. We cannot pursue all the ramifications of his work and its great extensions at the hands of his successors, because it was so fundamental and so widely accepted, but in Chapter 17 we take up an important topic that he had taken for granted: the nature of the real numbers. This was first explained by Richard Dedekind and Georg Cantor in different ways, each of which led them and their successors into the paradoxical world of set theory. We look at some of the attempts to base pure mathematics on nothing but logic, and at the struggle to reach a new consensus on the foundations of mathematics.

Algebra has the study of numbers at its heart, and in Chapter 18 we look at Gauss's reformulation of number theory, which put the subject at the centre of mathematics in Germany and beyond. We also conclude our story about whether every polynomial equation has as many real or complex roots as its degree, and look at the first proofs of what became known as the Fundamental Theorem of Algebra.

But a second question was also being asked, which is much deeper: Can every polynomial equation be solved by radicals — that is, by an explicit formula involving only addition, subtraction, multiplication, division and the extraction of roots? As we show in Chapter 19, Niels Henrik Abel proved that most equations of degree 5 or more cannot be solved by radicals, and in 1832 Évariste Galois gave a profound but obscure account of which equations of any degree can be solved in this way. We shall see how related ideas also helped to resolve two of the three classical problems inherited from Greek times, again in the negative. Attempts to understand Galois's work were among the motivating factors for the creation of group theory, one of the central domains of modern structural algebra.

Applied mathematics — in the form of rational mechanics — had many successes in the 18th century, and more were to come in the 19th century. In Chapter 20 we look at Joseph Fourier's account of heat diffusion in solids, and at his proposal that every function can be written as an infinite series in sines and cosines — later known as a Fourier series representation of the function: this is often regarded as one of the finest contributions of applied mathematics to pure mathematics.

We then turn to what is called potential theory. In the opening decades of the 19th century, magnetism and electricity began to reveal their secrets, and the appropriate mathematics was close to what was already in use in the study of gravitation. This encouraged Gauss and the English mathematician George Green to develop potential theory. Green's work, in particular, rested on claims that were plausible on physical grounds but lacked mathematical justification, and we look at two initial investigations into this issue: the Dirichlet problem and the Dirichlet principle.

In Chapter 21, we return to celestial mechanics, one of the ongoing concerns of mathematicians since Newton, and trace the analysis of its subtleties until Henri Poincaré showed that all known methods would ultimately fail. He discovered chaos lurking inside even relatively simple and physically realistic dynamical systems, and began to develop theories to deal with it.

13

The Profession of Mathematics

Introduction

Who were the mathematicians of the early 19th century? How were they educated, and what kinds of mathematics did they do? Was being trained as a mathematician in Paris the same as being trained as a mathematician in Berlin or London? We seek some answers to these questions by considering the structure of the mathematical communities in France, Germany, and Britain, as these were the most important for the development of mathematics during this period.

We shall describe how mathematics increasingly became practised within a mathematics 'profession'. We shall look at the development of the social structure within which mathematical activity took place, with an eye to the mathematical developments that will occupy us in later chapters. In doing so, we do not claim that they were linked in a causal way. These two developments were different in kind — one had to do with a mathematical subject, the other with social life — but seeing the two together suggests questions that are interesting to pursue.

13.1 The social context

The notion of a 'profession' is a complex one to pin down, and historians who have analysed such developments have often used different definitions. What we mean by the term here is that increasingly there were roles within society for people called 'mathematicians', and these mathematicians came to be seen as a specific group of people whose income derived from doing mathematics. Their roles differed from earlier ones (for example, 'mathematical practitioner') by the fact that objective criteria such as success in examinations became prerequisites for entry to the profession.

Professionalisation has various aspects: institutions in which training is received and through which qualifications are awarded; journals for communicating recent research results between members of the profession; and professional bodies and social structures to further the interests of the profession. Usually there were professional bodies that controlled membership and determined who was, and who was not, a professional. That was not quite the case for mathematicians in the 19th century, but there was a marked tendency for mathematicians to accumulate in universities, and by the end of the century they were usually in specially designated mathematics departments. We look at examples of these aspects in this chapter.

The rise of professionalism saw an uneven development of mathematics. Particular mathematical styles, approaches, and topics fared differently from one country or institution to another. To some extent it was always thus — many earlier examples of mathematical developments were associated predominantly with one country or another — but in the 19th century the role of universities and similar institutions as training grounds for future mathematicians, and as employers of mathematicians, became more significant than before. What the universities and colleges taught depended on priorities that emerged from political discussions within each country, as did the very existence of some of these institutions. We shall see something of this relation between mathematical styles and national political processes in Sections 13.2 and 13.3, where we describe developments in France and Germany at the beginning of the 19th century.

As it turned out, France and Germany had different priorities as to what kind of mathematics should be taught. Very broadly, there was a period in the early part of the 19th century when mathematics was taught in France for its applicability to the world (and so tended towards mathematical physics) and in Germany for its role in training the mind (and so tended towards the purer aspects of the subject). It is not surprising, for example, that the emergence of number theory as a major mathematical discipline was largely a German development.

In Section 13.4 we turn to another aspect of professionalisation, the growth of journals specifically devoted to mathematics. In surveying their origins and the kind of articles that they carried, we again find interesting differences on either side of the Rhine. In Section 13.5 we shall see that these differences persisted in the later 19th century, when other countries, such as Italy, America, and Britain, became more involved in the expanding world of mathematics and took the French and German communities as their examples.

13.2 Mathematics in France

In the years following 1800, Paris was the centre of the mathematical world. It had not been so for long: indeed, up to the 1780s the most important mathematical activities and developments seemed mostly to take place further east, in the Academies of Berlin and St Petersburg. But after Euler's death in 1783 there was no mathematician of international repute left in Russia, and in the late 1780s Lagrange ended his twenty years as Director of the Académie Royale in Berlin and moved to Paris — in effect, he was headhunted by the French government and the Académie des Sciences in Paris, much as the head of a large international corporation might be today. With the arrival of Lagrange in 1787, Paris became the mathematical focus of Europe. Over the next twenty years Lagrange worked hard to further the development of mathematics in France, perhaps all the more successfully because as a foreigner (he was Italian-born) he was somewhat detached from the revolutionary French politics of the next few years.

Within two years of Lagrange's arrival in France, the Bastille fell and the French Revolution was under way. Of the many changes during those momentous years, the most relevant here is that the French educational system totally collapsed. Many schemes were put forward in the 1790s for what should be done, but political agreement and financial support were hard to achieve.

The two most important educational developments of these years both involved Lagrange, as well as other mathematicians. A significant, though short-lived, institution was the École Normale de l'An III (the Normal School of Year 3 [of the Republic]) where teachers were to be trained.[1] In an endeavour to establish educational norms for the whole country, potential student–teachers from throughout France were assembled in Paris and received lectures from the leading mathematicians of France: Lagrange, Laplace, and Monge. For a variety of reasons this enterprise lasted for only a few months, but its influence lasted longer, because the lectures were copied by stenographers in the audience and printed. The École Normale illustrates, too, the strong centrally directive impulse of the educational reformers; the idea was that the educational system for all France should be determined and influenced by decisions and practices in Paris, and further, that there should be a strong mathematical component in the curriculum at all levels.

Figure 13.1. The original premises of the École Polytechnique

These aspects were to be seen again in the most famous and prestigious new institution of the early years of the Republic, the École Polytechnique (see Figure 13.1). Founded in 1794 as the École Centrale des Travaux Publics (the Central School for Public Works — meaning civil and military engineering), its name was changed to the École Polytechnique in the following year. This was the first institution devoted to the systematic training of large numbers of students in physics, mathematics, and engineering (its first-year intake was 400), and, as with the École Normale, some of the

[1] The French Republican calendar was introduced in October 1793 as part of a raft of measures designed to mark a break with the old regime. It was eventually abolished by Napoléon in 1806.

Figure 13.2. Gaspard Monge (1746–1818)

foremost mathematicians of France taught there. The founding of the École Polytechnique marks the transition between two ways of ordering mathematical and scientific activity within society: from that based on prestigious national academies staffed by a scientific elite, to a system focussing on educational institutions in which students, teachers, and researchers were more intimately intermingled.

Gaspard Monge. One of those most involved in planning and administering the school was Gaspard Monge. Lagrange taught there too, and although Laplace did not lecture there he held the even more influential position of examiner, and so he determined what would be examined, as well as what level of attainment the students ought to display.

Monge was almost a generation older than Lagrange, and a very different person in many ways. While still in his teens he had decisively simplified the study of what was called *descriptive geometry*: a technique for depicting three-dimensional configurations in a plane. It was used in two ways: to display accurately those configurations that already exist, such as buildings or fortifications, and to facilitate the design of new ones. Lengths had to be depicted systematically so that stresses along them could be calculated, and it had to be easy to decide which parts of the design could be seen from which other parts (for example, in the design of gun emplacements). Monge's method, consisted of an ingenious study of how a figure in three-dimensional space can be projected vertically onto a horizontal plane and horizontally onto a vertical plane: these give the 'plan' and 'elevation' of architectural drawings. Monge coupled this familiar idea to some simple algebra and developed the associated calculations algebraically until descriptive geometry became a truly flexible and useful tool that gave builders and designers a quick way of finding the information they needed. His new way of doing descriptive geometry also proved to be the start of a successful military career

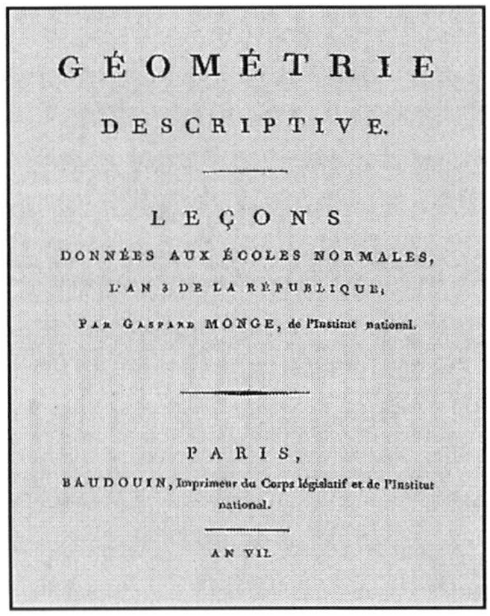

GÉOMÉTRIE

DESCRIPTIVE.

LEÇONS

DONNÉES AUX ÉCOLES NORMALES,

L'AN 3 DE LA RÉPUBLIQUE;

PAR GASPARD MONGE, de l'Institut national.

PARIS,

BAUDOUIN, Imprimeur du Corps législatif et de l'Institut national.

AN VII.

Figure 13.3. The title page of Monge's *Géométrie Descriptive* (1799)

— in fact, for many years his method was a military secret. His descriptive geometry was eventually published as a book in 1799, which was year 7 (An VII) in the French Republican calendar, see Figure 13.3.

As a loyal and patriotic Frenchman, Monge became involved in the French Revolution and for a time was Minister for the Navy. He prudently withdrew from public life during the Terror of 1793–1794 when it became dangerous, and then threw himself into the creation of the École Polytechnique, becoming its first director. For many years he was an influential Professor of Geometry there, and under his leadership descriptive geometry was taught to all students. Monge was aided by a marvellous capacity to visualise three-dimensional geometrical figures, and was apparently able to convey them into the minds of his students, as his former pupil Michel Chasles attested. Chasles eventually became a professor there in 1841 (and at the Sorbonne in 1846) and was the leading French geometer of the next generation. In his *Aperçu Historique* (1837) he wrote:[2]

Chasles on pure geometry.

> In recent times, after a rest of almost a century, pure Geometry has been enriched by a new doctrine, descriptive geometry, which is the necessary complement to the analytic geometry of Descartes and which, like it, must have immense results and mark a new era in the history of geometry. This science is due to the creative genius of Monge. It embraces two objects. The first is to represent all bodies of a definite form on a plane area, and thus to transform into plane constructions graphical operations which it would be impossible to execute in space. The second is to deduce from this representation of the bodies their mathematical relationships resulting from their forms and their relative positions.

[2] See (Chasles 1837, 189–190), in F&G 17.A2(a).

> This beautiful creation, which was initially intended for practical geometry and the arts which depend on it, really constitutes a general theory, because it reduces to a small number of abstract and invariant principles and to easy and always correct constructions, all the geometric operations which can be involved in stone cutting, carpentry, perspective, fortifications, gnomonics, etc, and which apparently can only be executed by mutually incoherent processes, which are uncertain and often scarcely rigorous. But besides the importance due to its first intention, which gives a character of rationality and precision to all the constructive arts, descriptive geometry has another great importance due to the real services which it renders rational geometry, in several ways, and to the mathematical sciences in general. [For this reason] geometry thus reaches a state where it can most easily lend its generality and its intuitive evidence to mechanics and the physico-mathematical sciences.

Monge's skill as a teacher and his considerable organisational ability and influence won him many disciples. To quote from Chasles again:[3]

> Monge gave us, in his *Traité de Géométrie Descriptive*, the first examples of the utility of the intimate and systematic alliance between figures in three dimensions and plane figures. It is by such considerations that he proved, with rare elegance and perfect evidence, the beautiful theorems which constitute the theory of poles of curves of the second degree; the properties of centres of similitude of three circles taken two by two whose centres lie three by three on a line, and various other figures of plane geometry. Since then, the pupils of Monge have cultivated this truly new kind of geometry with success, so that one has often, and with reason, given them the name of the school of Monge, and it consists, as we shall say, in introducing into plane geometry considerations of the geometry of three dimensions.

Chasles then went on to speak of Monge's geometry as being intuitive, general, and based on the idea of making transformations of figures, and said that Monge had regarded descriptive geometry as enriching rational geometry. This suggests that Monge was not exclusively interested in the practical uses of his descriptive geometry.

On the other hand, however useful it might have been (and it was to be taught at the École Polytechnique until the First World War), descriptive geometry did not offer much to interest research-minded mathematicians: there were seemingly no significant *mathematical* problems that it helped them to solve. Moreover, the whole subject was surely unattractive to someone as algebraically focussed as Lagrange, and this was to matter when the wheels of politics turned once again in France. But for as long as Napoléon reigned, Monge, who had identified himself strongly with the Napoleonic cause, was secure.

The third person who was involved in setting up the École Polytechnique was Pierre-Simon Laplace. As we saw in Chapter 11, he was not only the acknowledged expert on celestial mechanics but a smooth political operator within the scientific world with the ear of Napoléon.

[3]See (Chasles 1837, 191), in F&G 17.A2(b).

Box 32. Curricula at the École Polytechnique.

First year of study

	1801	1806	1812
Analysis	16 %	29%	25 %
Mechanics	10 %	17 %	18 %
Geometry	40 %	26 %	23 %
Chemistry	10 %	9 %	12 %
Total	76 %	81 %	78 %

Second year of study

	1801	1806	1812
Analysis	11 %	18 %	20 %
Mechanics	12 %	22 %	25 %
Geometry	4 %	4 %	2 %
Chemistry	20 %	9 %	11 %
Architecture	7 %	13 %	9 %
Other applications	29 %	14 %	13 %
Total	83 %	80 %	80 %

The precise role and function of the École Polytechnique changed somewhat as the years went by, for this vital educational institution was not to be independent of government policies. The teaching was reorganised in 1799, and the three-year course was converted to a two-year course preparatory to the training that graduates of the school could go on to obtain in one of the specialised Écoles (the Schools of Mining, Artillery, Military Engineering, Bridges and Roads, and so on). This change took place shortly after the coup d'état that brought Napoléon to power as First Consul in November 1799. In 1804, the year in which he declared himself Emperor, he ordered that the École Polytechnique be militarised, and its students treated like cadet soldiers. The students' reluctance to welcome his elevation to Emperor was probably one factor in his desire for a more disciplined regime at the school.

These years also saw a significant change in the balance of subjects within the curriculum. Statistics of this shift have been worked out by the historian Matthias Paul[4] (see Box 32).

From 1801 to 1806 the overall trend was for the proportion of time spent learning analysis and mechanics to increase, in both the first and second years, while the amount of time devoted to geometry dropped. One explanation for this might be the militarisation of the school, but it is not immediately clear why military needs should cease to value the kind of geometry promulgated by Monge, especially when one recalls that descriptive geometry was originally regarded as a military secret. The obvious place to start looking for an explanation is to see who controlled the syllabus; with the

[4]See (Paul 1980, 123–124), in F&G 17.A4.

importance of the École Polytechnique, decisions over what was taught were likely to be highly charged politically.

In the early years of the 19th century, the influence of Laplace was paramount in determining the direction of the syllabus at the École Polytechnique, and certainly the shift of balance was in accord with his own mathematical interests. Laplace also worked hard to extend Newtonian ideas to many other kinds of natural phenomena, although with less success than in celestial mechanics. As an exponent of rational mechanics he would have had little time for geometry, which might explain why the subject's place in the syllabus shrank, but he would have denied that his kind of mathematics was less useful than that of Monge. So it is important to be clear what the shift in the syllabus entailed.

The increase in the time devoted to analysis and mechanics, with a consequent decrease in the time allotted to geometry, was not what the modern sense of these terms might suggest. It was rather a shift away from a mathematical style designed for immediate use and applicability, and for use by engineers (as the proponents of Monge's geometry saw it), towards the purer style of algebra and analysis favoured by Lagrange and Laplace, with its roots deep in the 18th century. The result was to make mathematics at the École Polytechnique more abstract and formal, and less intuitive. For instance, the 'Mechanics' whose study time virtually doubled between 1801 and 1806 was not an earthy engineering subject, but rather the rational mechanics of the 18th century described in Chapter 10 — an abstract and highly theoretical branch of the calculus.

We can infer, then, that Laplace was trying to move mathematics away from being a useful subject for everyday needs, designed to equip the students with skills that could serve the immediate needs of the state, and towards a pure subject with its own cadre of trained personnel. This is typically what happens in the professionalisation of any discipline, as its leading figures shape its future practitioners by providing the training without which it is increasingly difficult to obtain a job. Laplace saw the abstract formal training that he advocated as particularly appropriate to the role of the École Polytechnique as a general institution, preparatory not only to a number of more specialist Écoles but also to the careers of researchers in mathematics and science. In his view, once you had learned elementary rational mechanics you could then go on to study the design of machines. His view may also have accorded with the dignified status that attached to the intellectual, rather than the practical, sphere of life to which the new generation of professionals might reasonably also aspire.

Laplace's intentions received a yet stronger boost with the further reorganisation of the École Polytechnique that took place with the restoration of the Bourbon monarchy in 1816. Unlike Monge, Laplace had the gift of tempering his coat to the political wind, and when he was put in charge of setting up the new royalist institution, he consolidated his approach to mathematics by determining the courses of studies for students. But Monge was too prominently associated with the defeated dictator, and he was harassed politically by the new regime, which expelled him from the Institut de France and the Académie des Sciences. When he died in July 1818, the Bourbon monarchy even opposed his funeral being made into a major occasion, but many current and former students attended and paid tribute, as did Laplace and Legendre and former officers of all the public services.

Some indication of the acrimony that these changes in the syllabus caused among those who had adhered to the earlier ideals of mathematics may be seen in the vituperative recollections of one of Monge's pupils, Théodore Olivier, over thirty years later.[5]

> The theoreticians who take the name of pure scholars regard themselves as forming an aristocratic corps, having the right to command and dominate practical men, who are treated by them like helots or peasants ... it was under the fatal inspiration of Laplace, Poisson and Cauchy that the École Polytechnique was reorganised in 1816 ... these men incapable of understanding anything other than algebra and incapable of rendering services to their country other than in algebra, destroyed from top to bottom the old organisation of studies of the École Polytechnique ... the reformers ought to have changed the name from École Polytechnique to École Monotechnique.

As we shall see, the École Polytechnique was influential across Europe, and indeed further afield. In particular, it directly inspired the United States' military academy of West Point, which was set up in 1802, two years before the École Polytechnique became a military school. One reason for its great influence, both at home and abroad, was the multiplicity of textbooks written by teachers at the school. Following the practice at the short-lived École Normale, lectures at the École Polytechnique were written down and published — indeed, this was made a legal requirement. There were various reasons for this, some concerned with the style of teaching: a break was made with the older tradition of teaching in which lessons were simply dictated, and it was also thought useful for students to have an opportunity to study the content of a lesson in advance, as they could if a printed textbook existed. But a further important factor in the spread of textbooks was a judgement by Monge and his colleagues about how to raise the level of mathematics teaching throughout France. This is explained by the historian Jean Dhombres:[6]

> The idea which guided the 'founding fathers' of the École Polytechnique ... was to impose uniformity on an advanced level of organised knowledge in mathematics all over France. As around 150 students entered the École Polytechnique each year and many more prepared for the entrance examination from everywhere in France, and sometimes from outside France, teachers in charge of the classes eagerly tried to provide their students with the prerequisites and more by reading the textbooks used at the École.

The sales figures for some of these textbooks have been calculated by Dhombres, and were remarkably high. In 1812, to take a year at random, the two main texts on geometry each came out in their ninth editions, in runs of 3000 copies. One was by Sylvestre Lacroix, a protégé of Monge, who was Lagrange's successor as professor at the École Polytechnique and a prolific writer of high-quality textbooks; the other was by Legendre, who had succeeded Laplace as examiner. Other books on geometry reached similar numbers: for example, 2000 copies of Monge's work on descriptive geometry (a transcript of his lectures to the École Normale) were issued in that year. Lacroix's book on algebra reached 3000 copies, and a text by Jean-Guillaume Garnier for aspiring polytechnic students sold 1500 copies in 1811. Lagrange's book on the theory of functions ran to 1500 copies in 1813, as did Lacroix's more elementary textbook on the calculus. This flurry of mathematics texts was remarkable, as may be seen from another statistic: in the period from 1795 to 1830, more books were published on mathematical subjects than on physics, chemistry, pharmacy, and mineralogy combined.

---

[5]See (Olivier 1847), quoted in (Grattan-Guinness 1981, 665–666).

[6]See (Dhombres 1984, 158).

It seems clear, then, that mathematics played a fuller and stronger cultural role than the immediate needs of École Polytechnique students. As a result, the school curriculum throughout the country was dominated by mathematics. Various broad educational and social justifications attended this predominance. Lacroix's textbooks were a marvellous jumble of approaches, and he consciously espoused the 18th-century Enlightenment spirit of inviting students to make a rational choice among a multiplicity of views. Other writers, such as the astronomer Joseph Lalande, were explicitly atheist in their views, and saw mathematics as a model of clear, rigorous, non-religious reasoning, exemplifying how one could argue without relying on authority. Many people at the time would have seen the dominant place of mathematics in the curriculum as justified by much the same qualities as Frederick the Great (the patron of Euler and Lagrange) had seen in it, when he commended geometry by saying, 'That science is the only one which has not produced sects'.[7] Even when Napoléon, for political reasons, reincorporated Christianity into French public policy in 1801, the religious neutrality of mathematics was a strong point in its favour.

Several of the themes that we have considered — the fame and attractiveness of the École Polytechnique, the influence of Lagrange, the way that textbooks influenced what was studied throughout France, the role of examiners, and the zeal of young people to become part of the mathematical community — may be seen in an interesting autobiographical account. The physicist François Arago rose from a village childhood in the Pyrenees to become, eventually, the influential permanent secretary of the Académie des Sciences. In his autobiography he described how youthful ambition drove him, in the early 1800s, to apply to the École Polytechnique:[8]

Arago applies to the École Polytechnique.

> I saw at a glance that [my school] lessons would not be sufficient to secure my admission to the Polytechnic School; I therefore decided on studying by myself the newest works, which I sent for from Paris. These were those of Legendre, Lacroix, and Garnier ... I increased my library with Euler's *Introductio in Analysin Infinitorum*, with the *Resolution des Equations Numériques*, with Lagrange's *Traité des Fonctions Analytiques* and *Mécanique Analytique*, and finally with Laplace's *Mécanique Céleste*.
>
> At last the moment of examination arrived ... Monge [the entrance examiner — a nephew of Gaspard Monge] put to me a geometrical question, which I answered in such a way as to diminish his prejudices. From this he passed on to a question in algebra, on the resolution of a numerical equation. I had the work of Lagrange at my fingers' ends; I analysed all the known methods, pointing out their advantages and defects ... all were passed in review; the answer had lasted an entire hour. Monge, brought over now to feelings of great kindness, said to me, 'I could, from this account, consider the examination at an end. I will, however, for my own pleasure, ask you two more questions. What are the relations of a curved line to a straight line tangent to

[7] Frederick the Great, *Oeuvres de Frédéric le Grand*, Vol. 7, 100.

[8] See (Arago 1855, 6–8).

> it?' I looked upon this question as a particular case of the theory of osculations which I had studied in Lagrange's *Fonctions Analytiques*.
>
> 'Finally', said the examiner to me, 'how do you determine the tension of the various cords of which a funicular machine is composed?' I treated this problem according to the method expounded in the *Mécanique Analytique*. It is clear that Lagrange had supplied all the resources of my examination. I had been two hours and a quarter at the table.

Arago proved to be a brilliant student at the École Polytechnique, and in 1809 he was appointed Monge's successor as Professor of Descriptive Geometry. But his interests lay more in astronomy and physics, and he is not generally thought of as having played much part in the history of mathematics.[9]

By contrast, Arago's contemporary Augustin-Louis Cauchy was to become the most productive and influential mathematician of his generation, and stamped his personality irrevocably on the development of mathematics in France and throughout Europe. Cauchy entered the École Polytechnique in 1805 at the age of 16, moving on to the more specialised training of the School of Bridges and Roads two years later. After a few years as an engineer he began to produce mathematical papers of high quality. These attracted attention in the right quarters, and Cauchy was duly appointed a member of the Académie des Sciences and Professor of Analysis and Mechanics at the École Polytechnique in 1816, the year in which he won a prize contest of the Académie (on the subject of hydrodynamics). In the next few years he held the Chair of Mechanics at the Paris Faculty of Sciences and was simultaneously the assistant to the Chair of Mathematical Physics at the Collège de France, besides continuing to produce a ceaseless flow of mathematical publications on an extraordinary range of subjects. In this period, and for some time afterwards, mathematicians had to do several jobs at once to earn enough money and to become noticed; their collection of positions was known as the *cumul* (see Box 33).

Cauchy's reformulation of the foundations of the calculus, which we examine in Chapter 16, was very thorough and hard to understand, so that not only the students but also his fellow professors protested. But Cauchy's energy carried the day. He consolidated the direction in which Laplace had pointed the syllabus, of levering the pure analytical style away from geometrical connotations, and did so with such success that the new style survived the next revolutionary trauma, the overthrow of the corrupt Bourbon monarchy in 1830. Cauchy, however, felt that he owed his allegiance to the Bourbons, and viewed the events of 1830 with such alarm that he followed them into exile in Italy, eventually winding up as a tutor to the Dauphin when the Court settled in Prague. Cauchy's self-imposed exile markedly weakened his impact on contemporary mathematicians for most of the 1830s.

A good impression of the French mathematical community during the 1820s is given in a letter from the young Norwegian, Niels Henrik Abel, who had travelled to Paris to learn the latest mathematics and to meet mathematicians. On his arrival, in the summer of 1826, Abel wrote home to a friend:[10]

---

[9]Arago did important work on the nature of light and on electromagnetism. He was also active in left-wing politics: as Minister of War during the 1848 revolution, he was responsible for abolishing slavery in the French colonies.

[10]Quoted in (Ore 1957, 146–147).

Box 33. The *cumul*.
For a young person wishing to teach and research into higher mathematics the career situation was chaotic. At the École Polytechnique, only the Director's salary was enough to live on; other teachers had to look elsewhere for supplementary income. One might put in a spell of teaching at one of the specialised higher Écoles (the School of Mining, the School of Artillery, and so on), or be an examiner, or serve on an advisory board or committee at one of the Écoles or the preparatory institutions that were established all over France. Some teaching institutions remained from the old regime, such as the Collège de France, and there were specialist state institutes for particular purposes, such as the Bureau des Longitudes (founded in 1795 with Lagrange and Laplace as its chief mathematicians) which had astronomical, metrological, and geodesic concerns.

The result was that a young man (no provision existed anywhere for the higher education of women) would graduate from the École Polytechnique, attend a specialist school for higher training, and — if he spurned a career in the army — look for a mixture of teaching jobs that were sufficient to make a living. This was his *cumul*. He would hope to accumulate positions which, taken together, might involve a thirty-hour week of teaching and academic business at a variety of institutions throughout Paris. This did not leave much time for research, although successful independent investigations would be desirable for further advance, and might bring one to the notice of the right people. He might hope gradually to acquire some influence of his own, until a more prestigious and remunerative appointment would at last allow him to relax and dispense some of his old positions to younger men whom he wished to help in turn.

The Parisian mathematical community, then, was a hothouse of intrigue and competition, gossip, and skulduggery; only the cleverest and most ruthless survived. The priority disputes, the charges of plagiarism, and other quarrels of which historical records survive, reflect a world where professional survival as well as serious intellectual issues were at stake.

Abel writes home from Paris.

> To this moment I have only met Legendre, Cauchy, Hachette, and a few lesser but quite clever mathematicians: Monsieur Saigey, editor of the *Bulletin*, and Herr Lejeune-Dirichlet, a Prussian who called on me one day, taking me for a fellow countryman.[11] That is a very ingenious mathematician; with Legendre he has shown that the equation $x^5 + y^5 = z^5$ is impossible in integers, and other beautiful things.
>
> Legendre is an extremely amiable man, but unfortunately hoary with age. Cauchy is mad, and there is no way to get along with him, although he is at present the mathematician who knows best how mathematics ought to be treated. His things are excellent, but he writes

[11] Jacques Frédéric Saigey was a mathematician and astronomer who edited the mathematical and physical sections of Ferussac's *Bulletin*.

Figure 13.4. Niels Henrik Abel (1802–1829)

very obscurely ... Cauchy is immoderately Catholic and bigoted, a very strange thing for a mathematician. Otherwise he is the only one who at present works in pure mathematics; Poisson, Fourier, Ampère, etc., are exclusively occupied by magnetism and other physical theories. Laplace presumably does not write any more ... I have seen him often in the Institute. He is a small, lively man ...

Everyone works for himself without concern for others. All want to instruct, and nobody wants to learn. The most absolute egotism reigns everywhere.

I have completed a large memoir on a class of transcendental functions to be presented to the Institute; it will be done on Monday. I showed it to Cauchy but he would hardly cast a glance at it.

We return later to the life and work of Abel, whose time in Paris was not very stimulating, as this letter suggests. Abel was not the only foreigner to journey to Paris to explore the mathematical community. His Prussian acquaintance, Peter Gustav Lejeune Dirichlet, later to be one of the leaders of the 19th-century revival of German mathematics, had discovered as a Cologne schoolboy of 17 that the mathematical instruction he sought was unavailable in Germany, so he went to study in Paris, where, in 1822, he enrolled at the Collège de France. Others also made the journey to Paris, which held the liveliest and richest mathematical community; there was then no mathematical community to speak of in Germany.

Dirichlet is, however, our link to the other French mathematician whose influence matches that of Cauchy: Joseph Fourier, whom Dirichlet came to know and admire during his years in Paris.

Figure 13.5. Joseph Fourier (1768–1830)

Joseph Fourier was born in Auxerre in central France in 1768. Orphaned at the age of 9, he was placed in the town's military school where he discovered both a passion for mathematics and a calling to civic and military duty. At 21 the French Revolution found him in his home town as a schoolteacher, and he became involved in politics. He was arrested in 1794, the year of the Terror, and a personal appeal to Robespierre failed, but when the Terror turned on one of its leading architects and Robespierre was himself executed on 28 July 1794, Fourier was able to go to the École Normale. He must have made an excellent impression there in a short time, for in 1795 he was appointed an assistant lecturer at the École Polytechnique, working under Lagrange and Monge. Soon he was arrested again, on a charge of having supported Robespierre, but his colleagues got him released. In 1798 Monge selected Fourier to go with him on Napoléon's expedition to Egypt, which lasted from 1798 to 1801. In military terms this was an attempt to exclude the British from the eastern Mediterranean, but it also had imperialist and cultural dimensions, and French engineers made impressive surveys of Egyptian pyramids and other remains.

In 1801, Fourier returned to France. Napoléon had noticed his impressive organisational talents and sent him to be the Prefect (or Governor) of the Department of Isère, which at that time extended from Grenoble to the French border. He was so successful there that Napoléon made him a Baron in 1808, and in 1809 he finished his contribution to the *Description de l'Égypte*, a massive account and glorification of ancient Egypt based on the surveys; this work marks the start of the celebration of ancient Egypt in the modern world.[12]

The turbulent events around the eventual defeat of Napoléon, his short-lived return in 1815 (the Hundred Days), and his final defeat and exile to Elba were the lowest points of Fourier's life, but in 1816 a former colleague on the Egypt expedition rescued

[12]The first edition, in 23 large volumes, was published between 1809 and 1813. There are nine volumes of text, the rest are plates and maps. Cleopatra's Needle and other obelisks began their journeys at this time, as did more and more fanciful statements about ancient wisdom, secret knowledge, and so forth, none of which can be held against Fourier.

him and gave him a job as Director of the Bureau of Statistics for the Department of the Seine, a position that left him good time for research. His association with Napoléon was, however, to cost him dear. It delayed his appointment to the reformed Académie des Sciences for a year (1816–1817), but, after more protests, he became its permanent secretary in 1822. Even more prestigiously, he was elected to the Académie Française in 1827 at the age of 59. He died in 1830 as the result of complications from an illness that he had caught years earlier in Egypt.

Fourier's name is nowadays inseparable from the representation of a function by a trigonometric series, which he came to as a result of a profound study of the diffusion of heat through a solid body. This work has variously been hailed as one of the few major contributions to both pure and applied mathematics, and as one of the earliest pieces of 'non-Newtonian physics' (that is, physics that is not based on the idea of atoms attracting one another at a distance).

Fourier wrote this work in several stages. The first version was finished in 1807 and submitted to the Académie des Sciences in Paris. Laplace, Lacroix, and Monge were in favour of publishing it, but Lagrange opposed it because its treatment of trigonometric series differed markedly from the way that he, in a paper written in the 1750s, had said they ought to be handled. In 1810 the Académie des Sciences announced a prize competition on heat diffusion, and Fourier revised his memoir and entered it for the competition. It won, but was nonetheless criticised, presumably at the instigation of Lagrange, for its lack of rigour and generality. Fourier thought the criticism unfair, but revised the work for publication, and the resulting book, *Théorie Analytique de la Chaleur* (*The Analytical Theory of Heat*), was published in 1822.

In this work, Fourier first obtained a partial differential equation that describes the flow of heat through a body.[13] This equation gives the temperature of each part of a body at any instant of time, and in order to solve it, the investigator needs to know the temperature of the body at every point on the surface.

This was already a significant achievement, but what gave Fourier's work its broader significance — and generated all the controversy — was his claim that any function can be written in a particular form, known today as its *Fourier series*. More precisely, Fourier claimed that any function f defined on the interval $[-\pi, \pi]$ can be written as an infinite series of sines and cosines in the form

$$f(x) = {}^1\!/\!_2\, a_0 + \sum_{n=1}^{\infty} (a_n \cos nx + b_n \sin nx).$$

As Fourier discussed, the coefficients of the series are given by the formulas

$$a_0 = \frac{1}{\pi} \int_{-\pi}^{\pi} f(x)\, dx,$$

$$a_n = \frac{1}{\pi} \int_{-\pi}^{\pi} f(x) \cos nx\, dx, \qquad b_n = \frac{1}{\pi} \int_{-\pi}^{\pi} f(x) \sin nx\, dx.$$

Two aspects of this claim are striking: it was made about *any* function; and it was essentially without proof. A considerable amount of difficult and important mathematics was to be done in marrying these claims to the rigorous theories of functions

[13] For this part of Fourier's work, see Section 20.1.

that Cauchy, Dirichlet, and their successors were shortly to develop.[14] We cannot pursue that story here, but a good over-simplification is that Fourier's claim is correct for functions with intuitively obvious properties but may be false for others.[15]

A third observation is that the idea of a Fourier series suggests a general moral: if you have a periodic function, write down its Fourier series. This approach was to be productive right across mathematics and beyond, from number theory to physics. When mathematicians found functions with other types of regular behaviour, the moral could be extended further, with functions replacing the trigonometric functions in the new setting. This aspect of mathematics is far from exhausted today.

13.3 Mathematics in Germany

The German situation could scarcely have differed more from the French. The country was divided into several independent principalities, each with its own attitudes towards education. Chief among these was Prussia, whose capital was Berlin, and which could boast of the man generally considered to be the greatest living mathematician — Carl Friedrich Gauss.

Gauss had been born into a labouring family on 30 April 1777, and soon displayed a precocious talent at mental arithmetic and languages that brought him to the attention of teachers and benefactors at an early age. Many stories grew up around him.[16]

At his first school Gauss was fortunate to find in Martin Bartels, the teacher's assistant, a young man who was a competent mathematician willing to give him special attention. Bartels not only stimulated Gauss to further study — indeed, it seems that the influence was mutual and helped Bartels to commit himself to a career in mathematics — he helped to bring the young Gauss to the attention of the Duke of Brunswick. In 1791 the Duke paid the 14-year old Gauss a regular stipend so that he could attend the nearby Collegium Carolinum, a newly founded, progressive, science-oriented academy of a type that was then springing up all over Germany. Liberal rulers quite often used their patronage to encourage bright children to stay in education, knowing that otherwise the family would keep them in the family business where their talents could be lost.

At the Collegium Carolinum Gauss read the works of Newton, Euler, and Lagrange, and when he left in 1795 to go to the University of Göttingen he was already doing original research; his discovery of a straight-edge and compasses construction for the regular 17-sided polygon dates from this period (see Section 19.1). In line with our earlier remarks about the poor state of universities in the 18th century, it is interesting to note that[17]

> In 1800 Göttingen University was an almost unique example of a well-functioning German university at a time when most were in a state of decline made worse by student rowdiness and drunkenness. It largely freed the Faculties of Arts and Sciences from the control of the Theology Faculty, and fostered research. Wilhelm von Humboldt studied

[14] See (Bottazzini 1986).

[15] For example, Dirichlet proved that if a function is defined on a finite number of possibly overlapping intervals, and is continuous and either increasing or decreasing on each interval, then it agrees with its Fourier series.

[16] For a debunking of some of these stories, see (Martinez 2012).

[17] See (McClelland 1980, 39).

Figure 13.6. Carl Friedrich Gauss (1777–1855)

there briefly, and took it as an inspiration for his reforms of the German University system and the creation of the University of Berlin.

In 1796, while he was still 18, Gauss began to keep a mathematical diary.[18] This startling object came to light only when Gauss's collected works were being prepared for publication at the end of the 19th century. It was published for the first time in 1901 when, such was its profundity, it was provided with lengthy commentaries by several leading mathematicians. We can form a good impression of the young Gauss from a quick look at some of its opening entries:

Gauss's mathematical diary.

[1] The principles upon which the division of the circle depend, and geometrical divisibility of the same into seventeen parts, etc. March 30 Brunswick.

[2] Furnished with a proof that in case of prime numbers not all numbers below them can be quadratic residues. April 8 Brunswick.

[3] The formulae for the cosines of submultiples of angles of a circumference will admit no more general expression except into two periods. April 12 Brunswick.

[4] An extension of the rules for residues to residues and magnitudes which are not prime. April 29 Göttingen.

[5] Numbers which can be divided variously into two primes. May 14 Göttingen.

[6] The coefficients of equations are given easily as sums of powers of the roots. May 23 Göttingen.

...

[18]An English translation of Gauss's diary appears in Dunnington (2004, 449–496); see F&G 15.A1 for a selection, from which the above is taken.

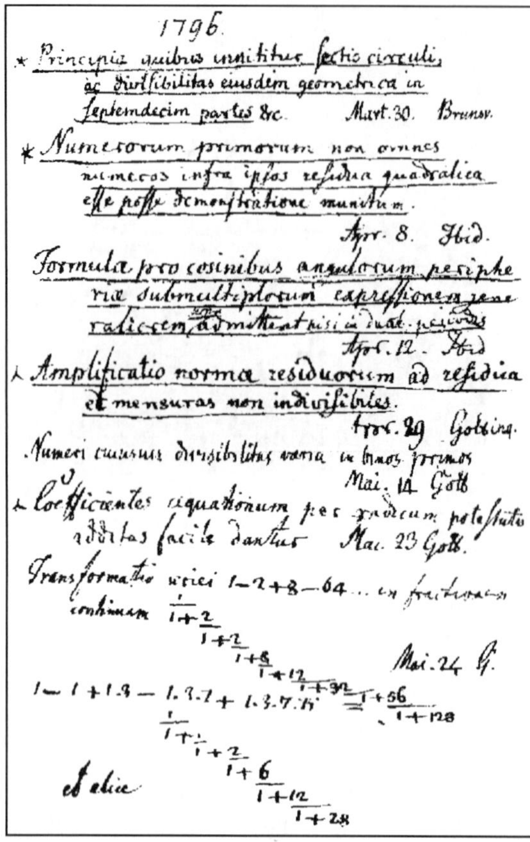

1796.

* Principia quibus innititur sectio circuli, ac divisibilitas eiusdem geometrica in septemdecim partes &c. Mart. 30. Brunsv.

* Numerorum primorum non omnes numeros infra ipsos residua quadratica esse posse demonstratione munitum. Apr. 8. Ibid.

Formula pro cosinibus angulorum peripheriae submultiplorum expressionem generaliorem non admittent nisi in duas periodos. Apr. 12. Ibid.

* Amplificatio normae residuorum ad residua et mensuras non indivisibiles. Apr. 29 Gotting.

Numeri cuiusvis divisibilitas varia in binos primos Mai. 14 Gott.

* Coefficientes aequationum per radicum potestatum summas facile dantur Mai. 23 Gott.

Transformatio seriei 1−2+8−64 ... in fractionem continuam Mai. 24 G.

et aliae

Figure 13.7. The opening page of Gauss's diary, 1796

[12] The sum of the periods when all numbers less than a [certain] modulus are taken as elements: general term $[(n+1)a - na]a^{n-1}$. June 5 Göttingen.

[13] Laws of distributions. June 19 Göttingen.

[14] The sum to infinity of factors $= \frac{\pi^2}{6}$ sum of the numbers. June 20 Göttingen.

[15] I have begun to think of the multiplicative combination (of the forms of divisors of quadratic forms). June 22 Göttingen.

[16] A new proof of the golden theorem all at once, from scratch, different, and not a little elegant. June 27

...

[18] EUREKA, number $= \triangle + \triangle + \triangle$. July 10 Göttingen.

...

[23] I have seen exactly how the rationale for the golden theorem ought to be examined more thoroughly and preparing for this I am ready to extend my endeavours beyond the quadratic equations. The discovery of formulae which are always divisible by primes: $\sqrt[n]{1}$ (numerical). August 13 Göttingen

[24] On the way developed $(a+\sqrt{-1})^{m+n\sqrt{-1}}$. August 14.

[25] Right now at the intellectual summit of the matter. It remains to furnish the details. August 16 Göttingen.

We might start by observing that Gauss found 25 things worthy of note in less than five months! Next, we may well think that most of its entries border on the unintelligible, but then Gauss was writing only for himself. Even so, it is clear that Gauss was interested in all sorts of topics, among them: numbers, infinite series, questions about equations, and trigonometric formulas.

We might recognise a few details. The first entry is obviously connected to his discovery of the construction of the regular 17-gon (see Section 19.1). The second one must have something to do with quadratic reciprocity (see Section 18.1), so he was already looking for a proof. Entry 6 has to do with the way that the roots of a polynomial equation are related to its coefficients, which is handled, as Gauss would have known, by Newton's rules.[19]

That said, the diary is most striking for its profusion and its opacity. It is profuse: there are so many original discoveries being broached here. For example, Entry 16 records his pleasure at the discovery of a second proof of the Golden Theorem (the theorem of quadratic reciprocity, see Box 55 in Section 18.1) — this one is 'not a little elegant'. Entry 23 announces the intention of going further, with success coming, apparently, three days later (one hopes that the confidence was rewarded). Entry 18, which records that every number is a sum of three triangular numbers resolves an old, and previously unproved conjecture of Fermat.[20]

The diary is opaque: knowing the mathematics to which Gauss could be referring is often of little help in understanding the entries. Indeed, Gauss himself admitted in later life that he could not always recapture the meaning of his own cryptic remarks.

In 1799 Gauss graduated from Göttingen University with a doctoral thesis on the Fundamental Theorem of Algebra; we shall return to his proofs of this theorem in Section 18.3. He was by then immersed in writing a book on the theory of numbers, the *Disquisitiones Arithmeticae* (Arithmetical Investigations), which appeared in 1801. This book subsumes much of what had been done before by Euler and Lagrange, notably in the theory of quadratic forms, and gave it a new direction, and it was to earn Gauss the title of the 'Prince of Mathematicians'. Lagrange wrote to him in 1804 that the *Disquisitiones* immediately put Gauss in the top rank of mathematicians.[21] As we shall see, this book was to prove decisive in putting number theory at the heart of the German mathematical community's interests, and indeed in defining what the subject was taken to be.

As if this were not enough, Gauss also established himself among the leading astronomers of Europe. On the very first day of the 19th century, 1 January 1801, the Italian astronomer Giuseppe Piazzi had discovered the first new body in the solar system since Herschel's discovery of Uranus twenty years earlier (it was the asteroid Ceres). But Piazzi was able to observe it for only 42 days before it vanished behind the Sun. Where and when would it re-appear? How could the position of such a small, faint object be determined from such a limited set of observations?

---

[19] Consider, for example, the cubic polynomial x^3+ax^2+bx+c, with roots α,β,γ. We have $\alpha+\beta+\gamma=-a$, $\alpha\beta+\beta\gamma+\gamma\alpha=b$, and $\alpha\beta\gamma=-c$. It follows that $\alpha^2+\beta^2+\gamma^2=a^2-2b$, and $\alpha^3+\beta^3+\gamma^3=3(ab+c)-a^3$.

[20] A triangular number is one of the form $\frac{1}{2}n(n+1)$, and is the sum of the first n positive integers.

[21] Quoted in (Weil 1984, 313).

Piazzi's data were published in June, and several astronomers began to make their predictions. So too did Gauss, and although his prediction was widely separated from the others it was clear by the end of the year that Gauss was correct. The foundations of his success were his willingness to investigate elliptical orbits for Ceres, and his use of innovative statistical tests of his own devising to refine Piazzi's data. Such was the excitement caused by this new 'planet' that Gauss briefly tasted European fame.

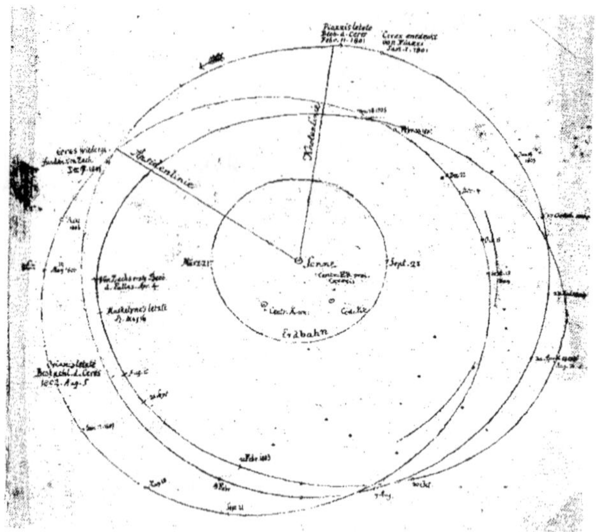

Figure 13.8. Gauss's diagram of the orbits of Ceres and Pallas (the second asteroid to be discovered)

For the rest of his life Gauss was to enjoy astronomical research and to prefer the company of astronomers to that of mathematicians. Three factors came together here. Astronomy was congenial to Gauss because it made particularly good use of his talents (including his remarkable ability at mental arithmetic); it enabled Gauss to repay his debt, as he felt it, to his benefactors; and it was there as a profession to belong to. There was no comparable position in mathematics in Germany for someone of Gauss's ability — he had become the director of the Observatory at Göttingen in 1807 — and in later life Gauss resisted attempts to bring him to Berlin, securing instead a pay rise that kept him in Göttingen and free of the heavy teaching loads that would have eroded his opportunities for research.

This all helps to explain the apparent paradox that an individual rightly remembered today as a great mathematician could spend so much of his time being paid to do something else. For example, in the late 1810s and early 1820s Gauss worked intensively on a survey of Hannover: he went on the surveying expeditions and carried out the lengthy calculations — one historian has estimated that over a million measurements were taken. But in those years he also made a remarkable discovery in differential geometry (the geometry of curves and surfaces using the methods of the calculus) that was to reformulate the subject completely (see Section 14.3).

However, because Gauss kept himself aloof, his influence on the mathematicians of his day was a complicated one. His work greatly stimulated others, not least in the theory of numbers, but it was a daunting edifice. Gauss would publish only when he felt that he understood something completely, prompting Abel to say of him that 'he is like the fox, who erases his tracks in the sand with his tail', because he concealed the ways in which he had been led to discover his theorems.[22] According to his first biographer, Sartorius von Waltershausen, who knew Gauss personally, Gauss said that when a building is finished you take down the scaffolding.[23] On non-Euclidean geometry, a subject he thought deeply about but chose not to publish on, it can even be argued that his influence was negative, as we discuss in Chapter 14.

The mid-19th century. German universities had not been vital centres of intellectual life in the 18th century, and did not have vigorous, established, mathematical traditions. The one possible asset that Germany possessed was the incomparable Gauss. But much of his best mathematics lay in his desk drawers, and was not published until after his death in 1855, so he did not provide the stimulus for other mathematicians that one might expect. Rather, we should perhaps see Germany as fallow ground, ripe to bear new influences. As we shall see in Chapter 15, German mathematicians were to take to projective geometry with a will, perhaps because it could be started easily from scratch, unlike such contemporary developments as Cauchy-style analysis. But they also took to this geometry in a different way from the French — a significant point that we dwell upon later.

Aside from Gauss, the overall level of mathematical training and attainment in Germany was low: mathematics participated in a more general educational and social structure. As the historian Herbert Mehrtens has emphasised, professors of mathematics existed, but played a different role from what their title might suggest to us now. Speaking of the general features of university mathematics in Germany around 1800, Mehrtens has written:[24]

> There was no strict delimitation of subjects in teaching assignments and personal combinations of chairs within the philosophical faculty were frequent. The equation of mathematical professorship with mathematical research is valid (with exceptions) only when 'mathematics' is taken in the very broad sense of the day... Performance was not that of a disciplinary specialist but rather either that of a scientific universalist ...or it had to be performance in the field of civil administration with a combination of technological, administrative, and maybe juridical expertise, which might all be centred around a basic 'mathematical' education and achievement.

In the early years of the new century, however, things began to change. The shock of Prussia's defeat in 1806 by Napoléon's troops, at the battle of Jena, led to an upsurge of patriotism and to what was later seen as the moral regeneration of the country. Intellectual life was re-evaluated and a flowering of national culture was promoted by educational reform, new institutions, and new social and professional structures. The University of Berlin was founded in 1810 by the philosopher and distinguished linguist Wilhelm von Humboldt, and it developed during the 19th century into the leading institution embodying the new, research-orientated, professional approach to academic subjects. The mathematical developments of the time are reflections or particular instances of a more wide-ranging impetus to professionalisation, which was felt in all

---

[22] Abel, writing to Christopher Hansteen, quoted in (Bjerknes 1885, 92).

[23] See (Sartorius 1856, 82).

[24] See (Mehrtens 1981, 410–411).

subjects. The 19th-century developments, here described by the scholar Joseph Ben-David, applied to many other subjects — history, linguistics, philosophy, and so on — as well as to the sciences and mathematics:[25]

> The transformation of science into a status approaching that of a professional career and into a bureaucratic, organized activity took place in Germany between 1825 and 1900. By the middle of the nineteenth century, practically all scientists in Germany were either university teachers or students, and they worked more and more in groups consisting of a master and several disciples. Research became a necessary qualification for a university career and was considered as part of the function of the professor ... The first step towards this transformation was the establishment in 1809 of a new type of university — the University of Berlin.

In fact, the institutional and professional setting of mathematics, in something close to the form that we know today, was created in Germany in the early 19th century. Because the immediate context for the renewal of German intellectual life was the shock of being defeated by Napoléon's armies, it is striking that those responsible for this movement did not copy the French and set up a university on the model of the École Polytechnique. Instead, they created something that became much more clearly the model for other places and nations to follow — the modern university. Military education was conducted outside the elite places of higher education, as happened increasingly in America too, and in Britain and Italy.

The Germans pursued a variety of styles of higher education, some openly in pursuit of higher learning, and others, later in the century, more deliberately tied to technical education and the needs of the emerging German industries. As in France, the impulse to produce a new group of specialists, operating at a higher level than the previous generation, was felt in Germany. These would be university graduates requiring, accepting, and administering a system of training and qualifications that would largely be funded by government. But the difficult and detailed decisions that a profession has to take were taken differently in France and Germany, with rather different consequences, as we shall see.

One of the factors influencing the nature of the new developments was a widely-felt philosophical attitude that is called *neo-humanism* — one of those somewhat vague social and educational philosophies that are no less potent for being difficult to pin down. Neo-humanism emphasised the expansion of knowledge through research, but saw the pursuit of pure knowledge as more commendable than utilitarian or applied research. The attitudes that guided the new developments are here characterised in the words of Ben-David:

> Neither natural science, nor any other kind of philosophically important knowledge, had to be directly useful for political or economic purposes either in the short or in the long run. Learning and knowledge were ends in themselves. Their importance derived from providing a spiritual justification for society and from their educational effects of shaping the mind.

Neo-humanism was most actively promoted in Prussia. It was expected that most university graduates would go on to teach in the higher classes of schools, and in that way promote the agenda for modernising Germany, and the qualification to become

[25] See (Ben-David 1971, 108–109).

one of these teachers was highly prized, more so than the Ph.D. degree, because it guaranteed a respected and well-paid job.[26]

In view of the guiding neo-humanist philosophy, it is not surprising that the mathematics that came to be developed in the new and reformed German universities was predominantly what we think of as pure mathematics. This contrasts with the greater French emphasis on mathematical physics and mathematics with some applicability to the empirical world that we saw alluded to in Abel's letter (although Abel had overstated the case somewhat).

The marriage of mathematics and neo-humanism was to prove successful, as we can see if we briefly look ahead to the mid-19th century. In his inaugural address as a professor in Tübingen in 1874, when he succeeded Hermann Hankel, the mathematician Paul du Bois-Reymond described mathematical life in these terms. Quoting from a letter that Dirichlet had written to his friend Jacobi, du Bois-Reymond said that mathematics 'was a science filled with cries of agony from mathematicians as they wrestled to uncover truth'. In the words of a later account:[27]

> Yet mathematics was measured not in the anguish of the creative process but in the joy and exhilaration of discovery. In some ways, it was a sport, like mountain-climbing. As an independent branch of learning, mathematics had to be allowed to follow its own course without being constrained to provide applications. Echoing Hankel, du Bois-Reymond did not oppose practical application, but he saw mathematics serving a goal higher than that of handmaiden to the sciences. His mathematics was a positive philosophy and a profound art. In the view of Hankel and du Bois-Reymond, engineers and practically inclined scientists calculated numerical solutions to problems, while pure mathematicians avoided practical considerations in favor of abstract, symbolic relationships. Two different communities were supposed to practice mathematics at the same time.

One subject that flourished was number theory, where the influence of Gauss was particularly strong. The leaders of the mathematical revival in Germany, Jacobi and Dirichlet, were to receive little personal support from Gauss, although both had a strong interest in the topic. But they helped to communicate an enthusiasm for it to the next generation, for it was through their teaching and research that Gauss's achievements gradually became the source of inspiration for a strong national school with a commitment to number theory. In particular, Dirichlet's lectures on that subject, aimed at simplifying and extending Gauss's work and so bringing it to a larger audience, were still valuable when they were published as a book in 1863, after Dirichlet's death.

Dirichlet and Jacobi also had a complicated relationship with the large and vigorous community based in Paris. Jacobi had acquired immediate fame in 1829 by decisively reformulating some ideas of Legendre. Legendre, by then in his 70s, had been highly impressed with what Jacobi had done and saw to it that it was well received. But his response to Dirichlet was less generous. Although Dirichlet was the first person since the 1760s to get anywhere with the case $n = 5$ of Fermat's Last Theorem,

---

[26] See (Pyenson 1983, 17).

[27] See (Pyenson 1983, 20), an account that draws on E. Lampe's version of an address that was published in 1910.

Legendre managed only an awkward acknowledgement to one 'Lejeune Dieterich' before going on to extend Dirichlet's work with some results of his own.[28]

Dirichlet and Jacobi were close friends who gathered around them most of the next generation of German mathematicians. As well as being a number-theorist Dirichlet was an analyst — that is, someone who worked with the calculus as it had been rigorised by Cauchy. One of his earliest results was to show that, under certain circumstances, the Fourier series representation of a function is equal to the function itself: Fourier had simply assumed this. Jacobi was also a number-theorist and a pioneer in the theory of elliptic functions, but he worked more in the older formal and algebraic style of Euler; indeed, he had Euler's formidable ability to marshal huge formulas. An indication of the difference between the two men, and of the strength of Jacobi's regard for Dirichlet, may be gained from Jacobi's remark:[29]

> If Gauss says he has proved something, it seems very probable to me; if Cauchy says so it is about as likely as not; if Dirichlet says so, it is certain. I would gladly not get involved in such delicacies.

Dirichlet was a professor at Berlin for 27 years, from 1828 to 1855, before moving to Göttingen to succeed Gauss. Jacobi was a professor at Königsberg for 18 years, from 1825 to 1843, before moving to Berlin. He lectured tirelessly, and introduced a novelty that did much to bring him followers: the research seminar. Dirichlet followed his example, and in this way the two fostered a generation of graduate students, thus bringing some organisation to the previously chaotic situation confronting the young would-be researcher.

This is indicative of one way in which German professors were to surpass their French rivals: the systematic production of the next generation of scholars. Absolute numbers would remain small throughout the 19th century because they depended on the slow and uneven growth of universities, but in mathematics and in other subjects German professors applied themselves to training at the highest level. The research seminar, at which the professor would discuss current work, was soon to be supported by a seminar library where advanced journals could be read. Mathematicians trained in this way might not get jobs in a university, but they would go on to contribute to raising the levels of mathematics teaching in the better German schools, and in this important way they raised the standards of the profession.

The French preferred the apprenticeship model for training researchers, and were cavalier about who survived and who did not. They also expected graduates of the École Polytechnique to move on to careers as engineers, and the École Normale, where teachers were trained, remained very much the junior institution until the last two decades of the 19th century. This largely explains why the leading centres for mathematics became those in Germany before the century was over.

---

[28] Dirichlet split his argument that there are no solutions in non-zero integers to the equation $x^5 + y^5 = z^5$ into two cases, but could deal with only one. In July 1825 Legendre presented this partial result to the Académie des Sciences in Paris, on Dirichlet's behalf (only full members of the Académie could present papers to it), and went on to dispatch the remaining case by an involved and artificial argument. In November, Dirichlet gave a much simpler account of the second case, which he published the following year in the new *Journal für Mathematik*. But in 1830, when Legendre gave his complete proof in the third edition of his *Théorie des Nombres*, he did not mention Dirichlet at all.

[29] Letter from Jacobi to A. von Humboldt, 21 December 1846, quoted in (Biermann 1959, 53). Alexander von Humboldt was a Prussian geographer, naturalist, and explorer, known for his investigations in Latin America, and a prominent supporter of science. He was the elder brother of Wilhelm, the founder of the University of Berlin.

Dirichlet and Jacobi also wrote works on applied mathematics, but here their example did not catch on so well: applicable work was not as obviously in the spirit of neo-humanism.[30] Nowhere was this clearer than in the unavailing proposal to create a version of the École Polytechnique in Berlin. Attempts to create a polytechnic that would train engineers foundered repeatedly, not least because of a distaste for the low level of mathematics that the students would need to be taught. Influential politicians who tried to bring Gauss to Berlin saw their efforts collapse in a morass of financial and administrative disagreements, and Gauss stayed in Göttingen. In the 1830s, both Dirichlet and Jacobi were drawn into discussions, but nothing came of them. In the 1840s Jacobi was offered the directorship of the proposed Polytechnic, but he withdrew on learning that the system of preparatory schools that were to be set up to feed the polytechnic would not teach Greek — 'the language of Euclid', as it was said. The result was that higher education became firmly established with universities at the top, and within each university the professors dominated their departments. Professors were free to choose what was taught, although the Ministry of Education had a strong say on who was appointed and how much they should be paid. It was a system more like the situation in the United States and Britain today than the strongly centralised French one.

It was also highly successful. By 1855 German (notably, Prussian) mathematics had begun to blossom, and the University of Berlin was becoming the centre of the higher mathematical world. Not only German students visited there to take part in the active seminar life; Scandinavians, Russians, and Italians also passed through.

If we compare the mathematical scenes in France and Germany in the period from 1820 to the 1850s, we see that there were obvious social differences between the two countries. Throughout the 19th century, France was a unified nation with a highly centralised intellectual life — most of the main academic institutions were in Paris. It also possessed a vigorous mathematical culture: Lagrange, Laplace, and Legendre were active throughout the Napoleonic Wars, and the eventual defeat of Napoléon seems to have had little effect on the development of mathematics. The subsequent generation of Cauchy, Fourier, and Poisson attracted mathematicians from all over Europe (Abel from Norway and Dirichlet from Germany, for example) and produced illustrious French successors such as Liouville, Hermite, and Jordan.

Germany, on the other hand, was a varied collection of separate principalities, which were not to be unified in anything like their present form until Bismarck waged wars to do so in the 1860s, culminating in the Franco–Prussian War of 1870–1871, which incorporated large parts of Alsace and Lorraine into Germany. German universities were small and varied. They were also short of money, and German academics were inclined to look nervously abroad whenever questions of status arose. A foreign reputation was widely believed to be of more use than a domestic one if one wanted to make a career in research. There was no single centre of German intellectual life, although, with the rise of Prussia, the University of Berlin came more and more to occupy such a position.

What was harder to see at the time, although many historians see it in retrospect, was a gradual shift of power and influence from France to Germany, and from Paris to

[30]What succeeded was first experimental, and then theoretical, physics which grew up outside the mathematics departments. See (Jungnickel and MacCormach 1986).

Berlin. This will occupy us in Section 13.5, but next we look at another feature of the early 19th century: the emergence of specialist journals.

13.4 Journals and publishing

Throughout history, mathematicians have tried various ways to inform others of their work. Archimedes and Apollonius wrote individual copies for distribution to their friends; Euclid's *Elements* may well have been professionally copied by scribes as part of a more systematic attempt to teach others. In the 16th and 17th centuries, the secrecy of people such as Tartaglia or Roberval can be contrasted with the greater openness of Bombelli or Descartes. But the publication of books and articles was far from being the norm in the early 17th century. Private circulation, through the good offices of Mersenne in Paris or Oldenburg and Collins in London, was much more usual. This made sense when the total number of people expected to be interested in what one had to say might not exceed a dozen or so.

Scientific journals were the creation of the late 17th century. These journals were often the property of the various national Academies, such as the Royal Society of London, which was founded in 1662 and sent out the first volume of its *Philosophical Transactions* in 1665, the earliest journal to be strictly devoted to science. Even though Leibniz, an editor of the *Acta Eruditorum* of Leipzig (founded in 1682), decided to publish his new calculus in its pages, Newton did not see fit to disseminate his own ideas in a similar way.

Figure 13.9. Joseph Diaz Gergonne (1771–1859)

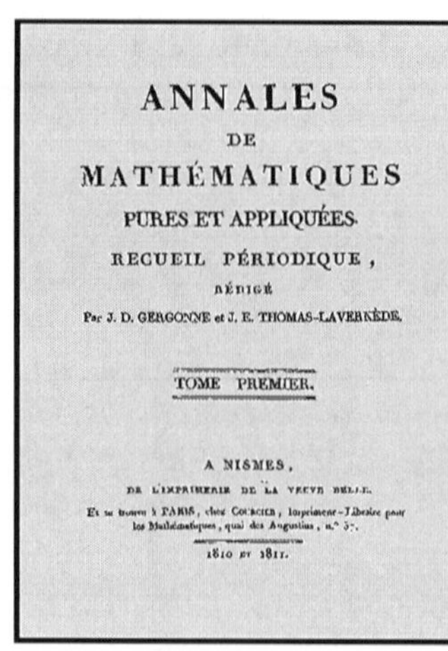

ANNALES
DE
MATHÉMATIQUES
PURES ET APPLIQUÉES.
RECUEIL PÉRIODIQUE,
RÉDIGÉ
Par J. D. GERGONNE et J. E. THOMAS-LAVERNÈDE.
TOME PREMIER.
A NISMES,
DE L'IMPRIMERIE DE LA VEUVE BELLE.
1810 ET 1811.

Figure 13.10. The title page of the first issue of Gergonne's *Annales*

As we discussed in Chapter 7, a change in publishing habits came with the growth of the profession in the 18th century, inspired by the example of Euler. By the end of that century, there were a number of journals in which mathematicians could expect to publish their work, although then (as now) there might well be a delay of several years between the submission of an article and its publication; matters had been even worse than this in the middle of the 18th century. We now pursue the story of books and journals through the 19th century, and see what they can tell us about the professionalisation of mathematics and the image of their subject that mathematicians chose to put forward.

As we might expect, there was an explosion of journal publishing in France between 1800 to 1820. At least eleven new journals appeared for those with an interest in mathematics or its applications. The École Polytechnique had already founded its own journal in 1795, and in 1824 there appeared Baron de Ferussac's small but important *Bulletin*, which was primarily a survey journal but also carried original work, so mathematicians were well served in France. Most of these journals adhered to the 18th-century custom of printing articles on mathematics and every branch of science, from astronomy and biology to zoology, and were a useful way of getting into print as the backlog for the journals of the Académie increased.

But one journal did specialise in mathematics: the *Annales de Mathématiques Pures et Appliquées* (Annals of Pure and Applied Mathematics). It was edited by the geometer Joseph Diaz Gergonne, and was strong on geometry and education.[31] Its editor was also unusual in eschewing the delights of Paris for the relative quiet of the provinces — in this case Nîmes, near Avignon in southern France. The journal lasted from 1810, when Gergonne was 39, to 1832, when he became Rector of the University of Montpellier (southwest of Nîmes) and could no longer find the time to run it. During that time it had acquired an international reputation and it was sorely missed.

Figure 13.11. August Leopold Crelle (1780–1855)

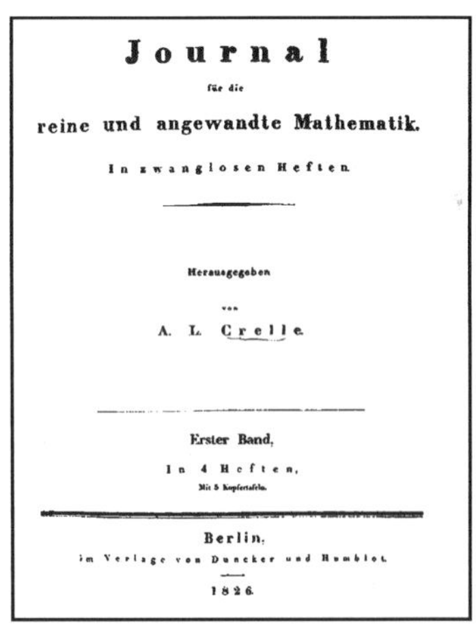

Journal

für die

reine und angewandte Mathematik.

In zwanglosen Heften.

Herausgegeben

von

A. L. Crelle.

Erster Band,

In 4 Heften,

Berlin,

im Verlage von Duncker und Humblot.

1826.

Figure 13.12. The title page of the first issue of Crelle's *Journal*

The first new journal that we look at in detail was not French, however, but German. It was founded by August Leopold Crelle, who exercised a considerable influence on mathematics without himself being a mathematician of the highest calibre. He was

[31] Although Gergonne's journal is often said to be the first journal devoted exclusively to mathematics, it was preceded by two short-lived journals edited by Carl Friedrich Hindenburg: the *Leipziger Magazin für reine und angewandte Mathematik* (The Leipzig Magazine for Pure and Applied Mathematics) (1786–1789) and the *Archiv für reine und angewandte Mathematik* (The Archive for Pure and Applied Mathematics) (1795–1799).

born in 1780, and although he wrote several books on mathematics he soon realised that his talents did not lie exclusively in that direction. He became a Privy Councillor and an accomplished construction engineer, responsible from 1820 to 1830 for the building of highways in Prussia, and later the railway between Berlin and nearby Potsdam. He was keen to revitalise German mathematics, and in 1821 and 1822 he published the first two volumes of an encyclopedia of mathematics for which he also wrote most of the articles. He then realised that this was not the ideal vehicle with which to carry out his task, and he abandoned this project in favour of initiating a journal. To this end he recruited his contacts in Berlin as authors and set about attracting more.

In 1825 Crelle met Abel, then on his way to Paris, and once the young Norwegian had explained his ideas as well as he could in his halting German, Crelle was happy to commission him too. Abel profited greatly from his acquaintance with Crelle. As he wrote home in glowing terms to his former professor:[32]

Abel writes home.

> When I expressed surprise over the fact that there existed no mathematical journal, as in France, he said that he had long intended to edit one, and would presently bring his plan to execution. This project is now organised, and that to my great joy, for I shall have a place where I can get some of my articles printed. I have already prepared four of them, which will appear in the first number. Since they are written in French, Crelle will oblige me by translating them. In this manner my little French serves me in good stead.
>
> Crelle has an excellent mathematical library which I may use as my own; it is particularly profitable because it contains all the newest material, which he obtains as soon as it is available. Among other things he subscribes to the *Bulletin Universel des Sciences et de l'Industrie*, which appears in France under the editorship of the Baron Ferussac; this is useful for me since it announces all new books and discoveries in mathematics.
>
> The young mathematicians in Berlin and, as I hear, all over Germany almost worship Gauss; he is the epitome of all mathematical excellence. But even if he is a great genius, it is also certain that he has a bad presentation. Crelle says that all Gauss writes is gruel since it is so obscure that it is almost impossible to understand.

In due course, the first issue of 'Crelle's *Journal*' (as it soon became called) appeared. Crelle himself wrote a preface, in which he observed that whereas almost every significant part of knowledge was now treated in some German journal or other,[33]

> only capacious and unbounded mathematics, that sublime science above time and place, opinion and passion, that above all is perhaps the most closely related to truth, is not served in this way.

He noted that the French had Gergonne's journal, and declared that he sought the broadest possible audience for his own offering. That said, what he then presented to

[32] Quoted in (Ore 1957, 90–91).

[33] *Journal für Mathematik* 1, 1826, 1.

his readers is interesting. In its first year the journal, which was conventionally called the *Journal für die reine und angewandte Mathematik* (Journal for Pure and Applied Mathematics) carried 15 articles on analysis (broadly speaking, advanced calculus and algebra), 13 on geometry, 4 on mechanics, and only 4 on applied mathematics (defined as comprising the theories of light, heat, sound, probability, hydraulics, and machines). Of these 36 articles, no fewer than 7 were by Abel, including his proof that the general quintic equation is unsolvable by radicals, and his proof of the binomial theorem in which he correctly criticised Cauchy's earlier attempt. This shows how seriously Abel took Cauchy's strictures about rigour — but that said, Abel proceeded to make much the same mistake as Cauchy later in his proof, which shows just how difficult being rigorous was proving to be. Five of the geometry articles were by Jakob Steiner, who was based in Berlin — among them his generalisation of Desargues's theorem on triangles in perspective to a theorem about two tetrahedra in perspective.

Crelle's choice of articles sheds an interesting light on his interpretation of the title he gave to his journal, and suggests that the bias was tilted heavily towards the pure side of things. This might reflect the preferences of mathematicians in Berlin, in which case poor Crelle was stuck with what he could get; but 4 or at most 8 articles on applications out of a total of 36 is perhaps indicative of what he wanted to include. His lofty paean to mathematics could have been just part of the new humanist rhetoric suitable for such occasions, or it could have been from the heart.

Subsequent issues continued this pattern. The second one, in 1827, carried the first of several articles by Abel on his new elliptic functions (see Box 34), and as the months went by, a stream of articles on the same subject by Jacobi also appeared. In that year Jacobi managed 9 and Abel 3; in subsequent issues each might manage half a dozen. In the third volume (1828) Dirichlet published his proof that there are no non-zero integer solutions to $x^5 + y^5 = z^5$ — the case $n = 5$ of Fermat's Last Theorem which Legendre completed — and other researches on rigorous Cauchy-style analysis and number theory, and the French mathematician Jean-Victor Poncelet contributed the first of two long articles introducing the projective geometry of his *Traité*. It is small wonder that wits quickly dubbed Crelle's *Journal* 'the journal of purely inapplicable mathematics'! Matters quietened down with Abel's death in 1829, but tables published in the 50th issue, when Crelle lay dying, showed that scarcely a quarter of the articles he had chosen to publish over thirty years were on applied mathematics, however it might be defined.

The tables in Crelle's *Journal* also showed quite explicitly that 111 Prussian authors had been published, contributing 664 articles between them and amounting to over 10,000 pages. Other Germans, 41 in number, had managed 193 memoirs in just over 3,000 pages, but even counting them in with the rest of the non-Prussians (mostly French, Russian, German, Scandinavian, English, and Italian) the Prussians still outnumbered the rest by a ratio of 2 to 1. Similarly, there were nearly 12,000 pages in German, about 4,500 in French, 2,400 in Latin, 112 in English, and 88 in Italian.

Apart from the fact that somebody in the editorial offices seemingly had a mania for counting, this tells us a great deal. Evidently, there were many people around with something mathematical to say (about 200), and Crelle had no trouble finding his native Prussians willing and able to contribute to the cause of reviving mathematics. It seems that he and his assistants were concerned to demonstrate their success to those

Box 34. Elliptic integrals and elliptic functions.
Elliptic integrals occur when one tries to evaluate the arc length of an ellipse. Because planets travel along elliptical orbits, problems in astronomy lead naturally to the attempt to evaluate such integrals. However, it proved impossible to evaluate them explicitly — that is, to express the answers in terms of known functions. More generally, integrals involving the square root of a quartic polynomial, such as the arc-length integral for an ellipse, or, more simply,

$$s = \int_0^t \frac{dx}{\sqrt{1-x^4}},$$

were equally difficult. They came to be called *elliptic integrals*, and people studied them in the hope of extending the power of the calculus.

One way forward, advocated most energetically by Legendre, was to say that such an integral *defines* a new function of its upper endpoint t, which is therefore the answer. Legendre found ways of compiling tables for these functions. There is an analogy here with the logarithm function. If we define $\log t$ as $\int_1^t dx/x$, then we have evaluated this integral if we obtain values of the log function (say by considering it as an area, or by using infinite series).

Gauss, Abel, and Jacobi found a neater way forward, by observing another analogy. The integral for the arc-length around a circle is

$$s = \int_0^t \frac{dx}{\sqrt{1-x^2}}.$$

If we write $x = \sin\theta$, we obtain the familiar answer $s = \sin^{-1} t = \arcsin t$. They all observed that $t = \sin s$ is an easy function to understand, so they took an elliptic integral like

$$s = \int_0^t \frac{dx}{\sqrt{1-x^4}}$$

and *defined* a function analogous to the sine function. In Gauss's notation this function is $t = sl(s)$.

Such a function is called an *elliptic function*, and it turned out to be possible to exploit the analogy with the trigonometric functions to obtain a rich and remarkable theory. Indeed, this step is analogous to regarding $y = \int_1^t dx/x$ as defining not $y = \log t$ but the simpler inverse function $t = e^y$.

who ran Prussian education. It also seems that the *Journal* acquired a good international reputation, with contributors from nearly all of the major countries of Europe.

We note that Latin was still in use, not least by Jacobi, which suggests that some Germans considered it a better medium for reaching an international audience than their own tongue. Finally, just compiling such tables indicates a degree of awareness that the *Journal* could itself be an object of interest and something to be proud of. It suggests that the editors considered it a success, and also that statistical arguments such as these were a reasonable way to demonstrate that success.

It is unlikely that Crelle could have created all this enthusiasm for pure mathematics single-handedly. In fact, it grew naturally out of the spirit of neo-humanism that dominated German intellectual life at the time. Crelle's priorities were those of the leading mathematicians around him, to whom he naturally turned for advice. Chief amongst these were two men whom we have already encountered, Dirichlet and Jacobi. When they in their turn joined the ranks of the profession, and wrote up their research, it was to Crelle's *Journal* that they would submit it. The crude page count above tells us that their preferences were also for pure mathematics.

Crelle's career helped him to become a member of five Royal Societies — those of Berlin, St Petersburg, Naples, London, and Stockholm — as well as the American Society for Promoting Useful Knowledge, in Philadelphia. He also played a major role in the Berlin Academy, to which he was elected in 1827 on the strength of his work in creating his *Journal*, and over the next fifteen years he set about proposing the leading Prussian mathematicians for ordinary membership and the best foreign ones as corresponding members. His nominations show clearly how seriously the Academy took its self-appointed task of stimulating research and how it (and Crelle too) interpreted the aims of that research. Only two of the fifteen he proposed were applied mathematicians. The others included Dirichlet, Jacobi, and Kummer (a geometer and number-theorist), the geometers Steiner and Möbius, and from France the geometers Poncelet and Gergonne.[34]

Another way in which the Academy sought to encourage research was by continuing the old habit of prize competitions. Crelle organised the competition for 1836, and the topics he proposed and the terms in which he expressed himself tell us a lot about his preferences in mathematics.

> Firstly, to give an elementary but rigorous proof, suitable for teaching, that the roots of a polynomial equation of degree n, cannot generally be expressed algebraically in terms of the coefficients as soon as n is greater than 4, in so far as this shall really be the case after the work of Riccati,[35] Abel, and others to resolve it.

The next two proposals concerned the uses of elliptic functions in this connection.

> Fourthly, to give a general account of the extent to which equations of degree higher than the fourth are algebraically solvable, and to give conditions on the coefficients and the degree, or between the roots, in these cases.

The remaining problems were on the convergence of infinite series, the theory of polyhedra, the calculus of variations, the solution of cubic equations in integers, and the theory of transcendental functions. Once again, we see that Crelle's priorities for mathematical research were heavily weighted in favour of pure mathematics — here was one engineer at least who was not worried about the consequences of pure research. Only the calculus of variations could have had anything to do with the 'real world', if that term was narrowly understood as meaning the world of industry and money making. His proposed topics fitted happily with a neo-humanist view of the world. They also fitted very well with the view of Crelle's friend Abel, and with that of the geometers that Crelle put forward to the Academy.

The next important new journal which, like Crelle's, continues to be important today was French: Joseph Liouville's *Journal de Mathématiques Pures et Appliquées*

[34]We consider the work of these geometers in Chapter 15.

[35]Presumably, Crelle meant Ruffini here, see Section 19.1.

Figure 13.13. Joseph Liouville (1809–1882)

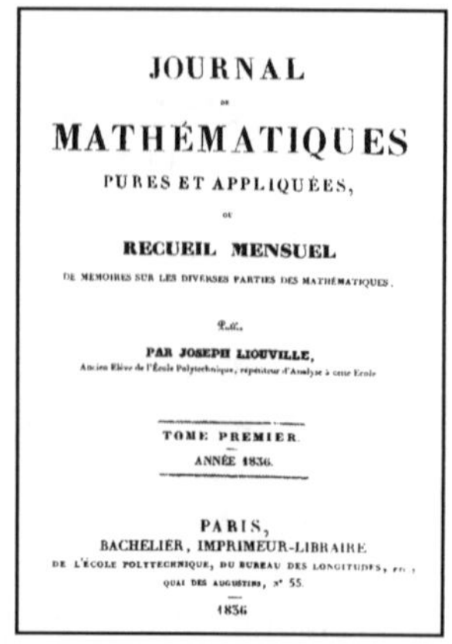

JOURNAL

DE

MATHÉMATIQUES

PURES ET APPLIQUÉES,

OU

RECUEIL MENSUEL

DE MEMOIRES SUR LES DIVERSES PARTIES DES MATHÉMATIQUES.

PUBLIÉ

PAR JOSEPH LIOUVILLE,

Ancien Élève de l'École Polytechnique, répétiteur d'Analyse à cette École

TOME PREMIER.

ANNÉE 1836.

PARIS,

BACHELIER, IMPRIMEUR-LIBRAIRE

DE L'ÉCOLE POLYTECHNIQUE, DU BUREAU DES LONGITUDES, ETC.,

QUAI DES AUGUSTINS, N° 55.

1836

Figure 13.14. The title page of the first issue of Liouville's *Journal*

(Journal of Pure and Applied Mathematics). There are some obvious similarities between the two journals. Liouville gave his journal the same straightforward title that Crelle (and Gergonne) had chosen, which is remarkable only for acknowledging, even insisting, that there are two kinds of mathematics. He also nodded in the direction of Gergonne's journal, by then defunct, and hoped for as large a readership as possible, for whom he would provide both elementary and advanced articles. But Liouville interpreted his title quite differently from Crelle.

The first volume, which appeared in twelve monthly parts in 1836, contained several papers on differential equations, as well as two on geometry (one by Chasles and one by Julius Plücker, another geometer whose work we look at in Chapter 15). Five papers were by Liouville himself, including two on celestial mechanics, and he was the co-author of two more with his friend, the Swiss mathematician Charles-François Sturm, on the theory of differential equations with applications to the theory of heat conduction.

Even this limited information suggests something about the editorial policy of Liouville's *Journal*. Celestial mechanics and differential equations lay in the overlap between pure and applied mathematics, and this is in keeping with what we have learned about French priorities. Unlike Crelle, Liouville was actually an author in his own journal — indeed, a prolific one. This might suggest that he could have been much firmer in propelling it in his preferred direction, should the need arise.

Liouville was born in St Omer, near Calais, and was admitted into the École Polytechnique in 1825 when he was only 16. Among the courses he took was one on analysis and mechanics, given alternately by Cauchy and André-Marie Ampère.[36] In Liouville's year the course was given by Ampère, but since Ampère agreed with Cauchy on the need for rigour in analysis, the course was Liouville's first meeting with this new spirit of mathematics. It is likely that he found it a shock — indeed, students in its first year (1821) so disliked the course that the École ordered Cauchy to water it down

[36]André-Marie Ampère was born in 1775, and began teaching himself advanced mathematics at the age of 12. He took up a position at the École Polytechnique in 1804, becoming a professor there in 1809 without ever having acquired a professional qualification, and was appointed a professor of experimental physics at the Collège de France in 1824. In the 1820s he did fundamental work on the new science of electromagnetism, for which his name is remembered — the ampere (or amp) being a unit of electric current.

in favour of things that the students could actually do! Still, Liouville soon set about getting hold of Cauchy's textbooks, and Ampère was to look favourably on Liouville in later life. Another useful contact that Liouville made in his two years at the École Polytechnique was François Arago, who was to help him build up his cumul.

Liouville graduated from the École Polytechnique and went on to the École des Ponts et Chaussées (School of Bridges and Roads) in November 1827. But in October 1830, in a bold and unusual move, he resigned from there to pursue a career exclusively in mathematics. By then he had submitted seven memoirs on the theory of electrodynamics, differential equations, and the theory of heat to the Académie des Sciences. In 1831, on the strength of these memoirs, he was appointed 'répétiteur' at the École Polytechnique — this involved running tutorials, seminars, and problem classes on a professor's lectures. Liouville was appointed répétiteur to the man who had been Arago's répétiteur, who in his turn was taking up a position that had been relinquished by Ampère.

Some of Liouville's memoirs were published in Gergonne's *Journal*, and so his decision to go for a career in mathematics paid off quickly. His career was further advanced by his later decision to launch a journal to fill the void left after the disappearance of Gergonne's. In the years between 1831 and 1836 he had made a number of good acquaintances, which put himself in a strong position to drum up authors. He had also continued to develop as a mathematician, and so he was in a good position to edit articles on a variety of subjects.

In 1836 Liouville, his friend Sturm, then in his early 30s, and one other were the candidates for election to the Académie des Sciences in the contest for the seat left vacant by the death of Ampère. In what must have been an unheard-of feat of altruism, Liouville virtually campaigned on his older friend's behalf, presenting a paper praising Sturm's work to the Académie and then withdrawing in his favour. Sturm was duly elected. In 1838 Liouville was appointed Professor of Analysis at the École Polytechnique, and in that year he began a long-running dispute with an Academician, the remarkable Guglielmo Libri, over the originality and accuracy of various people's work. Despite this feud, Liouville was elected to the Academy in 1839, where he continued to quarrel with Libri. In 1848 — the 'year of revolutions' as it is sometimes called — political turmoil in France led to Libri's influential patrons losing power, and the liberal Arago, another of Libri's enemies, gained power (he was briefly the Minister for War). Old charges against Libri were now more actively investigated and, feeling the heat, he absconded to England in 1850, taking many valuable papers with him, and was expelled from the Académie.

In his absence, Libri was charged with the theft of books from many libraries across France. Since 1840 he had been in charge of a royal commission for registering old manuscripts, and it was alleged that he had taken advantage of the trust placed in him, and the low standards of record keeping, to remove a considerable number of valuable documents. He vigorously denied the charges from the safety of England, but he was convicted in his absence and given the maximum prison sentence of ten years.

Among the papers that Libri came to be accused of stealing was the manuscript of one of Abel's most important papers, on a generalisation of the theory of elliptic functions. This paper had disappeared into the bowels of the Académie des Sciences in Paris, some time after Abel had submitted it in 1826. In 1829, Abel had died of tuberculosis at the age of 26; Crelle had just managed to obtain for him a professorship at Berlin, but the news arrived only after the young Norwegian had died. Hearing that a significant unpublished work of Abel's still existed, Jacobi led an international outcry demanding that it be found. Eventually it was, and Libri, who had never known the young Norwegian but who had written a sympathetic, if rather fanciful, biography of him, was put in charge of printing it. It was printed, in 1841, but then, strange to say, it disappeared again. Only in 1952 was the manuscript discovered — in Florence, in a collection of Libri's papers!

Liouville was a successful editor, and his *Journal* blossomed from the first. He was prominent in its pages, as he was entitled to be, being a prolific mathematician at the height of his powers, and he wrote on a wide variety of topics: in 1840 he proved that the number e is irrational, and in the next issue he returned to his interest in the theory of differential equations. But he also sought out authors. He sometimes obtained permission to translate articles that had appeared in German in Crelle's *Journal*, and when this was denied he would write a one-page abstract of a paper that had caught his interest, and try to solicit a second paper from the author for his own journal. In this way he helped to bring some of the latest German work to French eyes. He brought the work of the German geometers across the Rhine, and in 1839 he published Dirichlet's proof that every arithmetic progression

$$a,\ a+b,\ a+2b,\ldots,a+nb,\ldots,\ \text{where } a \text{ and } b \text{ are relatively prime}$$

contains infinitely many primes. Dirichlet's novel treatment of this important result, previously stated without proof by Legendre, did much to confirm the German's reputation. Liouville's own interest in the theory of numbers led him in 1840 to publish Gabriel Lamé's proof of another case of Fermat's Last Theorem (the case $n = 7$), and his interest in applications of mathematics led him to publish Jacobi's work on dynamics.

These achievements would have been impressive enough, but he also published articles on such diverse topics as astronomy, probability and, gratifyingly, the history of mathematics. Mathematicians and historians of mathematics also owe him a great debt for a short series of papers that he had published for the first time in 1846, some fourteen years after they had been written: the memoirs on the solution of equations by Évariste Galois (see Chapter 19).

13.5 The later 19th century

From a social–historical standpoint, the three most important political events for the development of mathematics in the second half of the 19th century were the unification of Italy, the unification of Germany, and the Franco–Prussian War of 1870–1871.

At the start of the 19th century Italy was a patchwork of principalities. Some, such as Naples, were under the influence of France, while others, perhaps more, were under the influence of Austria, and some were independent. The Napoleonic invasion brought several of the more liberal ideas of the French Revolution to France, including the ending of the ghettoes and a reduction in the laws against Jews in public life, but the collapse of Napoléon's empire in 1815 saw the old divisions mostly restored by the Treaty of Vienna. The movement for Italian liberation largely took on the cast of a struggle against Austria, and it lasted until the 1860s when Garibaldi's army and Cavour's political skills brought all of Italy, except for Venice and Rome, into Italian hands.

In 1866 Austrian power, already on the wane, was further weakened when Austria lost a war with Prussia. This war was fought by the Prussians to enhance their influence in the German-speaking territories of Europe, but the Italians had allied themselves with the Germans and benefitted from the terms of the Treaty that ended the war with the acquisition of Venice and its surrounding territory. Further gains followed with the Franco–Prussian War, which the Prussians fought in order to enlarge and secure their border in the west with France. This caused Napoléon III to withdraw his troops from Rome, and the Holy See, which had steadily opposed unification, fell into Italian hands. Rome was declared the capital of Italy in 1871, with the Vatican City as an independent state within it.

For Italy, unification meant freedom for individuals to travel around the country. Several Italian mathematicians, such as Luigi Cremona, had been prominent in the liberation struggle, and they now emerged onto the scene in considerable numbers. They were strongly attracted to geometry, not least because of the influence of Cremona, who was based in Rome and played much the same role that Monge had played earlier in France.

For Germany, the wars against Austria to the south-east, Denmark to the north, and France to the west, established the unification of several principalities into a state dominated by Prussia, and the emergence of a single powerful regime in the centre of Europe. The new country was stronger than Austria and the weakening Austro–Hungarian Empire, and became a powerful rival to France. Germany was respected, as France had been two generations before, for its political will and military might, and also for the institutions, including its universities, that had made this possible.

For France, the Franco–Prussian War, which saw German troops lay siege to Paris and annex large swathes of Alsace and Lorraine, including the military fortress of Metz, was a national disaster. Profound soul-searching ensued. The reign of Napoléon III was replaced by the Third Empire, which lasted until the Second World War, and the institutions of the state were restructured. Among the changes were the creation of new universities in the provinces — the French were acutely aware that there were twenty-two universities in Germany, many more than they possessed — and an increase in the pay of professors.

Also at this time the École Normale, which had been an institution for training teachers, grew into an independent rival to the École Polytechnique, and more and more gifted students who passed the demanding entrance exams chose it. The syllabus there was less oriented towards engineering, and the hold of engineers on French mathematics was weakened by men with broader, and more theoretical, interests. New journals were founded — notably, the *Bulletin des Sciences Mathématiques et Physiques* by the geometers Gaston Darboux and Jules Hoüel — which carried both research papers and extensive reviews of the literature, the better to acquaint French mathematicians with what was happening in their subject. The Société Mathématique de France was established in 1871, and it too produced a journal for research.

Among the nations that did not read the signs was Great Britain. Foreign policy continued to keep hold of the world-wide Empire and to play off one Continental power against another in a shifting web of treaties, but not enough was done to reform the universities. Cambridge University concentrated on applied mathematics and physics and produced a succession of major figures, among them George Gabriel Stokes, George Biddell Airy, James Clerk Maxwell, and John William Strutt (later Baron Rayleigh).

Scotland, under the influence of William Thomson (later Lord Kelvin) and Peter Guthrie Tait, promoted technology. Oxford University spent decades embroiled in theological debates, and quite failed to appreciate the presence of Henry J.S. Smith in its midst. Such was the force of the Thirty-nine Articles of the Church of England on Oxford and Cambridge that Jews, Roman Catholics, and Dissenters were forced to work elsewhere. The result was that men like Augustus De Morgan and William Kingdon Clifford taught at University College, London, which even so was not much more than a feeder for Cambridge, and James Joseph Sylvester, a Jew, eventually went off to Baltimore in the United States, where he made a great success of the Mathematics Department of the newly founded Johns Hopkins University.[37] Only Smith, Sylvester, and Arthur Cayley can be counted as pure mathematicians of truly international standing among British mathematicians. Cayley became the first Sadleirian Professor of Mathematics at Cambridge in 1863, at the age of 42, and remained there until his death in 1895. Prior to that he had been a successful barrister, and his friend Sylvester a successful actuary.

In Britain, the London Mathematical Society was founded in 1865 with De Morgan, Sylvester, and Cayley as its first three presidents. Despite its name, it was the first national society for mathematics in Britain, and its journal, the *Proceedings*, provided an important outlet for the publications of British mathematicians.

A feature of the new Europe was that it was less French-oriented and more German. The dominant university became the one in Berlin, the home of Crelle's *Journal für Mathematik*, but in 1869 this journal acquired a rival, the *Mathematische Annalen*. It was set up by Rudolf Clebsch in Göttingen and Carl Neumann in Leipzig in 1869, almost deliberately in opposition to the Berlin-based journal, and it was used by many mathematicians outside Berlin as a repository for their research.[38] The *Annalen* was at first difficult to obtain, but it gradually became well established, and it certainly blossomed when Christian Felix Klein became its Editor-in-Chief in 1876.

Klein saw to it that the *Mathematische Annalen* attracted the best writers from abroad, as well as those German mathematicians whose interests were not congenial in Berlin. In particular, he cultivated good contacts with many Italian geometers, and encouraged them to submit their articles to the *Annalen*. In the 1890s, he was appointed the senior Professor of Mathematics at the University of Göttingen. A formidable organiser, he dominated the institutional side of German mathematics, and built up the University of Göttingen until it was the leading mathematical centre in the world (as it remained, until destroyed by the Nazis in 1933).[39] He was also a motive force behind the international conferences in mathematics that began around the turn of the century, and directed a team of Germans, Austrians, and Italians towards the production of a multi-volume *Encyklopädie der mathematischen Wissenschaften* (Encyclopedia of the Mathematical Sciences), which was published between 1904 and 1920.

---

[37]Sylvester had been able to study at Cambridge, but not to take a degree. He returned from America, to Oxford, in 1883.

[38]Clebsch started in hydrodynamics, but turned to geometry and complex function theory around 1870 when he felt that no-one was attending sufficiently to the works of Riemann, who had died in 1866. He himself died unexpectedly of diphtheria in 1872 at the age of 39.

[39]The Department, but not the University, was progressive in its attitude to women mathematicians; it is where Grace Chisholm Young became the first British woman to obtain a Ph.D. degree in mathematics — see Figure 15.22, and (Rothman 1996, 89).

A measure of the rise of Germany is that Italian mathematicians travelled to Germany to learn advanced mathematics. Looking back in 1900 Vito Volterra, one of the leading Italian mathematicians of his day, spoke of Italian unification as being responsible for the 'scientific existence of the nation'. He described how Italian mathematicians travelled to German universities, chiefly Berlin and Göttingen, as soon as they felt confident to learn the newest mathematics, and how they became better able to conduct research when they returned home.

More remarkably, a generation later, American mathematicians did much the same. That Americans chose Germany is perhaps unexpected; one might have supposed, if only on linguistic grounds, that they would have gone to England. Their choice of German universities shows that they believed that the best instruction, and the most advanced mathematics, were to be found in Germany.

One issue that mathematicians confronted was intellectual and definitional. What should mathematics be taken to be? How pure? How close to contemporary physics? How obedient to the demands of education? In no country was there a Government capable of imposing its will on the professors, and so it was left to each country, and each university, to decide these issues; by and large they did so by looking at the German example, which was overwhelmingly pure. The Prussian ideology of neo-humanism emphasised subjects for their own sake, and for as long as German industry continued to prosper — which it energetically did — there would be no pressure to rethink the issue. The French tried to get every doctoral student to read and comment on a significant recent piece of research done in Germany, but otherwise hoped, unsuccessfully, that the tradition of science-oriented mathematics would continue to flourish. The Italians also took to pure mathematics in considerable numbers. The British, as we noted, made no organised response at all.

Consequently, for the first time, mathematics came to seem almost synonymous with pure mathematics. At the same time — and again German universities led the way — physics ceased to be a primarily experimental discipline and became a theoretical one as well. As this disciplinary divide grew, it reinforced the view that mathematics for its own sake belongs in mathematics departments, while mathematics that is applied might be done in science departments.

This was not everyone's opinion, of course. Individual mathematicians — most notably Henri Poincaré in France — stood out against this trend. Many of the best moved across disciplines as they saw fit. But the view in Berlin was strongly pure, and the graduates of the École Normale likewise seldom occupied themselves with applied or applicable mathematics. This trend will be clearly visible in the following chapters.

13.6 Further reading

Ben-David, J. 1971. *The Scientist's Role in Society*, Prentice-Hall. This remains one of the most interesting attempts at the difficult task of analysing science in a social context.

Bühler, W.K. 1981. *Gauss: A Biographical Study*, Springer. This book provides a good short introduction that punctures some myths about Gauss while still describing several of his remarkable achievements.

Dunnington, G.W. 2004. *Gauss: Titan of Science*, Mathematical Association of America. This is a reprint of the author's biography of 1956 (with a new introduction and appendices by Jeremy Gray), which usefully tells us more about Gauss as a scientist and astronomer than as a mathematician

Lützen, J. 1990. *Joseph Liouville, 1809–1882, Master of Pure and Applied Mathematics*, Springer. This is one of the best biographies of any mathematician, and is a rich account of the world in which he moved.

Merzbach, U. 2018. *Dirichlet: A Mathematical Biography*, Birkhäuser. This thorough and readable account of the life and work of this influential mathematician was completed by Jeanne LaDuke and Judy Greene on the basis of an almost full-length manuscript by its author, who died in 2017.

14

Non-Euclidean Geometry

Introduction

The history of geometry in the 19th century can be followed by tracing two separate stories until they come together, and then seeing how the implications of that unification were drawn.

The first story concerns the resolution of questions about Euclid's parallel postulate, which had previously occupied Greek and Islamic mathematicians.[1] As we shall see in this chapter, a novel approach to the study of parallels around 1830 led to the discovery of a wholly new geometry, significantly different from Euclid's but also capable of representing the true geometry of space: it is known today as *non-Euclidean* or *hyperbolic geometry*. But this new geometry almost died at birth, and by the middle of the century it was almost forgotten. Then in the 1860s a new generation of mathematicians rediscovered it, with dramatic results.

Our second story, which we shall study in Chapter 15, concerns what came to be called *projective geometry*. This can be regarded as the study of those properties that geometrical figures share with their shadows—for example, if a line is tangent to a curve, then its shadow touches the shadow of the curve. Put in this way it may not sound very deep or interesting, but we shall see that many 19th-century mathematicians considered it the most basic and important of all geometries. It too has a curious history: a flurry of interest in France in the 1820s and 1830s proved short lived, but it was taken up enthusiastically in Germany, where it became the central domain of geometry and an active area of research.

The rediscovery of non-Euclidean geometry in the 1860s led some geometers to feel that, where once there had been just 'Geometry', there were now Euclidean geometry, two new geometries called projective geometry and non-Euclidean geometry, and perhaps others, and consequently the very growth and diversity of geometrical research had made the subject almost impossible to grasp as a whole. In the next chapter we shall see how a movement to re-unify geometry led to projective geometry playing

[1] See *Volume* 1, Chapters 2 and 7.

a central role. Finally, influenced by the discovery of non-Euclidean geometry, mathematicians looked again at the whole question of geometry as a deductive science, and were led to propose new axiomatic approaches that were to be widely copied elsewhere in mathematics.

Greek and Islamic investigations. David Hilbert was the German mathematician who, with Henri Poincaré, dominated the world of mathematics in the years around 1900. In a speech that helped to set the agenda of 20th-century mathematics, he said that the discovery of non-Euclidean geometry was the most suggestive and notable geometrical achievement of the 19th century.[2] Historians have tended to agree: the historian Morris Kline, for example, wrote in 1972 that:[3]

> Amidst all the complex technical creations of the nineteenth century the most profound one, non-Euclidean geometry, was technically the simplest,

and that its discovery was

> the most consequential and revolutionary step in mathematics since Greek times.

So it is a dramatic tale we have to tell. To gauge its implications, ask yourself whether you believe that Euclidean geometry is *true*, and if so, why? Consider, too, what you know about the role of Euclidean geometry in anchoring mathematics to a bedrock of certainty. All the judgements about geometry made by mathematicians over two millennia were called into question by the discovery of non-Euclidean geometry, and subjected to the most drastic revision.

The story in some ways resembles a mystery story. It begins in the time of Euclid with a tidy, logical, deductive edifice in which only one small piece did not seem to fit. The first attempts to make it do so all failed, and gradually ever bolder attempts were made, more of the structure was called into question, and finally the building was brought down. The obstinate piece turned out to be a clue pointing in a different direction altogether.

The parallel postulate. We begin our account by recalling some of the earliest investigations into the nature of parallel lines, and then quickly revisit the later story, bringing it up to the 18th century.

Geometry was codified in Greek times by Euclid, whose *Elements* is justly considered a deductive masterpiece. The role of the *Elements* in Greek, Islamic, and Western intellectual life has already been described in *Volume* 1; here we concentrate on one, seemingly tiny, problem.

The fifth of Euclid's postulates asserts (see Figure 14.1) that[4]

> if two straight lines l and m cross a line n at angles α and β, and if $\alpha+\beta < 180°$, then l and m meet — to the right of n in this case.

The fifth postulate is often called the *parallel postulate*; we shall sometimes call it the PP for short, although, as stated, it says nothing about lines being parallel. But we can show that parallel lines exist by modifying Figure 14.1 so that $\alpha+\beta = 180°$. Euclid

[2]See his address to the Paris International Congress of Mathematicians, 1900, reprinted in Hilbert's *Gesammelte Abhandlungen*, Vol. 3 (1935), 290–329. For a reprint of the English translation, see *Bulletin of the American Mathematical Society*, Vol. 37 (1902), 407–436, the quoted remarks are on p. 409.

[3]See (Kline 1972, 867, 869).

[4]See *Volume* 1, Sections 2.4 and 7.2, and F&G 3.B1.

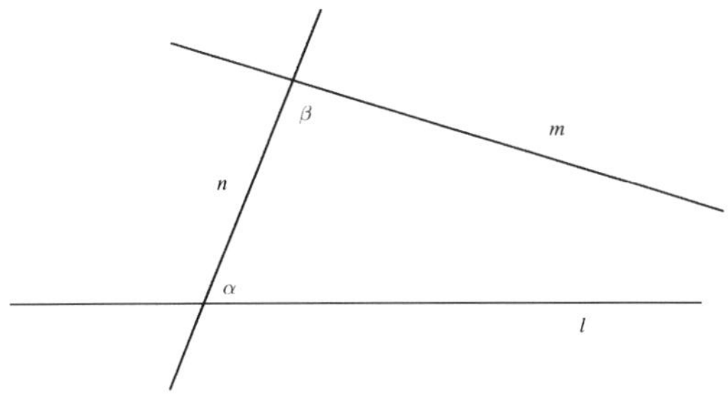

Figure 14.1. The fifth or parallel postulate

showed (in *Elements* I, 17) that the sum of any two angles in a triangle is always less than 180°. So when $\alpha + \beta = 180°$, the three lines cannot form a triangle and l and m cannot meet: therefore they are parallel. On the other hand, if $\alpha + \beta < 180°$, then the postulate asserts that the straight lines l, m, and n form a triangle, thus guaranteeing the existence of a figure satisfying certain conditions.

The remarkable and useful thing about the PP is that it can be used, with the other postulates in the *Elements*, to show that:

> In a plane, given a point P not on the line l, the line m through P parallel to l is unique.

The proof is straightforward, and was pithily summed up by Proclus in his commentary on the *Elements*:[5]

> For if two parallels to a straight line can be drawn through the same point, there will be parallels intersecting one another at the given point, which is impossible.

For future reference, we call this result the Uniqueness of Parallels Property:

> In a plane, given any line l and a point P not on it, there is a unique line parallel to l that passes through P.

In some 19th-century editions of Euclid's *Elements* the PP was replaced with the Uniqueness of Parallels Property, which was usually called 'Playfair's Postulate', after the Edinburgh mathematician John Playfair, who used it in his edition of 1795.

So Euclid showed first that parallels exist — without assuming the PP, but only his other postulates and definitions — and then used the PP to show that parallels are unique. It is really the Uniqueness of Parallels Property that is used in the *Elements*, rather than Postulate 5 as it stands; for example, the Uniqueness of Parallels Property is used to show that the sum of the angles of any triangle is 180° and to prove theorems such as these:[6]

- parallel lines are everywhere equidistant (I, 33)
- the Pythagorean theorem (I, 47)
- all of the theorems on similar figures (Books V and VI).

---

[5]See (Proclus, ed. Morrow, 1970, 376).

[6]See *Elements* I, 32, in F&G 3.B4.

On the other hand, without the PP it seemed that very little plane geometry could be proved rigorously. In fact, no-one in classical times doubted either its utility or its correctness, but it seems to have worried people to have to postulate it. This worry derived in part from its lack of obviousness; if, in Figure 14.1, l and m are aligned so that they only just fail to ensure that $\alpha + \beta = 180°$, then the PP asserts that they meet, but a very long way away. Plainly one cannot verify empirically an assumption about intersections that may be a million miles distant.

In Proclus's view the PP should not be classified as a postulate, because, he said, it is a theorem that can be proved once you have the right underlying definitions and theorems (see Box 35).[7] But he urged caution: a hyperbola and one of its asymptotes are a curve and a straight line that draw ever closer, but never meet. So, must two straight lines meet just because they seem to be heading towards each other? Not only is the implication insufficiently clear, but also the 'inference' if applied to any two lines (or curves) may even be false. However, Proclus may have been making the startling metamathematical claim that the PP must be true because it is the converse of a true theorem (one proved by Euclid in the *Elements*), or he may have been making the weaker assertion that the converse of something that you can prove is itself capable of proof or disproof, and is not something that you have to assume or deny a priori.

Proclus's openly made assumption, which he says he took from Aristotle, is that the distance between two intersecting lines increases indefinitely as you move outwards from their point of intersection. He used it to establish (see Box 35) that the point G moves arbitrarily far from the line AFB. The tacit assumption is much harder to spot. It was made when Proclus spoke of '*the* distance between parallel lines'. Why should there be a *fixed* distance between parallel lines, as this implies? Why should they not diverge, growing ever further apart? If they did, then G might diverge from AB without ever meeting CD. Proclus's proof assumes that the distance between parallel lines is constant, an assumption that is itself equivalent to the PP. (We have seen above that it is a consequence of the PP; the reverse implication turns out also to be valid.)

Proclus's attempt must stand for many down the ages, including some splendidly vigorous ones by Islamic mathematicians such as Ibn al-Haytham, Omar Khayyām, and Nasīr-al-Dīn al-Ṭusī.[8] Gradually the PP became notorious as one of the two so-called blots on Euclid (the other was the definition in Book V of proportion between magnitudes). But centuries of investigations led only to the discovery of further assumptions that, by judicious use of the other assumptions in the *Elements* both implied and were implied by the PP, were in this sense equivalent to it. But none of these investigations led to a means of deriving it as a theorem from the other Euclidean postulates alone without any extra assumptions. Here we recall two Islamic assumptions of this kind:[9]

- Ibn al-Haytham: The equidistant curve to a straight line is itself straight.
- Nasīr-al-Dīn al-Ṭusī: The angle sum of a triangle is 180°.

[7] See *Volume* 1, Section 2.4, and F&G 3.B1.

[8] We looked briefly at their contributions in *Volume* 1, Chapter 7.

[9] We discussed these in *Volume* 1, Section 7.2.

Box 35. Proclus on the parallel postulate.

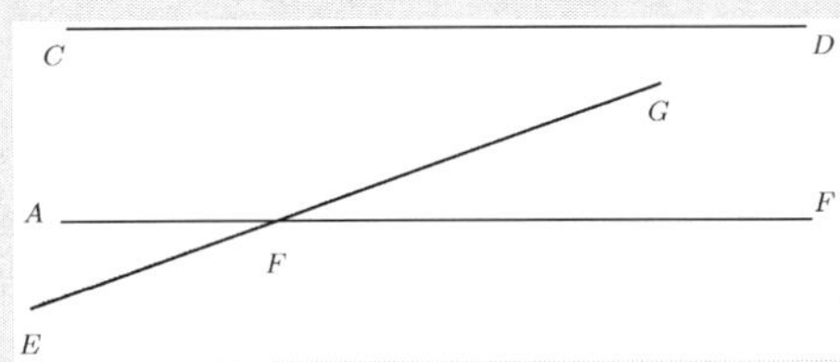

Figure 14.2. If a line crosses one of two parallel lines, must it cross the other?

The lines *AFB* and *CD* are parallel, and the line *EFG* crosses *AB* at *F*.

Proclus claimed that:

1. As *G* moves away from *F* along *EFG*, so it moves further from the line *AFB*;
2. The distance between the lines *AB* and *CD* is constant.

From these he deduced that the line *EFG* must eventually cross *CD*. So any straight line through *F*, other than the parallel line *AFB*, must meet *CD* and so cannot be a parallel. So the parallel to *CD* through *F* is unique. This argument is a proof of the Uniqueness of Parallels Property that purports not to rely on the PP, and so (other things being equal) Proclus hoped to have achieved his goal of demonstrating the PP as a theorem.

14.1 The first Western attempts

When mathematics was revived in the West in the 16th and 17th centuries the PP drew renewed attention. While Federigo Commandino and Christoph Clavius in their editions of Euclid had regarded equidistance simply as equivalent to parallelism, others attempted to derive the PP from the other assumptions of Euclid.[10]

In a lecture at Oxford on the evening of 11 July 1663, John Wallis, the Savilian Professor of Geometry, showed that the parallel postulate can be deduced from the assumption that similar copies of given figures can have different sizes. Figures of different sizes cannot be congruent, because two figures are congruent if and only if each can be superimposed exactly on top of the other.

Wallis showed that the existence of similar non-congruent figures in geometry is equivalent to the PP. He found this assumption very plausible, but the necessity of making it weakens any claim that the PP follows from the other axioms of Euclid. It does, however, imply a remarkable result:

[10]See *Volume* 1, Chapter 9, and F&G 16A.1.

In any geometry in which the PP does not hold, but the other postulates of the *Elements* do, similar figures must be identical in size as well as shape, and so scale copies cannot be made.

This is doubtless bizarre (and it would give architects a difficult time!), but it is not logically impossible.

Gerolamo Saccheri. All these earlier attempts were eclipsed by the attempt of Gerolamo Saccheri, a Jesuit logician who was Professor of Mathematics at Pavia (in northern Italy) from 1697 until his death in 1733.

In the final year of his life, Saccheri published a book entitled *Euclides ab Omne Naevo Vindicatus* (Euclid Freed of Every Flaw) attacking the two 'failings' in the *Elements* (the status of the PP, and proportion theory, which had also been singled out by Islamic mathematicians). His assault on the PP was most ingenious. Saccheri considered the system of the *Elements* with the PP removed and one of the following assumptions put in its place.

1. The hypothesis of the obtuse angle: the angle sum of a quadrilateral is greater than 360°.

2. The hypothesis of the right angle: the angle sum of a quadrilateral is exactly 360°; this yields Euclid's geometry, and is equivalent to the PP.

3. The hypothesis of the acute angle (HAA): the angle sum of a quadrilateral is less than 360°.

These names derive from the fact that, on taking each hypothesis in turn, we can construct a quadrilateral, two of whose angles are right angles, while the remaining two angles are obtuse, right, or acute, respectively (see Figure 14.3).

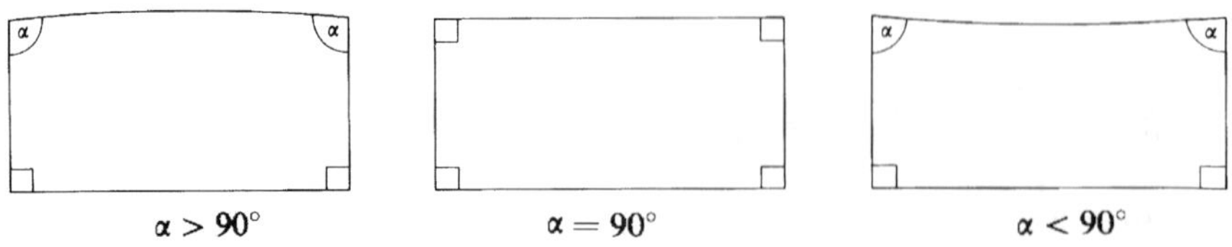

Figure 14.3. Three types of quadrilateral

So Saccheri made three hypotheses, each of which could potentially yield theorems when taken together with Euclid's other assumptions (excluding the parallel postulate). We may say that he had three potential geometries, but could they be actual geometries?

He first proved that making each assumption for just one quadrilateral allows one to prove it for *all* quadrilaterals. He then took each case in turn. He hoped to show that the hypothesis of the obtuse angle and the HAA each separately destroys itself, which would leave only the hypothesis of the right angle, Euclidean geometry. By 'self-destruction' he meant entailing a contradiction; reductio ad absurdum was his favourite form of argument, and the principal theme of his earlier book, his *Logica Demonstrativa* (Demonstrated Logic) of 1697.

The hypothesis of the obtuse angle does indeed yield a contradiction: Saccheri showed by a lengthy argument that this assumption both implies and contradicts the PP. Saccheri said of it that it 'is absolutely false, because it destroys itself'.[11] This left the HAA.

The hypothesis of the acute angle. This hypothesis proved much harder to destroy. Saccheri began the second half of his book with the words 'And here begins a lengthy battle against the hypothesis of the acute angle, which alone opposes the truth of that axiom [the PP]'. He was led to several novel results. For example, his Proposition XXXII (see Figure 14.4):[12]

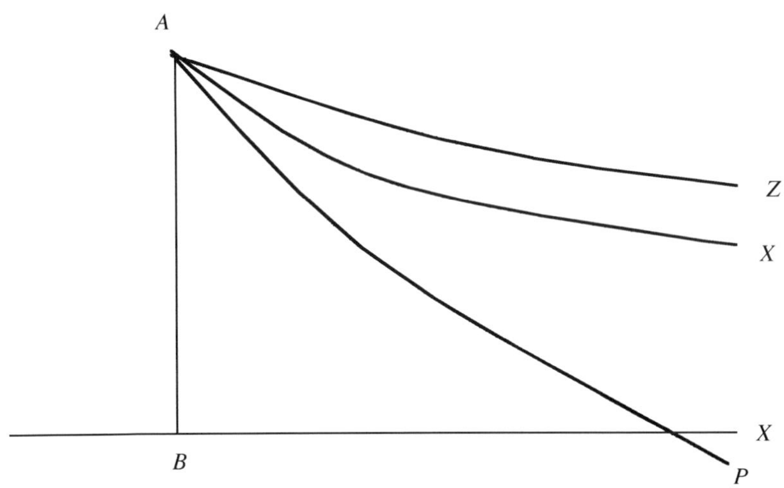

Figure 14.4. The three types of line through a point A with respect to a line BX

Saccheri on the hypothesis of the acute angle.

Proposition XXXII

Now I say there is (in the hypothesis of acute angle) a certain determinate acute angle BAX drawn under which AX only at an infinite distance meets BX, and thus is a limit in part from within, in part from without; on the one hand of all those which under lesser acute angles meet the aforesaid BX at a finite distance; on the other hand also of the others which under greater acute angles, even to a right angle inclusive, have a common perpendicular in two distinct points with BX.

We see that in this passage Saccheri divided the lines through A into three types: those, like AP, that meet the horizontal line BX; those like AZ, that never meet BX but have a common perpendicular with it; and the singular line AX, which he says meets BX 'at an infinite distance' — where it also has a common perpendicular with it.[13]

[11]See (Saccheri 1733, Prop. XIV), (Saccheri 1920, 59), or (Saccheri 2014, 93).

[12]See (Saccheri 1920, 169) or (Saccheri 2014, 153). Note the use of the 18th-century convention that a line denoted AX is the line through the point A in the direction of X.

[13]See (Saccheri 1733, 3–9), (Saccheri 1920, 169–173), or (Saccheri 2014, 153–157), in F&G 16.A2.

The construction of a common perpendicular to two divergent lines is important in Saccheri's argument. In a more modern notation, Saccheri established the following consequences of the HAA (see Figure 14.5).

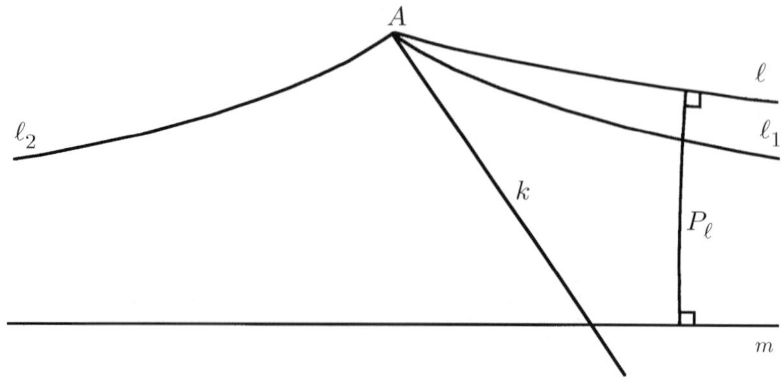

Figure 14.5. A common perpendicular to two lines

For a point A and a line m, the lines through A are of three kinds:

- lines that meet m, such as k
- lines that do not meet m and eventually diverge from it, such as ℓ
- lines that do not meet m but are asymptotic to it; there are exactly two of these, one in each direction, which are shown as ℓ_1 and ℓ_2.

Each line of the second kind, such as ℓ, shares a common perpendicular P_ℓ with m. Moreover, this common perpendicular is unique, because if there were two such perpendiculars then, together with m and ℓ, they would make a quadrilateral whose angle sum is four right angles, contradicting the HAA.

Conversely, if a line through A has a common perpendicular with m then it is of the second kind.

Lines near to ℓ are of the same sort as ℓ, provided that they lie above ℓ_1, and lines near to k are likewise similar to k, provided that they lie below ℓ_1.

To establish the contradiction that he wanted, Saccheri considered the common perpendicular to two lines of the second kind and showed that, if ℓ and ℓ' are lines of the second kind and ℓ is lower than ℓ' then the common perpendicular of ℓ' and m is further from A than the common perpendicular of ℓ and m (see Figure 14.6).

Thus far, Saccheri's arguments are entirely valid.

For the contradiction that he sought, Saccheri considered the limiting position of these lines, and his argument fails because of an interesting flaw that it worth trying to spot.[14]

Proposition XXXIII

The hypothesis of acute angle is absolutely false; because repugnant to the nature of the straight line.

[14]See Saccheri (1733, 3–9, 169–173) and F&G 16.A2.

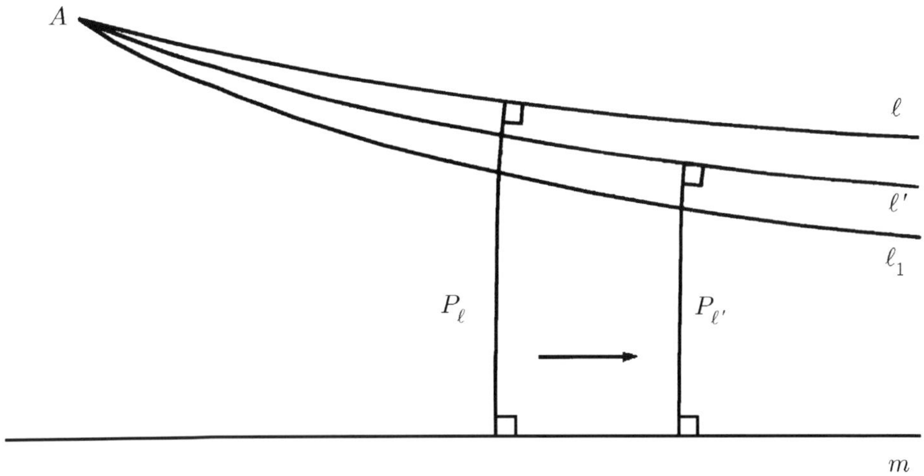

Figure 14.6. The common perpendicular moves away

> Proof: From the foregoing theorem may be established, that at length the hypothesis of acute angle, inimical to the Euclidean geometry, has as outcome that we must recognise two straights AX, BX, existing in the same plane, which produced in infinitum toward the parts of the points X must run together at length into one and the same straight line, truly receiving, at one and the same infinitely distant point a common perpendicular in the same plane with them.
>
> But since I am here to go into the very first principles, I shall diligently take care, that I omit nothing objected almost too scrupulously, which indeed I recognise to be opportune to the most exact demonstration.

Here Saccheri claimed that two straight lines $\ell = AX$ and $m = BX$ that have a common point X at infinity also have a common perpendicular at the point X. It is indeed true that two different straight lines cannot have a common perpendicular at a common point in their plane, but — alas for Saccheri — the point under consideration is an infinitely distant point. He now imagined the lines ℓ and ℓ_1 that he had just considered to coincide, ℓ_1 to meet m at infinity, and the common perpendicular still to exist and to pass through the common point. His argument is invalid, because such a common point does not exist.

Did Saccheri go wrong because these arguments were difficult in 1733, because he was ill (he died that year), or because he so badly wanted to prove the PP? We do not know.

Johann Heinrich Lambert. With Saccheri's death we enter a phase of more critical investigations. Foremost amongst the investigators was the Swiss polymath Johann Heinrich Lambert: cartographer, linguist, theorist of heat, investigator of magnetism and optics, philosopher, astronomer, and self-taught mathematician. In the late 1750s, Lambert was amongst the first, with Immanuel Kant, to interpret the Milky Way as evidence that we live in a galaxy or 'Island of Stars', and was the first to prove that π is an irrational number. In mathematics he particularly liked to establish sound foundations — for example, in his study of mechanics and perspective drawing.[15] He was a member of the Académie Royale in Berlin, to which he contributed papers in every one of its sections.

Figure 14.7. Johann Heinrich Lambert (1728–1777)

In the 1760s Lambert's attention was drawn to the PP by his friend Abraham Kaestner, whose student Georg Klügel had just written a thesis in which some thirty attempts on the PP were analysed and found to be faulty. One of these was Saccheri's. Intrigued, Lambert took up the problem, and followed Saccheri's three-pronged approach, although without mentioning him by name. Like Saccheri, he found the hypothesis of the obtuse angle to be self-contradictory, but in studying the HAA he discovered two further noteworthy results to which we now turn.

An absolute measure of length. First, Lambert found that if you accept the HAA then there is an absolute measure of length: this means that lengths, like angles, can all be measured by a fixed quantity or unit which can be defined without arbitrariness. For angles, this unit is the 'whole angle' — the angle subtended by the circumference of any circle. It is equal to 2π radians, 4 right angles, or 360° — the choice of unit is arbitrary, as we can define each of these angle measures in terms of the 'whole angle'. The crucial thing is that two speakers can agree on the size of angles even if they cannot see each other and are, for example, communicating by telephone.

This is not the case for length in Euclidean geometry: the foot or the metre have no such absolute reference. But, Lambert claimed, it *is* the case for a geometry based on the HAA. To see this, take a length d, and construct an equilateral triangle whose three sides have length d. It has three equal angles α. Now, as Wallis had shown, similar

[15]See Lambert's *Freye Perspektive* (Free Perspective) (1759).

figures are congruent in any geometry in which the PP is false, so d is the unique side-length of *every* equilateral triangle with angle α. So the concept of length reduces to that of angle: lengths can be specified purely in terms of angles.[16]

Lambert found the existence of an absolute measure of length in a geometry based on the HAA most remarkable. It was something that no-one had thought possible, and had seemingly been shown to be impossible by the philosopher Christian Wolf.[17] But Lambert decided in favour of mathematical rigour over philosophical arguments, remarking that the third hypothesis may yet be true of space, despite its unpalatable consequences. To accept or reject it on the grounds of the desirability of its implications would be to use 'arguments drawn from love and hate', which geometers must set aside entirely.[18]

Area and angle sums. The second consequence of note that Lambert observed is that, under the HAA, the area of a triangle is proportional to the difference between π and the sum of the angles in the triangle, where the angles are measured in radians and this sum is less than π. This led Lambert to invoke a mysterious analogy with an 'imaginary sphere'.

To understand what Lambert was saying, we first recall some properties about the geometry on a normal sphere (which Lambert had discussed earlier in his work). On a sphere, a 'triangle' is a figure whose three sides are arcs of great circles — circles of maximal radius cut out by planes through the centre of the sphere (see Figure 14.8). If these three arcs meet each other at angles α, β, and γ, then we can show that the area of such a 'triangle' is proportional to $\alpha+\beta+\gamma-\pi$. In fact, the area is $R^2(\alpha+\beta+\gamma-\pi)$, where R is the radius of the sphere.

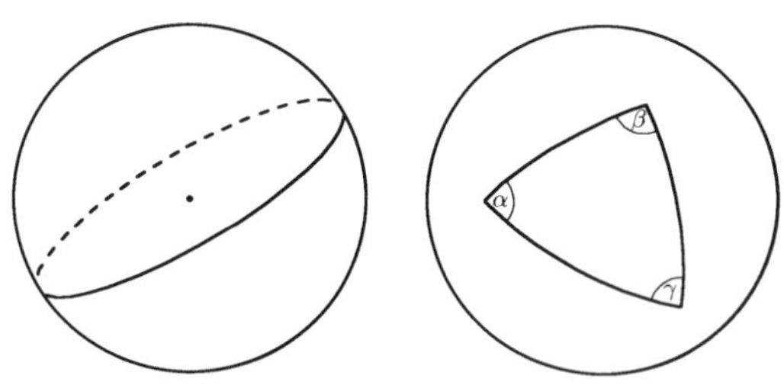

Figure 14.8. A great circle on a sphere (left), and a triangle composed of arcs of great circles arcs (right)

Lambert did not explain what he meant by the phrase 'imaginary sphere', so we have to make a conjecture of some kind about it. We get a clue on seeing Lambert's claim that, in any triangle on an imaginary sphere with angles α, β, and γ, the angular defect $\pi - (\alpha + \beta + \gamma)$ is proportional to the area of the triangle. Lambert claimed that he had a proof, but he did not supply one. He pointed out, however, that exactly the opposite is true on a (genuine) sphere. Indeed, as we have just remarked, the area of a triangle with angles α, β, and γ on a sphere of radius R is given by the formula

[16]Lambert's argument is like this one, but is not exactly the same; see (Engel and Stäckel 1899, §§79–82), and F&G 16.A5.

[17]Christian Wolf (or Wolff) was a philosopher in the first half of the 18th century who was much influenced by Leibniz.

[18]See F&G 16.A5, p. 518.

area $= R^2(\alpha + \beta + \gamma - \pi)$. Lambert presumably observed that if you replace R by $\sqrt{-1}R$, this formula becomes

$$\text{area} = R^2(\pi - (\alpha + \beta + \gamma));$$

the precise formula for area in a geometry satisfying the HAA. Accordingly, he said, the HAA is valid for the surface of an imaginary sphere. We could suppose this to be a sphere with an imaginary number for its radius, but Lambert did not explain what such a curious thing might be, and it remained a purely formal algebraic and verbal flourish in his work.

In this connection, we note that Lambert was not afraid of imaginary numbers. A few years later, in 1770, he wrote to the philosopher Immanuel Kant that 'The sign $\sqrt{-1}$ represents an unthinkable non-thing. And yet it can be used very well in finding theorems.'[19] Kant had a great respect for Lambert, ever since their independent realisation of the nature of the Milky Way, and he would have dedicated his *Critik der reinen Vernunft* (*Critique of Pure Reason*) to him had not Lambert died unexpectedly of pneumonia in 1777.

However, Lambert did not publish his work on the theory of parallels. He seems to have abandoned it in September 1766, perhaps because he could not reach a conclusion that satisfied him, and it was published only in 1786, after his death.[20] As a result, it was seldom cited in the burgeoning literature on the parallel postulate in the 1780s and 1790s.[21] Mathematicians, especially German ones, knew of it by the start of the next century, but most investigators, whether British, French, or German, proceeded rashly where Lambert had known better. The result was that the PP was under continuing, if mediocre, investigation. It is not necessary to survey these attempts, but it is worth noticing that they were made by many people — there were about three attempts a year between 1780 and 1820; a survey published in 1838 lists 91 of them.[22]

France: Legendre's attempts. The situation in France was paradoxical. By the end of the 1780s, France had effectively become the centre of the mathematical world, and although French mathematicians revitalised almost every other subject they touched, they did not take to foundational questions in the study of geometry. Of the major mathematicians, only Legendre was to take up the challenge of sorting out the basic assumptions of geometry with real vigour.

Legendre's approach to Euclid's *Elements* was part of a 'back-to-basics' movement in teaching, which went against one hundred years of believing that what was obvious need not be proved. That view in turn had derived from Descartes's belief in the essential veracity of what was clear and distinct to the mind. Legendre, in contrast, favoured rigorous deductions from a small list of initial assumptions, and sought to improve the teaching of mathematics in schools and colleges by giving a fresh account of geometry along Euclidean lines. This led him to confront the vexed question of parallels, and

---

[19]See (Zweig 1967, nr. 61), and F&G 16.A3. Lambert and Kant exchanged several letters about geometry.

[20]It was published in a short-lived journal edited by Johann III Bernoulli, who had bought most of Lambert's unpublished papers from the Berlin Academy on condition that he publish most of them. See M. Bullynck's article, 'Johann Heinrich Lambert: History of the Editions so far', `http://www.kuttaka.org/~JHL/JHLHistory.html`.

[21]See (Engel and Stäckel 1895, 299–307).

[22]See (Sohncke 1838, 383–384).

he made several attempts at proving the PP.[23] They were of uneven merit, and in successive editions of his *Éléments de Géométrie* (which ran to twelve editions between 1794 and 1823) he had to retract each one in favour of the next. All of them, however, established only the equivalence between the PP and some other assumption. Despite this lack of success, Legendre continued to believe in the truth of the PP until the end of his life.

Germany: Schweikart and Taurinus. It is interesting to compare the French experience with the contemporary situation in Germany. With the notable exception of Gauss, none of the major German mathematicians who worked after the Napoleonic War was interested in the PP. However, two investigators, Ferdinand Karl Schweikart and his nephew Franz Adolf Taurinus, thought the problem was worthwhile.

Schweikart was a professor of Jurisprudence at the University of Marburg and a colleague of Christian Ludwig Gerling, who taught mathematics there and was a friend of Gauss. In 1818 Schweikart communicated a remarkable theorem to Gauss via Gerling. The theorem was based on the HAA, which was known to him from Lambert's work. He wrote:[24]

Schweikart's astral geometry.

> There are two kinds of geometry — a geometry in the strict sense — the Euclidean; and an astral geometry.
>
> Triangles in the latter have the property that the sum of their three angles is not equal to two right angles.
>
> This being assumed, we can prove rigorously:
>
> (a) that the sum of the three angles of a triangle is less than two right angles;
>
> (b) that the sum becomes ever less, the greater the area of the triangle;
>
> (c) that the altitude of an isosceles right-angled triangle continually grows, as the sides increase, but it can never become greater than a certain length, which I call the *Constant*.
>
> Squares have, therefore, the form [shown in Figure 14.9]. If this Constant were for us the Radius of the Earth, (so that every line drawn in the universe from one fixed star to another, distant 90° from the first, would be a tangent to the surface of the earth), it would be infinitely great in comparison with the spaces which occur in daily life.
>
> The Euclidean geometry holds only on the assumption that the Constant is infinite. Only in this case is it true that the angles of every triangle are equal to two right angles: and this can easily be proved, as soon as we admit that the Constant is infinite.

In his letter, Schweikart showed a striking degree of confidence in his ability to build on a geometrical assumption that was not one of Euclid's. He was not looking for a contradiction in his new geometry; rather, he accepted the new assumption and

[23]Two of these can be found in F&G 16.A6.

[24]See Gauss, *Werke* 8, 175–176, and F&G 16.B1.

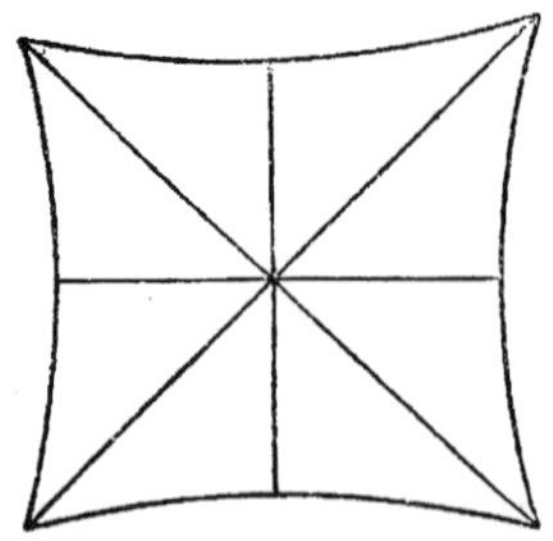

Figure 14.9. Schweikart's quadrilateral

explored its consequences. He even speculated that it might provide a suitable geometry for studying physical space. Gauss agreed with Schweikart's theorem, adding that he could solve all the problems in this 'astral geometry' once the Constant spoken of by Schweikart was given.

In 1825 and 1826 his nephew Taurinus tiptoed even further, only to withdraw. He produced a thorough study of geometry based on trigonometric formulas, but, because he always believed in the truth of Euclid's geometry and the PP, his endeavours were ultimately directed to understanding geometry better in order to prove the PP. Using his new formulas, Taurinus studied a geometry in which triangles with sides a, b, c, and corresponding angles A, B, C, are related by the final two formulas of Box 36. He showed that this geometry agreed with the 'astral geometry' of his uncle, and he even found Schweikart's Constant as a function of the radius k. Taurinus called his geometry 'log-spherical', however, because of the role played by the hyperbolic functions cosh and sinh (see Box 37).

But then, in the first of his two books, Taurinus gave ten reasons why what he was describing could not be a geometry of space. His reasons are far from convincing; for example: 'Were the log-spherical geometry true, Euclidean geometry could not be, the possibility of which cannot be doubted'. He simply did not want to accept it. In his second book, however, he decided that such a geometry could exist; it was that of some surface or other, but he could go no further with it. In 1824 Taurinus tried to get Gauss to endorse his findings, but Gauss prudently declined to lend his name to work that was not only muddled but also contrary to his own beliefs. Taurinus's first book, which he had printed at his own expense, was a failure, and finally he had all the unsold copies burned.

Gauss. Carl Friedrich Gauss published almost nothing on the topic of non-Euclidean geometry, but it was a lifelong interest of his. One of the first things that he read on his arrival in Göttingen in 1795 was Lambert's treatment of the theory of parallels. We shall argue that non-Euclidean geometry was one discovery that the 'Prince of Mathematicians' missed, but other historians disagree. Following the Italian mathematician and historian Roberto Bonola's lead, Morris Kline praised Gauss for being the first to realise that Euclidean geometry was not necessarily true of physical space. We shall present arguments in support of the contrary opinion that Gauss was not the first, but note that there is evidence for both sides.

Kline quoted from a letter written in 1799 from Gauss to his friend Farkas Bolyai, where Gauss says of his investigation that 'It seems rather to compel me to doubt the

Box 36. Taurinus's analogy.
It had been known for many years that the following formulas connect the angles A, B, C on the surface of a sphere of radius k with the angles $a/k, b/k, c/k$ subtended at the sphere's centre by the lengths a, b, c:

$$\cos a/k = \cos b/k \cos c/k + \sin b/k \sin c/k \cos A,$$

$$\cos A = \cos B \cos C + \sin B \sin C \cos a/k.$$

Taurinus asked what would happen to these fundamental formulas of spherical trigonometry if one considered a sphere of imaginary radius (such as Lambert had hinted at, fifty years earlier) or (to put it another way) if one made the purely formal substitution of replacing k by ik.

The effect is rather startling:

$$\cosh a/k = \cosh b/k \cosh c/k - \sinh b/k \sinh c/k \cos A,$$

$$\cos A = \cos B \cos C + \sin B \sin C \cosh a/k,$$

where cosh and sinh are the hyperbolic equivalents of our usual cosine and sine — see Box 37.

Thus, just as the spherical trigonometric formulas are used to study the properties of triangles on spherical surfaces, so Taurinus's new formulas could be used to study the properties of triangles on surfaces with a new geometry, akin to that hinted at by Lambert.

truth of geometry itself'. It seems that Gauss had discovered statements equivalent to the PP, but knew that such discoveries were not a proof of the PP. Kline also quoted from a letter that Gauss wrote in 1817 to Heinrich Wilhelm Olbers, a fellow astronomer, in which he said:[25]

> I am becoming more and more convinced that the [physical] necessity of our [Euclidean] geometry cannot be proved, at least not by human reason nor for human reason. Perhaps in another life we will be able to obtain insight into the nature of space, which is now unattainable. Until then we must place geometry not in the same class with arithmetic, which is purely a priori, but with mechanics.

Kline could also have quoted some other fragmentary pieces of evidence, such as Gauss's 1829 letter to his friend the mathematician and astronomer Friedrich Bessel, in which he wrote that he had been dragged through the mud for publishing a book review in 1822 in which he admitted that the question of the PP was still open, adding[26]

> I fear the howl of the Boeotians if I speak my opinion out loud.

Kline concluded: 'By 1817 Gauss was certain not only that the axiom could not be proved but that a logically consistent and physically applicable non-Euclidean geometry could be constructed.'

---

[25]See Gauss, *Werke* VIII, 177. Kline's quotes appear in (Kline 1972, 872).

[26]See Gauss, *Werke* VIII, 1900, 200–201, in F&G 15.B2(a) — the inhabitants of rural Boeotia in classical times were supposed by sophisticated Athenians to be dull, boring, and stupid.

Box 37. The hyperbolic trigonometric functions.
The two functions cosh and sinh are defined for all x by

$$\cosh x = \frac{e^x + e^{-x}}{2}, \quad \sinh x = \frac{e^x - e^{-x}}{2}.$$

Their graphs look like this:

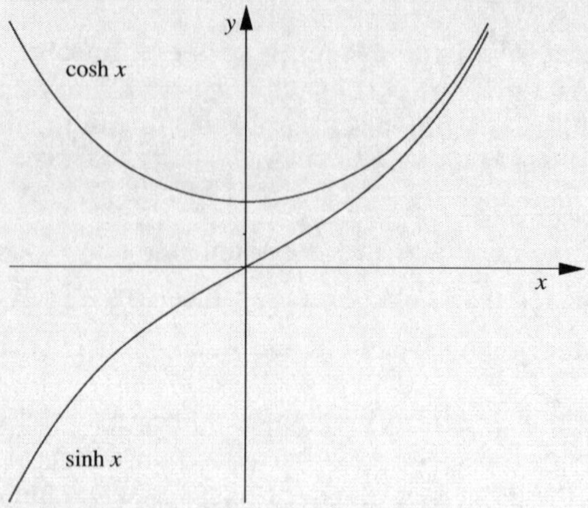

Figure 14.10. Graphs of $y = \cosh x$ and $y = \sinh x$

Their names arose because Lambert showed that they give a simple parametrisation of the hyperbola $x^2 - y^2 = 1$. Indeed, since

$$(\cosh t)^2 = {}^1\!/\!_4\left(e^{2t} + 2 + e^{-2t}\right) \quad \text{and} \quad (\sinh t)^2 = {}^1\!/\!_4\left(e^{2t} - 2 + e^{-2t}\right),$$

subtraction gives $(\cosh t)^2 - (\sinh t)^2 = 1$. So for all values of t, the point $(x, y) = (\cosh t, \sinh t)$ lies on the hyperbola $x^2 - y^2 = 1$.

This is a gratifyingly close analogy with the usual 'circular' trigonometric functions $\cos t$ and $\sin t$, which satisfy $(\cos t)^2 + (\sin t)^2 = 1$, so the point $(x, y) = (\cos t, \sin t)$ lies on the circle $x^2 + y^2 = 1$.

Lambert exploited this similarity to the full when he introduced the functions cosh and sinh into mathematics.

It is possible to argue that all the available evidence falls short of supporting this view. Gauss's views in 1799, and even in 1817, were those of the profound sceptic. He felt himself compelled 'to doubt the truth of geometry', and was 'more and more convinced ...[that] geometry cannot be proved'. By 1829 he was 'even firmer' in his views that geometry is not completely a priori.[27] But it is something of a leap to say that Gauss was convinced that a new geometry could be constructed; the best evidence we have for this is that Gauss did not refute the comments of Taurinus. For example,

[27] See Gauss's letter to Bolyai of 1799, Gauss's letter to Olbers of 1817, and Gauss's letter to Bessel of 1829, in Gauss, *Werke* VIII, 1900, 159, 177, 200.

Kline did not cite any investigations of a new geometry which are not hypothetical; Gauss's ideas were all couched in the vein of exploring new ideas to see whether they held up or whether they collapsed.

One crucial observation is that Gauss did not transform the subject utterly, unlike number theory, celestial mechanics, the theory of elliptic and complex functions, differential geometry, statistics, and probability theory — all of which were completely reconstituted by him. That he all but left non-Euclidean geometry alone suggests that he could not fundamentally accept it. He may have thought it to be true, in some sense, but it was alien to him. Certainly he rejected the false 'proofs' of the PP, and embraced the proposals of Schweikart instead, but it is nonetheless possible to imagine that had a suitable proof turned up, Gauss could have decided without anguish that the PP was true after all. Gauss spoke only of his 'persuasion' or 'conviction' that a new geometry was possible, and never of a proof. This is true of the passages that Kline cited, and we know of no better. Gauss possibly knew that he had not got to the bottom of the matter, and this suspicion only deepens when the whole of Gauss's work on the question is examined, because it is so fragmentary.[28] Kline's claim seems to allow Gauss too much.

The crucial point here is that space is three-dimensional, and any geometry that purports to describe it must be a three-dimensional geometry. It seems very likely that, at some stage in the 1820s, Gauss came to accept the two-dimensional geometries of Lambert, Schweikart, and Taurinus as valid geometries that are different from Euclid's. But this does not mean that there is a three-dimensional equivalent, and nothing in Gauss's writings suggests that he could describe a three-dimensional non-Euclidean geometry mathematically. Gauss would have known perfectly well that a two-dimensional object need not have a three-dimensional analogue, but he might well have considered that space could be non-Euclidean, even though a mathematical description of it was lacking.

Here we enter a continuing controversy in the history of mathematics. Later scholarship has disagreed with Kline over his specific claim that Gauss used a survey of mountains in the state of Hannover in the late 1820s to check on the physical nature of space. This claim is often made, but it was rejected by Breitenberger and then partially defended by Scholz.[29] The truth seems to be a compromise. Gauss used the data from the largest triangle in his survey to show that the question of the nature of space was too delicate to be resolved by terrestrial measurements, even with the state-of-the-art equipment that he was employing, but he knew this early on and did not devise his survey with that question in mind. Gauss knew that Euclidean geometry was true to within the limits of the best observational error of the time, but he also believed that space might be non-Euclidean. His confidence was that of the scientist who knows of no objection to an idea, but not that of the mathematician who can offer a mathematically valid description.

If the test of leadership is to enter the promised land of a geometry other than Euclid's, and not merely to see it from afar, then the credit must go to Nikolai Ivanovich Lobachevskii in Russia, and János Bolyai in Romania/Hungary, whose work in the late 1820s and early 1830s we now consider in detail.

---

[28] See (Reichardt 1976).

[29] See (Breitenberger 1984) and (Scholz 1992).

14.2 Lobachevskii and Bolyai

Figure 14.11. Nikolai Ivanovich Lobachevskii (1792–1856)

Figure 14.12. János Bolyai (1802–1856)

Nikolai Ivanovich Lobachevskii was born on 2 December 1792 in Nizhni Novgorod (now Gorky) in Russia. His family were not well off, but soon after they moved further east to Kazan, Lobachevskii and his brother were enrolled at the Gymnasium on public scholarships. At the age of 15, Lobachevskii entered the recently founded University of Kazan, which had been formed from the old Gymnasium. Here he studied under Martin Bartels, who had once been Gauss's teacher, and who had become a professor there in 1814. Lobachevskii went on to earn the respect of his colleagues for being not only a dedicated teacher and administrator, but a man of immense energy who helped to protect the University from cholera in 1830 by stringently isolating it from the community. He was Rector of the University from 1827 to 1846, and a leading figure in rescuing the library and astronomical instruments from a fire that engulfed the University in 1842. In 1846 he had to surrender his position as Rector, in accordance with a 20-year rule then in force. Refusing to be made a special case, he was thereupon made a director of education for the whole of Kazan province. He became blind in 1855, and died the following year.

Lobachevskii tried several times to convey his momentous discovery of a non-Euclidean geometry, which was nothing less than a new description of space itself. He published an account of it in Russian in 1829, and in a longer Russian article of 1835 he again outlined how a new geometry could be defined in which the PP does not hold. He also showed how to derive the basic formulas for areas and volumes of figures in their new setting, and observed that his formulas closely resembled those of spherical trigonometry, and that his geometry, like spherical geometry, was well approximated by Euclidean plane geometry over small patches. Accordingly, only astronomical observations would be able to distinguish his 'imaginary geometry' (as he called it) from Euclid's — indeed, in his first publication on the subject he described (inconclusively) how empirical observation on stellar parallax could distinguish between his geometry and the familiar Euclidean one.

Lobachevskii tried yet again in the article 'Imaginary geometry' that he published in French in 1837 in Crelle's *Journal*, where he said that:[30]

> After developing a new theory of parallels, I undertook to show that nothing except direct observation empowers us to suppose that the sum of the angles in a triangle is two right angles, and to show that a geometry can exist, if not in nature then at least in mathematics, in which the angle-sum is assumed to be less than two right angles.

He then tried for a fourth time in a booklet of 1840 written in German.[31] Here he began his attempt to reach a Western European audience as follows:

Lobachevskii on the foundations of geometry.

In geometry I find certain imperfections which I hold to be the reason why this science, apart from transition into analytics, can as yet make no advance from that state in which it has come to us from Euclid. As belonging to these imperfections, I consider the obscurity in the fundamental concepts of the geometrical magnitudes and in the manner and method of representing the measuring of these magnitudes, and finally the momentous gap in the theory of parallels, to fill which all efforts of mathematicians have been so far in vain.

For this theory Legendre's endeavours have done nothing, since he was forced to leave the only rigid way to turn into a side path and take refuge in auxiliary theorems which he illogically strove to exhibit as necessary axioms. My first essay on the foundations of geometry I published in the Kasan *Messenger* for the year 1829. In the hope of having satisfied all requirements, I undertook hereupon a treatment of the whole of this science, and published my work in separate parts in the *Gelehrten Schriften der Universitaet Kasan* for the years 1836, 1837, 1838, under the title 'New Elements of Geometry, with a Complete Theory of Parallels'. The extent of this work perhaps hindered my countrymen from following such a subject, which since Legendre had lost its interest. Yet I am of the opinion that the Theory of Parallels should not lose its claim to the attention of geometers, and therefore I aim to give here the substance of my investigations, remarking beforehand that contrary to the opinion of Legendre, all other imperfections — for example, the definition of a straight line — show themselves foreign here and without any real influence on the theory of parallels.

He then set off on 'the only rigid way', with a new definition of parallels (see Figure 14.13).

All straight lines which in a plane go out from a point can, with reference to a given straight line in the same plane be divided into two classes — into cutting and not-cutting.

The *boundary lines* of the one and the other class of those lines will be called *parallel to the given line*. From the point A let fall upon the line

[30]See (Lobachevskii 1837, 295).

[31]*Geometrische Untersuchungen zur Theorie der Parallellinien* (*Geometrical Researches on the Theory of Parallel Lines*), Berlin, 1840.

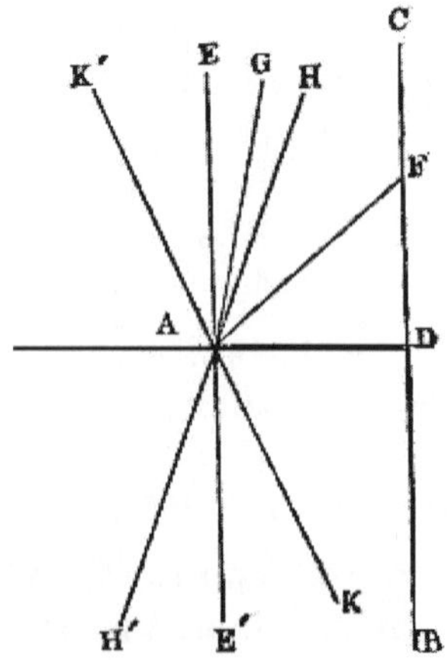

Figure 14.13. Lobachevskii's figure of lines through a point A

BC the perpendicular AD, to which again draw the perpendicular AE. In the right angle EAD either will all straight lines which go out from the point A meet the line DC, as for example AF, or some of them, like the perpendicular AE, will not meet the line DC. In the uncertainty whether the perpendicular AE is the only line which does not meet DC, we will assume it may be possible that there are still other lines, for example AG, which do not cut DC, how far soever they may be prolonged. In passing over from the cutting lines, as AF, to the not-cutting lines, as AG, we must come upon a line AH, parallel to DC, a boundary line, upon one side of which all lines AG are such as do not meet the line DC, while upon the other side every straight line AF cuts the line DC. The angle HAD between the parallel HA and the perpendicular AD is called the parallel angle (angle of parallelism), which we will here designate by $\Pi(p)$ for $AD = p$.

If $\Pi(p)$ is a right angle, so will the prolongation AE' of the perpendicular AE likewise be parallel to the prolongation DB of the line DC, in addition to which we remark that in regard to the four right angles, which are made at the point A by the perpendiculars AE and AD, and their prolongations AE' and AD', every straight line which goes out from the point A, either itself or at least its prolongation, lies in one of the two right angles which are turned toward BC, so that except the parallel EE' all others, if they are sufficiently produced both ways, must intersect the line BC.

If $\Pi(p) < \pi/2$ then upon the other side of AD, making the same angle $DAK = \Pi(p)$ will lie also a line AK, parallel to the prolongation DB of the line DC, so that under this assumption we must also make a distinction of *sides in parallelism*.

Lobachevskii concluded that lines through the point A cut the line BC if they lie inside the angle KAH (facing the line BC), that lines through the point A do not cut the line BC if they lie outside the angle KAH, and that the lines AK and AH that separate these two families of lines may be called the two parallels to BC through A.

Poor Lobachevskii! All this was to no avail. He was careless in his French article, and lost his readers when he said 'as I have shown' of passages that had been proved only in his Russian article of 1829. His German booklet of 1840 was wildly misunderstood by the only reviewer it received. In 1842 Lobachevskii sent Gauss a copy of his book — Gauss had heard of Lobachevskii's work, and was appalled by the review. Indeed, he thought so highly of Lobachevskii that he successfully proposed him for corresponding membership of the Göttingen Scientific Society, but this was the only international recognition that Lobachevskii ever received. At home his work was persistently attacked in St Petersburg by Mikhail Ostrogradsky (in his day the better-known mathematician), and Lobachevskii seems to have withdrawn into administration. Finally, in 1855, the year before his death, he published his *Pangéométrie*, but this last attempt to win support for his views was also unsuccessful. He was never to enjoy proper credit for his momentous discovery.

János Bolyai's story is scarcely more encouraging. Born in 1802 in Kolozsvár, he was the son of Farkas (Wolfgang, in German) Bolyai, whom we met fleetingly in the previous section.[32] Farkas had studied at Göttingen with Gauss, and they kept in touch when Farkas returned to his native Hungary. Farkas had a lifelong interest in the PP, and kept trying to prove it; his best result was that the PP is false if and only if there are three points in a plane that do not lie on a straight line or a circle.

His son János was apparently a precocious mathematician (and violinist), and in due course he took up the vexed question of the PP. Like Lobachevskii, he first thought that he had derived the postulate as a theorem, but then realised that his 'proof' was flawed. He then set to work to prove that the PP could be false. A marvellously vivid exchange of letters between father and son in 1823 was published by the mathematician and historian of mathematics Paul Stäckel, which sheds light on János Bolyai's view of the PP.[33]

The Bolyais on geometry.

> In Spring 1820 Wolfgang Bolyai wrote to his son: You must not attempt this approach to parallels. I know this way to its very end — I too have traversed this bottomless night, all light and every joy of my life was extinguished in it — I beg you before God, leave the science of parallels in peace.
>
> ...
>
> I was ready to sacrifice myself for the truth, I was ready to become a martyr if only in that way I could rid geometry of this flaw of human making. I accomplished monstrous, enormous labours, far better than those of others, but I have not achieved complete satisfaction, for here it is true that si paullum a summo discessit, vergit ad imum [if one departs a little from the summit, one turns to the depths]. I turned

[32] Kolozsvár (German, Klausenberg) in former Transylvania is now called Cluj, in Romania.

[33] See (Stäckel 1913, 76–77, 82, 85, 86). For a different translation from ours, see (Meschkowski 1964, 31–34), in F&G 16.B4.

> back when I saw that no man can reach the end of this night on the Earth. I turned back unconsoled for myself and all mankind.
>
> ...
>
> It is unbelievable that this stubborn darkness, this eternal eclipse, this flaw in geometry, this eternal cloud on virgin truth can be endured.
>
> I admit I expect nothing from the deviation of your lines. It seems to me that I too have visited these regions; that I have travelled past all the cliffs of this infernal Dead Sea and have always come back with broken mast and torn sail. The ruin of my disposition and my fall date to this time. I thoughtlessly risked my life and happiness — aut Caesar aut nihil [either Caesar or nothing].

In 1823 János Bolyai could tell his father, who had tried so hard to make him give up his interest in the problem, that he was succeeding:

> My fixed intention is I shall publish a work on parallels as soon as it is in order, I have completed it, and I have the opportunity. At the moment it is still *not* thoroughly worked through, but the path which I have taken almost certainly allows me to attain my goal, provided it is possible at all. I do not yet have it but I have found things so magnificent that I astonished myself and it would be an eternal pity if these things were lost; as you, my dear father, are bound to recognise when you see them. All I can say no more now, only *that I have created a new and different world out of nothing.* All that I have sent you so far is like a house of cards compared with a tower.

His father advised him to publish his results as soon as possible. János Bolyai's comment follows:

> He gave me the opinion that if I was really successful, I should quickly make a public announcement and that for two reasons. First because the idea might easily pass to someone else who would then publish it; second because, and this is another truth, several things ripen at the same time and appear in different places in the manner of violets coming to light in early spring, and since all scientific striving is like a great war in which one does not know when peace will come, one must win, if possible; for here preeminence comes to him who is first.

János Bolyai seems to have discovered something new, but he was not yet sure. He was 'almost certain' that he had 'created a new and different world out of nothing'. However, he first wanted to put everything in order, for the path he had found would lead to the goal only if reaching it were possible.

These letters, written at the height of the Romantic era and in splendidly Romantic prose, give a wonderful impression of the psychological pressures involved in trying to do truly original mathematics; plainly, father and son were not merely dabbling in the subject. Farkas's remarks about simultaneous discovery were strikingly prescient, given that both men were unaware of Lobachevskii's work at the time. Perhaps he was wondering about Gauss, or perhaps he was reflecting that discoveries seem to be made when the time is ripe, so speed was of the essence. But he could not have expected

Lobachevskii's work to be so very similar — their work on non-Euclidean geometry was to prove a most striking example of simultaneous discovery, as we shall see.

In 1823, however, the two Bolyais were still uncertain about exactly what it was that János had discovered. The difficulty was a constant that cropped up in all their formulas. It was unexpected and unexplained. Did it mean that the formulas would yield a self-contradiction (as would happen if one could show that this constant had simultaneously to take two different values), or did this constant have a geometrical meaning that would lead to a deeper understanding of the new geometry? Eventually they agreed to stop worrying, and János's work was published as a 24-page appendix to a two-volume work on geometry that was published by his father in 1831. A copy was sent to Gauss, and they waited nervously for his reply. But the copy was lost in the post. Another was sent, and this time Gauss replied.[34]

Gauss to Farkas Bolyai.

> If I commenced by saying that I am unable to praise this work, you would certainly be surprised for a moment. But I cannot say otherwise. To praise it, would be to praise myself. Indeed the whole contents of the work, the path taken by your son, the results to which he is led, coincide almost entirely with my meditations, which have occupied my mind partly for the last thirty or thirty-five years. So I remained quite stupefied. So far as my own work is concerned, of which up till now I have put little on paper, my intention was not to let it be published during my lifetime. Indeed the majority of people have not clear ideas upon the questions of which we are speaking, and I have found very few people who could regard with any special interest what I communicated to them on this subject. To be able to take such an interest it is first of all necessary to have devoted careful thought to the real nature of what is wanted and upon this matter almost all are most uncertain. On the other hand it was my idea to write down all this later so that at least it should not perish with me. It is therefore a pleasant surprise for me that I am spared this trouble, and I am very glad that it is just the son of my old friend, who takes the precedence of me in such a remarkable manner.

Farkas Bolyai professed himself comfortable with this letter when he forwarded it to his son, saying that it was 'very satisfactory and redounds to the honour of our country and of our nation', but János was deeply angered by this chilling reply. Gauss further compounded the offence by giving a delightfully simple proof of Lambert's claim that in the new geometry the area of a triangle with angles α, β, and γ is proportional to $\pi - (\alpha + \beta + \gamma)$. János never published again, and took a lifelong dislike to Gauss.

János Bolyai's work was published, and the news was out; but the result was a stony silence — the scientific world simply did not notice. Indeed, the Bolyais heard of Lobachevskii's work only in 1848, and although János then made a thorough study of it, he published nothing in response.

The work of Lobachevskii and Bolyai. We must now try to form an impression of what Bolyai and Lobachevskii actually did. In what ways had they advanced beyond

[34]See Gauss, *Werke* III, 220–221, (Bonola 1912, 100), and F&G 16.B2.

some of the earlier studies we have seen? Since their discoveries are remarkably similar, we ignore the slight differences and take the work of Lobachevskii to stand for both.

The first thing to notice is that Lobachevskii built on earlier ideas — in particular, he followed the strategy of Saccheri's 'hypothesis of the acute angle', but did so more doggedly and with clearer definitions. Lobachevskii then set out to follow the consequences of non-Euclidean assumptions about a straight line BC and a point A that does not lie on BC. In particular, he considered the straight lines through A that do meet BC and those that do not, as we saw in Figure 14.13.

There are infinitely many lines going through A that never meet BC. Lobachevskii called these 'not-cutting' lines; they are all the lines further away from BC than the 'boundary lines' AH and AK (see Figure 14.14).

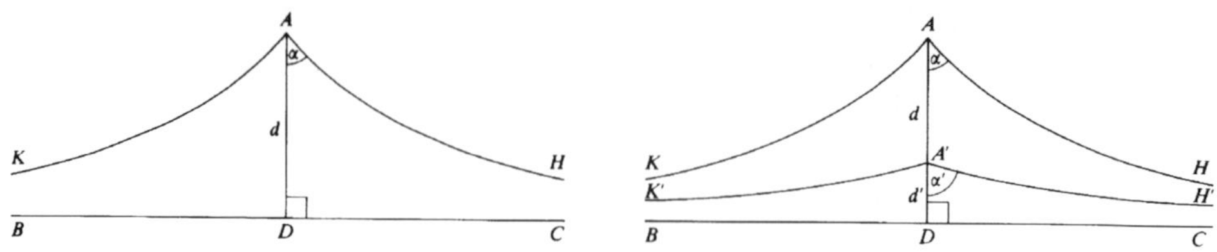

Figure 14.14. The angle of parallelism

For him, the lines parallel to BC through A are the boundary lines — the two lines passing through A with the property that a line nearer to BC cuts BC, and a line further away does not. In effect, he narrowed down his definition of a line parallel to a given line. For him, it was a line that approaches ever closer to the given line in one direction, but never meets it — the kind of line that Saccheri had thought of as asymptotic.

Lobachevskii went on to investigate the angle between a line parallel to BC and the perpendicular AD to BC.[35] If the parallel AH made a right angle with AD, then the geometry would be Euclidean geometry, for this would be Saccheri's 'hypothesis of the right angle'. In the geometry that Lobachevskii investigated, the angle $\alpha = HAD$ was less than a right angle.

This angle α, the 'angle of parallelism' as he called it, depends on how far A is from BC: the further away, the smaller the angle. In the right-hand picture in Figure 14.14 this appears plausible, for when the distance d' is short, α' must be virtually a right angle.

The precise formula that Lobachevskii calculated to show the relationship between α and d is

$$d = k \log \cot(\alpha/2), \text{ or } \alpha = 2 \arctan e^{-d/k}.$$

We do not need this formula as such, so we shall not concern ourselves with the details, but that Lobachevskii derived it is an important clue to the significance of his approach. With Lobachevskii we see a further move away from the classical Euclidean style of geometrical argument, and towards a more serious consideration of trigonometric formulas as an integral part of the argument. Recall that Taurinus in the 1820s had derived trigonometric formulas for his 'log-spherical' geometry by what amounted

[35] See (Lobachevskii 1840, 11–15, 44–45) in (Bonola 1912, Appendix), and F&G 16.B3(a).

to the formal trick of replacing k in ordinary spherical formulas by $\sqrt{-1}k$ (see Box 36). Taurinus's results were pretty, but the way in which he had reached them gave no special confidence that his formulas actually applied to anything. But, after making the above assumption about parallels, Lobachevskii followed a lengthy and complex argument to derive the relationships between the sides and angles of triangles in his new geometry, and he arrived at the same formulas as Taurinus's.

In order to do this, Lobachevskii took the vital step of working in three dimensions, as we describe in Box 38. His insight was to see how to use his formula for the angle of parallelism in the setting of Figure 14.15 to obtain the new trigonometric formulas that he needed.

Bolyai's work is similar to Lobachevskii's. He called his results 'theorems in absolute geometry', whereas Lobachevskii had spoken of an 'imaginary geometry'; this was rather underselling it, for Lobachevskii took seriously the possibility that his geometry might be the actual geometry of space, and analysed observations on stellar parallax (inconclusively) to try to resolve the matter.

The reception of the work of Bolyai and Lobachevskii. Why should Lobachevskii and Bolyai have fared so badly in terms of the public response? Part of the answer lies with the novelty of what they had to say, and part with the way in which they said it. Because Lobachevskii published more, his case is more interesting to discuss, so let us concentrate on that. Some of the problems that he created (by the way he described his work) seem to derive, paradoxically, from the fact that he had little choice about how to describe his results. As we said, Lobachevskii was able to deduce trigonometric formulas for his geometry: crucially, without these he could prove nothing about his imaginary geometry.

To convince his readers, Lobachevskii had to persuade them of the validity of his formulas. But in his 1837 paper he conjured up the trigonometric formulas by some arbitrary definitions, and the derivation of these from the basic formula for the angle of parallelism was missing. Perhaps he thought that his Russian audience would consult his 1829 article, but his French and German readers had no such opportunity. So why did he not incorporate the early discussion at this point? His other publication of 1835 is 170 pages long, but even in that paper the arguments are not entirely conclusive. They deal with the derivation of the formulas from the original assumption about parallel lines, but do not defend that assumption. Rather, Lobachevskii's defence of his geometry was precisely that it was described by formulas. It is this view that he presented in his German version of 1840, which was self-contained and intelligible.

By any standards, this justification was a weakness in his approach. No one has to agree that a hypothesis is plausible just because its consequences are attractive. Rather, it is possible to refute a hypothesis by finding that it entails a contradiction. No amount of trouble-free deduction can prevent a contradiction from existing around the next corner, and Lobachevskii knew this perfectly well. But somehow the formulas encouraged him to affirm the existence of the new geometry, and this is why he was willing to start with them in the first paper that he wrote for Western eyes. To see why they convinced him, we must consider what novel perceptions he brought to the very idea of geometry itself.

Box 38. Lobachevskii's trigonometric formulas.

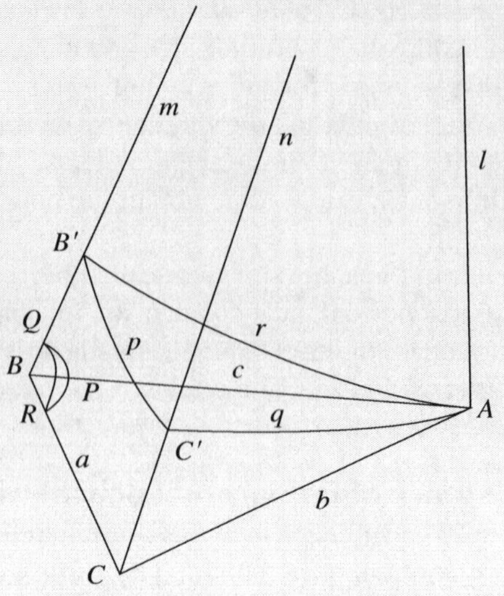

Figure 14.15. Three types of triangle in non-Euclidean 3-dimensional space, adapted from (Lobachevskii 1840)

First, we set the scene for what will be a tale of three triangles. Figure 14.15 shows three parallel lines l, m, and n in three-dimensional space. The plane ABC is perpendicular to l at A, and the lines m and n meet this plane in the points B and C. Each pair of lines, lm, mn, and nl, defines a plane, and these planes cut the plane ABC in the three sides of the triangle ABC.

Lobachevskii considered a small sphere around B, and the spherical triangle whose vertices P, R, and Q are the points where the sphere cuts the lines BA, BC, and m.

He also showed that there is a surface that passes through the point A and meets at right angles all the lines parallel to l (including m and n). He called this surface the *horosphere*. The three planes defined by the pairs of lines lm, mn, and nl cut the horosphere in three arcs r, p, and q, respectively.

He then showed that the angles in the triangle ABC, the spherical triangle PQR, and the triangle $AB'C'$ on the horosphere are all determined by the positions of the lines m and n if l is kept fixed (much as the strings of a puppet determine its shape). He next supposed that the lines m and n are such that the angle ABC is a right angle in the plane. Under these conditions he could show that the angles in any one of these three triangles determine the angles in any of the others.

(continued on next page)

Box 38. Lobachevskii's trigonometric formulas — continued

The scene was now set. Lobachevskii had already obtained some deep results that he could now apply. He had shown that the trigonometric formulas relating the sides and angles of the spherical triangle PQR are precisely the same formulas that hold in Euclidean geometry, even though this triangle is sitting in a space with a different concept of parallel lines.

He had also shown that the sum of the angles in the triangle $AB'C'$ on the horosphere is π — remarkably, the geometry on the horosphere is Euclidean!

These results allowed him to deduce from the trigonometric formulas for the triangles PQR and $AB'C'$ the corresponding formulas for the triangle ABC, and to show that they were the analogues of the sphere but with hyperbolic functions replacing the usual trigonometric ones. This was what Taurinus had claimed, but Lobachevskii had proved it.

In Lobachevskii's view, the crucial point about geometry was that it is grounded in measurement. However, to resolve the problems that necessarily arise when passing from the measurement of straight things to curved things, he needed to resort to axioms. It follows from the first point that geometrical statements are numerical — they are about quantities and their measurements — and so they must lead to formulas. In this connection, it is worth noting that Lobachevskii was one of the first to propose a very general conception of a mathematical function. So it was entirely natural for his geometrical researches to be based on formulas and functions.

As for the point about axioms, it was the very arbitrariness in his choice of axioms that enabled him to defend his own choice of a parallel postulate that was different from Euclid's. This combination was very potent, because it gave Lobachevskii a novel starting point in the long debate about space and geometry. Legendre, by contrast, had adhered throughout the 1820s to the old view of geometry as a matter of rigorous deduction from true axioms, and so had denied himself the opportunity to develop formulas that were suggestive, however obscure the material that was actually referred to. Consequently, Legendre's work fell prey to sundry errors, whereas Lobachevskii's had the rigour of analysis.

The upshot of all this was that, for Lobachevskii, geometry was naturally described by formulas, and that, conversely, a plausible set of formulas would be geometrical, so his new formulas would describe a new geometry. However, to the more traditional mathematician, geometry was about lines, angles, planes, and so forth: no formulas, however plausible, could describe a geometry unless they could be given a geometrical interpretation. When Lobachevskii's formulas were subjected to such an analysis, they turned out to rest on an unproved assumption about lines. If Euclid's geometry were to be correct after all, there would still be nothing wrong with Lobachevskii's formulas — but his assumption would have been wrong and his formulas would have ceased to carry any geometrical meaning.

So the views of Lobachevskii and Bolyai about the nature of geometry were implicit, yet profound, and differed greatly from the prevailing orthodoxy. On the other hand, that orthodoxy was seemingly unable to resolve the problem of the parallel postulate. The only way out was to take a radically new view, and we shall shortly look at its discovery.

The 'standard' historical accounts usually make a break here, and look only vaguely at the new ideas (which have to do with what is called differential geometry and the work of Riemann and Beltrami). So it is sensible to look at what we have learned, and at the criticisms that can be levelled at the 'story so far'. Before doing so, however, we describe how the tide finally began to turn in favour of Bolyai and Lobachevskii.

One argument for this change was put forward by Bonola.[36] In his opinion, the obstacles to its initial reception were not just their obscure places of publication, but the facts that Bolyai and Lobachevskii were otherwise little known, and that the prevailing Kantian conceptions seemed to establish that space is necessarily Euclidean. What turned the situation around were the efforts of the mathematicians Christian Ludwig Gerling, Heinrich Baltzer, and (above all) Guillaume-Jules Hoüel, who translated works that showed that Gauss, the 'Prince of Mathematicians', had looked favourably on the new geometry.

This argument is not entirely compelling. It is true that Bolyai and Lobachevskii published obscurely, so Baltzer and Hoüel's writings surely helped. But Lobachevskii's booklet of 1840 was self-contained and accessible in German, so the story must be more complicated. It is also true that Bolyai and Lobachevskii were both unknown. The relevance of this is hard to evaluate, but we can agree that Gauss's posthumous support seems to have helped. As for Kant, no-one cited by Bonola seems to have argued against him. All that Gauss or Baltzer did was to state a disagreement with Kant, and to present a new geometry in a mathematical way. This seems not to be confronting the Kantian hegemony, so much as slipping past it. On the other hand, a remarkable feature in the work of Bolyai and Lobachevskii — and one that Gauss had missed — was their use of a novel three-dimensional geometry. It played a crucial role in the derivation of the trigonometric formulas that they found for triangles. Even more importantly, it established that there was a novel geometry of space — specifically *three-dimensional* space — and not just that there was a novel geometry on some two-dimensional surface. Even if they had reached that conclusion by impeccable reasoning — and we have seen that their work was open to criticism — the existence of a strange two-dimensional geometry does not establish the existence of a three-dimensional geometry of the same kind. So the sheer novelty of their accomplishment may have worked against its easy acceptance when it was first published, and then it gradually became forgotten.

Gauss died in 1855, Lobachevskii in 1856, and János Bolyai in 1860, and none of them knew that the new non-Euclidean geometry would soon be accepted. In their lifetimes the writings of Bolyai and Lobachevskii fell dead from the press. Had Gauss not recorded his own ideas on the subject and the high regard he had for their work, it is entirely likely that their work would have remained obscure for even longer than it did. To see how these ideas achieved their posthumous rehabilitation, we must return to Göttingen, and to other ideas of Gauss that were to be triumphantly extended.

[36] See (Bonola 1955), and F&G 16.C1.

14.3 The reformulation of metrical geometry

From the time of Gerard Mercator in the 16th century, if not Ptolemy in the 2nd century, spherical geometry had attracted the interest of terrestrial map-makers, just as it had already captured that of astronomers. Experts had struggled with the way that maps of the known world could be made, for they involve depicting the curved surface of the Earth on flat pieces of paper. When this is done, what can be saved, and what, if anything, must go wrong?

Mercator's projection turned curves of constant compass bearing on the globe into straight lines on the map.[37] It is possible to map curves of shortest length on the globe to straight lines on a map, and it is even possible to map a region of the earth onto a page while preserving angles. But you cannot do all of these simultaneously. This is because of a fundamental feature of spherical geometry attributed to Albert Girard (c.1640):

The area of a spherical triangle is proportional to its angle sum $-\ \pi$.

Euler seems to have been the first to deduce that there cannot be a perfect map — one that represents great circles by straight lines, and angles by equal angles — and we can see that this must be the case because the angles of a spherical triangle (whose sides are parts of great circles) always add up to more than π, whereas those of a plane triangle (with straight sides) always add up to exactly π. So any depiction of a sphere onto a plane must involve distortions.

Let us look at the map that Lambert called 'central projection'. This is obtained as follows (see Figure 14.16). Place a glass sphere on a table, and shine a light from its centre: a copy of the southern hemisphere appears on the (infinite) table. Since great circles on the sphere lie on planes through the centre, their shadows are cast as straight lines, and so the map sends geodesics (curves of shortest length) on the sphere to geodesics on the plane.[38] On the other hand, it greatly distorts some distances, because arcs near the equator cast much longer shadows than arcs of the same length near the South Pole.

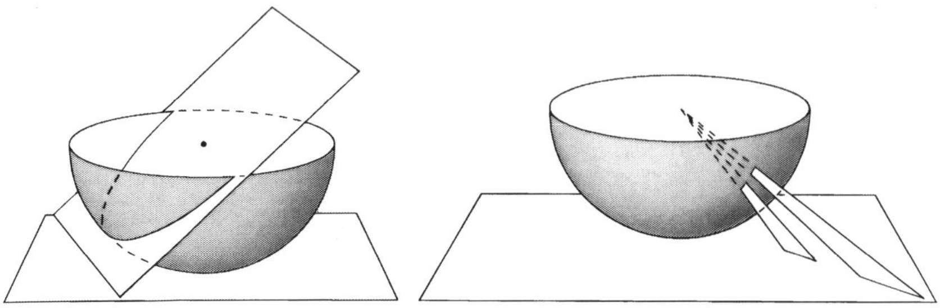

Figure 14.16. A geodesic projection

This leads to the question: Why not simply say that spherical geometry is a geometry different from Euclid's? In fact, despite the apparently axiomatic style of his work, Saccheri had spoken of things being 'repugnant to the nature of the straight line', and

[37] Discussed in *Volume* 1, Chapter 7.

[38] It is now called geodesic projection for that reason, and is used by airlines and shipping companies.

both Lambert and Taurinus had mentioned spherical geometry.[39] Lambert had not dismissed it, but Taurinus immediately ruled it out as irrelevant, because in spherical geometry two lines (great circles) can enclose an area, unlike straight lines in Euclidean space.[40]

Indeed, the valid refutations of the hypothesis of the obtuse angle may have diverted mathematicians away from spherical geometry. Spherical geometry is inconsistent with other postulates of Euclid as well, for it denies not only the existence of parallels but also a simple conclusion drawn from his Postulate, that allows straight lines to be extended indefinitely. This is impossible in spherical geometry because every great circle on a sphere has the same finite length. The conclusion we must draw is that investigations into the parallel postulate were not investigations into an axiom system, but into the correct geometry of space.

Apparently, geometers interested in this question knew some things about honest-to-goodness straight lines — but that only made them try to say precisely what a straight line is. The issue was raised quite sharply by an otherwise minor Cambridge mathematician, Perronet Thompson, who had no answer to it. Referring to a triangle with circular sides and an angle sum that was much less than $2R$, he put the point this way in 1830:[41]

> And if it was urged that these are curved lines and the statement was made of straight; then the answer is by demanding to know, what property of straight lines has been laid down or established which determines that what is not true in the case of other lines is true in theirs.

We have seen that most attempts on the PP foundered in claiming that a straight line is simultaneously the curve that is the shortest path between any two of its points, and that it has some (usually undefined) property of straightness.

The surprising resolution of this tricky conundrum began with the work of Gauss. It turned out that the 'straightness' of a straight line can be defined in differential geometry only by first considering surfaces and deciding what is flat about a plane and round about a sphere. There are essentially two different ways to persuade someone that the world is round like an orange. The simple answer makes use of the fact that the world sits in three-dimensional space: by leaving the surface and travelling in this third dimension, we can *see* that it is round. This is an example of *extrinsic* reasoning — it makes essential use of something outside the surface. The subtle answer is that we can tell by making accurate measurements on the surface itself. Such an answer is *intrinsic*, because it could be given even if the surface of the Earth were all that existed. For example, we can tell intrinsically that a sphere is round by observing that the angles of a triangle whose sides are arcs of great circles add up to more than π, and that the area of such a triangle is proportional to the amount by which this sum exceeds π.

Gaussian curvature. In the 1810s Gauss discovered that intrinsic answers can be given. Since then his arguments have been simplified, but they are essentially a sophisticated version of the way that one defines the curvature of a curve by looking at how the tangent (or the normal) to the curve varies from point to point. When a curve

---

[39] See (Saccheri 1733), and F&G 16.A2.

[40] Euclid had failed to state that two distinct lines cannot meet in two points, and used it without comment in *Elements* I, 4, prompting later commentators to interpolate an explicit claim.

[41] Perronet Thompson, quoted by Halsted in (Bonola 1912, Appendix, 67).

is only slightly curved, the variation in direction of the tangent is slight, but when the curve is tightly curved, the variation can be considerable.

In 1827 Gauss published his ideas in a book that changed the way that mathematicians came to think about geometry: his celebrated *Disquisitiones Generales Circa Superficies Curvas* (*General Investigations of Curved Surfaces*). He began his analysis of how to define the curvature of a surface by considering a way of mapping the surface onto a sphere. As you will see, this map was familiar to him from his work as an astronomer.[42]

Gauss on a map from a surface to a sphere.

> In researches in which an infinity of directions of straight lines in space is concerned, it is advantageous to represent these directions by means of those points upon a fixed sphere, which are the end points of the radii drawn parallel to the lines. The centre and the radius of this auxiliary sphere are here quite arbitrary. The radius may be taken equal to unity. This procedure agrees fundamentally with that which is constantly employed in astronomy, where all directions are referred to a fictitious celestial sphere of infinite radius.
>
> ...If we represent the direction of the normal at each point of the curved surface by the corresponding point of the sphere, determined as above indicated, namely, in this way to every point on the surface, let a point on the sphere correspond; then, generally speaking, to every line on the curved surface will correspond a line on the sphere, and to every part of the former surface will correspond a part of the latter. The less this part differs from a plane, the smaller will be the corresponding part on the sphere. It is, therefore, a very natural idea to use as the measure of the total curvature, which is to be assigned to a part of the curved surface, the area of the corresponding part of the sphere. For this reason the author calls this area the *integral curvature* [Gauss's italics] of the corresponding part of the curved surface.

Later in Gauss's paper, it became clearer that what he meant by the 'integral curvature' of a surface at a point P is the limiting value of the ratio

$$\frac{\text{area of the image of a piece of surface containing } P}{\text{area of a piece of surface containing } P}$$

as the piece of the surface containing P shrinks to the point P. With this definition, a sphere of radius R has the same curvature R^{-2} everywhere, whereas the plane has curvature zero at every point (because the entire plane is mapped to a point).

Gauss next discussed how to define a surface in space by introducing a system of coordinates on it — as we do with latitude and longitude coordinates on a sphere. If we denote longitude by θ and latitude by φ then a point on the sphere has coordinates

$$(\cos\theta\cos\varphi, \sin\theta\cos\varphi, \sin\varphi).$$

More precisely, we can think of the sphere as the image of the plane with its x-coordinate denoted by θ and its y-coordinate denoted by φ. Under this map, the point in the plane

[42]See the Anzeige (Abstract), (Dombrowski 1979, 85).

with coordinates (θ, φ) corresponds to the point with coordinates

$$(\cos\theta\cos\varphi, \sin\theta\cos\varphi, \sin\varphi).$$

More generally, Gauss thought of the plane as having (u, v) coordinates and a surface in space as having the coordinates

$$(x(u, v), y(u, v), z(u, v)),$$

where x, y, and z are functions of u and v. He then had to show how to define lengths and areas on a surface in terms of a system of coordinates on the surface. He supposed that two nearby points had coordinates

$$(x(u, v), y(u, v), z(u, v)) \quad \text{and} \quad (x(u+du, v+dv), y(u+du, v+dv), z(u+du, v+dv)),$$

and applied a three-dimensional version of the Pythagorean theorem to work out the distance ds between them. He showed in this way that

$$ds^2 = E(u, v)du^2 + 2F(u, v)dudv + G(u, v)dv^2,$$

where E, F, and G are expressions in x, y, z and their derivatives with respect to u and v (see Figure 14.17). He called this expression for ds^2 the line or *linear element* of the surface.

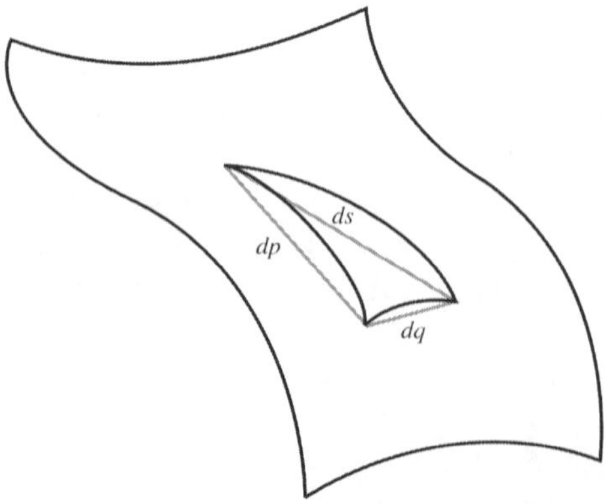

Figure 14.17. An infinitesimal triangle with curved sides on a surface, and its linear approximation

When one surface can be mapped upon another without any distortion of length, the term used in Gauss's day was that one surface is *developed* on the other. Gauss now observed that:[43]

Gauss on curvature.

> The new expression for the measure of curvature mentioned above contains merely these magnitudes [that is, E, F, and G] and their partial differential coefficients of the first and second order. Therefore we notice that, in order to determine the measure of curvature, it is necessary to know only the general expression for a linear element; the expressions for the coordinates x, y, z are not required. A direct

[43]See (Dombrowski 1979, 89, 91).

> result from this is the remarkable theorem: If a curved surface, or a part of it, can be developed upon another surface, the measure of curvature at every point remains unchanged after the development. In particular, it follows from this further: Upon a curved surface that can be developed upon a plane, the measure of curvature is everywhere equal to zero.

He then turned to discuss systems of coordinates on a surface that would be easy to use.

> If upon a curved surface a system of infinitely many shortest lines of equal lengths be drawn from one initial point, then will the line going through the end points of these shortest lines cut each of them at right angles. If at every point of an arbitrary line on a curved surface shortest lines of equal lengths be drawn at right angles to this line, then will all these shortest lines be perpendicular also to the line which joins their other end points. Both these theorems, of which the latter can be regarded as a generalization of the former, will be demonstrated both analytically and by simple geometrical considerations. *The excess of the sum of the angles of a triangle formed by shortest lines over two right angles is equal to the total curvature of the triangle.*
>
> ...Evidently we can express this important theorem thus also: the excess over the two right angles of the angles of a triangle formed by shortest lines is to eight right angles as the part of the surface of the auxiliary sphere, which corresponds to it as its integral curvature, is to the whole surface of the sphere. In general, the excess over $2n-4$ right angles of the angles of a polygon of n sides, if these are shortest lines, will be equal to the integral curvature of the polygon.

Let us comment on these results. Gauss showed that the curvature of a surface can be positive or negative, as shown in Figure 14.18.

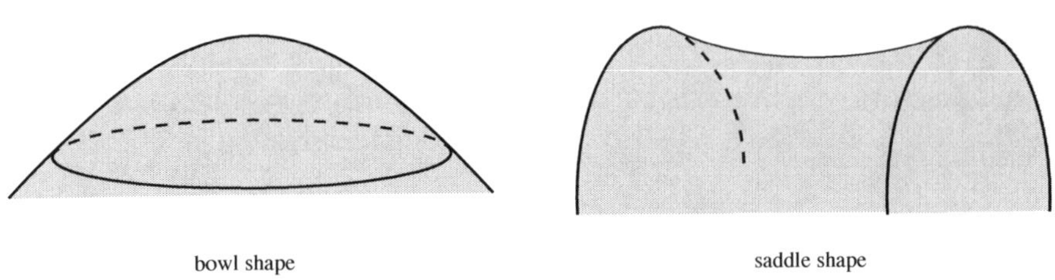

Figure 14.18. Regions of positive curvature (left) and negative curvature (right)

Gauss even found (but did not publish) a surface with constant negative curvature: he called this the analogue of the sphere. He took a curve called the tractrix (see Section 4.1) and rotated it about its asymptotic line, as shown in Figure 14.19. The result is a

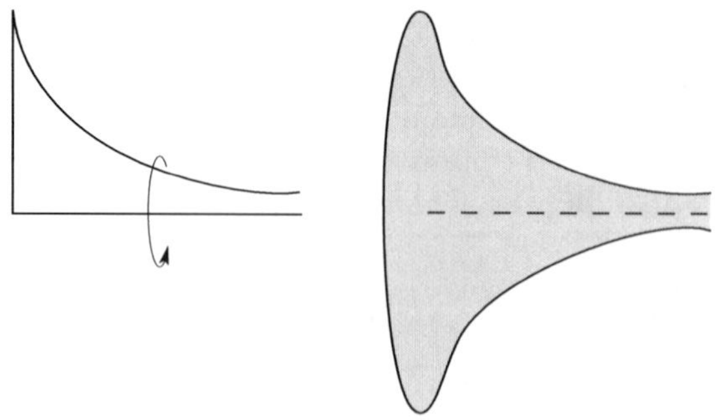

Figure 14.19. A surface of constant negative curvature

'bugle-shaped' surface which, Gauss showed, has constant negative curvature. The resemblance of this figure to Saccheri's (Figure 14.5) is striking.

Gauss was also concerned about the definition of a plane, which is inadequately defined in Euclid's *Elements*.[44] In 1827 he showed that a plane is a surface of zero curvature,[45] and that this characterises the surface in the following sense: Any small piece of a surface of zero curvature is geometrically indistinguishable from a piece of the plane. Thus a cylinder, which is a surface of zero curvature, is (locally) geometrically indistinguishable from a plane, which is why printing from a cylinder is possible.

He also showed that curvature can vary from point to point. As an example of a surface of variable curvature, consider the surface of a pear. Imagine two equilateral triangles on it, each of the same area, but with one triangle at the sharply curved top, and the other at the gently curved bottom. Although equal in area, these triangles cannot be put on top of one another, because one is more curved, and their angles differ. If all equal figures are to be superimposable, then the surface must have constant curvature.

Most importantly, as Gauss showed with a long and somewhat intimidating proof, curvature is an intrinsic property: it can be ascertained independently of the space containing the surface. He did this by showing that the expression for curvature involves only the functions E, F, G, and their derivatives with respect to the coordinates u and v, but not the explicit functions x, y, and z that determine the embedding of the surface in space.

Gauss's discovery of intrinsic curvature entirely revolutionised differential geometry, as this study of surfaces is called. But its full significance for non-Euclidean geometry eluded Gauss.

The work of Riemann. In 1854, the year before he died, Gauss was an examiner of a thesis by a brilliant young student at Göttingen, Bernhard Riemann. This thesis was part of Riemann's final examination in mathematics (the Habilitation, which earned him the right to lecture at German universities). Gauss, by now in his late 70s, was much impressed.

Riemann gave a novel formulation of geometry based on Gauss's idea of curvature. Applied to surfaces, his proposals went like this. A surface is a two-dimensional set

[44]See Gauss, *Werke* VIII, 200–201, and F&G 15.B2(a).

[45]See (Gauss 1827, 83–93), and F&G 15.A5.

Figure 14.20. Approximating the length of a curve

of points. Given a curve in it, try to measure its length by using a finite ruler, as in Figure 14.20. Since this usually gives you only an approximation to the length, you try smaller and smaller rulers. You can measure the length correctly only in the limit, and by means of an infinitesimal ruler — Riemann gave a precise description in formulas. Once you have a concept of length you can find the geodesic between two points and compute the curvature of the surface as Gauss had shown. If the surface has constant zero curvature a geodesic is a straight line, and on a sphere it is a great circle.

All that Riemann said about the content of non-Euclidean geometry was that the angle-sum of a triangle on a surface of constant negative curvature is always less than π.[46] He never mentioned the subject by name, nor any investigator of it after Euclid except Legendre, as we can see in this introductory passage from the published version of the lecture in which he first set out his views, and which is given here as translated, rather unclearly, by his leading English contemporary, W.K. Clifford.[47]

Figure 14.21. Bernhard Riemann (1826–1866)

[46]He also gave the part of a torus lying near the central axis as an example of a surface of constant negative curvature.

[47]See (Riemann 1867), (Clifford 1882, 55–72), and F&G 16.C2.

Riemann on the nature of geometry.

> It is known that geometry assumes, as things given, both the notion of space and the first principles of constructions in space. She gives definitions of them which are merely nominal, while the true determinations appear in the form of axioms. The relation of these assumptions remains consequently in darkness; we neither perceive whether and how far their connection is necessary, nor, a priori, whether it is possible.

Riemann was asking whether the definitions and axioms of geometry are mutually consistent ('possible') and, if so, whether they define a unique geometry ('necessary'). To analyse the matter he proposed to introduce systems of coordinates for spaces of any dimension (the 'multiply extended magnitudes' that he introduces below) and in particular coordinates for three-dimensional space ('triply extended magnitudes'):

> From Euclid to Legendre (to name the most famous of modern reforming geometers) this darkness was cleared up neither by mathematicians nor by such philosophers as concerned themselves with it. The reason of this is doubtless that the general notion of multiply extended magnitudes (in which space-magnitudes are included) remained entirely unworked. I have in the first place, therefore, set myself the task of constructing the notion of a multiply extended magnitude out of general notions of magnitude. It will follow from this that a multiply extended magnitude is capable of different measure-relations, and consequently that space is only a particular case of a triply extended magnitude. But hence flows as a necessary consequence that the propositions of geometry cannot be derived from general notions of magnitude, but that the properties which distinguish space from other conceivable triply extended magnitudes are only to be deduced from experience.

With these words, Riemann claimed that many different three-dimensional spaces are mathematically possible, and so determining the correct mathematical description of space would involve some experimental observations.

> Thus arises the problem, to discover the simplest matters of fact from which the measure-relations of space may be determined; a problem which from the nature of the case is not completely determinate, since there may be several systems of matters of fact which suffice to determine the measure-relations of space — the most important system for our present purpose being that which Euclid has laid down as a foundation. These matters of fact are — like all matters of fact — not necessary, but only of empirical certainty; they are hypotheses. We may therefore investigate their probability, which within the limits of observation is of course very great, and inquire about the justice of their extension beyond the limits of observation, on the side both of the infinitely great and of the infinitely small.

The crucial point here is that Riemann asserted that all geometry is based on intrinsic measurement, whereas Gauss had made only the more limited claim that one can do geometry on a surface without referring to the surrounding (three-dimensional) Euclidean space. Recall that in their work on cartography Euler and Lambert had shown that you cannot have a perfect map of a sphere on a plane, so the sphere and the plane must have different geometries. Following Gauss, Riemann said that any two surfaces have different geometries — that is, different mathematical theorems are true for them — if they have different curvatures anywhere.

But Riemann did much more. He gave a wholly novel answer to the question: *What is geometry*? To him, and to people who agreed with him, geometry was to do with concepts such as length and angle that can be intrinsically defined on a surface or space of some sort. So there are *infinitely many* geometries, one for each kind of surface and each definition of distance: a geometry arises from anything in which it makes sense to talk of a distance between two points, and it then has a set of theorems associated with it.

This was a major step. We have gone from having only one true geometry to having infinitely many different geometries, none of which has any special status. So, if we have a three-dimensional set of points which possess some sense of distance, then it carries a geometry. As an example, consider physical space. Do Riemann's ideas mean that space is Euclidean? Not at all, since we have not mentioned Euclidean geometry. Far from being the origin of geometrical properties, Euclidean space now becomes just one candidate for physical space. To discover whether it is the lucky winner it would be necessary to make measurements and calculate the three-dimensional analogue of the curvature. Euclid's postulates are completely subverted — no longer can they be regarded as unproblematically true assumptions about physical space!

The work of Beltrami. Riemann's thesis remained unpublished until 1867, the year after his early death. But in 1857, the Italian mathematician Delfino Codazzi had already investigated the above surface of constant negative curvature and had shown that the appropriate trigonometry on it is hyperbolic trigonometry. However, he was unaware of (or uninterested in) the controversy surrounding the PP. Then, in 1868, his fellow Italian Eugenio Beltrami brought it all together by exhibiting a space of constant negative curvature in which all the axioms of Euclid hold except for the PP, which is replaced by the HAA. From then on non-Euclidean geometry never lacked supporters among mathematicians, although some philosophers continued to reject it.

Beltrami had the inspired idea that a geometrical space exists mathematically if a map of it can be drawn. He took Lambert's central projection as his starting point. Beltrami wanted to construct a similar map of a non-Euclidean space, but he did not have a surface from which to start, such as the sphere in Euclidean space. So he took the mathematical description of central projection, which allows one to calculate lengths on the sphere given paths on the plane, and identified (in the relevant formula) what makes it describe lengths on a surface of positive curvature. He then changed that part of the formula so that it then describes lengths on a surface of negative curvature. In this way he obtained a formula that makes sense only when the points lie inside a disc of unit radius. But he showed that inside this disc is a map of an entire two-dimensional surface of constant negative curvature, in which curves of shortest length appear as segments of Euclidean straight lines, and everything corresponds to geometry based on the HAA with Lobachevskii and Bolyai's trigonometric formulas being valid. In

Figure 14.22. Eugenio Beltrami (1835–1900)

particular, a constant r that appears in those formulas turns out to be closely related to the curvature of the space.

Beltrami's paper of 1868 portrays a rather pleasing picture of him, then around 32, earnestly setting out to discuss the 'fruits of conscientious and sincere investigations' in a spirit of calm, as he said was his duty.[48] The whole passage is full of that 19th-century European sense of optimism and progress, although he did say that it would be impossible to proceed further than the planar (two-dimensional) case of non-Euclidean geometry. (As it happens, when he read Riemann's thesis he realised that he was wrong about this, and that the analysis was not so restricted.)

Beltrami based his geometry on two principles. The first was that figures should be capable of movement without distortion, so that two equal figures can be moved on top of one another: this requires that the surface be of constant curvature. The second was that there should be a *unique* geodesic through any two points: this rules out the sphere, on which infinitely many geodesics join any two antipodal points. It is unclear from the passage exactly what Beltrami thought about surfaces of constant negative curvature; in fact, as he showed in his paper, surfaces of constant negative curvature do obey the second principle.

[48]See also the extract in F&G 16.C3.

Box 39. Geodesics.

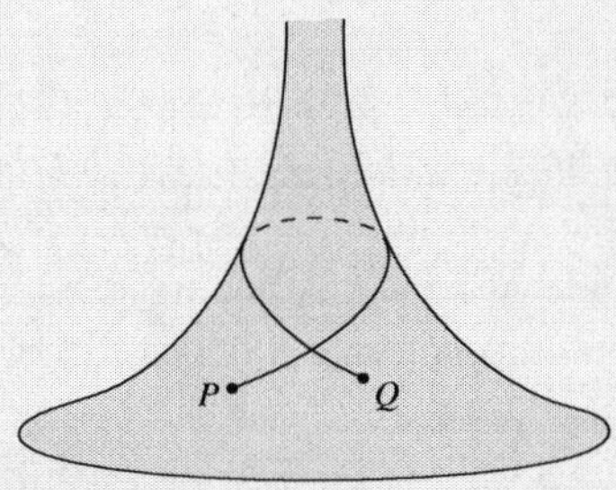

Figure 14.23. A geodesic on a surface of constant negative curvature

To find a geodesic between two points P and Q on a surface, we imagine them joined by a piece of elastic that is constrained to lie in the surface. When stretched between the points, the elastic will lie along a geodesic. This being the case, we can 'trick' the elastic into yielding unexpected examples of geodesics.

Suppose that the surface chosen is Gauss's bugle-shaped surface, and that the elastic is wrapped once around it before being stretched between P and Q (as shown in Figure 14.23). The elastic cannot unwrap itself, so it lies instead along the shortest path that goes from P to Q subject to its wrapping itself once around the surface. This path is shorter than any other path into which the elastic can be pushed, so in this sense it is still a geodesic. Such geodesics are never to be found in the plane, so they highlight one way in which the bugle-shaped surface is not an acceptable model of two-dimensional space.

Beltrami's achievement was not a trivial one. Gauss's bugle-shaped surface cannot be an acceptable description of a two-dimensional version of physical space because there are geodesics on it that intersect themselves (see Box 39); this is not the sort of behaviour that one likes in curves that are meant to be the equivalent of straight lines. Indeed, as Hilbert was to show in 1901, no surface can be found in space that has constant negative curvature and no 'bad' points (such as the rim of the bugle), so Beltrami's task was harder than perhaps he suspected.

How did matters stand in 1870? Only then did mathematicians have a conceptual framework to rescue Bolyai and Lobachevskii, for, after the publication of Riemann's work, it had become possible to say what a straight line is — namely, a geodesic in a surface of zero curvature. Other surfaces, even of constant curvature, have other geodesics — for example, the sphere's geodesics are great circles. In a surface of constant negative curvature we can map them onto straight lines, but at the cost of distorting angles. Beltrami's map was a map of a world of constant curvature. Upon seeing it, mathematicians began to claim that space could be negatively curved, like a bugle. It was Beltrami's map that swung mathematicians decisively behind non-Euclidean geometry. Gauss's posthumous remarks opened the way for what became a triumphal procession, but Riemann's new ideas about the fundamental notions of geometry were what finally enabled mathematicians to accept the creation of 'a whole world out of nothing'.

14.4 Further reading

Bonola, R. 1912. *Non-Euclidean Geometry*, transl. H.S. Carslaw, Open Court Publications, repr. Dover 1955. This classic text is now over a hundred years old. Rather formal, but highly informative on many issues, it has appendices containing Halsted's translations of Bolyai's *Appendix* (1831) and Lobachevskii's *Geometrical Researches* (1840).

Braver, S. 2011. *Lobachevski Illuminated*, Mathematical Association of America. This stimulating, yet leisurely and thorough, account of what Lobachevskii did works its way through his 1840 booklet in detail.

Gray, J.J. (ed.). 2004. *János Bolyai, Non-Euclidean Geometry and the Nature of Space*, Burndy Library, M.I.T. Press. This is a reprint of Bolyai's 1831 Appendix with a fresh commentary.

Gray, J.J. 2011. *Worlds out of Nothing; A Course on the History of Geometry in the 19th Century*, 2nd edn., Springer. This is a much more detailed and wide-ranging account of the material in this chapter and the next.

Trudeau, R.J. 1987. *The Non-Euclidean Revolution*, Birkhäuser, Boston. This enjoyable book by an exciting teacher covers much of the material in this chapter clearly and in depth.

15

Projective Geometry and the Axiomatisation of Mathematics

Introduction

In this chapter we consider the creation of projective geometry in the 19th century, the role that it subsequently played in the re-unification of geometry, and the moves to axiomatise geometry at the end of the century.

15.1 The rediscovery of projective geometry in France

The central figure in the re-birth of projective geometry was Gaspard Monge's former student Jean-Victor Poncelet.[1] Monge had always had practical and military issues at heart and, as befits a true disciple, Poncelet spent most of his life as the Professor of Technical Mechanics at the Military Engineering School at Metz, investigating machines and mechanical structures, but his first interest was in geometry.

Poncelet gave the following striking account of the circumstances in which he was led to his discovery of projective geometry during the months following his capture at the age of 25 by enemy troops, at the end of Napoléon's ill-fated invasion of Russia in 1812.

> This book is the result of researches which I undertook in the spring of 1813 in the prisons of Russia: deprived of every kind of books and help, and the proper facilities, above all distracted by the misfortunes of my country, I was unable to give it all the perfection desirable. However, I had at the time found the fundamental theorems in my work: that is to say the principles of central projection of figures in general and conic sections in particular, the principles of secants and tangents common to those curves, those of polygons which are circumscribed or inscribed to them, etc.

[1] We briefly discussed the earlier history of projective geometry and the work of Girard Desargues in *Volume* 1, Section 13.3.

Figure 15.1. Jean-Victor Poncelet (1788–1867)

This account was published in the preface to his *Traité des Propriétés Projectives des Figures* (Treatise on the Projective Properties of Figures) of 1822. The phrase 'central projection' alerts us to the projective element in Poncelet's work.

The idea is to imagine a figure drawn in one plane and projected onto another plane from a point source of light — the *centre* of the projection. Suppose that the first figure is a conic section (for example, an ellipse). The point source of light causes the figure to cast a conical shadow, so the image on the second plane is a conic section, but by tilting the second plane appropriately the image can be a circle, an ellipse, a parabola, or a hyperbola — in short, any conic section (see Figure 15.2). This means that any two conic sections are equivalent under central projection. In the same way, any triangle can be projected onto any other, so all triangles are equivalent. Projective geometry is the study of properties of geometrical figures that are unaltered by a sequence of central projections.

Central projection therefore offers a way to prove certain properties of all conics by proving them for one (such as a circle) and appealing to this sense of equivalence. It was for this reason that Girard Desargues had invoked the same idea a century and a half earlier. But what properties could there be that cannot distinguish an ellipse or a circle from a hyperbola? They cannot involve lengths, because lengths in a figure are usually changed by projection. Nor can these properties involve angles, for the same reason. But they can involve a line crossing a conic in two points, because the image is also a line crossing a conic in two points. And if the line is a tangent to the conic, the same is true of the image.

Poncelet found that central projections have several properties. They allow one to project any conic into a circle and also to send any specific point not on the conic to the

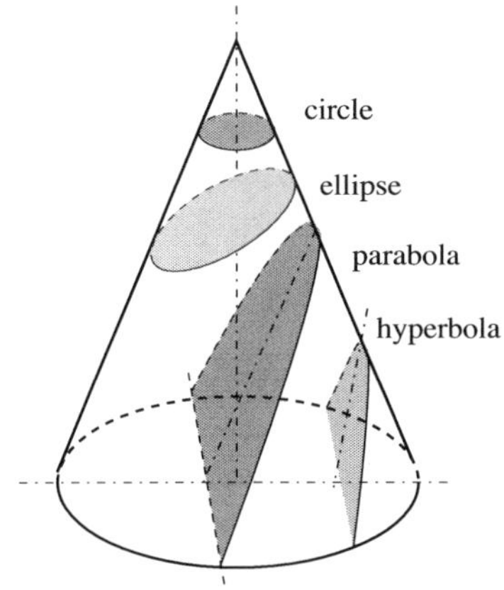

Figure 15.2. The different conic sections are all equivalent under central projection from the vertex of the cone

centre of the circle. So the promise is that some general properties of conics and points can be established by studying the simpler case of a circle and its centre. The challenge is to find interesting properties. Poncelet hinted at what those properties might be with his passing remark about inscribed polygons, and we shall come back to that later. But first we look at his aims in writing his *Traité*:[2]

> The point of this book, voluminous as it may appear, is less to increase the number of properties [of figures] than to indicate the route by which they are found. In a word, I have sought above all to perfect the method of proof and discovery in elementary geometry.

He explained his ideas more fully a few pages later on. He first praised analytical geometry for its generality, and then wrote:[3]

Poncelet on generality in geometry.

> In ordinary geometry, which one often calls *synthetic*, the principles are quite otherwise, the development is more timid or more severe. The figure is described, one never loses sight of it, one always reasons with quantities and forms that are real and existing, and one never draws consequences which cannot be depicted in the imagination or before one's eyes by sensible objects. One stops when those objects cease to have a positive, absolute existence, a physical existence. Rigour is even pushed to the point of not admitting the consequences of an argument, established for a certain general disposition of the objects of a figure, for another equally general disposition of those objects which has every possible analogy with the first. In a word, in this restrained geometry one is forced to reproduce the entire series of primitive arguments from the moment where a line and a point have passed from the right to the left of one another, etc.

[2]See (Poncelet 1822, v–vi).
[3]See (Poncelet 1822, xix–xxi).

> Now here precisely is in fact the weakness; here is what so strongly puts it below the new geometry, especially analytic geometry. If it was possible to apply implicit reasoning having abstracted from the figure, if only it was possible to apply the consequences of that kind of reasoning, this state of things would not exist, and ordinary geometry, without needing to employ the calculus and the signs of algebra, would rise to become in all respects the rival of analytic geometry, even if, as we have said already, it was not possible to conserve the *explicit* form of the reasoning.

We see that Poncelet objected to the need to give two different arguments in geometry when only one is necessary in algebra. By contrast with algebra, which can handle negative and even imaginary magnitudes, synthetic geometry 'is more timid or more severe'. For instance, configurations like those in Figure 15.3 must often be treated differently, depending on whether the perpendicular AD from A to BC falls inside or outside BC. If it falls inside, then BD is positive; if it falls outside, then DB is positive, so BD is negative.

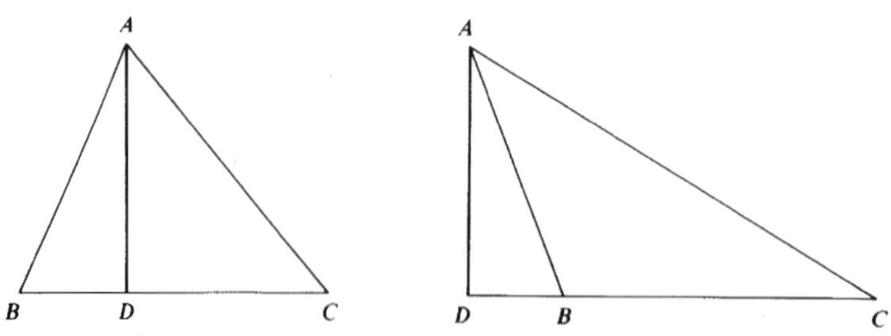

Figure 15.3. Working with triangles

Poncelet then proposed a remedy, and it is here that all later readers were to find that his book divided into claims and methods that were fairly rigorous and easy to understand, and others that were obscure. We shall start with the obscure ones.

Here, first of all, is the proposed remedy.

> Let us consider an arbitrary figure in a general position and indeterminate in some way, taken from all those that one can consider without breaking the laws, the conditions, the relationships which exist between the diverse parts of the system. Let us suppose, having been given this, that one finds one or more relations or properties, be they *metric* or *descriptive*, belong to the figure by drawing on ordinary explicit reasoning, that is to say by the development of an argument that in certain cases is the only one one regards as rigorous. Is it not evident that if, keeping the same given things, one can vary the primitive figure by insensible degrees by imposing on certain parts of the figure a continuous but otherwise arbitrary movement, is it not evident that the properties and relations found for the first system, remain applicable to successive states of the system, provided always that one has

> regard for certain particular modifications that may intervene, as when certain quantities vanish or change their sense or sign, etc., modifications which it will always be easy to recognise *a priori* and by infallible rules? ...
>
> Now this principle, regarded as an axiom by the wisest mathematicians, one can call the *principle* or *law of continuity* for mathematical relationships involving abstract and depicted magnitudes.

So, in Poncelet's opinion, if one figure can be obtained from another by changing it by insensible degrees — say, by an arbitrary continuous movement — then some properties of the first figure would obviously persist through these changes to the final figure, provided that one took note of the fact that certain quantities (which could be specified in advance) may change in size, vanish, or become negative. He called this proposal the *principle* or *law of continuity*. It is, to be frank, somewhat vague. It was never made rigorous by Poncelet, and he was strongly attacked as soon as it was published, as we shall see. He admitted that it led to paradoxes; for example, where are the common points — which he called *ideal points* — of the pair of circles on the left in Figure 15.4? According to his principle of continuity these points should be obtained by a continuous movement of the intersecting circles on the right in Figure 15.4. But, he said, the paradoxes do not go away if you use algebraic, rather than geometrical, analysis. So the problem was to explain them directly, rather than to let them halt progress.

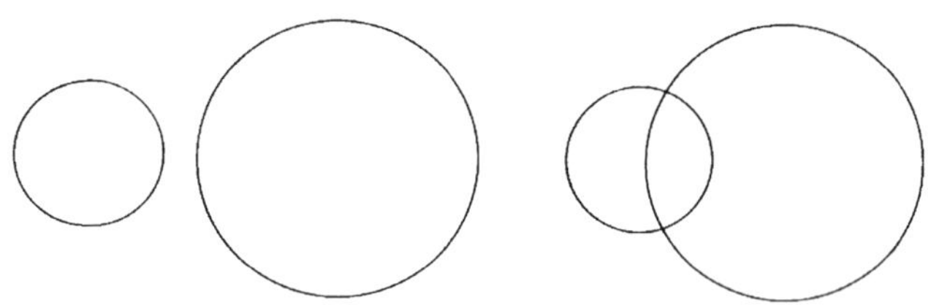

Figure 15.4. Working with circles

Poncelet's account of these mysterious points of intersection was not equivalent to what a geometer relying on algebraic methods would say — namely, that such circles meet in points with complex coordinates. In this respect at least Poncelet's presentation of his ideas, however 'geometrical', was not likely to replace the algebraic formulations of the previous hundred years, and indeed it did not. The principle of continuity had its supporters, but it remained obscure and finally lapsed from the subject.

Poncelet was on surer ground with the idea of projective transformations and properties of figures that are preserved by projections. Here he acknowledged the influence of Monge, and he also praised Desargues highly (although he had not read his main work), calling him 'the Monge of the 17th century'.[4] Another influence on Poncelet

[4]At this stage Desargues's work was known only at second hand through the work of Abraham Bosse, who had published what is nowadays called 'Desargues's theorem' on triangles in perspective. See *Volume* 1, Section 11.1, and F&G 11.D6.

was Lazare Carnot's study of the properties of pairs of intersecting curves, published in 1806 as the third of his books aimed at revising and extending the science of geometry.[5]

Poncelet also made considerable use of the concepts of the 'pole' of a line with respect to a conic and of the 'polar' of a point with respect to a conic. If a line ℓ meets a conic at two points Q and Q', then the tangents to the conic at Q and Q' meet at a point P, which is called the *pole* of ℓ (see Figure 15.5).

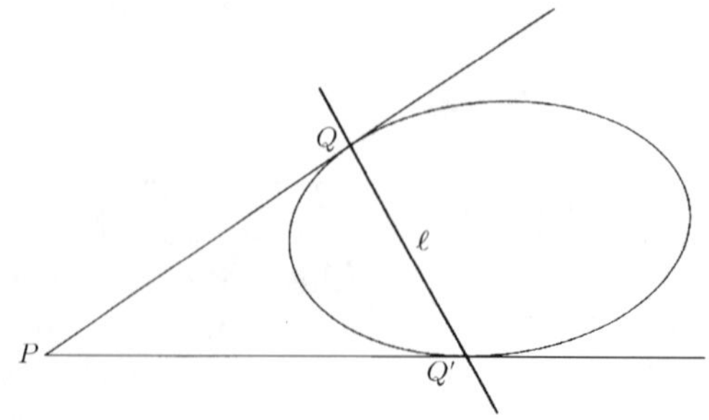

Figure 15.5. P is the pole of ℓ and ℓ is the polar of P with respect to the given conic

The converse is also true: given a point P outside a conic, one can draw through P two tangents to the conic. Label the points where they touch the conic Q and Q'; then to P one can associate the line QQ', which is called the *polar* of P.[6] Note that all the terms so far (point, conic, line, tangent) refer to properties of figures that are unchanged by a projection; this will remain the case in what follows.

In this way, given a conic, Poncelet could associate to any point outside the conic a line that cuts the conic, and to any line that cuts the conic he could associate a point outside the conic. It is also clear that starting with a point P outside the conic, the pole of the polar of P is P itself, and starting with a line ℓ that cuts the conic, the polar of the pole of ℓ is the line ℓ itself.

But what if the point P is *inside* the conic, or the line ℓ does not cross the conic? To answer this question, Poncelet invoked an idea that went back to Philippe de la Hire in the 17th century.

De la Hire's theorem (see Figure 15.6) says:

1. Let P be a point inside a conic, let QPQ' be a line through P meeting the conic at Q and Q', and let the tangents to the conic at Q and Q' meet at P'. Then for any position of the line QPQ' the corresponding points P' lie on a common line ℓ.

2. Let ℓ be a line outside a conic, let P' be a point on ℓ, and let the tangents from P to the conic touch it at the points Q and Q'. Then for any position of the point P' on ℓ the corresponding lines QQ' meet in a common point P.

The first result allowed Poncelet to define the polar of a point P inside the conic as the line ℓ.

[5]Lazare Carnot was another mathematician and scientist who was actively involved in politics. His defence of Paris in 1794, when he was in charge of the revolutionary army, had earned him the popular title of 'Organiser of the Victory'.

[6]The names 'pole' and 'polar' come from their earlier use in spherical geometry, where the polar circle of a point is the equator for that point.

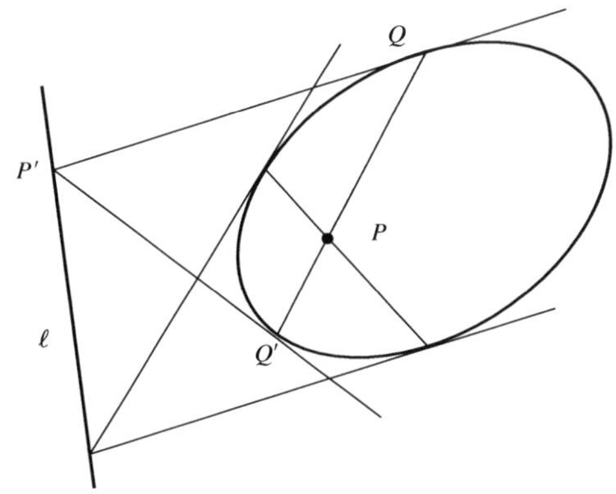

Figure 15.6. La Hire's theorem

The second result allowed Poncelet to define the pole of a line ℓ outside the conic as the point P.

We note also that the first result starts with a family of concurrent lines and produces a family of collinear points, whereas the second result starts with a family of collinear points and produces a family of concurrent lines.

Poncelet was pleased to find also that in this case the processes of going from a point to its polar line, and going from a line to its pole, are inverse to one another. So he could say quite generally that if one starts with a line ℓ, constructs its pole (the point P) and then constructs the polar line of P, the result is just the original line ℓ. Equally, if one starts with a point P, constructs its polar line ℓ, and constructs the pole of ℓ, the result is the original point P.

Poncelet remarked that the two cases, where the pole is inside the conic and where it is outside the conic, are equivalent if one admitted the principle of continuity, and spoke of the ideal intersection points of the polar and the conic when the polar lay entirely outside the conic. At least in this application of the principle of continuity, later mathematicians had a more comprehensible collection of insights to work with.

Poncelet also commented on two special cases. First, the polar of a point on the conic is the tangent to the conic at that point. Second, when the point P lies at the centre of the conic, the points Q and Q' in the above construction are diametrically opposite and the tangents at Q and Q' are parallel. In this case, he said that the polar line of P is the 'line at infinity' — so we see that Poncelet's geometry, like that of Desargues, involved points at infinity. Considerations of this kind were forced on him by his heavy reliance on central projection.

The duality controversy. All these properties of pole and polar led Poncelet to proclaim a principle of *duality*: given a conic, if you replace each line in a figure by its corresponding pole, and each point by its polar line, you obtain a new figure in which concurrent lines are replaced by collinear points and vice versa. The original theorem becomes a theorem about the new figure on exchanging the words 'point' and 'line', 'collinear' and 'concurrent', and the phrases 'lying on a line' and 'passing through a point', throughout. (Such a pair of theorems are called *dual theorems*.) The simple act of dualising a theorem yields another theorem, this time about the dual figure. This is

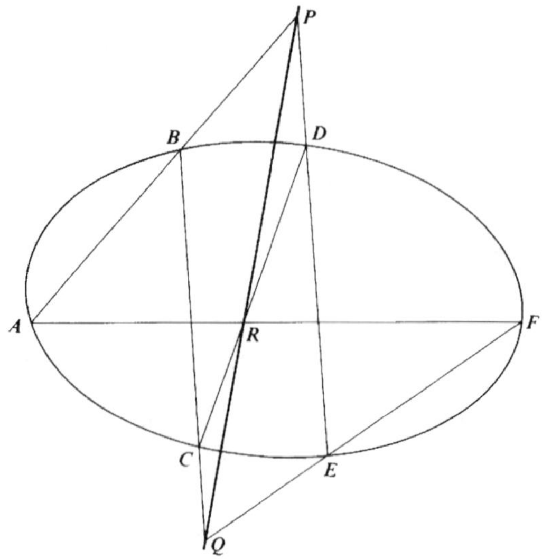

Figure 15.7. Pascal's Theorem

not the case in Euclidean geometry, because, although it is true that any two points lie on a line, it is true that two lines meet in a point only when they are not parallel.

In fact, the idea of duality had been around for some time. One can, for example, dualise *Pascal's Theorem* (see Figure 15.7). This theorem says that:

> *Pascal's Theorem*: If six points A, B, C, D, E, F lie on a conic, and if AB and DE meet at the point P, BC and EF meet at the point Q, and CD and FA meet at the point R, then the points P, Q, and R lie on a straight line.[7]

Charles-Julien Brianchon, a pupil of Monge, had already dualised this result in 1806, to get this beautiful result about any six tangents to a conic (see Figure 15.8):

> *Brianchon's Theorem*: Six tangents to a conic meet in six points which lie in pairs on three concurrent lines. More precisely: if a, b, c, d, e, f, are six tangents to a conic, and a and b meet at A, b and c meet at $B, \ldots, f$ and a meet at F, then the lines AD, BE, and CF meet in a point.

But it is one thing to realise that dualising a figure is a good way to obtain new theorems, which is what Poncelet did, and quite another to claim that points and lines are interchangeable concepts that must logically be treated on a par. This was the view that Gergonne put forward in 1825. Interpreted in such generality, Gergonne's principle of duality is one of the most profound and simple ideas to have enriched geometry since the time of the Greeks. Gergonne hoped that it would be a guiding principle that would enable one to see how certain geometrical ideas are related, and to deduce new ones, so unifying the study of geometry.

This generated a typically Parisian controversy. Poncelet disagreed forcefully with Gergonne, believing that there was nothing more to duality than the exchange of pole and polar with respect to a conic. In the resulting dispute, which concerned priority as

[7]See *Volume* 1, Chapter 13.

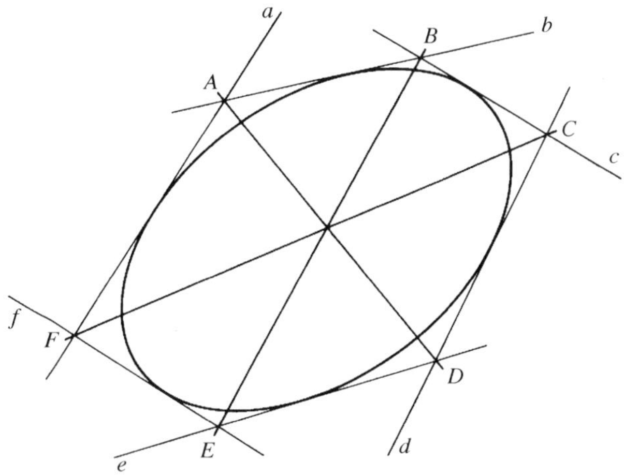

Figure 15.8. Brianchon's Theorem

well as mathematical depth, little justice was done to either side. As is often the case in such matters, each side benefitted from mistakes made by the other that had nothing to do with the fundamental issue.

Gergonne had initiated this study of duality, applied to curves other than lines and conics, in papers that he printed in his own journal, the *Annales des Mathématiques*, in 1827–1828. His idea was this. Suppose you have a curve. Use duality to replace each point on the curve by a line. When you do this for every point on the curve you get a family of lines; Gergonne asked what this family looks like. It turns out that it envelops a curve, called the *dual curve* to the original one (see Figure 15.9).

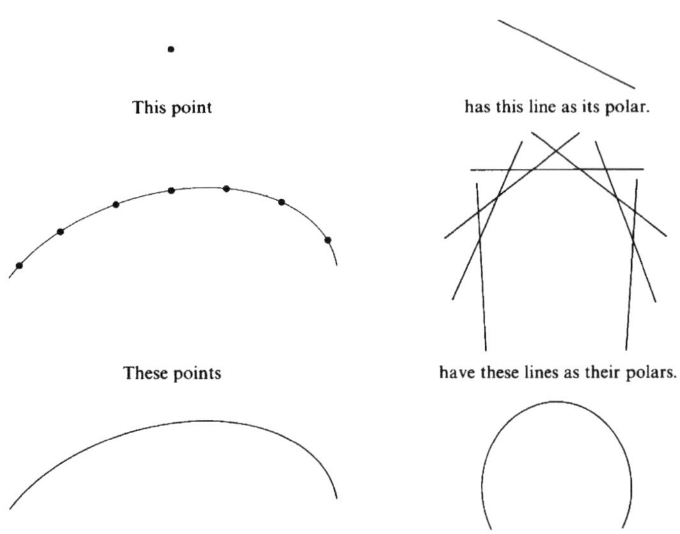

Figure 15.9. Constructing a dual curve

Box 40. An example of duality.
In an article of 1826, Joseph Gergonne dualised Desargues's Theorem on triangles in perspective.

Desargues's Theorem: Let the points O, A, A' lie on a line, the points O, B, B' lie on a line, and the points O, C, C' lie on a line; let the lines AB and $A'B'$ meet at the point R, the lines BC and $B'C'$ meet at the point P, and the lines CA and $C'A'$ meet at the point Q; then the points P, Q, and R lie on a line.

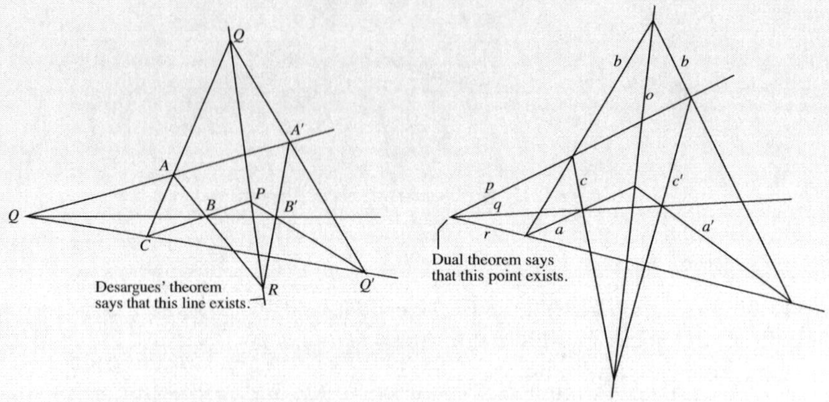

Figure 15.10. Dualising Desargues's theorem

We dualise this as follows. The duals of the collinear points O, A, A' are the lines o, a, a' meeting in a point, and similar statements apply to O, B, B' and O, C, C'. What about the lines AB and $A'B'$ meeting at R? The dual statement must say that the point where the line a and the line b meet and the point where the lines a' and b' meet define a line r (the dual of the point R). In the same way the lines q and p are obtained as the duals of the points Q and P. Finally, the original theorem concludes that the points Q, P, and R are collinear. The dual statement is that the lines p, q, and r are concurrent. So we obtain the dual theorem as follows.

Dual Theorem: Let the lines o, a, a' meet in a point, the lines o, b, b' meet in a point, and the lines o, c, c' meet in a point; let the points where the lines a and b meet and the lines a' and b' meet lie on a line r, the points where the lines b and c meet and the lines b' and c' meet lie on a line p, and the points where the lines c and a meet and the lines c' and a' meet lie on a line q. Then the lines p, q, and r meet in a point.

This is the converse of Desargues's theorem.

As an example of what can be done, Gergonne wrote down two pairs of theorems about a curve and its dual. Here he rather spoiled the effect by making a serious mistake. Poncelet pointed out the mistake, but was unable to see how to put it right.[8] It

[8]See F&G 17.A3(a).

is true for conics, but for curves of higher degree it implies that the dual of a cubic is another cubic, which is true only in special cases.

The remedy was simply to interchange two words, 'degree' and 'order' — the order of a curve is the number of tangents that can be drawn to it from a point.[9] Seldom can so grave a mistake have been put right so deftly — but the dispute left Poncelet looking the better of the two.

Ironically, it was to turn out that as far as plane geometry is concerned there is no difference in mathematical terms between the two ideas of duality. In plane geometry, if you can exchange collinear points and concurrent lines then you can define a conic with the property that in this exchange:

- the line corresponding to a given point is the polar line of the given point with respect to this conic
- the point corresponding to a given line is the pole of the given line with respect to this conic.

So Gergonne duality in the plane is a special case of Poncelet duality. The other way round is trivial: Poncelet duality is a special case of Gergonne duality.

The problem that Gergonne's erroneous claim had highlighted was not to be solved by either of the protagonists, and until it was settled, a question mark hung over the whole topic of duality. It was to be resolved by a German geometer, Julius Plücker, but in an entirely algebraic way that surely did not please Poncelet.[10]

The reception of Poncelet's ideas. Using the precise concepts of pole and polar, and his rather vague ideas of ideal points and continuous deformations of figures, Poncelet went some way towards meeting his aim of giving a unified and powerful treatment of geometry that did not rely on algebra for its proofs. He also obtained some interesting new results, the most remarkable of which is called *Poncelet's porism*.

Poncelet's porism starts with two conic sections — for simplicity, we will take one inside another (see Figure 15.11). Pick a point P on the outer conic, draw from it a tangent to the inner conic, and extend this tangent until it meets the outer conic again.

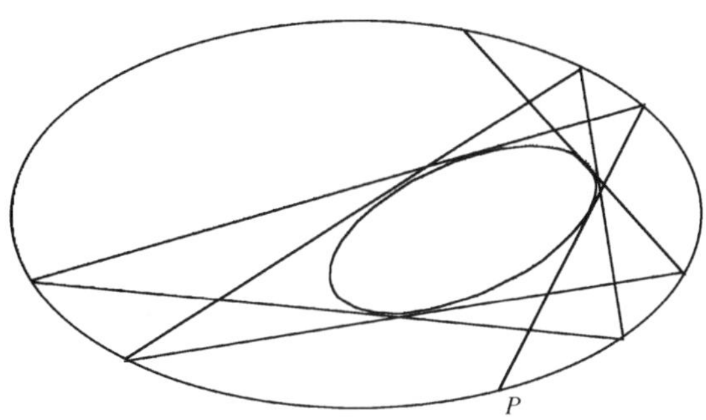

Figure 15.11. Poncelet's porism

[9]Counting these tangents is as tricky as calculating the points of intersection of a line with a curve, which we discussed in Section 8.4.

[10]The details would take us beyond the limits of this book, see (Gray 2011).

Now repeat the construction, starting at the point just obtained. The remarkable theorem is this: Either the lines you are drawing never return to the initial point, or they return for *every* choice of the initial point.

The porism tapped into a line of earlier results about inscribed and circumscribed figures, but it greatly generalised them. It is a thoroughly projective theorem, concerning only lines crossing and touching conics. Poncelet began its proof by describing how any pair of conics can be regarded as equivalent under central projection to a pair of conics, one inside the other. This remarkable claim made full use of his theory of conics under central projection. The result was that people were impressed by the result and were inclined to believe it, but were not inclined to accept his proof. They waited for more convincing proofs to come along, as later happened, and regarded Poncelet's account as just another original idea from a man whose work could not be said to be wholly convincing.[11]

Accordingly, responses to Poncelet's *Traité* were of two kinds. It was welcomed by other former students of Monge at the École Polytechnique, such as Dupin, Brianchon, Olivier, and Lamé, and by other geometers, such as Gergonne and Chasles. They appreciated it for its generality and because it grounded its arguments on objects that you could imagine as physical objects: points, lines, etc.

Others, not in this tradition, were less persuaded. In a review of an earlier version of Poncelet's work, Cauchy was extremely critical of the principle of continuity, remarking that it could hold only within certain limits. He made an analogy with mathematical analysis, pointing out that power series often made sense only for restricted ranges of the variable. We might say that on Poncelet's side was a way of doing mathematics that was intuitively plausible, but whose degree of rigour was obscure, while on Cauchy's side was a style that no one doubted was rigorous, but which had attained that state by eliminating all intuitive props. Perhaps unsurprisingly, those who liked Poncelet's geometry were teachers of engineers at the advanced specialist Écoles, whereas those who opposed it taught mathematics in the more general École Polytechnique but were dominant in the prestigious Académie des Sciences. Poncelet's way of dealing with Cauchy's criticism was to reprint it in his *Traité* and not to act on it at all — a rather ostentatious way of showing that he disagreed with it.

Although in the later 1820s Cauchy's rigorous analysis was regularly censured for its lack of geometry by the external examiner (Gaspard de Prony) at the École Polytechnique, Cauchy's influence was the more powerful. With the exception of Chasles, Monge's pupils taught in the engineering schools and found themselves unable to keep geometry sufficiently alive. It was the teachers at the École Polytechnique, Cauchy among them, who determined the syllabus, not former pupils who were by then elsewhere. When Cauchy took himself into voluntary exile in 1830 with the collapse of the Bourbon regime, the syllabuses did not change. Pure mathematics, as he had defined it, reigned supreme, and geometry was gone. Poncelet's projective geometry had proved to be neither applicable nor rich in its own right; nor was it adequately rigorous, whereas Cauchy's style of calculus was agreed to be more rigorous than earlier ones (as we shall see in Chapter 16).

The successes of descriptive geometry turned out to be confined to techniques of use in engineering; the subject did not become a tool of great conceptual power. Even

[11]For more information about Poncelet and the early years of projective geometry, see the essays in (Bioesmat-Martigon 2010).

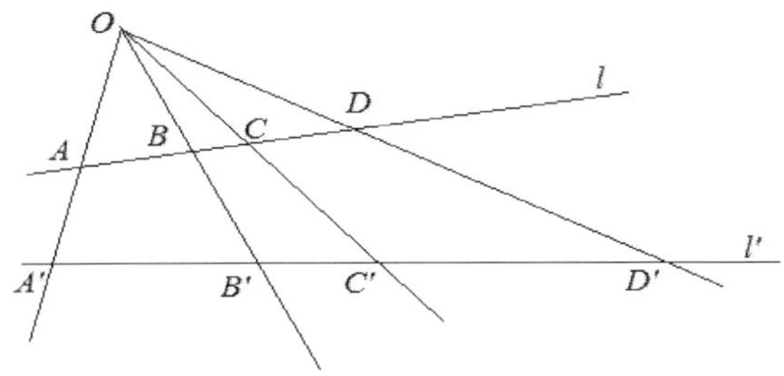

Figure 15.12. Cross-ratio is invariant under projection from O

Monge's own work lacked good examples of what descriptive geometry could achieve in theoretical geometry, whereas pure geometry, such as Poncelet produced, never generated much in the way of applicable mathematics. It could have flourished as a pure subject — the revitalised heir of Euclidean geometry — but even this did not happen. Poncelet virtually abandoned it in the 1830s, as did others, like Gergonne.

Projective geometry's stoutest defender was Michel Chasles, but he did not attempt to retain any of Poncelet's dubious ideas. He dropped the law of continuity, and never talked about ideal or imaginary points. Instead, he based his account of the new geometry on the following property of central projection that Poncelet had noticed but not sufficiently appreciated.[12]

It involves what came to be called the *cross-ratio* of four points on a line. The cross-ratio of the four points A, B, C, and D on a line ℓ is defined to be $\frac{AB}{BC}\Big/\frac{AD}{DC}$. This expresses it as a ratio of ratios, the ratios in which the points B and D are said to divide AC internally and externally. It is easier to work with if it is written in the form $\frac{AB.CD}{AD.CB}$.

The cross-ratio of four points is a projective invariant; the cross-ratio of four points is not altered by a projection. To see this, take four points A, B, C, D on a line ℓ, and project them by an arbitrary central projection from a point O onto four points A', B', C', D' on another line ℓ'. Then the cross-ratios of A, B, C, D and A', B', C', D' are the same:

$$\frac{AB.CD}{AD.CB} = \frac{A'B'.C'D'}{A'D'.C'B'}.$$

It follows that although a central projection can map any three distinct points on a line to any three distinct points on another line, its effect on any other point is then determined. Chasles coupled this insight with the simple fact that a central projection maps collinear points to collinear points and concurrent lines to concurrent lines, and found that he had a simple and transparently rigorous account of the new geometry. Its only defect was that it was much better suited to the study of conic sections than to that of any other type of curve.

For whatever reason, in his *Aperçu Historique* (Historical View) of 1837, the book in which he first set out his new ideas, Chasles gave no more credit to Poncelet than was strictly necessary, and was noticeably warmer in his praise for Monge. Poncelet had already abandoned the subject in favour of the engineering and military work that

[12]It was known much earlier, for example to Pappus in the 3rd century AD.

secured his reputation in his day, and Chasles had eventually to admit that he had lived to see it pass across the Rhine to Germany, where it was written in a language that he could not understand.[13] It is time for us to follow the subject there.

15.2 Projective geometry in Germany

The German story makes an interesting comparison with the French one. As we noticed earlier, Germany was a collection of principalities that were not unified until Bismarck's time in the 1860s, and they lacked a mathematical centre capable of matching Paris until the 1850s. Gauss was German, but although he was interested in differential geometry he was not interested in projective geometry. So the rise of this new geometry in Germany is a story of the piecemeal revival of mathematics in Germany in the hands of just a few mathematicians.

We begin with a fascinating essay of 1872, 'Julius Plücker — in Memoriam', written by a man who was in the first rank of German geometers of the next generation, Rudolf Friedrich Alfred Clebsch.[14] In his essay Clebsch suggested that mathematics had once been at a low ebb in Germany. He mentioned Gauss — but pointed out how isolated he was, and how immersed in astronomy — and considered the crucial changes, which took place in the 1820s, to be the work of Jacobi, Abel, and Dirichlet, and the appearance of three geometers: Möbius, Plücker, and Steiner.

Figure 15.13. August Ferdinand Möbius (1790–1868)

It is clear from this profusion of names that mathematics was reviving in Germany with a considerable degree of success. Two main streams can be distinguished. The dominant one, led by Dirichlet and Jacobi, was mainly interested in analysis and number theory. The other stream, however, is the one that interests us here, for it was more geometrically inclined. It was arguably more parochial, but for a generation it produced most of the important work done in geometry at the time.

[13] See (Chasles 1837, 215).

[14] See the reprint in (Clebsch 1895, 1, 10, 14), and F&G 17.B3.

If we look more closely at this school of geometers, we find that it divides neatly into two. The first and larger faction was represented by August Möbius and Julius Plücker, who pioneered the study of projective geometry by algebraic methods. These, they argued, were essential both in the pursuit of new results and in the rigorous establishment of geometrical theorems. In preferring such algebraic methods to the synthetic methods of Poncelet, they were thereby continuing the earlier work of Euler and Cramer (as described in Section 8.3). The smaller faction, which held the contrary view, can be identified with one man, Jakob Steiner.

At stake in such debates was a range of views about geometry. What is it, and why should it be studied? Is geometry simply an aid to other mathematical investigations, or is it a major subject in its own right? As we have seen in earlier chapters, mathematicians of the 18th century had increasingly become inclined to use algebraic methods to study curves — Euler had even gone so far as to define the conic sections by means of equations. From this point of view, it can be hard to feel that geometrical properties are truly basic to one's study of mathematics. The new geometers, whether algebraic or synthetic in their preferences, agreed that one does geometry, in part, to find out properties of curves. But they differed over what should be counted as a good proof, and over what sort of proof conveys understanding. They differed, too, over what methods would be needed in order to make new discoveries.

We start with the algebraists. Möbius was educated at Leipzig and Göttingen, where he studied theoretical astronomy under Gauss for two terms, and went on to become a Junior Professor of Astronomy at Leipzig in 1816 when he was 25 — he became a Senior Professor only in 1844. He was an original, but not a widely read, mathematician, and on several occasions he rediscovered results that were already in the published literature. His habit of publishing slowly nonetheless meant that he had priority on a number of discoveries that are often credited to others. Like Gauss, he ranged over astronomy, number theory, and geometry. His name is still attached to the Möbius band, a surface with the unexpected property that it has only one side and only one edge (see Figure 15.14).

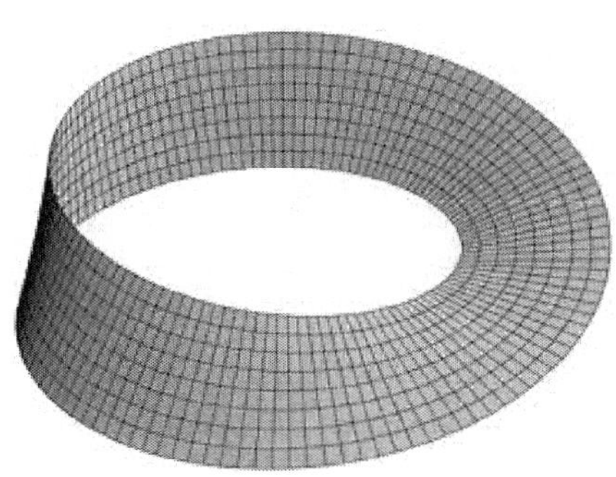

Figure 15.14. A Möbius band

In 1827 Möbius published *Der Barycentrische Calkul* (The Barycentric Calculus). The name of this book derives from the novel idea (summarised in Box 41) that he presented for assigning coordinates to points of the plane.

Using his new coordinate system, Möbius was able to treat simply and algebraically most of the topics in projective geometry that we have met in the course so far. He could describe lines and conics, and find tangents to curves. In particular, lines have equations of the form $ax + by + cz = 0$, giving rise to a duality between the point with coordinates $[a, b, c]$ and the line $ax + by + cz = 0$.

Möbius could also describe algebraically all transformations of figures by shadow projection, without any equivocation about points and lines being sent to infinity, such as had complicated most synthetic accounts. So he too had a unified theory of conics, albeit one that dealt with them algebraically.

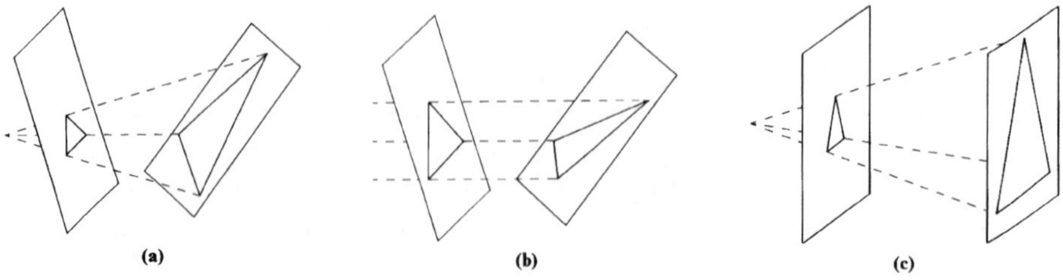

Figure 15.15. Various geometrical transformations

Möbius described his transformations in a less algebraic and more visual way, by supposing that lines were drawn through a fixed point and connected one plane to another; he wrote, 'a specified point [in the image] is produced by drawing a line' — it will help to think of such lines as rays of light. When the planes are arbitrarily inclined (as in Figure 15.15 (a)) we have a picture of projective transformation of figures in one plane onto figures in the other. When the light source is at infinity (that is, the rays are parallel, as in Figure 15.15 (b)) the transformation is called an *affine transformation*. When the two planes are parallel (as in Figure 15.15 (c)) then the picture describes a similarity transformation in which figures have the same shape, but not necessarily the same size. Finally, keeping the planes parallel, but putting the source either at infinity or midway between the planes, describes a Euclidean transformation (a congruence).

After Möbius, Plücker independently began his study of projective geometry. Plücker had an interesting career. He studied mathematics at various universities in Germany, as custom then allowed, and he obtained a teaching position at the University of Bonn in 1825, the year that he turned 24. He then proceeded to publish three books and several articles on geometry. But in 1846 he abandoned mathematics in favour of experimental physics, where he was much influenced by the British scientist Michael Faraday and studied the magnetic properties of gases and crystals. In 1868, the year of his death, Plücker was awarded the Royal Society of London's prestigious Copley Medal for his work. However, his scientific work was not appreciated in Germany, which may explain why he switched back to the study of projective geometry in 1864.

In such a wide-open field as geometry, a basic question is: What is there to prove? We saw in Section 8.3 that Euler, following Newton, had classified cubic curves, and also quartics (curves of degree 4). A few other mathematicians, such as MacLaurin, de Gua, and Cramer, had also considered these curves, but little had been done beyond their classification. So Plücker had to decide what the basic problems were. (This is different from the situation in the theory of conic sections, where the basic geometrical properties of the curves were known, and an improved method of organising the theory was being sought.) The properties that Plücker chose to emphasise were projective

Box 41. Barycentric coordinates.
Möbius's elegant idea was this. Let A, B, C be the three vertices of a triangle, and attach weights a at A, b at B, and c at C. Then the triangular system has a centre of gravity or *barycentre* P, which Möbius labelled $[a, b, c]$. The same point could equally be labelled $[ka, kb, kc]$ for any non-zero k, as he pointed out, because multiplying each weight by the same factor does not move their barycentre — the coordinates are called *homogeneous* for this reason. (The triple $[0, 0, 0]$ cannot occur.)

Departing a little from physical intuition, one can even let the weights be negative (in which case their barycentre lies outside the triangle ABC). Conversely, given a fixed triangle ABC, every point of the plane can be assigned a label (that is unique up to multiplication by any non-zero k) — namely the number triple $[a, b, c]$ (its *barycentric coordinates*), whose barycentre is the point P in question.

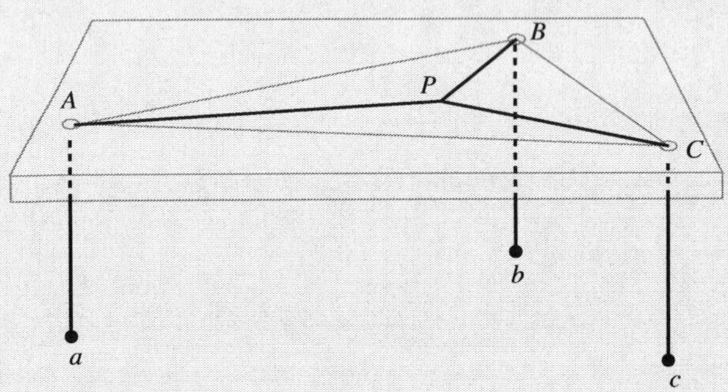

Figure 15.16. The barycentre P of three weighted points

ones, and the result was a series of articles and two important books, published in 1835 and 1839, in which he revived the theory of cubic and quartic curves.

One of his most important results was his theorem that a quartic curve in the plane can have at most 28 bitangents.[15] It proved very suggestive, and many similar results were found about the meeting points of cubics and conics, quartics and cubics, and so forth. They provided a novel kind of theorem that was not original to the 19th century, but which was much more developed in those years than hitherto: theorems for which the raw ingredients are continuous objects (generally curves) but where the results are something finite, a specific number of possibilities.

But all this work was done at a price — that of having to move quite far from geometrical intuition. The new developments were increasingly algebraic, and complex numbers were often found to be essential. The method of barycentric coordinates proved particularly amenable to such developments, and led to the emergence of a battery of increasingly sophisticated algebraic techniques. These methods were known

[15] A bitangent to a curve is a line that is a tangent to the curve at two points.

Box 42. Pestalozzi, Steiner, and mathematics education.
The Swiss teacher and writer Johann Heinrich Pestalozzi was one of the most influential educationalists of the late 18th and early 19th centuries. He believed that love was the vital force in education, and sought to develop a child's whole being through gradual encouragement at a natural pace. In mathematics he fought against the baneful effects of teaching by rote, and advocated learning by experience so as to develop true inner understanding.

He wrote of the children who were taught mathematics on his principles that:[a]

> They were perfectly aware, not only what they were doing, but also of the reason why. They were acquainted with the principle on which the solution depended; they were not merely following a formula by rote; the state of the question changed, they were not puzzled, as those are who only see as far as their mechanical rule goes, and not farther.

From 1814 to 1818 Steiner was a pupil, and then a teacher, at Pestalozzi's school at Yverdon, and was deeply influenced by Pestalozzi's ideas and approach. Consider, for example, Steiner's synthetic conception of mathematics as an organic unity, which he described in 1826 in these words:[b]

> Just as related theorems in a single branch of mathematics grow out of one another in distinct classes, so I believed, do the branches of mathematics itself. I glimpsed the idea of the organic unity of all the objects of mathematics.

It was this perception that guided Steiner's mathematical researches throughout his life.

[a]Quoted in (Lawrence 1970, 195).
[b]See (Burckhardt 1976, 13).

collectively as those of *invariant theory*.[16] Their spread was quite vigorously opposed, notably by Steiner, and we conclude this section by looking briefly at this remarkable man.

Jakob Steiner came from a very poor background and learned to write only when he was 14. When he was 18 he overrode his parents' wishes and went to Pestalozzi's school at Yverdon in Switzerland (see Box 42). There he impressed the mathematicians on his arrival by being able to solve immediately, and in his head, such questions as: 'Divide a regular pentagon into two equal pieces by means of a line parallel to one side'. Those who taught him later, at the University of Berlin, marvelled at how this gruff young man with his nearly unintelligible Swiss accent could perform such complicated manipulations of geometrical figures in his head. This was his greatest gift and, inspired perhaps by the example of Pestalozzi, he lectured to students at the University of Berlin in the dark, the better to encourage their mind's eye! Throughout his life he had a dislike of using algebraic methods in geometry.

[16]The term 'invariant' refers to an algebraic expression, derived from an equation defining a curve, which remains unaltered when the curve is projected into another curve described by a different equation.

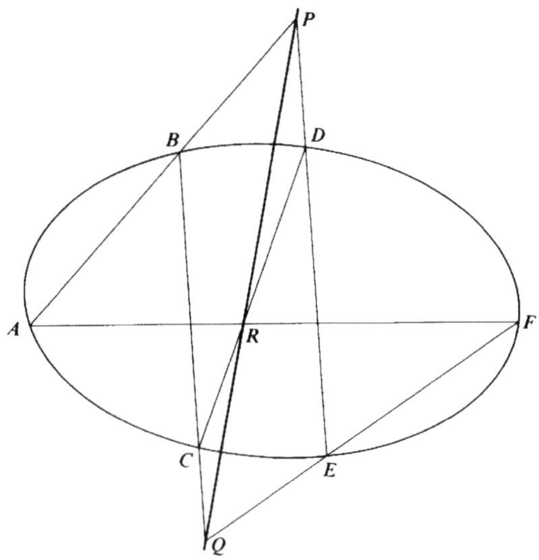

Figure 15.17. The hexagon $ABCDEF$ can be labelled in 120 different ways, which give rise to 60 different Pascal lines (such as PRQ)

We can get an indication of Steiner's mathematical ability from an extraordinary early result of his, announced without proof in his *Systematischer Entwickelung der Abhängigkeit geometrischer Gestalten von einander* (Systematic Development of the Dependence of Geometrical Forms on one Another) of 1832. Savour it purely as an example of his remarkable gift for visualising geometrical configurations — it generalises Pascal's theorem about six points on a conic (see Figure 15.17):

> Six points on a conic may be relabelled in 60 different ways so as to be the vertices of 60 inscribed hexagons; the 60 corresponding Pascal lines meet in threes at 20 points which lie in fours on 15 lines in such a way that through each of the 20 points run three of these 15 lines.

The 60 hexagons arise from the 120 possible labellings of the vertices which give rise to different lines; we count two lines as different if they are either in different places or in the same place but labelled in different ways.

Imagine seeing this in your mind's eye! Even when the diagram is drawn on paper, the theorem is hard to understand. The proof, which he gave in his lectures in Berlin, was published posthumously and makes considerable use of transformations of the basic figure to establish the desired collinearities and concurrences.

Steiner's book, which he dedicated to Wilhelm von Humboldt, was well received. Jacobi commented that:[17]

> Steiner had tried to obtain from a few spatial properties, and by means of a simple scheme, a clear overview of an entire multitude of clever geometrical theorems ...to bring order to what had been chaos ... He has discovered an organisation which ...not only offers a geometric synthesis, but also provides a roll-call of a complete method for all other branches of mathematics.

[17]Quoted in (Kollros 1947, 7).

As a result of his book's success, Steiner was appointed to a professorship at the University of Berlin, where Pestalozzi's methods were then in favour; Humboldt even sent several young teachers to study at Yverdon.

Steiner remained true to his geometrical gifts throughout his life. He published several ingenious proofs (not all of them completely valid) of the old claims that, of all figures with a given area the circle has the smallest perimeter, and of all solids with a given volume the sphere has the smallest surface area. In 1841 he presented a French translation of his work on the subject to the Académie des Sciences in Paris, where a panel of members led by Cauchy gave it a good report.

In 1852 Steiner took up the question of the bitangents to a quartic curve and announced a number of theorems about how they are interrelated (they can be arranged in groups of six in 63 ways such that these sixes are related in various ways!) — but he did not publish proofs of his theorems. Such proofs were speedily given by geometers working in the algebraic tradition. When Clebsch turned to the matter, a decade later, he found it remarkable that Steiner could have discovered them without the full range of the recently discovered algebraic and function-theoretic methods.

Sadly, Steiner's gifts began to desert him in old age, and his increasingly irascible manner alienated his friends. It came to be felt that the 'celebrated sphinx', as Luigi Cremona, the Italian synthetic geometer of the 1860s, had called him, had been mixing his synthetic methods with algebraic ones. There was even a suggestion of outright plagiarism on occasion. Whatever the truth might be, no one was ever to match the strange and brilliant Swiss geometer at his chosen subject. Synthetic geometry gradually slipped from the prominent position that Steiner had enabled it to occupy, until by the 1930s one informed observer could pronounce it 'pretty much worked out'.[18]

How do we summarise the twin schools pioneered by Plücker and Steiner? In the end, the blend of algebra and geometry seemed to be more vigorous than the synthetic geometry of Steiner: it would seem that algebra contributed something by way of power or rigour that was essential to the growth of projective geometry. But it is less important to stress their differences of method, as their common emphasis on the importance of the sensual. In their hands, geometry of whatever hue began to be appreciated internationally as a source of many beautiful results — hence a profusion of plaster-of-Paris models of surfaces, testimony to the problems inherent in teaching the subject and also to the visual wonders that it contained. Geometry enjoyed a new lease of life. As we shall see in the next section, it once again became a candidate for the role of provider of certainty in mathematics — but above all it became a thing of beauty. Even Clebsch, one of the more algebraic of German geometers, was moved to observe: 'it is the joy of form in the highest sense that makes the geometer'.[19]

15.3 The establishment of projective geometry

Poncelet's strange non-metrical geometry had been tamed by Chasles, who pruned it of its visionary elements. Chasles in France, and Steiner in Germany, then developed a theory of geometry that made fundamental use of central projections and the invariance of cross-ratio. This was a curious hybrid of novel but intuitively attractive ideas

[18]See (Coolidge 1934, 217–228, esp. 227).

[19]See (Clebsch 1872).

on the one hand and largely elementary results, known in many cases since the time of Apollonius, on the other. The subject thus created was ideal material for a university mathematics course, but it was not clear what contributions it could make to research. At a time when research was emerging as the way to build a career in the major universities of France and Germany, this was an issue that any ambitious young mathematician had to confront.

Where could research in projective geometry be done? One area was in the geometry of plane curves of higher degree (cubics, quartics, and beyond), and much was done, especially in Germany, as we have seen. Another area was the geometry of surfaces. Here a single discovery did more than anything else to bring people into the field.

In 1847 the 26-year-old English mathematician Arthur Cayley discovered that a general cubic surface necessarily has straight lines in it. He communicated this unexpected fact to his friend, the Irish mathematician and theologian George Salmon, who found that in general there are precisely 27 straight lines on a cubic surface (see Figure 15.18).

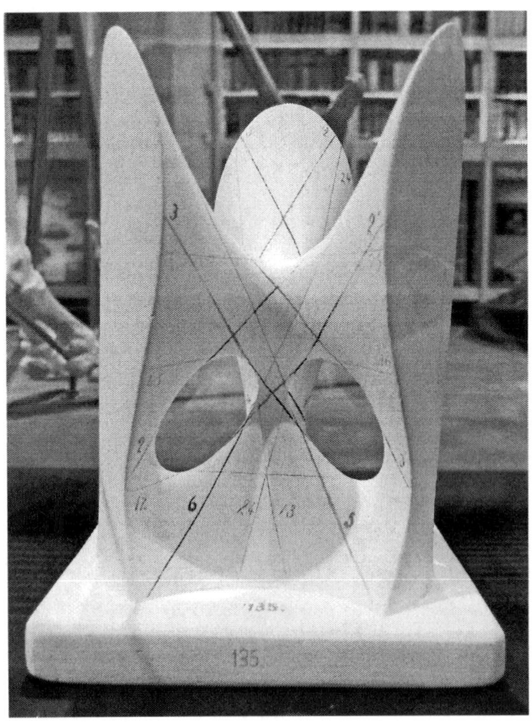

Figure 15.18. The 27 straight lines on a cubic surface

This made the discovery even more attractive, because 27 is a manageable number: one could hope to discover which lines meet which others, and ultimately to give an account of the configuration that they form. Cubic surfaces were made the object of a prize competition for a new Steiner Prize, which was administered in Berlin on money bequeathed by Steiner, who had died in 1863. Other mathematicians joined in, and eventually the prize was awarded in 1866 jointly to Luigi Cremona and the German

Richard Sturm. In 1870 the 27 lines were studied by the group-theorist Camille Jordan, who described the group of symmetries of the lines.[20]

Once again the successful treatment of a new geometrical problem had been found by treating it algebraically. The synthetic approach of Steiner was not obliterated by the successes of the algebraic method. It retained its fundamental charm, but despite Steiner's best efforts it could not keep up. The future of the geometry of central projections looked increasingly hybrid: synthetic foundations and applications to the theory of conic sections, upon which the professional mathematician could base more advanced algebraic investigations.

By 1860 it was clear that the subject had a rosy future. With the success of these investigations came a deepening acceptance of the new subject, and in 1873 it acquired an agreed name: *projective geometry*. This was bestowed upon it by Cremona, who called his book *Elementi di Geometria Projettiva* (*Elements of Projective Geometry*).[21] The book was a success: it was quickly translated into English, for example, and it inspired the work of Salmon, who saw his task as that of writing the books that made the research of his friend Cayley intelligible and attractive to students. His books on the subject, notably his *Higher Plane Curves* of 1852, were translated into German by Wilhelm Fiedler. The result, known as Salmon/Fiedler because of the quality of Fiedler's additional notes, became the standard German introduction to the subject. These books, and others like them, mark the emergence of the new domain in geometry.

What gave projective geometry an added attraction was its fundamental nature: it involves just points, lines, and planes that satisfy simple rules. Naively, the subject of geometry is about figures, and one studies lengths and angles. The subject might be written up in the manner of Euclid, or in the manner of Descartes as coordinate geometry, but its essentially metrical character was always apparent. The new projective geometry dispensed with these metrical features, and spoke only of incidence, cross-ratio, and of curves by type (line, conic, cubic curve, etc.). It considered collinear points and concurrent lines. Any true statement in projective geometry is automatically true in Euclidean geometry, but many Euclidean theorems (such as the isosceles triangle theorems) cannot even be stated in projective geometry. This meant that projective geometry was in some sense more fundamental than Euclidean geometry, and this gave it a prestigious position in mathematics. Just as Cauchy's analysis became the right foundation for the calculus, and discoveries such as Galois's were uncovering the right way to think about polynomial equations, so too were geometers uncovering a new and more fundamental way to think about geometry — and geometry, after all, was the study of space.

15.4 The re-unification of geometry

By the end of 19th century there had been a great deal of new geometrical research. There had been considerable growth in projective geometry, and this was beginning to be matched by the eruption of non-Euclidean geometry, resulting in the work of Riemann and Beltrami published in 1867 and 1868. One who feared that this very

[20] We discuss the origins of group theory in Chapter 19.

[21] Felix Klein also spoke of projective geometry in his Erlangen Programme of 1872, which we discuss shortly.

growth imperilled the true understanding of geometry was the German geometer Felix Klein.

Klein set himself the task of re-unifying geometry, which he saw as breaking up into separate disciplines. He knew that this was ambitious: he regarded it as his first major mathematical goal, and he made sure that he was well placed to reach it. As a student at Bonn he had learned projective geometry from Plücker, and (when only 19) he had helped to produce the second volume of Plücker's last book on the subject, which came out after Plücker's death. Then he had set about learning from Clebsch in Göttingen the more algebraic side of projective geometry. In 1870 he went to Paris to learn the new theory of groups from its master, Camille Jordan, although his studies there were interrupted by the Franco–Prussian War.

On his way to Paris, Klein made the then-obligatory trip for any ambitious German mathematician to Berlin, where his views on geometry led him into several arguments. For by now he was hardening his view that Euclidean and non-Euclidean geometries are somehow special cases of projective geometry. This clashed with the view of non-Euclidean geometry held by Karl Weierstrass, the senior professor at Berlin, who argued that these geometries are fundamentally distinct, because Euclidean and non-Euclidean geometries are essentially metrical (having a concept of distance), whereas projective geometry has no such concept. It took Klein a year to sort the matter out to his satisfaction, and he was able to present his findings when, at the remarkably early age of 23, he became a professor at Erlangen (whence the usual German name of his account, the *Erlangen Programme*). At his inauguration he circulated a written version of these ideas — the audience for his inaugural address were perhaps relieved to hear the young man discourse instead on the tasks of mathematical education.

What took Klein a year to resolve was a problem that projective geometers had already encountered lurking in the very definition of cross-ratio. If you define the cross-ratio of the four points A, B, C, D to be $\frac{AB.CD}{AD.CB}$, then what are these individual items AB, CD, AD, and CB? They cannot be lengths, because there is no concept of length between two points in projective geometry. You could start from Euclidean geometry and base projective geometry on it, but there was a widespread feeling that projective geometry was more basic than Euclidean geometry — because, as we noted earlier, any projective property (say, that of three lines meeting in a point) is also Euclidean, but the converse is not true. In any case, because Klein considered projective geometry to be more basic than either Euclidean or non-Euclidean geometry, he certainly could not take that tack. So what was he to do?

He found part of his answer in a paper written by Cayley in 1859, with the honest if unappealing title 'A sixth memoir on Quantics'.[22] In this paper Cayley had shown how to deduce Euclidean geometry from projective geometry, by showing how one can systematically discard two points from a cross-ratio of four points and be left with something that defines the Euclidean distance between the remaining two. Since the details of this need not concern us, we note only that Klein had in Cayley's work a precedent that helped to confirm him in his belief that projective geometry is truly the basic geometry. He found, moreover, that a simple modification of Cayley's argument could enable him to define the non-Euclidean distance between two points in terms of

[22]A *quantic* was Cayley's name for a homogeneous polynomial.

Figure 15.19. Felix Klein (1849–1925)

the cross-ratio of four—see Box 43 for details.[23] Klein called this concept of distance in non-Euclidean geometry a projective metric, to indicate both that it was a concept of distance (a metric) and that it derived from projective considerations.

This potentially still left Klein with a vicious circle in the definition of cross-ratio. His way out, based partly on refinements of some old ideas of Möbius, is worth examining because it shows how profoundly Klein thought through his ideas about geometry.

A fundamental fact about Euclidean geometry is that any point can be mapped by a congruence to any other point, but there are no congruences that map a line segment onto a subset of itself, or onto a larger segment that strictly contains it. If a segment AB is moved around in the plane (or space) by the transformations allowed in Euclidean geometry, and if it ends up lying with A in its original position and with the segment pointing in its original direction, then the point B has also returned to its starting position. As a result, we can speak about length in Euclidean geometry. We pick a segment as a unit of length, and say that a segment has length m if it is congruent to m copies of the unit segment laid end to end in a straight line. This is possible because the unit segment is never congruent to either a subset of, or a superset of, itself.

Klein argued that much the same was true for configurations of four collinear points in projective geometry. It is a fundamental fact of projective geometry that any set of three points on a line can be mapped onto any other set of three points on a line by a sequence of projective transformations, but if a sequence of projective transformations maps three collinear points to themselves then it also maps any fourth point

[23]Throughout his life Cayley was hostile to non-Euclidean geometry; he was almost unique among major mathematicians in this respect, and he may well have missed the opportunity to extend his discovery to include non-Euclidean geometry because his heart was not in it.

to itself. He had learned this idea from the work of his predecessor at Erlangen, Karl Georg Christian von Staudt, who had written a substantial — but forbidding and little read — presentation of projective geometry based on this idea, his *Geometrie der Lage* (Geometry of Position) of 1847. It built systematically on earlier arguments by Möbius, who had shown how a particular set of four points on a line can be taken for reference and provide the projective analogue of a ruler. What the reference set of four points then measures is the cross-ratio of any set of four collinear points, but it had taken von Staudt much effort, and many pages of his book, to establish how this works in detail.

It is instructive to see how Klein expressed these ideas in more mathematical language, because in that way we can see how he unified geometry. He argued that in any geometry one studies points that make up figures, and hence that the geometry is the study of a space of points. He then argued that you can always move figures around in any geometry, but that different geometries have different allowable motions.

For example, in Euclidean geometry the allowable motions are the congruences (translations, rotations, and reflections) and they preserve the lengths of line segments. In projective geometry, the allowable transformations are the projective transformations, and they preserve the cross-ratio of any four collinear points.

Klein noted that in each case these motions form a group. By this he meant only that if one motion is followed by another then the result is a third motion of the same kind. The result of following one Euclidean transformation by a second is another Euclidean transformation. The result of following one projective transformation by a second is another projective transformation.[24] Each geometry has its own group of allowable motions — those that preserve the basic properties of that geometry. So in Klein's view, every geometry involves both a space and a group, and different geometries can be compared by studying relationships between their spaces and their groups.

When Klein did this, he found he could reach the goal he had set himself of unifying geometry. Projective geometry had the largest space and the largest group. Non-Euclidean geometry could be described on a subset of the points in the space of projective geometry (those inside a conic, as it turned out) and had a smaller group (those projective transformations that map the interior of the conic to itself). Euclidean geometry became a special case of projective geometry in somewhat the same way, as Cayley had indicated. If anything, the situation became clearer when non-Euclidean geometry was at hand than when Euclidean geometry had been the sole issue.

So Klein had achieved his goal: all known geometries were shown to be 'subgeometries' of projective geometry.[25]

What impact did Klein's ideas have on the mathematical community? To some extent, this was affected by the way that he chose to present them. In 1871 he presented a paper to the Göttingen Scientific Society at the invitation of Clebsch.[26] Klein began:

Klein on geometry.

> The present discussion relates to the so-called non-Euclidean geometry of Gauss, Lobachevskii, Bolyai and the related considerations which

[24] If you look ahead to Chapter 19, Box 60, you will see that this falls short of the definition of a group that later became standard, as Klein himself later acknowledged.

[25] In fact, Klein initially missed one geometry, Möbius's affine geometry, which has the ratio of lengths as fundamental, but it also fits nicely into his programme, as he was able to show when he became aware of it.

[26] See (Klein 1871), and F&G 16.C4.

Box 43. Distance derived from cross-ratio.

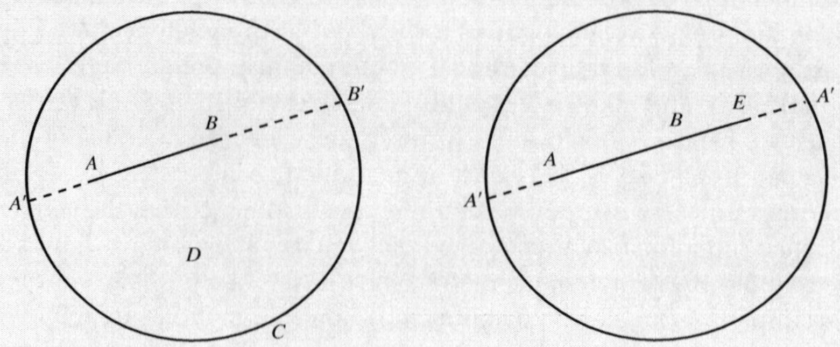

Figure 15.20. Distance derived from cross-ratio

Klein took a conic and considered the points inside it as the points of non-Euclidean space, rather as Beltrami had done. He wanted to be able to say of any two points A and B inside the conic that they are a certain distance $d(A,B)$ apart. This means that no allowable motion can map the segment AB to a subset of itself. He had the concept of cross-ratio available, but it requires four points.

The solution that occurred to Klein was to join A and B by a line meeting the conic in two more points, C and D, and to consider the cross-ratio of A, B, C, D, which we write as: $\mathrm{cr}(A,B;C,D)$. It might seem sensible to define this cross-ratio to be the distance $d(A,B)$, but he wanted the distance to satisfy the usual additive rule that if A, B, and E lie on a line then

$$d(A,B) + d(B,E) = d(A,E).$$

But cross-ratio obeys a product rule:

$$\mathrm{cr}(A,B;C,D) \times \mathrm{cr}(B,E;C,D) = \mathrm{cr}(A,E;C,D),$$

so Klein took logarithms, and defined distance by

$$d(A,B) = \log(\mathrm{cr}(A,B;C,D)).$$

This distance function now satisfies the additive rule.

Riemann and Helmholtz have put forward concerning the foundations of our geometric ideas. Nothing in it will pursue the philosophical speculations which have been attached to the works mentioned; the purpose is much more to put the mathematical results of those works, insofar as they relate to the theory of parallels, in a new, intuitive way and to make them accessible to a clear and general understanding.

> The route to this leads through projective geometry, whose independence from the question of the theory of parallels will be explained. One can now, after the start made by Cayley, construct a general projective metric which belongs on an arbitrarily chosen surface of the second degree taken as the so-called fundamental surface. This projective metric yields, according to the way in which the surface of the second degree is used, a picture for the different parallel theories of the above works. But it is not only a picture of them, it also shows their inner meaning.

If we put on one side material that seems familiar and on the other side material that is incomprehensible, just as we do when listening to a talk, the impression that we form of Klein's approach and ideas is that non-Euclidean geometry can be made to reveal its inner mathematical (if not its philosophical) meaning by being related to projective geometry — although what a projective metric is may not yet be clear.

Klein then gave a quick review of the history of the subject. Of Gauss's contribution, he remarked:

> A similar reflection seems to have been the starting point for Gauss's researches into this question. Gauss took the view that it would in fact be impossible to prove the theorem that the angle sum would be two right angles, and rather that, as a consequence, one could construct a geometry in which the angle sum was less. Gauss called this geometry non-Euclidean; he occupied himself a lot with it, but sadly, apart from a few remarks, published nothing about it.

One might wonder whether Gauss's importance was not a little exaggerated, but Klein quickly managed to mention several other names, including those of Bolyai and Lobachevskii, who would not have been known to everyone in the audience at the time. Indeed, as Klein remarked, these names were unknown until Gauss's correspondence with Schumacher was published in 1862, since when 'the view has spread that the theory of parallels is completely sorted out, i.e. that it is known to be truly undecidable'.

Klein then briefly described the ideas of Riemann and the distinguished German scientist Hermann von Helmholtz about geometry on a sphere, before concluding his address.

As he surely intended, Klein, who was barely 22, came across as widely read, energetic, and ambitious. Moreover, he followed this presentation of his views to a learned society with a printed account in the Society's journal.

The likely impact of Klein's next publication on this theme, the Erlangen Programme, is much harder to determine. As we noted, it was distributed as a pamphlet on the occasion of Klein's inaugural lecture as a professor in 1872, and much depends upon whether the pamphlet was widely distributed by post or, at the other extreme, just left on a few chairs in the lecture hall. It depends, too, on whether Klein also chose to re-cycle its contents in other publications, and to promulgate it in lectures and seminars. We should not be too optimistic about the impact of the pamphlet — the history of mathematics has many examples of good ideas falling dead from the press — and indeed it seems that it made little impact at first, being hard to come by. As noted, however, Klein reworked his Göttingen address as an article in 1871 in the newly founded *Mathematische Annalen*, and followed it with one on n-dimensional non-Euclidean geometry from a projective point of view in 1873, before moving on to other matters to do with the connection between groups and geometry. It was probably these two articles that carried his message most publicly to the world, and the rest had to be done by word of mouth.

A revealing and poignant illustration of the way that Klein operated in those years is afforded by the way that he responded to the emergence of the French mathematician Henri Poincaré, only five years his junior. In 1880 Poincaré had just finished his doctorate and was working at the École des Mines in Caen, when the French Académie des Sciences announced a prize competition on the theory of differential equations. This was a topic dear to Poincaré's heart (as perhaps the judges knew), but it was also one of widespread interest. In May of that year Poincaré sent in an entry in which, for reasons that we need not discuss, he considered nets of triangles in which each triangle is obtained by a sequence of transformations from an initial one. Over the summer he sent in three supplementary parts that dramatically improved on his original essay. The way in which he came to some of the insights in these supplements is surprising. He tells us that:[27]

Poincaré on discovery.

> I left Caen, where I then lived, to take part in a geological expedition organized by the École des Mines. The circumstances of the journey made me forget my mathematical work; arriving at Coutances we boarded an omnibus for I don't know what journey. At the moment when I put my foot on the step the idea came to me, without my being prepared for it, that the transformations I had made use of [in my work to date] were identical with those of non-Euclidean geometry. I did not verify this, having no time for it, since scarcely had I sat down in the bus than I resumed a conversation already begun, but I was entirely certain at once. On returning to Caen I verified the result at leisure to salve my conscience.

In this unexpected way Poincaré reached a crucial insight that inspired him to write not only the three supplements, but also a steady stream of papers emphasising the important role of non-Euclidean geometry in his work. He also expressed a challenging view on the nature of geometry, as the next quotation (taken from the first supplement) makes clear:[28]

Poincaré on groups and geometry.

> There are close connections with the above considerations and the non-Euclidean geometry of Lobachevskii. In fact, what is a geometry? It is the study of the group of operations formed by the displacements to which one can subject a body without deforming it. In Euclidean geometry the group reduces to the rotations and translations. In the pseudogeometry of Lobachevskii it is more complicated.

Poincaré was studying the consequence of rotating a particular figure through rotations he called M and N, and he went on:

---

[27]See Poincaré, *Science et Méthode* (Science and Method, 1908), 51–52.

[28]Quoted in (Poincaré 1997, 11).

> Indeed, the group of operations formed by means of M and N is isomorphic to a group contained in the pseudometric group. To study the group of operations formed by means of M and N is therefore *to do the geometry of Lobachevskii.* Pseudogeometry will consequently provide us with a convenient language for expressing what we have to say about this group. [Emphasis in the original].

This makes it look as though Poincaré was familiar with Klein's work. But was he? One can give arguments for and against this view. There might, for example, have been another source, such as Beltrami's ideas on geometry.[29] Poincaré's assertion that a geometry is a group of motions of figures in a space seems very Kleinian, but we have seen reason to doubt that the Erlangen Programme was widely distributed. The *Mathematische Annalen* was easier to come by, but Poincaré was not a widely read mathematician, and so he may well not have known of it. It is clear that he had learned of non-Euclidean geometry from somebody, which is where the Beltrami piece may come in. Beltrami spoke of geometry as being about the 'superimposability of equal figures', which is the same sort of idea as Poincaré's displacements that do not deform a figure (although Beltrami did not mention the concept of a group). So it is possible that Poincaré drew his inspiration from Italian work in geometry, rather than from Klein. Another source may have been the essays on non-Euclidean geometry by Helmholtz, who was responding to some of Beltrami's ideas.

In fact, it is rather hard to decide what Poincaré knew of Klein's work at this stage, but a close reading of Poincaré's paper suggests that he did not know it at all, and was relying entirely on Beltrami's work together with his own interest in groups of transformations — a topic that he had probably learned from Camille Jordan in Paris. Be that as it may, we see that by 1880 Poincaré's views were strikingly similar to those of Klein on the subject of groups and geometry, except that Poincaré never displayed any interest in the notion that all geometries are sub-geometries of projective geometry. Ignorance of the literature was no obstacle to Poincaré, and once the Académie competition was over — he came second, behind the author of a much more polished piece of work, now largely forgotten — he began to publish his research and so came to Klein's attention.

The contrast is a fascinating one. Klein prided himself on his breadth of reading, and was actively pursuing a career as the bright young man of German mathematics. He had probably come across no-one of his age who was his equal, and may even have initially underestimated Poincaré. And while Klein's own work was sometimes criticised in Berlin for its lack of rigour, it was certainly as rigorous as anyone else's in the domain of geometry. The work of Poincaré, on the other hand, was full of unproved assertions, and even remarks that were plainly wrong, mixed with wonderful insights and enough of an argument to make it clear that something important was going on.

Moreover, Klein's own work was currently at an impasse. He saw at once that Poincaré was doing what he was trying to do, and wrote to him accordingly. A correspondence began in which the scholarly and well-educated Klein, somewhat to his surprise, had to fill some remarkable gaps in Poincaré's education, while the younger man sent back a daunting mixture of naive questions and fresh observations. In due course, Klein asked Poincaré for an article on his work for the *Annalen*, explaining that

[29] See (Beltrami 1868), and F&G 16.C3.

he would add a preface to it showing how it related to his own work. (One wonders who he thought would gain more from this comparison.) Poincaré obliged, and the article, thus enriched, appeared a year later, in 1882.

The topic that Poincaré had discovered, and which, through their correspondence, Klein also sought to master, broke into two halves — one just within reach, but the other still a source of great difficulty to this day. The one that lay within reach was non-Euclidean geometry in two dimensions, the one that was beyond their powers was non-Euclidean geometry in three dimensions. Klein threw himself into the work, but although he completed a fine paper upon it, his health broke down completely and he was never able to work again at the same high level. It took him a year to recover, by which time Poincaré was securely in possession of the high ground, and thereafter Klein's work was never so original.

Throughout this time Klein remained devoted to geometry, which he viewed in a most general way. Recalling in 1923 his hopes on taking up a professorship in Leipzig in 1880, he wrote:[30]

> I did not conceive of the word 'geometry' one-sidedly as the subject of objects in space, but rather as a way of thinking that can be applied with profit in all domains of mathematics.

As Klein's influence grew, so did the respect in which his work was held, and in 1892 he reprinted his Erlangen Programme in the *Annalen*. It had already been translated into Italian in 1890, and appeared in English in 1892. In the 20th century generous claims were made for it: Klein's successor at Göttingen, Richard Courant, called it 'perhaps the most influential and widely read memoir of the last sixty years'.[31] This seems to be an over-statement. The idea of geometry as the subject of groups and spaces was indeed a growing one, but that might well have had more to do with Poincaré and others who came after him than with the Programme itself. Moreover, the truly Kleinian idea of a hierarchy of geometries, with projective geometry as the fundamental geometry and the other geometries as special cases, seems not to have caught on. What is true is that the Erlangen Programme, being both early and programmatic, was a convenient peg on which to hang a philosophy and was as good a way as any of marking the debut of one of the most influential mathematicians of the late 19th century.

15.5 The axiomatisation of geometry

In the 17th century a tension had been introduced into mathematics with the work of Descartes; it is now time to see how it worked out. The history of geometry can be viewed as a struggle between the algebraic and synthetic styles. One might argue, with Newton, that synthetic geometry is the true geometry because it reasons directly about lines, angles, tangents, and other obviously geometrical concepts; on this view algebra is at best a crutch. But then, as even Newton found, algebraic methods make results possible which synthetic methods find hard to match. On this alternative view algebraic methods more than compensate, by their success, for introducing into geometry techniques of arithmetic that are alien to the study of shape. Yet, algebra can blind one

[30] See (Klein 1923, 20).
[31] See (Courant 1925, 200).

to geometrical questions — this was to some extent the 18th-century experience. Duality in projective geometry is surely more geometrical than manipulating equations, however much the latter subject began through considering the intersections of curves.

The classic statement of the synthetic point of view was Euclid's *Elements*. How did it stand in the 1880s and 1890s? The principle of duality (which can be treated synthetically, although we have stressed the algebraic approach) was an eminent addition. On the other hand, the existence of non-Euclidean geometry was a grievous blow, because now Euclidean geometry could no longer in any simple way be true — that is, a correct description of the physical world. Its truth would have to be maintained empirically, not on a priori grounds. The belief that Euclidean geometry was true had rested on the inevitability of its derivation from simple postulates that were apparently true. Now one would have to be more careful, and geometers would have to recognise the existence of at least five logically valid geometries (projective, affine, Euclidean, non-Euclidean, and spherical).

How could Euclidean geometry have deceived people for so long? As we shall see, it came to be felt that the 'mistake' had been to rely on the implicit meanings of terms like 'line' and 'angle', which had concealed the novel geometries from overly Euclidean eyes. People began to think more abstractly about axiomatic reasoning, making a small, diplomatic retreat from synthetic geometry — or so they may have thought.

It will be helpful if we first make clear what is meant by an 'abstract axiomatic system'. As Aristotle had pointed out, the most basic terms in any discussion cannot be defined.[32] We must be initiated into their meanings in some other way, perhaps by being shown them in action (like a game with its rules, we do not ask about what the rules mean, we just accept them and play the game). Instead, the undefined terms in an axiomatic system are provided with rules that govern their use, but not with an account of what they might mean. For example, the most successful modern axiomatisation of geometry, David Hilbert's, begins:[33]

> Consider three distinct sets of objects. Let the objects of the first set be called points and denoted $A, B, C \ldots$; let the objects of the second set be called lines and be denoted by $a, b, c \ldots$ [and the third be called planes] ... The points, lines and planes are considered to have certain relations ... . The precise and mathematically complete description of these relations follows from the axioms of geometry. [For example]
>
> I, 1 For every two points A, B, there exists a line a that contains each of the points A, B.
>
> I, 2 For every two points A, B, there exists no more than one line that contains each of the points A, B.

Hilbert did not say what these 'points' and 'lines' are, but merely stated that any two points specify a unique line. In reasoning axiomatically about points and lines, he did not appeal to the 'meanings' of the terms, but argued that we must go strictly by the book of rules. It was said that when Hilbert became interested in the axiomatic method, he used to joke that beer mugs would do just as well as points — provided that they satisfy the same axioms!

One of the first to move towards rethinking geometry on axiomatic lines was another German, Moritz Pasch. Encouraged by Felix Klein he published his *Vorlesungen*

[32]See Aristotle, *Posterior Analytics*, $76^a31 - 77^a4$, extract in F&G 2.H1.

[33]See Hilbert, *Grundlagen der Geometrie* (*Foundations of Geometry*), 1899, p. 3.

Box 44. Pasch on a gap in Euclid's reasoning.

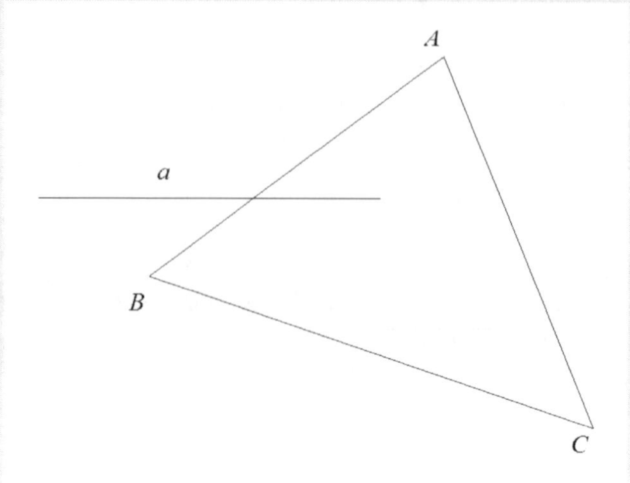

Figure 15.21. Pasch's axiom

Pasch introduced into Euclidean geometry axioms for the concept *A lies between B and C*, which Euclid had used without defining it. He found it necessary to state explicitly that if A, B, and C are three non-collinear points, and if a is a line that passes between A and B (that is, goes through a point on the straight-line segment joining A and B) and does not meet A, B, or C, then a passes between either A and C, or B and C.

über neuere Geometrie (Lectures on Modern Geometry) in 1882, when he was 39. The book is an interesting mixture of the conservative and the original. He wrote:[34]

> By introducing numerical concepts, relations between geometrical concepts become manifest so that they can be recognised by observation. The standpoint is thus indicated which is assumed in the following, according to which geometry is part of natural science.

In Pasch's opinion, Euclid did not explain the basic concepts of the *Elements*, nor need he have done so. Indeed, the basic concepts of point and line had, in Pasch's view, to be leached of meaning, since the proofs of two dual theorems are essentially the same, and their validity cannot therefore depend on the meanings of the terms involved. But this meant that the ways in which the proofs were found became all the more important, and Pasch was therefore led to examine Euclid's rules of inference. He found that he needed to supplement them with some more — such as the example in Box 44.

[34]Quoted in (Nagel 1939, 194).

In his book, Pasch wrote:[35]

> Indeed, if geometry is to be really deductive, the deduction must everywhere be independent of the meaning of geometrical concepts, just as it must be independent of the diagrams; only the relations specified in the propositions and definitions employed may legitimately be taken into account. During the deduction it is useful and legitimate, but in no way necessary, to think of the meanings of the terms; in fact, if it is necessary to do so, the inadequacy of the proof is made manifest.

It is not easy to see how such views fit with his opinion that geometry 'is part of natural science'. For, on the one hand, if geometrical relations are amenable to observation, and as such are part of science, then they are necessarily meaningful — how else would we know what to observe? On the other hand, the logical validity of a mathematical argument is here presented as being the more certain the more meaningless the terms in it are. It is possible to assert both propositions at once, because logical deduction plays a role in scientific thought, but it is a complicated position to maintain.

Pasch's views on the nature of geometrical reasoning were well received, notably in Italy, where there was quite a debate about the nature of geometry. In 1889, for example, Pasch's book was reworked by Giuseppe Peano as an illustration of Peano's notation for mathematical logic.[36] Peano, who had just turned 30, was a truly original thinker, and it is impossible to do him justice in the space available to us. He was a first-rate mathematician, a logician, and a propagandist for his own artificial international language. He was an enthusiast for the axiomatic approach to all areas of mathematics, and by presenting Pasch's ideas in his severely formal way he did much to emphasise their logical side. Indeed, Pasch's view that geometry is a natural science is invisible in Peano's treatment.

But Peano's views were too severe for many Italian geometers, whose work often relied on a vigorous intuitive sense of the meaning of geometrical terms. For example, in 1894 the 40-year-old Giuseppe Veronese argued that one should not reduce mathematics to a system of conventions for the manipulation of signs. For Veronese, geometry was the most exact experimental science, and he deliberately introduced terms with a strong intuitive sense. He did this even in the work with which he made his name in 1880 (a profound extension of the ideas of projective geometry to spaces of n dimensions). This accorded with Klein's views, in whose journal, the *Mathematische Annalen*, Veronese published his work, for in Klein's opinion 'what a geometer thinks, he sees'.

Veronese's views were also the views of the Italian mathematician Federigo Enriques, as presented in his long contribution 'On the principles of geometry' to Klein's *Encyclopedia*.[37] In the period from 1890 to 1914, Enriques was a leading algebraic geometer, who later devoted himself to the history and philosophy of mathematics, and his arguments were both considered and influential. Geometrical terms, he argued, are defined implicitly by the true things that can be said about them. Truth is a matter of proof and logic, but the terms themselves, and the statements involving them, may well be analogous to the familiar terms of elementary geometry.

---

[35] Quoted in (Nagel 1939, 197).

[36] See (Peano 1889). We discuss Peano's work on foundational issues in Chapter 17.

[37] The Encyklopädie der mathematischen Wissenschaften (Encyclopedia of Mathematical Sciences) surveyed mathematics and its applications, including physics and statistics, in 23 large volumes, and Klein recruited a roster of mathematicians from across Europe to write it. A French update, in which many essays were rewritten and new topics introduced, continued for a while after the First World War.

It is difficult to compare the views of Pasch, Peano, and Enriques with those of Euclid, for although we have seen that Pasch identified some quite specific logical gaps in the *Elements* — and although the *Elements* is more overtly logical and formal than many subsequent texts — we simply do not know Euclid's philosophical views. It is likely, though, that Euclid's basic terms were intended to be more meaningful than Pasch, and certainly Peano, would have allowed. Before non-Euclidean geometry was discovered, the physical meaning of the term 'straight line' was surely unproblematic.

However, the decisive figure in all the debates about the foundations of geometry was that of David Hilbert, and it is his work on the foundations of mathematics that brought world-wide recognition to the man already regarded by his German colleagues as the leading mathematician of his generation.

Hilbert first came to fame in the late 1880s by resolving the hardest outstanding question in the theory of invariants (by now a highly algebraic topic), with such success that for a time further work in the field seemed superfluous. He then turned his attention to the algebraic theory of numbers, and in 1896 contributed a 371-page summary that reformulated almost the entire subject to the *Journal* of the newly founded Deutsche Mathematiker-Vereinigung (German Mathematical Society). This work, which Hermann Weyl called 'a jewel of the mathematical literature', typified Hilbert's way of working.[38] Hilbert would submerge himself in a new topic until he had mastered it completely. From the first, Klein had been eager to bring such a gifted man to Göttingen, but jobs could not be created at will, and it was not until 1894 that a post was found for him. Klein stimulated Hilbert's interest in elementary geometry (an interest that had lain hidden under his remarkable algebraic work), and Hilbert presented his novel ideas, first in lectures and then in his book of 1899, *Grundlagen der Geometrie* (*Foundations of Geometry*).

Initially, Hilbert had agreed with Pasch that geometry was about the facts of experience, but he came to feel that the task was to codify each geometry in a system of axioms. Although this was a step in Peano's direction, it seems that Hilbert was still unaware of the Italian work. In his lectures he took seriously the problem of axiomatising non-Euclidean geometry as well as Euclidean geometry. He also investigated what axiom systems give rise to projective geometry, and what geometrical consequences follow from other axiom systems. But in his *Grundlagen der Geometrie* he contented himself with axiomatising only elementary geometry. In 1903 he revised his book, which was already a great success, to include a discussion of non-Euclidean geometry. Here he took his axiomatisation of Euclidean geometry, removed his version of the parallel postulate from the list of axioms, and replaced it by an axiom equivalent to the Bolyai–Lobachevskii definition of parallels. He then showed that the resulting list of axioms describes non-Euclidean geometry.

So we can say that Hilbert and Pasch agreed that deduction in geometry should be independent of meaning, and that Hilbert inclined more to Peano's formalism than to the residually intuitive character of geometry as presented by Veronese and Enriques. Indeed, Hilbert made explicit what they said was done implicitly.

Hilbert's eminence, and his natural affability, which made him a relatively easy mathematician to talk to, soon made his approach to the foundations of geometry into a thriving branch of mathematics. The backing of the influential but much more intimidating figure of Klein also helped.

---

[38]Quoted in (Reid 1970, 254).

Hilbert's work raised, but left open, two different kinds of problem:

1. For any given geometry, find the simplest set of axioms that describe it
2. For any given set of axioms, determine whether they are mutually consistent.

Around 1910 the American mathematician Oswald Veblen and his colleague John Wesley Young successfully tackled projective geometry in this spirit. Their two-volume *Projective Geometry* (1910, 1918), carried through the entire Kleinian Erlangen Programme in this axiomatic setting, thereby providing axioms for every classical geometry and explaining how they were interrelated, as well as giving a sound axiomatic base for the study of ideal points. Meanwhile, others, such as Veronese and Hilbert's student Max Dehn, explored the geometries that arise as more and more of the axioms that generate Euclidean geometry are replaced.

However, it is less important that axioms be independent of each other, as that they be mutually consistent, for if an axiom system has contradictory axioms then it is void. In order to establish the consistency of a set of axioms it became customary, following Hilbert, to exhibit a model — a system that obeys all the axioms. Such models need not be physical objects. For Hilbert, a model of plane Euclidean geometry was the usual coordinate plane: a point is modelled by a pair of real numbers, and a line by a set of pairs of real numbers satisfying an equation of the form $ax + by = c$. The idea was a simple one: if something can be found that obeys all the axioms of a system, then that system cannot be inconsistent. But once mathematicians had begun to doubt the truth of geometry, it became difficult for them to know what to trust. For example, what are the real numbers? How can we be sure that they exist?

As you will see in Chapter 17, the real numbers also came to be axiomatised. A long chain came to be built up, whereby:

- geometry was axiomatised, and the axioms were shown to be consistent by being provided with a model involving real numbers
- the real numbers were axiomatised, and the axioms were shown to be consistent by being provided with a model involving the integers
- the integers were axiomatised, and the axioms were shown to be consistent by being provided with a model involving the concept of a set

Thus the axiomatising movement in geometry now joined a broad stream of debates about the foundations of the calculus. Nor was this all. The novel objects in abstract algebra, such as the concept of a group, were likewise supplied with axioms. In the 20th century many mathematical gadgets old and new were given axiomatic treatments. This move was essential if the mathematical arguments that were needed to deal with the new concepts were to have any rigour, but it should not be taken to mean — as Hilbert's 'formal' approach might imply — that mathematics ceased to be about anything in particular. What 20th-century mathematicians found interesting proved to be a rather complicated dialectic between their formalism and their intuitions.

By about 1910 it was possible to talk about points, lines, and other geometrical objects, secure in the formal rigour of abstract axiomatics; in that sense, synthetic geometry had found bedrock. However, the price had been high, for the sense that geometry was about real things (objects intelligible to the senses, as Poncelet had put it) was precisely what had to be put aside when passing to the abstract systems.

Figure 15.22. The Mathematics Club of Göttingen, 1902, with Hilbert and Klein in the front row. Also in the front row is Grace Chisholm Young, a former doctoral student of Klein's.

How the consistency of the various geometries was established sheds an intriguing light on the question of what geometry is about. Because consistency was achieved via the passage to coordinate models, one is led to ask what the difference in rigour was between Möbius and Plücker (say) on the one hand, and Peano and Hilbert on the other. One answer is that Hilbert's and Peano's work rested on deeper foundations: they could invoke an axiomatisation of the real and complex numbers, whereas earlier generations could appeal only to obscure intuitions. Another answer is that 20th-century mathematicians could be much more sophisticated in their discussions of what mathematics is about.

The predominant view came to be that every system of geometry must be abstract: a *geometry* was a set of logical deductions from a list of initial assumptions or axioms. These assumptions were to be taken on trust for the purpose of discussion, and were not necessarily to be regarded as true statements about the physical world. Hilbert's formulation was the starkest expression of this point of view, and for that reason it was the easiest to use. On this approach, the deductions in any system of geometry had to be valid independently of any meanings that might be attached to the terms. Indeed, the very meaninglessness of the terms was to be the spur that drove one to check that the deductions were always made legitimately. Anyone interested in applying geometry to the real world was then free to shop around among the various systems, secure in the knowledge that each separate system was coherent. Such a person would give interpretations to the basic terms of the geometry, and proceed with the new system of terms and their interpretations as scientific hypotheses — that is, as items of belief that were held until shown to be false. Some geometries do not suggest themselves as plausible candidates for the geometry of space, because they lack the concept of distance. But one could investigate whether space is Euclidean or non-Euclidean, for example, by interpreting the term 'straight line' as 'the path of a ray of light'.[39]

Geometry, then, is a family of abstract deductive systems, whose coherence is guaranteed by the existence of some model which obeys the axioms. The idea that projective geometry has a kernel of coordinate geometry would have been a bitter truth to Poncelet, but the use of models (he might have replied) not only makes an axiom system consistent, it also makes it intelligible. On such a view, mathematics is saved from complete formality by the vividness of its examples.

We have reached the end of our study of the history of geometry. Debates about the foundations of mathematics did not cease, but that is another story. It may not be fanciful, though, to see many analogies between what we can perceive of the situation in ancient Greece and the situation today. Then, as now, there were times when such debates raged and others when they were quiet. Then, as now, their resolution was not something that was destined to last for all time. Then, as now, there are mathematicians for whom foundational questions are unimportant, and who prefer to invest their energies in finding new results and reliable new methods of discovery. The nature of mathematics itself, the scope of its applications, the reason for its efficacy, and even the moral constraints upon its use, were all themes of interest then, as they are now, and are matters upon which no firm and final conclusions have been reached. But then, it can be argued that the next best thing to a good answer is a really good question.

15.6 Further reading

Coolidge, J.L. 1940. *A History of Geometrical Methods*, Oxford University Press. Dover reprint, 1963. Coolidge was Harvard's Professor of Algebraic Geometry for many years, and this was reflected in the choice of topics, and indeed methods, that he emphasised in this book. It is partisan for projective geometry, but austere in its account. If you have a serious interest in the mathematics of the 19th century then this is a book that you will ultimately want to consult.

Fauvel, J., Flood, R., and Wilson, R. (eds.) 1993. *Möbius and his Band: Mathematics and Astronomy in Nineteenth-Century Germany*, Oxford University Press. A multi-sided treatment of many topics that covers Möbius's work on geometry, mechanics, and topology (the eponymous band), it includes an interesting account by Gert Schubring of the rise of pure mathematics and neo-humanism in Germany.

Klein, C.F. 1983. *Lectures on the Development of Mathematics in the Nineteenth Century*, MathSci Press. (Original German edition *Vorlesungen über die Entwicklung der Mathematik im 19. Jahrhundert*, 2 vols. R. Courant and O. Neugebauer (eds.) Springer, 1926. Repr. Chelsea, 1967.) A classic by a master of the subject, this is a well-informed and highly personal account of the development of the subject by one who was often intimately involved in the stories he had to tell.

[39]The modern answer is that gravity is best understood as a curving of space, and because gravity varies from point to point, so does the curvature. The fact that the curvature is not constant then prevents either geometry from being true.

Reid, C. 1970. *Hilbert*, Springer. This biography gives a vivid impression of Hilbert's tremendous range of interests. The author concentrates on the story of his life, which has allowed her to paint a spirited picture of the people that Hilbert knew, and of the University of Göttingen which he and Klein made the centre of the mathematical world. Included as an Appendix is a masterly essay by Hermann Weyl on Hilbert's mathematics.

16

The Rigorisation of Analysis

Introduction

At the start of the 19th century, two mathematicians pursued programmes for making the calculus rigorous. One was Bernard Bolzano, whose standpoint is often said to have been philosophical, and the other was Augustin-Louis Cauchy, one of the most prolific mathematicians of all time. Both perceived certain inadequacies in Lagrange's work, and, in order to make progress with the foundations of the calculus, both turned to an *algebra of inequalities*.

Of the two, it was Cauchy's work on the foundations of the calculus that became the basis of further developments, because it was widely available in Paris. Bolzano's work had little immediate influence, because it was obscurely published in Prague, outside the mainstream of mathematics, and became better known only after Cauchy's writings had been published and studied widely by the mathematical community. Nevertheless, the sharpness of his criticisms, and his priority over Cauchy, have rightly earned Bolzano a permanent place in the history of the calculus.

Cauchy's contributions to mathematics were extensive: his collected mathematical papers run to 31 volumes and touch on almost every branch of the subject. Here we look at his attempts to base the calculus on the arithmetic of inequalities. We shall see that he was profoundly insightful — although not always correct — and we go on to consider how he drew implications for the existence of solutions to differential equations.

16.1 Bolzano, Cauchy, and continuity

Bolzano. Bernard Bolzano was born in Prague in 1781 and worked there for many years; he was thus somewhat isolated from the European mathematical community at large. He was not only a mathematician and a philosopher, but also a Roman Catholic priest whose popular Sunday sermons were eventually suppressed by the authorities

Figure 16.1. Bernard Bolzano (1781–1848)

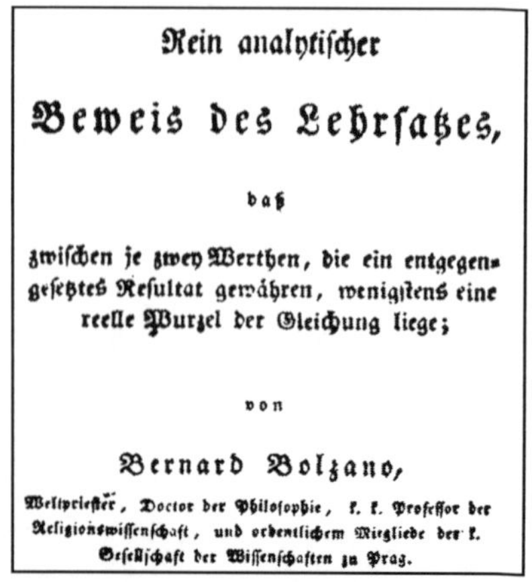

Rein analytischer

Beweis des Lehrsatzes,

daß

zwischen je zwey Werthen, die ein entgegengesetztes Resultat gewähren, wenigstens eine reelle Wurzel der Gleichung liege;

von

Bernard Bolzano,

Weltpriester, Doctor der Philosophie, k. k. Professor der Religionswissenschaft, und ordentlichem Mitgliede der k. Gesellschaft der Wissenschaften zu Prag.

Figure 16.2. The title page of Bolzano's 1817 pamphlet

for being too liberal. One biographer, Jaroslav Folta, recorded that, as a direct result of the Church's influence, Bolzano was attacked[1]

> not only for his rationalism and replacing religion by ethics, but also for his adopting modern scientific results and, last but not least, especially for his theses that had an anti-feudal, mostly bourgeois, but also socialist character; the assertions on perpetual progress and on the equality of all people, the rejection of hereditary aristocracy with all their privileges and possessions, the exposure of the origins of private property ...

Bolzano criticised his predecessors for leaving logical gaps in their work when they made intuitive assumptions about the continuity of functions. He was led to recognise and fill such gaps when he sought to provide a rigorous proof of the Fundamental Theorem of Algebra (see Section 7.3). As we shall see in Chapter 18, Gauss had given a proof of this theorem in 1799 which resorted to geometrical concepts. Bolzano wished to eliminate this reliance on geometrical intuitions, and he showed how to do so in a pamphlet, published privately in Prague in 1817, with a long title that translates as 'Purely analytical proof of the theorem that between any two values which give a result of opposite sign there lies at least one real root of the equation' (see Figure 16.2).

The Fundamental Theorem of Algebra would seem to be very different from the last-mentioned theorem; the former makes a claim about polynomial equations, the latter about continuous curves. To understand how they are related, we need to look at Bolzano's theorem in some detail.

The theorem seemed plausible, as a look at the graph of a suitable polynomial $p(x)$ confirms (see Figure 16.3). As we saw in Section 7.3, many writers had observed that the values of a polynomial of odd degree and with real numbers as coefficients have one sign when x is large and negative, and the opposite sign when x is large and positive. Accordingly, they said, the values of the polynomial must change sign as x runs through its values from large negative ones to large positive ones. At a point where it changes sign it must be 0, thus confirming the existence of a solution of the equation $p(x) = x^n + a_{n-1}x^{n-1} + \cdots + a_0 = 0$ when n is odd. More elaborate arguments were applied to the cases of polynomials of even degree and polynomials with complex coefficients.

[1] Quoted in (Jarník 1981, 14).

The argument about the graph of a polynomial is plausible. But to make it rigorous, Bolzano said (as Gauss had earlier) that the claim about continuous functions changing sign had to be established. This, it seemed to him, was a general claim about continuity, quite independent of polynomials, and he treated it as such. It is the claim that we today call the *Intermediate Value Theorem*

> If a continuous function defined on an interval takes a negative value at one end and a positive value at the other, then it must take the value 0 at least once somewhere in between.

Consider this claim for the function $f(x) = x^2 - 2$ on the interval from $x = 0$ to $x = 2$ (see Figure 16.4). When $x = 0$, $f(x) = -2$, which is negative. When $x = 2$, $f(x) = 2$, which is positive. So, according to the Intermediate Value Theorem, $f(x)$ must be 0 somewhere in between. It is 0, of course, at the value $x = \sqrt{2}$ — but is $\sqrt{2}$ actually a number on the line? Even this simple example shows that we have to be quite careful about numbers when using this theorem. A rigorous proof of the Intermediate Value Theorem requires that we can say what $\sqrt{2}$ is.

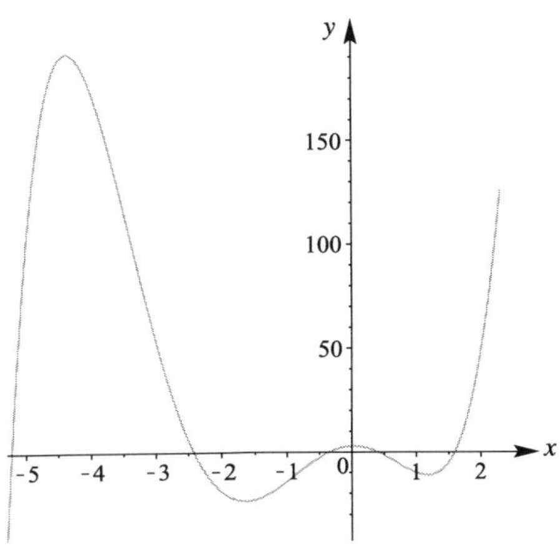

Figure 16.3. The graph of a polynomial of odd degree

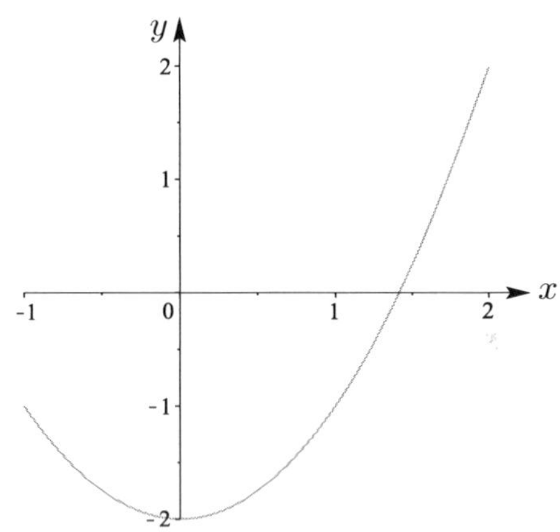

Figure 16.4. The graph of $y = x^2 - 2$

It might seem from the agreement between the approaches of Bolzano and Gauss that Bolzano was a true mathematician, and indeed, in view of the thoroughness of his arguments in the 'Purely Analytical proof', Bolzano was capable of original mathematics. In what sense, then, is Bolzano often said to be more of a philosopher than a mathematician? The point rests on a distinction between arguments that were more concerned with correct reasoning, wherever such reasoning might be found, and those that advance mathematics as such, and it does indeed seem that Bolzano's comments are largely those of a philosopher. He complained that, in order to establish proofs in '*pure* (or general)' mathematics, his predecessors had recourse to considerations taken from geometry — a 'merely *applied* (or special) part' of mathematics. This is described

Box 45. The algebra of inequalities.
The symbolic form of Bolzano's definition illustrates how inequalities enter the discussion of continuity.

First, some necessary definitions. The *absolute value* or *modulus* of a numerical expression X, denoted by $|X|$, is defined to be X if X is positive or 0 and $-X$ if X is negative — for example, $|-31| = 31$. So $|X|$ is always either 0 or positive.

To restrict ω to a range of values we say that there is a positive quantity δ such that $-\delta < \omega$ (thus putting a lower bound on ω) and such that $\omega < \delta$ (thus putting an upper bound on ω). We can write these requirements as

$$-\delta < \omega < \delta \quad \text{or} \quad |\omega| < \delta.$$

Similarly, the restriction on f emerges as the claim that there is a positive number ε such that

$$-\varepsilon < f(x+\omega) - f(x) < \varepsilon,$$

that is,

$$|f(x+\omega) - f(x)| < \varepsilon.$$

The definition of continuity proposed by Bolzano (and, later, by Cauchy) is, in effect, that for each x in the specified range and for any given number $\varepsilon > 0$ there is a number $\delta > 0$ such that

$$|\omega| < \delta \text{ implies } |f(x+\omega) - f(x)| < \varepsilon.$$

With this definition we can show that familiar functions are indeed continuous when their graphs look continuous, although the details can be surprisingly intricate. This may account for why Bolzano and Cauchy gave so few examples of the criterion in practical operation. Conversely, functions whose graphs have obvious gaps in them do indeed turn out to be discontinuous. So the Bolzano–Cauchy definition provides a firm foundation for our intuitive hopes about continuous functions.

as 'an intolerable offence against *correct method*'.[2] Bolzano regarded explaining continuity in terms of time as the illegitimate replacement of a general term by a particular case, and fallacious because it supplants a proof with an example.

How, then, did Bolzano define the continuity of a function without appeal to intuition? He said that a function $f(x)$ varies continuously for all values of x in a certain interval if, for any x in that interval, the difference

$$f(x+\omega) - f(x)$$

can be made smaller in absolute value (see Box 45) than any given quantity by insisting that ω be taken as small as we please. The idea is that if ω lies within certain limits,

[2]These quotes come from the extract of Bolzano's 'Rein analytischer Beweis' (Purely analytical proof) in F&G 18.B1.

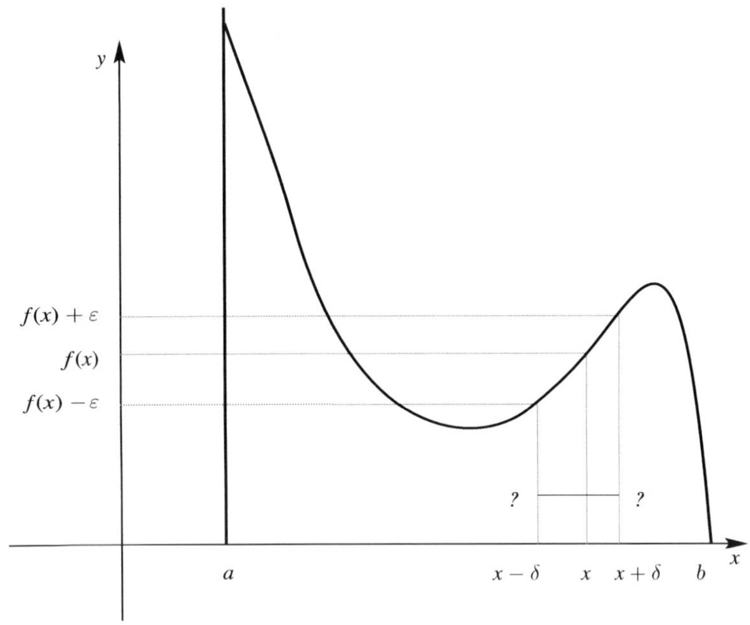

Figure 16.5. How $f(x)$ varies as x varies

then $f(x + \omega)$ differs from $f(x)$ by less than a certain amount. If the latter amount is stipulated in advance, then $f(x + \omega)$ can be kept within those bounds by suitably restricting ω.

Figure 16.5 shows the graph of a function f on an interval $a \leq x \leq b$. To each point x in the interval, there corresponds a point $f(x)$ on the y-axis that is the value of the function f at x.

Continuity is concerned with the relationship between nearby values of x and the corresponding values $f(x)$. As ω varies, so do the values $f(x+\omega)$. The test of continuity of the function $f(x)$ at the point x is whether, for each value of $\varepsilon > 0$, we can restrict the values of ω so that the values $f(x + \omega)$ lie between $f(x) - \varepsilon$ and $f(x) + \varepsilon$, however small ε is chosen to be.

From the figure, we see that, for the indicated value of ε there is a number $\delta > 0$ such that if $-\delta \leq \omega \leq \delta$ then $f(x) - \varepsilon \leq f(x + \omega) \leq f(x) + \varepsilon$.

Cauchy and real analysis. We saw in Chapter 13 how Cauchy emerged to set his stamp on the mathematics of the École Polytechnique. In his *Cours d'Analyse* (A Course in Analysis) of 1821, published four years after Bolzano's pamphlet, Cauchy aimed for 'geometrical rigour' and disavowed the 'generality of algebra' — he thus openly disagreed with Lagrange. The generality of algebra displeased him because it worked only 'in general' — that is to say, most of the time but not all of the time. As we shall see, Cauchy devised explicit examples that showed how Lagrange's methods would sometimes fall into error.

But this still leaves open the question: Why did Cauchy's method for rigorising the calculus differ so radically from Lagrange's? This is a vexed question, which historians have debated from many points of view. It is entangled with an even more imponderable question: Why did Cauchy devote such energetic attention to the foundations of analysis?

Sooner or later, every teacher of advanced mathematics has to explain why the calculus works. Euler and D'Alembert had done so, and presumably were satisfied enough with their explanations. Among Cauchy's contemporaries, the prolific textbook writer Sylvestre Lacroix also wrote at length on the question from a variety of points of view (see Chapter 13). But it was Cauchy's contribution to tackle the problem with vigour, and to get all the way from the foundations to the crucial theorems and principal applications of the calculus, thereby satisfying all of Lagrange's strictures of 1784. The methods of Cauchy and Lagrange certainly differed, but they were at one in seeking rigour, clarity, and an explanation of why all the known results of the calculus were valid.

Some have argued that it was Cauchy's position as lecturer at the École Polytechnique that forced him to sort the matter out. Put this baldly, such a claim is almost certainly false, because others in similar situations did not respond in the same way. But it is possible that having to lecture on analysis may have given Cauchy the spur to think the matter through from start to finish. Others have argued that Cauchy had been confronted with problems in his research that required a re-think of the foundations, and have speculated that problems in the theory of differential equations drove him on. Such speculations, if they could be confirmed, would return us neatly to the question of why Cauchy departed from Lagrange.

This question has been considered by the historian Judith Grabiner, who points out that anyone working with infinite series must confront the problems of convergence. Because we cannot literally add up infinitely many terms, the best we can do is to add up finitely many terms and estimate the value of the rest: if the value of the 'infinite tail' is sufficiently small, then we can ignore it. Now, it so happened that Lagrange was interested in calculating these estimates — not when he was trying to make the calculus rigorous, but elsewhere in his mathematics. Grabiner suggests that Cauchy saw in these questions, as Lagrange had not, a way to make the calculus rigorous by basing it on an algebra of inequalities, rather than on formal algebra. On this view, Cauchy was forced to adopt such an analysis by his study of the power series that arise in the study of differential equations.

Are these speculations valid? We cannot yet say. Research among Cauchy's papers may reveal matters upon which the voluminous published records speak only enigmatically: there is still much that we do not know. But if the motivation were pedagogical, then we should note that Cauchy's students and some of his colleagues found his new ideas so hard to understand that the director of the École Polytechnique eventually ordered him to teach old-fashioned, unrigorous, but comprehensible calculus instead![3]

> I must no longer leave unknown that numerous warnings have been given, for five years, to Mr. Cauchy to undertake to simplify his methods and to conform exactly to the programmes [of lectures] ... [by] letting himself go in his imagination beyond all measure, he employed a luxury of analysis without doubt appropriate for papers to be read at the Institut; but superabundant for the teaching of Students at the School and even harmful; that he neglected to train the Students in applications ... and that there

[3]Quoted in (Grattan-Guinness 1990, Vol. II, 712).

> has sometimes been as a result of this habit of devoting himself to scientific development a lack of clarity in his lectures which has even been expressed with justice by the Students.

In any case, there is a further puzzle about Cauchy's work that these speculations do not address. As we shall see, Cauchy took up the topic of rigorising the calculus, which is about differentiation and integration, and did so in terms of the novel concept of continuity. This is no idle concept. Cauchy found that so much needed to be said about continuity, and the idea of limits on which he based it, that he needed a whole lecture course to say it, and got round to the theories of differentiation and integration only in his next course, published as the *Résumé* in 1823.

We can see the germ of the idea of limits in Newton's first and last ratios, in Leibniz's infinitesimals, and in D'Alembert's writings on mathematical analysis, but that does not help us to understand the great weight that Cauchy placed on the concept. To do that we must recognise that Cauchy had realised that the calculus could not be based on a limit concept that drew on analogies from motion or geometry if the calculus was then going to be applied to those subjects. To do so would be to risk creating a vicious circle, as difficult concepts in geometry are elucidated by appeals to the calculus, whose difficulties are then explained by appeals to geometry. New foundations must be found, and Cauchy sought them in the domain of fixed and variable quantities.

Cauchy's *Cours d'Analyse*. In his lectures at the École Polytechnique in 1821, promptly published as his *Cours d'Analyse* in accordance with the rules of the institution, Cauchy defined and discussed a number of concepts with a remarkably novel degree of precision. These include finite quantities, real functions, infinitely small and infinitely large quantities, the continuity of functions, singular values of functions, real convergent and divergent series, rules for convergence, summation of some convergent series, and the binomial theorem.[4]

Numbers and quantities (by which Cauchy meant various kinds of magnitudes, such as lengths), said Cauchy, are represented by letters, some of which stand for variables and some for constants.

In the introduction to his *Cours d'Analyse*, Cauchy defined the concept of limit in these words:[5]

> When the values successively attributed to the same variable indefinitely approach a fixed value, in such a way as to finish by differing from it by as little as one wishes, this last is called the *limit* of the others.

In particular, an irrational number is a limit of a sequence of rational numbers.

He defined an infinitely large positive number, written $+\infty$, to be a variable that takes successive numerical values and increases indefinitely in such a way that it becomes larger than any given number.[6] (He defined $-\infty$ similarly.)

The concept of the *infinitely small* was particularly important for Cauchy — it is how he expressed many arguments involving limits. He defined it in Chapter 2 of his *Cours* as follows:[7]

---

[4]See Cauchy, *Cours d'Analyse*, and the extract in F&G 18.B2. For a detailed study of Cauchy's approach to analysis in the context of his place and his time, see (Schubring 2005).

[5]Cauchy, *Cours d'Analyse*, p. 4.

[6]Cauchy, *Cours d'Analyse*, p. 5.

[7]Cauchy, *Cours d'Analyse*, p. 26.

COURS D'ANALYSE

DE

L'ÉCOLE ROYALE POLYTECHNIQUE;

PAR M. AUGUSTIN-LOUIS CAUCHY,

Ingénieur des Ponts et Chaussées, Professeur d'Analyse à l'École polytechnique, Membre de l'Académie des sciences, Chevalier de la Légion d'honneur.

I.re PARTIE. *ANALYSE ALGÉBRIQUE.*

DE L'IMPRIMERIE ROYALE.

Chez DEBURE frères, Libraires du Roi et de la Bibliothèque du Roi, rue Serpente, n.° 7.

1821.

Figure 16.6. The title page of Cauchy's *Cours d'Analyse* (1821)

> One says that a variable quantity becomes *infinitely small* when its numerical value decreases indefinitely in such a way as to converge to the limit zero.

By the 'numerical value' of a quantity, Cauchy meant the absolute value (or modulus) of a quantity (see Box 45). He then went on to give rules for handling an infinitely small quantity α, raised to various powers α^n, that would play a role in his study of the continuity of functions.

Cauchy explained what he meant by the term 'function' at the start of Chapter 1 of his *Cours*:[8]

> When variable quantities are so related to each other that, the value of one being given, one can deduce the values of all the others, one usually thinks of the various quantities being expressed in terms of one of them, which is then called the *independent variable*; and the other quantities expressed in terms of the independent variable are what one calls functions of this variable [Cauchy's emphasis].

Cauchy then singled out the crucial concept of the *continuity of a function*, without completely realising what the problems were going to be.[9] Nonetheless, his was a dramatic innovation in a mathematical environment that took differentiability for granted.

Cauchy on continuous functions.

> Among the objects that attach themselves to the consideration of the infinitely small, one must place notions relating to the continuity or discontinuity of functions. Let us first of all examine functions of a single variable from this point of view.

[8] Cauchy, *Cours d'Analyse*, p. 19.

[9] Cauchy, *Cours d'Analyse*, pp. 34–35.

> Let $f(x)$ be a function of the variable x, and let us suppose that, for each value of x between two given limits, this function always takes a unique and finite value. If, starting from a value of x between those limits, one gives to the variable x an infinitesimally small increase α, the function itself increases by the difference
>
> $$f(x+\alpha)-f(x),$$
>
> which depends at the same time on the new variable α and the value of x. This done, the function $f(x)$ will be, between the assigned limits of the variable x, a *continuous* function of this variable, if, for each value of x between these limits, the numerical value of the difference
>
> $$f(x+\alpha)-f(x)$$
>
> decreases indefinitely with that of α. In other terms, *the function $f(x)$ remains continuous with respect to x between the given limits, if, between these limits, an infinitely small increase in the variable always produces an infinitely small increase in the function itself.*
>
> One also says that the function $f(x)$ is, in the neighbourhood of a particular value attributed to x, a continuous function of that variable every time it is continuous between two limits for x, even very close, which enclose the value in question.
>
> Finally, when a function $f(x)$ ceases to be continuous in the neighbourhood of a a particular value of the variable x, one says that it then becomes *discontinuous*, and that there is for this particular value a *solution of continuity*.

We see that Cauchy defined f to be *continuous between certain limits* if, for each x within those limits, the value of $|f(x+\alpha)-f(x)|$, where α is infinitely small, decreases indefinitely with α; equivalently, an infinitely small increase in x produces an infinitely small increase in $|f(x)|$. Notice that this is not continuity at a point, a concept that Cauchy never defined, but continuity on an interval.

He then gave many examples of continuous functions, such as $f(x) = a + x$, ax, a^x, $\sin x$, $\arcsin x$, and x^a, where a is a constant. These were exhibited without proof, although particular values of x (such as 0) were sometimes explicitly noted.

He next proved some theorems, such as this result on the composition of functions:

> If f and g are continuous within certain limits, and the range of values of g is included in the range of values for which f is defined, then $f(g(x))$ is continuous.

We can give a flavour of Cauchy's methods by looking at how he proved the Intermediate Value Theorem, which states that:[10]

> If $f(x)$ is continuous between x_0 and X, and b lies between $f(x_0)$ and $f(X)$ then there is at least one x between x_0 and X such that $f(x) = b$.

Cauchy's proof in the body of the lectures is nothing but an appeal to the geometrical idea that the curve $y = f(x)$ must cross the line $y = b$ because the curve is

---

[10] Cauchy, *Cours d'Analyse*, pp. 43–44.

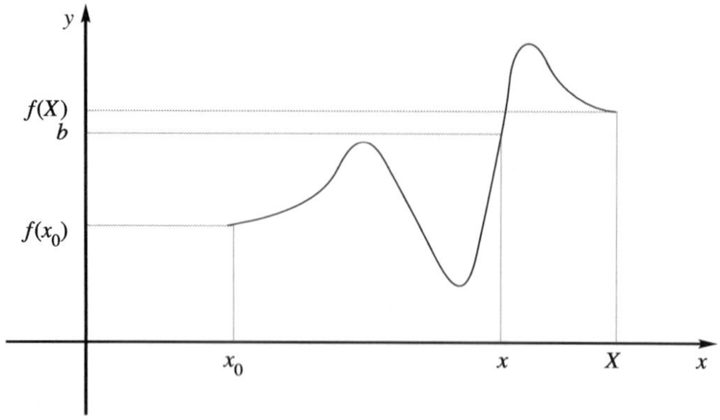

Figure 16.7. Cauchy's Intermediate Value Theorem

continuous. However, in a Note added to the lectures he gave a much better argument, which we paraphrase in Box 46.[11]

What did Cauchy mean by what he called 'geometrical rigour'? He cannot have meant 'geometrical' as in geometry (in the literal sense of Euclidean or coordinate geometry) without risking a vicious circle. Rather, he intended it in the metaphorical sense that ancient mathematicians had been rigorous in their presentation of geometry. Lagrange had tried to base the calculus on algebra, but as we have seen, this did not satisfy Cauchy. But if algebra is ruled out as insufficiently secure, what is left?

Cauchy found a solution in terms of his algebra of approximations based on inequalities. The calculus was to be saved by guaranteeing that it always makes numerical sense — this is one reason why his work laid such stress on the idea of continuity — and because elementary geometry is always helpful in such matters, it was reasonable on these grounds for Cauchy to appeal to geometrical rigour also.

As we have seen, Cauchy's approach began by taking up and refining the notion of a limit. Using it as a cornerstone, he provided a coherent foundation for the concepts of continuity, function, and convergence. Although Cauchy's definition of a limit looks somewhat like D'Alembert's earlier definition, it was markedly more sophisticated, especially in the way that it was used. Cauchy dropped the requirements that the limit must be approached from only one side, and more importantly, he no longer appealed to geometry but relied on the algebra of inequalities. The validity of the fundamental process of the calculus then no longer depended on geometrical or dynamical intuition, but on the formal language of algebra and number. For the first time, the intuitions that made the calculus so powerful and easy to use were given a coherent and (almost) rigorous foundation. To be precise, Cauchy's algebra of inequalities gave adequately sound foundations to the calculus, until mathematicians began to ask deep questions about the nature of numbers.

We can compare Cauchy's definition of continuity with Bolzano's by looking at the concluding paragraphs of Bolzano's 'Rein analytischer Beweis' and the opening pages of Cauchy's *Cours*.[12] When we do this, we see that the two definitions are fundamentally equivalent. Both mathematicians considered variations in the value of the

[11]Cauchy, *Cours d'Analyse*, Note 3.

[12]See (Bolzano 1817) and F&G 18.B1; (Cauchy 1821) and F&G 18.B2(b).

Box 46. Cauchy's proof of the intermediate value theorem.
Without loss of generality we can assume that $f(x_0) < f(X)$ and $b = 0$. Cauchy divided the interval $[x_0, X]$ of length h into m equal parts and considered the sequence of values

$$f(x_0),\ f(x_0 + h/m),\ f(x_0 + 2h/m),\ \ldots,\ f(X - h/m), f(X).$$

A comparison of successive values shows that there must be a consecutive pair of opposite signs, unless one of these values is zero in which case there is nothing more to prove. Let $f(x_1)$ and $f(X_1)$ be such a pair, so

$$x_0 \le x_1 < X_1 \le X \text{ and } f(x_1) < 0 < f(X_1).$$

Repeat the argument, obtaining a new consecutive pair x_2 and X_2 where f takes opposite signs (unless one is zero and the theorem is proved), and note that

$$x_0 \le x_1 \le x_2 < X_2 \le X_1 \le X \text{ and } f(x_2) < 0 < f(X_2).$$

Continue in this fashion and two sequences are obtained:

$$x_0 \le x_1 \le x_2 \le \ldots \le x_n \le \ldots$$

and

$$X \ge X_1 \ge X_2 \ge \ldots \ge X_n \ge \ldots.$$

The first sequence is non-decreasing, the second non-increasing, and their successive terms differ by no more than

$$h = X - x_0,\ h/m,\ h/m^2,\ \ldots,$$

so their terms 'will differ by as little as desired and therefore must converge to a common limit'. Let a be that limit. Since the values of f on

$$x_0, f(x_1), f(x_2),\ \ldots\ , f(X_2), f(X_1), f(X)$$

always remain of opposite sign (or are zero) 'it is clear that the quantity $f(a)$, which must be finite, cannot differ from zero'. Therefore the value a is such that $f(a) = 0$ as required, and the theorem is proved.

function in relation to small variations in the independent variable. While Bolzano was explicit in stating that we must first specify that the difference $f(x + \omega) - f(x)$ should be smaller than a given quantity, this was only implicit in Cauchy's formulation, which spoke of 'infinitely small increases' and so of limits.

Cauchy's definition of the continuity of a function, based as it was on his concept of a limit, underpinned his whole theory of functions and of the calculus. This emphasis on the concept of continuity was unexpected. Previously, most mathematicians had taken the calculus to be about differentiation and integration, tangents and quadrature, and continuity was not clearly distinguished from differentiability (see Figure 16.9). Cauchy's perspicuity, like Bolzano's, is notable in this regard, and is partly to be explained by its utility. For, with his definition of continuity, Cauchy was now able to prove the Intermediate Value Theorem, thereby giving a logical underpinning to the

Figure 16.8. Augustin-Louis Cauchy (1789–1857)

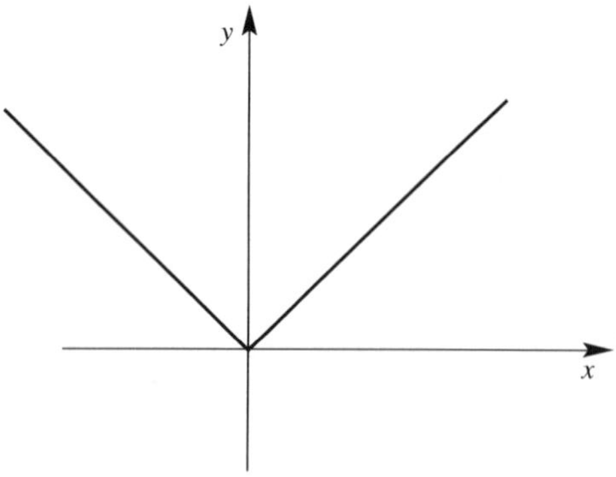

Figure 16.9. A graph of the function $y = |x|$, which is continuous but not differentiable at the origin

calculus that helped to make this important and successful branch of mathematics rigorous. In Cauchy's hands, mathematical analysis began to be the study of continuous functions.

But Cauchy also overhauled the theories of differentiation and integration. He defined integration independently of differentiation — that is, not as an anti-derivative, but as a mathematical operation in its own right, related to finding areas. So he needed to *prove* the inverse relationship between differentiation and integration enshrined in the Fundamental Theorem of the Calculus. As a glance at the above passages by Cauchy shows, Cauchy's definitions of limit, continuity, and convergence lie at the heart of his development of the subject.

Grabiner gave this helpful summary of Cauchy's achievements.[13]

> Cauchy's work established a new way of looking at the concepts of the calculus. As a result, the subject was transformed from a collection of powerful methods and useful results into a mathematical discipline based on clear definitions and rigorous proofs.

[13] In (Grabiner 1981, 164–165), see F&G 18.B4.

> His views were less intuitive than the old ones, but they provide a new set of interesting questions. His definition of limit and elaboration of the associated method of proof by the inequalities are the basis for modern theories of continuity, convergence, derivative, and the integral. And many of the important consequences of these theories — in the study of convergence, existence proofs for the solution of differential equations, and the properties of definite integrals — were pioneered by Cauchy himself.
>
> Moreover, Cauchy's rigorisation of the calculus was much more than the sum of its separate parts. It was not merely that Cauchy gave this or that definition, proved particular existence theorems, or even presented the first reasonably acceptable proof of the fundamental theorem of calculus. He brought all these things together into a logically connected system of definitions, theorems, and proofs.

This is dramatic enough, but she went further:

> The implications of this achievement go beyond the calculus. In a very important sense, it may be said that Cauchy brought ancient and modern mathematics together. He cast his rigorous calculus in the deductive mould characteristic of ancient geometry. And unlike his predecessors, he did this successfully; that is, he not only gave his work a Euclidean form but presented definitions that generally are adequate to support the desired results, proofs that basically are valid, and methods that were fruitful sources for later mathematical work. Cauchy, then, brought together three elements : the major results of analysis, most of which he could now prove; some fruitful concepts and techniques from algebra (particularly algebraic approximations) and analysis; and the rigor and proof structure of Greek geometry. For a long time Greek geometry had been considered the model for all of mathematics. If the origins of modern mathematics are traced to the Renaissance, then the rigor and structure characteristic of Greek geometry first effectively became part of modern mathematics only with Cauchy's work. Of course the late-19th-century idea that mathematics is the science of abstract logical systems in general is absent from Cauchy's work. But Cauchy's rigorisation of the calculus was an indispensable first step in that direction.

It is not only the foundation of the calculus that is Cauchy's great achievement, but the restoration of high standards of rigour through the meticulous way in which the *Cours d'Analyse* is structured. Grabiner's point is not that Cauchy's predecessors were not rigorous — Lagrange certainly aspired to complete rigour — but that Cauchy's use of explicit definitions, theorems, and proofs began the move towards the explicit formal style of modern mathematics and away from the leisurely manner of 18th-century authors. That said, Grabiner then rightly concluded:

> Cauchy left some unfinished business, as subsequent history shows. Some gaps in specific proofs had to be filled; some assumptions had to be proved, or at least explicitly stated; some crucial distinctions had yet to be made. But there is little in nineteenth-century analysis that was not marked, directly or indirectly, by his ideas. The basic logical structure Cauchy erected provides the framework in which we still think about rigorous calculus.

16.2 Cauchy's mistake

Cauchy reformulated every aspect of the calculus. His investigations were largely successful, but we shall now follow him in one direction where he went astray, because it tells us much about the limits of his understanding, and also what others made of his approach.

Ever since the time of Newton, power series of the form

$$a_0 + a_1x + a_2x^2 + \cdots + a_nx^n + \cdots$$

had been part of the calculus. In the 18th and early 19th centuries mathematicians had begun to consider trigonometric series of the form

$$a_0 + a_1 \sin x + a_2 \sin 2x + \cdots + a_n \sin nx + \cdots.$$

Cauchy was interested in when such infinite sums make sense, but we shall see that his intuition was strongly guided by his experience of dealing with power series.

To give one of his examples, the series $1 + x + x^2 + \ldots$ has a finite sum only when $-1 < x < 1$. This shows that whether an infinite sum of this sort makes sense can depend on the range of the variable x. (Note that, by the binomial theorem, this sum is equal to the value of the function $(1-x)^{-1}$, which is defined for all $x \neq 1$.)

To handle the concept of a convergent sequence, Cauchy fell back on his concept of a limit. He said that a sequence of numbers $a_0, a_1, a_2, \ldots, a_n, \ldots$ converges to a number a if the differences $|a - a_n|$ become arbitrarily small as n increases.[14] For any fixed value of x, a sequence of functions $s_n(x)$ is simply a sequence of numbers, and so the concept of convergence also applies to a sequence of functions. It also applies to an infinite series of functions

$$u_1(x) + u_2(x) + \cdots + u_n(x) + \cdots$$

if we say that a series converges if the sequence $s_n(x)$ of its partial sums

$$s_n(x) = u_1(x) + u_2(x) + \cdots + u_n(x)$$

converges. As the above example shows, there can be values of x for which a series of functions converges, and others for which it does not.

With this in place, the question that Cauchy confronted was to determine what properties of the functions in a sequence $(s_1(x), s_2(x), s_3(x), \ldots)$ are shared with the limit of that sequence, when this limit exists. For example, if the individual functions $s_n(x)$ are all continuous for all n, must the limit function also be continuous?

Although the binomial theorem had been claimed by Newton, as we discussed in Section 4.2, Cauchy felt that Newton's and later proofs relied too much on intuition, and so he set out to prove it. His proof took the form of a general theorem, which was then applied to the binomial theorem. The application to the binomial theorem was sound; here we concentrate on his general theorem, which was flawed.

Cauchy considered a series of functions

$$u_1(x) + u_2(x) + \cdots + u_n(x) + \cdots,$$

and asserted that:[15]

> when the terms of a series are continuous functions of a variable x in the neighbourhood of a particular value of x for which the series converges, the sum of the series is also a continuous function in the neighbourhood of that particular value.

Here, the clause 'in the neighbourhood of a particular value of x for which the series converges' is far from clear. It means that the individual functions $u_n(x)$ are all continuous functions on the same neighbourhood of a point x at which the series converges. So the theorem asserts that when a series of continuous functions converges, the function to which it converges is also continuous.

---

[14] Cauchy also showed that if the limiting value is not known in advance, then the series converges if the differences $|a_n - a_{n'}|$ become arbitrarily small as n and n' increase.

[15] See Cauchy, *Cours d'Analyse*, pp. 131–132.

To prove it, Cauchy let $s(x)$ be the sum of the series

$$s(x) = u_1(x) + u_2(x) + \cdots + u_n(x) + \cdots,$$

where the functions $u_1(x), u_2(x), \ldots$ are functions of x that satisfy the conditions of the theorem. He let $s_n(x) = u_1(x) + u_2(x) + \cdots + u_n(x)$ be the sum of the first n terms, and $r_n(x) = u_{n+1}(x) + u_{n+2}(x) + \cdots = s(x) - s_n(x)$ be the remainder term. He then argued that:

> When x increases by an infinitely small value α, the increase in s_n will also be infinitely small, for all values of n, and the increase in r_n will become insensible with r_n when n is large. So the increase in s will also be infinitely small.

This argument is quite convincing when you first hear it, but it is invalid. The first to comment on it was Abel, who observed that 'it seems to me that the theorem admits exceptions'.[16] He gave this example, the series

$$s(x) = \sin x - 1/2 \sin 2x + 1/3 \sin 3x - \cdots.$$

(To compare it with Cauchy's series, set $u_n(x) = (-1)^{n+1} \frac{1}{n} \sin nx$.) Fourier had shown that this function is equal to the function $y = x/2$ between $-\pi$ and π, and is periodic with period 2π (see Figure 16.10 and Section 20.1). As the variable x approaches π from below, the value of the function increases towards $\pi/2$. As the variable x approaches π from above, the value of the function decreases towards $-\pi/2$. So the function is discontinuous at all odd multiples of π, and it is likely that its occurrence in Fourier's work is what alerted Abel. It had indeed been known to Euler, who discussed it in 1783, and it had been picked up and commented on by Lacroix in his book of 1810.

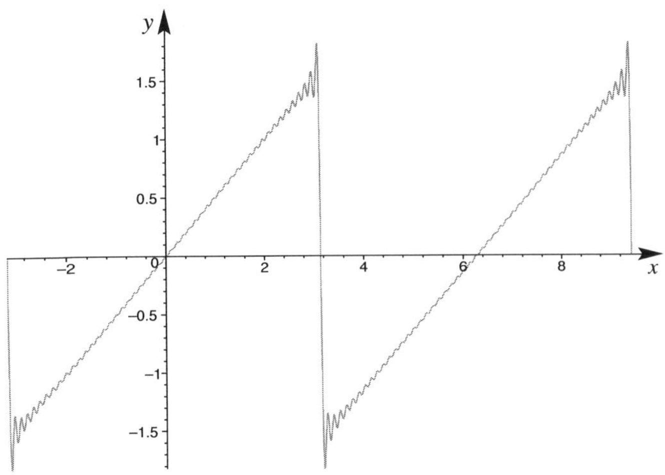

Figure 16.10. The sum of the first 50 terms of
$\sin x - \frac{1}{2} \sin 2x + \frac{1}{3} \sin 3x - \cdots$
(the vertical lines are not part of the graph of the function)

[16]See (Abel 1826, 225).

So this Fourier series provides an example of a series of functions, each of which is continuous everywhere but which has a sum that is discontinuous at certain points.

How should we understand Cauchy's error? One possibility is that it was a simple mistake — but then one might expect Cauchy, the architect of rigour in the calculus, to find an opportunity to correct it. However, Cauchy not only knew Fourier's work, he did not react when Abel published his remarks. Plainly, he did not consider them to be relevant to his theorem, whereas Abel did think that this series was an exception.

One interesting possibility is to note that, in the 1820s, Cauchy never expressed the concept of continuity *at a point*, but only continuity *on an interval*. It is possible to use the definition of continuous on an interval in such a way that Cauchy's theorem becomes true — even though this is not the modern definition of continuity at a point that mathematicians now attribute to Cauchy! Could it be that Cauchy was confused about the difference between being continuous at a point and continuous on an interval? If so, then we must conclude that Abel and Cauchy wrote the same words and meant something different by them. We can note, for example, that Cauchy's notation suppressed the variable x, whereas Abel's example did not.

Cauchy did, however, return to the topic, but only in 1853 when prompted by some remarks made by Charles Briot and Claude Bouquet, two young French mathematicians whom he wished to support. He now admitted that he had been wrong, and gave the above Fourier series example to show the failure of his theorem. Somehow, this series had become a counter-example in the intervening years — but what had changed to make this apparent?

In his article of 1853, Cauchy also showed exactly what was wrong with the purported proof of the theorem as he had originally stated it: the remainders do not necessarily behave as he had supposed. More precisely, he now noted that, for the theorem to be true, the difference of the two remainders, $r_n - r_{n'}$, for $n' > n > N$, must approach 0 as N increases indefinitely, in such a way that the value of N needed to make this difference arbitrarily small does not depend on x. His notation was no better, but his modification of the theorem spoke explicitly of the role of the variable.

Let us consider what it means for a sequence of remainders $r_n(x)$, $n \geq 1$, to become arbitrarily small, when x lies in a given interval. Which of the following two statements does it mean?

1. For each $\varepsilon > 0$ and for each x in the interval, there is a number N such that $n > N$ implies $r_n(x) < \varepsilon$.

2. For each $\varepsilon > 0$, there is a number N such that $n > N$ implies $r_n(x) < \varepsilon$ for all x in the interval.

In the first case the number N may depend on x, whereas in the second case the number N must hold simultaneously for all x in the interval.[17] In the case of Cauchy's purported theorem, he showed in 1853 that the theorem is true if the second interpretation of the behaviour of the remainder terms is chosen.

It would be helpful if the Fourier series in question gave a simple illustration of what can go wrong, but it does not. Instead, and to illustrate the point, we offer a simpler example of what is involved in Box 47.

[17] Compare the claims about a competition that 'for each entrant there is a prize' and that 'there is a prize that everyone gets'. In the first case the prize may vary from person to person; in the second case everyone gets the same prize.

Box 47. A counter-example to Cauchy's theorem.
Consider the function v_n, which is defined as follows, for $n > 0$,

$$v_n(x) = \begin{cases} -1, & \text{if } x \le -1/n \\ nx, & \text{if } -1/n \le x \le 1/n \\ +1, & \text{if } 1/n \le x. \end{cases}$$

Define $u_1(x) = v_1(x)$, and $u_n(x) = v_n(x) - v_{n-1}(x)$.

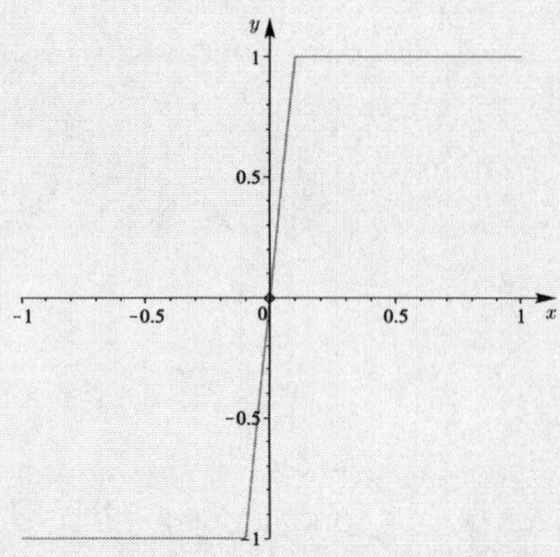

Figure 16.11. The graph of the function $v_n(x), n = 10$

Then

$$s_n(x) = u_1(x) + u_2(x) + \ldots + u_n(x)$$
$$= v_1(x) + (v_2(x) - v_1(x)) + (v_3(x) - v_2(x)) + \cdots + (v_n(x) - v_{n-1}(x)) = v_n(x).$$

In the limit, $s(x) = \lim_{n\to\infty} s_n(x)$ is the function

$$s(x) = \begin{cases} -1, & \text{if } x < 0 \\ 0, & \text{if } x = 0 \\ +1, & \text{if } x > 0. \end{cases}$$

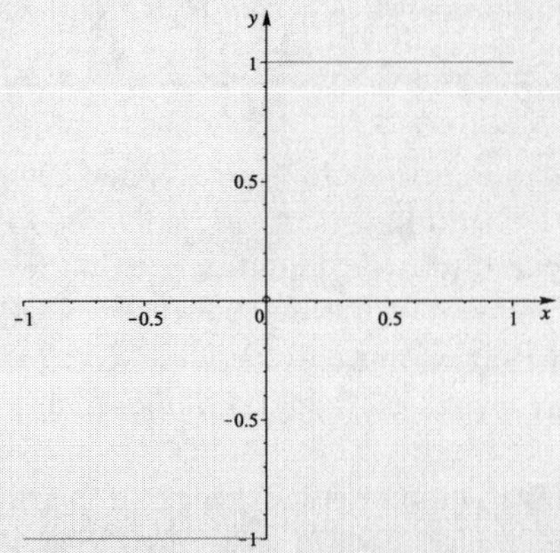

Figure 16.12. The graph of the limit function $s(x)$

(Continued on the next page)

Box 47. A counter-example to Cauchy's theorem (continued)
So the limit function is not continuous at $x = 0$, and Cauchy's purported theorem is incorrect.

Where is Cauchy's mistake? Consider the remainder $r_n(x) = s(x) - s_n(x)$. Outside the interval $-1/n \leq x \leq 1/n$, the remainder is 0. Inside this interval it is defined by

$$r_n(x) = \begin{cases} -1 - nx, & \text{if } -1/n \leq x < 0 \\ 0, & \text{if } x = 0 \\ 1 - nx, & \text{if } 0 < x \leq 1/n. \end{cases}$$

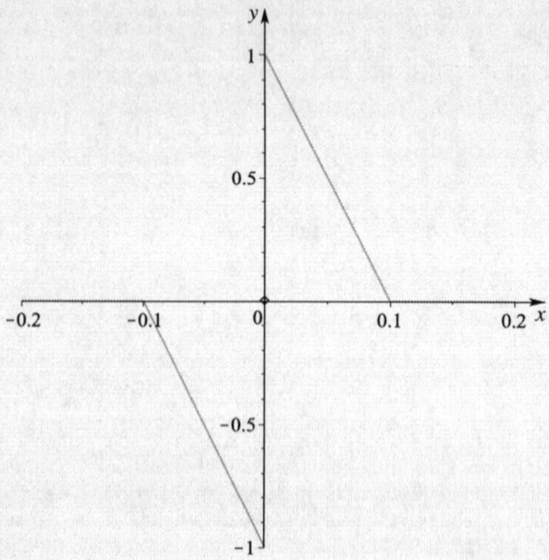

Figure 16.13. The graph of the remainder term $r_n(x)$, $n = 10$

So although for each individual value of x the remainder term approaches 0 as n becomes indefinitely large, it is not the case that the remainder term becomes 0 for all values of x *simultaneously* as n becomes indefinitely large.

One likely explanation of these developments is that Cauchy initially believed that a function might either be continuous on an interval or fail to be continuous at a single point. This could have given him undue confidence that the way that the remainder terms $r_n(x)$ behave is independent of x, once they are all small. On this interpretation, Cauchy did not suspect in 1821 what he was compelled to recognise in 1853, that the way that these subtleties are expressed in mathematics (using the apparatus of ε, δ, and N) might vary from point to point.[18]

Cauchy's formulation was a major step forward in the rigorisation of analysis, but clearly it contained surprises, even for him. These surprises grew out of his difficulties in adapting his intuitions about continuity to what his methods could establish. His

[18]It is worth noting (although we do not have the space to prove this here) that the behaviour of convergent power series is much simpler than the behaviour of general sequences of functions.

mistaken theorem is an indication of his relative naivety, despite all his successes; more importantly, the later correct proofs of the theorem show that the way forward was to adapt one's intuitions to what the use of inequalities would allow one to prove.

16.3 Cauchy on differentiation and integration

In his *Résumé* of 1823, Cauchy broke decisively with the formulation of the derivative that Lagrange had given, and also revived and made more rigorous the former conception of an integral as an area. We shall take each idea in turn.

Differentiation. Cauchy began his *Résumé* by restating the definitions of limit, infinitesimal, and continuity from the *Cours d'Analyse.* He then defined differentiation in these terms:[19]

Cauchy on differentiation.

> When the function $y = f(x)$ remains continuous between two given limits of the variable x, and one assigns to this variable a value included between the two limits in question, an infinitely small increase in the variable produces an infinitely small increase in the function itself. As a consequence, if one then sets $\Delta x = i$, the two terms of the *ratio of differences*
> $$\frac{\Delta y}{\Delta x} = \frac{f(x+i) - f(x)}{i}$$
> will be infinitely small quantities. But, while these two terms will indefinitely and simultaneously approach the limit zero, the ratio itself can converge towards another limit, either positive or negative. This limit, *when it exists,* has a determinate value for every particular value of x, but it varies with x ... The form of the new function that will serve as the limit of the ratio
> $$\frac{f(x+i) - f(x)}{i}$$
> will depend on the form of the proposed function $y = f(x)$. To indicate this dependence, we give the new function the name of derived function, and designate it, with the aid of an accent, by the notation y' or $f'(x)$.

In his next lecture Cauchy explained why the first derivative can be written as dy/dx — that is, as the ratio of the differential of the function and the variable.

Cauchy's thoroughgoing rewrite of the foundations of the calculus is based on the next result:[20]

> If, the function $f(x)$ being continuous between the limits $x = x_0, x = X$, we designate by A the smallest and by B the largest of the values that the derived function $f'(x)$ assumes in this interval, the ratio of increments
> $$\frac{f(X) - f(x_0)}{X - x_0}$$
> will necessarily be included between A and B.

[19] Cauchy, *Résumé*, pp. 22–23, in (Bottazzini 1986, 120).

[20] Cauchy, *Résumé*, p. 44, in (Bottazzini 1986, 120).

In the course of proving this theorem, Cauchy made the following remark:[21]

> Designate by δ and ε two very small numbers: the first being chosen in such a way that, for numerical values of i less than δ, and for any value of x between x_0 and X, the ratio of $(f(x+i) - f(x))/i$ always remains greater than $f'(x) - \varepsilon$ and less than $f'(x) + \varepsilon$.

Grabiner has observed that this is the first appearance of the 'delta–epsilon' notation familiar to all modern students of analysis, although it is modelled on more verbal expressions of the same kind in the *Cours d'Analyse*. The formal notation therefore reached print some years after Bolzano's work, but from then on it was a growing presence in Cauchy's work.[22]

Cauchy next stated and proved a Mean Value Theorem,[23] which says that:
If, the function $f(x)$ being continuous between the limits $x = x_0, x = X$, one denotes by A the smallest and by B the largest of the values that the derived function $f'(x)$ assumes in this interval, then the ratio of finite differences

$$\frac{f(X) - f(x_0)}{X - x_0}$$

will necessarily be contained between A and B.

He gave a proof that looked good in its day, but has slightly crumbled with time. In the course of it, he made assumptions that hold only if the derivative is assumed to be continuous in the interval $[x_0, X]$. Contrariwise, when Cauchy did assume the continuity of the derivative (in a corollary) he need not have done so. So Cauchy's genuine insights left a delicate muddle to be sorted out, but that should not obscure the magnitude of his achievement in rewriting the foundations of the calculus in a much more rigorous form that had ever been achieved before, and one that showed how to replace verbal reasoning with mathematical inequalities.

Cauchy went on to study the Taylor series representation of a function,

$$f(0) + xf'(0) + \frac{x^2}{2!}f''(0) + \dots,$$

and he concluded with a remarkable observation about the Taylor series expansion of a function, 'If it is convergent', he said, albeit obscurely:[24]

> then one might think that its sum is the function $f(x)$, and in particular that if all the terms of the Taylor series vanish, then the function itself vanishes — but this is not necessarily the case. But to be certain of the contrary it is sufficient to observe that the second condition will be fulfilled if we suppose $f(x) = e^{-(1/x)^2}$, and the first if we suppose $f(x) = e^{-x^2} + e^{-(1/x)^2}$. However, the function $e^{-(1/x)^2}$ is not identical to zero, and the series derived from the last supposition does not have the binomial $e^{-x^2} + e^{-(1/x)^2}$ as its sum, but its first term $e^{-(1/x)^2}$.

Cauchy here observed that the function $f(x) = e^{-1/x^2}$ and its derived functions all vanish at the point $x = 0$, but the function is not zero everywhere.[25] However, according to Lagrange's theory, if all the derived functions vanish then the function

[21] Quoted in (Grabiner 1981, 115).

[22] As Grabiner also noted, Cauchy was claiming the uniform continuity of the quotient.

[23] Cauchy, *Résumé*, pp. 44–46.

[24] Cauchy, *Résumé*, pp. 229–230.

[25] This claim requires more care to establish than Cauchy gave it, but it is correct.

itself must vanish everywhere that the Taylor series converges, which means in this case that

$$e^{-1/x^2} = 0 + 0.x + 0.\frac{x^2}{2!} + 0.\frac{x^3}{3!} + \cdots.$$

But this is not true — so in this case Lagrange's conceptions were wrong. This is one reason why Cauchy abandoned the Lagrangian foundations for the calculus.

It is difficult to overestimate the significance of this example. The prior belief of every mathematician had been that every function can be expanded as a Taylor series, except in trivial cases, and that the task of the mathematician was to explain this in a rigorous way. Cauchy here showed that it is possible to define a function that does not agree with its Taylor series. This not only destroyed the foundations of the Lagrangian calculus, it opened up the question of how, if at all, a function can agree with a representation of it. In the case at hand, the representation was as a power series, but Cauchy may have known that much graver problems were to present themselves in the theory of Fourier series. Beyond that was the hint, for the first time, that functions may not, after all, naturally submit to the operations of the calculus. Cauchy's formulation was to give mathematicians the means to explore that new and uncomfortable territory.

The definite integral. In the second part of his *Résumé* of 1823, Cauchy gave new foundations for the integral calculus, starting with his definition of the integral as a limit of sums that has since become known as the *Cauchy integral*.

Cauchy considered a continuous function $y = f(x)$ in a given interval, divided the interval up into n not necessarily equal subintervals $[x_{i-1}, x_i]$, and considered the sum

$$S = \sum_i (x_i - x_{i-1}) f(x_{i-1}).$$

He claimed that if S tends to a finite limit as the number of subintervals increases indefinitely then the limiting value of S can be taken as the integral of the function f over the given interval, and furthermore that this limit exists when the function f is continuous.[26]

He also observed that when f is continuous and n is large, the individual terms make very small contributions to the sum, and he offered a proof that the limiting value of the sum is consequently independent of the choice of subintervals. His argument drew on earlier results in the *Résumé* and the *Cours*, and he concluded:[27]

> if we let the numerical values of these elements decrease indefinitely by increasing their number, the value of S will end by being sensibly constant or, in other words, it will end by attaining a certain limit that will depend uniquely on the form of the function $f(x)$ and the extreme values x_0, X attributed to the variable x. This limit is what we call a definite integral.

Once again, Cauchy made a claim in the course of his proof that required a justification that he did not give, and probably did not realise was necessary, but this time his claim was one that later mathematicians were able to establish.

---

[26]See Leçon 21, pp. 122–127. Riemann later modified the definition by taking the sum

$$S = \sum_i (x_i - x_{i-1}) f(x_{i-1} + \varepsilon_{i-1}(x_i - x_{i-1})), \text{ where } 0 \le \varepsilon_{i-1} < 1,$$

so the function is evaluated at an arbitrary point within each interval.

[27]Cauchy, *Résumé*, p. 125, in (Bottazzini 1986, 144).

Cauchy went on to investigate whether certain kinds of discontinuous functions also have integrals. He looked at functions that become infinite at one or more points of the interval (such as $y = x^{-2}$). We shall not follow him here, but we note that he was interested in defining the integral of a function that is not continuous.

The indefinite integral. This work led Cauchy to the concept of the indefinite integral. In the 'Avertissement' to the *Résumé* Cauchy had written:[28]

> it seemed to me necessary to demonstrate the existence of the integrals or primitive functions in general before making their various properties known. To this end, it was first necessary to establish the idea of integrals taken between given limits or definite integrals.

To do this, Cauchy returned to Leibniz's original conception of the integral as the sum of infinitesimal elements and made it rigorous. This was a decisive break with the common practice in the early 19th century of assuming the existence of the indefinite integral and deriving the definite integral from it, according to the classic formula

$$\int_a^b f(x)dx = F(b) - F(a),$$

where $F'(x) = f(x)$.

To obtain this formula, which is one form of the *Fundamental Theorem of the Calculus*, Cauchy considered what can be regarded as the definite integral of a function $f(x)$ on the interval $[x_0, X]$, where the upper endpoint X is allowed to vary. He set $T(X) = \int_{x_0}^X f(x)dx$, and proceeded to show that the function T is differentiable and its derivative $T'(x)$ is the function $f(x)$. This established the theorem.

He then showed that if a differentiable function defined on a given interval has a zero derivative everywhere then it is constant on that interval. From this he deduced that

> the general value of y, the solution of the equation $dy = f(x)dx$, is given by $y = \int_{x_0}^x f(x)dx + constant$.

This is what he called the 'indefinite integral'. It is interesting that he regarded it as the solution of a differential equation.[29]

Cauchy concluded his *Résumé* with a discussion of differentiation under the integral sign.

16.4 Conclusion

Our account of the rigorisation of the calculus has concentrated mainly on Cauchy, because his work marked a turning point. Cauchy showed how the calculus could be made into rigorous mathematics, although he left much to be done and the standards of his own work sometimes left much to be desired, as Jacobi's comments to Alexander von Humboldt (quoted above in Section 13.3) make clear.

After the publication of the *Cours d'Analyse* and the *Résumé*, mathematicians had no excuse for lapsing back into their old ways, and gradually they renounced them. Prominent among those promoting and extending the new analysis were the German

[28] *Résumé*, p. 10.

[29] See Cauchy, *Résumé*, p. 154.

mathematicians Dirichlet and Weierstrass, who placed the algebra of inequalities at the heart of the rigorous calculus. We look at two of Dirichlet's contributions in Section 20.1, but otherwise we shall not describe that long process, because of the mathematical technicalities involved. Instead, we next look at the deepest area that Cauchy left out of consideration: the nature of the real numbers and the foundations of mathematics.

16.5 Further reading

Anderson, M., Katz, V., and Wilson, R. (eds.) 2009. *Who Gave You the Epsilon? And Other Tales of Mathematical History*, Mathematical Association of America. This is a very readable book with many stimulating articles at about the level of this book, including one by Judith Grabiner on Cauchy.

Bolzano, B. 2004. *The Mathematical Works of Bernard Bolzano*, transl. and ed. by S.B. Russ, Oxford University Press. This book presents Bolzano's major writings and some of his later ideas with short helpful commentaries.

Bottazzini, U. 1986. *The Higher Calculus, A History of Real and Complex Analysis from Euler to Weierstrass*, transl. W. van Egmond, Springer. This is still the best treatment of the subject, and the only one to include complex analysis, but it is not an easy read if the mathematics is unfamiliar.

Bradley, R.E. and Sandifer, C.E. 2009. *Cauchy's Cours d'Analyse: An Annotated Translation*, Springer. This is the first English translation of Cauchy's classic textbook of 1821 — one of the most influential texts in the history of mathematics.

Grabiner, J.V. 1981. *The Origins of Cauchy's Rigorous Calculus*, MIT Press. This is a pioneering study of the way in which Cauchy used the work of his predecessors to rewrite the calculus and create real analysis.

Jahnke, H.N. (ed.) 2002. *A History of Analysis*, American and London Mathematical Societies. This is a collection of essays on various aspects of the history of analysis by many of the present-day experts in the field. It is likely to be the best survey of its subject for a number of years, but one that naturally makes some assumptions about the erudition of its readership.

17

The Foundations of Mathematics

Introduction

In the previous chapter we saw how Cauchy's ideas helped to rigorise the calculus. But new problems in the subject were continually thrown up as the calculus advanced, and they served to show how much still needed to be clarified. Important among these topics was the exact nature of the real numbers: How should they be defined, and how can the most basic theorems in the calculus then be rigorously proved?

As we shall see in Section 17.1, Richard Dedekind gave one of the most successful answers to the first question. His foundational interests fitted well with the interests of his friend Georg Cantor in defining infinite sets, as we see in Section 17.2, but these ideas raised problems in their turn when people's interests turned to providing adequate foundations for the whole of mathematics, as we discuss in Section 17.3. One possibility, explored by Gottlob Frege and Bertrand Russell among others, was to try to define mathematics in terms of logic, but this approach was to fail dramatically, as we describe in Section 17.4. Instead, David Hilbert proposed that a solution might be found by extending both logic and set theory simultaneously, and in Section 17.5 we look at the start of that chain of ideas.

17.1 Dedekind's definition of the real numbers

Throughout the first half of the 19th century, mathematicians assumed that an independent variable, and any functions of that variable, always take numerical values that can be represented by geometrical magnitudes. The rigorous approach that had replaced geometrical intuition by the algebra of inequalities remained dependent upon assumptions about the nature of the real numbers — in particular, that there is an absolute equivalence between magnitudes and real numbers. One of the first to appreciate the unsatisfactory nature of such assumptions, made without adequate investigation or proof, was Dedekind, when he was a young professor at the Zürich Polytechnic.

Figure 17.1. Richard Dedekind (1831–1916)

Dedekind grew up in the city of Brunswick. In 1850 he entered the University of Göttingen to study mathematics, and he obtained his doctorate under the direction of Gauss. He continued at Göttingen as a lecturer, being only the second person (after Liouville) to teach Galois theory (see Section 19.2). At the same time he extended his mathematical education by attending the classes of Riemann and of Gauss's successor, Dirichlet. Riemann became a friend of his, and Dedekind adapted Riemann's highly conceptual approach to the more algebraic topics that interested him particularly.

In 1858 Dedekind accepted a post at the polytechnic in Zürich, where he had to lecture to students on the elements of the differential calculus, but four years later he returned to Brunswick to teach at its newly established polytechnic, where he spent the rest of his life in teaching and research. In 1880 he was elected to the Berlin Academy of Sciences because of his work in number theory. In 1894 he was made an emeritus professor at Brunswick, but continued to give occasional lectures. In 1899 he was able to refute a premature report of his death, which had appeared in the *Mathematicians' Calendar*, by informing the Editor that he had spent that day in the company of Cantor. He survived his reported death by a further seventeen years.

Dedekind's decision to remain in comparative isolation in Brunswick, rather than to move to a more famous centre of mathematical activity, no doubt contributed to a lack of immediate recognition of the value of his work. He was a modest man who attributed his achievements to hard work and the influence of others, rather than to any outstanding talent of his own. Indeed, he once wrote that he was reluctant to publish his account of the real numbers, which he had worked out in 1858. His essay describing this work, *Stetigkeit und irrationale Zahlen* (*Continuity and Irrational Numbers*) was not published until 1872.

In the introduction to his essay Dedekind said that the study was taken up because of his teaching duties at Zürich in 1858.[1] His aim was to give a purely arithmetic foundation for analysis and, in particular, to define the concepts of 'limit' and 'continuity' in a rigorous manner, in contrast to those who argued geometrically, and despite the pedagogical advantages of the intuitive approach. He also wanted to prove rigorously that every quantity that is increasing but bounded approaches a finite limit.[2] He decided to publish his essay because he discovered that others — notably, the mathematicians Cantor and Edouard Heine — were thinking along similar lines, and he regarded his own presentation as simpler and clearer.

Once Dedekind had identified 'numbers' as the area of mathematics that could support analysis, he began his essay by reviewing the properties of the rational numbers and the four basic arithmetical operations on them, together with the concept of order (the relation 'less than', written $<$). He noted the familiar point that the rational numbers and magnitudes (defined as lengths of segments of a straight line) are not the same.[3]

In what sense, however, is there a discrepancy between the set of rational numbers and the number line? Dedekind observed that there are infinitely many points on the line (for example, $\sqrt{2}$ and π) that have no corresponding rational number — this is, in effect, the Greek problem of incommensurables. It is therefore necessary to create new numbers to extend the set of rational numbers and 'fill in the gaps', so that numbers shall have the same *continuity* (or *completeness*) as the line.[4] Just as the rational numbers can be created from the integers alone, so these new numbers must be defined by means of the rational numbers alone.[5]

Dedekind's solution to this problem was one of those ideas that seems obvious once you have seen it, but is not at all obvious in advance. The straight line ℓ and the set of rational numbers R differ because the set of rational numbers has 'gaps' in it. The aim is to plug the gaps with numbers. Here is how he reflected on his ultimate success:[6]

Dedekind on the real numbers.

> By vague remarks upon the unbroken connection in the smallest parts obviously nothing is gained; the problem is to indicate a precise characteristic of continuity that can serve as the basis for valid deductions. For a long time I pondered over this in vain, but finally I found what I was seeking. This discovery will, perhaps, be differently estimated by different people; the majority may find its substance very commonplace. It consists of the following. In the preceding section attention was called to the fact that every point p of the straight line produces a

[1] See (Dedekind 1872), and F&G 18.C1(a).

[2] 'Quantity' was the all-purpose terms for numbers, magnitudes, and possible values of a variable; Dedekind's work on real numbers can be read as a way of making this vague term precise.

[3] See (Dedekind 1872), and F&G 18.C1(b). Note that Dedekind used R to denote the set of *rational* numbers.

[4] Dedekind's use of the term 'continuity' as a property of the line can be confusing. It had nothing to do with the continuity of a function, and it may be helpful to replace it throughout with the word 'completeness' or 'connectedness'.

[5] There were also other definitions of the real numbers. Cantor, for example, produced a definition of a real number that formalised the intuitive idea that a real number is the limit of converging sequences of rational numbers. See (Cantor 1872), and F&G 18.C2.

[6] See (Dedekind 1872), and F&G 18.C1(c).

separation of the same into two portions such that every point of one portion lies to the left of every point of the other. I find the essence of continuity in the converse, i.e., in the following principle: *If all points of the straight line fall into two classes such that every point of the first class lies to the left of every point of the second class, then there exists one and only one point which produces this division of all points into two classes, this severing of the straight line into two portions.*

His solution is brilliant: each gap is a number! It is defined by all the rational numbers less than it, and by all the rational numbers greater than it. (A gap, after all, is 'stuff to the left', 'stuff to the right', and a hole in between!) To be precise, what Dedekind called a *cut* (*Schnitt*), and denoted by (A_1, A_2), is a pair of subsets of R such that every member of A_1 is less than every member of A_2, and every rational number is in either A_1 or A_2, as the next extract explains.[7]

From the last remarks it is sufficiently obvious how the discontinuous domain R of rational numbers may be rendered complete so as to form a continuous domain. [Earlier] it was pointed out that every rational number a effects a separation of the system R into two classes such that every number a_1 of the first class A_1 is less than every number a_2 of the second class A_2; the number a is either the greatest number of the class A_1 or the least number of the class A_2. If now any separation of the system R into two classes A_1, A_2, is given which possesses only *this* characteristic property that every number a_1 in A_1 is less than every number a_2 in A_2, then for brevity we shall call such a separation a *cut* and designate it by (A_1, A_2). We can then say that every rational number a produces one cut or, strictly speaking, two cuts, which, however, we shall not look upon as essentially different; this cut possesses, besides, the property that either among the numbers of the first class there exists a greatest or among the numbers of the second class a least number. And conversely, if a cut possesses this property, then it is produced by this greatest or least rational number.

But it is easy to show that there exist infinitely many cuts that are not produced by rational numbers. Dedekind went on to call such cuts 'irrational numbers', and the set of all cuts he called the 'real numbers', which we denote by $\mathbb{R}$.

The crucial point in the italicised passage in the first extract is that there *exists* a point that produces this division. Every division in ℓ (of the kind he is considering) is made in this way — the line has no gaps. Dedekind's hope was that the gaps in R can be plugged with his cuts. Some cuts would correspond to, or be produced by, rational numbers, of course, but others would not. He then said:[8]

Whenever, then, we have to do with a cut (A_1, A_2), produced by no rational number, we create a new, an *irrational* number α, which we regard as completely defined by this cut (A_1, A_2).

The set of all cuts, however obtained, he called the *real numbers*. It may not be obvious that his real numbers satisfy the desired property, until you start thinking about

[7]See (Dedekind 1872), and F&G 18.C1(d).

[8]See (Dedekind 1888, 15).

it. A theorem, which Dedekind provided, is required to show that his set of all real numbers does not in turn admit gaps.[9]

Dedekind next considered the standard arithmetical operations on the real numbers. He discussed only the operation of addition in detail, because the other three basic operations can be derived from it or handled similarly. He then passed on to powers, roots, and logarithms, pointing out that only now could one justify operations on radicals, such as $\sqrt{2} \times \sqrt{3} = \sqrt{6}$. Such operations had been accepted without challenge in algebra, but had been justified only for geometrical *magnitudes.*

Dedekind was also clear that the real numbers that he was defining were not part of a description of actual lines in space, but were part of a formal structure. After apologising for what he worried might strike some of his readers as the commonplace nature of his proposal, he spelled this out:[10]

> I may say that I am glad if every one finds the above principle so obvious and so in harmony with his own ideas of a line; for I am utterly unable to adduce any proof of its correctness, nor has any one the power. The assumption of this property of the line is nothing else than an axiom by which we attribute to the line its continuity, by which we find continuity in the line. If space has at all a real existence it is not necessary for it to be continuous; many of its properties would remain the same even were it discontinuous. And if we knew for certain that space was discontinuous there would be nothing to prevent us, in case we so desired, from filling up its gaps, in thought, and thus making it continuous; this filling up would consist in a creation of new point-individuals and would have to be effected in accordance with the above principle.

The point of his definition was to enable crucial theorems in analysis to be proved — in particular, the theorem that 'every magnitude which grows continually, but not beyond all limits, must certainly approach a limiting value'.

Dedekind next showed how this theorem could now be proved, as follows:[11]

> I prove it in the following way. By hypothesis there exists one and hence there exist infinitely many numbers α_2 such that [a magnitude] x remains continually $< \alpha_2$; I designate by A_2 the system of all these numbers α_2, by A_1 the system of all other numbers α_1; each of the latter possesses the property that in the course of the process x becomes finally $\geq \alpha_1$, hence every number α_1 is less than every number α_2 and consequently there exists a number α which is either the greatest in A_1 or the least in A_2 [by the theorem above]. The former cannot be the case since x never ceases to grow, hence α is the least number in A_2. Whatever number α_1 be taken we shall have finally $\alpha_1 < x < \alpha$, i.e., x approaches the limiting value α.

He then concluded his essay with a section relating his theorem on the continuity of real numbers to infinitesimal analysis.[12]

It might be claimed that Dedekind completed missing elements in Cauchy's attempt to make the calculus rigorous, because his work established the continuity or completeness of the real numbers, and thus laid a firm foundation for what Cauchy had assumed about numbers. As to whether all references to motion and geometry ought to be avoided entirely in analysis, Dedekind thought that references to time, motion, and space can be entirely eliminated, but he condoned their use for pedagogical reasons.

---

[9]See (Dedekind 1872, 20), and F&G 18.C1(e).

[10]See (Dedekind 1872, 11), and F&G 18.C1(e).

[11]See (Dedekind 1872, 25), and F&G 18.C1(f).

[12]See (Dedekind 1872, 25–27), and F&G 18.C1(f).

Dedekind's approach is one of many that collectively formed what was called the 'arithmetisation of analysis', a process that gave arithmetical foundations to the calculus in all its forms. Many others contributed, most notably Weierstrass, his pupil Heine, and Cantor.

In Cantor's view an irrational number was usually studied by means of a suitable converging sequence of rational numbers (such as the decimal approximations to $\sqrt{2}$ or π). It is possible to say that such a sequence converges without there being a rational number to which it converges: it is enough that the terms become arbitrarily close to each other. In Cauchy-style formalism:

> the sequence a_n converges if, for all $\varepsilon > 0$, there is a number N such that $m, n > N$ implies $|a_n - a_m| < \varepsilon$.

Can we define a real number as such a sequence? The answer is 'no'. There are technical problems, because two different sequences a_n and a'_n may converge to the same number: this happens when the as and the a's become arbitrarily close to each other. Cantor considered such sequences to be equivalent, and defined a real number to be the whole equivalence class of sequences of rational numbers converging to the same limit. He then showed that this gives exactly the required properties of the real numbers. In particular, repeating the construction gives no new numbers, and so all the gaps in the rationals are filled simultaneously by this construction.

We add that Dedekind did not stop with defining the real numbers as sets of rational numbers, but went on to define the rational numbers in terms of the integers (which is easy) and then to define the integers in terms of sets alone. This was a very radical step, and it is indicative of Dedekind's desire to define everything in mathematics and to take nothing from intuition. He published it in his *Was sind und was sollen die Zahlen?* (*The Nature and Meaning of Numbers*) in 1888. His approach was a complicated one, and we shall not enter into the details.

It is more interesting here to look at a letter that Dedekind wrote in 1888 to his friend, the mathematician Heinrich Weber, in which he urged him not to accept Cantor's definition of the concept of number as fundamental, but as resting on a more primitive idea of counting — that is to say, we should not explain what 'one', 'two', 'three' and so on mean, via some concept of size, but as the outcome of going 'first', 'second', 'third' and so on.

We explain Cantor's approach in the next section, but it attempts to formalise the idea that the one thing that all the various sets with, say, six elements have in common is that they have the same number of elements (in this case, six) *because* they can be put into a one-to-one correspondence with each other but with no other sets; so the collection of all such sets can be taken as the size or *cardinality* of the set. For this reason, Cantor is said to have regarded the basic concept of number as that of *cardinal numbers*, whereas Dedekind defined them as *ordinal numbers* ('ordered numbers'). Dedekind's objection can now speak for itself:[13]

> But if one were to take your route — and I would strongly urge that it be explored once to the end — then I would advise that by number (*Anzahl*, cardinal number) one understand not the class itself (the system of all finite systems that are similar to each other) but something new (corresponding to this class) which the mind creates. We are a divine race and undoubtedly possess creative power, not merely in material things (railways, telegraphs) but especially in things of the mind.

[13]See (Dedekind 1930–1932, Vol. 3, 488–490); English transl. in (Ewald 1996, Vol. 2, 834–835).

This emphasis on mathematics as a free activity of the human mind was one that Dedekind shared with many leading mathematicians of his day. It is a key feature of the shift at that time towards modern pure mathematics.

After a short digression, Dedekind went on to complain that on Cantor's approach:

> one will say many things about the class (e.g. that it is a system of infinitely many elements, namely, of all similar systems) that one would apply to the number only with the greatest reluctance; does anybody think, or won't he gladly forget, that the number four is a system of infinitely many elements? (But that the number four is the child of the number three and the mother of the number five is something that nobody will forget.)

Dedekind's work was not always received enthusiastically by his contemporaries — his letter to Weber carries the remark 'I am delighted that you take such an interest in my article on numbers; not many do so'. In particular he was opposed by the influential Berlin mathematician Leopold Kronecker, who believed that the integers were fundamental and need not (perhaps could not) be defined in terms of more basic objects. Moreover, there is a price to be paid for accepting either Dedekind or Cantor's definition of the real numbers: the admission into mathematics of infinite sets. This was a momentous step, the full implications of which were not clear at first and were soon to prove controversial. Indeed, the very work that had made the calculus rigorous soon seemed to threaten the essential nature of mathematics. This alarming discovery arose out of further work by Cantor, to which we now turn.

17.2 Cantor, sets, and the infinite

Georg Cantor was born of Danish parents living in St Petersburg in 1845. The family moved to Frankfurt in Germany when he was 11. He began his university studies at the Polytechnic in Zürich when he was 17, but transferred the following year to Berlin University, attracted there by its high reputation. In Berlin, Cantor was taught by some of the most outstanding mathematicians of the day, notably Karl Weierstrass, Ernst Edouard Kummer, and Leopold Kronecker. In 1869 he became a teacher at the University of Halle, where he was to remain for the rest of his life, always hoping that a post at Berlin would be offered to him. Here we concentrate upon his two most notable achievements: the development of set theory, and his theory of transfinite numbers.

Problems and paradoxes of infinity. Problems relating to the infinite had been raised as far back as the time of the Greek philosopher Zeno and his school (5th century BC). That there were paradoxes associated with the idea of infinity was recognised anew in the 17th century by Galileo. In his *Two New Sciences* (1638), he made one of his characters, Salviati, raise the paradox involved in attempting to count the number of perfect squares. The principle of counting involves putting a collection of things into *one-to-one correspondence* with the natural numbers. Thus, when counting the perfect squares, we establish the following correspondence

$$\begin{array}{ccccccccc} 1 & 2 & 3 & 4 & 5 & 6 & 7 & 8 & \dots \\ \updownarrow & \updownarrow & \updownarrow & \updownarrow & \updownarrow & \updownarrow & \updownarrow & \updownarrow & \\ 1 & 4 & 9 & 16 & 25 & 36 & 49 & 64 & \dots \end{array}$$

There is no problem if the collection of things is finite, since the lower sequence ends, and we can answer the question 'how many things are there?'. If, however, the

Figure 17.2. Georg Cantor (1845–1918)

collection is infinite, as here, then the sequence of natural numbers in the correspondence does not end. This seems quite clear and straightforward — until we enquire whether there are as many perfect squares as natural numbers. The one-to-one correspondence seems to give an affirmative answer to this, but our common sense tells us that most of the natural numbers are missing from the sequence of perfect squares. As Salviati says:[14]

> If I should ask how many squares there are, one might reply truly that there are as many as the corresponding number of roots, since every square has its own root and every root its own square ... This being granted, we must say that there are as many squares as there are numbers ... yet there are more numbers than squares, since the larger portion of them are not squares.

This led Galileo to proclaim (through Salviati) that:

> Infinities ... transcend our finite understanding ... The attributes 'equal', 'greater', and 'less' are not applicable to infinite, but only to finite quantities.

It seemed as though infinite collections cannot be discussed; the finite human mind can grasp only finite things. Following Aristotle, generations of mathematicians had accepted the concept of the infinite as something *potential* rather than *actual*: there are always more numbers than in any finite set you can imagine, but you cannot contemplate *all* numbers at once.

Not until the 19th century did mathematicians come to grips with *actually infinite* collections. Bolzano, in his posthumously published *Paradoxien des Unendlichen* (*Paradoxes of the Infinite*) of 1851, a work strongly grounded in theology, defended the actually infinite, noting that it was a property of infinite collections that (contrary to the fifth common notion of Euclid's *Elements*) the whole is not necessarily greater in

[14]Galileo, *Dialogues Concerning Two New Sciences*, pp. 26, 32.

number than its parts. Later in the century, Dedekind took the Galilean paradox to be the *defining property* of infinite collections:[15]

> A system S is said to be *infinite* when it is similar to a proper part of itself; in the contrary case S is said to be a *finite* system.

Here Dedekind used the term *similar* to mean 'in a one-to-one correspondence with' and a *proper part* to be a part of a collection that is not the whole collection itself. Thus, a set is infinite if and only if it can be put in a one-to-one correspondence with a proper subset of itself.

Cantor's earliest work, published in Crelle's *Journal*, was concerned with the role of sets of real numbers in analysis — specifically in the theory of Fourier series. In the course of his investigations he was led to ask questions about the size of these sets — for example, how many real numbers are there? As we shall see, he also came across some highly unexpected correspondences between infinite collections of points of very different kinds. This led him to investigate infinite point sets in their own right.

Cantor's 'set theory' was presented in a number of papers appearing from 1874 onwards. We quote here from a later presentation of his work, in which he explained his idea of a set (*Menge* in German) and of the size (power or cardinal number) of a set. It is interesting to see how he thought of these objects as being made accessible through the power of our thought, as well as to see what he thought he had proved and what still needed proof.[16]

Cantor on cardinal numbers.

> By a 'set' we understand any collection M of definite and separate objects of our intuition or our thought (which will be called the 'elements' of M) gathered into a whole ...
>
> We call any other set M_1 whose elements are also elements of M a 'subset' or 'partial subset' of M. If M_2 is a subset of M_1 and M_1 is a subset of M then M_2 is also a subset of M.
>
> Every set M has a definite 'power', which we will also call its 'cardinal number'.
>
> *We call the 'power' or 'cardinal number' of M the general concept that arises from M by means of our capacity for active thought when we abstract the properties of its different elements and the order in which they appear.*
>
> The result of this twofold act of abstraction, the cardinal number or power of M, we denote by
>
> $$\overline{\overline{M}}.$$
>
> Because, if we abstract its nature from each element it becomes a 'unit', the cardinal number $\overline{\overline{M}}$ itself becomes a definite set composed of these units that exists as an intellectual image or projection of the given set M in our minds.

[15]See (Dedekind 1888, 64).

[16]See (Cantor 1895, 481).

We call two sets M *and* N *'equivalent' and denote this by*

$$M \sim N \quad \text{or} \quad N \sim M$$

if it is possible to relate them to each other in such a way that each element of each one corresponds to one and only one element of the other. Then, to each subset M_1 of M there corresponds a definite equivalent subset N_1 of N and conversely ...

Every set is equivalent to itself:

$$M \sim M.$$

If two sets are equivalent to a third, then they are equivalent to each other:

$$M \sim P \text{ and } N \sim P \text{ implies } M \sim N.$$

In fact, by the above definition of the power, the cardinal number of $\overline{\overline{M}}$ remains unchanged if any, or even all, of its elements are replaced by other things. If now $M \sim N$ then there is a correspondence between M and N by means of which the element m corresponds to the element n. We can now replace each element m by the corresponding element n, and this transforms M into N without altering its cardinal number, consequently

$$\overline{\overline{M}} = \overline{\overline{N}}.$$

[Cantor then showed that two sets are equivalent if and only if they have the same cardinal number. His argument hinged on the equivalence $M \sim \overline{\overline{M}}$]

'Greater' and 'lesser' for powers

Let the two sets M and N with cardinal numbers $a = \overline{\overline{M}}$ and $b = \overline{\overline{N}}$ satisfy the following conditions:

1. There is no subset of M that is equivalent to N,
2. There is a subset N_1 of N such that $N_1 \sim M$.

It is then evident that these conditions are still satisfied when M and N are replaced by equivalent sets M' and N', and so they express a definite relationship between the cardinal numbers a and b. Moreover, the equivalence of M and N, and so the equality of a and b, is impossible. For, if $M \sim N$ then, because we have $N_1 \sim M$ also $N_1 \sim N$ and therefore, if $M \sim N$ it must be the case that there is a subset M_1 of M such that $M_1 \sim M$ and so $M_1 \sim N$ which contradicts condition 1 ...

We express the relationship between a and b characterised by (1) and (2) above by saying that a is smaller than b or b is greater than a, in symbols:

$$a < b \quad \text{or} \quad b > a.$$

One easily shows that if $a < b$ and $b < c$ then always $a < c$. Likewise, it follows without more work that if P_1 is a subset of a set P then $a < \overline{\overline{P_1}}$ implies $a < \overline{\overline{P}}$ and $\overline{\overline{P}} < b$ also implies $\overline{\overline{P_1}} < b$.

We have seen that of the three conditions

$$a = b, \quad a < b, \quad b < a,$$

each one excludes the others.

This in no way implies, what cannot be proved at this point in our process of thought, that given any two cardinal numbers $\mathfrak{a}$ *and* $\mathfrak{b}$ *precisely one of these three conditions must necessarily hold.*

This definition of a set is quite abstract: it allows sets to comprise very different kinds of things. Thus the letters of the alphabet constitute a set, as do the natural numbers, the rational numbers, the real numbers, functions with a particular property, and sets themselves. Two sets are said to be 'equivalent' or, informally, 'of the same size', if there is a one-to-one correspondence between their respective elements, in which case they are also said to have the same 'power' or, in Cantor's later terminology, *cardinal number*.

But what, precisely, is the cardinal number of a set? For finite sets, it is the number of elements in the set, and there is no problem in principle in determining whether the cardinal number of one set is greater than, equal to, or smaller than that of another. But what about infinite sets? Can there be infinite sets of different sizes? It is not clear what such a question means, and intuition suggests conflicting answers. To understand what is going on, a set is said to be *countable* if it can be put into a one-to-one correspondence with the set of all natural numbers. The correspondence then 'counts' the set by labelling the elements of the set as the first, second, third, and so on. Our question then becomes: Is every infinite set countable? If the answer to this question is 'no' then the next questions are: Does every infinite set have a cardinal number, and if so, how can it be determined?

The above extract shows that, even after over twenty years' work, Cantor's concept of cardinal number was still vague. It was something like a collection of all sets that can be put in one-to-one correspondence with each other — so there was a collection of sets each with precisely one element, another for all the sets with precisely two elements, another for all the sets with precisely three elements, and so on, and these collections formed the cardinal numbers 'one', 'two', 'three', etc. This approach was later shown to be incoherent — it was destroyed by paradoxes that involve defining phrases of the form 'the set of all sets with such-and-such a property' that we discuss below — but even if there were no such flaw, one can agree with Dedekind's remark that this is an unwieldy way to define a very natural concept that everyone learns.

Cantor called the cardinal number of the natural numbers $\aleph_0$.[17] It is called a *transfinite number* because it is larger than any finite cardinal $1, 2, 3, \ldots$.

What about the infinite set of rational numbers? Is it countable? Intuitively, this seems unlikely. For instance, between any two distinct rational numbers r and s there is always another (we can take their average, $\frac{1}{2}(r+s)$). It follows that there are infinitely many rational numbers between any two rational numbers, and indeed between any two natural numbers. But between any two natural numbers m and n, with $m < n$ there are only $n - m - 1$ natural numbers — for example, no natural number lies between 2 and 3. Does this disparity mean that there are far more rational numbers than natural numbers? This would mean that the set of rational numbers would have

[17] This symbol is pronounced 'aleph-zero'; aleph is the first letter of the Hebrew alphabet.

Box 48. The set of positive rational numbers is countable.
We first arrange the positive rational numbers in a grid, as shown.

1/1	2/1	3/1	4/1	...
1/2	2/2	3/2	...	...
1/3	2/3	...	...	...
1/4	...	...	...	...

Then we establish a one-to-one correspondence with the natural numbers by threading our way diagonally through the grid, following the line that goes:

1/1, 2/1, 1/2, 1/3, 2/2, 3/1, 4/1, 3/2, 2/3, 1/4, and so on.

We omit any numbers equal to those already obtained (such as 2/2), and the correspondence is

1	2	3	4	5	6	7	8	9	...
↕	↕	↕	↕	↕	↕	↕	↕	↕	
1/1	2/1	1/2	1/3	3/1	4/1	3/2	2/3	1/4	...

so, for example, the rational number 2/3 corresponds to the natural number 8.

a cardinal number greater than $\aleph_0$, or, to put the point another way, that the set of rational numbers is not countable.

This intuitive guess is wrong, however. Cantor gave two proofs, in 1874 and 1895, that the set of rational numbers is in fact countable. The second proof, the one most widely used today, is carried out by arranging all the positive rational numbers m/n in an array, as shown in Box 48. It establishes a one-to-one correspondence between the set of natural numbers and the set of positive rational numbers; it is then straightforward to show that there is a one-to-one correspondence between the set of natural numbers and the set of all rational numbers.

In his 1874 paper, Cantor proved another surprising result: The set of all algebraic numbers is countable. This set comprises all the numbers that are solutions of polynomial equations with integer coefficients:

$$a_n x^n + a_{n-1} x^{n-1} + \cdots + a_1 x + a_0 = 0.$$

The set of all algebraic numbers includes all the rational numbers, since the rational number m/n is a solution of the equation $nx - m = 0$. But it also includes many irrational numbers: for example, $\sqrt{2}$, which is a solution of $x^2 - 2 = 0$, and $\sqrt{2} + \sqrt{3}$, which is a solution of $x^4 - 10x^2 + 1 = 0$. The set of rational numbers seems to form a rather small subset of the set of all algebraic numbers, so it is perhaps surprising that the set of algebraic numbers is also a countable set.[18]

The countability of the set of all rational numbers, and of the set of all algebraic numbers, might suggest that every infinite set is countable. Indeed, in a letter to Dedekind of 1873, Cantor asked whether the set of real numbers is also countable. A few weeks later he was able to write and say that it is not, and so he had discovered an infinite set

[18]We omit the proof — see (Cantor 1874a), and F&G 18.C4(a).

Box 49. The set of real numbers between 0 and 1 is uncountable. Cantor proved this by a reductio ad absurdum argument. Assume that the set of real numbers between 0 and 1 is countable. Write each of these numbers as an infinite decimal, putting, for example, 0.5 in the form 0.5000 ... rather than 0.4999 ... to avoid ambiguities. Because, by assumption, this set is countable, the real numbers between 0 and 1 can be put into a list and each assigned to a natural number. We would thus have a list of the following form:

1. $0.a_1a_2a_3a_4\ldots$
2. $0.b_1b_2b_3b_4\ldots$
3. $0.c_1c_2c_3c_4\ldots$
4. $0.d_1d_2d_3d_4\ldots$

Cantor now defined a new real number $0.abcd\ldots$, where a differs from a_1, b differs from b_2, c differs from c_3, d differs from d_4, and so on, and no $a, b, c, d, \ldots$ is 0 or 9. This number differs from each number in the list because it differs from the nth number in the nth decimal place. But the list was assumed to include *all* the real numbers between 0 and 1.

This contradiction tells us that the original assumption must have been false and so the set of real numbers between 0 and 1 is uncountable. It follows that the set of all real numbers is uncountable.

larger than $\aleph_0$. He published one proof in his 1874 paper and a simpler proof in 1891.[19] We outline the later argument in Box 49.

Clearly, the uncountability of the set of all real numbers must arise from non-algebraic irrational numbers. These last numbers were termed 'transcendental' because they 'transcend' the operations of algebra. Note that they have been proved to exist because of the uncountability of the set of real numbers. This is a *non-constructive existence proof* — that is, a proof that something exists without displaying it or showing how it may be constructed. Cantor's proof does not exhibit even one transcendental number explicitly, let alone an uncountable infinity of them! That transcendental numbers must exist had been shown earlier, by Liouville in 1844, who constructed one explicitly.[20] It was to prove much harder to show that the familiar numbers e and π are transcendental.

So Cantor's surprising conclusion was that there are infinite sets of different sizes. He next argued that a set A is *smaller than* a set B if A can be put into a one-to-one correspondence with a proper subset of B, but B cannot be put into a one-to-one correspondence with a proper subset of A. So the set of natural numbers and the set of rational numbers are both smaller than the set of real numbers. We also say that the cardinal number, or cardinality, of the former set is *smaller than* that of the latter.

---

[19]See (Cantor 1874a), and F&G 18.C4(b).

[20]See (Liouville 1844).

The cardinal number of the real numbers is usually denoted by c (the first letter of 'continuum'). Because the set of reals is not countable, c is a transfinite number larger than $\aleph_0$. But is c the *next* cardinal after $\aleph_0$, or is there a set that is smaller than the set of real numbers but larger than the set of rational numbers? The conjecture that c is the next cardinal greater than $\aleph_0$ is known as the 'continuum hypothesis', which Cantor stated in this way:[21]

Cantor's 'continuum hypothesis'.

> Since we are led in this way to an extraordinary rich and broad domain of manifolds with the property that they can be put into a unique and complete correspondence with a line or a part of a line ... the question arises of how the different parts of a continuous straight line, i.e. the thinkable different manifolds of points, relate to their respective powers. If we clothe this problem in its geometric dress and understand ... by a linear manifold of real numbers any thinkable collection of infinitely many distinct real numbers, then one can ask into how many and what classes the linear manifolds fall, if manifolds of the same power are put into one and the same class and manifolds of different powers into different classes. By an inductive process, which we will not present here, we are led to the theorem that the number of classes which arise according to this principle of classification is equal to *two*.

Cantor's idea was that the real line certainly contains a countable infinite set (for example, the set of rational numbers) but is itself uncountable. If its power (cardinality) were larger than that of the smallest uncountable set, then presumably it would contain a proper subset that had the same power as the smallest uncountable set, and so the number of classes that Cantor was counting would be at least 3.

Despite great efforts, Cantor was never able to prove his 'inductive process', and in the end it turned out to be invalid. But it was not until 1963 that his continuum hypothesis was shown to be independent of Cantor's other assumptions: it can be neither proved nor disproved.[22]

Cantor also turned his attention to equivalences between the set of all points on a line and the set of all points in a plane. Surely these infinities are of different sizes? Does not the whole of geometry assume a distinction between a line or curve on the one hand and a plane on the other? Cantor and Dedekind corresponded about this in 1877, in the course of which Cantor came to believe that he had found a one-to-one correspondence between the points of the unit interval and those of the unit square — a conclusion about which he memorably said 'I see it, but I don't believe it'.[23] As he pointed out, if this conclusion were correct then we would have to reject making the number of independent coordinates the basis of a definition of *dimension*.

He waited anxiously for Dedekind's reply, which was broadly supportive.[24] In the first letter, Cantor raised the question of mapping a surface onto a line by a one-to-one correspondence. The intuitive answer is that this is not possible. However, in his

---

[21]See (Cantor 1874b), and F&G 18.C5.

[22]After Cantor's informal ideas had been axiomatised, this was shown by the American mathematician Paul Cohen in 1963.

[23]See the letter of 29 June 1877 in *Georg Cantor Briefe*, p. 44.

[24]See the first two extracts from the Cantor–Dedekind correspondence, in (Purkerts and Ilgauds 1985, 32–35), and F&G 18.C3.

Box 50. A one-to-one correspondence between the unit interval and the unit square.

We give Cantor's original version of the proof. As Dedekind pointed out to him, it is not quite rigorous, but can straightforwardly be put right.

Cantor's idea was to make the point of the square with coordinates

$$(0.a_1a_2a_3a_4\ldots,\ 0.b_1b_2b_3b_4\ldots)$$

correspond to the point on the line with coordinate

$$0.a_1b_1a_2b_2a_3b_3a_4b_4\ldots.$$

So, for example, the point of the square with coordinates

$$(0.1357924\ldots, 0.2468045\ldots)$$

corresponds to the point on the line with coordinate

$$0.12345678902445\ldots.$$

Conversely, to each point on the line, there corresponds a point of the square, according to the rule that $0.c_1c_2c_3c_4\ldots$ corresponds to

$$(0.c_1c_3c_5c_7\ldots, 0.c_2c_4c_6c_8\ldots).$$

Thus each point of the square corresponds to a *unique* point of the line, and vice versa — so the correspondence is one-to-one. Therefore the unit square and the unit interval have the same cardinality.

second letter, Cantor accepted that there is such a one-to-one correspondence. The implication is that the intuitive response to the question raised in the first letter is wrong.

Cantor had indeed proved that there is a one-to-one correspondence between a line and a plane, and more generally between a one-dimensional space and an n-dimensional space. He was himself so surprised that he waited for Dedekind to confirm the validity of his proof. In Box 50 we describe a one-to-one correspondence between the unit interval and the unit square — that is, between a one-dimensional figure and a two-dimensional figure.

Dedekind's letter to Cantor of 2 July 1877 indicated that such correspondences are likely to be discontinuous.[25] This is the case with the correspondence in Box 50 — points near to each other in the square do not necessarily correspond to points near to each other on the line. Indeed, all such correspondences have to be discontinuous, as will be seen below. This insistence on continuity rescues geometry: there is no one-to-one correspondence between a line and a plane that is continuous in each direction.

We conclude this section by looking at an Italian mathematician whose work extended the paradoxical conclusions of Cantor. Giuseppe Peano was a professor of mathematics at Turin who had a particular interest in rigorous analysis. In 1890 he produced a remarkable curve: one that passes through every point of a square — a space-filling curve! Because he was a sharp critic of covert appeals to geometry, Peano's

[25]See (Purkerts and Ilgauds 1985, 32–35), and Dedekind's letter in F&G 18.C3.

own definition of the curve was entirely analytical — that is, it involved formulas but made no appeal to visual intuition. However, other mathematicians felt less inhibited and gave a pictorial definition of the curve, as we shall now see.

The approach of the American mathematician Eliakim Hastings Moore in 1890 was to draw successive approximations to the space-filling curve, as follows. In Figure 17.3 we start on the left with the square and a diagonal: this diagonal is the first approximation to the Peano curve. The second approximation, shown in the middle, is obtained by dividing the square into nine equal squares, and replacing the original diagonal by the nine new diagonals as shown: these nine diagonals, traced as numbered, form the second approximation to the curve. To form the third approximation (shown on the right), replace each square and diagonal in the second approximation by nine smaller squares and diagonals, following the same basic pattern. The Peano curve is the limit (in a sense that Peano made precise) of this succession of approximations.

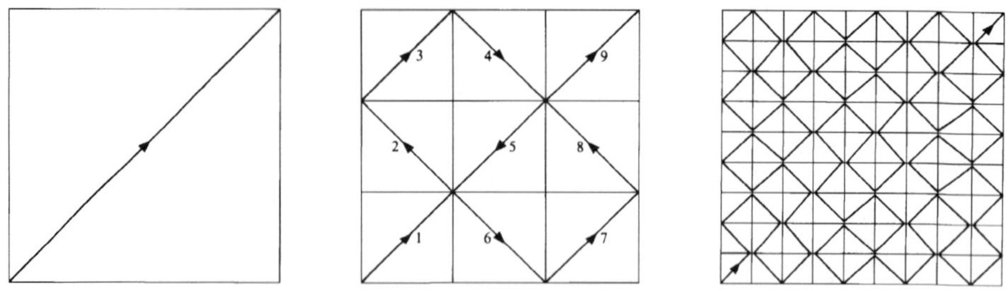

Figure 17.3. Peano's space-filling curve, as described by E.H. Moore

Although it is not obvious that this limiting process produces anything of consequence, the result turns out to be a *continuous* map from the unit interval onto the unit square. This again raised the spectre that there might be no significant mathematical distinction between a curve and a surface, and that Dedekind's conjecture of 2 July 1877 was false, but Peano pointed the way out: some points of the square are visited more than once by the curve, so the map is not one-to-one. Later it was shown that there are no continuous one-to-one maps from a curve onto a surface, and so the intuitive idea of dimension can indeed be made precise.

17.3 Foundational questions

Cantor's work contained many novel and disturbing conclusions. His claims, and the questions that they raise, need to be clarified if we are to understand the reception of his ideas over the subsequent fifty years.

- What is a set? How is the concept to be defined, or is it an undefinable basic notion?
- There are finite and infinite sets, and among each kind are sets that are 'numbers', which yield finite and *transfinite* arithmetic.
- Can mathematicians reason validly about infinite sets?
- There is the claim that mathematics is basically about sets, and it is these objects about which true mathematical statements are made.

- Without Cantor's theory of infinite sets, can certain problems in mathematics, notably in analysis, ever be resolved (or even stated in the first place)?

Cantor thought long and hard about these issues. He was convinced that sets have a real existence, and that his set theory was therefore absolutely true — that is, it makes valid statements about existing objects. A deeply religious man, he believed that his ideas came from God, and that the transfinite sets exist as ideas in the mind of God. J.W. Dauben, a biographer of Cantor, summarised Cantor's views by saying that, for Cantor:[26]

> Consistency alone was the determining factor in any question of mathematical existence, since God could realise any 'possibility', and by possibility Cantor meant that the ideas capable of realisation *in concreto* be only consistent. In this respect he seems very much a latter-day Leibnizian, believing that it would have contradicted God's omnipotence had he been unable to realise any possible, that is, consistent idea. But Cantor did not go so far as to insist that such possibilities actually had a physical existence somewhere in the phenomenal world. He would only say that if ideas were consistent, then they were possibilities, and as possibilities they had to exist in the mind of God as eternally true ideas; this was sufficient to confer upon them the right to mathematical existence.

Like any sincere religious person, Cantor was sensitive to the problems caused by our imperfect human minds with their limited understanding. He hoped that his ideas about the infinite would help the Church to reach a better understanding of God, but he also struggled to improve his own grasp of the infinite. The problems came with his infinite numbers and transfinite arithmetic, and Cantor changed his mind about what a number actually is. Up to the mid-1890s he had defined the cardinal number of a set in this way:[27]

> We will call by the name 'power' or 'cardinal number' of M the general concept which, by means of our active faculty of thought, arises from the aggregate M when we make abstraction of the nature of its various elements M and of the order in which they are given.

This is somewhat vague, and when writing to Dedekind in 1899, Cantor proposed another definition:[28]

> When a set is presented, I call the general idea which it and only sets equivalent to it give rise to, its cardinal number or its power.

On this approach, the *cardinal number* of a set is the set of all sets that can be put into a one-to-one correspondence with it — so the cardinal number 5 is the set of all sets with five elements in them. The fate of this idea will concern us later, when we shall see that its inadequacies imperilled much of Cantor's whole scheme.

Cantor did have his supporters: Klein was happy to publish papers that the Berlin authorities did not appreciate in his *Mathematische Annalen*, and in 1891 Cantor was elected the first president of the German Mathematical Union. But his work was not accepted by everyone. The most forceful dislike for it was expressed by Leopold Kronecker.

---

[26]See (Dauben 1979, 229).
[27]See (Cantor 1895, 481–512).
[28]See (Cantor 1932, 444).

Kronecker had been taught at school by Kummer, and graduated in 1841 from the University of Berlin, where he had studied under Dirichlet and Steiner. His interests lay in the fields of number theory and elliptic functions, and although the family business was sufficiently successful for him not to need to seek a university position, he became a member of the Berlin Academy of Sciences in 1861 and actively helped to propose other mathematicians for membership. His position in the Academy allowed him to lecture at the University, which he did, confining himself to advanced topics that few could follow. To those who kept up, however, he was an inspiring and supportive teacher. In 1880 he became an editor of Crelle's *Journal*, and in 1883 he succeeded Kummer as a professor (a position Kummer had held in Berlin since 1855, when he succeeded Dirichlet) and became a very influential figure. This was distressing to the other senior professor, Weierstrass, for by then Kronecker was moving more and more to the position that mathematicians should concern themselves only with objects that can be defined in finitely many steps. These included the integers but, despite his prowess as an analyst, Kronecker seemed increasingly to reject the real numbers, because they require infinite processes for their definition. Such a stricture applied equally to Cantor's sequences and to Dedekind's cuts. Kronecker's attitude is often summed up in his famous dictum:[29]

> The good Lord made the integers; all the rest is the work of man.

And writing to Ferdinand Lindemann about the latter's proof that π is transcendental, he said:[30]

> What use is your beautiful investigation regarding π ... since irrational numbers do not exist?

There is no doubt that Cantor found Kronecker's hostility hard to bear, and felt that Crelle's *Journal* was unreceptive to him. In his biography of Cantor the mathematician Arthur Schoenflies wrote:[31]

> Kronecker's attitude inevitably conveys the impression that Cantor, in his capacity as a researcher and teacher, was a corrupter of the youth.

This was an extraordinary thing to say. It went so far beyond the bounds of normal academic politeness, and was surely libellous if false, that it has to be taken as evidence of hostility, however much it was also the opinion of its author.

We also know, from a letter that the Swedish mathematician Gösta Mittag-Leffler wrote to Cantor on 17 January 1884, that Kronecker had let it be known that he 'was thinking of submitting a work [to *Acta Mathematica*, the journal edited by Mittag-Leffler] in which he would show that the results of modern function theory and set theory are of no real significance'. He added, somewhat sarcastically, that he hoped that 'I would publish his work with same impartiality as I publish my 'friend Cantor's'.[32] It is not unknown for one mathematician to publish a paper attacking the work of another, but it would have been a serious blow if it had landed. Mediocre work is usually left to wither with time.

It is often said that Kronecker referred to Cantor as a 'scientific charlatan'. The historical situation is somewhat more complicated. In 1891 Cantor complained that

---

[29] See (Weber 1891, 19).

[30] See (Weber 1891, 19). It has been suggested that this remark was intended in jest.

[31] See (Schoenflies 1927, 2).

[32] See (Cantor 1999, 166).

Kronecker had criticised his work in front of impressionable students and called it 'sophistry'. Kronecker's lectures have recently been published, and in them Kronecker complained of sophistry in the current fashion for mixing philosophy and mathematics, which is exactly what Cantor was accused of.[33] Kronecker mentioned no-one by name in his strictures on philosophy at this point, and Cantor was not the only one doing this, but his remarks are far from exonerating Cantor. There was no reason for Kronecker to name names when his target would have been widely recognised, and to this day lecturers deplore certain activities but withhold names, for a variety of reasons. Finally, a 'sophist' is a deliberately confusing or fallacious reasoner — not much different from a scientific charlatan, who is a pretender to scientific knowledge.

For all these reasons, the intellectual disagreements between Kronecker and Cantor hovered on the brink of an open feud.[34] There were partial reconciliations, but Kronecker's central position in Berlin, compared with Cantor's marginal one in Halle, often left Cantor feeling that the world was against him. This, coupled with the strains of pursuing his highly novel and often counter-intuitive research, increased the pressure that he was under. Sadly, Cantor's mental health broke in 1884, again in 1899 (not long after the death of his youngest son, aged only 13), and increasingly often from 1904 until his death in 1918.

It is not easy to say what caused Cantor these personal problems. The historian Ivor Grattan-Guinness gave careful consideration to Cantor's personal history, and found that he was always liable to bouts of disabling depression.[35] Accordingly, he considered that the academic conflicts that Cantor experienced were 'little more than the clap that starts the avalanche'. It is interesting to note that other mathematicians of the period also suffered nervous collapses, occasioned in part by the pressure of work — for example, Klein and Weierstrass. Plainly many factors were involved, Cantor's high sense of his own mission among them. Dauben even goes so far as to suggest that Cantor saw himself as being 'God's messenger to mathematicians everywhere', which would have raised the stakes considerably: inspiring Cantor to his most original ideas, but also making any inadequacies in his work impossible to concede.[36]

Other mathematicians also had ideas about the infinite. Paul du Bois-Reymond, who had studied under Weierstrass in Berlin and had discovered some highly counter-intuitive facts about Fourier series, had been led to propose a theory of infinitesimals. Although his work grew from the same starting point as Cantor's, Cantor was scathing about such numbers. According to Cantor, they could not exist; they were a 'cholera bacillus' infecting mathematics.[37] It would be churlish to point out that something that does not exist can hardly be a bacillus, but Cantor's invective is instructive. Cantor's theory of infinite sets does not generate a theory of infinitesimals, but that does not mean that such a theory cannot exist — yet Cantor does not sound as if he was merely upbraiding a colleague for making a mathematical mistake.

If we ask what feature of Cantor's attitude towards sets might account for the intensity of his rejection of infinitesimals, one answer would be his belief that all consistent ideas co-exist in the mind of God, which allowed him to feel that his theories

[33]See (Boniface and Schappacher 2001).

[34]See (Dauben 1979, 1).

[35]See (Grattan-Guinness 1971, 378).

[36]See (Dauben 1979, 291).

[37]See (Dauben 1979, 131). The cholera bacillus had been discovered by Robert Koch ten years earlier.

were true. This would incline him to reject theories that differed from his own. Such theories might not have appeared to Cantor as possible, but simply as false.

17.4 The philosophy of mathematics

Others were also at work on foundational questions about mathematics. It was the coming together of their ideas with Cantor's that plunged mathematicians into what some felt was the biggest crisis in the history of their subject.

Figure 17.4. Giuseppe Peano (1858–1932)

Giuseppe Peano was a crucial transitional figure in this enterprise. His work on logic was directed towards refining its role in mathematical deduction, and not towards showing how it could provide foundations for mathematics. As Peano wrote to Klein in 1894, 'The purpose of mathematical logic is to analyse the ideas and reasoning that especially figure in the mathematical sciences'.[38]

Peano helped to provide mathematicians with a system of notation that could express ideas based on a theory of sets, and also helped to cast mathematics in an axiomatic form. In 1891 he founded a journal, the *Rivista di Matematica*, with which to promulgate his views — *rivista* means review, in the sense of 'looking again' — and in 1895 he began work on a project to rewrite all the theorems and proofs of mathematics in his system of mathematical logic. This project was eventually completed in 1908, when some 4200 theorems had been rewritten in this way. Peano attracted many Italian mathematicians to his cause.[39]

Peano's earliest work on logic was much influenced by Dedekind's analysis of the concept of number. One way in which this influence is apparent is in Peano's decision to give an axiomatic treatment of arithmetic — his work also displays his new notation to good advantage.

[38]See Giuseppe Peano, letter to Felix Klein, dated 29 August 1894, Niedersächsische Staats- und Universitätsbibliothek Göttingen Handschriftenabteilung, in (Kennedy 1974, 443).

[39]We mentioned some of their work on the foundations of geometry in Chapter 15.

ARITHMETICES PRINCIPIA.

§ 1. De numeris et de additione.

Explicationes.

Signo N significatur *numerus (integer positivus).*
» 1 » *unitas.*
» $a+1$ » *sequens a, sive a plus 1.*
» $=$ » *est aequalis.* Hoc ut novum signum considerandum est, etsi logicae signi figuram habeat.

Axiomata.

1. $1 \in \mathrm{N}$.
2. $a \in \mathrm{N} . \supset . a = a$.
3. $a, b, c \in \mathrm{N} . \supset : a = b . = . b = a$.
4. $a, b \in \mathrm{N} . \supset \therefore a = b . b = c : \supset . a = c$.
5. $a = b . b \in \mathrm{N} : \supset . a \in \mathrm{N}$.
6. $a \in \mathrm{N} . \supset . a + 1 \in \mathrm{N}$.
7. $a, b \in \mathrm{N} . \supset : a = b . = . a + 1 = b + 1$.
8. $a \in \mathrm{N} . \supset . a + 1 -= 1$.
9. $k \in \mathrm{K} \therefore 1 \in k \therefore x \in \mathrm{N} . x \in k : \supset_x . x + 1 \in k :: \supset . \mathrm{N} \supset k$.

Definitiones.

10. $2 = 1 + 1$; $3 = 2 + 1$; $4 = 3 + 1$; etc.

PEANO, *Arithmetices principia.* 1

Figure 17.5. Peano's axioms for the natural numbers, 1889

As his brief text explains (see Figure 17.5), Peano began by asserting the existence of a set N, containing an element that he suggestively called 1, and such that every element in N has a successor. Axioms 1–5 give an abstract characterisation of the natural numbers, for, as the third one assures us, any set that obeys these axioms certainly contains N, so the set of natural numbers can be said to be the 'smallest' set satisfying these axioms. The third axiom also enshrines the 'principle of induction', which asserts that *if*

- a proposition about a number n is true for 1, and
- it is true for the integer n then it is true for the number $n + 1$,

then the proposition is true for every natural number.

Peano's axioms exactly capture what we need to specify the natural numbers: there is a 1, and every number has a successor: 2, 3, 4, and so on. A modern analogy would be with what is needed in order to write a computer program. But it does not base

Figure 17.6. Gottlob Frege (1848–1925)

the concept of a natural number on anything more fundamental. The search for foundations of mathematics was, however, already under way, personified by the forceful figure of Gottlob Frege.

Frege studied mathematics at Jena and Göttingen before becoming Professor of Mathematics at Jena, where he remained all his working life. He took his early training in mathematics with him when he started his researches into logic, and he shared with Cantor a hostility towards the idea that mathematics is based on psychological considerations. Neither of them regarded mathematics as being about feelings, sensations, or physical perceptions, and consequently it was not a branch of empirical science. It was, they agreed, about objects of thought that were to be presented to the mind clearly and distinctly. But Frege felt that the clarity of definitions in Cantor's work left much to be desired, and in an unpublished article of 1891 he caustically satirised Cantor's approach:[40]

> If, for example, one finds a property of a thing upsetting, one abstracts it away. If one wants to order a stop, however, to this destruction, so that properties which one wants to see retained are not obliterated, then one reflects upon these properties. Finally, if one painfully misses properties of the thing, one adds them back by definition. Possessing such magical powers, one is not very far from omnipotence.

Can Cantor's first definition of the cardinality of a set be criticised on these lines? The answer is surely 'yes'. The cardinality (power, or size) of a set is supposed to be the only property of a set that is left when all its other properties have been abstracted away — but how can we be sure that we have not abstracted away the crucial property we want in the process, since it would seem that we have no means of recognising it other than by its mysterious ability to reside in an object after all the other properties have left.

[40]Cited in (Dauben 1979, 221–222).

Frege had his own ideas about how to proceed, which are clear from the preface to his book, *Begriffsschrift*, of 1879.[41] As he wrote:

> The most reliable way of carrying out a proof, obviously, is to follow pure logic, a way that, disregarding the particular characteristics of objects, depends solely on those laws upon which all knowledge rests.

Frege set himself the task of showing how the concept of number could be *defined* entirely logically. He found it increasingly difficult to keep intuition at bay and produce a gapless chain of reasoning, and he devised his concept writing (see Figure 17.7) so that every deduction was explicit (much as one might write a computer program today). Inevitably, this led Frege to become emphatic on the need to subject logic to a rigorous analysis, and he also gave it a goal: to derive all of arithmetic.[42]

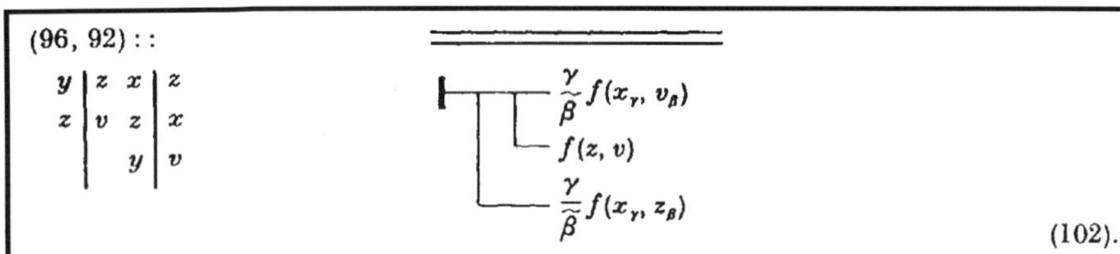

Let us here give the derivation of (102) in words.
If z is the same as x, then by (92) every result of an application of the procedure f to z follows x in the f-sequence. If z follows x in the f-sequence, then by (96) every result of an application of f to z follows x in the f-sequence.

Figure 17.7. An example in Frege's *Begriffsschrift*, and his own translation of the same passage into words

Figure 17.7 shows an impressive determination to rid mathematics of any trace of ambiguity. Each ideogram is a precise picture of a mathematical statement, but it is not easy to write, and Frege never returned to it. That said, he never relented on his drive for complete logical precision in his attempt to define every term used in arithmetic purely logically.

His criticisms of Cantor's approach meant that Frege could not define 'number' by a process of abstraction. So, to define what he meant by 'number', he first defined what he meant by 'equal'. He said that two concepts are equal when the objects corresponding to one could be correlated one-to-one with the objects corresponding to the other. Then he defined 'number' (in paragraph 68) as follows:

> The number which belongs to the concept F is the extension of the concept 'equal to the concept F'.

The *extension* of a concept can be thought of as all the things to which it applies, but even with this gloss one can only agree with Frege's own comment: 'That this definition is correct will perhaps hardly be evident at first'.

---

[41]The German title means 'concept writing', which neatly captures Frege's attempt to establish a truly logical form for presenting arguments.

[42]Frege never claimed that all mathematics is derivable from logic, because he felt that geometry is inescapably tied up with our perception of the real world, and he never accepted the discovery of a non-Euclidean geometry, because he felt that there is only one world and it is Euclidean.

Frege's definition of number proceeded as follows. To define the number 0 he noted that there are concepts for which there are no examples, such as a square circle. In Frege's more technical language, such a concept has no *extension* — that is, the list of examples of the concept is empty. Indeed, said Frege, there is only *one* such concept, for there are no objects available to distinguish between any two such concepts, and that makes the concepts indistinguishable. Frege said that 0 is the number which belonged to that concept, thereby formalising the observation that such a concept has a zero number of examples.

Then Frege defined the number 1 (in paragraph 77): '1 is the number which belongs with the concept 'Identical with zero'.' This is true; for there is exactly one concept with no examples. Like climbers inching their way up a rock face, Frege then defined each natural number in turn by defining what it is to be the next number in the sequence of numbers. He said that the number n followed the number m if

> there is a concept F, and an object falling under it, x, such that the number that belongs to the concept F and the number that belongs to the concept 'falling under F but not identical with x' is m.

So the number 2 is defined by taking the concept F that appears in the definition of the number 1, and forming the concept of having two non-identical objects each of which falls under F.

Frege went on to embrace Cantor's infinite numbers. In paragraph 85 he said that he heartily shared Cantor's 'contempt for the view that in principle only finite numbers ought to be admitted as actual', and he concluded by saying (in paragraph 87) that 'Arithmetic becomes simply a development of logic, and every proposition of arithmetic a law of logic, albeit a derivative one'.

Frege's work was neglected at first. Cantor gave it a welcoming review, although he did not like Frege's definition of number. But eventually Frege was followed in his work by Bertrand Russell, who went even further and saw no reason to stop at arithmetic. As Russell stated in the preface to his *The Principles of Mathematics* (1903), his aim was to prove

> that all pure mathematics deals exclusively with concepts definable in terms of a very small number of logical concepts, and that all its propositions are deducible from a very small number of fundamental logical principles [and to prove this with] all the certainty of which mathematical demonstrations are capable.

Russell was clearly another logicist, and, as we shall see, the logicist movement became important in the early years of the 20th century.[43]

Russell had been stimulated by Peano's work on logic. He met Peano at the sessions on logic and foundations of mathematics at the International Congress of Philosophy in Paris in 1900, which Peano and his followers dominated, an event that Russell called 'a turning point in my intellectual life'.[44]

Yet it was Russell who dealt the most grievous blow to Frege's whole programme. The story of one week in June 1902 almost tells itself, although you may also wish to consult Box 51 as well, which describes *Russell's paradox*. In that week, Russell wrote to Frege:

[43] Russell's best attempt to derive mathematics from logic was his three-volume *Principia Mathematica* (1910), jointly written with his Cambridge colleague A.N. Whitehead, which was generally considered to be magnificent but flawed.

[44] See (Russell 1967, 144).

Figure 17.8. Bertrand Russell (1872–1970)

Russell to Frege.

Friday's Hill, Haslemere, 16 June 1902

Dear colleague,

For a year and a half I have been acquainted with your *Grundgesetze der Arithmetik* [Foundations of Arithmetic], but it is only now that I have been able to find the time for the thorough study I intended to make of your work. I find myself in complete agreement with you in all essentials, particularly when you reject any psychological element in logic and when you place a high value upon an ideography (*Begriffsschrift*) for the foundations of mathematics and of formal logic, which, incidentally, can hardly be distinguished. With regard to many particular questions, I find in your work discussions, distinctions, and definitions that one seeks in vain in the works of other logicians. Especially so far as function is concerned (§9 of your *Begriffsschrift*), I have been led on my own to views that are the same even in the details. There is just one point where I have encountered a difficulty. You state that a function, too, can act as the indeterminate element. This I formerly believed, but now this view seems doubtful to me because of the following contradiction. Let w be the predicate: to be a predicate that cannot be predicated of itself. Can w be predicated of itself? From each answer its opposite follows. Therefore we must conclude that w is not a predicate. Likewise there is no class (as a totality) of those classes which, each taken as a totality, do not belong to themselves. From this I conclude that under certain circumstances a definable collection does not form a totality.

I am on the point of finishing a book on the principles of mathematics and in it I should like to discuss your work very thoroughly. I already have your books or shall buy them soon, but I would be very grateful to you if you could send me reprints of your articles in various periodicals. In case this should be impossible, however, I will obtain them from a library.

The exact treatment of logic in fundamental questions, where symbols fail, has remained very much behind; in your works I find the best I know of our time, and therefore I have permitted myself to express my deep respect to you. It is very regrettable that you have not come to publish the second volume of your *Grundgesetze*; I hope that this will still be done.

Very respectfully yours,

Bertrand Russell

Less than a week later, Frege replied to Russell:

Jena, 22 June 1902

Dear Colleague,

...Your discovery of the contradiction caused me the greatest surprise and, I would almost say, consternation, since it has shaken the basis on which I intended to build arithmetic. It seems, then, that transforming the generalisation of an equality into an equality of courses-of-values (§9 of my *Grundgesetze*) is not always permitted, that my Rule V (§20, p. 36) is false, and that my explanations in §31 are not sufficient to ensure that my combinations of signs have a meaning in all cases. I must reflect further on the matter. It is all the more serious since, with the loss of my Rule V, not only the foundations of my arithmetic, but also the sole possible foundations of arithmetic, seem to vanish. Yet, I should think, it must be possible to set up conditions for the transformation of the generalisation of an equality into an equality of courses-of-values such that the essentials of my proofs remain intact. In any case your discovery is very remarkable and will perhaps result in a great advance in logic, unwelcome as it may seem at first glance ...

Very respectfully yours,

G. Frege

To see how damaging Russell's intervention was, consider the consequences of trying to avoid his paradox. The weak point in his argument was the assumption that the set of all sets satisfying a given property *is itself a set*, something that one can talk meaningfully about. We have seen how useful such sets are in defining such basic concepts as number, even finite numbers. Russell's paradox therefore appeared to threaten the very definition of number, and with it Frege's programme of deriving arithmetic from logic. But the damage went further than this. If there are to be collections that are not sets, and reasonings that one cannot perform with sets, what are these anomalous collections and these invalid inferences? How can they be characterised, and how can the valid ones be defined? This problem had to be resolved before a theory of sets and logic could again hope to support arithmetic, and hence mathematics.

So the paradoxes seemed to undermine the whole of set theory and even of mathematics itself. Yet, as we said at the start of this section, set theory had grown out of the

Box 51. Russell's paradox.
We may put the paradox in the language of sets this way.

Let a set that does not contain itself be called 'normal', and a set that does contain itself be called 'abnormal'. For example, the set of all circles is normal, since it is not itself a circle; and the set of all sets with more than three members is an example of an abnormal set. (A normal set is our analogue of what Russell called 'a predicate which cannot be predicated of itself'.) Plainly, every set is either normal or abnormal.

Now consider the set S of all normal sets. If it is normal, then (by the definition of normal) it does not contain itself, and so (by the definition of S) it must be abnormal. But if it is abnormal, then (by the definition of abnormal) it contains itself, which (by the definition of S) makes it normal. So it is neither normal nor abnormal, yet every set must be one or the other. A most ingenious paradox!

real needs of mathematicians and seemed to be essential if many problems in analysis were to be solved. To Cantor the problem may not have appeared too grave, for since sets really exist (in the mind of God) the paradoxes could not imply that sets do not exist: they could be only a temporary local difficulty caused by our imperfect minds. But others had less confidence. Chief among these were the leading mathematicians of the day, Poincaré and Hilbert.

Hilbert had been in contact with Cantor since 1897, and in conversation and in correspondence he raised with him some ambiguities that he detected in Cantor's theory of sets.

He concentrated on Cantor's alephs. Cantor had defined the alephs as the sizes of sets: $\aleph_0$ was the size of the first infinite set, $\aleph_1$ was the size of the next infinite set, and so on. He had also shown that given a set of any size there is a set of a strictly larger size that has as its elements the subsets of the given set.[45] So there are infinitely many distinct alephs.

Hilbert asked whether the collection of all alephs could be a set. According to Cantor's naive rules it could, but then a paradox ensued: the aleph corresponding to this set would be a new aleph, but the set is supposed to contain them all. This paradox has a tangled history, and is usually known as the Burali–Forti paradox (see Box 52); we cannot pursue its history here, but simply note that it was raised.

In his reply, Cantor argued that the collection of all alephs was indeed a set, but not a 'completed' set. Hilbert rightly found this defence obscure. Nonetheless, Hilbert was impressed with Cantor's ideas, and sketched out a presentation of them for an Easter course in 1898 for school teachers on the infinite in geometry and arithmetic.[46] The aim of his course was to show how school mathematics could be made rigorous by ideas drawn from the latest research — no wonder that German universities could continue to set such high standards with such well-prepared and motivated school teachers!

[45]To give an example involving finite sets, the set $\{a, b, c\}$ has 3 elements, and the set of its subsets has 8 elements $\{\varnothing, \{a\}, \{b\}, \{c\}, \{b, c\}, \{a, c\}, \{a, b\}, \{a, b, c\}\}$.

[46]See (Moore 2002, 44).

Hilbert's course on set theory was the first one anywhere in Germany, and only the second in the world (after a course on the theory of functions given by Émile Borel in France the previous year which included a good deal of set theory). In the next few years, more courses exclusively devoted to set theory were given, all in Germany: by Ernst Zermelo, who was working with Hilbert in Göttingen, in 1900; by Felix Hausdorff in Leipzig in 1901; and by Edmund Landau in Berlin in 1902. This argues for a significant shift in accepting set theory as an acceptable foundation for mathematics, but disquiet about the paradoxes had not gone away.

In 1903 Frege sent a copy of the second volume of his *Grundgesetze* to Hilbert — by then they had been in correspondence for some time, but remained rather far apart intellectually. Hilbert thanked him, and observed that Russell's paradox had been known in Göttingen for some time. In fact, he wrote:[47]

> I believe Dr. Zermelo discovered it three or four years ago after I had communicated my example to him.

Quite what Hilbert might have discovered in 1898 or 1899 has long been a mystery to historians, because he never published anything about it, but Zermelo's contribution is clear; he may even have come upon it while preparing his lecture course. It concerns the paradoxical consequences of a set that contains all of its subsets as elements, and it seems that the Russell-type paradox to which it leads had already been discussed in Hilbert's circle around 1900.[48]

17.5 Set theory and logic

Hilbert had already queried the consistency of the usual axioms for arithmetic in his address to the International Congress of Mathematicians in Paris in 1900 (see Chapter 22). He returned to this theme at the next Congress (in Heidelberg) in 1904, spurred on by the newly discovered paradoxes which he recognised as threatening much that was valuable in Frege's work. But the problems also affected his own ideas, for in order to establish that his various axiomatic systems of geometry were consistent he had given them arithmetic models, and in so doing he had opened himself up to the charge that he did not know that arithmetic was itself consistent.

His proposal was to solve the matter in much the same way as he had dealt with problems in the foundations of geometry. He argued that difficulties in defining the basic terms (point, line, etc.) could be sidestepped by formalising the rules of inference instead. The basic terms were deliberately made meaningless, but in return one obtained a set of rules (the axioms) that could not lead to a state of self-contradiction.

Hilbert proposed to avoid the question of what a set is, and instead to make arithmetic secure by making it axiomatic. This approach was explicitly contrasted with the logicist approach of Frege:[49]

> Arithmetic is often considered to be a part of logic, and the traditional fundamental logical notions are usually presupposed when it is a question of establishing a foundation for arithmetic. If we observe attentively, however, we realise that in the traditional exposition of the laws of logic certain fundamental arithmetic notions are already used,

[47]See (Peckhaus and Kahle 2002, 157).

[48]See (Moore 2002) and (Peckhaus and Kahle 2002).

[49]Hilbert, in (Van Heijenoort 1967, 131).

Box 52. More paradoxes.

- *The paradox of the liar.* If a man says 'I always lie', is he telling the truth or telling a lie? If the truth, then on his own admission he has told a lie. If a lie, then his statement was false and he was therefore telling the truth.
- In 1908 Russell introduced 'the least integer not nameable in fewer than nineteen syllables' (he reckoned it was 111,777). There must be such a number, but the phrase in inverted commas defines it in eighteen syllables. This paradox had originally been introduced by the French writer Jules Richard.
- *Burali–Forti's paradox*: Consider the cardinal numbers 1, 2, 3, ..., and all the transfinite cardinals defined by Cantor. Cantor had shown that every set has a cardinality, and that the cardinality of the set of all subsets of a set S always exceeds the cardinality of the set S itself. But let S be the set of all cardinal numbers. What is its cardinality? The set of all cardinals must have a cardinality larger than any other cardinal, yet there can be no largest cardinal. (This is a contradiction that Cantor knew about.)
- *The paradox of the barber*: In a certain village, the sole (male) barber shaves every man who does not shave himself and no-one else. Who shaves the barber?

> for example, the notion of set and, to some extent, also that of number. Thus we find ourselves turning in a circle, and that is why a partly simultaneous development of the laws of logic and of arithmetic is required if paradoxes are to be avoided.

At the International Congress (in Rome) of 1908, Poincaré struck a more defensive note:[50]

> It has come about that we have run against certain paradoxes and apparent contradictions, which would have rejoiced the heart of Zeno of Elea and the school of Megara.[51] Then began the business of searching for a remedy, each man his own way. For my part I think, and I am not alone in so thinking, that the important thing is never to introduce any entities but such as can be completely defined in a finite number of words. Whatever be the remedy adopted, we can promise ourselves the joy of the doctor called in to follow a fine pathological case.

The difference between the views of Hilbert and Poincaré is a deep one. In the spirit of his own remark that 'in mathematics there is no 'we shall not know'' and with the support of many German mathematicians, Hilbert felt that the real difficulties confronting them could be solved; what was called for was a vigorous advance.[52] Poincaré, with the support of several young French mathematicians, felt that the difficulties were irresolvable, and that what was called for was a prudent retreat. This debate raged at

---

[50]See (Poincaré 1908, 1914, 45).

[51]There was a long-standing confusion between Euclid, the author of the *Elements*, and a philosopher known as Euclid of Megara.

[52]For the Hilbert quote, see Chapter 22.

many levels. It emerged that some operations with sets were unproblematic, whereas others (such as calling the collection of all sets with a certain property a set) were distinctly doubtful; other operations are valid when applied to finite sets but are less obviously valid when applied to infinite sets. In this way, the notion of a set ceased to be simple, and began to spawn a theory. The question here was: Were the difficulties in this theory to be resolved by improving the definition of a set, or by axiomatising the whole business?[53]

There was a more radical way out, hinted at by Poincaré and soon to be taken up most vigorously by the Dutch mathematician Luitzen Brouwer. Brouwer felt that the whole problem began with an uncritical acceptance of the idea that the human mind can deal with infinite sets in the way that it deals with finite ones. In principle at least, finite sets can be reviewed item by item, but infinite sets cannot be so reviewed. Consequently, Brouwer argued, we can say that a given proposition is true or false only when that proposition refers to a finite set of things. Some propositions referring to infinite sets may be neither true nor false, he claimed, because the human mind is incapable of deciding a question when it cannot review all of the evidence. So Brouwer's proposal clashed head-on with the so-called 'law of the excluded middle', which asserts that any proposition must be either true or false; there is no third possibility, intermediate between the two. This law underlies proof by *reductio ad absurdum*, for instance, so it was a serious criticism.

Brouwer's proposal to reject 'the law of the excluded middle', just where it is most useful, appalled many mathematicians. They feared that to adopt Brouwer's rigorous alternative, that of explicitly constructing or defining every item you need in a finite number of steps, could never reach far enough. Too much mathematics would be lost: quite possibly even the real numbers could not be defined in a way that would satisfy Brouwer, and most of the theory of functions would need to be discarded. The price, said Hilbert, was too high to pay.

Although logicism was the current that began the debate about the foundations of mathematics, it came to be felt that the paradoxes that arose from an uncritical use of the concept of a set could not be resolved from within the logicist framework. Hilbert's point that a simultaneous attack on logic and arithmetic would be necessary came to prevail, and an effort was made to establish mathematics upon an axiomatic basis. Because this approach stressed the importance of formal rules governing mathematics, which were to be adopted much as one adopts the rules of chess, it is called the *formalist* approach. The rival approach, deriving from Brouwer, stressed the role of intuition, the human mental capacity to engage with ideas, and was consequently called *intuitionism*. They differ so completely that it is not possible to see them as part of a debate, although each acted as a goad to the other, and so we shall treat them separately. But it is clear that each derived from important currents in 19th- and early 20th-century mathematical thought.

On the one hand, formalism grew out of successful attempts to make various parts of mathematics rigorous, and it appealed to those who valued mathematics for its strict deductive character. It endeavoured to present a vision of mathematics as something upon which one could rely with certainty. On the other hand, intuitionism appealed to one's sense of mathematics as a creative activity in which the mind intuits truths in a

[53]There are no grounds for the oft-repeated statement that Poincaré said 'Set theory is a disease from which later generations of mathematicians will say we have recovered'; see (Gray 1991, 19–22).

way that is not merely logical. It required mathematicians to recognise that, whatever lay outside them, what they actually dealt with were the ideas that they had in their own minds. On such a view, it was merely prudent and honest to admit that judgements about infinite collections could not be made.

Brouwer felt that the quest for formal rigour had led, in the hands of Hilbert and others, to no secure foundations at all. He wrote in 1927 that:[54]

> the formalist school should ponder the fact that in the framework of formalism nothing of mathematics proper has been secured up to now ... whereas intuitionism, on the basis of its constructive definition of set and the fundamental property it has exhibited for finitary sets, has already erected anew several of the theories of mathematics proper in unshakeable certainty.

He ironically concluded, in direct response to Hilbert (who had recently spoken of the 'modest' assertions of 'the recent doctrine called 'intuitionism'):

> If therefore, the formalist school ... has detected modesty on the part of intuitionism, it should seize the occasion not to lag behind intuitionism with respect to this virtue.

It is impossible to tell the full story of this clash in the space we have available — it could easily fill a whole book in the history and philosophy of mathematics! Both ideas had enough in them to survive to the present day as, in modified forms, they do. Intuitionism was driven back from its heyday in the 1920s, partly because mathematicians proved reluctant to give up so many of their hard-won theorems on what seemed to them to be merely philosophical grounds, and preferred to invest their hopes in an elimination of the paradoxes. Brouwer's aggressive personality, and his seeming indifference to the loss of so much mathematics, did not help. Only more recently have his questions been taken up again by people curious to see how much mathematics can survive without the law of the excluded middle. But the main reason for the decline of intuitionism was Hilbert's powerful and influential defence of formalism — although, ironically, his hopes for its complete success were to be dashed.

The principal paper in which Hilbert presented his views was his 'Über das Unendliche' (On the infinite), published in the *Mathematische Annalen* in 1926. He was forceful about what was at stake, and said of his theory that:

> Its aim is to endow mathematical method with the definitive reliability that the critical era of the infinitesimal calculus did not achieve.

The problem, he argued, is with the notion of the infinite:

> The definitive clarification of the *nature of the infinite* has become necessary, not merely for the special interests of the individual sciences, but rather for the *honour of the human understanding* itself.

Hilbert considered the various ways in which the infinite enters mathematics. There are 'ideal elements', such as points at infinity in projective geometry, and complex numbers. But above all there are the actual infinities of Cantor's transfinite numbers. 'Finally,' he said, 'through the gigantic collaboration of Frege, Dedekind, and Cantor the infinite was enthroned.' But then the paradoxes appeared, which 'had a downright catastrophic effect in the world of mathematics'. Hilbert continued:[55]

[54]See Brouwer, in (Van Heijenoort 1967, 492).
[55]See Hilbert, in (Van Heijenoort 1967, 370–371, 375, 376).

Box 53. The logical calculus.
In the logical calculus each statement is denoted by a letter, $A, B, C, \ldots$, and each logical operation by a symbol, such as & for 'and', $\vee$ for 'or', $\neg$ for 'not', $\Rightarrow$ for 'implies', and so on.

Criteria are given for recognising a correctly written formula — one that arises in a formal argument; thus '$A \,\&\, B$' is acceptable but '$A \,\&$' is not. Rules for manipulation of the symbols are given; these are rules of inference — for example, $\neg(A \vee B)$ is equivalent to $(\neg A) \,\&\, (\neg B)$. The logical calculus is the manipulation of these symbols according to agreed rules.

Hilbert on the infinite.

> Let us admit that the situation in which we presently find ourselves with respect to the paradoxes is in the long run intolerable. Just think: in mathematics, this paragon of reliability and truth, the very notions and inferences, as everyone learns, teaches, and uses them, lead to absurdities. And where else would reliability and truth be found if even mathematical thinking fails?
>
> But there is a completely satisfactory way of escaping the paradoxes without committing treason against our science. The considerations that lead us to discover this way and the goals toward which we want to advance are these:
>
> (1) We shall carefully investigate those ways of forming notions and those modes of inference that are fruitful; we shall nurse them, support them, and make them usable, wherever there is the slightest promise of success. No one shall be able to drive us from the paradise that Cantor created for us.
>
> (2) It is necessary to make inferences everywhere as reliable as they are in ordinary elementary number theory, which no one questions and in which contradictions and paradoxes arise only through our carelessness.
>
> Obviously we shall be able to reach these goals only if we succeed in completely clarifying the nature of the infinite.

Hilbert proposed to analyse the way that proofs are constructed, and to augment the propositions that govern our use of the finite with what he explicitly called 'ideal propositions' (by analogy with ideal points) that would govern our treatment of the infinite. He invoked a familiar analogy: one can go from statements about numbers to formulas and then take the further step of considering those formulas purely as abstract expressions, and so (the analogy ran) we can look at a proof, analyse it into its component statements and the inferences that hold it together, and then pass to considering the chain of inferences purely abstractly. Indeed, he pointed out, the algebraic expressions for deductions already exist in the form of what he called the *logical calculus* (see Box 53).

In this way, Hilbert suggested, a proof could be written down entirely symbolically. The formulas for valid inferences in the logical calculus, denuded of all meaning, were the ideal propositions that Hilbert required:[56]

> In this way we now finally obtain, in place of the contentual mathematical science that is communicated by means of ordinary language, an inventory of formulas that are formed from mathematical and logical signs and follow each other according to definite rules. Certain of these formulas correspond to the mathematical axioms, and to contentual inference there correspond the rules according to which the formulas follow each other; hence contentual inference is replaced by the manipulation of signs according to rules, and in this way the full transition from a naive to a formal treatment is now accomplished.

A fundamental question that one can ask about Hilbert's proposal is: Was Hilbert giving an account of why mathematics is *true*, or of why it is *valid*? The answer must be that truth as such does not enter into this formulation, in that the touchstone for correctness of a mathematical statement is not its correspondence with some other reality. For Hilbert, mathematics was meaningless and formal, but it was not a system where anything goes. Rather, statements are valid if they are deduced from agreed premisses by means of agreed rules.

What matters, by Hilbert's account, is whether certain formal statements are consistent with the original axioms, and, of course, that the axioms themselves are mutually consistent. Consistency — lack of contradiction — is crucial, for an inconsistent set of statements permits one to deduce anything, and so cannot have anything to do with an account of the validity of mathematics. In an inconsistent system, for example, one can prove both a proposition and its negation — not a happy state of affairs! To make the point vivid, there is an anecdote about Bertrand Russell. Asked to deduce from the contradiction '1 = 2' that he was the Pope, he replied: 'The Pope and I are two; therefore we are one, therefore I am the Pope'.

At a later stage, the formalist approach added two other requirements to that of consistency:

- the system should be *complete* — that is, given any validly formulated statement, either it or its negation should be derivable from the basic axioms.
- the system should be *decidable* — that is, given a validly formulated statement, one should be able to decide whether it is derivable from the basic axioms.

Hilbert's approach was highly influential. Brouwer himself was not convinced, and continued to castigate what he called Hilbert's thoughtless use of the logical principle of the excluded middle. And Hermann Weyl, the leading mathematician of the next generation at Göttingen and, like Hilbert, a man of immense intellectual range, fell quiet on foundational questions after several years of arguing for a modified version of Brouwer's intuitionism. Once again Hilbert seemed to have isolated the crucial features of a subject — in this case, the vexed and deep question of criteria for the validity of mathematics.

A way of thinking that is consistent, complete, and decidable would be a very powerful one indeed. All the more shocking, then, was the discovery that mathematicians were about to be driven from Hilbert's paradise. We cannot describe in any detail here

[56] See (Van Heijenoort 1967, 381).

how this came about, but in 1930 and 1931 the Austrian mathematician Kurt Gödel, then in his mid-20s, showed that no formal system capable of describing arithmetic can be both consistent and complete.[57]

What Gödel's result means is that mathematics, if it is consistent, is incomplete. Whatever basic axioms are chosen, mathematics must contain statements that are consistent with those basic axioms and whose negations are likewise consistent. That is, in any axiomatised system of mathematics there will be true statements that cannot be proved. Faced with such a statement, the mathematical community can choose to make either it or its negation an axiom, but it cannot hope to prove that the favourite candidate is the only validly deducible one.

In 1936 the 23-year-old British mathematician Alan Turing resolved the decidability question, again in the negative, when he showed that given a mathematical statement, there is no general method that can decide whether the statement is provable from the axioms, or whether it is independent of them. Matters were nothing like as simple as Hilbert had hoped.

The situation resembled that of geometry in the mid-19th century. The basic assumptions of Euclid were not in dispute, except for the parallel postulate. But both the parallel postulate and a contradictory postulate turned out to be consistent with the uncontested postulates, and consequently, neither could be derivable from the uncontested list. Considerations derived from the utility of mathematics might predispose people to prefer Euclidean to non-Euclidean geometry, or the other way round, but considerations internal to geometry could not. But since the whole question of the validity of mathematics had been separated from its utility, mathematicians of the 20th century could take no comfort there.

Hermann Weyl summarised his own and the mathematical community's response to Gödel's discovery in these words:[58]

> Since then [1931] the prevailing attitude has been one of resignation. The ultimate foundations and the ultimate meaning of mathematics remain an open problem; we do not know in what direction it will find its solution, nor even whether a final objective answer can be expected at all. 'Mathematizing' may well be a creative activity of man, like music, the products of which not only in form but also in substance are conditioned by the decisions of history and therefore defy complete objective rationalisation.

So the situation was not as bad as was at first feared. Debates continue over what the basic axioms should be, and what additions should be made to the list of simple ones in order to ensure the validity of advanced mathematics. But, as so often, making the best of a bad job turned out not to be so bad after all. If mathematics cannot have totally secure foundations, then let it be without them. There is a sense in which the subject grows both upwards and downwards, developing new branches and at the same time extending its roots. The unexpected discovery is that such a way of life can be as healthy and as vigorous as any other.

17.6 Further reading

Cantor, G. 1915. *Contributions to the Founding of the Theory of Transfinite Numbers*, Open Court, repr. Dover, 1955. Two of Cantor's most important papers are

[57] Gödel's paper was entitled (in translation) 'On formally undecidable propositions of *Principia Mathematica* and related systems, I', which is interesting testimony to what Russell and Whitehead had achieved.

[58] See (Weyl 1949, 219).

presented here in an English translation. This is not an easy read, but it is a fascinating one. The book also contains a good introductory essay by the translator, Philip Jourdain, who overcame severe physical handicap to become a leading exponent of Cantor's ideas.

Dauben, J.W. 1979. *Georg Cantor: His Mathematics and Philosophy of the Infinite*, Harvard University Press. This thorough study of one of the most remarkable mathematicians of the 19th century is sensitively attuned to both Cantor's mathematics and his troubled personal life. It is also very informative about the mathematical community of the day.

Dawson, J.W., Jr. 1997. *Logical Dilemmas, The Life and Work of Kurt Gödel*, A.K. Peters. This is a fine account of the man many consider to have been the greatest logician of the 20th century, by one who has been deeply involved in editing Gödel's collected works.

Ewald, W.B. 1996. *From Kant to Hilbert: A Source Book in the Foundations of Mathematics*, 2 vols., Oxford University Press. This rich collection is as fascinating for the many well-known items it includes as for the unexpectedly relevant ones that it sweeps up because of its astute interpretation of its theme (it runs from Berkeley in 1707 to Bourbaki in 1948).

Franzen, T. 2005. *Gödel's Theorem: An Incomplete Guide to Its Use and Abuse*, A.K. Peters. This clear and careful presentation of Gödel's theorem also explains what it does *not* say about mathematics and the human mind.

Kennedy, H.C. 1980. *Peano: Life and Works of Giuseppe Peano*, Reidel. This is a clear and well-researched account of a diverse and influential figure.

Monk, R. 1997. *Bertrand Russell: The Spirit of Solitude, 1872–1920*, Vol. I, Simon and Schuster. This is a richly documented account of the most productive years of Russell's life as a philosopher, as well as his principled pacifist stand in the First World War.

Nagel, E. and Newman, J.R. 1959. *Gödel's Proof*, Routledge and Kegan Paul. This is a clear and elementary presentation of this remarkable result in only 100 pages.

Shapiro, S. (ed.) 2005. *The Oxford Handbook of Philosophy of Mathematics and Logic*, Oxford University Press. This book provides a clear introduction to, and critique of, the main positions, including the philosophies of logicism, formalism, and intuitionism.

Van Dalen, D. 1999, 2005. *Mystic, Geometer, and Intuitionist; The Life of L.E.J. Brouwer*, 2 vols., Clarendon Press, Oxford. This is a rich account of the life and work of Brouwer up to 1925, paying equal attention to his mathematical work and the circumstances of his life.

Van Heijenoort, J. (ed.) 1967. *From Frege to Gödel: A Source Book in Mathematical Logic, 1879–1931*, Harvard University Press. The papers have been chosen with remarkable care and are often presented in their entirety. The result is richly enthralling, but is not for the beginner.

Vilenkin, N. Ya. 1968. *Stories about Sets*, transl. Scripta Technica, Academic Press. This classic of popularisation tells us much of what we need to know about the strange world of set theory. After a clear exposition of Cantor's ideas, meet the devil's staircase, a space-filling curve, and much more.

Weyl, H. 1949. *Philosophy of Mathematics and Natural Science*, Princeton University Press. This is a profound book by a man who was not only a leading mathematician, but also a physicist, a philosopher, and a fine writer.

18

Algebra and Number Theory

Introduction

As we indicated in Section 13.3, one of the first influences that Gauss exerted in the mathematical community was to reshape and deepen the appreciation of number theory. In Section 18.1 we indicate briefly how this was done by looking at the theorem of quadratic reciprocity, which occupies a central position in his book, the *Disquisitiones Arithmeticae* (Arithmetical Investigations), and at how it led him to give a new and more rigorous introduction to the subject. Then we turn from the algebraic side of number theory to the analytic, and look in Section 18.2 at questions about prime numbers and their distribution. Research in this area was to lead to what has become the most famous unsolved problem in mathematics, the Riemann hypothesis, and we examine how it came about.

Then we turn to two topics that belong firmly to the history of algebra. The first is the culmination of several decades of work to prove the Fundamental Theorem of Algebra, and in Section 7.3 we see how this was provided by the four proofs that Gauss published. The second was the emergence and eventual formalisation of the idea of a 'vector', a quantity with both magnitude and direction. The earliest to introduce and promote the idea of vectors were William Rowan Hamilton and Hermann Gunther Grassmann, and we compare their approaches in Section 18.4 before looking at how the work of James Clerk Maxwell and Josiah Willard Gibbs led to the introduction of vectors into contemporary mathematical physics.

18.1 Number theory

During the 19th century the theory of numbers grew from a collection of isolated results into a more coherent theory that formed one of the major branches of mathematics; indeed, it exemplifies the 19th-century tendency of mathematics as a whole towards increasing purity.

Opinions differ over the merits of this development. Mathematicians then and now can be found who applaud the theory of numbers and find in it a beauty, elegance, and profundity second to none. Others inveigh against the abstraction of the subject and what they detect as a turning-away from any concern with the natural world. Why, then, did some early 19th-century mathematicians feel attracted to number theory? In this section we look at the work of two men whose contributions were crucial in effecting the change from isolated results to systematic theory: Legendre in France and Gauss in Germany.

In Chapter 9 we saw that Lagrange went a long way to replace Euler's remarkable, but scattered, insights into number theory with a more systematic account. Lagrange's theory of quadratic forms (expressions of the form $ax^2 + bxy + cy^2$, for particular integers a, b, c, as x and y range over the integers), and his influence on the mathematicians of Paris, were powerful stimuli for an algebraic theory of numbers. His theory systematically studied questions about numbers using the algebraic techniques that had proved so powerful in other branches of mathematics during the 18th century. One mathematician who took up the challenge to extend such work was the geometer and textbook writer Adrien-Marie Legendre, whose work in geometry we looked at briefly in Chapter 14.[1]

Figure 18.1. Adrien-Marie Legendre (1752–1833)

Legendre was a self-effacing man who led a turbulent life. Born into a well-to-do family he made his living as a professional mathematician before the Revolution, first at the École Militaire and later at the Académie des Sciences in Paris, as successor to Laplace. The Revolution nearly bankrupted him, and in order to make a living he occupied various positions and served on numerous committees, including the commission chaired by Lagrange that brought in the metric system. He also led the section of analysts at work, under the overall direction of Gaspard de Prony, on calculating logarithmic and trigonometric tables. From 1799 to 1815 he was an examiner at the École Polytechnique, and from 1813 to 1833 he was Lagrange's successor at the Bureau des Longitudes. But his passions were for the theory of numbers and the theory of elliptic integrals (see Section 13.4, Box 34).

[1]For the fascinating story of why the above illustration of Legendre is the only one we have, and why the one frequently reproduced is of someone else, see (Duren 2009).

Quadratic reciprocity. Legendre was interested in Lagrange's work on quadratic forms, but in a lengthy paper on the subject which he published in 1788, he also raised other questions.[2] He commented that whereas some isolated results of Fermat had been partially worked up into a theory by Euler and Lagrange, others had not. He cited two in particular, which he said were related, although he did not explain how. Then he suggested that the way forward would be to prove a certain theorem about prime numbers, which did indeed turn out to be remarkably influential.[3]

> The theorems thus described are of great generality, but can all be included in the following statement: c and d being prime numbers, the expressions $c^{(d-1)/2}$ and $d^{(c-1)/2}$ are of different signs only when c and d are both of the form $4n-1$; in all other cases these expressions are of the same sign.

This is one form of the *theorem of quadratic reciprocity*, and we shall shortly make more sense of it by using a symbolism that makes it easier to grasp, but first we look ahead to see how the result was received. We shall be interested to see whether Legendre actually proved this theorem, for if he was theory-building it was plainly important that his theorems be proved. We note here that he concluded his paper in the following way, which surely raises at least a hint of a doubt:[4]

> It will perhaps be necessary to prove rigorously something which we have assumed at several places in this article, i.e. that there is an infinity of prime numbers contained in any arithmetic progression whose first term and increment are relatively prime, or, which comes to the same thing, are of the form $2mx+\mu$ where $2m$ and μ have no common divisor. This proposition is quite difficult to prove, however, one can assure oneself that it is true by comparing this arithmetic progression with the ordinary progression $1, 3, 5, 7$ etc. If one takes a considerable number of terms in these progressions, the same in both, and arranges them, for example, in such a way that the greatest terms shall be equal and in the same place on both sides; one will see that in omitting from each side the multiples of $3, 5, 7$, etc. up to a certain prime number p, the same number of terms must remain on each side or else there will be fewer remaining in the progression $1, 3, 5, 7$, etc. But as prime numbers necessarily remain in this one, they must also remain in the other. I content myself with outlining the means for proving this theorem, which it would be too long to give in detail, and besides this memoir has already exceeded the ordinary limits.

It seems that Legendre was confident of the truth of what he said, and although he admitted that he did not have a proof of it, he claimed that he could see how such a proof might begin.[5] Gauss, however, was dismissive.[6]

> The illustrious Legendre presented his demonstration again in his excellent work *Essai d'une Théorie des Nombres* but in such a way as to change nothing essential. So this method is still subject to all [my earlier objections]. It is true that the theorem (on which one supposition is based) which states that any arithmetic progression $l, l+k, l+2k$, etc. contains prime numbers if k and l do not have a common divisor, is given more fully here, but it does not yet seem to satisfy geometric rigour. But even if this theorem were fully demonstrated, the second supposition remains (that there are prime numbers

[2] See (Legendre 1788).

[3] See (Legendre 1788), and F&G 15.C1(a).

[4] See F&G 15.C1(a).

[5] Before him, Euler had also established a series of results that taken together imply the theorem of quadratic reciprocity, but state it in a very different form. He sent these results to Goldbach in a letter of 28 August 1742, which was first published only in (Fuss 1843), see (Euler 2015, Nr. 54).

[6] Gauss, *Disquisitiones Arithmeticae*, 1801, and F&G 15.C1(b).

> of the form $4n + 3$ for which a given positive prime number of the form $4n + 1$ is a quadratic non-residue) and I do not know whether this can be proven rigorously unless the fundamental theorem is *presumed* [Gauss's italics]. But it must be remarked that Legendre did not tacitly assume this last supposition, nor did he ignore it.

However politely, Gauss was arguing that not only did Legendre's result still lack a proof, but that Legendre's whole strategy was flawed. Gauss was actually saying (correctly, as it turned out) that Legendre had fallen into a vicious circle.

The result that Legendre strove so hard to prove was shortly proved by Gauss, who thought so highly of it that he called it his 'Golden Theorem' and went on to give seven more proofs of it. Where does it come from, and what does it say?

It concerns prime numbers, and Legendre had been led to it from his interest in the work of Lagrange and, above all, of Fermat. Fermat's little theorem (see Section 8.1) says that if p is a prime number, and a is a number that is not divisible by p, then a^{p-1} leaves a remainder of 1 when divided by p. For example, if $p = 7$ and $a = 4$, then

$$a^{p-1} = 4^6 = 4096 = (7 \times 585) + 1.$$

Legendre asked himself what, so to speak, the score was at half-time — that is, what remainder does the square root of a^{p-1} leave when divided by p? The square root is $a^{(p-1)/2}$, and its remainder must be either $+1$ or -1, since its square is 1—but which one is it? In our example, $4^3 = 64 = (7 \times 9) + 1$ leaves a remainder of 1. On the other hand $5^3 = 125 = (7 \times 18) - 1$ gives a remainder of -1.

Legendre found that he could answer his question when a is also a prime number, say $a = q$. However, what he discovered was not a result along these lines:

> If two prime numbers q and p are of such-and-such a kind, then the remainder on dividing $q^{(p-1)/2}$ by p is $+1$, and otherwise it is -1

but rather, of this kind:

> If two prime numbers q and p are of such-and-such a kind, *and* if the remainder on dividing $q^{(p-1)/2}$ by p is $+1$, then the remainder on dividing $p^{(q-1)/2}$ by q is also $+1$.

Because q and p are interchanged, the theorem is called a *reciprocity* theorem; the idea is that the two prime numbers in it play reciprocal roles. (It is called the *quadratic* reciprocity theorem because it is about squares and square roots.)

As it stands, the theorem does not answer the question originally asked, even when we explain what is meant by the phrase 'of such-and-such a kind'. It does not tell you what the sign of the square root is; all it does is trade one question in for another. To answer the original question (What is the remainder on dividing $q^{(p-1)/2}$ by p?) some further calculation is needed, but this turns out not to be too arduous (see Box 55 for a worked example).

Mathematicians were familiar with a short cut that arose naturally when they dealt with questions about remainders, and which is taught quite often these days under the heading of 'modular arithmetic' (see Box 54). This idea was codified by Gauss in the opening pages of his book of 1801, and has been further tidied up since then. It is worth looking at now, because it is good evidence for the rising abstraction of 19th-century mathematics. Moreover, its place of publication, in the first truly systematic presentation of number theory as a subject extending from elementary but precise foundations to advanced research topics, is emblematic of what professionalisation requires.

Box 54. Modular arithmetic.
The basic idea of modular arithmetic applies whenever we dealing with remainders on division by a fixed number p (not necessarily prime). The idea is to work with the remainders themselves, and not with the numbers, which can grow uncomfortably large.

To fix our ideas, let us again take the prime number $p = 7$. The remainders on dividing by 7 are $0, 1, 2, 3, 4, 5$, and 6. Let $a = 3$. The square of 3 is 9, and the remainder obtained on dividing 9 by 7 is 2. Gauss introduced the technical terms 'congruent', to express the fact that two numbers give the same remainder on division by a third number, and 'modulus' for the divisor. He also introduced a modified equality sign $\equiv$, which stands for the phrase 'is congruent to'. Because 9 and 2 give the same remainder on division by 7, we say that 9 is congruent to 2 (modulo 7), and write

$$9 \equiv 2 \pmod 7.$$

Fermat's little theorem, expressed in modular arithmetic, says that if p is a prime number and a is not congruent to $0 \pmod p$ then $a^{p-1} \equiv 1 \pmod p$.

It is now easy to verify Fermat's little theorem in the case when $a = 3$ and $p = 7$. We find that

$$3^6 = 9 \times 9 \times 9 \equiv 2 \times 2 \times 2 \equiv 1 \pmod 7,$$

as required.

Modular arithmetic is a good example of the abstractness of 19th-century mathematics, because it involves objects that are not numbers, but can be treated like numbers. As Gauss emphasised, the remainders (or 'residues', as he called them) modulo a prime can be added, subtracted, multiplied, and divided, although we cannot, of course, divide by the residue 0. Nonetheless, the residues themselves are novel quantities abstracted from numbers.

To see how modular arithmetic helps in the statement of the quadratic reciprocity theorem, we recall Legendre's own statement of that result above (where his c and d have been replaced by p and q). In his way of putting it, we must understand that $p^{(q-1)/2}$ means $p^{(q-1)/2} \pmod q$, and $q^{(p-1)/2}$ means $q^{(p-1)/2} \pmod p$. Using Gauss's notation, we can rewrite Legendre's statement of the theorem as:

> p and q being odd prime numbers, $p^{(q-1)/2}$ [mod q] and $q^{(p-1)/2}$ [mod p] are different only when p and q are both of the form $4n - 1$; in all other cases (when either or both of p and q have the form $4n + 1$) they are the same.

This is already easier to understand. But the theorem can be reformulated so that it deals with square roots directly. When this is done — we suppress the details — it is not concerned with whether $p^{(q-1)/2}$ is $+1$ or -1, but whether it makes sense to talk of $p^{1/2} \pmod q$. Only half the residues modulo a prime have a square root. For example, the non-zero squares (mod 7) are $1^2 = 6^2 = 1$, $3^2 = 4^2 = 2$, and $2^2 = 5^2 = 4$, as you can easily check. So only the residues 1, 2, and 4 (mod 7) have square roots; the other residues 3, 5, and 6 (mod 7) do not.

If it makes sense to say that $p^{1/2} = a \pmod{q}$, then raising it to the power $q-1$ gives

$$p^{(q-1)/2} \equiv a^{q-1} \equiv 1 (\mathrm{mod}\ q),$$

by Fermat's little theorem. On the other hand, if p does not have a square root (mod q) then $p^{(q-1)/2} \equiv -1 \pmod{q}$. So the distinction is between those numbers modulo a prime that have square roots — and so are themselves squares — and those that do not have square roots and are therefore not squares.

In Gauss's terminology, if p is a square (mod q) one says that p is a *quadratic residue* (mod q), or that p is a *square* (mod q). When this is not the case, we say that p is a *quadratic non-residue* (mod q). So we finally get the statement of the quadratic reciprocity theorem in a form close to the one that Gauss gave it in his *Disquisitiones Arithmeticae* (§131):

> p and q being odd prime numbers, p is a quadratic residue (mod q) if and only if q is a quadratic residue (mod p), unless p and q are both of the form $4n-1$; in this case p is a quadratic residue (mod q) if and only if q is a quadratic non-residue (mod p).

All that remained was to determine when p is a quadratic residue (mod q). This was answered in principle by Legendre himself, when he gave a way of finding this by a method that was astonishingly quick to use (see the details in Box 55). With this refinement the theorem of quadratic reciprocity was both remarkable and extremely useful — to number theorists at least.

But while some mathematicians found it useful to be able to find whether a given number is a square by modular arithmetic, they were more excited by the quadratic reciprocity theorem itself. In the course of a monumental report on the subject of number theory that he wrote for the British Association for the Advancement of Science between 1859 and 1865, the English mathematician Henry Smith said of the theorem that it was: 'without question, the most important general truth in the science of integral numbers which has been discovered since the time of Fermat'.[7]

There are various reasons for this. Coupled with Legendre's algorithm, the reciprocity theorem helped to resolve many outstanding problems in the theory of numbers. It was also a surprising theorem: Why should the study of squares or square roots modulo a prime p have anything to do with squares or square roots modulo a different prime q? It invited generalisations to cube and higher roots, and it was hard to prove. Legendre, as we saw, failed to secure his result, and Smith said of Gauss's first proof that it was 'very repulsive to any but the most laborious students'.[8] Consequently the result attracted a lot of attention; some 196 proofs of it are given in the reference section of Lemmermayer's history of reciprocity theorems.[9]

It was not the difficulty that was so intriguing, but the reason for the difficulty. It emerged, as Gauss said, that the theorem seems connected to others that do not lie close at hand. Legendre himself had noticed a surprising connection between the theorem of quadratic reciprocity and the claim that any arithmetic progression of the form $a, a+b, a+2b, a+3b, \ldots$, in which a and b are relatively prime integers, contains infinitely many

[7] See (Smith 1859, 56). The British Association for the Advancement of Science was founded in 1831 'to give a stronger impulse and more systematic direction to scientific inquiry, to obtain a greater degree of national attention to the objects of science, and a removal of those disadvantages which impede its progress, and to promote the intercourse of the cultivators of science with one another, and with foreign philosophers'.

[8] See (Smith 1859, 59).

[9] See (Lemmermayer 2000).

Box 55. Quadratic reciprocity.
We give examples of all three types of behaviour (mod p) and (mod q).

Let $p = 5, q = 13$; both are of the form $4n + 1$.
Mod 5, the squares are 0, 1, and $4 \equiv -1$, so $13 \equiv 3$ is not a square (mod 5);
Mod 13, the squares are $0, 1, 4, 9, 16 \equiv 3$, and $25 \equiv -1$, so 5 is not a square (mod 13).

So p is not a square (mod q) and q is not a square (mod p).

Let $p = 7, q = 29$; p is of the form $4n - 1$, q is of the form $4n + 1$.
Mod 7, we have $29 \equiv 1$ (mod 7), so 29 is a square (mod 7).
Mod 29, we have $7 \equiv 36 \equiv 6^2 \equiv 7$ (mod 29), so 7 is a square (mod 29);

So $p = 7$ is a square (mod q) and q is a square (mod p).

Let $p = 7, q = 11$; both are of the form $4n - 1$;
Mod 7, we have $11 \equiv 4$ (mod 7) , so 11 is a square (mod 7);
Mod 11, the squares are $0, 1, 4, 9, 5 \equiv 4^2$, and $3 \equiv 5^2$, so 7 is not a square (mod 11).

So p is not a square (mod q) but q is a square (mod p).

In 1798 Legendre introduced a notational device that makes the quadratic reciprocity theorem easy to use. Assume that a is not divisible by p. He wrote

$$(a/p) = 1 \text{ if } a \text{ is a square (mod } p)$$
$$\text{and } (a/p) = -1 \text{ if } a \text{ is not a square (mod } p).$$

He then proved that $(ab/p) = (a/p)(b/p)$. The theorem of quadratic reciprocity says that if p and q are odd primes, then (p/q) and (q/p) are equal unless p and q both are of the form $4n - 1$.

So to investigate, say, whether 61 is a square (mod 137), he could calculate as follows:

$$\begin{aligned}(61/137) &\equiv (137/61), \text{ by the quadratic reciprocity theorem}\\ &\equiv (15/61), \text{ since } 137 \equiv 15 \text{ (mod 61)}\\ &\equiv (3/61)(5/61), \text{ by his new result}\\ &\equiv (61/3)(61/5), \text{ by the quadratic reciprocity theorem}\\ &\equiv (1/3)(1/5) = 1, \text{because } 61 \equiv 1 \text{(mod 3) and } 61 \equiv 1 \text{(mod 5)}.\end{aligned}$$

So 61 is a square (mod 137). Note that Legendre's method does not find either of its square roots, which are 46 and $-46 \equiv 91$ (mod 137).

primes. It is not obvious what this claim has to do with quadratic residues. It is not even clear that it is true, nor is it easy to see how it can be proved. Unfortunately for him, as we saw, Legendre based part of his proof of the quadratic reciprocity theorem on the truth of this claim.

Box 56. Unique factorisation.
Not only can every integer be factorised into primes, but the factorisation is unique apart from the order of the factors. For example, $168 = 2^3.3.7$, so 168 is a product of the prime numbers 2 (taken three times), 3 and 7, but it is not the product of any other set of primes.

When the new 'integers' are introduced, the list of primes changes. For example, when we use 'integers' of the form $m + n\sqrt{-1}$, the integer 5 is no longer prime, because it can be written as the product $(2 + \sqrt{-1})(2 - \sqrt{-1})$. But it turns out that each of these factors is irreducible (that is, it cannot be factorised further) and that the factorisation of every 'integer', including 5, into irreducibles is unique.

However, when we use 'integers' of the form $m + n\sqrt{-5}$ and the concept 'irreducible' is defined as before, factorisation into irreducibles is no longer unique. For example, $6 = 2.3 = (1 + \sqrt{-5})(1 - \sqrt{-5})$, and all these factors are irreducible.

For this reason, mathematicians did not call irreducible numbers 'prime'.

In due course Legendre's claim about primes in arithmetic progressions was proved — but only in 1837, by Dirichlet. The proof makes essential use of the quadratic reciprocity theorem, which forcefully bears out Gauss's feeling that Legendre had things back to front. No simpler proof than Dirichlet's of this seemingly elementary result has ever been found, and so this may stand as our illustration of Gauss's remark that number theory is full of the most unexpected interconnections.

Dirichlet's work on Fermat's Last Theorem had emphasised what Euler had suggested earlier, that progress seemed to depend on the introduction of certain numbers which, although not integers, could be treated as such. In Box 17 we saw that Euler had used 'integers' of the form $m + n\sqrt{-3}$, where m and n are ordinary integers; Dirichlet similarly used 'integers' of the form $m + n\sqrt{5}$. It gradually came to be felt that the way to tackle every case of Fermat's Last Theorem might be to introduce such 'integers'. It turned out that Gauss's *Disquisitiones Arithmeticae* offered a natural definition of these 'integers', but it was not clear that they obeyed all the rules that the familiar integers $(1, 2, 3, \dots)$ obey. The precise rule at stake was unique factorisation (see Box 56).

What happened next says much about the status of number theory within the Parisian and the German mathematical communities in the 1840s.

There seems to have been some interest in Fermat's Last Theorem in Paris. Three Academicians, Liouville, Lamé, and Cauchy, became involved, and Lamé and Cauchy even claimed to have ideas about proving it in general. By contrast, the Germans seem to have been highly motivated to take up number theory, but not to have had much interest in Fermat's Last Theorem. In this, as in so many things, they took their lead from Gauss, who regarded it as a mere curiosity, writing to Olbers in 1816 that even if he were to develop his ideas much further, Fermat's theorem would appear only as one of the least interesting corollaries.[10] In particular Kummer, inspired by Jacobi's work on

[10]See Gauss, *Werke*, X.1, p. 76.

a generalisation of the quadratic reciprocity theorem, was working out a fully-fledged theory of some of these unusual 'integers'.

The 'integers' that Lamé introduced into the study of Fermat's Last Theorem are Gauss's 'cyclotomic integers' (from the Greek words for circle division), which are expressions of the form

$$a_0 + a_1 z + a_2 z^2 + \cdots + a_{p-1} z^{p-1},$$

where the a_j are integers, p is a prime number, and z is a complex number with the property that $z^p = 1$ but no smaller power of z equals 1. He believed that he had shown that there are no solutions in these integers to Fermat's Last Theorem for the exponent p, and therefore that there are no solutions in ordinary integers either. However, his argument relied on the claim that there is a unique factorisation theorem into prime 'integers' for these 'integers', and when he announced his result to the Académie des Sciences in Paris on 1 March 1847, Liouville expressed doubts about it. Lamé and Cauchy agreed that a unique factorisation theorem for these integers was required, but then spent the rest of the month trying to make the idea work.

On 24 May, Liouville, who alone of the three seems to have had any interest in reading the literature on the subject, presented a letter to the Académie from Kummer, in which he pointed out that he already published the discovery that unique factorisation can fail for 'integers' of this kind.[11] Kummer enclosed a copy of his paper, which had been published rather obscurely, and Liouville promptly published it in the journal that he edited. Kummer added that he was aware that the new 'integers' might still yield results in the study of Fermat's Last Theorem, and over the next few years he was to obtain deep and difficult results that extended mathematicians' understanding of that problem and put it back on the agenda. In particular, they extended the results that Sophie Germain had already obtained, which established the truth of Fermat's Last Theorem in a substantial number of cases. Unfortunately, there is no space here to pursue the matter further.[12]

Sophie Germain is interesting.[13] She was born in 1776, and educated at home, where she taught herself Latin and Greek and set about becoming a mathematician by reading works by Newton and Euler. When she was eighteen, she was able to get hold of students' notes of courses at the École Polytechnique, at a time when women were not admitted, and, as students were expected to do, she submitted an end-of-term report on Lagrange's lectures (under the pseudonym of Monsieur Le Blanc). Her report, which was a paper on analysis, impressed Lagrange greatly, and when he found out the true identity of its author he became her supporter and adviser.

Germain became interested in the theory of numbers, which she had learned from Legendre's book, *Théorie des Nombres* (1799), and in correspondence with him. Later she read Gauss's *Disquisitiones Arithmeticae*, and corresponded with him also. When he in turn found out the identity of his correspondent in 1807, he wrote to her to say:[14]

> How can I describe my astonishment and admiration at seeing my esteemed correspondent M. LeBlanc metamorphosed into this celebrated person, yielding a copy so brilliant

---

[11] It would take us too far afield to prove this result, but the first prime p for which the theorem fails is $p = 23$.

[12] See (Edwards 1977), Chapter 4, for a rich account of these matters. For a discussion of Germain's manuscripts on Fermat's Last Theorem, which show that she had made much more progress on the Theorem than revealed in the published footnote by Legendre, see (Laubenbacher and Pengelley 2010).

[13] For a biography of Sophie Germain, see (Bucciarelli and Dworsky 1980).

[14] Quoted in (Bucciarelli and Dworsky 1980, 25).

> it is hard to believe? The taste for the abstract sciences in general and, above all, for the mysteries of numbers, is very rare ... But when a woman, because of her sex, our customs and prejudices, encounters infinitely more obstacles than men in familiarising herself with their knotty problems, yet overcomes these fetters and penetrates that which is most hidden, she doubtless has the most noble courage, extraordinary talent, and superior genius.

Germain went on to do important work in the study of Fermat's Last Theorem. She then turned her attention to elasticity theory and worked on the oscillations of membranes, for which she was awarded a prize by the Académie des Sciences in Paris in 1809, the first woman to be so honoured.

It seems that the leading mathematical communities of the early 19th century differed in the enthusiasm with which they took to the theory of numbers, and it is tempting to attribute the stronger German attraction to their tradition of neo-humanism. The influence of Gauss was also a factor: he continually returned to the subject, and in various ways let it be known that he considered it to be full of deep and difficult connections with other branches of mathematics that merited attention. The fact that his *Disquisitiones Arithmeticae* was obviously central to this endeavour, but yet very difficult to read, did no harm either, because Dirichlet took it upon himself to produce a more comprehensible, almost textbook, version of the subject, thus making it easier to learn, and even to teach.

In both Germany and France we can see the emerging characteristics of a new mathematical style: a livelier appreciation of the need for careful definitions of novel objects, and for more rigorous proofs. The rise in abstraction associated with modular arithmetic, quadratic reciprocity, and the emerging theory of algebraic 'integers', came with an increasing interest in rigour. The 18th-century appeals to the generality of algebra were beginning to give way to a more critical, questioning spirit; this is also an aspect of professionalisation. An emerging profession set standards and 'policed' them, if that is not too strong or anachronistic an expression. In the institutions responsible for producing the next generation of professionals, professors determined which skills should be possessed by these people and also which people actually possessed them. Among these qualities were the abilities to argue correctly, to recognise valid proofs, and to be able to produce them. These had always been requirements placed on a mathematician, but now they carried extra weight. Originality was wanted, and the ability to get by with merely plausible arguments and simple assertions was henceforth to diminish.

18.2 Prime numbers

Prime numbers have some claim to be the 'atoms' of numbers: Every integer greater than 1 is a product of primes in an essentially unique way. But it is strangely difficult to prove claims about them. Currently, the codes used to make financial transactions over the internet depend on it being impossible for even a very powerful computer to factorise large integers quickly into primes. Only in 2002 was a reasonably rapid test discovered that detects whether a given number is prime, but it gives no information about what the factors of a given composite number are.

In this section we consider some of the questions about prime numbers that have animated mathematicians.

The Prime Number Theorem. In his inaugural lecture at the University of Bonn in 1975, the distinguished number-theorist Don Zagier said:[15]

> There are two facts of which I hope to convince you so overwhelmingly that they will permanently be engraved in your hearts. The first is that the prime numbers belong to the most arbitrary objects studied by mathematicians: they grow like weeds, seeming to obey no other law than that of chance, and nobody can predict where the next one will sprout. The second fact is even more astonishing, for it states just the opposite: that the prime numbers exhibit stunning regularity, that there are laws governing their behaviour, and that they obey these laws with almost military precision.

Let us first document some of the random behaviour of the primes. Prime numbers are used in the multiplication and factorisation of numbers, and we can get into difficulties when we try to study their additive properties. In fact, some of the most intriguing puzzles in mathematics arise in this way; for example, all primes other than 2 are odd, so if we add two primes other than 2, then we always get an even number — but can we obtain all even numbers (greater than 2) in this way? The belief that we can is called the *Goldbach conjecture*, after Christian Goldbach who posed the question in 1742, but no-one knows for sure. It has been checked for all even numbers up to 4×10^{18} — but this is not good enough for mathematicians who need absolute proof, true in every case. The best results today are that:

- Every large enough even number is the sum of a prime and another number which is either a prime or a product of two prime factors. This was proved by the Chinese mathematician Chen Jingrun in 1966. His proof was purely an existence proof — it did not provide an estimate of what is meant by the phrase 'large enough' — but it was recently shown that the theorem applies to every even number greater than $e^{e^{36}} \approxeq 1.7 \times 10^{1\,872\,344\,071\,119\,348}$.
- Every odd number is a sum of at most five primes, which was shown by Terence Tao in 2012. In December 2013 Harald Helfgott announced a proof of the stronger result that every odd number greater than 5 is a sum of three primes.

Let us now subtract prime numbers. We cannot find two primes that are just 1 apart (except for 2 and 3), since beyond 2 all primes are odd, but there are many pairs of primes differing by 2. These are called *twin primes*, such as 3 and 5, 11 and 13, 41 and 43, 107 and 109, or 1,000,037 and 1,000,039, and they are not scarce: up to 10^{11} there are 224 million such pairs. The *twin prime conjecture* is that there are infinitely many such pairs. This is another conjecture that is easy to state, and yet no-one has found a method for proving it.

The best result known at the end of 2020 was that there are infinitely many prime numbers p and $p + h$ where $0 < h \leq 246$. The twin prime conjecture would follow if the number 246 could be replaced by 2. This result follows from a breakthrough by the mathematician Yitang Zhang in 2013. He did not use the bound 246 but a bound of 70,000,000, which simplified his proof.[16] However, it seems agreed that new ideas will be required to reduce this bound to 2.

---

[15] See (Zagier 1977, 7).

[16] See (Granville 2015) for a brief biography of Zhang and explanations of his work and its subsequent refinements, and also (Neale 2017).

Although twin primes seem to occur forever, we can still find arbitrarily large gaps between primes. For example, the numbers from 90 to 96 form a string of seven non-primes, and if we want a string of 100 non-primes, we consider the number $101! = 101.100.\cdots.2.1$, and look at the numbers $101!+2$ (which is divisible by 2), $101!+3$ (which is divisible by 3), and so on, up to $101!+101$ (which is divisible by 101). This simple idea shows that there are arbitrarily large gaps between primes.

This leads us to one of the main questions in number theory: How are the prime numbers distributed? As Zagier noted, they do not seem to occur regularly, as we can see if we look at the numbers just below and just above ten million. The hundred just below it contain nine primes, whereas the hundred just above it contain just two. However, as Zagier also remarked, there are striking regularities in the distribution of the primes.

To see what he meant by this, we follow Gauss and introduce the prime-counting function $\pi(x)$, which counts the number of primes up to any number x (it is nothing at all to do with the 'circle number' π). So $\pi(10) = 4$, because there are exactly four primes (2, 3, 5, 7) up to 10, and $\pi(20) = 8$, because there are then a further four (11, 13, 17, 19). Continuing, we find that

$$\pi(100) = 25, \ \pi(1000) = 168, \ \pi(10{,}000) = 1229,$$

and so on. If we plot the values up to 100 on a graph we get a jagged pattern — each new prime creates a jump. But if we stand further away and view the primes up to 100,000, as in Figure 18.2, we get a lovely curve that is almost smooth — the primes do indeed seem to increase very regularly.

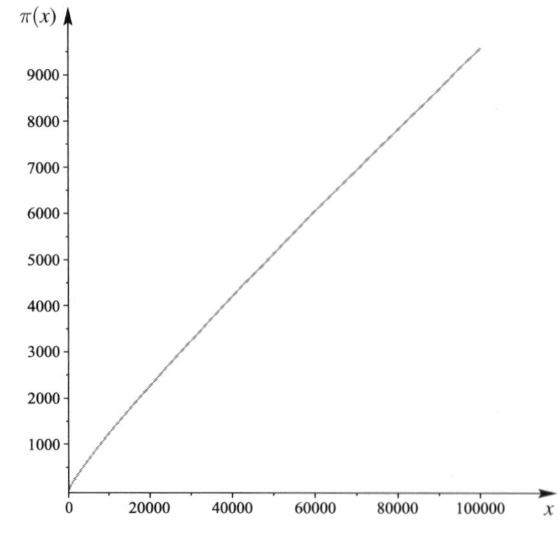

Figure 18.2. A graph of the number of primes up to 100,000

We can describe this more precisely by comparing the values of x and $\pi(x)$ as x increases. We obtain the table in Box 57 which lists x, $\pi(x)$, and their ratio $x/\pi(x)$. So up to 100 one-fourth of the numbers are prime, up to 1000 one-sixth of them are prime, and so on: they are gradually thinning out. But can we express this more precisely?

Box 57. The growth of $\pi(x)$.

x	$\pi(x)$	$x/\pi(x)$
10	4	2.5
100	25	4.0
1000	168	6.0
10,000	1229	8.1
100,000	9592	10.4
1,000,000	78,498	12.7
10,000,000	664,579	15.0
100,000,000	5,761,455	17.4
1,000,000,000	50,847,534	19.7

Notice that whenever x is multiplied by a factor of 10, the ratio $x/\pi(x)$ seems to increase by a constant amount, around 2.3. Now the mathematical device that turns multiplication into addition is the logarithm, and this number 2.3 is the natural logarithm of 10. So if we introduce the natural logarithm function $\log x$, we can summarise this phenomenon by saying that as x increases, $\pi(x)$ behaves like $x/\log x$, in the sense that the ratio between them approaches 1 as x increases without limit. This celebrated result is known as the *Prime Number Theorem*; it was guessed at by Gauss, when he was playing around with prime numbers in 1793 at the age of 15. Around 1851 the Russian mathematician Pafnuti Chebyshev proved that if the expression $x/\log x$ approaches a fixed number as x gets larger, then this limiting number must be 1 — but he was unable to prove that such a limiting number exists.

If we now compare the graphs of $\pi(x)$ and $x/\log x$, we see that they agree fairly well, but the agreement is not perfect and various mathematicians have tried to improve on it. Legendre had suggested introducing an adjustment of about 1.08, so that $\pi(x)$ ultimately behaves like $x/(\log x - 1.08)$, while Gauss proposed an improved estimate for $\pi(x)$ in terms of an integral, called the *logarithmic integral*, which is defined as

$$Li(x) = \int_0^x \frac{dt}{\log t}.$$

This is another slowly increasing function, so we can appreciate it better by comparing it with the prime-counting function, as we do below.

It was not until 1896 that the Prime Number Theorem was proved, independently, by the French mathematician Jacques Hadamard and the Belgian mathematician Charles de la Vallée Poussin. Their method was to use sophisticated ideas from calculus (specifically, the subject known as complex analysis).[17] From then on, most work on the prime number theorem used techniques from complex analysis, until the 1940s

[17] Complex analysis is the study of differentiable functions of a complex variable. It was largely created in the 19th century through the work of Cauchy, Riemann, and Weierstrass.

when Atle Selberg and Paul Erdős produced the first proof that relies on purely 'elementary' (though technically very difficult) techniques; Selberg was awarded a Fields Medal in 1950 for his work in this area.[18]

The Riemann hypothesis. Let us now leave prime numbers for a while and see how the celebrated (and still unproved) Riemann hypothesis arose. As we saw in Section 8.2, Euler's zeta function

$$\zeta(k) = 1 + \frac{1}{2^k} + \frac{1}{3^k} + \frac{1}{4^k} + \cdots$$

is defined for any real number $k > 1$. But can we extend the definition to other numbers?

By way of analogy, we know by the binomial theorem that

$$1 + x + x^2 + x^3 + \cdots \text{ is equal to } 1/(1 - x).$$

But the left-hand side makes sense only when x lies between -1 and 1, whereas the right-hand side makes sense for any x, apart from 1. So we can extend the formula on the left-hand side to all values of x (other than 1) by redefining it using the formula on the right-hand side.

In the same way, it is possible to extend the definition of the zeta function to almost all of the complex plane, which consists of all symbols of the form $a + bi$, where a and b are the x- and y-coordinates and i represents the 'square root of -1'. With the use of contour integration, a method in complex analysis that had been used to great effect by Cauchy, the definition of the zeta function can be extended to the whole plane, apart from the point 1 (since $\zeta(1)$ is not defined when $k = 1$). When $k > 1$, we get the same value as before — for example, $\zeta(2) = \pi^2/6$. For all other values of k, including complex numbers, we get a value of the zeta function.

It was Bernhard Riemann who extended the zeta function to the complex plane, and the function is now known as the *Riemann zeta function*. In a memoir of 1859 'On the number of primes less than a given magnitude' he linked the properties of prime numbers to those points in the complex plane where the zeta function takes the value 0, and stated the conjecture that is now known as the *Riemann hypothesis*, as we see below.

We have seen that Gauss attempted to explain how the primes thin out, by proposing the estimate $x/\log x$ for the number of primes up to x, and we discuss this further below. Riemann's great achievement was to obtain an excellent approximation to the number of primes up to x, and his formula involved the zeros of the zeta function — the solutions of the equation $\zeta(z) = 0$ — in a crucial way.

It turns out that the Riemann zeta function has zeros at the points $-2, -4, -6, -8, \ldots$ (these are called the *trivial zeros*), and that all the other zeros (the non-trivial ones) lie within a vertical strip between 0 and 1 (the so-called critical strip), beginning at the following points:

$$\tfrac{1}{2} \pm 14.1i, \quad \tfrac{1}{2} \pm 21.01i, \quad \tfrac{1}{2} \pm 25.01i, \quad \tfrac{1}{2} \pm 30.4i, \quad \ldots.$$

But these all have the form $\frac{1}{2}$ + (something times i). The question arises:

> Do all the zeros in the critical strip have this form? — that is, do they all lie on this central vertical line?

[18] The Fields Medals are awarded every four years at the International Congress of Mathematicians and are generally regarded as the highest honour that a mathematician can receive. They were first awarded in 1936, and eligibility is restricted to mathematicians under the age of 40.

This is the big question, and the Riemann hypothesis is the conjecture that the answer is 'yes'. It has been proved that the zeros in the critical strip are symmetrically placed, both horizontally and vertically, and that as one progresses vertically up and down the line, the first several zeros do lie on this critical line 1/2. In fact, the first hundred billion zeros lie on this line! Moreover, the English mathematician G. H. Hardy proved in 1914 that there are infinitely many zeros inside the critical strip, and that at least 40% of them lie on the critical line. But do they *all* lie on this line? Or are the first hundred billion just a coincidence?

It is generally believed that all the non-trivial zeros do lie on the critical line. Indeed, the appearance of just one of these zeros off the critical line would cause problems in number theory and throughout several areas of mathematics where there are theorems that assume the truth of the hypothesis. But no-one has been able to prove it, even after a century and a half — the Riemann hypothesis, which is now one of the Millennium prize topics of the Clay Mathematics Institute, remains (in 2021) one of the most celebrated unsolved problems in the whole field of mathematics.

What is the connection between the Riemann hypothesis and the Prime Number Theorem? If the Riemann hypothesis is true then the theorem follows fairly straightforwardly. If the Riemann hypothesis were markedly false, and zeros strayed considerably from the critical line, then the Prime Number Theorem would be false. This cannot be the case, and what de la Vallée Poussin proved was that the zeros lie in a region around the critical line which is such that the Prime Number Theorem holds.

The Riemann hypothesis acquired its present importance because it fits into a very general family of ideas, and because it has resisted many attempts at proving it. There are now many situations in mathematics where a variant of the Riemann hypothesis can usefully be formulated, and in most of these cases it has also been established. Mathematicians differ as to whether this suggests that the Riemann hypothesis is likely to be true, and we shall all have to wait for a proof. So it is perhaps surprising that Riemann himself set little store by it, writing only:

> A rigorous proof of this would certainly be desirable; however, after a few brief and fruitless attempts to find one, I have put this on one side for the time being, because it did not seem essential to the immediate object of my investigation.

The distribution of the primes. The problem that Riemann was interested in was the detailed distribution of the primes. There are intervals where primes are more numerous than the Prime Number Theorem would suggest, and regions where the primes are less numerous, and Riemann wanted to know how they alternate. This thickening and thinning of the primes is quite clear in small ranges. For example, there are no primes between 890 and 910 or between 7255 and 7282. Other regions, say between 800 and 1000, show the thickening quite markedly.

Riemann considered two error estimates. The first, $\pi(x) - (x/\log x)$, is a direct comparison of the counting function with the estimate provided by the prime number theorem (note that $x/\log(x)$ is an increasing function with a slowly decreasing slope).

The graph of $\pi(x) - (x/\log x)$ in the range $[100, 10{,}000]$ is generally *increasing*, rising from about 20 to just over 140, but it oscillates quite markedly and it sometimes decreases. Because the graph of $x/\log x$ increases smoothly, the dips correspond to regions where $\pi(x)$ increases very slowly; we interpret this as saying that the gaps between the primes here tend to be large. Likewise, the steeper parts of this graph cor-

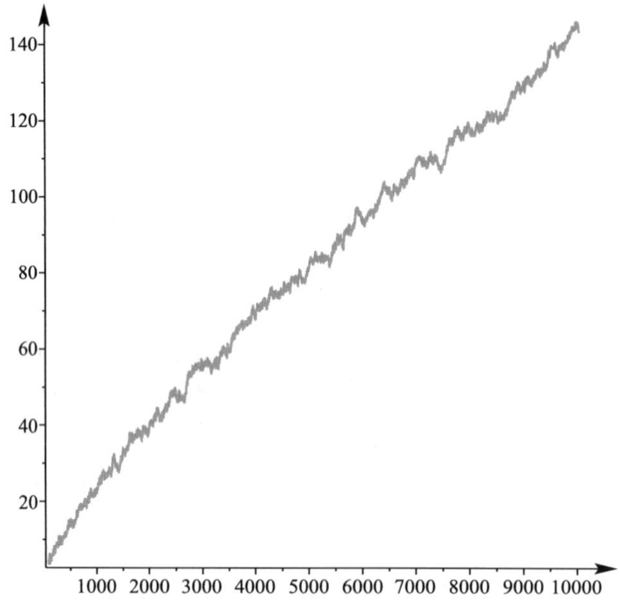

Figure 18.3. A graph of $\pi(x) - (x/\log x)$ in the range [100, 10,000]

respond to regions where $\pi(x)$ increases rather fast and the gaps between the primes here tend to be small.

Riemann also considered what was thought to be a better estimate provided by Gauss's logarithmic integral $Li(x)$. This function is also increasing with a slowly decreasing slope, and as an indication of its merits,

$$Li(1{,}000{,}000{,}000) = 50{,}849{,}234.96,$$

a strikingly good approximation to the number of primes less than 1,000,000,000, which is 50,847,534.

The graph of the function $\pi(x) - Li(x)$ is generally *decreasing* (see Figure 18.4), and as before, local peaks correspond to regions where the primes are close together, and local dips indicate regions where the primes are further apart. However, the range is much less, between -4 and -22, so on this evidence at least, $Li(x)$ is a better approximation. We can see that the thickening and thinning of the primes shows up well in the graph of $\pi(x) - Li(x)$. More precisely, $\pi(x) - (x/\log x) = 6115.58$ and $\pi(x) - Li(x) = -129.54$ when $x = 1{,}000{,}000$.

The delicate question that Riemann addressed was that of representing the difference $\pi(x) - Li(x)$. He concluded that this difference could be written as an infinite series of the form[19]

$$\frac{Li(x^{1/2})}{2} + \frac{Li(x^{1/3})}{3} + \frac{Li(x^{1/5})}{5} - \frac{Li(x^{1/6})}{6} + \frac{Li(x^{1/7})}{7} + \cdots.$$

What is this difference? Let us set

$$ri(x) = Li(x) - \left(\frac{Li(x^{1/2})}{2} + \frac{Li(x^{1/3})}{3} + \frac{Li(x^{1/5})}{5} - \frac{Li(x^{1/6})}{6} + \frac{Li(x^{1/7})}{7}\right).$$

[19]Notice the minus sign after the third term.

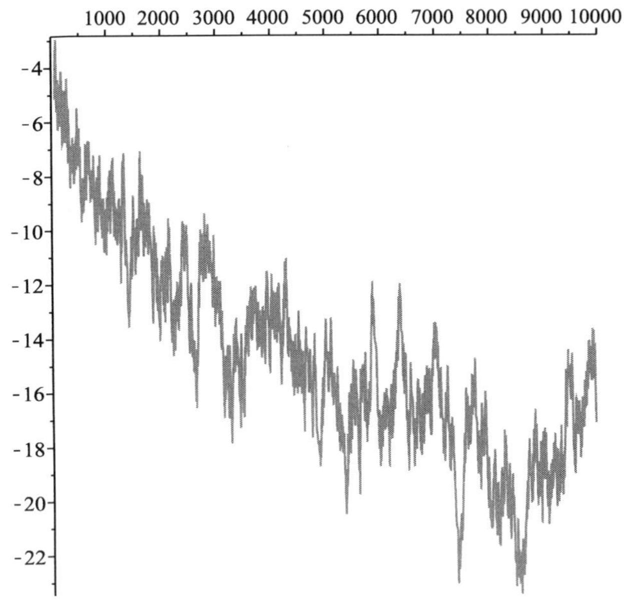

Figure 18.4. A graph of $\pi(x) - Li(x)$ in the range [100, 10,000]

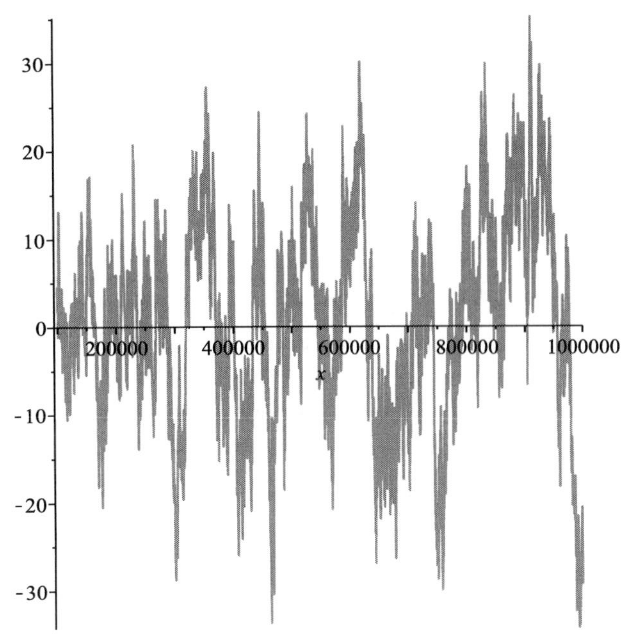

Figure 18.5. A graph of $\pi(x) - ri(x)$ in the range [100,000, 1,000,000]

In the range [100,000, 1,000,000] the graph in Figure 18.5 seems neither to increase nor decrease steadily, but to oscillate around the x-axis. The oscillations are slowly increasing, from around ± 12 to ± 30, which is also an improvement on the accuracy of the estimate. The error at 1,000,000 is -29.34. The same behaviour persists to

$x = 3{,}000{,}000$ (the range mentioned by Riemann as known to Gauss) and the error has grown only to ± 50.

So Riemann's new estimate is an improvement in two ways. First, the difference shows no trend of increasing or decreasing. Second, the difference is smaller and therefore the estimate is more accurate.

18.3 Complex numbers and quaternions

Complex numbers. We have mentioned complex numbers several times already, often with a hint of mystery about their nature. We have seen them discussed by Euler when he related the exponential and trigonometric functions (see Section 9.2), and we have seen that Euler spoke of novel kinds of integers that involved complex numbers (see Box 17). Work on Bézout's theorem led to the suggestion that points on curves might have complex numbers as coordinates, although the idea was left unexplored. We have even seen Lambert raise the idea of a sphere of imaginary radius in his investigations of the foundations of geometry.

Most often, we have seen complex numbers in connection with the Fundamental Theorem of Algebra, whether they were introduced or deftly avoided, and we shall shortly examine this topic in more detail. Indeed, it was in connection with the solution of polynomial equations that complex numbers first entered mathematics, when, in 1545, Cardano allowed them as solutions of the equation $x^2 - 10x + 40 = 0$, although he said that the arithmetic that gave rise to them was 'as refined as it is useless'.[20]

What we have not yet seen is any adequate discussion of what complex numbers *are*. From one point of view they are bizarre. Because the square of every number is positive or zero, how can it make sense to take the square root of a negative number? Indeed, if you believe that the natural numbers are what we count with, and numbers more generally are what we measure, then even an explanation of negative numbers is required.

John Wallis in Oxford was among the first to go beyond the formal rules governing the arithmetic of complex numbers and offer an explanation of their essence. In 1685 in *A Treatise of Algebra* he first explained negative numbers by invoking the idea of movement backwards and forwards along a straight line.[21] Then he said that the same ideas must also be admitted when talking about planes, and explained that $\sqrt{-1}$ can be regarded as a 'mean proportional' between -1 and 1. The mean proportional between two numbers a and b is the number h such that $a : h = h : b$, or $ab = h^2$, so Wallis's claim follows from the equation

$$-1 : \sqrt{-1} = \sqrt{-1} : 1, \quad \text{or } -1 \times 1 = \sqrt{-1} \times \sqrt{-1}.$$

To show that this was not mere arithmetic, he then explained how a mean proportional arises geometrically. In Figure 18.6, the line segment OC is the mean proportional between the segments OA and OB, because the right-angled triangles BOC and COA are similar, and so

$$BO/OC = OC/OA, \text{ and } OA \times BO = OC^2.$$

[20]See (Cardano 1968, Ch. 227). Complex numbers also arise in the Tartaglia–Cardano formula for the solution of a cubic equation. See *Volume* 1, Section 10.3.

[21]Wallis, *Algebra* (1685), Chapter 67. We discussed some of Wallis's work in *Volume* 1.

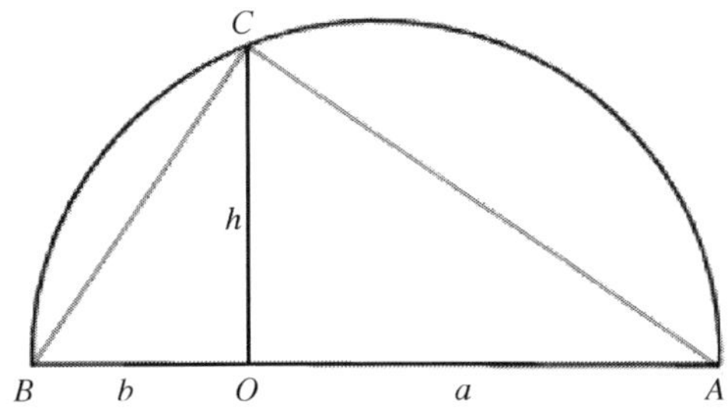

Figure 18.6. The segment OC is the mean proportional between OA and OB

Wallis then considered the case in which the segments OA and OB are taken to be of lengths a and b, except that the segment OB is measured in the negative direction, so b is negative. In this case their mean proportional OC has length $\sqrt{a \times (-b)}$. If we take the special case (which Wallis did not) where OA and OB are of unit length then the segment OC represents $\sqrt{-1}$. As Wallis put it 'we have shewed what in Geometry answers to a Root of a Negative Square in *Algebra*'.[22]

When faced with an apparently difficult question, the temptation is always to try to answer it directly, by accepting its premises. Quite often, however, the smart move is to contest the premises, and this is what Euler did in his *Algebra*, almost a century later.

Euler on imaginary numbers.

> 143. And, since all numbers which it is possible to conceive are either greater or less than 0, or are 0 itself, it is evident that we cannot rank the square root of a negative number amongst possible numbers, and we must therefore say that it is an impossible quantity. In this manner we are led to the idea of numbers, which from their nature are impossible; and therefore they are usually called *imaginary quantities*, because they exist merely in the imagination.
>
> 144. All such expressions as $\sqrt{-1}$, $\sqrt{-2}, \sqrt{-3}, \sqrt{-4}$, &c. are consequently impossible, or imaginary numbers, since they represent roots of negative quantities; and of such numbers we may truly assert that they are neither nothing, nor greater than nothing, nor less than nothing; which necessarily constitutes them imaginary, or impossible.
>
> 145. But notwithstanding this, these numbers present themselves to the mind; they exist in our imagination, and we still have a sufficient idea of them; since we know that by $\sqrt{-4}$ is meant a number which, multiplied by itself, produces -4; for this reason also, nothing prevents us from making use of these imaginary numbers, and employing them in calculation.

Euler here offered a different definition of number, as a concept that is in our minds, that we know how to handle — it is free of contradiction — and which embraces the usual numbers. To distinguish the new numbers from the counting and

[22]See Wallis, *Algebra* (1685), Chapter 68.

measuring numbers, he allowed that they can be called 'imaginary', but he plainly did not regard them as impossible in themselves — they are impossible only as quantities or measuring numbers. Indeed, he went on:

> 149. It remains for us to remove any doubt which may be entertained concerning the utility of the numbers of which we have been speaking; for those numbers being impossible, it would not be surprising if they were thought entirely useless, and the object only of an idle speculation. This, however, would be a mistake; for the calculation of imaginary quantities is of the greatest importance, as questions frequently arise, of which we cannot immediately say whether they include any thing real and possible, or not; but when the solution of such a question leads to imaginary numbers, we are certain that what is required is impossible.

This is another valuable insight. The price of declaring complex numbers to be logically impossible is high: the calculation of unknown quantities requires that complex numbers can be accepted as answers, even though in a problem concerning quantities the valid conclusion would be that 'what is required is impossible'.

We saw in Section 14.1 that in the same year, 1770, Lambert expressed a very similar view to Kant when he wrote: 'The sign $\sqrt{-1}$ represents an unthinkable non-thing. And yet it can be used very well in finding theorems'. In short, numbers are what one calculates with, and certain kinds of them arise in counting and measuring problems, but one should let go of the idea that every number is the measure of some object.

Difficult philosophical questions do not surrender easily — or perhaps one should say that people holding certain views are not easily persuaded to change them. What some mathematicians, such as Euler, seem to have believed is that there is really nothing to explain about complex numbers. Others may have thought that perhaps there was an issue, but it could be set aside. We shall see below that Gauss had no problem with assigning complex numbers as the coordinates of points in the plane in his *Disquisitiones Arithmeticae* of 1801, and mathematicians working with complex numbers after Euler generally seem not to have worried about what they are or could be. But around 1800 there was a last stand by those who thought that reliance on formal arithmetic was a shallow way out of a deep problem, and raised the question again.

We focus our attention on two people: Caspar Wessel in Denmark–Norway, and Jean Robert Argand in Paris and Geneva.

Wessel was a surveyor, and he had no doubt that complex numbers could be used to represent points in the plane. What he set out to do was to explain the utility of vectors by constructing algebraic operations for directed line segments.[23] He expressed his directed line segments in terms of a unit segment that he denoted by 1 and another unit segment of the same length and at right angles to the first that he denoted by ε, and showed that his rule for multiplication gave him $\varepsilon \times \varepsilon = -1$. He then remarked that 'from this it follows that ε becomes $\sqrt{-1}$'. As the historian Kirsti Andersen remarked in her study of Wessel's mathematics:[24]

> This observation shows quite clearly that he was aware that he had given a geometric interpretation of $\sqrt{-1}$, but he did not make any point out of this achievement.

[23]See (Andersen 1999).

[24]See (Andersen 1999, 74).

Wessel then applied his methods to the study of plane and spherical polygons, which involved extending his ideas of directed line segments into a third dimension.

Surprisingly little is known about Argand. Indeed, the historian Gerd Schubring has argued that even Argand's correct first name may be unknown, and the attribution to one Jean Robert Argand made by Hoüel, the editor of the reprint of Argand's paper in 1874, may be wrong. Schubring also questions Hoüel's claim that Argand was born in Geneva, and suggests that Argand may have been active in clock-making circles in Paris, and that he wrote an anonymous 'Essai' on complex numbers (possibly in 1806) and showed it to Legendre. Then sundry misunderstandings and confusions, exacerbated by Argand's bashfulness, resulted in his Essai being published only in 1813, following an article by Jacques Frédéric Français in Gergonne's *Annales*. Français's article referred to an unnamed author whose ideas had impressed Legendre, and urged him to come forward and identify himself, which is what Argand did. His Essai was then published in Volume 4 of Gergonne's *Annales* and, in an altered form, as a pamphlet.[25]

Argand began by explaining negative numbers as weights in the second pan of a balance, or as debts. Then he interpreted complex numbers as points in the plane. He discussed how this representation gives rise to the usual rules for addition and multiplication of complex numbers, deduced De Moivre's formula, and concluded with a sketch of a proof of the Fundamental Theorem of Algebra.[26] So we see that although Wessel's contribution of 1799 and Argand's contribution of 1813 are sometimes regarded as almost the same, they are diametrically opposite in their approaches.

For whatever reason, Wessel's work remained unknown until it was republished in 1897 in a French translation.[27] It cannot therefore be said to have changed the debate about the nature of complex numbers, which rumbled on in Europe for several more decades. The split gradually widened between those who held to a hard-line traditional view that numbers are necessarily counting and measuring numbers (and cannot, therefore, be complex numbers) and the mathematicians who preferred to stretch the concept of number in order to accommodate the needs of algebra, in which the traditionalists were not so much refuted as marginalised.

Two papers are generally taken to mark the consolidation of the mathematicians' point of view. In the first of these — an important paper on number theory in 1831 — Gauss introduced 'integers' of the form $m + in$, where m and n are ordinary integers (they are called *Gaussian integers* today). His principal point was to establish the utility of these new 'integers' in the theory of numbers, but he gave a very careful explanations of them in algebraic and geometrical terms. There he remarked that:[28]

> In this way the metaphysics of quantities that we call imaginary, is presented in a clear light ... The difficulties that one has usually believed attend the theory of imaginary quantities are largely based on their unfortunate names (some indeed refer to them by

---

[25]His paper was published under the name Argand with no initial given, and this has been a source of the subsequent confusion. For a full and recent discussion of the French debate, see (Schubring 1999). Almost all historical scholarship has relied on the republication of Argand's work by Hoüel in 1874.

[26]One interesting difference between the two publications is that the pamphlet omits a wholly unsuccessful attempt by Argand to regard $\sqrt{-1}^{\sqrt{-1}}$ as representing a third dimension. See (Argand 1813, paragraph 11).

[27]An English translation was published in 2001 to mark Wessel's bicentenary.

[28]See (Gauss 1831, §§31 and 38).

> the ill-sounding name of impossible quantities). Had one started from the representation provided by manifolds of two dimensions … simplicity would have followed instead of perplexity, and clarity instead of darkness.

The second of these two papers is William Rowan Hamilton's presentation of complex numbers as ordered pairs of real numbers, with appropriate definitions of addition and multiplication.[29] Hamilton was the Professor of Astronomy in Trinity College, Dublin, and Royal Astronomer of Ireland, and had been appointed to these positions in 1827, when still an undergraduate.

In his approach, ordered pairs (a, b) and (c, d) are added and multiplied as follows:

$$(a, b) + (c, d) = (a + c, b + d) \sim (a + ib) + (c + id) = (a + c) + i(b + d)$$

$$(a, b) \times (c, d) = (ac - bd) + (ad + bc) \sim (a + ib) \times (ci + d) = (ac - bd) + i(ad + bc).$$

The only novelty here is that the concept of an ordered pair can be understood directly, without the need for any appeal to geometry, thereby avoiding any confusion with the idea of measuring numbers. This helped to allay suspicions about complex numbers.

What we see in these papers are philosophical ideas about mathematics being adapted to fit new mathematical practices. It is striking to see that as mathematicians learned to accept complex numbers as reliable, their anxieties about what they might be subsided. The problem that the best mathematicians, such as Gauss and Cauchy, faced was not to use or understand complex numbers, but to understand the nature of functions of a complex variable. That is another story, which we cannot tell here, but from that point of view the attention that focussed narrowly on debates about the nature of complex numbers may be said to be excessive.

That said, it remained for mathematicians to put their confidence in complex numbers to the test and to solve important problems, such as proving the Fundamental Theorem of Algebra, and that is the topic to which we now turn.

The issue now is not the philosophical or foundational nature of complex numbers, but their utility in solving polynomial equations. None of the above discussion had any bearing on the algebra of complex numbers, which had been well understood since the time of Bombelli if not before.[30]

Gauss's proofs of the Fundamental Theorem of Algebra. As we saw in Section 7.3, Gauss first swept aside all known proofs of the Fundamental Theorem of Algebra in his Ph.D. dissertation of 1799, and considerably raised the standards by which attempts on the theorem are to be judged.

He then offered his own proof. He took a polynomial

$$f(z) = z^m + a_{m-1}z^{m-1} + \ldots + a_0$$

with real coefficients, and although he had promised in §2 of his essay not to use imaginary quantities he looked for its roots in an infinite plane whose points are specified by polar coordinates r, ϕ: this is to use complex numbers in polar form without saying so. His strategy was now to write

$$z = re^{i\phi} = r(\cos\phi + i\sin\phi),$$

[29] See (Hamilton 1834).

[30] See *Volume* 1, Chapter 9.

and to use the mathematical identity

$$z^k = r^k e^{ki\phi} = r^k(\cos k\phi + i\sin k\phi),$$

to separate the real part U and the imaginary part T of the equation. The result is that the roots of the equation are the common points of the equations

$$U = r^m \cos m\phi + a_{m-1}r^{m-1}\cos(m-1)\phi + \ldots + a_1 r\cos\phi + a_0 = 0,$$

$$T = r^m \sin m\phi + a_{m-1}r^{m-1}\sin(m-1)\phi + \ldots + a_1 r\sin\phi = 0.$$

These are the curves defined by the real and imaginary parts of the equation $f(z) = 0$: that is, Re $f(z) = 0$ and Im $f(z) = 0$, as z varies in the complex plane. (Note that real roots of a polynomial lie on the horizontal axis in the plane of complex numbers.)

Gauss now argued that these curves meet in precisely m points. We give the following example.

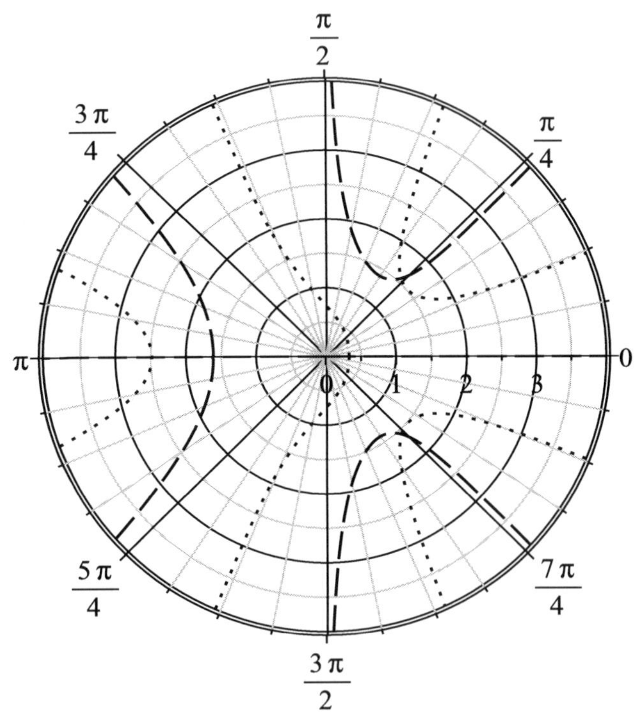

Figure 18.7. The real and imaginary parts of $f(z) = z^4 - 3z^2 + 7z - 2$

In this example (see Figure 18.7), the real and imaginary parts of the complex polynomial $z^4 - 3z^2 + 7z - 2$ are shown: the dotted curve shows where the real part vanishes, the dashed curve shows the where the imaginary part vanishes — this includes the real axis. Each of these curves breaks into several pieces. There are four dotted pieces, and four dashed pieces (one of which is the real axis) — four pieces because the degree of the polynomial is 4.

The four dotted and the four dashed pieces meet in four points that correspond to the roots of $f(z) = 0$. Two of these are complex, and two are purely real and lie on the horizontal axis. Notice also that the four pieces of each kind are so arranged that they meet the circle that bounds the figure in alternating points (one dotted, then one dashed, then one dotted, and so on).

Gauss's argument breaks into two parts. In the first part he observed that outside a suitably large circle of radius R centred on the origin, each of these curves meets a concentric circle of radius $r \geq R$ in two disjoint sets of $2m$ distinct points, so these curves consist of $2m$ arcs going off to infinity in the plane. Moreover, when r is suitably large, the z^m term is so dominant that the curve $\mathrm{Re}\,(f(z)) = 0$ meets the circle of radius r approximately at the points where $\cos m\phi = 1$. Likewise, the curve $\mathrm{Im}\,(f(z)) = 0$ meets the circle of radius r approximately at the points where $\sin m\phi = 1$, so we see that these two sets of points alternate. It now follows that the curves $\mathrm{Re}\,(f(z)) = 0$ and $\mathrm{Im}\,(f(z)) = 0$ never cross, and so there are no common zeros outside the circle of radius R.

For future use, let us call the $2m$ points where the curve $\mathrm{Re}\,(f(z)) = 0$ meets the circle of radius R the U_R points, and the $2m$ points where the curve $\mathrm{Im}\,(f(z)) = 0$ meets the circle of radius R the U_T points.

Gauss now set to work to show that inside the circle of radius R, the curve $\mathrm{Re}\,(f(z)) = 0$ is made up of m disjoint parabola-shaped pieces, which join up pairs of the U_R points, and likewise the curve $\mathrm{Im}\,(f(z)) = 0$ is made up of m disjoint parabola-shaped pieces, which join up pairs of the U_T points.

Intuitively, it is clear that because the U_R and U_T points alternate, corresponding to the vanishing real and imaginary parts means that these pairs can be joined only by parabola-shaped pieces that cross, and furthermore, because there are m pairs of each kind, there will be m such crossing points. At such a crossing point, the polynomial equation has a root, and so the Fundamental Theorem of Algebra is proved.

In general, Gauss argued that the curves $\mathrm{Re}\,(f(z)) = 0$ and $\mathrm{Im}\,(f(z)) = 0$ are real algebraic curves, so they consist of m pieces that, as it were, come from and go to infinity; they cannot stop, break apart, or spiral to a point in the fashion of some transcendental curves (Gauss gave the example of $y = 1/\log x$). As he put it, if an algebraic curve enters a bounded region of the plane, then it also leaves it.[31] This is true, but it is surely no easier to prove than the Fundamental Theorem of Algebra itself, and to that extent Gauss's proof is also defective.

On the other hand, his argument was systematic in that it dealt with polynomial equations of any degree. Moreover, the topological nature of Gauss's proof is attractive, and that was what Gauss saw as the heart of the matter. (We say that the proof was topological because it relied on the general shape of the curves involved but not on any special property of them.) This was a remarkable insight for 1799, even if neither Gauss nor anyone else could have provided a clear account of the completeness of the real numbers and a proper distinction between the real numbers and the rational numbers.

Gauss's second proof (from late 1815) is a skilful use of symmetric functions and does not involve complex variables, so we pass it over.[32] His third proof (from January 1816) used a theory of complex integrals that Gauss had been developing; the proof rests on the insight that when a double integral is replaced by a repeated integral, the order of integration may matter when the integrand becomes infinite.

[31] See (Gauss 1799, §21, footnote).

[32] An English translation can be found in Smith's *A Source Book*, pp. 292–306.

The fourth and last of Gauss's proofs was published in 1849 and was produced on the occasion of the 50th anniversary of his first proof, which was marked by a celebration of Gauss's distinguished career. Gauss declined to repeat his criticisms of the 18th-century proofs, noted that Cauchy had given a proof more recently, and then took up his first proof again. This time he was more careful about the case of multiple roots and the possible configurations of curves that can arise, and showed that all roots occur in the way that he described.

Quaternions. The study of complex numbers in the early 19th century was productive in other, more surprising, ways, and for the best example we must return to William Rowan Hamilton.

Figure 18.8. William Rowan Hamilton (1805–1865)

As we saw above, in 1834 he had published a paper in which he explained carefully how complex numbers can be regarded as ordered pairs of real numbers; specifically the complex number $a + ib$ is written as the ordered pair (a, b). Complex numbers are a convenient notation for pairs, and so for points in the plane, but space is three-dimensional, and his paper sparked a rather British search for a similar notation for triples. What was required was a way of adding and multiplying two triples so as to obtain a third, in such a way that subtraction and division are also possible.

We pause to recall how division is done with complex numbers. Given a non-zero complex number $a + ib$ we must find another, $a' + ib'$, with the property that

$$(a + ib) \times (a' + ib') = 1.$$

This is accomplished by setting

$$a' + ib' = \frac{a - ib}{a^2 + b^2} = \frac{a}{a^2 + b^2} - i\frac{b}{a^2 + b^2},$$

which is possible precisely because $a + ib$ is non-zero and so $a^2 + b^2$ does not vanish. The consequence is that $a + ib$ has what is called a 'multiplicative inverse',

$$(a + ib)^{-1} = \frac{a - ib}{a^2 + b^2},$$

and so it is always possible to divide a complex number by a non-zero complex number.

The search for triples with the required properties always failed when it came to defining multiplication is such a way that division was possible. Later, it was proved that no such algebra of triplets can exist. But in 1843 Hamilton surprised himself by discovering an algebra of quadruples in which quadruples can be added, subtracted, multiplied, and (if non-zero) divided. Hamilton himself left this account of their discovery in the form of a letter to his son Archibald, written more than twenty years later, in 1865. He recalled that his children Archibald and William Edwin used to ask him at breakfast if he could multiply triples, 'Whereto I was always obliged to reply, with a sad shake of the head: 'No, I can only add and subtract them'.' Hamilton then went on:[33]

Hamilton discovers quaternions.

> But on the 16th day of the same month — which happened to be a Monday, and a Council day of the Royal Irish Academy — I was walking in to attend and preside, and your mother was walking with me, along the Royal Canal, to which she had perhaps driven; and although she talked with me now and then, yet an under-current of thought was going on in my mind, which gave at last a result, whereof it is not too much to say that I felt at once the importance. An electric circuit seemed to close; and a spark flashed forth, the herald (as I foresaw, immediately) of many long years to come of definitely directed thought and work, by myself if spared, and at all events on the part of others, if I should even be allowed to live long enough distinctly to communicate the discovery. Nor could I resist the impulse — unphilosophical as it may have been — to cut with a knife on a stone of Brougham Bridge, as we passed it, the fundamental formula with the symbols, i, j, k; namely
>
> $$i^2 = j^2 = k^2 = ijk = -1,$$
>
> which contains the Solution of the Problem, but of course, as an inscription, has long since mouldered away. A more durable notice remains, however, on the Council Books of the Academy for that day (October 16th, 1843), which records the fact, that I then asked for and obtained leave to read a Paper on Quaternions, at the First General Meeting of the Session, which reading took place accordingly, on Monday the 13th of the November following.

Hamilton introduced three symbols i, j, and k, and gave them simple but unexpected rules for their multiplication

$$i^2 = j^2 = k^2 = -1,$$
$$ij = k; \quad jk = i, \quad ki = j,$$
$$ij = -ji, \quad jk = -kj, \quad ki = -ik.$$

The first rule makes i, j, and k behave as the symbol i does in complex numbers — they are all square roots of -1. The second rule is a simple way of combining i, j, and k. But the third rule was a shock to almost everyone who saw it, because it says that the multiplication of the new symbols depends on the order of the symbols. Everyone

[33]See Hamilton, *Mathematical Papers* 2, 435–436.

who had been looking for an algebra of triples up to this point had assumed they must be *commutative*: the order of multiplication does not matter ($ab = ba$). In the context in which Hamilton was working, the idea of a non-commutative system of multiplication was unheard of: one might even say that people believed that multiplication of numbers was *meant* to be commutative.

Hamilton then considered quadruples, which he called 'quaternions' and studied the algebra that resulted; these are expressions of the form $w + xi + yj + zk$, where w, x, y, z are ordinary numbers.

From their definition, and various properties that he assumed for them, the formula for a product is as follows:

$$\begin{aligned}(w + xi + yj + zk)&(w' + x'i + y'j + z'k)\\ &= (ww' - xx' - yy' - zz') + (wx' + w'x + yz' - y'z)i\\ &\quad + (wy' + w'y + zx' - z'x)j + (wz' + w'z + xy' - x'y)k.\end{aligned}$$

Notice that

$$(w + xi + yj + zk)(w - xi - yj - zk) = w^2 + x^2 + y^2 + z^2,$$

which is the key to division:

$$(w + xi + yj + zk)^{-1} = \frac{w - xi - yj - zk}{w^2 + x^2 + y^2 + z^2}.$$

Also, when $w = w' = 0$,

$$(xi+yj+zk)(x'i+y'j+z'k) = (-xx'-yy'-zz')+(yz'-y'z)i+(zx'-z'x)j+(xy'-x'y)k.$$

Hamilton called the real number w the *scalar* part of the quaternion $w+xi+yj+zk$, and the purely imaginary part $xi + yj + zk$ its *vector* part (from the Latin meaning 'carrier'). The term 'vector' in modern mathematics derives from this. In the next section we shall see how the vector part of a quaternion was made to stand on its own and the modern theory of vectors was created. The last formula above shows that when two 'vectors' are multiplied, the scalar part of the product is the negative of the modern dot product of two vectors, and the vector part is the modern vector (or cross) product of the two vectors.

Hamilton's enthusiasm for quaternions drove him to produce 'at least thirty-four papers in five different journals' on quaternions by the end of 1847, to which he later added a further 75 papers, his *Lectures on Quaternions* (1853), a book of 873 pages, and his *Elements of Quaternions* (1854), a book of a mere 762 pages.[34] Initial hostility to the idea of non-commutative 'numbers' died down, and other authors took up the subject and found applications for quaternions in geometry.

The first of Hamilton's books can perhaps be regarded as a work of reference — it is certainly tedious to read — and the second is a poor response to what even Hamilton perceived as a need for a good introduction to the subject. But Hamilton had a true supporter in Peter Guthrie Tait, Professor of Natural Philosophy in Edinburgh, whose *Elementary Treatise on Quaternions* came out in 1867 and proved to be exactly the book to promote the cause.

[34]See (Crowe 1985).

18.4 Vectors

In mathematical physics many three-dimensional quantities such as forces, accelerations, and velocities have both size and direction. For example, to specify a force at a point (x, y, z), mathematicians may write something like this:

$$(f(x, y, z), g(x, y, z), h(x, y, z)).$$

In particular, electric and magnetic fields exert forces at each point that are represented by vectors. If Newton's laws of motion can claim to be the most important statements made in 17th- and 18th-century physics, then the equations relating electricity and magnetism can equally claim to be the most important statements made in 19th-century physics. They are called *Maxwell's equations*, after the Scot James Clerk Maxwell, whose *Treatise on Electricity and Magnetism* of 1873 became the definitive book on the subject.

Maxwell's equations are written these days as follows. Let **E** denote the electric field strength, and **B** denote the magnetic field strength. Let ρ denote the electric charge density and **j** denote the electric current density (the rate at which charge flows through a unit area per second), let ε_0 be a constant determined by the medium, and let c be the velocity of light in a vacuum. Then

$$\nabla.\mathbf{E} = \frac{\rho}{\varepsilon_0}, \quad \nabla \times \mathbf{E} = -\frac{\partial \mathbf{B}}{\partial t}, \quad c^2\nabla \times \mathbf{B} = \frac{\partial \mathbf{E}}{\partial t} + \frac{\mathbf{j}}{\varepsilon_0}, \quad \nabla.\mathbf{B} = 0.$$

The ∇ notation involves differentiation and is explained below. For the moment, it is enough to know that if **V** is a vector then $\nabla.\mathbf{V}$ is a scalar and $\nabla \times \mathbf{V}$ is another vector.

As we shall see, just as Newton's equations evolved over time, so too did Maxwell's equations — and just as equations involving gravitational forces, accelerations, and velocities can rapidly become complicated and burdensome, so too can the mathematics in electromagnetic theory. It is in this context that the vector notation for such quantities emerged, evolved, and became accepted as a significant simplification of the theory.

There are two principal sources for vectors: work by Hamilton and work by the German mathematician (and later linguist) Hermann Grassmann. We have already looked at Hamilton's initial contribution. A further example, which shows how he extended these ideas to the calculus, is described in Box 58, and was to catch the attention of Maxwell.

The other key author was Hermann Günther Grassmann. Paradoxically, his ideas about vectors were much clearer and deeper than Hamilton's but were much less successful for many years. One reason is that Grassmann presented a very ambitious system of ideas in a style so complicated and philosophical in his *Die lineale Ausdehnungslehre* (Theory of Linear Extension) (1844) that he was forced to try again. This he did in his completely revised *Ausdehnungslehre* (*Extension Theory*) of 1862, which was written in a less elaborate manner that made the mathematical content stand out, but even then it was hard to grasp what he was trying to say.

Grassmann had entered Berlin University in 1827, when he was 18, where he studied philology and theology for six semesters before returning to his native Stettin (Szczecin) near the Baltic coast (then in Germany, now in Poland). His father taught theology, mathematics, and science at the Gymnasium there, and Hermann

Box 58. Hamilton, quaternions, and partial differentiation

In 1846 and 1847 Hamilton began to advocate the use of quaternions in partial differential expressions such as

$$\lhd = i\frac{\partial}{\partial x} + j\frac{\partial}{\partial y} + k\frac{\partial}{\partial z}.$$

This allowed him to write

$$-\lhd^2 = \left(\frac{\partial}{\partial x}\right)^2 + \left(\frac{\partial}{\partial y}\right)^2 + \left(\frac{\partial}{\partial z}\right)^2,$$

an expression known as the *Laplacian* and crucial to the study of potential theory.

He wrote:[a]

$$\lhd f = i\frac{\partial f}{\partial x} + j\frac{\partial f}{\partial y} + k\frac{\partial f}{\partial z},$$

where f is a function. Then if f, g, and h are functions

$$\lhd(if + jg + kh) = -\left(\frac{\partial f}{\partial x} + \frac{\partial g}{\partial y} + \frac{\partial h}{\partial z}\right)$$
$$+i\left(\frac{\partial h}{\partial y} - \frac{\partial g}{\partial z}\right) + j\left(\frac{\partial f}{\partial z} - \frac{\partial h}{\partial x}\right) + k\left(\frac{\partial g}{\partial x} - \frac{\partial f}{\partial y}\right).$$

This is not quite the notation used for Maxwell's equations at the start of this section, but it does show the effect of applying $\lhd$ first to a function and then to a vector. Moreover, the departure from the form presented at the start hints at how later mathematicians were to develop this notation.

[a] See (Hamilton 1847, 292).

Grassmann trained to become a teacher himself. In 1834 he obtained his first teaching position, at a technical school in Berlin, where he succeeded Steiner. But in 1836 he returned to Stettin, and taught in schools there for the rest of his life.

In 1840 he submitted a 200-page paper as part of his examination to obtain a teaching certificate. It was on the theory of the tides (*Theorie der Ebbe und Flut*) and it contains the earliest expression of his ideas about vectors, some of which went back to the early 1830s. Here he introduced what he called the 'geometrical product' of two vectors: it has a size equal to the area of the parallelogram defined by the vectors, with a sign determined by the order in which the vectors are taken. Its size is that of the modern vector or cross product of two vectors, but Grassmann's product is not a vector but a signed area. He also introduced what he called the 'linear product' of two vectors, which is the modern dot product, and described the algebra of these products. However, the work was published only posthumously, in 1911, in the third volume of his *Gesammelte Werke*.

In 1844 he set down an expanded version of his theory of vectors in his *Die lineale Ausdehnungslehre*. He took the opportunity to write about Euclidean space and proof in mathematics, none of which helped with the book's reception.

Figure 18.9. Hermann Günther Grassmann (1809–1877)

We can get a hint of what went wrong, even in the non-philosophical passages, by looking at just one small part. Grassmann had become convinced that his ideas were more general than Euclidean geometry, which appeared to him as but one application, and so he couched his theory in very general terms. After some philosophical pages he introduced his forms purely symbolically. They come equipped with what he called a 'synthetic connection' $\wedge$, such that

$$a \wedge b = b \wedge a \quad \text{and} \quad (a \wedge b) \wedge c = a \wedge (b \wedge c).$$

They also have what he called an 'analytic connection', $\vee$, such that

$$(a \vee b) \wedge b = a.$$

Grassmann proved that these definitions imply various other results, but only after a lengthy account did he admit that $\wedge$ could have been called 'addition' and $\vee$ 'subtraction'.

The earlier geometrical product became a species of what Grassmann now called an 'outer product', but it has been estimated that his theory contains no fewer than sixteen different kinds of product. With some justice, George Sarton, one of the founders of modern history of science, compared it with Hamilton's *Lectures* — but Grassmann's ideas are more original. Sadly for Grassmann, however, his book was even more unreadable, as his few friendly readers had to point out when they failed to penetrate it. Möbius, for example, wrote to say that he had failed to understand the philosophy.

So Grassmann tried again in 1862. Out went the philosophy, as did the applications to physics. Grassmann now made it clear that his exposition worked with any finite number of basic units, or (as later mathematicians would say) in a vector space of any dimension. And in came a manner of exposition deliberately modelled on Euclid's

Elements, with definitions and theorems in a numbered sequence, a style that Grassmann had used a year earlier in a book on arithmetic. However, as the mathematician Friedrich Engel commented:[35]

> Without a doubt this was a disastrous mistake. What was perfectly in place in the treatment of a subject so commonplace for all mathematicians as arithmetic, at least for readers who pursue mathematics as a science, was the most unsuitable form of presentation for a subject to which the reader was for the first time being introduced. Though this form of presentation led to an admirable codification of the new concepts and laws of his theory of extension, still it was not a presentation likely to win followers for his ideas, let alone convert those who had not wished to read his first *Ausdehnungslehre*.

What was eventually to save Grassmann was the interest of other mathematicians who could see, more clearly than he could, what was valuable in his work. Among these were Hermann Hankel, who incorporated Grassmann's ideas in his book *Theorie der complexen Zahlensysteme* (Theory of Complex Number Systems) (1867) about complex numbers and their generalisations (including quaternions) and the English mathematician Clifford, who extended them to novel kinds of algebra. Possibly contrary to Grassmann's hopes, what was most appreciated in his work was the way that it suggested generating novel algebraic systems.

Moreover, by 1862 a number of mathematicians had come up with similar ideas, and there was something of a priority dispute going on about vectorial systems and their properties. We do not have space to describe this here, but it kept Grassmann's name in the public eye. After his death in 1877 Grassmann's ideas found more enthusiasts, among them Felix Klein and the American physicist Josiah Willard Gibbs.

How then did the modern vectorial system emerge? While Tait could see nothing but good in quaternions, James Clerk Maxwell, a better mathematician and a much better physicist, gave them only a cautious welcome in his *Treatise on Electricity and Magnetism*. This book redefined the subject and is comparable in its field to Newton's *Principia*, so the exposure was on balance positive — certainly, quaternions were seen to do good work in this very complicated and exciting subject. In particular, Maxwell took the twenty equations in which he formulated his theory and showed how they could be written to some advantage in quaternion notation.

Maxwell praised Hamilton's $\triangleleft$ operator, but wrote it in what became the standard 'nabla' or 'del' notation, ∇ and ∇^2, and noted the many uses that Tait had found for it. He stressed the importance of regarding some quantities as vectors — quantities with a magnitude and a direction. But he held off from writing his *Treatise* in quaternions throughout, not just because he judged (rightly) that it would make it harder to read, but because he was not convinced that quaternions made it easier to think. There was always a risk of a sign error, and the scalar and vector parts jostled uneasily for attention.

Enter Josiah Willard Gibbs, who was to become one of the great architects of the theory of thermodynamics. He learned the theory of electricity from Maxwell's book, and with it about quaternions, but all he knew of Grassmann's work was a paper that Grassmann had written in 1877 on electricity. With this background Gibbs set about proposing a system of notation that he thought was right for physics. He introduced his *Elements of Vector Analysis* this way:[36]

---

[35]See (Crowe 1985, 90).

[36]Quoted in (Crowe 1985, 155).

Figure 18.10. James Clerk Maxwell (1831–1879)

Figure 18.11. Josiah Willard Gibbs (1839–1903)

> the object of the writer does not require any use of the conception of the quaternion, being simply to give a suitable notation for those relations between vectors, or between vectors and scalars, which seem most important, and which lend themselves most readily to analytical transformations, and to explain some of these transformations. ... In this connection, the name of Grassmann may also be mentioned, to whose system the following method attaches itself in some respects more closely than to that of Hamilton.

Gibbs used the symbols i, j, and k to denote three mutually perpendicular unit vectors (if i points East and j North then k points upwards, as he put it), and he denoted vectors with Greek letters. He defined the dot product $\alpha.\beta$ and the vector (or cross) product $\alpha \times \beta$ of two vectors α and β in the modern way, noting that

$$i \times j = k, \;\; j \times k = i, \;\; k \times i = j, \;\text{ and } \alpha \times \beta = -(\beta \times \alpha).$$

He defined

$$\nabla f = i\frac{\partial f}{\partial x} + j\frac{\partial f}{\partial y} + k\frac{\partial f}{\partial z},$$

where f is a function, and when f, g, and h are functions,

$$\nabla.(if + jg + kh) = \left(\frac{\partial f}{\partial x} + \frac{\partial g}{\partial y} + \frac{\partial h}{\partial z}\right)$$

and

$$\nabla \times (if + jg + kh) = i\left(\frac{\partial h}{\partial y} - \frac{\partial g}{\partial z}\right) + j\left(\frac{\partial f}{\partial z} - \frac{\partial h}{\partial x}\right) + k\left(\frac{\partial g}{\partial x} - \frac{\partial f}{\partial y}\right).$$

Gibbs's little book of 83 pages was in no major way original — that was not his purpose. He arranged for it to be privately printed and distributed it in two parts in 1881 and 1884 to many leading scientists in Britain and continental Europe. The quaternionists hoped to dismiss his work — Tait called it a 'hermaphrodite monster' in 1890 — but they reckoned without one of the strangest people ever to take up physics, the brilliant but eccentric English physicist, Oliver Heaviside.[37]

Heaviside was born in London in 1850. His formal education ended at 16, and he never took a university degree. He got a job as a telegraph operator, and became

[37] Quotation from (Crowe 1985, 185).

interested in electrical problems, but he retired at age 24 to live with his parents on not very much money and pursue his interest in physics. In that year he read Maxwell's *Treatise*, which was his introduction to quaternions. From Tait's *Elementary Treatise* he learned how to use them, but he found them inconvenient and cut them down to vectors and scalars. Only in 1888 did he hear that Gibbs had already had similar ideas.

Heaviside developed his own ideas in a succession of papers that showed both his mastery of Maxwell's ideas and his own independence of mind. His approach was most conveniently and forcefully set out at the end of his life, in his three-volume *Electromagnetic Theory* (1893, 1899, 1912).

Heaviside was a powerful polemicist, as befitted his position as an outsider. In these books he explained that he found vector analysis was easy once he had 'thrown off the quaternionic old-man-of-the-sea ... Prof. Tait's Quaternions'. Indeed, 'the quaternion was not only not required but was a positive evil of no inconsiderable magnitude'.[38] He praised Gibbs's work, but not Gibbs's notation — Heaviside advocated a bold font that has since become standard — and he noted that, unlike vectors, truly quaternionic quantities are not to be found in physics. All of this would have been less significant had Heaviside not been the first to reduce Maxwell's twenty equations to the four presented earlier, and known today as 'Maxwell's equations', one of the finest applications of vector methods.

German physicists had followed Maxwell's ideas from the first. They found them difficult, and the leading German physicists Hermann Helmholtz and Heinrich Hertz both rewrote his theory in their own ways. It was Hertz who experimentally verified one of the most important of Maxwell's suggestions, that light may be an electromagnetic phenomenon. But Hertz died young of cancer, as Maxwell had, and it was one of his successors who brought Heaviside's approach and notation to a German audience. August Föppl's name is known only to experts these days, but his *Einführung in die Maxwell'sche Theorie der Elektricität* (Introduction to Maxwell's Theory of Electricity) (1904) proved to be one of the most important single works to make vector methods the preferred approach for mathematicians and physicists alike. Klein was to call it 'one of the most frequently used textbooks in electricity'.[39]

There never was a decisive confrontation between quaternions and vectors. As Gibbs and Heaviside saw sooner and more clearly than anyone else, many physical quantities are vectorial in nature, and the algebra of vectors deals naturally with them. Quaternionic quantities barely occur in physics, and Grassmann's ideas about algebra had almost no role to play in 19th-century physics. The triumph of vector methods over quaternionic ones is two-fold: the streamlining of notation originally produced for other reasons, and a clash of personalities and egotistical interests.

18.5 Further reading

Crowe, M.J. 1967. *A History of Vector Analysis*, University of Notre Dame Press, Dover reprint, 1985. This highly readable book is full of stimulating quotations from original sources on which the above section on vectors is based.

---

[38] See (Heaviside 1893, 27).

[39] See (Klein 1926–1927, Vol. 2, p. 47), quoted in (Crowe 1985, 227).

Dittrich, W. 2018. *Reassessing Riemann's Paper: On the Number of Primes less than a given Magnitude*, Springer. This book is rather technical, but will interest people with the appropriate background.

Hankins, T.L. 1980. *Sir William Rowan Hamilton*, Johns Hopkins University Press. This readable biography looks at every aspect of Hamilton's life and work.

Katz, V. and Parshall, K.H. 2014. *Taming the Unknown: A History of Algebra from Antiquity to the Early Twentieth Century*, Princeton University Press. As the title suggests, this is a sweeping history of the subject, equally valuable for its breadth and its detail.

Mazur, B. and Stein, W. 2018. *Prime Numbers and the Riemann Hypothesis*, Cambridge University Press. This highly acclaimed and accessible account covers the history and enters into many related subjects with clarity and an abundance of stimulating diagrams.

Neale, V. 2017. *Closing the Gap: The Quest to Understand Prime Numbers*, Oxford University Press. This very readable book looks at recent breakthroughs in the Twin Prime Conjecture, and provides a glimpse of how the international mathematical community works.

Wilson, R. 2020. *Number Theory: A Very Short Introduction*, Oxford University Press. This book introduces several of the topics that are discussed in this chapter.

Yavetz, I. 1995. *From Obscurity to Enigma: The Work of Oliver Heaviside, 1872–1889*, Birkhäuser. This is the definitive biography of this scientific outsider.

19

Group Theory

Introduction

In this chapter we address a significant question in the history of mathematics: Why is the term 'algebra' as used today so very different from what was meant two hundred years ago (and is still meant by 'algebra' at high school)? We answer this question with one example of the transformation that took place; a full answer would cover many different domains of what is called 'modern algebra' today, including more on number theory than we can cover here. However, the emergence of group theory is in many ways typical of what took place, and moreover the story involves one of the most colourful people in the history of mathematics — as you will see.

The story is a highlight in a process of enquiry that is as old as mathematics itself: the desire to find numbers that satisfy certain conditions. Italian mathematicians of the early 16th century had shown that polynomial equations of degrees 3 and 4 can be solved algebraically — that is, there is a formula for their solutions that involves the coefficients of the equation, and only the familiar operations of arithmetic (addition, subtraction, multiplication, and division), and the extraction of roots (square roots and cube roots in these cases). But, as Lagrange had discussed at length (see Section 7.3), there seemed to be no similar formula for the general equation of degree 5. Nor was there to be — and one interesting question that we investigate here is: How can it be proved that something is impossible?

19.1 Solving polynomial equations

We begin slightly to one side, with Gauss's *Disquisitiones Arithmeticae* of 1801, because the book that placed number theory centre stage in mathematics also began a wholesale change in the nature of algebra.

In Section 6 of his book, Gauss was interested in the particular kind of complex numbers that arise when solving an equation of the form $x^n - 1 = 0$, for $n > 1$; these are called the nth roots of unity.[1] The equation $x^n - 1 = 0$ has $x = 1$ as a solution, so it can be written as

$$x^n - 1 = (x-1)(x^{n-1} + x^{n-2} + \cdots + x + 1) = 0,$$

and it is the complex solutions, the solutions of

$$x^{n-1} + x^{n-2} + \cdots + x + 1 = 0,$$

that are interesting. They are the complex numbers $e^{2ki\pi/n}$, for $k = 0, 1, \ldots, n-1$, because $(e^{2ki\pi/n})^n = e^{2ki\pi} = 1$ for each integer k, so geometrically they form the vertices of a regular n-gon in the plane of complex numbers.

For example, when $n = 3$ this equation is

$$x^2 + x + 1 = 0,$$

whose roots are

$$x = {}^1\!/\!_2 \pm i\,{}^1\!/\!_2\sqrt{3}$$

—that is,

$$e^{2i\pi/3} = {}^1\!/\!_2 + i\,{}^1\!/\!_2\sqrt{3} \quad \text{and} \quad e^{4i\pi/3} = e^{-2i\pi/3} = {}^1\!/\!_2 - i\,{}^1\!/\!_2\sqrt{3}.$$

Together with $x = 1$, they form the three vertices of an equilateral triangle in the plane of complex numbers.

When $n = 4$ the original equation factorises as

$$x^4 - 1 = (x^2 - 1)(x^2 + 1) = (x+1)(x-1)(x^2+1) = 0.$$

Its four solutions are the complex numbers $1, i, -1$ and $-i$, which form the vertices of a square.

What about when $n = 5$? The solutions of $x^5 - 1 = 0$ form the vertices of a regular pentagon in the plane of complex numbers. Can we find these solutions algebraically? And can we connect the algebra with the fact, known to Euclid, that the regular pentagon is constructible by straightedge and compasses? This suggests that the solutions will involve only square roots.

The answer to both these questions is 'yes'. To solve the equation

$$x^4 + x^3 + x^2 + x + 1 = 0,$$

the trick is to divide through by x^2, to get the equation

$$x^2 + x + 1 + x^{-1} + x^{-2} = 0,$$

and then to introduce the new variable $y = x + x^{-1}$. Because $y^2 = x^2 + 2 + x^{-2}$, the original quartic equation in x becomes this quadratic equation in y:

$$y^2 + y - 1 = 0.$$

The solutions of this equation are

$$y_+ = -{}^1\!/\!_2 + {}^1\!/\!_2\sqrt{5} \quad \text{and} \quad y_- = -{}^1\!/\!_2 - {}^1\!/\!_2\sqrt{5}.$$

To find the four values of x that we want, we now solve the two quadratic equations obtained from $y = x + x^{-1}$ (which is equivalent to $x^2 - yx + 1 = 0$) and which are

$$x^2 - y_+x + 1 = 0 \quad \text{and} \quad x^2 - y_-x + 1 = 0.$$

[1]When n is prime, these are the cyclotomic integers we met in Chapter 18.

Box 59. Powers of 2 (mod 19).

n :	1	2	3	4	5	6	7	8	9
2^n:	2	4	8	16	13	7	14	9	18

n :	10	11	12	13	14	15	16	17	18
2^n:	17	15	11	3	6	12	5	10	1

After some algebra, we find

$$x = \frac{1}{2}\left(y_+ + \sqrt{y_+^2 - 4}\right), \quad \frac{1}{2}\left(y_+ - \sqrt{y_+^2 - 4}\right),$$
$$\frac{1}{2}\left(y_- + \sqrt{y_-^2 - 4}\right), \quad \frac{1}{2}\left(y_- - \sqrt{y_-^2 - 4}\right).$$

As expected, the solution involves only square roots, although the original equation was of degree 4. Moreover, a straightforward calculation shows that these four solutions are powers of

$$\frac{1}{2}\left(y_+ + \sqrt{y_+^2 - 4}\right) = \frac{1}{4}\left(-1 + \sqrt{5} + \sqrt{-2\left(5 + \sqrt{5}\right)}\right),$$

as required.

The solution process involved solving one quadratic equation with rational coefficients, and using the solutions, which are not rational numbers, as the coefficients in a further pair of quadratic equations.

Now for the breakthrough. Gauss investigated the equation $x^n - 1 = 0$ for higher values of n, concentrating on the case when n is a prime number (which we denote by p). He presented a general theory in his *Disquisitiones Arithmeticae*, as well as a detailed discussion of two special cases, when $p = 17$ and when $p = 19$.

The equation

$$\frac{x^p - 1}{x - 1} = x^{p-1} + x^{p-2} + \cdots + x + 1 = 0$$

is of degree $p - 1$, and Gauss found that the theory depends on the prime factors of $p - 1$. For example, when $p = 19$, $p - 1 = 18 = 2 \times 3^2$, and Gauss found that the equation can be solved by first solving one cubic equation, using its solutions to form three more cubic equations, and then using the solutions of these equations to form nine quadratic equations. The eighteen solutions of those equations were the solutions (other than $x = 1$) of $x^{19} - 1 = 0$. The difficult part was to find a way of constructing the intermediate equations, which Gauss handled by a very ingenious method. It made use of the fact that the 18 non-zero numbers (mod 19) are all powers of a single number (2 in this case, see Box 59). Indeed, Gauss was the first to prove that, for any prime p, the $p-1$ non-zero numbers (mod p) are always powers of a single number, which depends on p, although Euler and others had asserted it.

Gauss used the existence of this number, which he called a *primitive element*, to label the solutions and find patterns among them that led him to group the solutions together and so obtain the intermediate equations.[2]

In the case when $p = 17$, $p - 1 = 16 = 2^4$, Gauss showed that the equation $x^{17} - 1 = 0$ can be solved by solving a sequence of quadratic equations. The payoff was that the regular 17-gon is therefore constructible by straightedge and compasses, the first discovery of this kind since the Greeks. On discovering it, Gauss told his mathematics professor at Göttingen, Abraham Kästner, who was then nearly 80. Sadly, Kästner failed to appreciate what Gauss was trying to tell him, and merely observed that very accurate approximate constructions were available for polygons of any number of sides, but Gauss proceeded to publish the result, though not the method, in the local newspaper, the *Allgemeine Literaturzeitung*, in April 1796, when he was still 18. It was his first publication, and he remained proud of it all his life, even suggesting to his friend Farkas Bolyai that a regular 17-gon should be inscribed on his tombstone.[3]

Gauss went further. When he presented the general theory of equations of the form $x^p - 1 = 0$ in his *Disquisitiones Arithmeticae*, he went beyond what he had proved and in §359 claimed of

> the search for a general solution of equations of higher than the fourth degree ... [that] there is little doubt that this problem does not so much defy modern methods of analysis as that it proposes the impossible ....

Section 6 of the *Disquisitiones Arithmeticae* was easier to understand than the monumental rewriting of quadratic forms in Section 5, and was the first to be widely adopted. Numbers that solve an equation of this form are rather like the roots of the equation $x^2 + 1 = 0$ that give rise to the square roots of -1. Gauss showed that integer combinations of the roots of the equation $x^p - 1 = 0$ for any given prime p have interesting properties, and because of their association with the regular polygons with p vertices these numbers became known as the *cyclotomic integers.*[4]

Ruffini and Abel. By Lagrange's time, it was well known that there were formulas for solving quadratic, cubic, and quartic equations (see Section 7.3). Lagrange had shifted attention away from the formulas onto certain expressions, which he called *resolvents*: these were polynomials in the solutions of the original equation (the solutions now being treated as variables themselves) with particular properties under permutations of the solutions.

The first mathematician after Lagrange to publish a major investigation of the problem of whether polynomial equations can be solved by radicals was the Italian Paolo Ruffini. Ruffini was clear how the matter stood; he began his book *Teoria Generale delle Equazioni* (General Theory of Equations), published in 1799, by stating:

> The algebraic solution of equations of degree greater than 4 is always impossible. Behold a very important theorem which I believe I am able to assert (if I do not err); to present the proof of it is the main reason for publishing this volume. The immortal Lagrange, with his sublime reflections, has provided the basis of my proof.

---

[2]These patterns correspond to the subgroups of the cyclic group of order $p - 1$ — but Gauss did this before group theory had been created!

[3]See (Dunnington 2004, 28), where a translation of Gauss's announcement appears.

[4]We mentioned their use by Lamé in Section 18.1.

Figure 19.1. Paolo Ruffini (1765–1822)

Historians agree that Ruffini's proof of his claim is long, obscure, and fails to secure a key theorem; there is some debate about how easily Ruffini could have proved it, had he noticed. In their opinion that his work is flawed, historians are joined by Ruffini's contemporaries. Lagrange never responded directly to Ruffini, although he was sent copies on two occasions. Disappointed by its reception, Ruffini tried again, writing new versions in 1806 and 1813. These papers met with slightly greater success: the rather vague approval of the Royal Society of London, and a remark that Ruffini had attacked the problem is recorded in Jean Delambre's report to Napoléon on progress in mathematics since 1789. This marks the official French judgement on Ruffini's work.[5]

What is the problem with Ruffini's work? Ruffini accepted Lagrange's analysis that an algebraic solution to the general quintic equation depended on the existence of a certain resolvent — in this case, one taking only five distinct values. He tried to show that such a thing could not exist, by an exhaustive (and exhausting) study of the possible permutations of the five roots. The gap in his 1799 proof is here, and it was only imperfectly plugged in his later work. A fair assessment would be that Ruffini, by pushing Lagrange's ideas hard in the direction of impossibility, came very close to establishing that the general quintic equation is not solvable by radicals.

Abel and Cauchy. The next person to attack the problem was Niels Henrik Abel. At first he thought he had found a method for solving the general quintic by radicals, but he became convinced of its inadequacy when asked to apply it to numerical examples. Then he changed his approach, and started instead to prove unsolvability. In 1824, at

[5]The *Report* was submitted 'to his Majesty the Emperor and King' on 6 February 1808 and published in 1810, see (Delambre 1810, 64). Much more enthusiasm was expressed for Gauss's work on cyclotomy.

his own expense, he published a proof of the unsolvability, which was rather sketchy because he was about to leave for Paris, and his arguments were not well received; a copy sent to Gauss was found unopened amongst Gauss's papers after his death in 1855. Abel then rewrote the piece, and it was published in 1826 in the first volume of Crelle's *Journal*, as we saw in Section 13.4.

In this second account, Abel agreed with Lagrange that for the general quintic to be solvable by radicals, there would have to be a resolvent that takes only 5 distinct values under the 120 possible permutations of the roots. Turning to Ruffini's work, which he does not appear to have known in 1824, Abel said: '[his] memoir is so complicated it is difficult to judge the validity of his reasoning. It seems to me that his reasoning was not always satisfactory'. Abel then established the general form of the resolvents and finally showed that the required resolvent cannot exist in general. He also showed that certain specified types of quintic are solvable by radicals, but not all.

It is no easy matter to write down all the polynomials in the several variables involved, and to see how they vary as the variables are permuted. It was hard to be sure that one had written down every possible candidate for the (non-existent) resolvent in five variables taking only five values as the variables are permuted. One ingredient used by Abel (but not by Ruffini) was a study of permutations that Cauchy had made in 1815. The study of permutations in their own right not only made the question of solvability by radicals easier, but opened the door into other domains of mathematics. Many years later, in 1845, a clear explanation of permutations was given by Cauchy, writing in the knowledge that Galois's work had recently been re-discovered — Liouville in 1843 had announced his intention of publishing a thorough explanation of it.[6]

Cauchy began with the Lagrangian idea of a function — thought of as a formal expression — taking different values as its variables are permuted.

Cauchy on permutations.

> One calls a permutation or substitution an operation which consists in moving the variables and substituting them for each other in a given value of a function Q, or in the corresponding arrangement. To indicate this substitution we shall write the new arrangement which is produced above the first and enclose the system of the two arrangements in parentheses. So, for example, given the function $Q = x + 2y + 3z$, where the variables x, y, z occupy the first, second, and third positions respectively, and consequently follow one another in the order indicated by the arrangement xyz, if one interchanges the variables y, z which occupy the last two places one obtains a new value Q' for Q, which will be distinct from the first, and determined by the formula $Q' = x + 2z + 3y$. Moreover the new arrangement corresponding to this new value will be xzy, and the substitution by which one passes from the first value to the second will be found to be represented by the notation $\begin{pmatrix} x & z & y \\ x & y & z \end{pmatrix}$, which is a sufficient indication of the way in which the variables have been moved.

[6]See (Cauchy 1845, 280–283), F&G 15.D3, and (Stedall 2008, 366–373).

Cauchy then proceeded to consider how permutations can be combined.[7]

> The *product* of a given arrangement xyz by a substitution $\begin{pmatrix} x & z & y \\ x & y & z \end{pmatrix}$ will be the new arrangement xzy which one obtains by applying this substitution to the given arrangement.
>
> The *product* of two substitutions will be the new substitution which always furnishes the result to which the application of the first two, operating one after the other on an arbitrary arrangement, would lead. The two given substitutions shall be the two *factors* of the product. The product of an arrangement by a substitution or of a substitution by another will be indicated by one of the notations which serves to indicate the product of two quantities, the multiplicand being placed, following custom, on the right of the multiplier. So one finds, for example,
>
> $$\begin{pmatrix} x & z & y \\ x & y & z \end{pmatrix} xyz = xzy \quad \text{and} \quad \begin{pmatrix} y & x & u & z \\ x & y & z & u \end{pmatrix} = \begin{pmatrix} y & x \\ x & y \end{pmatrix}\begin{pmatrix} u & z \\ z & u \end{pmatrix}.$$
>
> There is more: one can, in the second term of the last equation, exchange the two factors with each other without inconvenience, in such a way that one still has
>
> $$\begin{pmatrix} y & x & u & z \\ x & y & z & u \end{pmatrix} = \begin{pmatrix} u & z \\ z & u \end{pmatrix}\begin{pmatrix} y & x \\ x & y \end{pmatrix}.$$
>
> But this exchange is not always possible, and often the product of two substitutions will vary when one exchanges the two factors with each other. So, in particular, one will find
>
> $$\begin{pmatrix} y & x \\ x & y \end{pmatrix}\begin{pmatrix} z & y \\ y & z \end{pmatrix} = \begin{pmatrix} y & z & x \\ x & y & z \end{pmatrix}$$
>
> and
>
> $$\begin{pmatrix} z & y \\ y & z \end{pmatrix}\begin{pmatrix} y & x \\ x & y \end{pmatrix} = \begin{pmatrix} z & x & y \\ x & y & z \end{pmatrix}$$
>
> We will say that the two substitutions are permutable with each other when their product is independent of the order in which the two factors occur.

Cauchy's explanation of permutations and how to combine them should be clear. If you want to satisfy yourself that you have understood it, first check that there are 24 different permutations of the symbols x, y, z, and u. Then ask yourself: how many different values does the Lagrange resolvent $xy + uz$ take under these 24 permutations, and which permutations leave $xy + uz$ unaltered? For example, it is unaltered when x and y are interchanged, but the permutation

$$\begin{pmatrix} x & y & u & z \\ y & u & x & z \end{pmatrix}$$

[7]Notice that in a product of two permutations the permutation on the left is performed first.

that replaces x by y, y by u, u by x, and leaves z unchanged, sends $xy + uz$ to $yu + xz$, which is different. In fact, $xy+uz$ takes a total of three distinct values (the other values are $xz+yu$ and $xu+yz$) and there are eight permutations that leave $xy+uz$ unaltered.

Abel's point that some equations of higher degree are solvable by radicals is not a kind of irritating quirk. It is instead the sort of detail that Galois, who knew Abel's work well, realised was crucial to the whole question of solvability by radicals. Abel had got the idea from carefully reading Gauss's account of the cyclotomic polynomials in the in the *Disquitiones Arithmeticae*, and may have been struck by the impossibility claim in the last paragraph of that influential book.

19.2 Galois and Galois theory

Of all the transformations that mathematics underwent in the 19th century, one of the most momentous is associated with the work of a young Frenchman, Évariste Galois, who died as a result of a duel at the age of 20. His remarkable ideas not only reformulated the theory of equations, but introduced into mathematics the concept of a *group*, which was to initiate wholly new lines of research. His contributions can be compared in their effects with those of Descartes: he made it possible to think of algebra in a new way, and so opened the door to completely novel kinds of discovery.

Figure 19.2. Évariste Galois (1811–1832)

Galois made himself completely familiar with the published works of Gauss and Abel. The question that he addressed was, therefore, to explain which polynomial equations are solvable by radicals, and which are not. His solution was technical, but we can learn much by patiently studying the frankly obscure letter that he wrote on 29 May 1832 to his friend Auguste Chevalier to accompany his account of it.[8]

Galois on groups.

> Here is a summary of the most important things.
>
> (1) Following propositions II and III of the first memoir one sees a great difference between adjoining to an equation one of the roots of an auxiliary equation or adjoining all of them.

[8]Taken from (Neumann 2011, 85–87, 97).

In both cases the group of the equation is partitioned by the adjunction into groups such that one passes from one to the other by means of the same substitution; but the condition that these groups should have the same substitutions does not necessarily hold except in the second case.[9] That is called a 'proper decomposition'.

In other words, when a group G contains another H, the group G can be partitioned into groups each of which is obtained by operating on the permutations of H with one and the same substitution, so that

$$G = H + HS + HS' + \cdots$$

and also it can be decomposed into groups all of which have the same substitutions so that

$$G = H + TH + T'H + \cdots .$$

These two kinds of decomposition do not ordinarily coincide. When they coincide, the decomposition is said to be 'proper'.

It is easy to see that when the group of an equation is not susceptible of any proper decomposition one may transform the equation at will, and the groups of the transformed equations will always have the same number of permutations.

When, on the contrary, the group is susceptible of a proper decomposition, so that it is partitioned into M groups of N permutations, then one will be able to solve the given equation by means of two equations: the one will have a group of M permutations, the other one of N permutations.

Therefore, when one has effected on the group of an equation all the possible proper decompositions on this group, one will arrive at groups which one will be able to transform, but in which the number of permutations will always be the same.

If each of these groups has a prime number of permutations, the equation will be solvable by radicals; if not, not.

The smallest number of permutations that can have an indecomposable group, when this number is not prime, is $5.4.3$.

(2) The simplest decompositions are those which arise by the method of M. Gauss's. Since these decompositions are obvious, even in the actual form of the group of the equation, it is useless to pause for long on this topic

...You will have this letter printed in the *Revue Encyclopédique*.

Often in my life I have risked advancing propositions of which I was not sure. But all that I have written here has been in my head for almost a year and it is not in my interest to make a mistake so that someone could suspect me of having announced theorems of which I did not have the complete proof.

[9]What Galois called a group is today often called a subset, and need not be a group in the modern sense of the term.

You will publicly ask Jacobi or Gauss to give their opinion not on the truth but of the importance of the theorems.

After this, there will, I hope, be people who will find profit in deciphering all this mess.

As we shall see, Galois's letter is one of the most remarkable in the history of mathematics, and a contemporary reading it would have to decide how strong a claim Galois was putting forward, and how much confidence one could put in his claim. To judge by the letter alone, it would seem that Galois was claiming to have gone beyond the simple question of whether a given equation is solvable by radicals, and to have insights into equations that are not solvable.

What does a first reading of the letter tell us? Galois claimed to have a worked-out theory in three parts, one of which was written, so this does not sound entirely like a pipe-dream. His first crucial insight apparently involved something called 'the group of the equation', whatever that might be — the concept was novel, and Galois did not define it in the letter, although he did give examples elsewhere. This 'group' is something that comes in two types — those that can be properly decomposed, and those that cannot. We may have no idea what these words can mean, but producing a dichotomy sounds like the kind of thing that could explain why some equations of degree 5 are solvable by radicals whereas others are not. Equally obscure, at first glance, is Galois's second insight, which says something about the crucial importance of the permutations being prime in number.

Evidently Galois was claiming, with whatever justification one cannot decide, to have solved the problem completely. There is a cryptic reference to some work of Gauss that makes it seem that Galois knew the work of the best people in the field, and the bold claim that he wanted Gauss and Jacobi to pronounce not on the *truth* but on the *importance* of the claims. This could be the work of a crank — but it could be the work of someone who knew what he was talking about, knew the current state of the art, and was content to let his work be judged by the best in the field.

Galois's claims make more sense when seen in context. Cauchy had observed in 1815 that it was worth while studying permutations in their own right, and Abel had shown that, in this way, Lagrange's question about solvability by radicals could be answered in the negative.

Galois now claimed that the set of all permutations of the roots of any given equation has a structure that enables one to tell whether the equation is solvable by radicals. It is not the degree of the original equation that conveys this information, but the collection of permutations that it brings in its wake, which depends not only on the degree of the equation, but also the particular coefficients. It can be argued that the step that Galois took beyond Lagrange was to adapt Lagrange's theory so that it was sensitive to the coefficients of an equation and not just to its degree.

To understand the reception of Galois's profound but extremely difficult work, we must look at the circumstances of his life and tell the most famous, colourful, and tragic biography in all of mathematics.[10]

Born on 25 October 1811 in a suburb of Paris to a family that had prospered under Napoléon (his father was mayor of Bourg-la-Reine during Napoléon's brief return from

[10]Our account follows T. Rothman, *Science à la Mode* (1989), 148–194, which demolishes some of the more romantic myths about this young man.

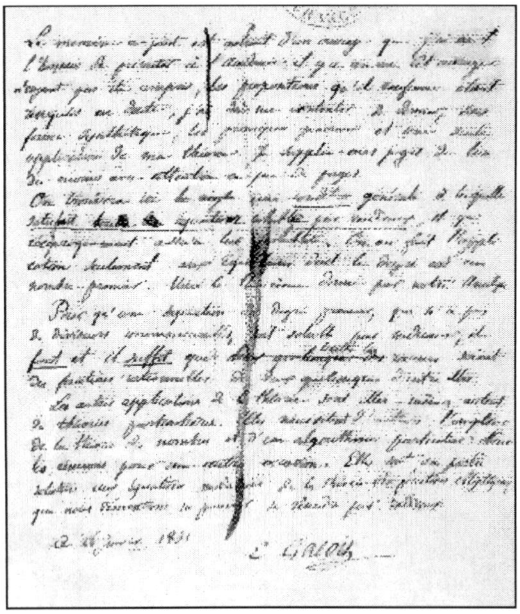

Figure 19.3. A manuscript by Galois

exile), Évariste Galois was educated at home by his mother and then, from 1823, at the Collège Royal de Louis-le-Grand. He was introduced to serious mathematics when he was 15 by his teacher H.J. Vernier. It immediately became his consuming passion. He read Legendre's *Géométrie* and Lagrange's 'The Resolution of Algebraic Equations' to the neglect of his other studies, and although Vernier was delighted with his work, other teachers reported that he was becoming increasingly withdrawn, 'singular', and even 'bizarre'. He insisted on taking the entrance exam to the École Polytechnique a year early, and failed, which hardened his growing dislike of authority. He then enrolled in the advanced class at the Collège de Louis-le-Grand, where he had the good fortune to be taught by L.P.E. Richard. Richard immediately recognised Galois's brilliance, and called on the École Polytechnique to accept the young man without examination. They did not, and failed him when he sat the examination, but Galois was sufficiently encouraged by Richard to publish his first paper (on number theory) in the 1828 issue of Gergonne's *Annales de Mathématiques*. A year later, on 25 May and 1 June 1829, he submitted two papers to the Académie des Sciences on the solvability of equations of prime degree. These are the first papers on 'Galois theory', and Cauchy was invited to report on them.

Suddenly tragedy struck. Galois's father was the innocent victim of a politically inspired plot against him, organised by the local priest, and unable to bear the calumnies directed at him, he committed suicide on 2 July 1829. Within days of this blow, Évariste re-sat the examination for the École Polytechnique and again failed. Now Galois had no choice but to sit for what was then the lesser college, the École Normale Supérieure, which he did in November 1829. This time he passed. Cauchy then failed to report on Galois's work, in circumstances that are unclear, but it is possible that Cauchy thought that Galois should re-work his papers as an entry for the Grand Prize in Mathematics organised by the Académie, for which the closing date was 1 March 1830. At all events,

Galois submitted an entry in February. But in May Fourier, the permanent Secretary of the Académie and chief judge of the competition, died and Galois's paper could not be found. Galois was convinced that he was the victim of an establishment conspiracy, and given the traditional viciousness of Parisian academic life and the political passions of the time his opinion was quite reasonable (even if wrong, as it may well have been). From then on, Galois moved steadily towards the revolutionary left.

Although three more articles by Galois soon appeared in Ferussac's *Bulletin*, the political events of 1830 became more important for him. The July revolution, as it is called, saw the end of the Bourbon regime of Charles X, who fled from France after three days of riots in Paris, and the installation of the Orleanist King Louis-Philippe. Cauchy followed Charles X voluntarily into exile in September. The students of the École Polytechnique played a prominent role in these affairs, but those of the École Normale, Galois included, were simply locked in to the École by Guigniault, the head of the school. Galois retaliated by taking up with the revolutionaries Blanqui and Raspail, and by the end of the year Guigniault expelled him.

Figure 19.4. The 1830 Revolution, 'Prise de l'Hôtel de Ville: le Pont d'Arcole', by Amédée Bourgeois

Throughout 1831 Galois was immersed in the revolutionary politics that had returned to Paris almost with the fervour of the 1790s. On 9 May Galois was at a militantly republican dinner called to celebrate the acquittal of nineteen republicans on conspiracy charges. About two hundred people were there, including the novelist Alexandre Dumas, who later wrote:[11]

> It would be difficult to find in all Paris two hundred persons more hostile to the government than those to be found re-united at five o'clock in the afternoon in the long hall on the ground floor about the garden [of the Vendanges des Bourgogne].

[11] Quoted in (Rothman 1989, 165–166).

Dumas went on:

> Suddenly, in the midst of a private conversation which I was conducting with the person on my left, the name Louis-Philippe, followed by five or six whistles caught my ear ... A young man who had raised his glass and held an open dagger in the same hand was trying to make himself heard. He was Évariste Galois ... All I could perceive was that there was a threat and that the name of Louis-Philippe had been mentioned; the intention was made clear by the open knife.

Dumas prudently made his escape through an open window. Galois was arrested the next day, and put on trial on 15 June for threatening the King's life, but was acquitted. Within a month he appeared on Bastille Day, dressed in the uniform of the banned Artillery Guard and accordingly carrying several weapons. This was interpreted as an extreme act of defiance, and Galois was again arrested. This time he was sentenced, on 23 October, to six months in the prison of Sainte-Pélagie. He spent some of this time with the eminent botanist and fellow republican François-Vincent Raspail, who was to spend 27 months in prison between 1830 and 1836, but lived to receive the Cross of the Legion of Honour from Louis-Philippe.[12] Raspail has left us this haunting and prophetic account of a prison scene on 25 July 1831, when Galois, taunted by his fellow prisoners, became drunk on emptying a bottle of brandy at a single draught.[13]

> You do not get drunk, you are serious and a friend of the poor. But what is happening to my body? I have two men inside me, and unfortunately I can guess who is going to overcome the other ... And I tell you I will die in a duel on the occasion of a worthless coquette. Why? Because she will invite me to avenge her honour which another has compromised.

In a final twist, the Académie now rejected another manuscript by Galois, who responded by rejecting them. He arranged to have his work published by his friend Auguste Chevalier, who belonged to the proto-socialist Saint-Simonian movement inspired by Claude Henri de Rouvroy, Comte de Saint-Simon. On 29 April 1832, Galois left prison. On 29 May he wrote the above letter to Chevalier sketching what he had achieved in mathematics, and went the next morning to his fatal duel. Shot in the abdomen, he died in hospital the day after, still only 20 years old. Although politics marked his life, it seems not to have caused his death, but his funeral was the occasion for a republican demonstration that sparked off a week of rioting.

It would be pleasant to record that, even posthumously, French mathematical society soon recognised that it had had a genius in its midst, but for fourteen years, and despite Chevalier's ensuring that some of Galois's work was published, the response was silence. It was only with the publication of more of Galois's work in Liouville's *Journal* in 1846 that the tide began to turn significantly.

It must be said that Galois's style of writing did not help — as you may have already begun to suspect. It was excessively terse, omitting both definitions of novel concepts and quite significant steps in the proofs. Reporting on one of the works that Galois had submitted to the Académie des Sciences, Poisson said of it:[14]

> We have made every effort to understand Mr. Galois's proof. His arguments are not clear enough, nor developed enough, for us to be able to judge their correctness ... in the state in which it is now submitted to the Academy, we cannot recommend that you give it your approval.

---

[12] He is now honoured by a boulevard and a Metro stop in Paris.

[13] See (Rothman, 1989, 94).

[14] Quoted in (Edwards 1984, 44).

So it is only with an element of hindsight that we recognise just how vital his contributions were. As to what Galois did that was so profound, one account was given by Camille Jordan, a leading French mathematician of the second half of the 19th century. Jordan stressed that Galois 'put the theory of equations on a definitive footing' by means of the concept of a 'group of substitutions', and introduced what is apparently a crucial distinction between simple and compound groups.[15]

Accordingly, we have two tasks: to explain what a 'group' is (for the modern definition, see Box 60), and to explain the distinction between simple and compound groups. Galois did not in fact define a group, he merely presented examples of them, and they are groups of permutations. As an example, Galois gave the set of all 6 permutations of three objects. If we label the objects x, y, and z, then the permutations are (in modern notation, rather than Cauchy's):

$$\begin{pmatrix} x & y & z \\ x & y & z \end{pmatrix}, \quad \begin{pmatrix} x & y & z \\ x & z & y \end{pmatrix}, \quad \begin{pmatrix} x & y & z \\ z & y & x \end{pmatrix},$$

$$\begin{pmatrix} x & y & z \\ y & x & z \end{pmatrix}, \quad \begin{pmatrix} x & y & z \\ y & z & x \end{pmatrix}, \quad \begin{pmatrix} x & y & z \\ z & x & y \end{pmatrix}.$$

The convention is that, in each case, the arrangement at the top is replaced ('permuted') by the arrangement underneath. What makes this set of six permutations into a 'group' is that they can be combined two at a time and the result is always another permutation in the set. For example, in our way of writing things, the result of following the second of these permutations by the third permutation is the last of the six. The reason is that the second permutation sends x to x, and then the third permutation sends x to z, so the combination sends x to z, as the sixth permutation says. Likewise, y goes to z which then goes to x, and finally, z goes to y which then stays as y.

Other examples are the groups of all permutations of four, five, or any other number of objects. There are yet other types of groups, some of them introduced by Galois, and all share the feature that any two objects they contain combine to yield another element within the group. As a result, there are many examples of groups in mathematics (see Box 61), and Galois's concept is applicable in many contexts other than the theory of equations. And, as the concept spread throughout mathematics, other properties of a group emerged.

It is clear from Galois's memoirs that he knew that the theory of groups provides the deepest explanations for questions in the theory of equations. He located the precise point at which the theory of groups elucidates the question of solvability by radicals with his distinction between 'simple' and 'compound' groups, so let us now attend to that.

As Galois observed, there are several different types of groups, and this observation opened the way to what we have been calling the 'structural analysis' of groups. An illustration of such a structural property was provided by Cauchy's distinction between commutative and non-commutative permutation groups. A group is said to be *commutative* if the order in which elements are combined does not affect the result (that is, $a \circ b = b \circ a$ for all elements a, b in the group); otherwise it is called *non-commutative*. As Cauchy pointed out, the above group of permutations of three objects is not commutative.

[15]See (Jordan 1870, Preface), extract in F&G 15.D4.

Box 60. The modern definition of a group.
Today we define a *group* G to be a set with a binary operation, denoted by $\circ$ and called the *product*, that satisfies four properties (we use lower case letters $a, b, c, \ldots$ to denote the elements of G):

1. the product operation is closed: if a and b are in G, then $a \circ b$ is in G;
2. there is an identity element e such that $e \circ a = a$, for each element a in G;
3. each element a in G has an inverse a^{-1} in G such that $a^{-1} \circ a = e$;
4. the product operation is associative: if a, b, c are in G, then
$$(a \circ b) \circ c = a \circ (b \circ c).$$

A *subgroup* of G is a subset of G which is a group with the same binary operation.

Box 61. Modular arithmetic provides examples of groups.
For any fixed integer n, the integers (mod n) form a group under addition. Thus, when $n = 7$, the numbers 0, 1, 2, 3, 4, 5, 6 (mod 7) form a group, where for example
$$4 + 5 \equiv 2 \pmod 7.$$
Here, 0 is the identity element, and the inverse of k (mod 7) is $7-k$; for example, the inverse of 5 is 2.

If n is prime, then the non-zero numbers $\{1, 2, \ldots, n-1\}$ (mod n) also form a group under multiplication, where, for example
$$4 \times 5 \equiv 6 \pmod 7.$$
Here, 1 is the identity element, and the inverses are
$$2^{-1} = 4, \;\; 3^{-1} = 5, \;\; 4^{-1} = 2, \;\; 5^{-1} = 3, \;\text{ and }\; 6^{-1} = 6.$$

For each group the product is associative, because addition and multiplication of integers is associative.

Different polynomial equations can have different groups that arise from the permutations of its roots. In particular, where Lagrange had looked at the 'general' equation of degree 5, Galois looked at each possible type of equation of degree 5. In group-theoretic terms, Lagrange considered equations associated with the permutation group on five objects, and Galois looked at equations whose group might be that group or one of its subgroups. The group corresponding to the equation $x^5 - 1 = 0$ is such a group. Galois's remarkable conclusion was that an equation of degree 5 is solvable by radicals if and only if its group is a particular group with 20 elements that Galois described in detail, or one of its subgroups, This group has a special property that is not shared with the permutation group on five objects, and Galois claimed that an equation of whatever degree is solvable by radicals if and only if its group also has this special property.

Galois characterised the special property of those groups that come from equations that are solvable by radicals. These groups have to be 'compound' (to use Jordan's later term) or to 'admit a proper decomposition' (to use Galois's), and when compound they have to be composed of smaller groups in a particular way, which he also specified.

We do not need to know what these technical terms mean to appreciate the nature and size of the step that Galois invited mathematicians to take. Before Lagrange, mathematicians had started from the general polynomial equation and tried to solve it directly. Lagrange pushed them instead to look at resolvents. Cauchy and Abel then turned the problem into one about permutations. Finally, Galois transformed it into a problem about the structure of groups of permutations.

We may wonder whether this is the kind of answer that Lagrange had been looking for. A theorist himself, Lagrange would have accepted this resolution of the problem of solvability by radicals at a theoretical level. But he might nonetheless have agreed with those (like Poisson) who observed that it is not easy to determine the group of a given equation. In this sense, Galois's work was incomplete. It was also negative, for it endorsed Abel's bleak conclusion that polynomial equations of degree 5 or more are generally not solvable by radicals. It was only gradually that the fertile and positive side of his theory began to emerge, and we shall return to this point after we have indicated the technical details upon which Galois's theory rests. But first we look at some related results about what cannot be done in mathematics.

19.3 Impossibility theorems

The so-called 'three classical problems' in Greek geometry ask for the following constructions by straightedge and compasses alone:[16]

1. doubling a cube: finding a cube whose volume is twice that of a given cube.
2. trisecting an angle: dividing an angle into three equal parts.
3. Squaring a circle: constructing a square equal in area to a given circle [more accurately, a disc].

Once it became clear that allowable constructions can arise only from square roots (that is, from quadratic equations) and never from cube roots (that is, from cubic equations), much else became clear.

The first problem, if we set the volume of the given cube equal to 1, asks for a number x such that a cube of side x has a volume equal to 2, so $x^3 = 2$. No such x can be constructed by straightedge and compasses alone.

The second problem leads to a cubic equation because of the trigonometric identity

$$\cos 3\theta = 4\cos^3\theta - 3\cos\theta.$$

If the given angle is 3θ then the above equation, regarded as an equation for the unknown $x = \cos\theta$, says

$$4x^3 - 3x - \cos 3\theta = 0.$$

[16]See *Volume* 1, Sections 3.3 and 3.4.

This is a cubic equation for $\cos\theta$ and except in rare cases (such as $\theta = \pi/2$ and $\theta = \pi/3$) this equation cannot be suitably factorised. For example, if $\theta = 20°$, then $\cos 3\theta = 1/2$ and the equation cannot be factorised into factors with rational coefficients. This means that an angle of 20° cannot be constructed by straightedge and compasses alone.[17]

The third problem is solved by a square of side-length $\sqrt{\pi}$ because the area of that square is π, the area of a disc of radius 1, it was eventually shown that such a length cannot be constructed by straightedge and compasses.

The history of the classical problems is complicated. It is clear that they were only three of a number of unsolved problems around the time of Plato; another asked for the construction of a regular 7-sided polygon, for example. On the other hand, it was not clear what made them difficult, and that depended on the means that were allowed to tackle them.

It was known from early on that the cube can be duplicated by a construction involving the intersection of a hyperbola and a parabola.[18] Later, Archimedes gave a construction for trisecting an angle, but it involves sliding a marked ruler around while ensuring that it passes through a given point. The restriction to the use of a straightedge and compasses alone is implied but nowhere absolutely stated; it gained weight from the absence of any other kind of 'instrument' in Euclid's *Elements*, but that may be irrelevant. Finally, no-one disputed that there is an angle that is one-third the size of a given one, or that there is a cube twice the size of a given one — the only question was how to construct them by allowable procedures. Squaring the circle was regarded as impossible from very early on, and remains to this day a synonym for impossibility.

The last of the three classical problems is very deep. What is involved, and how was it discovered that these problems cannot be solved?

These two classical problems ask whether quadratic methods can solve cubic equations (to phrase them in a way current since the 16th century). The consensus by 1800 was that the answer is 'no'. This is correct, and we can ask: Who first established this result?

The answer is Pierre Laurent Wantzel, a French mathematician about whom little was known until recently.[19] He published his demonstration in 1837, when he was 23, in Liouville's *Journal des Mathématiques*. The first historian of mathematics to mention Wantzel by name was Julius Petersen, a Danish mathematician, in 1877, but no-one picked up the mathematical argument except Felix Klein, who dropped the reference to Wantzel, until Florian Cajori mentioned Wantzel in 1918.[20]

What did Wantzel do, and why did it fall dead from the press? His paper was in four parts. First, he showed how to translate geometrical problems involving constructions into algebraic ones. Second, he showed that when a construction uses only straightedge and compasses, the solution of the corresponding geometrical problem appears as a solution of an equation whose degree is a power of 2. Third, he showed that under certain further assumptions, this equation is irreducible. Fourth, he showed that the

[17]A consequence of this result is that a regular 9-sided polygon is not constructible by straightedge and compasses alone.

[18]See *Volume* 1, Chapter 3.

[19]See (Lützen 2009).

[20]See (Cajori 1918).

problems of duplicating a cube and trisecting an angle lead in general to irreducible cubic equations—and therefore cannot be solved by a sequence of quadratic equations.

The first, second, and fourth of these points are unproblematic, but Wantzel's proof of the third step was deficient. Suppose that the problem of trisecting an angle had led to a reducible cubic equation of the form

$$(ax - b)(x^2 + cx + d) = 0.$$

The roots are $x = b/a$ and the roots of the *quadratic* equation $x^2 + cx + d = 0$, but these can be found by straightedge and compasses. So, for the trisection problem to be unsolvable by straightedge and compasses, it has to translate into an irreducible equation whose degree is not a power of 2. Wantzel did his best, but his argument left gaps for later mathematicians to fill, as Petersen and Klein succeeded in doing.

So why did Wantzel not become famous? He published an original result in a major journal while still a student in a mainstream place. He was even quite well-known, being the first student to come top in the entrance examinations for both the École Polytechnique and the École Normale Supérièure.

The historian Jesper Lützen has two interesting explanations. The first is that perhaps the result was already thought to be true. Gauss had claimed in his *Disquisitiones Arithmeticae* that the only polygons that are constructible by straightedge and compasses have $2^n p_1 p_2 \ldots$ sides, where the p_i are distinct Fermat primes.[21] This implies in particular that the regular 9-gon is not constructible, but its sides subtend an angle of 40° at the centre, which is one-third of the constructible angle of 120°. Therefore angles cannot, in general, be trisected by straightedge and compasses. Although Gauss's claim was not accompanied by a proof, there is no evidence that any of his readers doubted that Gauss knew how to prove the claim, and so they would think that Wantzel was saying nothing new; such readers would have found Wantzel's paper little more than a spelling-out of ideas and methods that had been published a generation before by Gauss.

Lützen's second explanation is that perhaps the result was simply not found to be interesting. Wantzel's readers could argue that this merely confirmed what they had suspected all along: two of the classical problems are insoluble. So what? What can be done with an impossibility proof? Nothing. On the other hand, the equations are solvable numerically, so why care about a requirement that they be solved in a particular (and old-fashioned) way? A number of mathematicians (Gauss himself, Abel, and on another topic Liouville) phrased the problem in a more positive way: look for a solution of such-and-such a kind; if you fail then prove that there is no solution of *that* kind and look for solutions of another kind.

Lützen also points out that Wantzel had a liking for negative results. Wantzel was one of the first to publish a simplification of Abel's proof that the quintic equation cannot be solved by radicals, and his simplification was adopted by Joseph Serret, a protégé of Liouville's, in his *Cours d'Algèbre* (1849). Wantzel was also the first to prove that if all the solutions of a cubic equation with rational coefficients are real, then any expression for the solutions in terms of radicals must involve complex numbers.

[21] Fermat primes are prime numbers of the form $2^{2^n} + 1$, such as 3, 5, 17, 257, and 65,537; these are the only ones known.

Our modern interest in this result is partly because it fits in nicely with our understanding of Galois theory, a topic unknown to Wantzel and his readers. More broadly, it fits into a whole range of topics where we are interested in solutions and solution methods of various kinds. Impossibility results have become positive results in mathematics — but they were not seen positively in 1837.

From the one article on his life by someone who knew him, as cited in Lützen's paper, it seems that Wantzel did not live up to his early promise. He was unwilling to stick at any one topic for long, was given a heavy teaching load, abused coffee and opium, and died in 1848 aged only 33, his work already forgotten.

19.4 Galois's theory of groups and equations

Galois's difficult and obscurely worded theory relies on a particular way of decomposing a group into a sequence of smaller groups contained within it. We now briefly outline the mathematics involved, using modern terminology.

If G is a group, recall that a subset H of G is a *subgroup* of G if it is a group in its own right. This means that the following conditions are met:

1. The identity element e of G is an element of H.
2. If h_1 and h_2 are elements of H, then so is their product $h_1 \circ h_2$.
3. If h is an element of H, then so is its inverse h^{-1}.

A subgroup of a finite group can be used to partition a group into disjoint subsets, as follows. If $H = \{e = h_1, h_2, h_3, \ldots, h_n\}$ is a subgroup of G and g is an element of G, then the set

$$gH = \{g, g \circ h_2, g \circ h_3, \ldots, g \circ h_n\}$$

is called a *left coset* of H in G, and the set

$$Hg = \{g, h_2 \circ g, h_3 \circ g, \ldots, h_n \circ g\}$$

is called a *right coset* of H in G. Notice that if g is an element of H then $gH = H = Hg$, but in general we cannot expect a relationship between gH and Hg when G is not a commutative group. It can be shown that any two left cosets either have no elements in common, or they are equal, and the same result holds for right cosets.

A *decomposition* of a group G with respect to a subgroup H presents it as the union of pairwise-disjoint cosets (all left cosets, or all right cosets). Galois called such a decomposition *proper* if every left coset is also a right coset and vice versa.[22] In other words, for every g in G there is an element g' such that $gH = Hg'$.

For example, let G be the group S_3 of all permutations on three objects x, y, z, and H be the subgroup A_3 that consists of the permutations e, xyz, and xzy.[23] Straightforward calculation shows that there are only two left cosets of A_3 in S_3: A_3 itself, and the coset consisting of the other three elements of S_3. The same is true of the two right cosets of A_3 in S_3. So every left coset is a right coset in this case and the decomposition is proper.

A more important example relates to the solvability of the quartic equation. It is provided by the group A_4, which consists of all the 12 permutations of 4 objects

[22] Today we say that H is a *normal* subgroup of G.

[23] The permutation xyz sends x to y, y to z, and z to x; the permutation xzy sends x to z, z to y, and y to x.

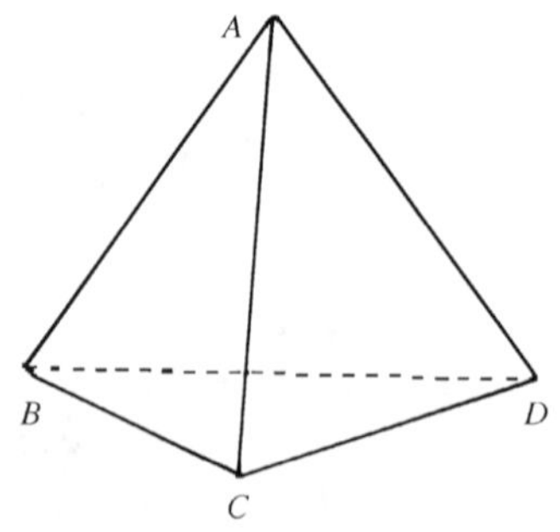

Figure 19.5. A regular tetrahedron

that can be thought of as rigid-body symmetries (rotations and reflections) of a regular tetrahedron with vertices A, B, C, D.

We count these symmetries as follows. The vertex A can be put at any position: 4 choices. With A in that position, the vertex B can be put in any remaining position: 3 choices. The positions of the remaining two vertices are now fixed.

So there are $4 \times 3 = 12$ symmetries of the tetrahedron that correspond to a rigid body motion of the tetrahedron — and they form a group, called the *alternating group* on four objects. If reflections are also allowed then there are $4 \times 3 \times 2 = 24$ symmetries, which form the group S_4 of all permutations of four objects.

The group of all permutations of three objects, which we denote by S_3, and whose six elements were listed earlier, is a subgroup of the group S_4— simply, fix one of the four objects. However, it turns out that the left and right cosets of S_3 in S_4 are different, and so they do not provide a proper decomposition of S_4.

The group A_4 is a subgroup of the group S_4, and it does provide a proper decomposition of S_4, because its two left cosets in S_4 are also its two right cosets.

Moreover, the group A_4 has a subgroup (often called K) with four elements.[24] These other than the identity, are:

$$\begin{pmatrix} A & B & C & D \\ B & A & D & C \end{pmatrix}, \quad \begin{pmatrix} A & B & C & D \\ C & D & A & B \end{pmatrix}, \quad \begin{pmatrix} A & B & C & D \\ D & C & B & A \end{pmatrix}.$$

It can be checked that every left coset of K in A_4 is a right coset, so the decomposition is proper.

Any element of K other than the identity e forms with the identity a subgroup H of K that provides a proper decomposition of K. So we have a chain of proper decompositions:

$$S_4 \rhd A_4 \rhd K \rhd H \rhd \{e\},$$

where $A \rhd B$ means that B provides a proper decomposition of A.

In Galois's view, the existence of such a sequence of proper decompositions partly explains why all polynomial equations of degree 4 are solvable by radicals. The remaining part of his explanation came from the fact that the sizes of the groups are 24, 12, 4, 2

[24]The group K is often called the 'Klein four-group', named after Felix Klein. Klein introduced it in his study of polynomial equations.

and 1, and so the quotients are

$$\frac{24}{12} = 2, \quad \frac{12}{4} = 3, \quad \frac{4}{2} = 2, \quad \frac{2}{1} = 2.$$

In Galois's theory, this corresponds to the successive use of a square root, a cube root, and then two more square roots when solving a polynomial equation of degree 4. That 2 and 3 are prime numbers is essential: it corresponds to the use of square and cube roots in solving the given equation.

Earlier we remarked that Galois had claimed that an equation is solvable by radicals if and only if its group is either a specific group with 20 elements, or a subgroup of that group. Galois had found such a subgroup of S_5, the group of all 120 permutations of five objects, and a proper decomposition of the required kind. It follows that any equation associated with this group and any of its subgroups, is solvable by radicals. Moreover, he also knew that this is the largest subgroup of S_5 that has a proper decomposition of the above kind.

On the other hand, most polynomial equations of degree 5 are not solvable by radicals, because there is no chain of subgroups of S_5, like the above chain for S_4. The group S_5 has only one proper decomposition, namely $S_5 \rhd A_5 \rhd \{e\}$, where the group A_5 has 60 elements; here, the corresponding quotients are $120/60 = 2$ and $60/1 = 60$ but 60 is not prime.

19.5 Group theory

In the years around 1870, group theory began to emerge as a subject in its own right, and as one with significant contributions to make to other fields. This has much to do with the work of Camille Jordan in France.

Jordan published the first paper to present Galois's ideas successfully, as part of an independently intelligible theory of groups in Liouville's *Journal*, in 1869. This was his 'Commentaire sur Galois' that he then incorporated into the book that established group theory as a subject in its own right in mathematics, his *Traité des Substitutions et des Équations Algébriques* (Treatise on Substitutions and Algebraic Equations) of 1870.

Mathematicians before Jordan, such as Kronecker, Cayley, Serret, and the Italian mathematician Francesco Brioschi, had filled in holes in Galois's presentation of the idea of a group. Serret in particular had incorporated Cauchy's presentation of the theory of permutation groups into the subject that Galois had so tantalisingly outlined.

But Jordan's systematic theory of permutation groups was much more abstract. Among many novel ideas, he singled out subgroups for which every left coset is a right coset; the modern term 'normal' for such subgroups came later. Indeed, he came close to presenting the idea of an abstract group. He wrote:[25]

> One will say that a system of substitutions forms a group (or a *faisceau*) if the product of two arbitrary substitutions of the system belongs to the system itself.

In the *Traité*, Jordan described a great many groups, and indicated that many finite groups present themselves as groups of matrices. He stated 'Lagrange's theorem' in the form it retains to this day, as a theorem about the number of elements in a group (called its 'order'): the theorem says that in a finite group the order of any subgroup divides the order of the group. And he described a wide range of situations in which groups could be found permuting geometrical objects.

[25] See (Jordan 1870, 22).

Figure 19.6. Camille Jordan (1838–1922)

When it came to Galois's work, Jordan proved what Galois had only asserted: a polynomial is solvable by radicals if and only if its group G admits a chain of proper decompositions,

$$G \rhd G_1 \rhd G_2 \rhd \ldots \rhd G_{k-1} \rhd \{e\},$$

where the quotients of the numbers of elements in successive subgroups are all prime. This last condition is the one that corresponds to the requirement that the polynomial equation be solved by radicals.

We began this chapter by asking why the term 'algebra' has such very different meanings in traditional and modern mathematics. Our answer is that problems about algebraic equations gradually became answered with reference to an increasingly elaborate theory of equations, couched in terms of a novel concept of a 'group', and that groups turned out to have structures that were of interest in its own right. So the topic of algebra moved from the study of equations to include that of groups—and in due course to other types of structures.

With Jordan's work the process of turning Galois's insights into a theory of groups was complete. The question of solvability by radicals was submerged in the new and exciting topic of group theory. The work of Jordan in France, of others initially in Germany, and, by the end of the century, of mathematicians in many countries, established group theory as a vital part not only of algebra but of many other parts of mathematics. The group of an equation became known as its 'Galois group', and the mathematics concerning these groups, which also turned up in the algebraic theory of numbers, became known as 'Galois theory'. It would seem after all that there were people who did 'find it to their advantage to decipher all this mess', as Galois had hoped.

19.6 Further reading

Neumann, P.M. 2011. *The Mathematical Writings of Évariste Galois*, Heritage of European Mathematics. European Mathematical Society. This is a parallel (French/English) text of all Galois's writings, and as such is an invaluable source.

Pesic. P. 2003. *Abel's Proof*, MIT Press. This is an exploration for the general reader of impossibility proofs in mathematics, with an account of what Abel did.

Richeson, D.S. 2019. *Tales of Impossibility: The 2000-Year Quest to Solve the Mathematical Problems of Antiquity*. Princeton University Press. This is a clearly written and thorough account of the attempts to solve the three classical problems of antiquity and other ancient mathematical problems.

Rothman, T. 1989. *Science à la Mode*, Princeton University Press. This set of sharp essays on the presentation of science, of which only the one on Galois is relevant here, is a fine antidote to the hyperbole that surrounds him.

Tignol, J.-P. 2001. *Galois' Theory of Algebraic Equations*, World Scientific. This is the most thorough and readable account available on the mathematics and its history, with much more on the work of Lagrange, Vandermonde, and Gauss than its title would suggest.

Weil, A. 1984. *Number Theory. An Approach through History from Hammurapi to Legendre*, Birkhäuser. This celebrated account by a 20th-century master of the subject is very readable and careful in its judgements.

Wussing H. 1984. *The Genesis of the Abstract Group Concept*, transl. A. Shenitzer, MIT Press. This is the standard account of the subject, covering the origins of group theory in geometry and number theory as well as the solution of equations, and taking the story through permutation groups to abstract groups.

20

Applied Mathematics

Introduction

At the beginning of the 19th century, in the age of the Industrial Revolution and of continued nation building, state funding and political support led (largely through teaching and journals) to a new level of recognition for mathematics as a discipline. The older bifurcation of pure and mixed mathematics was replaced by that of pure and applied mathematics, the latter meaning the part of mathematics that is concerned with the application of mathematical methods in practical contexts or in other subjects.

The term 'applied mathematics' seems to have appeared first in German in Wolf's dictionary of 1716, but it featured little during the 18th century, and made its first appearance in English only in 1798 in a translation of a German work by Immanuel Kant. The institutional confirmation of the term came with its appearance in the names of journals, such as those founded in the early decades of the 19th century by Gergonne, Crelle, and Liouville (as discussed in Chapter 13), although the inclusion of 'applied mathematics' in the names of these journals did not guarantee a strong representation of applied topics. But these journals were not the only outlets for articles on applied mathematics. Journals associated with national academies, such as the *Philosophical Transactions of the Royal Society*, carried articles on applied topics, while the *Philosophical Magazine* (launched in 1798) was the journal of choice for several leading 19th-century British applied mathematicians.

The early 19th century was also the period in which positions explicitly devoted to the applications of mathematics were created at universities. The first time this happened seems to have been in Norway, which had just introduced a constitution and was emancipating itself from Danish rule. Christopher Hansteen's position as 'Lecturer for Applied Mathematics' at the newly founded university in Christiania (Oslo) was justified in May 1814 by 'the broad scope of applied mathematics and its importance for Norway'. In 1815 Hansteen was promoted to 'Professor Matheseos Applicatae'.

As the century progressed, so did the interest in mathematical applications. In this chapter we show how mathematicians in the 19th century applied a variety of mathematical ideas to solve real-world problems, from the flow of heat down a pipe to the laying of the transatlantic cable.

20.1 The uses of Fourier series

Joseph Fourier reworked his prize-winning memoir of 1810 into the book *Théorie Analytique de la Chaleur* (The Analytical Theory of Heat), which he published in 1822. It is devoted to the topic of heat diffusion. Fourier imagined a solid block of a metal body, with a specified heat distribution on some or all of its surface, and asked: What is the temperature at each point of the block at each moment of time?

THÉORIE
ANALYTIQUE
DE LA CHALEUR,
PAR M. FOURIER.

A PARIS,
CHEZ FIRMIN DIDOT, PÈRE ET FILS,

Figure 20.1. Joseph Fourier's *Théorie Analytique de la Chaleur* (1822)

At this stage in the study of heat, it was reasonably well known that heat was a state of the hot material and not a fluid substance contained in the material, but little else was understood, and so Fourier made as few assumptions as possible about its nature. Instead, he formulated the way in which heat passes from a hot part of the body to an adjacent colder part in a very short interval of time, by supposing that the amount of heat that passes is proportional to the duration of the time interval, the infinitesimal temperature difference between adjacent parts, and a certain function of the distance between the parts. He gave his reasons for believing that this captured the essence of the problem, and then examined homogeneous bodies of simple shapes with simple temperature distributions, usually when the boundaries are kept at fixed temperatures. Before we look in detail at one of them, we note how confident he was of the importance of his achievement. After describing the problems he had solved, and remarking that they 'have never yet been submitted to calculation', he wrote:[1]

[1] See Fourier, *The Analytical Theory of Heat*, p. 6.

Fourier on the propagation of heat.

> If we consider further the manifold relations of this mathematical theory to civil uses and the technical arts, we shall recognize completely the extent of its application. It is evident that it includes an entire series of distinct phenomena, and that the study of it cannot be omitted without losing a notable part of the science of nature.
>
> The principles of the theory are derived, as are those of rational mechanics, from a very small number of primary facts, the causes of which are not considered by geometers, but which they admit as the results of common observations confirmed by all experiment.
>
> The differential equations of the propagation of heat express the most general conditions, and reduce the physical conditions to problems of pure analysis, and this is the proper object of the theory. They are not less rigorously established than the general equations of equilibrium and motion. In order to make this comparison more perceptible, we have always preferred demonstrations analogous to those of the theorems which serve as the foundation of statics and dynamics. These equations still exist, but receive a different form, when they express the distribution of luminous heat in transparent bodies, or the movements which the changes of temperature and density occasion in the interior of fluids ...
>
> The equations of the movement of heat, like those which express the vibrations of sonorous bodies, or the ultimate oscillations of liquids, belong to one of the most recently discovered branches of analysis, which it is very important to perfect. After having established these differential equations their integrals must be obtained; this process consists in passing from a common expression to a particular solution subject to all the given conditions. This difficult investigation requires a special analysis founded on new theorems, whose object we could not in this place make known. The method which is derived from them leaves nothing vague and indeterminate in the solutions, it leads them up to the final numerical applications, a necessary condition of every investigation, without which we should only arrive at useless transformations.

We also see that he claimed that his method solved completely every problem of this kind. Let us see how he justified this, before we look at the interest and criticism it was to inspire.

The problems that Fourier considered have two aspects. One concerns the flow of heat in the body, and he showed that this is described by a differential equation. The other concerns the temperature on the boundaries of the body, and although these can be arbitrary, they are easiest to handle if the boundary is made of simple shapes. Only if the boundary conditions are simple can explicit solutions be found.

We now turn to one of his examples to see how he formulated the mathematical equation that describes the flow of heat.

Fourier considered infinite horizontal bars with either circular or square cross-sections, in which one end is kept hot and the rest uniformly cool. The problem is to find how hot the bar becomes when it reaches a steady state.

He argued that at each point, heat (as measured by the temperature v) passes through the bar in each of the x, y, and z directions. In §98 he considered an infinitesimal cube in the bar, and stated that

- the amount that enters the face with sides dx and dy at a height z is $Kdxdy\frac{\partial v}{\partial z}$ evaluated at that face, where K is a quantity determined by the nature of the body
- what leaves the opposite face is $Kdxdy\frac{\partial v}{\partial z}$ evaluated at that face (where the height is now $z + dz$).

So what left the cube at the second face was found by replacing z by $z + dz$ in the expression $Kdxdy\frac{\partial v}{\partial z}$.

To understand what the transformed expression is, Fourier used the facts that the cube is infinitesimal and that the variation takes place in the z-direction only. So he could use the familiar expression

$$f(z+dx) - f(z) = \frac{df}{dz}dx,$$

where f is a differentiable function. He applied it in the case where $f = \frac{\partial v}{\partial z}$, so

$$\frac{\partial v}{\partial z}(z+dz) - \frac{\partial v}{\partial z}(z) = \frac{\partial^2 v}{\partial z^2}dz.$$

So the difference between what enters the cube and what leaves it in this direction is

$$Kdxdy\frac{\partial v}{\partial z} = Kdxdydz\ \frac{\partial^2 v}{\partial z^2}.$$

A similar argument that involves only switching the roles of x, y, and z, deals with each of the other pairs of faces of the infinitesimal cube. Because the temperature is in a steady state, heat neither accumulates nor drains from the cube, and so the sum of these quantities, taken over the three pairs of opposite faces of the cube, is 0. But the heat flow in and out of the cube is

$$Kdxdydz\ \frac{\partial^2 v}{\partial z^2} + Kdydzdx\ \frac{\partial^2 v}{\partial x^2} + Kdzdxdy\ \frac{\partial^2 v}{\partial y^2},$$

and so the resulting equation, the steady state equation, is

$$\frac{\partial^2 v}{\partial x^2} + \frac{\partial^2 v}{\partial y^2} + \frac{\partial^2 v}{\partial z^2} = 0.$$

The distribution of heat is given by the particular solution of this differential equation that satisfies the stated boundary conditions.

A similar argument allowed Fourier to obtain the even more important equation for a body that is not in a steady state but is, for example, a uniformly hot cube cooling down in a uniformly cool medium. Now the problem is to find the temperature at any point at any moment of time.

As before, the amount of heat leaving a cube in the z direction is $Kdxdydz\frac{\partial^2 v}{\partial z^2}$, but now the sum over the pairs of opposite faces equals the rate of change of temperature, which is $\frac{\partial v}{\partial t}$, where v is the temperature at the point with coordinates (x, y, z). The result (§128) is the *heat equation*:

$$K\left(\frac{\partial^2 v}{\partial x^2}+\frac{\partial^2 v}{\partial y^2}+\frac{\partial^2 v}{\partial z^2}\right)=\frac{\partial v}{\partial t},$$

As before, solutions of this partial differential equation are required that satisfy the given boundary conditions.

Fourier now argued (§142) that the heat equation holds for the flow of heat in any body, because the body can be regarded as made up of little cubes. The boundary conditions, however, generally forced him to suppose that the body has a simple shape, such as a cuboid or one of a limited range of other shapes that can be handled by finding suitable coordinate transformations.

It remained for Fourier to show how to solve the differential equation. He started (§166) with the simplest case, a semi-infinite strip of a given width. For convenience he chose the width to be π in suitable units. He supposed that the strip lay between $x = -\pi/2$ and $x = +\pi/2$, that the two parallel infinite sides were kept at a given temperature (which he took to be 0 in some units), and that its base had a temperature of 1. He chose coordinates (relabelled here to make them more familiar) in which y measures the height above the base and x measures the distance of a point from the mid-line of the strip. The differential equation of the steady state distribution, which is essentially the steady state equation but now involves only two variables, is

$$K\left(\frac{\partial^2 v}{\partial x^2}+\frac{\partial^2 v}{\partial y^2}\right)=0.$$

Fourier looked for a solution of the form in which the variables are separated.[2]

$$v(x, y) = f(x)g(y).$$

This led him to the differential equation

$$g''(y)/g(y) = -f''(x)/f(x).$$

But a function of y cannot equal a function of x unless both are constant, so both sides must be constant (say, m) which led him to solutions of the form

$$f(x) = \cos mx \text{ or } \sin mx \quad \text{and} \quad g(y) = e^{-my}.$$

So

$$v(x, y) = e^{-my}\cos mx \quad \text{or} \quad e^{-my}\sin mx.$$

The boundary conditions force m to be positive, since otherwise e^{-my} would become infinitely large, and if the solution is to vanish for $x = \pm\pi/2$ for all y, then m must be an odd integer. The solution $f(x) = \cos mx$ fits the boundary condition that the solution vanishes when $x = \pm\pi/2$, but the solution $f(x) = \sin mx$ does not, and so must be discarded.

[2]This was a familiar technique. D'Alembert had used it when solving the wave equation, see Section 10.1.

Fourier also knew that if $e^{-m_1 y}\cos m_1 x$ and $e^{-m_2 y}\cos m_2 x$ are both solutions, then so is their sum $e^{-m_1 y}\cos m_1 x + e^{-m_2 y}\cos m_2 x$, and this led him to contemplate solutions in the form of infinite sums, which he wrote in the form (§169):

$$v(x,y) = ae^{-y}\cos x + be^{-3y}\cos 3x + ce^{-5y}\cos 5x + de^{-7y}\cos 7x + etc.,$$

subject to the boundary condition at the base (where $y = 0$) that

$$1 = a\cos x + b\cos 3x + c\cos 5x + d\cos 7x + etc.$$

The infinitely many arbitrary constants $a, b, c, \ldots$ must now be determined from the boundary conditions. Fourier gave two methods, of which the first is so impressive in its virtuosity that he probably included it for purposes of display. But his second method is much easier and became standard in the subject, so we proceed directly to it.

In §220 Fourier observed that, when $j \neq k$,

$$\int_{-\pi/2}^{\pi/2} \cos jx \cos kx\, dx = \left(\frac{1}{k-j}\cos(k-j)x - \frac{1}{k+j}\cos(k+j)x\right)\Bigg|_{-\pi/2}^{\pi/2} = 0,$$

and that the integral is $\pi/2$ when $j = k$. He deduced that the coefficients of the series can be found by multiplying the original series by $\cos jx$, for each value of j, and then integrating. This gave him:

$$a = 1, \quad b = -1/3, \quad c = 1/5, \quad d = -1/7, \ \ldots\ .$$

So he found this solution to the heat distribution problem in the semi-infinite strip:

$$v = \frac{4}{\pi}\left(e^{-y}\cos x - \frac{1}{3}e^{-3y}\cos 3x + \frac{1}{5}e^{-5y}\cos 5x - \frac{1}{7}e^{-7y}\cos 7x + \cdots\right).$$

As he observed, this series, when evaluated at $y = 0$, represents the constant function 1, because that was one of the boundary conditions. On substituting the value $x = 0$, he obtained Leibniz's series for $\pi/4$:

$$\frac{\pi}{4} = 1 - \frac{1}{3} + \frac{1}{5} - \frac{1}{7} + \cdots.$$

Fourier now argued that his method was very general, and indicated how it applied to a variety of problems. This led him to series involving sines instead of cosines, to series involving both sines and cosines, and to other intervals such as $(-\pi, \pi)$. Specific choices of boundary conditions led him to expressions for functions specified on the boundary, and some of these were interesting. For example, he obtained this series for a function that Abel was to refer to in 1826, when questioning a theorem of Cauchy's:

$$f(x) = \sin x - \frac{1}{2}\sin 2x + \frac{1}{3}\sin 3x - \cdots.$$

The graph of the function f that is defined by this series is the graph of $x/2$ on the interval $[-\pi, \pi]$, and copies of this on successive intervals $((2k-1)\pi, (2k+1)\pi)$, for every integer k (see Figure 20.2).[3]

Importantly, it is more correct to say Fourier *claimed* that all this could be done. His claim quickly became a challenge to mathematicians to prove that it is correct — and this was to become a long and fascinating story that we can only begin to tell here.

Fourier's ideas generated quite some discussion in print. Poisson complained that Fourier's method for finding the coefficients in a Fourier series 'has not in fact been

---

[3] See (Fourier 1822, §182), and for Abel's use of it, see Section 16.3.

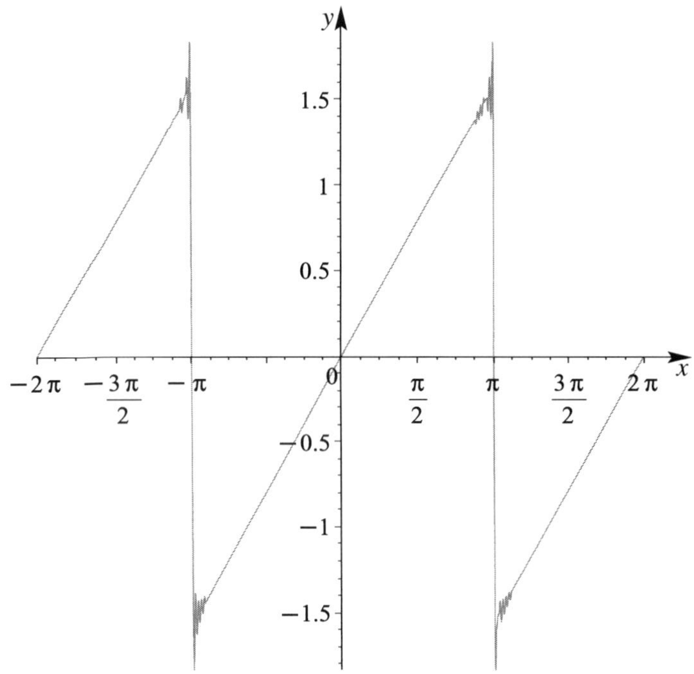

Figure 20.2. A graph of the first 100 terms of Abel's function

demonstrated in a precise and rigorous manner'.[4] A decade later, he objected that Fourier's fundamental assumption that an arbitrary function defined on an interval can be expanded as an infinite series of sines and cosines had not been proved, and Charles-François Sturm joined in, remarking that[5]

> Fourier and other geometers [mathematicians] seem to have misunderstood the importance and the difficulty of this problem, which they have confused with that of determining the coefficients.

By 1827 Cauchy had taken up the subject and offered his own argument, but this was brought down by a fallacious convergence argument. The mathematician who spotted this flaw, and went on to be the first to give a rigorous proof of special cases of Fourier's claim, was Dirichlet.

Dirichlet, who knew Fourier personally, said that he knew of no other attempt on the problem than Cauchy's. In his opinion, the Fourier series expansion of an arbitrary function enjoys the remarkable property of being convergent (so it seems that he doubted neither the existence nor the convergence of the series). But, he said, noone had given a satisfactory proof, and he proposed to establish the convergence of a Fourier series directly, and to show that the series and the function agree.

---

[4]See (Poisson 1823a, 46), quoted in (Bottazzini 1986, 188).

[5]See (Poisson 1835, 186), and (Sturm 1836, 400).

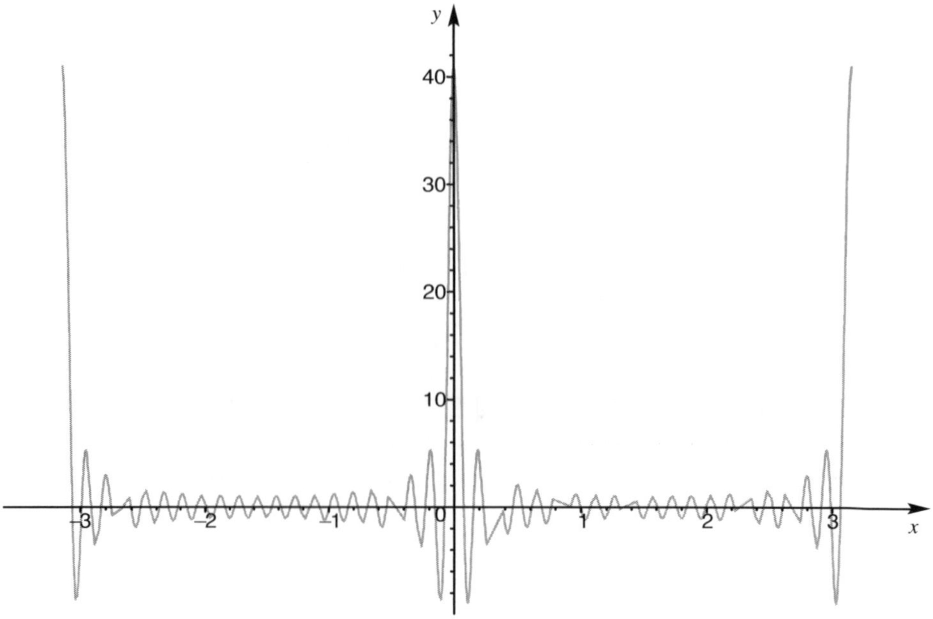

Figure 20.3. The function $\frac{\sin 41x}{\sin x}$ in the range $[-\pi, \pi]$

Dirichlet's argument is a classic in the application of Cauchy's own $\varepsilon - \delta$ analysis and the theory of convergence. He was able to show, using properties of the trigonometric functions, that the sum of the first n terms of the Fourier series of a function $f(x)$ in the interval from $-\pi$ to π is (in his notation):

$$\frac{1}{\pi}\int_{-\pi}^{\pi} \frac{\sin(2n+1)\frac{\alpha-x}{2}}{2\sin\left(\frac{\alpha-x}{2}\right)} f(\alpha)\, d\alpha.$$

So the Fourier series converges if and only if this integral (known today as a *Dirichlet integral*) converges to a finite sum as n tends to infinity.

There are two problems, and they are both illustrated in Figure 20.3 for the case $n = 20$. One problem is that the function $\sin(2n+1)\alpha$ has a period of $\frac{2\pi}{2n+1}$, and so it changes sign faster and faster as n increases. The other problem is that the limiting value of $\frac{\sin(2n+1)t}{\sin t}$ is $2n+1$ when $t = 0$. This number increases with n, and this potentially raises problems for the existence of the integral.

Two things happen as n increases. The rapidly oscillating part away from $x = 0$ diminishes in amplitude, and while the peak around $x = 0$ becomes ever higher, the base of the peak gets narrower. This means that the contribution of the part away from $x = 0$ should tend to 0 as n increases, and there is some hope that the contribution of the peak tends to a finite limit as n increases.

This is what Dirichlet was able to show. In particular, he proved that

$$\int_{-\pi}^{\pi} \frac{\sin(2n+1)t}{\sin t} dt = 2\pi.$$

To conclude his argument, Dirichlet was forced to make some simplifying assumptions about the function f. He could establish that the Fourier series converges, and does so to the value of the function at each point, only for functions that are continuous, positive, and monotonic increasing or decreasing in a finite set of intervals that cover $[-\pi, \pi]$. Dirichlet's argument was rigorous, and showed that it would be hard to prove Fourier's claim in any greater degree of generality.[6]

Dirichlet may be considered the mathematician who brought rigorous mathematics to Germany, if not indeed to the wider mathematical world, as his work on the convergence of Fourier series demonstrates. Another indication of how exceptional he was is gained from a remark once made by Dirichlet's wife, Rebecca Mendelssohn-Bartholdy, who wrote that Jacobi would spend hours with Dirichlet:[7]

> being silent about mathematics. They never spared each other, and Dirichlet often told him the bitterest truths, but Jacobi understood this well and he made his great mind bend before Dirichlet's great character.

In the next section we shall meet Dirichlet again, at work on potential theory, where, however, his rigour failed him in what turned out to be a productive way for others who came after him.

Thomson and the modelling of the tides. Many nearly periodic phenomena can be modelled by finite Fourier series — often the first two or three terms are enough. The best-known example of this concerns the Victorian debate over the age of the Earth. Charles Darwin, whose *The Origin of Species* was published in 1859, estimated that the Earth had to be at least 300 million years old to accommodate the very slow processes of evolution that he felt his theory required. This brought him into conflict with William Thomson, who objected that Darwin's geological speculations conflicted with the laws of physics. Thomson therefore undertook to arrive at a physically sound estimate of the age of the Earth, and used Fourier series methods to do so; Thomson had a lifelong appreciations for Fourier's work.[8]

Fourier himself had discussed how the Earth might have an internal source of heat, how it might be heated by the Sun, and how it might be heated by the universe. On the basis of certain assumptions about the Earth's core, and estimates of the amount of heat that is lost at the surface of the Earth, Thomson argued that the Earth was losing heat and could not be more than 100 million years old. The very public controversy remained unresolved until Thomson's argument was invalidated by the discovery that the Earth's core has a source of heat in the radioactive decay of radium.

In that debate the intellectual difficulties were more geological than mathematical.[9] A better example for our purposes concerns the modelling of the tides.

[6]Later generations of mathematicians would discover that Fourier's claim is false in general, and that it applies only to continuous functions that do not oscillate wildly.

[7]Quoted in (Scharlau and Opolka 1985, 148). Rebecca Mendelssohn-Bartholdy was a singer and a sister of the composer Felix Mendelssohn; Dirichlet was a good pianist himself.

[8]William Thomson was born in Belfast in 1824. The family moved to Glasgow in 1832 where he and his older brother James were educated before William went to Cambridge. To everyone's surprise he graduated only second in January 1845, before winning the Smith's Prize later that month. He then returned to Glasgow, becoming the Professor of Natural Philosophy there in September 1846, a position which he retained for over 50 years until he retired in 1899. He was knighted in 1866 and was elevated to the peerage in 1892, taking the title of Lord Kelvin.

[9]For a recent account, see (Jackson 2008).

Figure 20.4. William Thomson, later Lord Kelvin (1824–1907)

For navigational reasons, port authorities need to know the motion of the tides, and specifically the height $h(t)$ of the tide at any time t. As Thomson put it in 1879:[10]

> The object is to predict the tides for any port for which the tidal constituents have been found from the harmonic analysis from tide-gauge observations; not merely to predict the times and heights of high water, but the depths of water at any and every instant, showing them by a continuous curve, for a year, or for any number of years in advance.

For any port, mechanical tide-gauges continuously collect data about the rise and fall of the sea over time, tracing out complicated curves. Harmonic analysis separates these curves into their various components, called the tidal constituents. These are the multiple influences which affect the change in the height of the tide over time. They include the rotation of the Earth, the motion of the Moon, and the position of the Sun. It makes sense therefore to write

$$h(t) = h_1(t) + h_2(t) + h_3(t) + \cdots,$$

where, crucially, if Fourier series methods are to be used, h_1 is a periodic function with period $2\pi/\lambda_1$ (the period of the rotation of the Earth with respect to the Moon), h_2 is a periodic function with period $2\pi/\lambda_2$ (the period of the rotation of the Earth with respect to the Sun), and so on. Other terms are there to allow for other periodic influences, whatever they might be. Thomson listed ten factors altogether.

[10]We quote here from Appendix B′ to Vol. 1 of Thomson and Tait's *Treatise on Natural Philosophy*, 1879, repr. 1912, pp. 479–482. This book covered almost all aspects of mechanics and was affectionately referred to as T&T′ by two generations of readers; we shall follow that custom here. Thomson's partnership with Tait began in 1861.

Because each function h_j is periodic, it can be closely approximated by a finite Fourier series:

$$h_j(t) = \alpha_{j0} + \sum_{k=1}^{N} \alpha_{jk} \cos k\lambda_j t + \sum_{k=1}^{N} \beta_{jk} \sin k\lambda_j t.$$

Add these together and the resulting function $h(t)$ is a finite trigonometric Fourier series. The problem now is to find some way to estimate its coefficients.

Here, Thomson's method was to vary an argument of Fourier's for picking out the individual coefficients of a Fourier series. Thomson was able to prove that, as T tends to infinity:

$$\frac{2}{T}\int_S^{S+T} h(t)\cos\omega_k t\,dt \quad \text{tends to } \alpha_{jk} \text{ when } \omega_k = k\lambda_j,$$

$$\frac{2}{T}\int_S^{S+T} h(t)\sin\omega_k t\,dt \quad \text{tends to } \beta_{jk} \text{ when } \omega_k = k\lambda_j,$$

and

$$\frac{1}{T}\int_S^{S+T} h(t)\,dt \quad \text{tends to } \sum_{j=1}^{N} \alpha_{j0},$$

which is the sum of the constant terms. So each coefficient is now well approximated by an appropriate integral, which has next to be evaluated.

This integral involves a run of observations of $h(t)$ from an arbitrary starting time S over a period of time T that, ideally, is as long as possible. This period of time has to be much longer than any of the periods imposed by astronomy, of which the strongest is the Moon's, which is a lunar fortnight. The period of the Sun of half a solar year is longer, but the influence of the Sun is less.

Thomson and Tait recorded that it took even a skilled arithmetician 20 hours to calculate a single coefficient — and the data varies, of course, from port to port.[11] Thomson therefore designed and built a machine, the Tidal Harmonic Analyser, to do the job in only a few minutes — substituting 'brass for brain', as he put it in his Report (p. 280). This was a task, Thomson said (p. 304) that Airy, the Astronomer Royal, had believed was 'so complicated and difficult that no machine could 'master it'.'

It remained to draw the curve represented by the Fourier series for the graph of the data $h(t)$. This was done by the Tidal Predictor (see Figure 20.5). As Thomson said of the tidal predictor (p. 290): 'The machine may be turned so rapidly as to run off a year's tides for any port in about four hours'.[12]

Because the tidal predictor marks one of the successful introductions of automatic calculation into the world of mathematics, it is worth looking at its design and function:[13]

> This object requires the summation of the simple harmonic functions representing the several constituents to be taken into account, which is performed by the machine in the following manner:- For each tidal constituent to be taken into account the machine has a shaft with an overhanging crank, which carries a pulley pivoted on a parallel axis adjustable to a greater or less distance from the shaft's axis, according to the greater or less range of the particular tidal constituent for the different ports for which the machine

[11]See T&T′, Vol. 1, Appendix B′, p. 496.

[12]Thomson's Report is (Thomson 1882). His machine was in use for many years, and was used to calculate the tides for the Normandy landings on D-Day in 1944.

[13]See T&T′, Vol. 1, Appendix B′, p. 479.

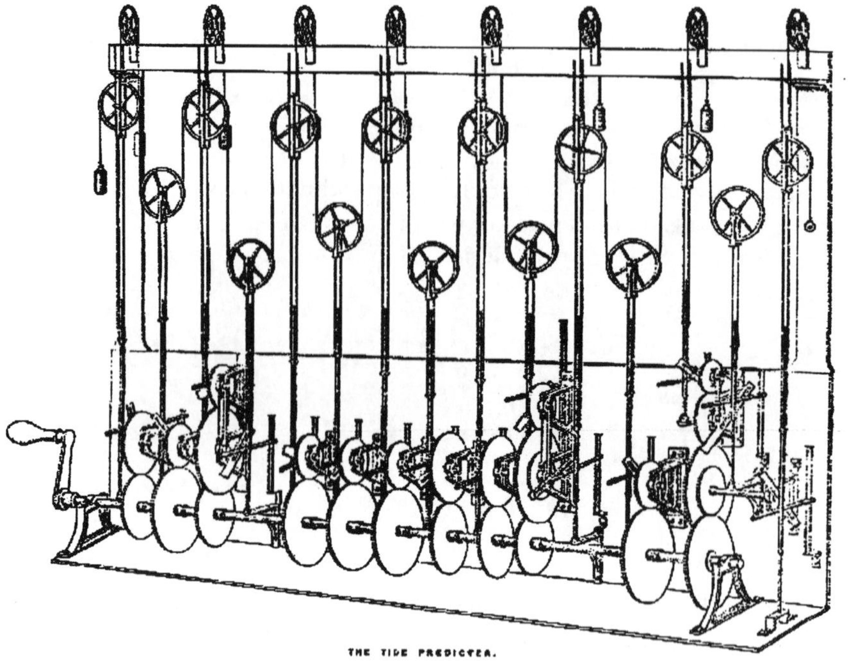

Figure 20.5. Thomson's machine for predicting the tides

> is to be used. The several shafts, with their axes all parallel, are geared together so that their periods are to a sufficient degree of approximation proportional to the periods of the tidal constituents. The crank on each shaft can be turned round on the shaft and clamped in any position: thus it is set to the proper position for the epoch of the particular tide which it is to produce. The axes of the several shafts are horizontal, and their vertical planes are at successive distances one from another, each equal to the diameter of one of the pulleys (the diameters of these being equal). The shafts are in two rows, an upper and a lower, and the grooves of the pulleys are all in one plane perpendicular to their axes.

It would be impossible to build a machine from this description alone — and Thomson's account continued for two more pages — but it is clear that each tidal constituent is modelled by shaft with a crank, and that the cranks are connected by gears that allow the shafts to turn with the correct speeds. It becomes clear from later in the account that the pulleys execute simple harmonic motion — the motion of an ordinary pendulum — and that when they are fastened together the machine can be made to draw the sum of these motions. In short, each shaft, crank, and pulley describes one term of the Fourier series, and the machine adds these together and draws the output.

Thomson explored a number of machines for automating various tasks, and in this he was far from alone; among the other people involved were Maxwell, and Thomson's brother James. Another paper by William Thomson gives a glimpse of his relationship with his brother. It concerns the development of his harmonic analyser, although the context in this paper is the task of evaluating integrals of products of functions, which is a routine but essential part of finding the Fourier series of a given function $f(x)$:[14]

[14]See T&T′, Vol. 1, Appendix B′, p. 493.

> In consequence of the recent meeting of the British Association at Bristol [1875], I resumed an attempt to find an instrument which should supersede the heavy arithmetical labour of calculating the integrals required to analyse a function into its simple harmonic constituents according to the method of Fourier. During many years previously it had appeared to me that the object ought to be accomplished by some simple mechanical means; but it was not until recently that I succeeded in devising an instrument approaching sufficiently to simplicity to promise practically useful results. Having arrived at this stage, I described my proposed machine a few days ago to my brother Professor James Thomson, and he described to me in return a kind of mechanical integrator which had occurred to him many years ago, but of which he had never published any description.[15] I instantly saw that it gave me a much simpler means of attaining my special object than anything I had been able to think of previously. An account of his integrator is communicated to the Royal Society along with the present paper.

Thomson was evidently pleased with the design of the machine, writing:[16]

> The machine thus described is immediately applicable to calculate the values H_1, H_2, H_3, etc. of the harmonic constituents of a function $f(x)$ in the splendid generalisation of Fourier's simple harmonic analysis, which he initiated himself in his solutions for the conduction of heat in the sphere and the cylinder, and which was worked out so ably and beautifully by Poisson, and by Sturm and Liouville in their memorable papers on this subject published in the first volume of Liouville's *Journal des Mathématiques*.

As well as being used in tidal analysis, Thomson's harmonic analysers were also used in meteorology, although these machines were smaller because there were fewer components in the curves to be analysed. In 1878 one was brought into use by the Meteorological Office for analysing the graphical records of daily changes in atmospheric temperature and pressure.

These few examples of the use of Fourier series methods must stand for many, because the method proved to be powerful in many different domains where periodic behaviour is at work. In the two centuries since they were introduced they have been applied to domains from celestial mechanics to number theory, and from probability theory to numerical computation. The idea of representing a function by its Fourier series has likewise led to many advances in the study of partial differential equations.

20.2 Potential theory

In this section we look at the introduction by Lagrange and Laplace of a potential function in problems involving gravity, and the creation in the 1830s of a theory of potential functions by Green and Gauss, who were motivated by the desire to advance the new theories of magnetism and electricity.

Potential theory in the late 18th century. Many problems in celestial mechanics and other areas deal with forces. These have magnitudes and directions, and therefore three components, and consequently can be hard to handle. In 1773, to simplify matters, Lagrange introduced the idea of a potential function, without giving it a

---

[15] James Thomson developed his integrator only after having read a paper by Maxwell from 1855 on an new idea for a planimeter. Maxwell wrote the paper after having seen the planimeters in the Great Exhibition in 1851 as a young student at Cambridge and being convinced he could do better, which obviously he did!

[16] See T&T′, Vol. 1, Appendix B′, p. 494.

Box 62. Lagrange's potential function.
The force $\mathbf{F}(\mathbf{p})$ exerted on a unit mass at a point $\mathbf{p}$ by a mass distribution $\rho(\mathbf{x})$, where $\mathbf{x} = (x, y, z)$, is given by the integral:

$$\mathbf{F}(\mathbf{p}) = \int \frac{\rho(\mathbf{x})(\mathbf{x}-\mathbf{p})}{|\mathbf{x}-\mathbf{p}|^3} d\mathbf{x}.$$

This is a complicated expression — especially so when written out in coordinates — that mathematicians of the 18th century found difficult to work with. Lagrange's potential function is the simpler scalar function

$$V(\mathbf{p}) = \int \frac{\rho(\mathbf{x})}{|\mathbf{x}-\mathbf{p}|} d\mathbf{x}.$$

Direct computation shows that $-\nabla V(\mathbf{p}) = \mathbf{F}(\mathbf{p})$.

Laplace showed that $\nabla^2 V = 0$ when $\rho = 0$, and Poisson showed that $\nabla^2 V = -4\pi\rho$ when $\rho \neq 0$.

name (see Box 62). It is a function $V(x, y, z)$ whose partial derivatives give the components of a force. These partial derivatives form what is today called the *gradient* ∇V of the function V:

$$\nabla V = \left(\frac{\partial V}{\partial x_1}, \frac{\partial V}{\partial x_2}, \frac{\partial V}{\partial x_3}\right).$$

This raised the question of how to determine the potential function for a given body, and thus its gravitational attraction? In 1782 Laplace, using some earlier results of Legendre, showed how to find the potential function V for an arbitrary spheroid by assuming that it satisfies the differential equation $\nabla^2 V = 0$ (called Laplace's equation today).[17] Here,

$$\nabla^2 V = \frac{\partial^2 V}{\partial x^2} + \frac{\partial^2 V}{\partial y^2} + \frac{\partial^2 V}{\partial y^2}.$$

In 1813 Poisson extended this work to the interior points of a solid, and obtained the equation $\nabla^2 V = -4\pi\rho$, where ρ is the density of the solid. He had been led to study this question in the course of mathematising the theory of electrostatics that had begun in the 1780s when the French military engineer Charles Coulomb proposed an inverse-square law for electrostatics. This law differs from Newtonian gravity only in that the force is one of repulsion, and so a minus sign enters the formulas.

Potential functions were first used in the study of the motion of the Moon, and then of the asteroids which began to be discovered in the years after 1801. But potential functions acquired a new importance when they were found to apply to the new, and more mysterious, processes of magnetism and electrodynamics.[18] In his masterpiece of 1826, the *Mémoire sur la Théorie Mathématique des Phénomènes Électrodynamiques, Uniquement Déduite de l'Expérience* (Memoir on the Mathematical Theory of Electrodynamics Solely Deduced from Experiment) the French mathematician André-Marie

[17]Two notations are commonly used, $\triangle$ and ∇^2.

[18]See (Darrigol 2000).

Ampère offered a unified mathematical theory of electricity and magnetism that seldom ventured into the realm of physical explanations about the nature of electric current. This recalled the way that Fourier had similarly refused to speculate about the nature of heat. Ampère explained magnetism in terms of electric currents in the magnetised body, an idea that he extended to the magnetic field of the Earth. His mathematical theory fitted in well with the contemporary and brilliant experimental work of the English physicist Michael Faraday, and for a time it seemed to offer a better explanation of the facts than Faraday's, and so it was widely taken up.

Gauss. In 1828 Alexander von Humboldt constructed a magnetic observatory on French principles for Berlin University. Gauss visited it, but was not impressed. He went back to Göttingen and in 1831 teamed up with the young physicist Wilhelm Weber, who had just been appointed to the chair of physics there. Together they successfully sought to develop new mathematics with which to penetrate the mysteries of electromagnetism, and to find applications in navigation and geodesy.

Gauss's most important work on the subject of potential theory is his *Allgemeine Lehrsätze* (General Propositions) of 1840. He began by giving rigorous proofs of results already published by the French — without mentioning any mathematicians by name — and then set out some new results of his own.

Gauss began with an attractive illustration of what can be done by starting with a simple system and progressively generalising it. The initial system consists of two masses: a single mass M at the point $\mathbf{P}$ with a potential function V, and a single mass M' at the point $\mathbf{P}'$ with a potential function V'. Then

$$MV'(\mathbf{P}) = \frac{MM'}{|\mathbf{P} - \mathbf{P}'|} = M'V(\mathbf{P}').$$

It follows that a similar expression holds for a sum:

$$\sum_i M_i V_i'(\mathbf{P}_i) = \sum_j M_j' V_j(\mathbf{P}_j'),$$

where the sums are taken over the two mass distributions, which may not be the same.

Gauss then supposed that the mass distribution is not discrete but lies on curves, surfaces, or in solids, and replaced the sum by an integral. The result is his *reciprocity theorem*:

> If μ and μ' are two mass distributions on regions R and R', respectively, and V and V' are their associated potential functions, then $\int_R V'\mu = \int_{R'} V\mu'$.

It can be helpful to think of this reciprocity theorem as an equality between two integrals taken over the whole of space, but for which μ and μ' are non-zero only where there is mass. That said, we shall see that a potential function cannot be zero everywhere in a region without vanishing altogether.

Using his reciprocity theorem, Gauss derived an expression for the potential $V(0)$ at the centre of a sphere S of radius R, due to a mass distribution μ outside the sphere and having a potential V:

$$V(0) = \frac{1}{4\pi R^2} \int_S V.$$

This says that the potential at any point is the average of the values that it takes on any sphere surrounding the point.

Gauss also investigated how the potential function V varies from point to point. It is a function of x, y, z, so he looked at its derivative $\left(\frac{\partial V}{\partial x}, \frac{\partial V}{\partial y}, \frac{\partial V}{\partial z}\right)$. He considered how this derivative varied over a closed surface S, and at each point he looked at the component of the derivative that is normal to the surface, which can be written as

$$\frac{\partial V}{\partial S}.$$

He then evaluated the integral of $\frac{\partial V}{\partial S}$ over the surface.

We can form a picture of this by supposing that the potential exerts its effect because it is a substance that flows, in somewhat the way that light flows out of the Sun. To measure the energy that flows out of the Sun, one imagines that it is surrounded by a large sphere, and estimates the energy flowing radially across the surface of the sphere.

Gauss found that the integral is non-zero for surfaces that enclose mass, and zero for ones that do not. More precisely, if V is a potential due to a mass distribution and if M is the total mass inside a closed surface S, then

$$\int_S \frac{\partial V}{\partial S} = -4\pi M.$$

This result is today called *Gauss's law*.[19] It follows, in particular, that if there is no mass inside the surface, then

$$\int_S \frac{\partial V}{\partial S} = 0.$$

George Green. The English mathematician George Green carried out his work before Gauss, but it received acclaim only after Gauss's, and so we deal with it now.

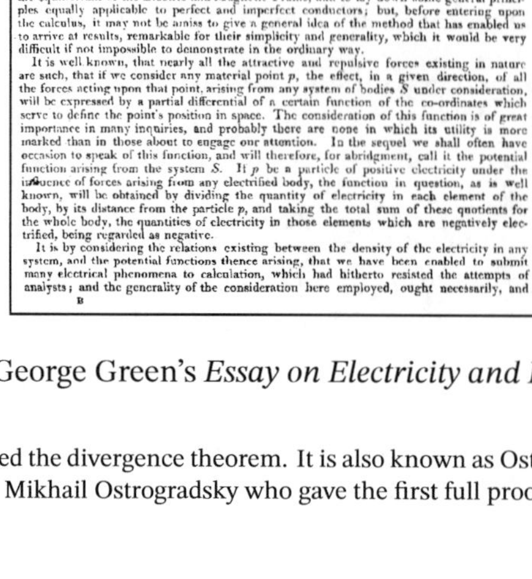
AN ESSAY

ON THE APPLICATION OF

MATHEMATICAL ANALYSIS TO THE THEORIES OF ELECTRICITY AND MAGNETISM.

INTRODUCTORY OBSERVATIONS.

THE object of this Essay is to submit to Mathematical Analysis the phenomena of the equilibrium of the Electric and Magnetic Fluids, and to lay down some general principles equally applicable to perfect and imperfect conductors; but, before entering upon the calculus, it may not be amiss to give a general idea of the method that has enabled us to arrive at results, remarkable for their simplicity and generality, which it would be very difficult if not impossible to demonstrate in the ordinary way.

It is well known, that nearly all the attractive and repulsive forces existing in nature are such, that if we consider any material point *p*, the effect, in a given direction, of all the forces acting upon that point, arising from any system of bodies *S* under consideration, will be expressed by a partial differential of a certain function of the co-ordinates which serve to define the point's position in space. The consideration of this function is of great importance in many inquiries, and probably there are none in which its utility is more marked than in those about to engage our attention. In the sequel we shall often have occasion to speak of this function, and will therefore, for abridgment, call it the potential function arising from the system *S*. If *p* be a particle of positive electricity under the influence of forces arising from any electrified body, the function in question, as is well known, will be obtained by dividing the quantity of electricity in each element of the body, by its distance from the particle *p*, and taking the total sum of these quotients for the whole body, the quantities of electricity in those elements which are negatively electrified, being regarded as negative.

It is by considering the relations existing between the density of the electricity in any system, and the potential functions thence arising, that we have been enabled to submit many electrical phenomena to calculation, which had hitherto resisted the attempts of analysts; and the generality of the consideration here employed, ought necessarily, and

B

Figure 20.6. George Green's *Essay on Electricity and Magnetism* (1828)

[19]This law is often called the divergence theorem. It is also known as Ostrogradsky's theorem, as it was the Russian mathematician Mikhail Ostrogradsky who gave the first full proof of it.

Green was born in Nottingham on 14 July 1793. He studied mathematics in local libraries and with a little help from local mathematicians who had passed through Cambridge, and he had sufficient leisure to do so because he had inherited a successful milling and bakery business from his father. He learned more from reading Laplace's *Mécanique Celeste* and some papers by Poisson, Biot, and Coulomb than he could have learned if he had gone to Cambridge, and in 1828 he wrote *An Essay on the Application of Mathematical Analysis to the Theories of Electricity and Magnetism*, which we consider below.[20] This remarkable work won him the support of Sir Edward Ffrench Bromhead, a local dignitary and Cambridge graduate, and in due course Green himself went to Cambridge University, first as a student and then for a brief time as a Fellow of Gonville and Caius college. At Cambridge he wrote a few other papers, also on magnetism, electricity, and hydrodynamics, and acquired a modest reputation, but finding little to stimulate him there he returned to Nottingham, where he died a few weeks short of his 48th birthday — not even, it seems, a local celebrity.[21]

Green had set himself the task of facilitating, as he put it, 'the application of analysis to one of the most interesting of the physical sciences'. Like Fourier before him, he saw that the new physics called for new mathematics, and that it was based on intuitions of a non-Newtonian kind.

Green coined the convenient term 'potential function' for the function U whose gradient $(\frac{\partial V}{\partial x}, \frac{\partial V}{\partial y}, \frac{\partial V}{\partial x})$ is a given force, and this name helped to move the concept centre stage. He had learned from Laplace and Poisson that the potential function satisfies two partial differential equations — one for points outside the body, and the other for points inside. He then formulated what today is called *Green's theorem*, which can be thought of as a reciprocity theorem. This theorem concerns a closed surface s enclosing a region S, and two potential functions, U and V, each in the variables x, y, z. The surface s is supposed to have an outward-pointing normal $\mathbf{n}$ at each point. The theorem, known today as 'Green's reciprocity theorem', states:

$$\int_S U\nabla^2 V - \int_S V\nabla^2 U = \int_s VU_{\mathbf{n}} - \int_s UV_{\mathbf{n}},$$

where the notation is explained in Box 63.

When V is the constant function 1, the theorem reduces to $\int_S \nabla^2 U = -\int_s U_{\mathbf{n}}$, which is Gauss's law. For, if we write $\nabla U = \mathbf{U}$, then the claim that $\int_S \nabla^2 U = -\int_s U_{\mathbf{n}}$ becomes the claim that

$$\int_S \nabla.\mathbf{U} = -\int_s \mathbf{U}.\mathbf{n}.$$

For a sketch of Green's ingenious proof of his theorem, see Box 64.

As we shall see, mathematicians see Green's theorem as generalising the Fundamental Theorem of the Calculus to several variables. Physicists see it as relating a flux across a surface to the quantity of material inside it. We can think of it this way: the integrals on the left-hand side are taken over the volume under consideration, and the integrals on the right-hand side are taken over the surface. The integrals on the left-hand side count what is inside the surface, and the integrals on the right-hand side

[20]Even the word 'analysis' in the title reminds us that Green was self-taught, for it refers to the calculus as it had become in France, and not to the sterile exercises in Newtonian methods preferred in England.

[21]See (Cannell 1993).

Box 63. Directional derivatives.
The *directional derivative* of a function, is defined as the component of the gradient of a function in a specified direction.

To find the component of a vector $\mathbf{v} = (x_0, y_0, z_0)$ in the direction of a unit vector $\mathbf{n} = (a, b, c)$, we compute

$$(a, b, c).(x_0, y_0, z_0) = ax_0 + by_0 + cz_0.$$

So to find the component of the vector $\mathbf{v}$ in the x-direction, we compute $(1, 0, 0).(x_0, y_0, z_0) = x_0$, because the vector $(1, 0, 0)$ is the unit vector in the x-direction.

To find the directional derivative of a function $U(x, y, z)$ in the direction of a unit vector (a, b, c), we compute

$$(a, b, c).\nabla U = (a, b, c).\left(\frac{\partial U}{\partial x}, \frac{\partial U}{\partial y}, \frac{\partial U}{\partial z}\right) = a\frac{\partial U}{\partial x} + b\frac{\partial U}{\partial y} + c\frac{\partial U}{\partial z}.$$

If $\mathbf{n}$ is a unit vector perpendicular (or normal) to a surface at a point, we write $U_{\mathbf{n}}$ for the directional derivative of U along the normal at a point (with a similar notation for V).

Box 64. How Green proved his reciprocity theorem.
Green proved his reciprocity theorem, apparently after some effort, in the following elegant way. He considered the integral $\int_S \nabla U.\nabla V$, and integrated it by parts (taking each of the three components of the sum separately). This gives

$$\int_S \nabla U.\nabla V = \int_S VU_{\mathbf{n}} - \int_S V\nabla^2 U.$$

Notice that if U is a potential function and there is no mass in the region S, then $\nabla^2 U = 0$ (Laplace's equation) and so Gauss's law is obtained. However, Green was working with arbitrary functions U and V, so in his case this term does not vanish. Instead, he noted that one could exchange U and V and the integral would not change, so

$$\int_S \nabla V.\nabla U = \int_S UV_{\mathbf{n}} - \int_S U\nabla^2 V.$$

On subtracting this from the first equation, he obtained the desired result:

$$\int_S U\nabla^2 V - \int_S V\nabla^2 U = \int_S VU_{\mathbf{n}} - \int_S UV_{\mathbf{n}}.$$

count what is entering or leaving it. Intuitively, there should be an equality of some sort, because, for example, to know how many cars there are in a car park, it is enough to know how many were inside when it opened, add to that the number that have since entered and subtract the number that have since left.

Another way to think of this theorem is as a generalisation of the Fundamental Theorem of the Calculus,

$$\int_a^b f'(x)dx = f(b) - f(a),$$

which compares an integral over a region (the interval $[a, b]$) with an integral (or sum) over its boundary (the two points a and b).

Green's deepest insight came with his introduction of what are called *Green's functions* today. His aim was to solve the differential equation satisfied by a potential function, and he began by considering extreme cases where the solution is easy to discover. He started with the case where the potential is caused by a single charge at an isolated point: here the potential function satisfies Laplace's equation in any region that does not contain the point. Green now assumed that the potential vanishes on the boundary of the body, and that it increases like $1/r$ as one approaches the point charge. (Notice that differentiating r^{-1} gives $-1/r^{-2}$, suggestive of the familiar inverse-square laws.)

He then showed how to solve the equation for such a curious function by using the theorem that he had developed in the earlier part of his *Essay*. He was aware that his treatment was not rigorous, but gave a plausible limiting argument to show how such potential functions can be defined. Such examples have tended to strike mathematicians as likely to make sense on physical grounds, and physicists (such as Maxwell) as likely to be amenable to rigorous mathematics.

It remained for him to show how to get from his solution to problems with extreme sets of boundary conditions (such as an infinite amount of electricity concentrated at a single point) to more plausible ones where the electricity is distributed through an entire body. He did this in two steps, reminiscent of those by which Gauss explained his reciprocity theorem. First, he solved the problem with finitely many point charges by adding the separate solutions. Then he solved a problem with infinitely many point charges distributed in some way, by replacing addition with integration.

Green's *Essay* was not appreciated in his lifetime. It became well known after his death in 1841 only because William Thomson had picked up a stray reference to it when he was at Cambridge and finally tracked the *Essay* down the day before taking his degree and leaving for Paris in January 1845. The essay entranced Thomson, and in Paris he showed it to Liouville and Sturm. They too were excited by it, as was Crelle, who immediately accepted it for his *Journal*; it was published there in three instalments between 1850 and 1854.

The elegance of Green's presentation, and his ability to be clear even when rigour lay out of reach, impressed mathematicians, and Green's name rapidly became securely attached to his discoveries, even when some of them had by then been independently discovered by others. Among the physicists, William Thomson and Maxwell remained staunch advocates, and Green's functions have played a prominent role in mathematical physics ever since.

To be accepted into mathematics, a rigorous proof of the existence of potential functions had to be found. Although we cannot take the story all the way to its successful conclusion, we can introduce the first important step in that direction — the introduction of Dirichlet's principle.

Figure 20.7. Gustav Lejeune Dirichlet (1805–1859)

Dirichlet on potential theory. In his lectures at Göttingen in 1856–1857 Dirichlet raised the following problem, now known as the *Dirichlet problem*:[22]

> Given a solid body and a continuous function V defined on its boundary, is there a function f that satisfies Laplace's equation on the interior of the body and agrees with V on the boundary?

To solve it, he introduced the contentious principle that the successful function exists because it is the function that minimises the integral

$$\int_{vol}\left(\left(\frac{\partial u}{\partial x}\right)^2+\left(\frac{\partial u}{\partial y}\right)^2+\left(\frac{\partial u}{\partial z}\right)^2\right)dxdydz$$

among all functions u that satisfy the conditions of the theorem. This claim is known as the *Dirichlet principle.*

Dirichlet then observed that the problem of *explicitly* finding such a function f 'cannot be solved; we can only speak of an existence proof for it. The latter presents no difficulty'. He then explained this comforting remark as follows:[23]

Dirichlet's principle.

> For every bounded connected domain T there are clearly infinitely many functions u continuous together with their first-order derivatives, for x, y, z which reduce to a given value on this surface. Among

[22]Dirichlet's lectures were edited and published posthumously in 1876.

[23]See (Dirichlet 1876, 127–128), quoted in (Bottazzini 1986, 300).

these functions there will be at least one which reduces the following integral

$$U = \int_{vol} \left[\left(\frac{\partial u}{\partial x}\right)^2 + \left(\frac{\partial u}{\partial y}\right)^2 + \left(\frac{\partial u}{\partial z}\right)^2 \right] dT$$

extended over the domain T, to a minimum; it is evident that this integral has a minimum since it cannot become negative. We can now show the following:

1. Every such function u which minimizes U, satisfies the differential equation $\frac{\partial^2 u}{\partial x^2} + \frac{\partial^2 u}{\partial y^2} + \frac{\partial^2 u}{\partial z^2} = 0$ everywhere in the domain T. This already makes it clear that there always exists a function u having the desired property, namely that function for which U becomes a minimum.
2. Every function u which satisfies the [above] differential equation within the domain T, minimizes the integral U.
3. The integral U can have only one minimum.

It follows from 2 and 3 that there is only one function u with the desired property.

When this 'principle' is applied in a genuine physical setting it seems entirely reasonable: nature itself tells us that there is an equilibrium distribution of charge, and so forth. But rigorous mathematics requires that this be proved, and Dirichlet's claim that 'It is evident that this integral has a minimum since it cannot become negative' is unconvincing (and surprising, coming from him), because it is easy to find examples of functions that are always positive but never take a minimum value. For example, the function $f(x) = x$ defined on the open interval $(0, 1)$, and the function $f(x) = 1 + \tanh x$ defined on the whole real line, are never negative and also never have a minimum.[24]

It was to turn out that Dirichlet's principle is false in the generality in which he stated it, even for functions of two variables rather than three. It is possible to define functions on a circle that oscillate so wildly that the 'Dirichlet integral'

$$U = \int_{vol} \left(\left(\frac{\partial u}{\partial x}\right)^2 + \left(\frac{\partial u}{\partial y}\right)^2 \right) dxdy$$

is infinite for all extensions to functions defined in the disc, and so there is no question of the integral having a minimum value. Nonetheless, arguments using Green's functions were found to show that the Dirichlet problem in the plane always has a solution, and that the Dirichlet problem in three dimensions has a solution for almost all boundaries. Furthermore, the Dirichlet principle can be made to work for a wide variety of regions and boundaries, as Hermann Amandus Schwarz in the 1870s was the first to show in two dimensions, and as Poincaré in the 1890s was the first to show in three dimensions.

---

[24]The function $\tanh x = \frac{e^x - e^{-x}}{e^x + e^{-x}} = \frac{\sinh x}{\cosh x}$.

20.3 Transatlantic cables

Here we look briefly at a dramatic 19th-century story: the struggle to connect Britain and America by a transatlantic cable.

The invention of the electric telegraph in the 1830s soon led to a network of cables across Europe and throughout the eastern seaboard of the United States. For the first time, information could be sent reliably much faster than by a rider on a horse. Typically, the message to be transmitted was transcribed letter by letter into a stream of short and long pulses, such as the dots and dashes of Morse Code.

Attempts to connect Britain to Continental Europe soon found unexpected problems, however, because of the increased capacitance of the cable under water. Capacitance is the ability of a body to store electric charge, and the increased capacitance caused the signal to spread out, so that the gap between one item and the next had to be increased, causing an increase in the time for a message to be transmitted. However, plans to run a cable across the Atlantic — a far more risky project — were regarded as extremely valuable and therefore attempts had to be made.

Thomson persuaded himself in 1855 that the one-dimensional heat equation,

$$\kappa \frac{\partial u}{\partial t} = \frac{\partial^2 u}{\partial x^2},$$

is the correct equation for describing the passage of electricity down a long wire, where κ is a constant determined by the wire.[25] As a result, he was able to show that an instantaneous pulse sent down a wire of length x is detected at the far end as a pulse lasting for a time of $T = kx^2$ seconds, where k is another constant determined by the properties of the wire, and so two separate pulses must be transmitted T seconds apart in order to be received as distinct signals at the far end. But Thomson's advice was ignored, and a cable was designed on the theory that the correct equation was a wave equation, and that the signal travels without distortion.

In 1857 the first attempt to lay a cable failed when it snapped after 338 miles and could not be recovered. A second attempt to lay the cable, in 1858, succeeded, and a 99-word message was sent from Queen Victoria to President Buchanan to mark the event, but it only signalled a greater failure: the message took 16½ hours to transmit. Attempts to improve performance by increasing the voltage only made things worse, and after a month the cable had to be abandoned. To quote the mathematician Thomas Körner, '2500 tons of cable and £350,000 of capital lay useless on the ocean floor'.[26]

The failure had been one of design. Thomson was now in a position to insist on his insights being implemented, including his version of the relevant equation. His analysis called for a much thicker cable, which would reduce the constant k in the equation $T = kx^2$. This would enable one to transmit pulses closer together, and also to use a much lower voltage, because varying the voltage had little effect on the signal. This time, to quote Körner again:[27]

> Half a million pounds was being staked on the correctness of the solution to a partial differential equation.

---

[25] See (Thomson 1856).

[26] See (Körner 2002, 334).

[27] See (Körner 2002, 336).

[28] *Illustrated London News*, 2 September 1865, p. 221.

Figure 20.8. Laying the transatlantic cable in 1865[28]

However, as before, the first attempt at laying the cable failed when the cable broke. This time, however, they were able to recover the end of the broken cable on the sea floor and reconnect it (see Figure 20.8), and by 8 September 1866 America and Europe were joined by two cables that worked as planned — a 99-word message now took only 8 1/2 minutes to transmit. Thomson received a knighthood and a considerable amount of money, some of which he used to buy a 17-year old, 126-ton, oak-built yacht, the *Lalla Rookh*. He was a keen sailor, and would usually spend much of the time between May and November on his yacht.[29]

The telegraphist's equation (see Box 65) was written down for the first time by the German physicist Gustav Kirchhoff in 1857, and was profoundly studied by Oliver Heaviside in 1876.[30] It was solved for the first time by Poincaré in 1893. In his short paper he showed that the general solution could be described as follows. If a pulse of some simple kind is transmitted between times $t = a$ and $t = b$ then, he wrote:[31]

> one sees first of all that the *head* of the perturbation will travel with a certain speed, in such a way that in front of this head the perturbation is zero, contrary to what happens in Fourier's theory of heat and in agreement with the laws of propagation of light or of plane sound waves deduced from the equation of the vibrating string. But there is an important difference with this latter case, because the perturbation, as it propagates, leaves behind a non-zero residue ... If $b - a$ is small ... the residue is negligible in front of the principal perturbation, but this is not the case if the perturbation lasts for a long time and if $b - a$ is finite. The residue can then disturb the observations ...

[29]See (McCartney 2008, 17). The name was taken from an oriental romance in verse by Thomas Moore, set in Mughal India and published in 1817. By 1870 it had been reworked as no fewer than three operas, and an episode from it had also been set to music by Schumann in 1843. According to Moore, 'Lalla Rookh' means 'tulip-cheeked' in Persian.

[30]See (Thomson 1856), (Kirchhoff 1857), and (Heaviside 1876).

[31]See (Poincaré 1893a, 1032).

Box 65. The telegraphist's equation.
The equation that describes the current u at any point x in a straight wire, at any time t during the transmission of electric signals down the wire, is

$$KL\frac{\partial^2 u}{\partial t^2} + (KR + LS)\frac{\partial u}{\partial t} + RSu = \frac{\partial^2 u}{\partial x^2},$$

where the constants that appear in the equation represent the capacitance K, the self-inductance L, the resistance R, and the leakage S of the wire. (Self-inductance is the induction of a voltage in a current-carrying wire as the current changes.)

The equations that were used in the two designs of the cable can be understood as follows. In the first case the equation used was of the form

$$KL\frac{\partial^2 u}{\partial t^2} = \frac{\partial^2 u}{\partial x^2},$$

which amounts to ignoring the leakage and the resistance.

In the second case (Thomson's) the equation used was of the form

$$KR\frac{\partial u}{\partial t} = \frac{\partial^2 u}{\partial x^2},$$

which amounts to ignoring the self-inductance and the leakage. Put another way, the effects of capacitance and resistance dominate the self-inductance in this case. So the first approach failed because it did not appreciate the effects of increased capacitance caused by the cable being under water, and Thomson's approach succeeded because it did.

Poincaré had shown that, when an attempt is made to transmit a periodic wave down a wire, the velocity and wavelength depend on the frequency, the waves undergo dispersion, and the head of the disturbance moves with a finite speed. This is also the case with the transmission of light, but not of heat, and the head, once it has passed, leaves behind a disturbance that never vanishes, unlike the case of the wave equation.

Poincaré was apparently unaware of a remarkable discovery that Heaviside had made in 1887, when he showed that the values of the physical constants can be so adjusted that the rate of dispersion is zero. This can be done both mathematically and physically; it requires merely that the leakage be non-zero. Far from being an inconvenience, this condition is necessary for the production of distortionless telephony. The signal becomes fainter over longer distances, but this can be corrected by fitting amplifiers. Long-distance telegraphy had dealt with distortion by accepting a low transmission rate, so as to separate the pulses. Telephony required much higher frequencies, but with some leakage and a deliberately high self-inductance it became distortion-free. Long-distance communication was reborn — although the money for the first successful patents went to the American electrical engineer Michael Pupin in 1901, and not to Heaviside.[32]

[32]See (Yavetz 1995).

It is intriguing to see that Poincaré in his 1894 lecture course, *Cours sur les Oscillations Électriques*, also failed to mention Heaviside's ingenious discoveries. The reason may have been a misplaced interest in the general case. Poincaré's analysis of the telegraphist's equation depended on the condition $KR \neq LS$, but the condition $KR = LS$ is exactly what Heaviside's insight depended upon. So the experimental work was given a theoretical twist, and the technological implications were not mentioned.

Conclusion. The three principal partial differential equations that we have considered — the heat equation, Laplace's equation, and the wave equation — became known as the partial differential equations of mathematical physics. It is a striking fact that between them they describe so many of the advances in applied mathematics made in the 19th century and into the 20th, and it is fortunate that in many cases they can be solved when appropriate boundary or initial conditions are specified — for, as the telegraphist's equation indicates, rigorous solution methods for general partial differential equations may be hard to find. It is remarkable how much of the modern world was made possible by the study of the calculus of functions of several variables.

20.4 Further reading

Cannell, M. 1993. *George Green, Mathematician and Physicist, 1793–1841. The Background to his Life and Work*, The Athlone Press. This labour of love by a local historian more than fills a gap in the literature, and says much about Green and mathematics in the Britain of his day.

Darrigol, O. 2000. *Electrodynamics from Ampère to Einstein*, Oxford University Press. Electrodynamics can claim to be the science that changed the world more thoroughly than any other, and this book does it full justice. Mathematically it is demanding, but the picture it paints is broad and dramatic.

Flood, R., McCartney, M., and Whittaker, A. (eds.) 2008. *Kelvin: Life, Labours, and Legacy*, Oxford University Press. This is an exceptionally readable book with many interesting essays on numerous topics.

Körner, T.W. 2002. *Fourier Analysis*, Cambridge University Press. This book concentrates on the mathematics and its applications, which are both well explained and are illuminated by numerous pertinent historical studies.

21

Poincaré and Celestial Mechanics

Introduction

In this chapter, we look at 19th-century work on the long-term behaviour of the solar system, and more generally at the contributions of Henri Poincaré. Poincaré was one of the two dominant mathematicians at the start of the 20th century (the other being David Hilbert). He originated new departures in several domains of mathematics; here we look at his reformulation of dynamics and its implications for astronomy. His wholly original study of the long-term behaviour of the motion of a particle governed by a system of differential equations initiated a new approach to mechanics that has since spread to many other branches of mathematics.

21.1 Late 19th-century celestial mechanics

As we saw in Chapter 10, Laplace's work in celestial mechanics had shown almost beyond doubt that the planets in the solar system obey Newton's laws, strongly suggesting that the solar system is stable and that the planets will go round in their orbits indefinitely, at least until some massive new event occurs. In 1808 the French mathematician Siméon-Denis Poisson, who had studied in Paris under both Lagrange and Laplace, further strengthened the general argument for stability. He improved Lagrange's approximation by showing that, when quadratic terms in the planetary masses are included, there are no secular variations in the mean motions of the planets. Poisson also proposed a new and broader definition of stability in general: he defined a system as 'stable' if it repeatedly returns close to its initial configuration; in other words, as long as the system returns, it does not matter how it behaves in between.

But an argument to the effect that the solar system is stable does not mean that every aspect of its behaviour is known. As we noted briefly in Chapter 13, there was great excitement in the astronomical community in 1801 with the discovery of Ceres, the asteroid whose orbit was calculated with such success by Gauss after it had first been spotted by Piazzi. Ceres was the first new object to be seen in the solar system since the discovery of the planet Uranus in 1781, and its detection meant that it was likely that there were other things to discover. In 1846 the French mathematician Urbain Le Verrier, puzzled by discrepancies between the observed orbit of Uranus and the orbit predicted by Newton's law of gravity, correctly predicted the existence of another planet. Armed with Le Verrier's calculations that told them where to point their telescope, observers at the Berlin Observatory were the first to see the new planet, later named Neptune. It soon transpired that John Couch Adams, a 28-year old Cambridge mathematician, had independently predicted Neptune's existence. But despite the fact that Adams's results were incomplete, and that Adams himself publicly credited Le Verrier with priority, the British were keen to champion their man, giving him equal status with Le Verrier, and a rather bitter Anglo-French controversy ensued.[1]

Priorities aside, the fact remains that both Le Verrier and Adams had deduced the existence of a new planet by calculation, using only the observed perturbations of one planet and assuming only Newton's law of gravity. It was the first time that a planet had been found by mathematics, rather than by observation, and it was a triumph for Newton's theory. The discovery also generated much popular interest and some mirth, as the French cartoons in Figure 21.1 show: in them, Adams is shown on the left looking in vain for Neptune, and on the right discovering it in Le Verrier's Notebook.[2]

Figure 21.1. The discovery of Neptune

There was also the important, but tedious, business of constructing nautical almanacs. These publications provide astronomical data, mainly in tabular form, for use in navigation at sea, as well as for general astronomical purposes. Although they are extremely useful for those requiring astronomical information in the short term, they serve no practical purpose for those wanting to look further into the future.

In England, the first almanac dedicated to data for finding the longitude at sea was *The Nautical Almanac and Astronomical Ephemeris*, which was first published by the Board of Longitude in 1766 (with data for 1767) and which was based on Mayer's tables.[3] For each hour of the year, it specified the positions on the Earth's surface at which the Sun, Moon, planets, and a point in the sky known as the first point of Aries

[1] It was only with the discovery in Chile in 1999 of the 'Neptune file', which had been missing from the Greenwich Royal Observatory since the mid-1960s, that a full account could be given of the British side of the story, which shows that Adams did not deserve credit for the discovery. See (Kollastrom 2006).

[2] See *L'Illustration*, 7 November 1846.

[3] We discussed Mayer's work in Section 11.4.

are directly overhead, as well as the positions of 57 selected stars relative to the first point of Aries.[4] Since voyages often took many years to complete, the *Almanac* came to be published annually with data for several years ahead.

Particularly important were the tables of lunar distances that were used to help mariners to determine longitude at sea from observations of the Moon. From 1862 the *Almanac* used the lunar tables of the Danish astronomer Peter Hansen, who had developed a new and powerful numerical method for computing the mutual perturbations of planets. He first applied it to Jupiter and Saturn (winning a Berlin Academy prize in 1830) and then to the motion of the Moon.[5] Although the lunar method was published in 1838, it was not until the 1850s that its efficacy was fully recognised when it received praise from both Jacobi and Cayley. In 1857 tables based upon it were published by the British Government. The construction of tables for these almanacs was painstaking work involving the use of power series in several variables, and the routine calculations were parcelled out to human computers often hundreds of miles apart. In 1883 Hansen's tables were also adopted by the Americans working in the Nautical Almanack Office (NAO) in Washington D.C. in its publication, *The American Ephemeris and Nautical Almanac*, first published in 1855. The two almanacs merged in 1891.

A second lunar theory, which vied with Hansen's theory although it was algebraic rather than numerical in nature, was that of Charles-Eugène Delaunay, an indefatigable computer and a colleague, and later rival, of Le Verrier. Delaunay, who earned the epithet 'Baron of the Moon' for his work, spent 20 years constructing his theory. He found a purely trigonometric series that formally satisfied the equations of motion, and used it to become the first to complete a total elimination of the secular terms in the lunar theory. His work appeared in two monumental volumes, each of over 900 pages, published in 1860 and 1867. As the historian David Aubin has observed:[6]

> In these books, Delaunay pushed to extreme the formal analytical expansion of a single function. He spent twenty years of his life developing it to seventh order (and sometimes even to ninth order), computing over 1259 terms in the expansion series for the moon's longitude and 1086 for its latitude. Although this extraordinary effort has sometimes been ridiculed, Delaunay's work is emblematic of the tremendous optimism invested both in the precision of measurements made in the observatory and in the precision of the analytical method.

Delaunay's series are polynomial approximations in trigonometric functions to infinite trigonometric series.

Among those who initially admired Delaunay's theory was the American mathematical astronomer George William Hill, a member of the staff of the NAO. Delaunay's method, which derived ultimately from Lagrange, was elegant but, as Hill discovered, it was also flawed. The higher-order approximations converged so slowly, and the complexity of the computations increased so dramatically, that Delaunay had been forced to resort to an element of guesswork in order to complete his computations, thereby compromising their reliability. He was exactly the right person to embark on such a

---

[4]The first point of Aries is one of the two points on the celestial sphere where the ecliptic and the celestial equator cross one another. When the Sun reaches the first point of Aries, an equinox occurs.

[5]For an account of how Hansen discovered his new method and its connections to Laplace's work, see (Wilson and Harper 2014).

[6]See (Aubin 2009, 293).

Figure 21.2. George William Hill (1838–1914)

project — Simon Newcomb (the head of the NAO)[7] described him as 'the greatest master of mathematical astronomy during the last quarter of the nineteenth century'. In 1877 Hill privately published a paper on the motion of the lunar perigee (the point of the Moon's orbit closest to the Earth) which included the first new periodic solutions to the three-body problem since Lagrange's discovery of equilateral triangle solutions in 1772.[8]

Hill's method, like Newton's, was to start with an approximate solution and use it to generate more accurate solutions. Hill had the original idea of finding a periodic solution (now called the 'intermediate orbit') of the differential equations that described the motion of the Moon and closely approximated the observed motion. Starting with this (rather than an ellipse, as Newton had done) led him to a complicated system of linear differential equations with periodic coefficients. He then simplified this to a single linear differential equation of the second order, although it was still difficult to work with.

After an extremely laborious calculation, Hill obtained a solution that enabled him to calculate the ratio of the mean motion of the lunar perigee to that of the Moon, accurate to 13 decimal places. His value was close to that given by observations, exceeding it by only 1.4%, a difference that he attributed to neglecting the inclination of the Moon's orbit. Although the value given by Delaunay's series also differs by the same amount, the convergence of this series is much slower — Hill estimated that it would be necessary to compute Delaunay's series as far as terms of order 27 to achieve the same degree of accuracy.

In the following year, 1878, Hill published a paper on the lunar theory that includes a more complete derivation of the periodic solutions, and shows that the motion of the planetoid (an infinitesimal body) in the three-body problem would be confined to certain regions of the plane. This caught the attention of Poincaré, who made no secret of his admiration for Hill. The British mathematician James Joseph Sylvester was in

[7] See (Newcomb 1903, 218–219).

[8] See Section 11.4.

Paris in 1891, shortly after the publication of Poincaré's great memoir on the three-body problem (described in Section 21.5), and recorded that:[9]

> They speak great things here of Poincarré's [sic] prize memoir in the Acta[10] — and he seems to have taken some of the most fruitful ideas from Hill of whom he speaks most highly both (as I noticed [?] in the memoir) and also in conversation as has been the case in talking with me. All the French mathematicians young and old bow their heads before Poincarré whom they regard as the greatest Mathematician in Europe and who is as simple and modest as he is eminent.

21.2 Henri Poincaré

Figure 21.3. Jules Henri Poincaré (1854–1912)

Born in 1854 in Nancy in Alsace-Lorraine, Jules Henri Poincaré came from an eminent professional family. His father, Léon Poincaré, was a professor of medicine at the University of Nancy, and his first cousin, Raymond Poincaré, was to become President of the French Republic during the First World War and serve as prime minister on four occasions. His sister Aline married Émile Boutroux, a well-known philosopher of science.

Poincaré distinguished himself at mathematics at school, and in 1873 entered the École Polytechnique. There his talent flourished and he began to display the qualities that were to characterise his working life. He seldom took notes, and when asked to solve a problem it was said that the answer came back with the swiftness of an arrow. Graduating second at the École Polytechnique — his weakness in drawing cost him first place — Poincaré enrolled at the prestigious École des Mines (School of Mining) and practised for a short time as a mining engineer. Meanwhile he was preparing for a

[9] J.J. Sylvester to S. Newcomb, 18 January 1891; see (Archibald 1936, 151–152).

[10] 'Acta' refers to the journal *Acta Mathematica*, which we discuss below.

higher degree in mathematics, and in August 1879 he was awarded his doctorate from the University of Paris for a thesis on partial differential equations.

In December of that year he was appointed to take charge of the analysis course at the University of Caen. In 1881 he returned to Paris to teach at the Sorbonne, and in 1886 he was elected to the chair of mathematical physics and probability. After ten years, on the death of his friend Felix Tisserand, he succeeded to the chair of mathematical astronomy and celestial mechanics, and remained in that position for the rest of his life. Poincaré also took on important roles in scientific public life. From 1902 he was Professor of Theoretical Electricity at the School of Posts and Telegraphs, and was twice President of the Bureau de Longitudes and once President of the Académie des Sciences in Paris. In 1908 he was elected to the Académie Française.

Many of Poincaré's lecture courses at the Sorbonne, particularly those on subjects in mathematical physics such as optics and electricity, were turned into books. These volumes were praised for their elegant literary style and, in contrast to some of his research papers, were clearly written and straightforward to follow. His writings on the philosophy of mathematics and science, which began in 1887 with an article on the foundations of geometry, are best known through four books of essays, *La Science et l'Hypothèse* (*Science and Hypothesis*) (1902), *La Valeur de la Science* (*The Value of Science*) (1905), *Science et Méthode* (*Science and Method*) (1908), and *Dernières Pensées* (*Last Essays*) (1913). These books, still available today, have been reprinted several times and translated into many languages. Throughout his career Poincaré was keen to make mathematics and physics accessible to a wide audience and articles by him can be found in a variety of popular journals of the day.

Poincaré created new fields of mathematics and transformed old ones, and his contributions to pure mathematics, mathematical physics, and the philosophy of science cover a wider range of knowledge than those of any other mathematician of his generation. A prodigious author, who wrote more than 30 books and almost 500 papers, he is known for his work in function theory, geometry, topology, differential equations, celestial mechanics, electromagnetic theory, and the foundations of science. That he was able to make so many important and unexpected mathematical discoveries stems in part from the range and depth of his scientific interests, as well as his capacity to apply the ideas or results from one subject area to another. By anyone's standards, the extent of his scientific production was remarkable — his ability to write profoundly and at such length on so many different subjects was unmatched in the 20th century. In fact, his output was so extensive that we may wonder how he managed to achieve so much. Pierre Boutroux, Poincaré's nephew, provides us with a clue:[11]

> [Poincaré] thought in the street on his way to the Sorbonne, while he was attending a scientific meeting, or while he was taking his customary long walks after lunch. He thought in his antechamber, or in the meeting room of the Institute, while he was strolling about, his face tensed, rattling a bunch of keys. He thought at table, during family reunions, in drawing rooms, often stopping suddenly in mid-conversation and leaving the other speaker stranded in order to follow a thought crossing his mind. All of his work of discovery was done mentally, usually without any need to check his calculations in writing or to commit his proofs to paper.

[11] See (Boutroux 1914/1921, 197–201).

21.3 Poincaré and differential equations

At the beginning of the 1880s, Poincaré began work on the qualitative theory of differential equations. He realised that it can be more productive to begin by studying the general properties of the solutions than trying to find the exact solutions themselves. In doing so he was echoing an established approach to solving polynomial equations — that is, equations of the form

$$a_n x^n + a_{n-1} x^{n-1} + \ldots + a_1 x + a_0 = 0.$$

In the polynomial case, qualitative information about the solutions, such as whether they are real or imaginary, can be helpful in the search for exact solutions.

Poincaré was not the first to work on the qualitative theory of differential equations, but his approach was radically different to anything that had been done before. What was particularly new and important was his idea of thinking of the solutions in terms of curves rather than as functions. Poincaré's method was to visualise the solutions as a 'flow' — that is, as trajectories (orbits) of points flowing through what we now call phase (or state) space. Each point in the phase space represents a particular state of the system, with the coordinates of the point taking on the numerical values of the variables at the moment that state occurs. In the case of a first-order differential equation, these are the values of x, y, and dy/dx, where x and y are real. As the system continuously unfolds over time, the solution flows as a trajectory through the phase space.

Poincaré's interest in differential equations was driven not only by an intrinsic interest in the equations themselves, but also by a special interest in some of the fundamental questions of mechanics, such as whether the solar system is stable. By using qualitative methods and focusing on the characteristics of solutions, rather than using quantitative methods and trying to find explicit formulas, Poincaré brought about a fundamental change in the way that mathematicians thought about the three-body problem and, in consequence, the stability of the solar system.

At the beginning of his first paper on the qualitative theory of differential equations, Poincaré made his motivation clear:[12]

> Moreover, this qualitative study has in itself an interest of the first order. Several very important questions of analysis and mechanics reduce to it. Take for example the three-body problem: one can ask if one of the bodies will always remain within a certain region of the sky or even if it will move away indefinitely; if the distance between the two bodies will infinitely increase or diminish, or even if it will remain with certain limits? Could one not ask a thousand questions of this type that would be resolved if one could construct qualitatively the trajectories of the three bodies? And if one considers a greater number of bodies, what is the question of the invariability of the elements of the planets, if not a question of qualitative geometry, since to show that the major axis has no secular variations shows that it constantly oscillates between certain limits.

Almost no-one before Poincaré had considered such questions in this way, as Jacques Hadamard, Poincaré's successor at the Académie des Sciences, later observed:[13]

> The most important of these [questions of analysis and mechanics] is well known, and by itself this example presents the whole spirit of the progress of astronomy: it is *the stability of the solar system*. The single fact that this question is essentially qualitative

[12] See (Poincaré 1881, 376).
[13] See (Hadamard 1921, 240).

> suffices to show the necessity of his [Poincaré's] point of view. However, this point of view was almost completely neglected and ignored by the predecessors of Poincaré.

The first-order case. Poincaré's initial researches centred on the simplest case: the construction of solution curves of the equation

$$\frac{dx}{X} = \frac{dy}{Y},$$

where X and Y are polynomials in x and y, and so dy/dx is a single-valued function of x and y. Unlike his contemporary pure mathematicians, but not astronomers, he considered only real values for x and y, as opposed to allowing complex values. Although this equation is too simple to have direct applications in celestial mechanics, it was to provide Poincaré with a basis from which he could extend and elaborate his results to more complicated systems.

At each point P of the plane, except those where both X and Y vanish, the differential equation gives a value for the slope dy/dx of the solution curve through P. Points at which both X and Y vanish, and so there is no definite value for dy/dx, are called *singular points*. They require special treatment, as does the behaviour of the solutions as x and y become very large.

Poincaré began by examining the behaviour of these curves in the neighbourhood of a singular point. He showed that there are four possible types of singular point and classified them by the behaviour of the nearby solution curves: *nœuds* (nodes) through which infinitely many solution curves pass; *cols* (saddle points) through which only two solution curves pass, the curves acting as asymptotes for neighbouring solution curves; *foyers* (foci) which the solution curves approach in the manner of a logarithmic spiral; and *centres* (centres) around which the solution curves are closed (see Figure 21.4).

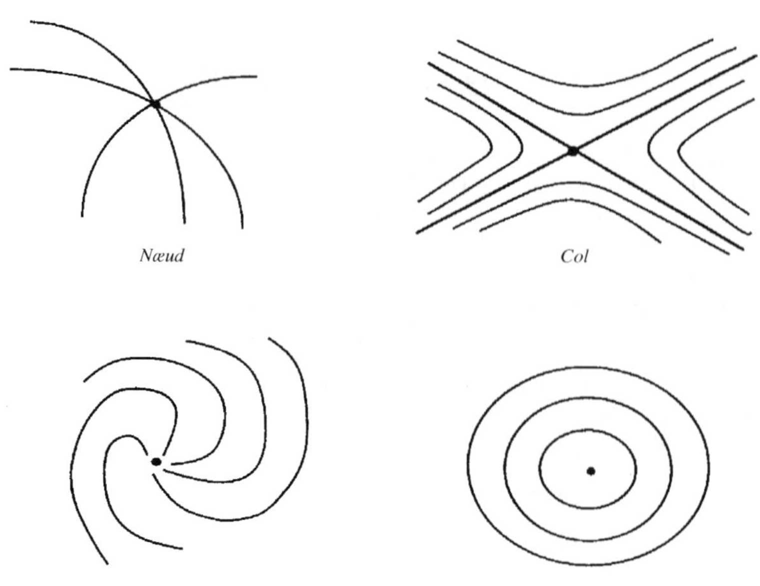

Figure 21.4. Flows near a node, a saddle point, a focus, and a centre

Poincaré next looked at the behaviour of solution curves beyond the neighbourhood of singular points. A type of closed curve which played an important role in his theory was one that he called a *limit cycle*. This is a closed solution curve with no singular points but which is asymptotically approached by other solution curves. These other solution curves spiral in towards a limit cycle but never actually reach it. He examined what happens when one follows a solution curve in one direction and looked at all the possible outcomes, and he found that a curve that does not asymptotically approach a limit cycle, and is not a closed curve, must end in a node or a saddle point.

By obtaining results about the distribution of limit cycles, Poincaré began to generate a qualitative description of the flow described by the differential equation. He showed how the functions X and Y determine the number of limit cycles within a given region of the plane, and found the particular regions in which a given number of limit cycles exist.

Poincaré also considered higher-order differential equations, such as those that describe the motion of celestial bodies. A detailed description of this aspect of Poincaré's work is beyond the scope of this book, but we describe some of his most important techniques in the next section.

21.4 Poincaré and celestial mechanics

The first Oscar. In July 1885 a notice appeared in mathematical and scientific journals in Europe and America announcing a mathematics competition that was being organised in Stockholm. It was sponsored by King Oscar II of Sweden and Norway, who had decided that a mathematics competition was an appropriate way to mark his forthcoming 60th birthday in 1889.

Figure 21.5. King Oscar II

King Oscar, who had studied mathematics at the University of Uppsala, was an active patron of the subject. He provided financial support for publishing enterprises, as well as making awards to individual mathematicians. He was well known within Scandinavian mathematical circles, and in her autobiography the Russian mathematician, Sonya Kovalevskaya, a professor of mathematics in Stockholm and the only female professor of mathematics in Europe, said of him:[14]

> King Oscar is a pleasant and cultivated person. As a young man he attended lectures at the university, and still today shows an interest in science, although I cannot vouch for the profundity of his erudition. He has no official contact with the university but is extremely sympathetic to it and very amicably disposed towards its professors in general and to myself in particular.

From its beginnings in 1884, the competition was organised by Gösta Mittag-Leffler, the professor of mathematics at the newly established Stockholm Högskola, soon to become the University of Stockholm. Like King Oscar, Mittag-Leffler had studied mathematics at the University of Uppsala, gaining his doctorate from there in 1872, and he was familiar with the leading mathematical centres in Europe, having studied with Hermite in Paris and Weierstrass in Berlin.

Inspired by Weierstrass, Mittag-Leffler's own mathematical interests lay in the field of complex analysis. He was an accomplished mathematician who published over a hundred papers, but his most valuable legacy to mathematics lies in his work as founder and Editor-in-Chief of the journal *Acta Mathematica*. This had been founded in 1882 with support from King Oscar, and Mittag-Leffler's duties as editor were his main preoccupation for forty-five years until his death. Under his editorship *Acta* became one of the first truly international mathematics journals. It was not only a showcase for Scandinavian mathematics, but mathematicians from many other countries, including France, Germany and Italy, as well as Great Britain and the United States, contributed to its pages.

Mittag-Leffler was also responsible for arranging Kovalevskaya's appointment in Stockholm. Like him, Kovalevskaya had been a student of Weierstrass, although she and Mittag-Leffler had not been in Berlin at the same time. In 1876 Mittag-Leffler visited St Petersburg where, on the bidding of Weierstrass, he went to see Kovalevskaya. It was the first time that the two of them had met, and she captivated him:[15]

> More than anything else in St. Petersburg what I found most interesting was getting to know Kovalevskaya ... As a woman, she is fascinating. She is beautiful and when she speaks, her face lights up with such an expression of feminine kindness and highest intelligence that it is simply dazzling. Her manner is simple and natural, without the slightest trace of pedantry or pretension. She is in all respects a complete 'woman of the world'. As a scholar she is characterised by her unusual clarity and precision of expression ... I understand fully why Weierstrass considers her the most gifted of his students.

Mittag-Leffler saw Kovalevskaya again in St Petersburg in 1880 when she delivered a paper at a conference, and from then on he devoted considerable energy to championing her as a mathematician. Eventually, in the face of substantial opposition from colleagues, he arranged a teaching position for her in Stockholm, where she arrived at the end of 1883, albeit to a somewhat mixed reception. One of the progressive Stockholm newspapers hailed her as a 'princess of science',[16] but the Swedish dramatist August Strindberg considered a female professor of mathematics to be a pernicious and

[14]See (Kovalevskaya 1978, 228).
[15]See (Mittag-Leffler 1923, 172).
[16]See (Koblitz 1983, 179).

Figure 21.6. Sonya Kovalevskaya (1850–1891)

Figure 21.7. Gösta Mittag-Leffler (1846–1927)

disagreeable monstrosity.[17] Kovalevskaya had not been in Stockholm for long before Mittag-Leffler invited her to join the editorial board of *Acta Mathematica*, and in accepting the position she became the first woman to occupy an editorial position on a major scientific journal.

Her most important mathematical work concerned the rotation of a solid body (such as a spinning top) — a problem to which Euler and Lagrange had previously made significant contributions — and for this she won the prestigious *Prix Bordin* of the Académie des Sciences in Paris in 1888. In 1889 she became a full professor of mathematics in Stockholm, the first woman anywhere to achieve such a position. She died suddenly in Stockholm in 1891 at the age of 41.

We do not know when the idea for the Oscar competition was first mooted, although a letter from Mittag-Leffler to Kovalevskaya tells us that by June 1884 it was already under discussion.[18] The first difficulty was to find the jurors who would run it. Although the King had wanted a panel of four or five jurors, he had to make do with a jury of only three: Hermite, Weierstrass, and Mittag-Leffler himself.

Karl Weierstrass had by then been a professor at the University in Berlin for twenty years. His research was focused principally on complex analysis, although he also lectured on the application of analysis to problems in mathematical physics, and he had a particular interest in the n-body problem. His influence was largely carried on by his former students, Mittag-Leffler and Kovalevskaya among them.

Charles Hermite, professor of mathematics at the Sorbonne, was one of the dominant figures of French analysis during the second half of the 19th century. Mittag-Leffler had studied with Hermite in Paris and a close friendship had developed between them. A leading exponent of Cauchy's complex analysis, he also actively promoted Weierstrass's ideas in France, and made no secret of the high regard in which he held the work of his German counterpart. As Mittag-Leffler later recalled, his earliest memory of Hermite was of being greeted by the words,[19]

---

[17] See (Cooke 1984, 109).

[18] The correspondence relating to the competition is preserved in the Institut Mittag-Leffler, the former home of Mittag-Leffler and now a mathematics institute.

[19] See (Mittag-Leffler 1902, 131).

> You have made a mistake, Monsieur, you should have taken the courses of Weierstrass in Berlin. He is the master of us all.

Once the jury was established, Mittag-Leffler's next (and formidable) task was to reach agreement on the questions to be set. If the competition was to be the success that he hoped, then it had to attract entries of the highest international calibre, and this depended crucially on the nature of the questions. There was an intensive correspondence among the members of the jury, who finally agreed that the competition would consist of four questions, the first three proposed by Weierstrass and the fourth by Hermite. Entrants could also submit an entry on a related topic of their own choice, but entries on one of the listed questions would be considered first. All four questions were on mathematical analysis, but the question of interest to us is the first one, which asked for a solution to the n-body problem:[20]

> A system being given of a number whatever of particles attracting one another mutually according to Newton's law, it is proposed, on the assumption that there never takes place an impact of two particles, to expand the coordinates of each particle in a series proceeding according to some known functions of time and converging uniformly for any space of time.
>
> It seems that this problem, the solution of which will considerably enlarge our knowledge with regard to the system of the universe, might be solved by means of the analytical resources at our present disposition; this may at least be fairly supposed, because shortly before his death Lejeune-Dirichlet communicated to a friend of his, a mathematician, that he had discovered a method of integrating the differential equations of mechanics, and that he had succeeded, by applying this method, to demonstrate the stability of our planetary system in an absolutely strict manner. Unfortunately we know nothing about this method except that the starting point for its discovery seems to have been the theory of infinitely small oscillations. It may, however, be supposed almost with certainty that this method was not based on long and complicated calculations but on the development of a simple fundamental idea, which one may reasonably hope to find again by means of earnest and persevering study.

In mid-1885 a notice about the competition was placed in *Acta Mathematica* and circulated to other journals. It stipulated the conditions of entry, included the names of the jury, a list of the questions, and details of the prize: a gold medal, together with the substantial sum of 2500 Crowns. (As a comparison, Mittag-Leffler's annual professorial salary in 1882 was 7000 Crowns.) The entries had to be sent to Mittag-Leffler before 1 June 1888 and, as was customary in such competitions at the time, they had to be sent in anonymously, identifiable only by a motto and accompanied by a sealed envelope bearing the motto and containing the author's name and address. Entries could not have been previously published and the winning entry would be published in *Acta Mathematica*.

The competition entries. The selection of topics for the competition was such that Poincaré could have submitted an entry on any one of them. This raises the question as to whether they had been chosen with Poincaré in mind. Hermite freely admitted that this was the case with his question, which included a reference to work by Poincaré, and perhaps Weierstrass too had designed his questions so as to appeal particularly to Poincaré. Although we cannot know for sure, the fact that Mittag-Leffler

---

[20] For a copy of the competition announcement (which originally appeared in *Nature* on 30 July 1885), including all four questions, see (Barrow-Green 1997, 229–231).

was a champion of Poincaré's work — he had secured important papers by him for publication in each of the first five volumes of *Acta Mathematica* — makes it seem probable.

By the closing date twelve entries had been received. Shortly afterwards a list of their titles, numbered in date order of submission, was published in *Acta* with the authors identified solely by their respective mottos. Five of the entrants, including Poincaré, had attempted the first question, one had attempted the third (on first-order non-linear differential equations), and the remaining six had chosen their own topics.

Within a fortnight of the closing date Mittag-Leffler had whittled the number of entries worth considering from twelve down to three, although none of the three provided a complete solution to any of the given questions. He then spent August in Germany studying the memoirs with Weierstrass. The following month he wrote to Hermite to tell him that they thought that Poincaré should win, with another French mathematician, Paul Appell, being given an honourable mention. Apart from the intrinsic quality of the two memoirs, Poincaré had an immediate advantage over Appell because he had attempted one of the set questions, although he limited his investigations to the restricted three-body problem, whereas Appell had chosen his own topic. Meanwhile Hermite, who had also been studying Poincaré's memoir, was equally convinced of its importance.

The jury rapidly reached a unanimous decision, but the hard part of their work had not yet begun. It was one thing to recognise the quality of Poincaré's work but quite another to understand it. Poincaré's entry was very long — when printed for *Acta Mathematica* it amounted to 158 pages — and it contained many new ideas and results. Moreover, as Hermite freely admitted in a letter to Mittag-Leffler, Poincaré's customary lack of detail made his work all the more difficult to follow. Hermite reported on the experience of Émile Picard, a young contemporary of Poincaré's, as follows:[21]

> But it has to be acknowledged in this work as in almost all of his research, Poincaré points the way and gives signs, but there is plenty to be done to fill the gaps and complete his work. Picard has often asked him for clarification and for explanations on very important points in his articles in the *Comptes Rendus*, without obtaining anything except the words: 'it is thus, it is like that', so that he seems like a seer to whom truths appear in a bright light, but largely to him alone.

Weierstrass, Hermite, and Mittag-Leffler all struggled with various parts of the memoir, but it was Mittag-Leffler, determined that the version submitted to the King should be as complete as possible, who entered into correspondence with Poincaré, asking for clarifications, despite the fact that he was not supposed to know who the author was. The gaps in Poincaré's arguments were not trivial: his response to Mittag-Leffler's questions resulted in a series of Notes which, when printed, added a further 93 pages to the original 158-page memoir. Mittag-Leffler may have had no qualms about his contact with Poincaré, but Weierstrass, a stickler for the rules, was distinctly unhappy about it and asked Mittag-Leffler not to mention the fact that he knew that Poincaré had entered the competition.

The competition result. On 20 January 1889, the day before the King's 60th birthday, Mittag-Leffler went to the palace to have the result sanctioned by the King. Only Weierstrass's report on Poincaré's memoir remained outstanding; although Weierstrass

[21] Hermite to Mittag-Leffler, 22 October 1888, Institut Mittag-Leffler.

Figure 21.8. Charles Hermite (1822–1901)

Figure 21.9. Karl Weierstrass (1815–1897)

Figure 21.10. Lars Edvard Phragmén (1863–1937)

had begun his report, he had been too ill to complete it, but he gave every indication that it would soon be finished. The news of the national double for France was well publicised in the French press, and both Poincaré and Appell were made Knights of the Legion of Honour in recognition of their achievement. The result was also good for Mittag-Leffler, for he too was similarly honoured for his role in promoting French mathematics.

But Mittag-Leffler's job proved to be far from over. Indeed, nothing so far was to compare with the problem that was about to arise.

Discovery of the error. In July 1889 Edvard Phragmén, whom Mittag-Leffler had charged with editing Poincaré's memoir for publication, alerted Mittag-Leffler to some passages in it that he found hard to follow. Thus prompted, Mittag-Leffler immediately wrote to Poincaré for clarification, but it was a few months before the scale of the problem became clear. Poincaré, while dealing with Phragmén's queries, realised that he had made a serious error elsewhere in the memoir. At the beginning of December, and making no attempt to conceal his distress, Poincaré wrote to Mittag-Leffler:[22]

> I have written this morning to M. Phragmén to tell him of an error I have made and doubtless he has shown you my letter. But the consequences of this error are more serious than I first thought. It is not true that the asymptotic surfaces are closed, at least in the sense in which I originally meant. What is true is that if both sides of this surface are considered (and which I still believe are connected to each other) they intersect along an infinite number of asymptotic trajectories … .
>
> I had thought that all these asymptotic curves having moved away from a closed curve representing a periodic solution, would then asymptotically approach the same closed curve. What is true, is that there are an infinity which enjoy this property.
>
> I will not conceal from you the distress this discovery has caused me. In the first place, I do not know if you will still think that the results which remain, namely the existence of periodic solutions, the asymptotic solutions, the theory of characteristic exponents, the non-existence of single-valued integrals, and the divergence of Lindstedt's series, deserve the great reward you have given them.
>
> On the other hand, many changes have become necessary and I do not know if you can begin to print the memoir; I have telegraphed Phragmén.

[22] Poincaré to Mittag-Leffler, postmarked 1 December 1889, Institut Mittag-Leffler.

> In any case, I can do no more than to confess my confusion to a friend as loyal as you. I will write to you at length when I can see things more clearly.

The effects of his error had turned out to be much more serious than Poincaré had originally thought. He was now questioning whether the memoir was still worthy of the prize.

Poincaré's news was most unwelcome for Mittag-Leffler, but it could have been worse. The volume of *Acta Mathematica* containing the memoir had been printed, but it had not yet been published. Nevertheless, a limited number of the printed copies had been circulated, and this meant that Mittag-Leffler's carefully cultivated reputation was in jeopardy. For if the error became public knowledge then everyone would know that neither he nor the other members of the jury had spotted it.

In order to reduce the risk of scandal, Mittag-Leffler suggested to Poincaré that they should keep everything concerning the error between themselves, at least until the new memoir was published.[23] He also gave detailed instructions to Poincaré about what he should write in the introduction to the reworked memoir to ensure that nothing about the error would be included. On top of everything else, he asked Poincaré to pay for the printing of the now-suppressed original version, despite the fact that the bill came to just over 3500 Crowns, some 1000 Crowns more than the prize that Poincaré had won! Without hesitation Poincaré agreed to make the payment.

By the beginning of January 1890 Poincaré had reworked the memoir and sent a copy to Phragmén for editing. As well as making substantial alterations to take account of all the corrections arising from the error, he also took the opportunity to incorporate the explanatory Notes — the ones that he had sent Mittag-Leffler during the judging of the competition — into the paper itself. Thus the memoir was not only different in content from its predecessor, it was also more coherent. The printers set to work at the beginning of April 1890, but due to a backlog of other work, the volume of *Acta Mathematica* that contained the memoirs by Poincaré and Appell, as well as Hermite's report on the latter, was not actually published until the middle of November. When Poincaré's memoir was printed, it was long enough to be a book in its own right. It ran to 270 pages, over 100 pages more than his original entry to the competition.

In November 1890, more than a year later than Mittag-Leffler had originally planned, the winning entries were finally published in *Acta Mathematica*. More than six years had elapsed since Mittag-Leffler had written optimistically to Kovalevskaya to tell her about the plans for the competition. Once the *Acta Mathematica* volume was in circulation, rumours of the error faded and the brilliance of Poincaré's memoir was acknowledged. Mittag-Leffler had succeeded: the competition had indeed brought forth important new mathematics, although just how important not even Poincaré himself could have guessed. Together Mittag-Leffler and Poincaré had ensured that King Oscar's 60th birthday was a royal birthday to remember.

---

[23]For the same reason, Mittag-Leffler also lost interest in Weierstrass's report on the memoir, which was promised for a future volume. In the event, Weierstrass's general introduction, which dealt with the n-body problem and makes no reference to Poincaré's paper, was eventually published in *Acta Mathematica* only in 1911, well after Weierstrass's death in 1897; see (Mittag-Leffler 1911) and (Weierstrass 1923).

21.5 Poincaré's memoir

For Poincaré the timing of the Oscar competition could not have been better. For several years he had been building up a battery of techniques to tackle the question of the stability of the solar system and the competition was just the spur he needed. Meanwhile, he had been profoundly influenced by Hill's paper of 1877 on the lunar perigee — which Mittag-Leffler had republished in *Acta Mathematica* in 1886 — and in particular by Hill's emphasis on periodic solutions.

In his introduction to the published memoir, Poincaré mentioned that his original paper had been revised for publication. Mindful of Mittag-Leffler's request not to reveal details of the error, he was careful to give no hint of the nature and extent of his alterations. In the event, some of the principal results for which the paper is most famous today were not in the original version but were added later. We can also see that these additions were not simply extensions of previously existing results but were a direct consequence of the discovery of the error. Most remarkably, it was to take almost 80 years before its full implications were uncovered — for it was in correcting his error that Poincaré discovered the behaviour that is the forerunner of what today is called *mathematical chaos*.

Poincaré divided his memoir into two parts, theory and application. The first part contains both analytic and geometrical theory, and the second part deals with the application of the theory to the restricted three-body problem. As we shall see, the error turned out to be so fundamental that it had serious implications throughout the memoir, implications that affected both the analytic and the geometrical methods, as well as their application. It is a measure of the novelty and brilliance of Poincaré's geometrical methods that, although it was to take some time before they became fully understood, they are now widely adopted, and today's mathematicians routinely represent dynamical problems geometrically.

Initially, Poincaré thought that the best strategy for tackling the n-body problem was to start with the general three-body problem and then try to extend his results. But the difficulties inherent in the general three-body problem were so great that he decided to focus his attention almost exclusively on the restricted version of the problem.

As we described in Chapter 11, the restricted problem is to ascertain the motion of an effectively 'massless' body (a planetoid), subject only to the gravitational attraction of two much larger bodies moving in planar circular orbits around their centre of mass. The motion of the planetoid can be described by a set of differential equations involving four variables, its position (x, y) — because the physical motion takes place in a plane — and its velocity $(\dot{x}, \dot{y})$. However, these four variables are connected by what is called the 'Jacobian integral', C:

$$C = n^2\left(x^2 + y^2\right) + 2\left(\frac{\mu_1}{r_1} + \frac{\mu_2}{r_2}\right) - \left(\dot{x}^2 + \dot{y}^2\right),$$

where the constant C is determined by the initial position and velocity of the planetoid, $n = 2\pi/T$ is the mean motion (where T is the orbital period), $\mu_1 = Gm_1$ and $\mu_2 = Gm_2$ for the two masses m_1 and m_2 (where G is the gravitational constant), and r_1 and r_2 are the distances of the planetoid from the two primary bodies.[24]

[24]This equation is called an 'integral' because it is obtained by solving (integrating) a second-order differential equation. See Box 31.

Poincaré framed the restricted three-body problem in the geometrical language of phase space, which in this case is four-dimensional: two dimensions of position and two of velocity. In the equation above, if the four numbers $x, y, \dot{x}, \dot{y}$ that describe the motion of the planetoid at any moment of time are thought of as the coordinates of a point in a four-dimensional space, then the 'point' $(x, y, \dot{x}, \dot{y})$ moves in a three-dimensional subset of that space. Poincaré could therefore regard the equations as defining flows in a three-dimensional space — but this three-dimensional space is not our ordinary real space. Mittag-Leffler was especially concerned about this aspect of Poincaré's work, fearing that astronomers would not understand it, and indeed Weierstrass found it difficult.

Poincaré did not believe that the equations describing the three-body problem could be solved explicitly (see Box 66) so, taking his cue from Hill, he focused on the periodic solutions which he considered to be of the utmost importance, and 'the only breach through which we have been able to approach a fortress hitherto considered inaccessible'.[25] In this case, a periodic solution corresponds to a closed curve in three-dimensional phase space, and Poincaré's discussion of them forms a central part of his memoir. He began by looking for solutions that start off close to a periodic solution.

To simplify the problem of understanding this part of the flow, Poincaré considered a surface S that crosses the periodic solution and all the nearby flow lines (this surface is called a two-dimensional cross-section transverse to the flow). He then considered the map that the flow induces on the surface S. For each point M_0 on S, the image (or iterate) of M_0 under the map is the point M_1 at which that trajectory through M_0 first intersects S again. A good way to think of this map is to think of athletes running round a track, and to take as the cross-section the line that marks the end of a lap. The 'first return map' notes where the athletes cross this line as they run round the track.

If M_0 is the point where the periodic trajectory crosses S, then eventually an image M_n of M_0 is M_0 itself.[26] This produces a picture of a piece of the three-dimensional flow in the form of a map of the two-dimensional surface S to itself that is easier to understand (see Figure 21.11).[27]

The first return map can also be run backwards: M_{-1} is the point that is mapped by the first return map to M_0, and generally M_{-k-1} is the point that is mapped to M_{-k}, for every integer k.

Poincaré had another idea that drew on Hill's approach. Recall that in the three-body problem the masses of the two planets are represented by the parameters $1 - \mu$ and μ. When $\mu = 0$, the problem reduces to a pair of two-body problems, where the motion is known. Poincaré therefore devised the strategy of starting with a particular solution for which $\mu = 0$, finding a periodic solution of the planetoid, and then varying μ to see whether periodic solutions still exist for small values of μ.

He found that in systems for which periodic solutions exist when $\mu = 0$, periodic solutions also exist for small values of μ. These new solutions, which are very close to the original ones, depend on certain numbers, which he called *characteristic exponents*, that are constants for each fixed value of μ. Importantly, the stability of these new

---

[25] Poincaré, *Les Méthodes Nouvelles de la Mécanique Céleste*, Vol. 1, p. 82.

[26] The simplest case, but not the only one, is when $M_1 = M_0$, and we shall restrict our account to this case.

[27] This map is now recognised as being an extremely powerful tool in the exploration of dynamical systems. In recognition of Poincaré's role in its development, it is often also referred to as the 'Poincaré map'.

Box 66. Can the equations of motion for the restricted three-body problem be solved?

It was widely believed that these differential equations cannot be solved explicitly, and so other ways of understanding them must be found. But in 1907 Karl Sundman, an otherwise little-known Finnish associate professor of astronomy at the University of Helsinki in Finland, showed that convergent series solutions to the three-body problem exist that express the x- and y-coordinates of the planetoid as functions of time.[a]

However, his solution answered no important questions about the motion of the planetoid. It was not a practical one because the rate of convergence of the series which he had derived was far too slow — it has been estimated that were the series to be used for astronomical observations then the computations would involve at least $10^{8,000,000}$ terms![b] Moreover, the solution provides no qualitative information, and for these reasons astronomers showed little interest in it. Mathematicians did take note — Sundman was awarded a prize by the Académie des Sciences in Paris — but the interest was limited.

[a] See (Sundman 1912).
[b] See (Beloriszky 1930).

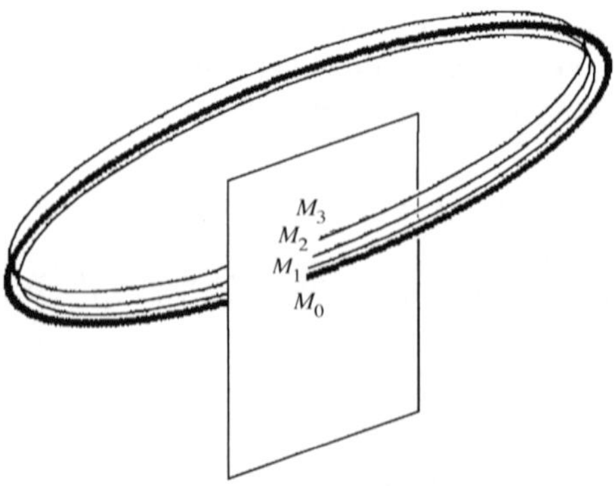

Figure 21.11. The first return map: M_0 is a periodic point, and M_1 returns successively as $M_2, M_3, \ldots$

periodic solutions is determined by the nature of these characteristic exponents. If they are imaginary, then the solution is stable; otherwise it is unstable. As will become clear shortly when we discuss stability, it is crucial to know whether a periodic solution is stable.

Apart from being easier to study in themselves, the periodic solutions also provide a natural starting point for studying and classifying other solutions, particularly for those solutions that are near by. By studying solutions differing only slightly from a given periodic solution, Poincaré was led to his remarkable discovery of an entirely new class of solutions, called *asymptotic solutions*. These are solutions that slowly approach or move away from an unstable periodic solution; they can get arbitrarily close to another periodic solution, but can never reach it.

Stability. There is a useful analogy with the behaviour of a simple pendulum in the form of a heavy bob which is suspended from a fixed point by a rod of negligible weight and swings freely in a vertical plane. This system has two states of equilibrium (or two equilibrium solutions). The obvious one is when the pendulum hangs at rest at its lowest point. The second is when the pendulum is balanced and at rest at its highest position — a theoretical, if not a practical, possibility. If we assume that we have a 'perfect' pendulum (one unaffected by other forces such as friction, air resistance, etc.), then once the pendulum has been disturbed from a position of equilibrium it will never again be at rest in that position. It may return to the same position, but not with the same velocity of zero.

If we now consider the slightest deviation from the equilibrium position, then there is a fundamental difference between the two states. In the first case the bob always remains close to equilibrium; in the second it descends to one side or the other and is carried far from its equilibrium position before returning close to it again. We describe the former as a *stable equilibrium*, and the latter as an *unstable* one. If an equilibrium solution is stable then nearby solutions remain nearby, but if it is unstable then at least one nearby solution leaves the neighbourhood of the equilibrium, even if it returns at some later stage.

When the solutions to a differential equation are represented as curves, an equilibrium solution appears as a single point — known as an 'equilibrium (or singular) point'. As we saw earlier in Poincaré's work on the qualitative theory of differential equations, these points can be classified according to the behaviour of nearby solutions. If all nearby solutions spiral in towards, or outwards away from, an equilibrium point, the point is called a 'spiral point'. If all nearby solutions tend towards, or outwards away from, an equilibrium point, these points are called 'nodal points'. Spirals and nodes are stable or unstable, depending on whether the trajectories tend towards the equilibrium point (stable) or away from it (unstable). If an equilibrium point is surrounded by closed trajectories, it is a centre. If two trajectories approach an equilibrium point, two leave it, and all others bypass it, then it is a saddle point. Centres are stable equilibrium points, and saddle points are unstable ones.

Armed with a notion of stability, we can now return to Poincaré's geometrical representation and his first return map. The simplest solutions are the periodic orbits, which are closed curves in the three-dimensional 'space' that he had introduced. Under the first return map they cut a plane section at a point; this section was later known as a 'Poincaré section'. If, moreover, the periodic solution is stable, then nearby solutions stay nearby, and cut the plane section in points close to the point P_0 where the periodic solution cuts the cross-section. But if the periodic solution is unstable, and if we take a point R_0, very close to a fixed point P_0 that corresponds to an unstable periodic solution, then the first return map for R_0 generates a sequence of image points R_1, $R_2, \ldots$, and Poincaré asked: Where do these image points appear on the cross-section?

He discovered that if the fixed point P_0 on the cross-section corresponds to an unstable periodic solution, then different things happen when the first return map is iterated.

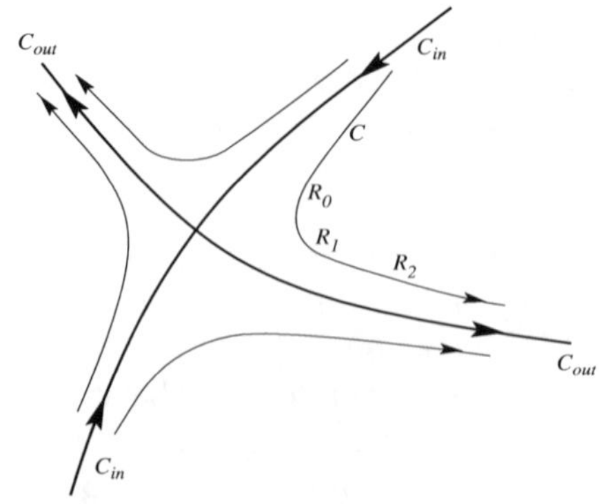

Figure 21.12. Two asymptotic curves on the Poincaré section, and the iterates of a typical point R_0

- If R_0 is a point on the two-part curve labelled C_{out}, then it moves away from P_0 to another point on C_{out} that is further from P_0.
- If R_0 is a point on the two-part curve labelled C_{in}, then it moves towards P_0 to another point on C_{in} that is nearer to P_0.
- If R_0 lies on neither C_{out} nor C_{in}, then its iterates lie on curves (such as C) that start off near C_{in} and finish up near C_{out}.

So the picture on the cross-section is of a two-part curve of points that approach the point P_0 and a two-part curve of points that move away from the point P_0; other points move away from the first curve and towards the second — the picture is of a saddle point (see Figure 21.12). Poincaré called the curves C_{out} and C_{in} *asymptotic curves*, and the flow lines that pass through them *asymptotic solutions* of the differential equations.

Poincaré now had to consider the full three-dimensional picture of the flow. His hope was that the asymptotic curves on the cross-section would generate asymptotic surfaces that divide the three-dimensional 'space' into regions (just as asymptotic curves divide flows in the plane into separate regions). If this were the case, then every flow line that was not on an asymptotic surface would be trapped in one of these regions.

It was in studying this problem that he made his major mistake. Recall that in his letter to Phragmén he had spoken of asymptotic surfaces that he had incorrectly believed to be closed, and that the correction established that these surfaces intersect along an infinite number of asymptotic trajectories. We cannot easily see the asymptotic surfaces, but we can see the asymptotic curves in which they meet the plane of section, and recognise Poincaré's error.

Poincaré failed to take proper account of the exact geometrical nature of the asymptotic curves. If the first return map has several fixed points, then there are several asymptotic curves. An inward-going asymptotic curve for one point P_0 cannot cross an inward-going asymptotic curve for a different fixed point Q_0, because the common point R would have to move steadily closer to two distinct points under iteration of the first return map, and this is impossible (see Figure 21.13).

Similarly, an outward-going asymptotic curve for one fixed point cannot cross an outward-going asymptotic curve for a different fixed point, because the common point

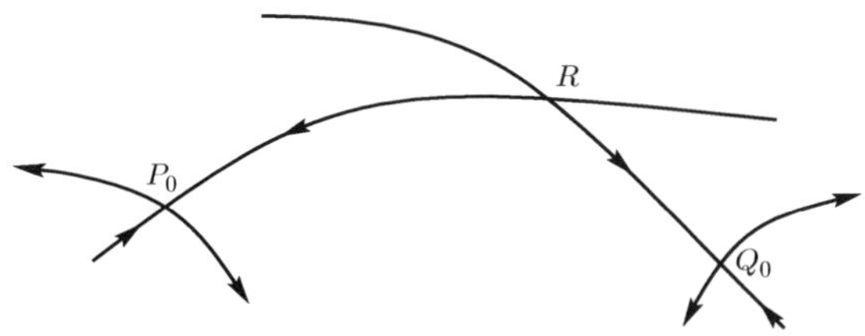

Figure 21.13. Two inward asymptotic curves on the Poincaré section, crossing at the point R

would have to move steadily closer to two distinct points under backward iteration of the first return map — simply reverse all the arrows in Figure 21.13.

But could an inward-going asymptotic curve for one point cross an outward-going asymptotic curve for another point? Poincaré believed that this too could not happen, although for more complicated reasons, and he offered a proof that showed that either an inward-going asymptotic curve for one point was an outward-going asymptotic curve for another point, or the two curves never met.

On being prompted to look at the matter again he found, to his evident distress, that inward and outward asymptotic curves can cross — but if they cross once then they cross infinitely often! To see this, suppose that an inward curve C_{in} for a point P_0 and an outward curve C_{out} for a point Q_0 cross at a point R, as in Figure 21.14.

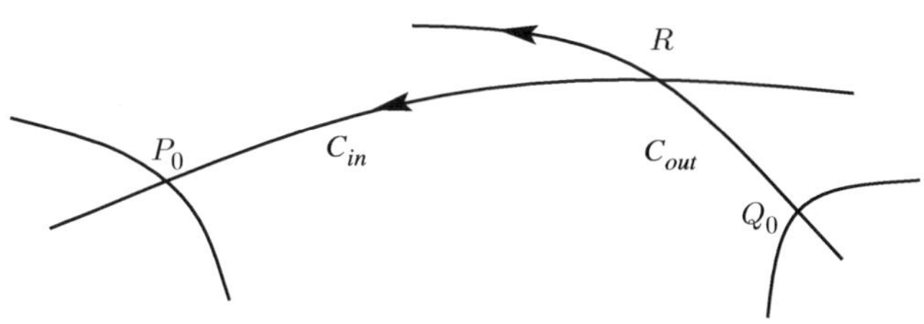

Figure 21.14. An inward asymptotic curve for P_0 and an outward asymptotic curve for Q_0 crossing at R

The image R_1 of R under the first return map must then lie on both curves C_{in} and C_{out}, and this argument can be extended indefinitely, establishing a sequence of points $R, R_1, R_2, \ldots$ common to both curves. But the point R has an antecedent R_{-1}, which is also common to both curves, and its antecedent R_{-2} likewise lies on both curves, and so on. Therefore there is a doubly infinite sequence of points common to C_{in} and C_{out}:

$$\ldots, R_{-2}, R_{-1}, R, R_1, R_2, \ldots.$$

Although he did not describe it as such, in discussing these curves Poincaré was providing the first description of mathematical chaos. Although he drew relatively little attention to the behaviour he had discovered and made no attempt to draw a diagram — a point we return to shortly — he was profoundly disturbed by what he had found, and almost a decade elapsed before he published anything further on the subject. He called the trajectories that pass through the points of intersection *doubly asymptotic trajectories*. Later he called them *homoclinic trajectories*, and the points of intersection are now known as *homoclinic points*.

Why did Poincaré make his mistake? Was it simply an oversight? As we have seen, he was renowned for paying scant attention to detail, and having a competition deadline would not have encouraged him otherwise. But a more convincing argument might be that he had a preconceived idea about how such a curve should behave. If he thought that he had found what he was expecting, then he might not have felt the necessity to scrutinise his results, especially if he felt pressed for time. The behaviour of the self-intersecting curve is extremely complex and quite unlike anything that Poincaré (or anybody else) had previously encountered. Indeed, when he discovered the mistake and its implications, it came as a complete shock to him.

He conveyed a sense of this to the readers of the newly founded *Revue générale des Sciences* in 1891 (this was one of a number of intellectual and popular scientific journals of the time to which Poincaré contributed). He described the types of orbits he had discovered, and wrote:[28]

Poincaré on the three-body problem.

> Firstly, there are the *periodic solutions.* These are solutions where the distances of the three bodies are periodic functions of time; at periodic intervals the three bodies therefore find themselves in the same relative positions. Periodic solutions are of several types. In those which I have called the first type, the inclinations are zero and three bodies move in the same plane; the eccentricities are very small and the orbits are almost circular; [and] the mean motions are not commensurable ... To this category belongs the first periodic solution to be discovered and which Hill, its discoverer, took as the starting point for his lunar theory.
>
> In the solutions of the second sort, the inclinations are again zero, but the eccentricities are finite; the motion of the perihelion is very slow; [and] the mean motions are almost commensurable ... In the solutions of the third sort, the inclinations are finite, the orbits are almost circular; [and] the motion of the perihelions is very slow ... I leave to one side the numerous categories of more complicated periodic solutions which it would take too long to enumerate.
>
> There are then the *asymptotic solutions.* In order to make it clear what is meant by these, allow me to give a simple example. First, let us imagine an Earth and a Sun isolated in space, thus moving according to Kepler's laws. Suppose, again for simplicity, that their motion is circular. Now give to the Earth two satellites L_1 and L_2 whose mass is infinitely small so they do not disturb the circular motions of the Earth and the Sun, and they do not disturb one another, each one moving as if it were alone. Choose the initial position of L_1 so that this satellite describes a periodic orbit, we can then choose the initial position of L_2 so that this satellite describes what we call an asymptotic orbit. At first quite far from L_1, it will indefinitely approach L_1 so that after an infinitely long time its orbit will differ infinitely little from that of L_1. Suppose there is an observer on the Earth and turning slowly so that

[28]See (Poincaré 1891).

> he is constantly facing the Sun. To him the Sun will appear immobile and the satellite L_1, whose motion is periodic will appear as a closed curve C. The satellite L_2 will appear to him as describing a sort of spiral with increasingly tighter turns that approach indefinitely closer to the curve C. There are an infinite number of similar asymptotic orbits. The set of these orbits form a continuous surface S which passes through the curve C and on which are drawn the spirals mentioned above.
>
> But there is another category of asymptotic solutions. It may happen, if the initial position of L_2 is chosen appropriately, that this satellite will move away from L_1 in such a way that at a very remote time in the past, its orbit differs very little from that of L_1. For our observer, this satellite will again describe a spiral whose turns bring it indefinitely closer to the curve C; but it will describe the spiral in the opposite direction, constantly moving away from C. The set of these new asymptotic orbits will form a second continuous surface S' passing also through the curve C.
>
> Finally, there is an infinity of doubly asymptotic solutions; this is a point I have had a great deal of difficulty in establishing rigorously. It may happen that the satellite L_2 is at first very close to the orbit of L_1, moves far away from it, and then approaches it again indefinitely closely. At a time in the remote past, this satellite was on the surface S' and while there described spirals moving away from C; it then moved far away from C but after a long time it will end up on the surface S and will again describe spirals approaching C.
>
> Let $L_2, L_3, \ldots, L_n$, be $n-1$ satellites describing doubly asymptotic orbits; at a time in the distant past, these $n-1$ satellites move following spirals on S'; by traversing this surface, we meet these $n-1$ orbits in a certain order. After a very long time, our satellites will be on S and will again describe spirals; but by traversing this surface S, we will meet the orbits of the $n-1$ satellites in a completely different order. This fact, as long as one takes the trouble to think about it, will seem a striking proof of the complexity of the three-body problem and the impossibility of solving it with the current methods of analysis.

Further evidence for the extreme novelty and complexity of Poincaré's asymptotic solutions can be gathered from the fact that, almost without exception, the contemporary commentators on Poincaré's memoir ignored their discovery. The situation was not helped by the fact that Poincaré was not the easiest of writers to follow, and that there were no diagrams.

It took the brilliant young German mathematician Hermann Minkowski seven pages to report on the memoir for the German reviewing journal, *Jahrbuch über die Fortschritte der Mathematik* (Yearbook on Advances in Mathematics), whereas most of the reviews carried by the *Jahrbuch* were no longer than a page, and many were considerably shorter.[29] Minkowski did not gloss over Poincaré's asymptotic solutions but freely acknowledged the difficulties that they raised, which probably indicates that he had a better grasp of Poincaré's discovery than most of his contemporaries who

[29] Minkowski later became well known for his work on number theory and the foundations of relativity.

abstained from passing comment. For example, in 1899 Edmund Taylor Whittaker, in his forty-page 'Report on the progress of the solution of the problem of three bodies' for the British Association, described the solutions as being 'approximately periodic when $t = -\infty$ and $t = +\infty$, but not periodic in the meantime'.[30] While this is certainly true, it hardly gives an adequate description of the behaviour of the solutions.

The stability of the solar system. When Poincaré wrote to Mittag-Leffler to apologise for his error, he remarked that the memoir still contained a number of good results. One of these concerned the stability of the solar system, and is another consequence of his discovery that the flow in phase space is volume-preserving. This is his *Recurrence Theorem*:[31]

> Suppose that the coordinates x_1, x_2, x_3 of a point P in space remain finite; then for any region r_0 in space, however small, there will be trajectories which traverse it infinitely often. That is to say, in some future time the system will return arbitrarily close to its initial situation and will do so infinitely often.

The Recurrence Theorem tells us that if we start with a region of phase space and a flow for which volume is preserved, then there are trajectories that traverse it infinitely often, however small the region is. Furthermore, there are infinitely many trajectories that pass through the region infinitely often. There may also be other trajectories that pass through it only a finite number of times, but these are the exceptions — their number is so small that they can be considered negligible. To put it more formally, the probability of a trajectory starting in a particular region and not returning to it infinitely often is zero. This means that at some point in the future (and it could be a very long time in the future) the system returns arbitrarily close to its initial state, and it does so infinitely often. The theorem may seem counter-intuitive, but Poincaré provided a proof that in essence goes as indicated in Box 67.

This theorem enabled Poincaré to establish that, under certain initial conditions, the restricted three-body problem has infinitely many solutions that possess what he called 'Poisson stability' — that is, they return infinitely often to positions arbitrarily close to their original position. The solutions that do not possess Poisson stability are the exceptions. As a consequence of this theorem, Poincaré deduced results about the motion of the Moon. Hill, in his paper of 1878 on lunar theory, had proved the existence of an upper bound for the radius vector of the Moon, so Poincaré concluded that the Moon has Poisson stability — that is, it returns infinitely often as close as desired to its initial position.

Poincaré also realised that his theorem does not apply to the general three-body problem, unless one assumes that the motion is bounded (which in general it is not) and that no collisions take place. Still less did it apply to the solar system. Although the results of his predecessors leant in favour of stability of the system, Poincaré himself did not discount the opposite possibility:[32]

> All persons who interest themselves in the progress of celestial mechanics, but can only follow it in a general way, must feel surprised at the number of times demonstrations of the stability of the solar system have been made ... The astonishment of these persons

[30] Whittaker later became Royal Astronomer for Ireland before taking up the chair of mathematics at Edinburgh.

[31] See (Barrow-Green 1997, 86).

[32] See (Poincaré 1898, 183).

Box 67. Poincaré's Recurrence Theorem.
Start with a region R filled with an incompressible continuously moving liquid with volume V. At an initial time t_0, consider a very small region r_0 of R with volume v. Because the liquid is moving, at a subsequent time t_1 the liquid that occupied the region r_0 occupies a new region r_1. Similarly, at time t_2, the liquid occupying r_1 has moved to r_2, at time t_3 it has moved to r_3, and so on, until at a time t_n it occupies a region r_n.

Because the liquid is incompressible, each region $r_0, r_1, r_2, \ldots, r_n$ occupies the same amount of volume, v. So, if n is greater than V/v, then at least two of the regions have a part in common, and if the regions are very small, some particles of the liquid will have returned very close to where they were at some previous time. Moreover, because the particles do not stop moving once they have returned close to where they were initially, the same argument can be applied infinitely often. This proves the theorem.

would doubtless be increased if they were told that perhaps some day a mathematician would show by rigorous reasoning that the planetary system is unstable. This may happen, however; there would be nothing contradictory in it, and the old demonstrations would still retain their value.

21.6 Poincaré's later work in celestial mechanics

Poincaré remained interested in celestial mechanics for many years, and his 1890 memoir formed the backbone to his acclaimed three-volume *Méthodes Nouvelles de la Mécanique Céleste* (New Methods in Celestial Mechanics), published between 1892 and 1899. Most of the ideas in the 1890 memoir can be found in the *Méthodes Nouvelles*, although in an expanded and elaborated form. The volumes included a greater number of applications of the theory, as well as a substantial amount of new material, with the emphasis as much on the general three-body problem as on the restricted problem.

In 1900, when he presented the Gold Medal of the Royal Astronomical Society to Poincaré, George Darwin, the Society's President and Plumian Professor of Astronomy at Cambridge, declared that, 'It is probable that for half a century to come it [*Méthodes Nouvelles*] will be the mine from which humbler investigators will excavate their materials'.[33]

The first volume of Poincaré's *Méthodes Nouvelles* includes an amplified treatment of periodic solutions, asymptotic solutions, and the non-existence of new uniform integrals for the problem. The second volume is devoted to the perturbation methods of other mathematical astronomers and their applications to the three-body problem. In the final volume, which is characterised by his geometrical ideas, Poincaré discussed invariant integrals and stability and, for the first time since the competition, returned

[33]See (Darwin 1900, 412). G.H. Darwin was the fifth child of Charles Darwin and Emma Wedgwood and so, to quote one of his brothers, George was 'born in the scientific purple', *The Times*, 9 December 1912, p. 9. He conducted extensive numerical work on the nature of periodic solutions in the three-body problem and corresponded with Poincaré about them.

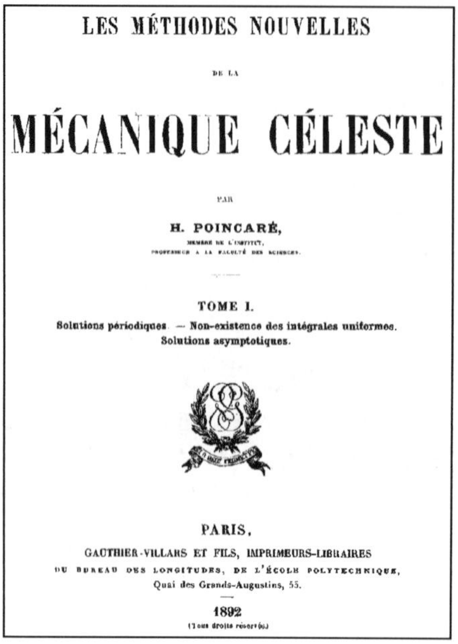

LES MÉTHODES NOUVELLES

DE LA

MÉCANIQUE CÉLESTE

PAR

H. POINCARÉ,

MEMBRE DE L'INSTITUT,
PROFESSEUR A LA FACULTÉ DES SCIENCES.

TOME I.

Solutions périodiques. — Non-existence des intégrales uniformes.
Solutions asymptotiques.

PARIS,

GAUTHIER-VILLARS ET FILS, IMPRIMEURS-LIBRAIRES
DU BUREAU DES LONGITUDES, DE L'ÉCOLE POLYTECHNIQUE,
Quai des Grands-Augustins, 55.

1892

(Tous droits réservés.)

Figure 21.15. Poincaré's *Les Méthodes Nouvelles de la Mécanique Céleste*, Vol. 1 (1892)

to the subject of doubly asymptotic solutions. Once again, he established the existence of homoclinic solutions in the restricted three-body problem, but this time he added an unequivocal statement about their bewildering complexity:[34]

> When one tries to depict the figure formed by these two curves and their infinity of intersections, each corresponding to a doubly asymptotic solution, these intersections form a kind of net, web, or infinitely tight mesh; neither of the two curves can ever intersect itself, but must fold back on itself in a very complex way in order to intersect all the links of the mesh infinitely often.
>
> One is struck by the complexity of this figure that I am not even attempting to draw. Nothing can give us a better idea of the complexity of the three-body problem ...

However, unlike the Oscar memoir, this time his discussion of doubly asymptotic solutions did not end with the discovery of homoclinic solutions. Remarkably, he had found an even more complex type of doubly asymptotic solution. These new solutions, which he called *heteroclinic solutions*, were associated with two unstable periodic solutions (rather than one), and thus were even more complicated than the homoclinic solutions. The fact that Poincaré published nothing on doubly asymptotic solutions in the decade between the publication of the Oscar memoir and that of the third volume of the *Méthodes Nouvelles*, despite publishing widely in related mathematical fields, is indicative of the difficulty that he (let alone anybody else) had with these solutions.

Poincaré also published papers of a general nature on the three-body problem and on the stability of the solar system.[35] These papers embraced a greater practical perspective than his 1890 memoir and were a well-judged response to the need for a more

[34] H. Poincaré, *Méthodes Nouvelles*, Vol. 3 (1899), p. 389.

[35] They are collected in various volumes of his *Collected Works*, notably Vol. 7.

popular presentation of his ideas. In one he provided a synopsis of the memoir, specifically designed to be accessible to astronomers and others whose interest in the three-body problem was motivated by practical considerations, while in another, his exposition of the results regarding the restricted three-body problem was almost completely descriptive. The concepts were illustrated with examples rather than theoretical mathematics, and not a single formula was used.

In his last paper on the three-body problem, Poincaré returned again to the question of periodic solutions, although the form of his attack was quite different from his original investigations. In this paper, which was published only a few weeks before he died, he announced a theorem which, if true, would confirm the existence of infinitely many periodic solutions for the restricted three-body problem for all values of the mass parameter. However, although he had been working on the theorem for two years, he had managed to find a proof only in certain cases. Uncharacteristically, he published it despite the fact that its proof was incomplete. One cannot but think that he had a portent of what was to come. He wrote:[36]

> It seems that in these circumstances, I should refrain from any publication for as long as I am unable to resolve the question; indeed after the useless efforts that I have made for so many months, it appeared to me that it would be wisest to leave the problem to mature, while resting for several years; that would be all very well if I was certain to be able to return to it one day; but at my age I cannot be sure. On the other hand, the subject is so important (and I will seek further to understand it) and all the results already obtained so considerable, that I am resigned to leaving them incomplete. I hope that the mathematicians who will interest themselves in this problem and who without doubt will be more successful than me, will be able to take advantage of them and use them to find the way in which they should proceed.

Shortly after Poincaré's death in July 1912, the young American mathematician, George D. Birkhoff was indeed successful, and in October of that year he supplied a brilliantly elegant proof, providing one of the mathematical sensations of the decade.[37]

21.7 Conclusion

Poincaré's geometrical approach to dynamics has been applied to many topics in mathematics in which a system evolves over time, ranging from Hadamard's investigations of geodesics on surfaces to modern weather forecasting. In the 1960s the National Aeronautics and Space Administration (NASA) had the *Méthodes Nouvelles* translated into English because they needed the mathematics for their satellite programme. Poincaré's books and papers can be seen as the start of a recognition that most problems involving differential equations do not have solutions that can be expressed in terms of well-known functions, and must be approached in more qualitative ways.

By the first decade of the 20th century, the long-term stability of the solar system seemed well established, and many astronomers moved their attention to the study of stars, nebulas, and galaxies. But with the advent of high-powered computing it became possible to revisit and extend the formerly time-consuming computations of Delaunay, Hansen, and Hill. Current estimates suggest that the planetary orbits, including those of the Earth and its Moon will remain much as they are for 10 to 20 million years, after

[36] See (Poincaré 1912, 376).

[37] See (Birkhoff 1913).

which they become too difficult to predict. This is partly because of certain resonances that will amplify small changes in the relative motions, and partly because small uncertainties in the calculations (as little as 1 metre in the position of Mercury) can cause radically different predictions for 200 million years hence.

21.8 Further reading

Barrow-Green, J.E. 1997. *Poincaré and the Three Body problem*, American and London Mathematical Societies, HMath 11. A thorough account of the prize competition and of Poincaré's work on differential equations and dynamics and its significance.

Barrow-Green, J.E. 2010. The dramatic episode of Sundman, *Historia Mathematica* 37, 164–203.

Diacu, F. and Holmes, P. 1996. *Celestial Encounters: The Origins of Chaos and Stability*, Princeton University Press. Starting with Poincaré, this book moves smoothly to the modern revival of his ideas.

Gray, J.J. 2013. *Henri Poincaré: A Scientific Biography*, Princeton University Press. An account of the life and work of one of the leading mathematicians, scientists, and philosophers of his day, that situates him in his milieu and discusses much of what he accomplished.

Verhulst, F. 2012. *Henri Poincaré: Impatient Genius*, Springer. This goes into more detail on differential equations and analysis than Gray's book, while making different choices about what to include and what to omit.

22

Coda

Introduction

We have chosen to conclude our history of mathematics around 1900, and to do so with a glimpse of three themes that were becoming increasingly important in the 20th century: the creation of a significant community of mathematicians in the United States; the emergence of national societies of mathematics in several countries; and the rise of the International Congress of Mathematicians. We shall see that Felix Klein and David Hilbert were involved in these in various ways, and this will deepen our sense of how German mathematicians were dominant figures in the subject as the new century began.

22.1 The international community of mathematicians

In 1893 the Americans celebrated the four-hundredth anniversary of Columbus's discovery of America a year later by mounting an exposition in Chicago, the fastest-growing city in the country at the time. Feeling that the anniversary marked a lofty cause, the organisers decided to play down its commercial aspects, and to stress the intellectual aspects of world culture. No fewer than 214 committees were formed to organise various parts of the World's Columbian Exposition, and one of these was charged with arranging a conference on mathematics.

The committee consisted of four mathematicians, three from the newly founded University of Chicago — Eliakim Hastings Moore, Oskar Bolza, and Heinrich Maschke (see Figures. 22.1, 22.2, and 22.3), and Henry S. White of Northwestern University in Evanston, Illinois. Moore was an American, born in Ohio in 1862 and educated at Yale University, where he took his Ph.D. degree in 1885 with a thesis on the geometry of n dimensions. At the suggestion of his supervisor, who had studied in Paris in 1855–1856, Moore then took himself to Europe, going not to France but to Germany — first to Göttingen and then to Berlin. On his return he taught for a while at Northwestern

Figure 22.1. Eliakim Hastings Moore (1862–1932)

Figure 22.2. Oskar Bolza (1857–1942)

Figure 22.3. Heinrich Maschke (1853–1908)

University before being lured to the embryonic University of Chicago in 1892 as its first professor of mathematics. There he was charged with creating a strong research department in a university that aimed to build up a tradition of pure research before attempting any applied research. This fitted well with Moore's own interests in algebra and the foundations of geometry, and he succeeded in building up a vigorous department.

Bolza and Maschke were his first hirings. They were Germans who had become friends whilst students at Berlin in 1875. Bolza, who turned 36 in 1893, had gone on to become a doctoral student of Felix Klein in Göttingen, where he was again joined by Maschke, who was four years older than him, but the stress of trying to keep up with the intense and demanding Klein in seminars, while simultaneously teaching in a high school, proved too much for Bolza. Fearing that a career in mathematics lay beyond him, he emigrated to the United States in 1888. Even there his doctorate, and a testimonial from Klein, did not at first bring him secure work. Eventually he was offered an appointment at the new University of Chicago. He agreed, on condition that his friend Maschke should be appointed as well. Maschke had also received his doctorate under Klein (in 1880), and then emigrated because he could not see a career for himself in Germany. Thus was the Chicago department created. The fourth man, White, who was born in 1861, had also been a student of Klein's in Germany.

The strength and influence of the German mathematical community is evident in this story. Moore had travelled to Germany to complete his education as a mathematician, and the rest of his faculty were Germans by birth, which suggests that at that time there may have been few qualified Americans to choose from. Even the problems that Maschke and Bolza encountered can be put down to the fact that German universities had found it easy to produce large numbers of mathematics postgraduates, but much harder to find jobs for them.

These four succeeded in putting a mathematical programme together for the Columbian Exposition and in getting an audience for it. Forty-five people attended, nearly all of them Americans. Many European mathematicians sent papers to be presented in their absence — in those days crossing the Atlantic, let alone a third of another continent, was not the easy journey that it later became. Of the thirty-nine papers presented, sixteen were by mathematicians in German universities, reflecting the strengths of German mathematics as well as the connections of the organisers.[1] Ten more were by various other Europeans (none by British mathematicians), and thirteen papers were by Americans. But perhaps even more important than this conference was its immediate consequence.

Alone of the foreign nations that took part in the Exposition, Prussia's Ministry of Culture sent an official representative — Felix Klein himself. His former students had been trying to lure him to Chicago permanently, and Klein was not uninterested, but that possibility had fallen through in 1892 when he had finally become the senior professor at Göttingen. However, he was keen to involve himself closely with any development that he could lead, and the opportunity to promote his own views in the name of the Ministry of Culture was too good to resist. It could do him no harm in the world of Prussian education, and besides, mathematics education was a topic in which he had a deep and abiding interest. So when Moore and the others asked him to speak at their conference, he not only accepted, but offered to give a further series of lectures free of charge.[2]

Klein's talk at the Congress stressed the essential unity of mathematics which, he said, had formerly been united in the great individual mathematicians, such as Lagrange, Laplace, and Gauss, but which was more lately threatened by the great growth of the subject. However, he felt that the developments of the previous two decades gave grounds for optimism, because the branches of the subject seemed now to be running in parallel directions. This new unity was made possible by the appearance of certain general conceptions — Klein singled out the concepts of a function and a group — and he cited the work of the American mathematical astronomer G.W. Hill on differential equations in astronomy, and that of the German mathematician Arthur Moritz Schoenflies, who was working in Göttingen on the use of group theory in crystallography. Klein concluded by praising the role of the newly formed national societies of

[1] See (Moore, Bolza, Maschke, White 1896).

[2] See (Klein 1893).

Box 68. National mathematical societies.
The formation of national societies of mathematicians was one way in which the emergence of a mathematics profession becomes clearer. These societies were professional bodies, with membership sometimes by election, set up to encourage and promote research in mathematics, offer expert advice, and represent the interests of mathematicians as a group when appropriate.

Such societies include the London Mathematical Society, founded in 1865, which mostly represented English mathematicians, and the Mathematical Society of France (Société Mathématique de France), founded in 1872. These societies, and others like them, usually established journals for the publication of research articles, and unlike previous learned scientific bodies were exclusively devoted to mathematics. The Germans followed suit in 1890 with the German Mathematical Society (Deutsche Mathematiker-Vereinigung, or DMV), whose annual reports carried both administrative items and extensive surveys commissioned several years in advance on specific branches of mathematics. The Americans established the American Mathematical Society (AMS) in 1894, when the New York Society and the Chicago group transcended (as they put it) local considerations, and started to publish its *Bulletin*.

Other national societies that emphasised the importance of education grew up in several countries. In the United States the Mathematical Association of America grew out of the AMS, and became an independent organisation in 1915. In the United Kingdom, the Association for the Improvement of Geometrical Teaching was formed in England in 1871, and was renamed as the Mathematical Association in 1894.

mathematicians, and advocating the formation of an international union, trusting, he said, that the present Congress at Chicago would be a step in that direction (see Box 68).

Klein then gave ten lectures (in English) at Northwestern University, Evanston, from 28 August to 9 September 1893. They were a great success and, when published the following year in book form,[3] they rapidly and deservedly came to enjoy a high reputation as a survey of the state of the art. In this book Klein lucidly described a number of topics in algebra and geometry, taking the opportunity to advance his own views on the nature and importance of various subjects, and stressing how seemingly different parts of the subject are interrelated. His lectures provide a fascinating glimpse into how a leading mathematician saw the state of mathematics at the time.

Klein began with the work of his former mentor, Clebsch, on geometry and algebra. He then turned to the work of another man he knew well, the Norwegian Sophus Lie, and discussed his contributions to geometry and his profound but difficult work on groups of transformations. Then he looked at questions about curves: How can significant properties of a curve be deduced from its defining equation? In other lectures he considered the theory of numbers, recent developments generalising the theory of elliptic functions, the solution of equations, non-Euclidean geometry, the way

[3]See (Klein 1894).

that mathematics was studied at Göttingen (for which, he said, some American graduate students seemed unprepared), and the relationship between intuition and proof in pure and applied mathematics. In almost every case, the topic of his lecture was one to which Klein had made an original contribution or had actively tried to encourage.

In his lecture on intuition he described a simple construction that, in theory, would produce a continuous curve that could never be drawn accurately because it was almost never smooth enough to have a tangent. Such curves severely stretched the understanding of contemporary mathematicians. Indeed, it has since been shown that the sketch that Klein provided of the curve was such a bad approximation as to fail to suggest any of its interesting properties![4] The moral that Klein drew from the existence of such curves was that intuition cannot discover everything in mathematics, and needs to be supplemented by the more rigorous and penetrating path of axiomatics — 'purely logical reasoning from exact definitions', as Klein put it.

The impression that one gets from Klein's lectures is a sense of the great vitality of the German mathematical scene, and of his position at the heart of it. To those such as Bolza and Maschke who knew him well, this must have seemed like a virtuoso performance by their former leader. Someone from Berlin or Paris would probably have discussed other topics, but Klein chose to emphasise those aspects of contemporary mathematics that he felt were most important: geometry, group theory, the role of intuition, and the essential unity of the subject.

Klein was a prolific writer, and it can be difficult to see beyond him to the views of other mathematicians. But while the gatherings at Chicago and Evanston were important for the growth of mathematics in America, they were also a spur to even grander events: truly international conferences at which many mathematicians from different countries could air their views.

The first International Congress of Mathematicians was held in Zürich in 1897 (see Figure 22.4); Klein headed the German delegation. The second Congress was held in Paris in 1900, and the topics that it addressed make interesting reading.

The Paris Congress opened with a lecture on the history of mathematics by Moritz Cantor (no relation to Georg) and one by Vito Volterra on the rise of the Italian school of mathematics. It then split into parallel sessions on different aspects of mathematics. The section on arithmetic and algebra was presided over by Hilbert and covered such topics as the theory of groups, the theory of equations, prime numbers, and the postulates of algebra. The section on analysis heard three lectures on various aspects of the theory of differential equations as well as other matters. Other sections discussed geometry and mechanics, and two general sections were devoted to bibliography, history, teaching, and methodology, including the foundations of arithmetic, the foundations of geometry, and the need for an artificial language (which provoked a vigorous, if inconclusive, debate). The Congress closed with a talk on the life of Weierstrass, who had dominated the Berlin school of mathematicians for forty years until his death in 1897, and one by Poincaré on the roles of intuition and logic in mathematics.

What can we infer from the Congress agenda about the image of mathematics that the organisers presented? Naturally enough, there was a certain amount of self-congratulation (or at least self-awareness), notably in the contributions on the recent history of mathematics, but that was to be expected at an international conference that

[4] Its definition is so straightforward, however, that it has long been available in computer curve-sketching packages.

Figure 22.4. A poster advertising the Zürich Congress, 1897

had been called after a century of unprecedented growth in the subject. The historical talks also mark an attempt to provide lectures that were intelligible to the non-specialist. That said, it was a conference aimed quite clearly at the new breed of specialist. Such people were offered three sections on pure mathematics and only one on applied, so it would seem that, even in Paris, the balance between pure and applied mathematics was shifting in the pure direction.

However much the French organisers may have busied themselves in advance of the Congress, the most important behind-the-scenes activity had taken place in Germany — once again in Göttingen. Shortly after becoming head of the Faculty, Felix Klein had succeeded in securing a professorship there for David Hilbert who, it was widely agreed, was the brightest German mathematician of his generation. Klein's acquisition of him was a coup that would help him considerably in his bid to make Göttingen the centre of the mathematical world.

David Hilbert was not the kind of mathematician who spread himself in several directions at once; his style was to work upon one branch of mathematics at a time. In this way he mastered and transformed algebraic invariant theory, then the theory of numbers, and then (as we saw in Section 15.5) the foundations of geometry. In 1900 Hilbert was between topics, and wondering what to speak about at the forthcoming Congress. It was the start of a new century: What could he say that would be appropriate to the event?

Hilbert wrote to Hermann Minkowski about his dilemma. Minkowski was a friend of Hilbert's from their student days together in Königsberg in the 1880s, a brilliant algebraist, geometer, and number theorist, who was later to cast Einstein's special theory of relativity into a simple geometrical form. In his letter, Hilbert remarked that he had read Poincaré's speech at the Zürich Congress, in which the eminent French mathematician had described what he saw as the mutually stimulating relationship of pure mathematics and mathematical physics, and he wondered whether he should not reply to it with a defence of mathematics for its own sake. But he was also attracted to individual problems in mathematics, which he thought kept the subject alive — perhaps he should present some of these, and offer some speculations about how the subject should develop. Minkowski replied that Poincaré's lecture had been unexceptional, and that it would be better to speculate on the future: 'With this, you might even have people talking about your speech decades from now. Of course, prophecy is a difficult thing'.[5]

Figure 22.5. David Hilbert (1862–1943)

Figure 22.6. Hermann Minkowski (1864–1909)

[5] Quoted in (Gray 2000, 57).

Despite his friend's encouragement, Hilbert remained undecided, and by June, when the programme of the Congress was announced, he had still not made up his mind. But in July he plumped for predicting the future, and so provided mathematics with the most talked-about and celebrated address in the entire history of the subject. For historians, his lecture is remarkable in presenting a rare glimpse not only of a leading mathematician's priorities before he or she can have any idea of how difficult they will be to carry out, but also of Hilbert's prescience — although, as with many who claim to predict the future, it is worth pointing out that opportunities existed for Hilbert to influence the subsequent course of events.

Hilbert sent Minkowski a draft of what he planned to say, and the two corresponded about it. Minkowski was excited by the idea, but worried that its execution would exceed the permitted hour. He thought that the problems themselves could be distributed separately to the delegates, and in due course Hilbert circulated an extract in French of the whole talk. The two friends met in Paris on 5 August, the day before the Congress was opened by Poincaré, and then they moved off with the 250 participants to the Sorbonne, the École Polytechnique, and the École Normale Supérieure, where the mathematical sessions were to take place. Hilbert was scheduled to speak in the Sorbonne on 8 August.

Hilbert's plan was not without its potential pitfalls. The problems should not be too general or too difficult, because it is not hard to think of such things. Nor should they be too easy, in case it be felt that you should simply have waited and worked until you had solved them. They must be reasonably spread out across the subject as a whole, challenging (but can you avoid giving away their solution, should you have an inkling of it?), and deep, rather than merely difficult, but not too philosophical, lest mathematicians be alienated.

The big day came, and, speaking clearly and slowly in German, Hilbert began:[6]

Hilbert on the future of mathematics.

> Who of us would not be glad to lift the veil behind which the future lies hidden; to cast a glance at the next advances of our science and at the secrets of its development during future centuries? What particular goals will there be toward which the leading mathematical spirits of coming generations will strive? What new methods and new facts in the wide and rich field of mathematical thought will the new centuries disclose?
>
> ... If we would obtain an idea of the probable development of mathematical knowledge in the immediate future, we must let the unsettled questions pass before our minds and look over the problems which the science of today sets and whose solution we expect from the future.
>
> ... We can ask whether here are general criteria which mark a good mathematical problem. An old French mathematician [Gergonne] said: 'A mathematical theory is not to be considered complete until you have made it so clear that you can explain it to the first man whom you meet on the street.' This clearness and ease of comprehension, here insisted on for a mathematical theory, I should still more demand

[6]Quoted in (Hilbert 1902, 437–479).

> MATHEMATICAL PROBLEMS
>
> *LECTURE DELIVERED BEFORE THE INTERNATIONAL CONGRESS OF MATHEMATICIANS AT PARIS IN 1900.*
>
> BY PROFESSOR DAVID HILBERT.
>
> WHO of us would not be glad to lift the veil behind which the future lies hidden; to cast a glance at the next advances of our science and at the secrets of its development during future centuries? What particular goals will there be toward which the leading mathematical spirits of coming generations will strive? What new methods and new facts in the wide and rich field of mathematical thought will the new centuries disclose?

Figure 22.7. The opening words of Hilbert's address[7]

> for a mathematical problem if it is to be perfect; for what is clear and easily comprehended attracts, the complicated repels us. Moreover a mathematical problem should be difficult in order to entice us, yet not completely inaccessible, lest it mock at our efforts.
>
> ... [The] conviction of the solvability of every mathematical problem is a powerful incentive to the worker. We hear within us the perpetual call: There is the problem. Seek its solution. You can find it by pure reason, for in mathematics there is no ignorabimus [we shall not know].

Hilbert argued that mathematics drew equally from problems in pure and applied mathematics. In pure mathematics there was the challenge posed by Fermat's Last Theorem; in applied mathematics there was the three-body problem, recently enriched, he said courteously, by the 'fruitful methods and far-reaching principles' introduced by Poincaré. These, he remarked, might seem like opposite poles, the one 'the free invention of pure reason', the other 'forced upon us by astronomy', but good problems find unlikely applications, and the interplay between the rigour of pure mathematics and the demands of the outer world, he suggested, had driven mathematics forwards.

Hilbert then presented ten of his list of twenty-three problems which he illustrated with brief remarks indicating their origins and significance.[8] The first was taken from the foundations of Cantor's theory of infinite sets,[9] which Hilbert hailed as a 'most suggestive and notable achievement'. Cantor had spent twenty years showing how the seemingly simple concept of an infinite set of objects led to a rich mathematical world seemingly capable of establishing a rigorous theory of the real numbers and of raising new and fundamental questions.

Hilbert's second problem asked for a set of axioms that would provide a rigorous basis for arithmetic. As we described in Section 15.5, there had been a widespread move towards formulating various branches of mathematics along axiomatic lines, and Hilbert followed his second problem with one asking for an axiomatisation of physics. There then followed two problems on the theory of numbers, one on the theory of functions, and one on the shape of curves. Finally, there was another on the theory

---

[7] *Bulletin of the American Mathematical Society* 8 (1902), p. 437.

[8] The full list of 23 problems was published in the Congress *Proceedings*.

[9] See Section 17.2.

of functions, one on the theory of differential equations, and one on the calculus of variations. But these labels may be deceptive: of these ten, only two could be said to be about applied mathematics. Hilbert ended with a statement of his belief in the unity of mathematics:[10]

> The organic unity of mathematics is inherent in the unity of this science, for mathematics is the foundation of all exact knowledge of natural phenomena. That it may completely fulfil this high mission, may the new century bring it gifted masters and many zealous and enthusiastic disciples!

The immediate response to the address on that hot and sultry morning in the Sorbonne was only a 'rather desultory discussion', to quote the reporter in the *Bulletin of the American Mathematical Society*. But it soon became clear that Hilbert had captured the imagination of the delegates, and in 1902 the same *Bulletin* carried an English translation of the entire address and all of the twenty-three problems.

Opinions differ even today, though, over exactly what some of the problems mean, and therefore over whether they have been solved. In 1974 the American Mathematical Society organised a conference specifically on Hilbert's problems, at which progress to date was reviewed and a new list of twenty-seven problems was drawn up. What one mathematician had felt able to do in 1900 now needed over twenty first-rate mathematicians (who, to be sure, also chose to report at much greater length on what had been done). This was impressive testimony to the influence that Hilbert's problems have continued to exert on 'zealous and enthusiastic disciples' ever since.

Among the problems in the full list that Hilbert did not present verbally were two more on geometry, one on Sophus Lie's theory of groups of transformations, four more on the theory of numbers (one being the Riemann hypothesis), and some that are harder to classify. We state three of these problems.

7. Are numbers of the form $2^{\sqrt{2}}$ or e^{π} transcendental, or at least irrational?

11. Attack successfully the theory of quadratic forms with any number of variables and any kind of numerical coefficients.

19. Has not every regular problem in the calculus of variations a solution, provided certain assumptions ... are met ... and provided also if need be that the notion of a solution shall be suitably extended?

It is not easy to say whether the problems mock us by their difficulty, especially when they ask us to 'attack successfully' a whole subject — evidently some of them are much more precisely formulated than others. But it is interesting to see how well represented was Hilbert's old love, the theory of numbers. He was also keen on giving axiomatic foundations for various branches of the subject, and while he was attracted to the theory of differential equations and the calculus of variations, he does not seem to have had any precise interest in physics.[11]

Overall, the problems do not sound like the sort of thing that one would expect to solve in six months — rather the opposite, if anything — but they do seem to be drawn from several of the concerns of mathematicians in the late 19th century. The range of problems was spread across the sphere of mathematics (more so, indeed, when you

[10] *Bulletin of the American Mathematical Society* 8 (1902), p. 479.

[11] Hilbert did, however, lecture extensively on physics at Göttingen in the years to come.

read the accompanying text in full), and even Hilbert cannot have had good ideas about how to solve them all. Their impact on mathematical life is hard to disentangle from that of Hilbert himself and the importance of Göttingen as a centre of mathematics, but it was considerable on any reckoning.[12]

Hilbert's problems provide as good a place as any to draw our story to a close. Hilbert's audience was recognisably modern in its avocation, and almost exclusively composed of people with professional training and qualifications in mathematics. Some looked towards the physical world for their inspiration, and probably more looked towards the problems of abstract mathematics, but almost all worked exclusively in universities where they set their own priorities for research. In that sense, the German neo-humanist model of the university, sustained as it was by the success of German universities in many fields, as well as its links with science and industry, had become the international standard.

We conclude our account of the history of mathematics on the brink of the 20th century. Much was to happen, which it will be the task of other historians to write.

22.2 Further reading

Gray, J.J. 2000. *The Hilbert Challenge*, Oxford University Press. This is a readable attempt to trace and explain the problems that Hilbert presented and to describe their influence on the mathematics of the 20th century.

Parshall, K.H. and Rice, A.C. (eds.) 2002. *Mathematics Unbound: The Evolution of an International Mathematical Community, 1800–1945*, American and London Mathematical Societies, HMath 23. Offering a range of topics and approaches, this is the first systematic attempt to survey its theme and with it the professionalisation of mathematics.

Parshall, K.H. and Rowe, D.E. 1991. *The Emergence of the American Mathematical Community, 1876–1900: J.J. Sylvester, Felix Klein, and E.H. Moore*, American and London Mathematical Societies. This rich, detailed, and very readable account traces the transformation of America from a mathematical backwater to a major presence during the last quarter of the 19th century.

Yandell, B.H. 2002. *The Honors Class: Hilbert's Problems and their Solvers*, A.K. Peters. This lucid account is particularly strong on the people involved, and on the problems to do with logic and the foundations of mathematics.

[12] For two accounts of their subsequent influence down the 20th century, see (Gray 2000) and (Yandell 2002).

23

Exercises

Advice on tackling the exercises

All our exercises take the form of essays, rather than exercises on the mathematics. We have not put a word length on these exercises, but you may wish to think in terms of 500–1000 words for each exercise in Part A, 1000–1500 words for each exercise in Part B, and 1500–2000 words for each exercise in Part C. Remember that keeping to the stated length is a skill that it is important to master.

When tackling a question from Part A, it is helpful to think in terms of the phrase 'content, context, and significance'. Being able to present the content of an extract, place it in a historical context, and then draw out its significance is a fundamental skill for a historian of mathematics. As a rough guide, you should allocate an equal amount of space to your discussion of each of the content, context, and significance passages in your answers, although sometimes you will find that there is rather more to say on each of the content and context than there is on the significance. There are also questions where the context and significance are quite distinct, and others where these categories merge into one another. A useful strategy is to go through the extract underlining all the words or phrases you intend to discuss (such as proper names, technical terms, etc.) and then, when you have finished writing your essay, go back and check that you have not omitted any of them.

The questions in Part B relate to a specific chapter (or chapters) in the book. Most of these questions consist of two equally weighted parts, and when answering the first part you should take care not to get so carried away that you forget to answer the second. You may, for example, be asked to describe a piece (or style) of mathematics and then to comment in a particular way upon it.

The Part C questions are on more wide-ranging themes and require you to use material from several chapters of the book. With these questions you should present a variety of evidence before reaching a conclusion in which you balance the weight and merits of the different arguments.

Writing an essay.

Essays are sometimes arguments in favour of a judgement. One way to think of how you should write an essay is to imagine that you are briefing your boss on the topic. It is your boss who will go into the meeting, who will present the arguments, and who will try to counter those on the other side. It is your job to tell your boss what he or she needs to win the arguments. Your boss will come out of the meeting and congratulate you or criticise you. You need to get it right, and remember, your boss cannot come out of the meeting halfway through to ask for clarification of what you wrote, or extra information. Everything has to be there in the essay, and it must not exceed the stated length (no-one wants to hear your boss droning on for far too long).

Or, to vary the metaphor, imagine that you are defending the teaching of the history of mathematics in a college. The syllabus can be filled many times over with topics, all of which have a legitimate claim on the students' attention. To argue for the subject is to give reasons why it matters (how will the students' education be enriched?), not to launch into an account of the life and work of Newton or Gauss, fascinating and important though they are. Of course, if the facts are relevant, as good accounts of the life and work of Newton or Gauss might well be, then they should be included, but the rule is: argument first, facts in support of the argument second.

So what does your boss want, and need, to know? First, what is the decision you are arguing for. Then, what are the good reasons for coming to that decision, what are the alternative decisions, and why are they not as good? Notice that it is a decision you are arguing for, a judgement that will help the company to perform better. We emphasise this point by including questions that explicitly call for a judgement. But it is not enough to state your opinion, however eloquently. You must have facts to back it up. With the facts clear in your mind, you can reach that judgement, discriminating between other judgements and refining your own in the process.

How to organise an essay. *Essays have a head, a long body, and a tail.*
There are many ways of writing an essay, but perhaps the main rules are these.

- Introduce the topic of your essay in a short paragraph at the start ('In this essay we argue that ...'; 'This essay describes ...').
- Fill the bulk of the essay with evidence in support of your initial claim.
- Each paragraph should make a single point.
- Organise your evidence clearly, and distinguish your position from others that have struck you as plausible but weaker than your own.
- In a short final paragraph, state your conclusion. This does not matter so much in a short essay, but in a lengthy article it is easy for the reader to lose the main point in a welter of detail.
- Diagrams/illustrations should be clearly captioned or referred to. If you discuss a diagram from a historical source and wish to refer to specific points on it, such as 'the point C on the line AB', a copy of the diagram should be included in your essay.
- If you transcribe historical mathematics into modern format/notation, make it clear that you are doing so and why.

Other rules are maybe less important, but they matter. Here are a few.

- Keep to the recommended length.
- Try to avoid writing in too personal a way, with an over-lavish use of the first person ('I') — you are supposed to be preparing a case that can be agreed by any reader.
- Use the past tense to refer to people who are now dead, but the present tense when writing about their surviving work, thus 'Newton showed that' but 'Newton's *Principia* establishes'. Of course, if the event happened in the past, then the past tense is required: 'Newton's *Principia* persuaded Halley'.

How to use sources. Always include citations for any information or opinion you use, so that readers can check them if they wish to. There are rules about using sources. If you want to quote anything you must say where the quotation come from. Citations that appear within the essay itself, are given in many different forms in the literature. We suggest that they should be in one of these forms:

- author, date of publication, page number(s), e.g. (Stedall 2004, 132–133)
- author, title of book or paper, page number(s), e.g. (Westfall, *Never at Rest: A Biography of Isaac Newton*, p. 279).

Full references, including publisher and date etc., should be given in the bibliography at the end of your essay. If the item is in this book, however, it is enough to give the author's name, followed by BGW and the page number. If you want to summarise somebody else's opinion, give the author's name, followed by the date (or title of book or paper) and page number.

These are the basic rules of evidence: any reader should be able to check that what you say is true. It follows that you must quote and summarise fairly, whether your source of information is one that you trust completely, or one that you have just pulled off a website. In this way you can also deal with errors and with conflicting opinions.

Examples of entries in a bibliography:

Article: Ferraro, G., 2007. Convergence and formal manipulation in the theory of series from 1730 to 1815, *Historia Mathematica* 34, 62–88.

Book: Meli, B., 1993. *Equivalence and Priority; Newton versus Leibniz*, Oxford University Press.

Book chapter: Pedersen, K. 1980. Techniques of the calculus, 1630–1660, in *From the Calculus to Set Theory, 1630–1910*, I. Grattan-Guinness (ed.), Duckworth, 10–48.

Website: Zach, R., Hilbert's Program, *The Stanford Encyclopedia of Philosophy*, `https://plato.stanford.edu/entries/hilbert-program/`.

Plagiarism. *Using other people's words as your own is theft*

You must not quote other people's work as if it was your own, and without giving proper reference to it. Presenting other people's words as your own is plagiarism. It is plagiarism if you send in a whole essay that somebody else wrote, and it is plagiarism if you just re-cycle a whole paragraph or sentence without reference. Plagiarism is theft. It is stealing somebody else's work, and even when it is not illegal it is an insult to the academic community because it denies the reader a chance to check your sources or to give due credit to the original author. The person who suffers most is yourself, as you are depriving yourself of the opportunity to test your own growing understanding of the subject through constructing and expressing your own arguments, which is integral to your learning experience.

Nor is there any reason to do it. If it turns out that you have found an essay that answers your question perfectly, just say so. Write 'This essay takes the view of [author] who in [give publication details] argued that ...' and then re-state that case in your own words. It is almost certain that you will then find that you have extra information to bring to bear, or different shifts of emphasis to make. The result will be your work, and that is what the reader (and ultimately you) want.

Sample exercises. The following is not a set of model answers — we do not believe there can be such things — but is an indication of how you might proceed when confronted with various types of question. They are sometimes longer than your answers should be, because we want to acquaint you with how you should proceed in general. They are intended to give hints about how to get started, and how to present your answer. They are meant to be helpful, not a straitjacket.

Question 1 is similar to a Part A type question, Question 2 is similar to a Part B type question, and Question 3 is similar to a Part C type question.

Question 1.

Proposition I. To find the differentials of simple quantities connected together with signs + and −. It is required to find the differentials of $a+x+y-z$. If you suppose x to increase by an infinitely small part, viz. till it becomes $x+dx$; then will y become $y+dy$; and z become $z+dz$: and the constant quantity a will still be the same a. So that the given quantity $a+x+y-z$ will become $a+x+dx+y+dy-z-dz$; and the differential of it (which will be had in taking from it this last expression) will be $dx+dy-dz$; and so of the others. From whence we have the following: *Rule I* For finding the differentials of simple quantities connected together with signs + and −. Find the differential of each term of the quantity proposed and which connected together by the same respective signs will give another quantity, which will be differential of that given.

Proposition II. To find the differentials of the product of several quantities multiplied, or drawn into each other.

The differential of xy is $ydx + xdy$: for y becomes $y + dy$, x becomes $x + dx$; and therefore xy then becomes $xy + ydx + xdy + dxdy$. Which is the product of $x + dx$ into $y + dy$, and the differential thereof will be $ydx + xdy + dxdy$, that is $ydx + xdy$: because $dxdy$ is a quantity infinitely small, in respect of the other terms ydx and xdy: For if, for example, you divide ydx and $dxdy$ by dx, we shall have the quotients y and dx, the latter of which is infinitely less than the former.

Whence it follows, that the differential of the product of two quantities, is equal to the product of the differential of the first of those quantities into the second plus the product of the differential of the second into the first.

(L'Hôpital, 1696)

Content

The extract, which is from l'Hôpital's 1696 textbook on the calculus, gives proofs of the sum rule and product rule for differentation.

Once you have stated the mathematical results in modern terms, you need to describe how l'Hôpital presented them. For example, are the two rules for differentiation stated and proved in complete generality or do examples come into play? What sort of quantities does l'Hôpital consider differentials to be? Notice how the two rules are presented: Proposition, proof, rule.

Context and significance

Whatever the extract, a good way to get started on its context and significance is to ask yourself what you know about the author and the circumstances in which he or she was working. In this case the author was a Frenchman, Guillaume de l'Hôpital (or the Marquis de l'Hôpital), who was a member of the philosopher Nicolas Malebranche's circle in Paris, and who, in 1691, commissioned the young Swiss mathematician Johann Bernoulli, who was visiting France, to give him lessons on the infinitesimal calculus. If you then ask yourself why l'Hôpital wanted lessons from Bernoulli, you will be led into considering issues relating to the wider context of the extract, such as Leibniz's invention of the calculus and its reception by the mathematical community in Paris.

L'Hôpital's text has a particular significance as the first published book on the calculus (remember to include the book in your bibliography). Moreover, although ostensibly authored by l'Hôpital, it consists largely of Bernoulli's results, many of which Bernoulli communicated to l'Hôpital by letter after he left Paris, a fact which l'Hôpital acknowledged, albeit rather briefly, in the preface of his book.

The source itself raises several questions. When was it written? (Always give the date of an extract.) What type of a source is it (book, letter, etc.)? What were the circumstances under which it was written? What can you say about the language/notation in which it was written? Who was the intended audience? Is it typical?

The information about Bernoulli's relationship with l'Hôpital prompts a consideration of Johann Bernoulli himself. While it is important not to get too side-tracked — several pages describing each member of the extended Bernoulli family of mathematicians is not required! — it is necessary to say something briefly about him and his status as a mathematician. We may also ask how he reacted to the publication of his work. Was it something he sanctioned, and when he was giving the lessons did he know that l'Hôpital intended to publish them? As it turns out, Bernoulli and l'Hôpital signed a contract giving l'Hôpital the right to do what he pleased with Bernoulli's results. One fall-out from this arrangement was that the well-known rule known as 'l'Hôpital's rule' was in fact due to Bernoulli, although concrete evidence about this did not come to light until 1922. It is worth noting that Bernoulli himself published a textbook on the integral calculus in 1742.

Returning to l'Hôpital's book as a whole, you may need to say something about its overall content. For example, are particular topics included or excluded? What can you say about the style of writing, the notation, or the way that the mathematics is presented? Are there other texts with which it can be compared? How was it received? What is its legacy? We are not going to answer these questions here, except to note that these questions are very general and would be adaptable to other sources.

If you now write up your answer, following the ideas suggested, you will see that you are in a good position to explain the importance of l'Hôpital's text in the early development of the calculus.

Question 2.
To what extent did the calculus develop between 1660 and 1760 in response to practical needs, and to what extent was it developed for other reasons?

Notes
Your answer should distinguish:

- practical problems that the calculus was explicitly developed to solve
- other kinds of problems that the calculus was explicitly developed to solve — for example, in geometry — but which were not of a practical nature
- problems, of whatever kind, that were raised with any hope of success only after the calculus was developed.

You may also wish your answer to reflect the growth and development of the calculus in the specified period.

You are asked to evaluate complementary claims about the development of the calculus. This will involve searching for evidence in support of each claim, weighing one against the other and attempting to balance your arguments in order to reach a conclusion. Although your final answer will be in the form of an essay, a good way to get started is to take the hint afforded by the *Notes* to the question and make a list of the different problems you want to discuss in your answer, identifying them as practical or otherwise. Notice the dates in the question: 1660 excludes, for example, the work of Kepler on the measurement of wine casks, whereas 1760 allows for the inclusion of some work of the Bernoullis and Euler.

Examples of a response to practical needs could include:

- navigation: cartography, and the shape of the Earth
- astronomy: the motion of the Moon; uses of differential equations in celestial mechanics (particularly by Clairaut)
- rational mechanics: mathematics was at the centre of 18th-century science, or at least the dominant part of that science — D'Alembert's support of this view
- motion: the early contributions of Euler (*Mechanica*).

Observations of a non-practical nature might include:

- a continuing Greek agenda — specifically exploring curves
- allowing analytical approaches
- the growth of algebra as a tool and a language.

These are not exhaustive lists, and there are also problems that do not fit snugly into either category. One such is Debeaune's problem which, although described by Debeaune in 1638 in terms of abstract analysis, had at its foundation a physical problem that required the language of the calculus to make apparent. What about Euler's *Institutiones Calculi Differentialis*? How does that fit in? On the one hand it was a textbook that gave mathematicians techniques that they could use to solve practical problems — his definition of a function in the introduction uses the example of the flight of a ball expelled from a cannon by the force of gunpowder! — while on the other it was an armoury of algebraic analysis.

Your job now is to weigh these up, with maybe other types of problems, bringing out their salient points so that you can make a coherent evidence-based argument. Finally, you have to come to a conclusion. And whatever your conclusion may be, it must be supported by the evidence!

Question 3.
How true is it that great mathematicians make bad writers? Give examples of mathematicians after 1700 whose work was easily understood and others whose work was not.

One way to begin would be to interpret some of the key words and phrases in the question, such as 'great mathematicians', 'bad writers'. When considering the latter, for example, you might want to discuss the audience — for whom is the mathematician writing? Also, the issue of writers (and writing) is one which naturally raises questions about terminology and notation.

Although you may have a clear idea about your overall answer to the question — 'very true', 'only partially true' — you need to be able to produce evidence on both sides. In this particular case, one strategy might be make a list of 'great mathematicians' (according to your interpretation) and divide it into two according as to whether or not the mathematician could be used to support the claim. So in support of it you might cite Leibniz (publications on the calculus), Galois (manuscripts on solving polynomial equations), and Lobachevskii (publications on non-Euclidean geometry). Opposing the claim you might include Euler (*Introductio in Analysin Infinitorum*) and D'Alembert (popular articles in the *Encyclopèdie*). You may also decide that certain 'great mathematicians' fall into both camps, such as Poincaré whose publications on technical mathematics were difficult for other mathematicians to understand, but whose publications on the philosophy of mathematics were accessible to a wider public.

Other mathematicians you might want to draw into your answer are the 'less than great' mathematicians who made 'good' writers, such as Nathaniel Bowditch, whose commentary on Laplace's *Mécanique Céleste* was widely admired. While you are not explicitly asked to give examples of such mathematicians, a short discussion of this category might help to inform your argument. As there are many texts from which you can quote, you will need to be disciplined in your answer in order to keep to the word limit. This is where judicious referencing to relevant passages can be very helpful. Remember that you need to end with a conclusion and that it must be supported by the evidence you have given.

Exercises: Part A

CHAPTER 2. The Invention of the Calculus.

1. This method [of tangents] never fails and could be extended to a number of beautiful problems; with its aid, we have found the centres of gravity of figures bounded by straight lines or curves, as well as those of solids, and a number of other results which we may treat elsewhere if we have time to do so.

(P. de Fermat, 1629)

2. Suppose two plane figures, or solids, are constructed to the same altitude; moreover having taken straight lines in the planes, or planes in the solids, parallel to each other in whatever way, with respect to which the aforesaid altitude is taken, if it is found that segments of the taken lines intercepted in the solids, are proportional quantities, always in the same way in each figure, then the said figures will be to each other as any of the former to the latter corresponding to it in the other figure.

(B. Cavalieri, *Geometria Indivisibilibus*, 1635)

3. Suppose there is given a series of quantities in duplicate ratio to arithmetic proportionals (or as a series of square numbers), continually increasing, beginning from a point or 0 (thus as $0, 1, 4, 9, 16$, etc.); it is proposed to enquire, what ratio does it have to a series of the same number of terms equal to the greatest? The investigation can be carried out by a means of induction, and we will have:

$$\frac{0+1=1}{1+1=2}=\frac{3}{6}=\frac{1}{3}+\frac{1}{6}, \qquad \frac{0+1+4=5}{4+4+4=12}=\frac{1}{3}+\frac{1}{12},$$

$$\frac{0+1+4+9=14}{9+9+9+9=36}=\frac{7}{18}=\frac{1}{3}+\frac{1}{18},$$

$$\frac{0+1+4+9+16=30}{16+16+16+16+16=80}=\frac{9}{24}=\frac{1}{3}+\frac{1}{24},$$

$$\frac{0+1+4+9+16+25=55}{25+25+25+25+25+25=150}=\frac{11}{30}=\frac{1}{3}+\frac{1}{30},$$

$$\frac{0+1+4+9+16+25+36=91}{36+36+36+36+36+36+36=252}=\frac{13}{36}=\frac{1}{3}+\frac{1}{36},$$

and so on. The ratio that arises is everywhere greater than one third, or $\frac{1}{3}$. Moreover, the excess decreases continually as the number of terms is increased; thus $\frac{1}{6}, \frac{1}{12}, \frac{1}{18}, \frac{1}{24}, \frac{1}{30}, \frac{1}{36}$, etc.; indeed the increased denominator of the fraction, of the consequent term of the ratio, is in each place a multiple of 6 (as is clear) which makes the excess above one third, of the ratio arising, one over six times the number of terms after 0.

(J. Wallis, *Arithmetica Infinitorum*, 1656)

4. Foreshadowings of the principles and even of the language of [the infinitesimal] calculus can be found in the writings of Napier, Kepler, Cavalieri, Pascal, Fermat, Wallis, and Barrow. It was Newton's good luck to come at a time when everything was ripe for the discovery, and his ability enabled him to construct almost at once a complete calculus.

(W.W. Rouse Ball, *History of Mathematics*, 1893)

CHAPTER 3. Newton and Leibniz.

1. I thinke it is almost a yeare since I accquainted the Reverend Doctor Barrow, that Mr James Gregory was by his owne Ingenuity falne into your methods of infinite Series, and that he had wrote to me to get his Booke de quadratura Circuli reprinted here with some new Additions, but the said Mr Gregory being since informed by me that you had taken much paines in that harvest, and invented the method some yeares before Mercators Logarithmotechnia was printed, hath laid aside his Intentions of publishing anything ...

(from a letter of J. Collins to I. Newton, 1671)

2. From all this it is to be seen how much the limits of analysis are enlarged by such infinite equations [infinite series]: in fact by their help analysis reaches, I might almost say to all problems.

   (from a letter of I. Newton to G.W. Leibniz, 1676)

3. It is the use of the Wallisian techniques of induction and interpolation that in the mid-1660s led a young Newton to the discovery of the binomial series, a tool that allowed him to deal with mechanical curves and solve quadrature problems that lay beyond the boundaries of the Cartesian canon.

   (N. Guicciardini, *Isaac Newton on Mathematical Certainty and Method*, 2009)

4. By the incomparable Leibniz was invented that famous calculus called differential to which all the questions that go beyond the common algebra are submitted and the curves that Descartes excluded from geometry are treated and expressed by their equations. Nevertheless, to that renowned scholar pleased to show to the mathematical community the beauty of his invention only through a fog curtain. Precisely in 1684 in *Acta Eruditorum* the monthly journal edited in Leipzig he wanted to display the first elements only in very few pages, without any explanation, but covered in enigma.

   (Johann Bernoulli, inaugural dissertation at Basel University, 1705)

5. Philosophically speaking, I no more believe in infinitely small quantities than in infinitely great ones, that is in infinitesimals rather than infinituples, I consider both as fictions of the mind for succinct ways of speaking, appropriate to the calculus, as also are the imaginary roots in algebra.

   (from a letter of G.W. Leibniz to B. Des Bosses, 1706)

6. One of the most precious documents of the Leibniz archive at Hannover is a set of mathematical manuscripts dated 25, 26 and 29 October, and 1 and 11 November, 1675. On these sheets Leibniz wrote down his thoughts more or less as they came to him, during a study of that most important problem of 17th-century mathematics: to find methods for the quadrature of curves. In the course of these studies he came to introduce the symbols '$\int$' and '*d*', to explore the operational rules which they obey in formulas, and to apply them in translating many geometrical arguments about the quadrature of curves into symbols and formulas.

   (H. Bos, 1980)

CHAPTER 4. The Development of the Calculus.

1. The chief Principle, upon which the Method of Fluxions is here built, is this very simple one, taken from Rational Mechanicks; which is, That Mathematical Quantity, particularly Extension, may be conceived as generated by continued local Motion; and that all Quantities whatever, at least by analogy and accommodation, may be conceived as generated after a like manner.

   (J. Colson, from the Preface to Isaac Newton, *The Method of Fluxions and Infinite Series*, 1736)

2. Those who have taken the measure of curvilinear figures have usually viewed them as made up of infinitely many infinitely small parts; I, in fact, shall consider them as generated by growing, arguing that they are greater, equal or less according as they grow more swiftly, equally swiftly or more slowly from their beginning. And this swiftness of growth I shall call the fluxion of a quantity. So when a line is described by the movement of a

point, the speed of the point — that is, the swiftness of the line's generation — will be its fluxion. I should have believed that this is the natural source for measuring quantities generated by a continuous flow according to a precise law, both on account of the clarity and brevity of the reasoning involved and because of the simplicity of the conclusions and the illustrations required.

(I. Newton, *Geometria Curvilinea*, c.1680)

3. To Mr Leibnitz's ingenious letter I have returned an answer which I doubt is too tedious. I could wish I had left out some things since to avoid greater tediousness I left out something else on which they have some dependance. But I had rather you should have it any way, then write it over again being at present otherwise incumbred. Sr I am in great hast

   Yours

   Is. Newton.

   I hope this will so far satisfy M. Leibnitz that it will not be necessary for me to write any more about this subject. For having other things in my head, it proves an unwelcome interruption to me to be at this time put upon considering these things.

(from a letter of I. Newton to H. Oldenburg, 24 October 1676)

4. Consider, that 'tis now about Thirty years since you were master of those notions about Fluxions and Infinite Series; but you have never published ought of it to this day, (which is worse than nonumque prematur in annum.) 'Tis true, I have endeavoured to do you right in that point. But if I had published the same or like notions, without naming you; & the world possessed of anothers Calculus differentialis, instead of your Fluxions: How should this, or the next Age, know of your share therein?

(from a letter of J. Wallis to I. Newton, 30 April 1695)

5. Proposition I, Problem 1
   Given any equation involving fluent quantities, to find the fluxions
   Let every term of the equation be multiplied by the index of the power of that fluent quantity it contains, and in each multiplication let a root of the power be changed into its fluxion, and the sum of all the terms under the proper sign will be a new equation.

(I. Newton, *Quadrature of Curves*, 1704)

CHAPTER 5. Newton's *Principia Mathematica*.

1. The Proof you sent me I like very well. I designed the whole to consist of three books, the second was finished last summer being short & only wants transcribing & drawing the cuts fairly. Some new Propositions I have since thought on which I can as well let alone. The third wants the Theory of Comets. In Autumn last I spent two months in calculations to no purpose for want of a good method, which made me afterwards return to the first Book & enlarge it with divers Propositions some relating to Comets others to other things found out last Winter. The third I now designe to suppress. Philosophy is such an impertinently litigious Lady that a man had as good be engaged in Law suits as have to do with her. I found it so formerly & now I no sooner come near her again but she gives me warning. The two first books without the third will not so well beare the title of Philosophiæ Naturalis Principia Mathematica & therefore I had altered it to this De motu corporum libri duo: but upon second thoughts I retain the former title. Twill help the sale of the book which I ought not to diminish now tis yours.

(from a letter of I. Newton to E. Halley, 20 June 1686)

2. *Book III, Proposition II, Theorem II*
   The forces by which the primary planets are continually drawn away from rectilinear motions and are maintained in their respective orbits are directed to the sun and are inversely as the squares of their distances from its center.
   (I. Newton, *Principia Mathematica*, 1687)

3. *Book III, Proposition XVIII, Theorem XVI*
   The axes of the planets are smaller than the diameters that are drawn perpendicular to those axes.

   If it were not for the daily circular motion of the planets, then, because the gravity of their parts is equal on all sides, they would have to assume a spherical figure. Because of that circular motion it comes about that those parts, by receding from the axis, endeavour to ascend in the region of the equator. And therefore if the matter is fluid, it will increase the diameters at the equator by ascending, and it will decrease the axis at the poles by descending. Thus the diameter of Jupiter is found by astronomical observations to be shorter between the poles than from east to west. By the same argument, if our earth were not a little higher around the equator than at the poles, the seas would subside at the poles and, by ascending in the region of the equator, would flood everywhere.

   (I. Newton, *Principia Mathematica*, 1687)

4. The Irregularity of the Moon's Motion hath been all along the just Complaint of Astronomers; and indeed I have always look'd upon it as a great Misfortune that a Planet so near to us as the Moon is, and which might be so wonderfully useful to us by her Motion, as well as her Light and Attraction (by which our Tides are chiefly occasioned) should have her Orbits so unaccountably various, that it is in a manner vain to depend on any Calculation of an Eclipse, a Transit, or an Appulse of her, tho never so accurately made. Whereas could her place be but truly calculated, the Longitudes of Places would be found everywhere at Land with great Facility, and might be nearly guess'd at Sea without the help of a Telescope, which cannot there be used.

   (I. Newton, *The Theory of the Moon's Motion*, 1702)

5. To the mathematician of the present century, however, versed almost wholly in algebra as they are, this synthetic style of writing [of the *Principia*] is less pleasing, whether because it may seem too prolix and took akin to the method of the ancients, or because it is less revealing of the manner of discovery. And certainly I could have written analytically what I had found out analytically with less effort than it took me to compose it. I was writing for Philosophers steeped in the elements of geometry, and putting down geometrically demonstrated bases for physical science. And the geometrical findings which did not regard astronomy and physics I either completely passed by or merely touched lightly upon.

   (I. Newton, *Mathematical Papers*, late 1710s)

CHAPTER 6. The Spread of the Calculus.

1. I must own myself very much obliged to the labours of Messieurs *Bernoulli*, but particularly those of the present Professor at *Groenengen*, as having made free with their Discoveries as well as those of Mr. *Leibnitz*: So that whatever they please to claim as their own, I frankly return to them. I must here in justice own, (as Mr. *Leibnitz* himself has done, in *Journal des Sçavans* for August, 1694) that the learned Sir *Isaac Newton* likewise discover'd something like the *Calculus Differentialis*, as appears by his excellent *Principia*, published first in the Year 1687 which almost wholly depends on the Use of the said *Calculus*. But the Method of Mr. *Leibnitz*'s is much more easy and expeditious, on account of the Notation he uses, not to mention the wonderful Assistance it affords on many Occasions.

   (E. Stone, *The Method of Fluxions, both Direct and Inverse*, 1730)

2. It is evident in general that Sir Isaac Newton's Figure of a flat Spheroid, and Mr Cassini's of a long one, will give very different Distances of Places that have the same Longitude and Latitude ... In a course of 100 Degrees Longitude, there might be a Mistake of more than two Degrees, if Sailing upon Sir Isaac Newton's Earth one should imagine himself to be upon Mr Cassini's. And how many Ships have perished by smaller Mistakes ...

(P. de Maupertuis, 1738)

3. This Determination would likewise be exceedingly useful in that important problem, *To Find the Parallax of the Moon*; which would greatly contribute to the completing of a Theory of this Satellite of our Earth; upon which the best Astronomers have always most reckoned for the discovery of Longitudes at Sea.

(P. de Maupertuis, 1738)

4. My present point is solely to bring to the notice of geometers who are interested in this question that having considered it anew from a viewpoint which no one had previously thought of I have been led to reconcile observations on the motion of the apogee of the moon with the theory of attraction without supposing any other attractive force than one proportional to the inverse-square of the distance.

(A.-C. Clairaut, May 1749)

5. A Frenchman arriving in London finds things very different, in natural sciences as in everything else. He has left the world full, he finds it empty. In Paris they see the universe as composed of vortices of subtle matter, in London they see nothing of the kind. For us it is the pressure of the moon that causes the tides of the sea; for the English it is the sea that gravitates towards the moon, so that when you think that the moon should give us a high tide, these gentlemen think you should have a low one. Unfortunately this cannot be verified, for to check this it would have been necessary to examine the moon and the tides at the first moment of creation.

   Furthermore, you will note that the sun, which in France doesn't come into the picture at all, here plays its fair share. For your Cartesians everything is moved by an impulsion you don't really understand, for Mr. Newton it is by gravitation, the cause of which is hardly better known. In Paris you see the earth shaped like a melon, in London it is flattened on two sides.

(F. Voltaire, 1734)

CHAPTER 7. The 18th Century.

1. I have heard with joy ... that you have been invited on behalf of the Prussian king to organize the new Academy in Berlin, and that you have already accepted this call, for the honour of which I sincerely congratulate you.

(from a letter of Johann Bernoulli to L. Euler, 1741)

2. We reproduce here a tabular survey (prepared by Adolf Pavlovich Yushkevich), ordered by decades, regarding the quantity of writings made ready for the press by Euler himself — without, to be sure, taking into account a few dozen works which could not yet be dated.

years	works	%
1725–1734	35	4
1735–1744	50	10
1745–1754	150	19
1755–1764	110	14
1765–1774	145	18
1775–1783	279	34

With regard to special topics, the respective shares in percentages look about like this:

algebra, number theory, analysis	40%
mechanics and the rest of physics	28%
geometry, including trigonometry	18%
astronomy	11%
naval science, architecture, ballistics	2%
philosophy, music theory, theology and what is not included above	1%

The listing does not include either c.3000 pieces of correspondence known so far, or the unedited manuscripts.

(E.A. Fellmann, *Leonhard Euler*, 1995)

3. It is a matter for considerable regret that Fermat, who cultivated the theory of numbers with so much success, did not leave us with the proofs of the theorems he discovered. In truth, Messrs Euler and Lagrange, who have not disdained this kind of research, have proved most of these theorems, and have even substituted extensive theories for the isolated propositions of Fermat. But there are several proofs which have resisted their efforts.

(A.-M. Legendre, 1785)

4. Finally, Lagrange has dealt with our theorem in the commentary *Sur la Forme des Racines Imaginaires des Equations*, 1772. This great geometer handed his work to the printers when he was worn out with completing Euler's first demonstration ... However, he does not touch upon the third objection at all, for his investigation is built upon the supposition that an equation of the *m*th degree does in fact have roots.

(C.F. Gauss, 1801)

CHAPTER 8. 18th-century Number Theory and Geometry.

1. Moreover recently in an unexpected manner, I have been able to deduce an expression for the entire sum of this series $1 + \frac{1}{4} + \frac{1}{9} + \frac{1}{16} + \frac{1}{25} +$ etc. which depends on the quadrature of the circle, thus, as if truly the sum of this series is obtained, then likewise the quadrature of the circle follows.

(L. Euler, 1734)

2. It was almost the same in every branch of mathematics until the invention of algebra: ingenious methods for reducing the problems to the most simple and easy calculations that the given question would admit. This universal mathematical key has opened the gate to many to whom it would always have been shut without its help. One can say that this discovery has produced a veritable revolution in all the sciences which depend on calculation.

   It therefore seems to be a whim and a sort of caprice, to despise so useful a method and to glory in only using the geometrical methods of the ancients. It has, I admit, the merit over algebra of being more apparent to the senses, and has a certain elegance which is infinitely pleasing, but it is nothing like so easy and so universal. So prefer it, if you wish, but do not exclude the other method. Mathematical truths are not so easy to find that it is worthwhile closing any of the routes that can lead to them.

   Above all it is in the theory of curves that one feels most forcefully the utility of a method as general as that of algebra. Descartes, whose inventive mind shines no less brilliantly in geometry than in philosophy, had not sooner introduced the way of expressing the nature of curves by algebraic equations than this theory changed its face. Discoveries multiplied with extraordinary ease: each line of the calculus gave birth to new theorems.

(G. Cramer, Preface to *Introduction to the Analysis of Algebraic Curves*, 1750)

3. Recently it occurred to me to determine the general properties of solids bounded by plane faces, because there is no doubt that general theorems should be found for them, just as for plane rectilinear figures, whose properties are

   (1) that in every plane figure the number of sides is equal to the number of angles and

   (2) that the sum of all the angles is equal to twice as many right angles as there are sides, less four.

   Whereas for plane figures only sides and angles need to be considered, for the case of solids more parts must be taken into account ...

   (from a letter of L. Euler to C. Goldbach, 1750)

4. It is impossible to find any two cubes, whose sum, or difference, is a cube.

   (L. Euler, *Algebra*, 1770)

5. In addition to that branch of geometry which is concerned with magnitudes, and which has always received the greatest attention, there is another branch, previously almost unknown, which Leibniz first mentioned, calling it the geometry of position. This branch is concerned only with the determination of position and its properties: it does not involve measurements, nor calculations made with them. It has not yet been satisfactorily determined what kind of problems are relevant to this geometry of position, or what methods should be used in solving them. Hence, when a problem was recently mentioned, which seemed geometrical but was so constructed that it did not require the measurement of distances, nor did calculation help at all, I had no doubt that it was concerned with the geometry of position — especially as its solution involved only position, and no calculation was of any use. I have therefore decided to give here the method which I have found for solving this kind of problem, as an example of the geometry of position.

   (L. Euler, The solution of a problem relating to the geometry of position, 1736)

CHAPTER 9. 18th-century Calculus.

1. A letter published in the year 1734, under the title of *The Analyst*, first gave occasion to the ensuing treatise, and several reasons concurred to induce me to write on this subject at so great length. The author of that piece had represented the method of fluxions as founded on false reasoning, and full of mysteries. His objections seemed to have been occasioned by the concise manner in which the elements of this method have been usually described, and their having been so much misunderstood by a person of his abilities appeared to me to be sufficient proof that a fuller account of the grounds of them was required.

   (C. MacLaurin, *A Treatise of Fluxions*, 1742)

2. The development of functions, generally considered, gives rise to derived functions of different orders; and the algorithm for these functions once being found, one can consider them in themselves and independently of the series from which they have been obtained. So, a given function being regarded as *primitive* one can deduce from it by simple and uniform rules other functions which I call derived; and, being given arbitrary equations in several variables, one can pass successively from these to derived equations and return from these to primitive equations. These transformations correspond to differentations and integrations; but in the theory of functions they only depend on purely algebraic operations based on the simple principles of the calculus.

   (J.-L. Lagrange, Discours sur l'objet de la théorie des fonctions analytiques, 1795)

3. We define as a *function* of one or several quantities any mathematical expression in which those quantities appear in any manner, linked or not with some other quantities that are regarded as having given and constant values, whereas the quantities of the function may take all possible values. Thus in a function we consider only the quantities which are supposed to be variables without regard to the constants it may contain.

   The term *function* was used by the first analysts to denote in general the powers of a given quantity. Since then the meaning of this term has been extended to any quantity formed in any manner from any other quantity. Leibniz and the Bernoullis were the first to use it in this general sense, which is nowadays the accepted one.

   (J.-L. Lagrange, *Théorie des Fonctions Analytiques*, 1797)

4. If a quantity was so small that it is smaller than any given one, then it certainly could not be anything but zero; for it were not $= 0$, then a quantity equal to it could be shown, which is against the hypothesis. To those who ask what the infinitely small quantity in mathematics is, we answer that it is actually zero. Hence there are not so many mysteries hidden in this concept as there are usually believed to be. The supposed mysteries have rendered the calculus of the infinitely small quite suspect to many people. Those doubts that remain we shall thoroughly remove in the following pages in which we shall explain this calculus.

   (L. Euler, *Institutiones Calculi Differentialis*, 1755)

5. In the same way that the differential calculus is called by the English the method of fluxions, so integral calculus is usually called by them the inverse method of fluxions, since indeed one reverts from fluxions to fluent quantities. For what we call variable quantities, the English more fitly call by the name of fluent quantities, and their infinitely small or vanishing increments they call fluxions, so that fluxions are the same to them as differentials to us. This variation in language is already established in use, so that a reconciliation is scarcely ever to be expected; indeed we imitate the English freely in forms of speech, but the notation we use seems to have been established a long time before their notation. And indeed since so many books are already published written either way, a reconciliation of this kind would be of no use.

   (L. Euler, *Institutiones Calculi Integralis*, 1768)

CHAPTER 10. Applied mathematics.

1. Mr. Bernoulli draws all his excellent reflections uniquely from the investigations made by the late Mr. Taylor on the motion of strings, and maintains against Mr. D'Alembert and me that the solution of Taylor is sufficient for the explanation of all motions to which a string can be subjected, in such a way that the curves which a string takes during its motion are always either a simple elongated trochoid, or a mixture of two or more curves of the same kind. Now although such a mixture can no longer be regarded as a trochoid, and the possibility of combining several of Mr. Taylor's curves already makes his solution insufficient in other respects as well, the motion of a curve could be such that it would be impossible to reduce it to the type of Taylor trochoids.

   (L. Euler, 1755)

2. But if the theory [of the vibrating string] leads us to a solution so general that it extends to all discontinuous as well as continuous figures, one must admit that this research opens to us a new road in analysis by enabling us to apply the calculus to curves which are not subject to any law of continuity, and if that has appeared impossible until now the discovery is so much more important.

   (L. Euler, 1766)

3. However sublime the researches on fluids that we owe to Messrs. Bernoullis, Clairaut, and D'Alembert may be, they derive so naturally from my two general formulas that one could not cease to admire this agreement of their profound meditations with the simplicity of the principles from which I have drawn my two equations and to which I have been immediately driven by the first axioms of Mechanics.

(L. Euler, 1757)

4. ... D'Alembert has tried to undermine [my solution to the vibrating strings problem] by various cavils, and that for the sole reason that he did not get it himself ... He thinks he can deceive the semi-learned by his eloquence ... He wished to publish in our journal not a proof, but a bare statement that my solution is defective ... From this you can judge what an uproar he would let loose if he were to become our president.

(from a letter of L. Euler to J.-L. Lagrange, 2 October 1759)

CHAPTER 11. 18th-century Celestial Mechanics.

1. To perfect the methods on which the lunar theory is based, to determine in this way those equations of this Planet which are still uncertain, and in particular to consider if one can explain, by this theory, the secular equation of the mean motion of the Moon.

(Prize competition announcement of the Paris Académie des Sciences, 1768)

2. I am now at the task of giving a complete theory of the variations of the elements of the planets due to their mutual action. What M. de la Place has done on this matter has pleased me a great deal, and I flatter myself that he will not be displeased with me for not keeping the sort of promise I made to abandon it entirely to him. I have not been able to resist the desire to occupy myself with it again, but I am not the less delighted that he should also work at it on his side. I am even very eager to read his further researches on this subject, but I beg him to send me nothing in manuscript, only the printed memoirs. I beg you to tell him this.

(from a letter of J.-L. Lagrange to J. D'Alembert, 1775)

3. The object of the author, in composing this work, as stated by him in his preface, was to reduce all the known phenomena of the system of the world to the law of gravity, by strict mathematical principles; and to complete the investigations of the motions of the planets, satellites, and comets, begun by Newton in his Principia. This he has accomplished, in a manner deserving the highest praise, for its symmetry and completeness; but from the abridged manner, in which the analytical calculations have been made, it has been found difficult to be understood by many persons, who have a strong and decided taste for mathematical studies, on account of the time and labour required, to insert the intermediate steps of the demonstrations, necessary to enable them easily to follow the author in his reasoning. To remedy, in some measure, this defect, has been the chief object of the translator in the notes. It is hoped that the facility, arising from having the work in our own language, with the aid of these explanatory notes, will render it more accessible to persons who have been unable to prepare themselves for this study by a previous course of reading, in those modern publications, which contain the many important discoveries in analysis, made since the time of Newton.

(N. Bowditch, *Mécanique Céleste by the Marquis De La Place, translated with a commentary*, 1829)

4. Simple as the law of gravitation is, its application to the motions of the bodies of the solar system is a problem of great difficulty, but so important and interesting, that the solution of it has engaged the attention and exercised the talents of the most distinguished

mathematicians; among whom La Place holds a distinguished place by the brilliancy of his discoveries, as well as from having been the first to trace the influence of this property of matter from the elliptical motions of the planets, to its most remote effects on their mutual perturbations. Such was the object contemplated by him in his splendid work on the Mechanism of the Heavens; a work which may be considered as a great problem of dynamics, wherein it is required to deduce all the phenomena of the solar system from the abstract laws of motion, and to confirm the truth of those laws, by comparing theory with observation.

(M. Somerville, *Mechanism of the Heavens*, 1831)

CHAPTER 13. The Profession of Mathematics.

1. Once our illustrious confrère [Cauchy] had been appointed to the École Polytechnique, he was not content to follow the programs and lectures that the eminent professors who had preceded him had devised. He restructured, so to speak, the material on algebraic and infinitesimal analysis along lines and by methods that he deemed appropriate. He was found to be — and why should it not be said here? For it is the only reproach that seems possible to make, and it is a reproach which is, as it were, a praise. He was found to be too learned, too brilliant, for the students who came to him in large numbers expecting to learn the practical material required for the public services ...

   (from a speech by Charles Dupin at the funeral of A.-L. Cauchy, 1857)

2. All the measurements in the world are not worth *one* theorem by which the science of eternal truths is genuinely advanced. However, you are not judge on the absolute, but rather on the relative value. *Such* a value is without doubt possessed by the measurements by which my triangle system is to be connected with that of Krayenhoff, and thereby with the French and English. And however low you estimate this work, in my eyes it is higher than those occupations which are interrupted by it. I am indeed far removed *here* from being master of my time. I must divide it between teaching (to which I have always had an antipathy, which is increased, though not caused, by the feeling of throwing my time away, an ever present concomitant of this activity) and practical astronomical work.

   (from a letter of C.F. Gauss to F.W. Bessel, 1820)

3. Monsieur Fourier was of the opinion that the principal aim of mathematics is to serve mankind and to explain natural phenomena; but a philosopher such as he ought to have known that the sole aim of science is the fulfilment of the human spirit, and that, accordingly, a question about numbers has as much significance as a question about the workings of the world.

   (from a letter of C.G.J. Jacobi to A.-M. Legendre, 1830)

4. Since our relations were interrupted, you will doubtless have remarked, Sir, the birth of two collections which imitate my own: the first is the Correspondence published in Brussels by Messers Quetelet and Garnier, and in which the latter has often copied me word for word without naming me. They have there among their collaborators a Mr. Dandelin who is of merit. The other collection is that which Mr. Crelle publishes in German in Berlin. I have just received the first three editions of which I have thus far read but the table of contents. Mr. Schmidten has reproduced therein, it seems to me, a work which he had already published in my annals; I also recognised a work by Mr. Louis Olivier on the general resolution of equations whose manuscript I have had in my hands for more than two years ago, and which I refused to publish at the time, because it leads to nothing and would not be able to go beyond the barrier one encounters beyond the 4th degree.

   (from a letter of J.-D. Gergonne to W.H.F. Talbot, 1826)

CHAPTER 14. Non-Euclidean Geometry.

1. Do not lose one hour on that. It brings no reward, and it will poison your whole life. Even through the ponderings of a hundred great geometers lasting for centuries it has been utterly impossible to prove the eleventh [fifth] without a new axiom. I believe that I have exhausted all imaginable ideas. ... I can prove in writing that [Gauss] racked his brains on parallels. He averred both orally and in writing that he had meditated fruitlessly about it.

   (from a letter of F. Bolyai to J. Bolyai, 1820)

2. The assumption that the angle sum is less than 180° leads to a geometry quite different from Euclid's, logically coherent, and one that I am entirely satisfied with. It depends on a constant, which is not given a priori. The larger the constant, the closer the geometry to Euclid's and when the constant is infinite they agree. The theorems are paradoxical but not self-contradictory or illogical ... All my efforts to find a contradiction have failed, the only thing that our understanding finds contradictory is that, if the geometry were to be true, there would be an absolute (if unknown to us) measure of length ... As a joke I've even wished Euclidean geometry was not true, for then we would have an absolute measure of length a priori.

   (from a letter of C.F. Gauss to F.A. Taurinus, 1824.)

3. Also, another theme which is almost 40 years old with me that I have been thinking about now and then in a few free hours; I mean the foundations of geometry. I don't know if I've told you my opinions about this. I have consolidated some things further here too, and my opinion that we cannot establish geometry completely a priori is, if possible, much firmer. Meanwhile I will not get around to it for some time and work up my very extensive researches for publication, and perhaps they will never appear in my lifetime, for I fear the howl of the Boeotians if I speak my opinion out loud.

   (from a letter of C.F. Gauss to F.W. Bessel, 1829)

4. Equations (1) [imply] that imaginary geometry passes over into the ordinary, when we suppose that the sides of a rectilineal angle are very small. The equations (1) attain for themselves already a sufficient foundation for considering the assumption of imaginary geometry as possible.

   (N.I. Lobachevskii, *Theory of Parallels*, 1840)

5. In recent times the mathematical public has begun to take an interest in some new concepts which seemed destined, if they prevail, to profoundly change the whole complexion of classical geometry. These concepts are not particularly recent. The master Gauss grasped them at the beginning of his scientific career, and although his writings do not contain an explicit exposition, his letters confirm that he had always cultivated them and attest his full support for the doctrine of Lobachevskii.

   (E. Beltrami, 1868)

CHAPTER 15. Projective Geometry.

1. The admission of the principle [of continuity] into geometry consists in supposing that, in the case where a figure composed of a system of lines or curves always has certain properties while the absolute or relative dimensions of its various parts vary in an arbitrary manner, between certain limits, the same properties necessarily persist when one goes beyond the dimensions which one had hitherto supposed were within these limits; and that, if some parts of the figure disappear on the second hypothesis those which remain

continue to enjoy, the ones with respect to the others, the properties that they had in the primitive figure. This principle, it should be said, is a bold induction, by means of which one can extend theorems, initially established with certain restrictions, to the case where these restrictions no longer hold. Applied to curves of the second degree, it leads the author to exact results. Nonetheless, we think it should not be admitted generally and applied indifferently to all sorts of questions in geometry, nor even in analysis.

(A.-L. Cauchy, 1820)

2. Let there be a plane figure composed in any way one wishes of points, lines, and curves. Let us conceive that having drawn arbitrarily in the plane of this figure, an arbitrary curve of the second order, one then constructs in the same plane another figure, all of whose points and all of whose lines are the poles and polars of all the lines and all the points of the first …

1. If there is a system of a certain number of points on a line, then in the other figure there will be a system of exactly as many lines meeting in a point.	1. If there is a system of a certain number of lines meeting in a point, then in the other figure there will be a system of exactly as many points lying on a line.

(J.-D. Gergonne, 1827–1828)

3. There was Brianchon, who in a memoir of 1810 presented new and extensive reflections on the subject to which, Poncelet tells us, he owes his own initial idea about the numerous beautiful geometric researches contained in his *Traité des Propriétés Projectives* [and] Gergonne who performed the useful service of writing his own works, always imprinted with profound philosophical insight, and of founding his *Annales de Mathématiques* for the productions of former pupils of the École Polytechnique.

(M. Chasles, 1837)

4. But simultaneously in Möbius, Steiner and Plücker there appeared three geometers of the greatest significance and innermost originality, who, advancing in different ways, are yet unified in their essential points of view, and whom one must thank in major part for the present form of our geometric understanding.

(from A. Clebsch's obituary of J. Plücker, 1895)

CHAPTER 16. The Rigorisation of Analysis.

1. The *most common* kind of proof depends on a truth borrowed from *geometry, namely: that every continuous line of simple curvature of which the ordinates are first positive and then negative* (or conversely), *must necessarily intersect the abscissae-line somewhere at a point lying between those ordinates.* There is certainly nothing to be said against the *correctness*, nor against the *obviousness* of this geometrical proposition. But it is also equally clear that it is an unacceptable breach of *good method* to try to derive truths of *pure* (or general) mathematics (i.e. arithmetic, algebra, analysis) from considerations which belong to a merely *applied* (or special) part of it, namely *geometry*.

(B. Bolzano, 1817)

2. As for methods, I have sought to give them all the rigor that one demands in geometry, in such a way as never to revert to reasoning drawn from the generality of algebra. Reasoning of this kind, although commonly admitted, particularly in the passage from convergent to divergent series and from real quantities to imaginary expressions, can, it seems to me, only occasionally be considered as inductions suitable for presenting the truth, since they accord so little with the precision so esteemed in the mathematical sciences.

(A.-L. Cauchy, 1821)

3. This work, undertaken at the request of the Committee of instruction of the École Royale Polytechnique, provides a summary of the Lectures on infinitesimal calculus which I gave at the École. It consists of two volumes corresponding to the two years of the course. Today I publish the first volume divided into forty Lectures, of which the first twenty concern the differential calculus, and the last twenty concern a part of the integral calculus. The methods which I have followed differ in several ways from those which can be found in works of the same genre. My main aim has been to combine the rigour, which I made a rule in my *Cours d'Analyse*, with the simplicity which results from the direct consideration of infinitely small quantities.

(A.-L. Cauchy, 1823)

4. On the whole divergent series is a devilry, and it is a shame that one dares to base any demonstration on them. One can deduce whatever one wants when one uses them, and they have done much harm and caused many paradoxes. Can you think of anything more terrible than saying that

$$0 = 1 - 2^n + 3^n - 4^n + etc.$$

where n is a positive integer. Risum teneatis amici [Friends, can you help but laugh]. My eyes have been opened in the most amazing way; indeed when you except the simplest cases, e.g. the geometric series, there hardly exists in all of mathematics a single infinite series the sum of which has been determined in a rigorous way. In other words the most important part of mathematics is without foundation.

(from a letter of N.H. Abel to his professor, C. Hansteen, 1826)

CHAPTER 17. The Foundations of Mathematics.

1. Of the greatest importance, however, is the fact that in the straight line L there are infinitely many points which correspond to no rational number.

(R. Dedekind, 1872)

2. Is it possible to map uniquely a surface (suppose a square including its boundaries) onto a line (suppose a straight line including its end points), so that to each point of the surface corresponds one point of the line and reciprocally to each point of the line corresponds one point of the surface? It seems to me at the moment that the resolution of this question — however much one is attracted to the answer 'no' that one may feel proof is almost superfluous — has very great difficulties.

(from a letter of G. Cantor to R. Dedekind, 5 January 1874)

3. For a time the influence of Kronecker's direction was very strong, and in my student years it was clearly disreputable to concern oneself with set theory. Through the mystical embellishment, which in the beginning Cantor gave to his theory, the antipathy to it was intensified. The first in the younger generation who seriously espoused Cantor's cause were Minkowski and myself...

(D. Hilbert, 1920)

4. Mathematics with Brouwer gains its highest intuitive clarity. He succeeds in developing the beginnings of analysis in a natural manner, all the time preserving the contact with intuition much more closely than had been done before. It cannot be denied, however, that in advancing to higher and more general theories the inapplicability of the simple laws of classical logic eventually results in an almost unbearable awkwardness. And the mathematician watches with pain the larger part of his towering edifice which he believed to be built of concrete blocks dissolve into mist before his eyes.

(H. Weyl, 1928)

CHAPTER 18. Algebra and Number Theory.

1. The doctrine of numbers, spite of [the works of previous mathematicians] has remained, so to speak, immobile, as if it were to stay for ever the touchstone of their powers and measure of their intellectual penetration. This is why a treatise as profound and novel as his *Arithmetical Investigations* heralds M. Gauss as one of the best mathematical minds in Europe.

   (L. Poinsot, 1807)

2. We shall probably have to wait [for a proof of the Prime Number Theorem] until someone is born into the world as far surpassing Tchebycheff [Chebyshev] in insight and penetration as Tchebycheff has proved himself superior in these qualities to the ordinary run of mankind.

   (J.J. Sylvester, 1881)

3. I have, moreover, in justice to declare that the substance of these new ideas do not belong to me. I found it in a letter from Mr. Legendre to my late brother, in which this great mathematician stated it (as something that was communicated to him, and as an object of pure curiosity)...

   I hope that the publicity that I give to the results which I have arrived at will cause the original author of these ideas to become known, and to update the work he has done on this subject.

   (J.F. Français, New principles of geometry of position, and geometrical interpretation of imaginary symbols, 1813)

4. Sir W. Hamilton, when I saw him but a few days before his death, urged me to prepare my work as soon as possible, his being almost ready for publication. He then expressed, more strongly than he had ever done before, his profound conviction of the importance of Quaternions to the progress of physical science; and his desire that a really elementary treatise on the subject should soon be published.

   (P.G. Tait, *Elementary Treatise on Quaternions*, 1876)

CHAPTER 19. Group Theory.

1. My dear friend, I have done several things new in analysis.
   Some concern the theory of equations; others, integral functions.
   In the theory of equations I have found out in which cases the equations are solvable by radicals, which has given me occasion to deepen the theory and to describe all the transformations admitted by an equation, even when it is not solvable by radicals.

   (from a letter of É. Galois to A. Chevalier, 29 May 1832)

2. It seems to us that it has not been demonstrated rigorously until now that these problems, so famous among the ancients, are not capable of a solution by the geometric constructions they valued particularly.

   (P.L. Wantzel, 1837)

3. There is nothing else on the subject [group theory] in English except a masterly little paper by Cayley in the Phil. Mag. and by this reprint [submitted], readers will have a better chance of understanding both Cayley there, & me in a short paper in the Brit. Assoc. Report for 1860. It has hitherto been quite a French question.

   (from a letter of T.P. Kirkman to W. Francis (a publisher), 1861)

4. These beautiful results [of Lagrange and Abel] were however only the prelude to a much greater discovery. It was reserved for Galois to put the theory of equations on its definitive footing and to show that to each equation there corresponds a group of substitutions, in which are reflected its essential characters, notably those that have to do with its solution by auxiliary equations [e.g., by radicals]. According to this principle, given any algebraic equation it suffices to know one of its characteristic properties to determine its group, whence, reciprocally one can deduce its other properties ...

   The point of this work is to develop Galois's methods and shape them into a body of theory, by showing how easily they permit the solution of all the principal problems in the theory of equations.

   (C. Jordan, *Traité des Substitutions et des Équations Algébriques*, 1870)

CHAPTER 20. Applied Mathematics.

1. In order to lay down the foundations for this theory [of heat], it would first of all be necessary to distinguish and to define with precision the elementary properties that determine the action of the heat. I have recognized then that all the phenomena dependent on this action separate into a very small number of general and simple facts; and in this way, any question in physics of this kind [concerned with heat phenomena] is reduced to a research in mathematical Analysis.

   (J. Fourier, *Théorie Analytique de la Chaleur*, 1822)

2. Fourier's theory of heat is one of the first examples of the application of analysis to physics. Starting from simply hypotheses, which are nothing but generalised facts, Fourier deduced from them a series of consequences which together make up a complete and coherent theory. The results which he obtained are certainly interesting in themselves, but what is still more interesting is the method which he used to arrive at them and which will always be a model for all those who wish to cultivate any branch of mathematical physics.

   (H. Poincaré, 1895)

3. In series of this type [Fourier series], the coefficients of the different terms are ordinarily definite integrals that include sines and cosines; and when the integrations can be made, by reason of the particular form attributed to the function that it is necessary to expand, it is easily seen that the series obtained are convergent. Nevertheless it is always desirable that this convergence be demonstrated in a general manner, independently of the functions.

   (A.-L. Cauchy, 1827)

4. First to observe the facts while varying the circumstances as much as possible, then to accompany this initial work with precise measurements in order to deduce from them general laws based solely upon experience and to deduce from these laws independent from any hypothesis on the nature of the forces that produce the phenomena, the mathematical value of these forces, that is, the formula that represents them, such is the procedure Newton followed. In general, it has been adopted in France by those savants to whom physics owes the immense progress it has made in recent times, and it is what has served me as a guide in all my research on electrodynamic phenomena.

   (A.-M. Ampère, 1827)

5. I was quite unacquainted with the contents of Green's memoir and was quite unable to procure it at Cambridge till Jan 1845 when through the kindness of my former tutor, Mr. Hopkins, I obtained two copies. I was of course very surprised to find that it contained the general theorem on Attraction besides other interesting matter and I have since endeavoured as much as possible to make the work known. Your kind cooperation will now insure its being generally known amongst scientific men ...

   (from a letter of W. Thomson to A.L. Crelle, 11 February 1846)

6. An application of the theory of the transmission of electricity along a submarine telegraph wire, which I omitted to mention in the haste of finishing my letter on Saturday, shows how the question raised by Faraday as to practicability of sending distinct signals along such a length as the 2000 or 3000 miles of wire that would be required for America, may be answered. The general investigation will show exactly how much the sharpness of the signals will be worn down, and will show what maximim strength of current through the apparatus in America, would be produced by a specified battery action on the end in England, with a wire of given dimensions etc.

(from a letter of W. Thomson to G.G. Stokes, 30 October 1854)

CHAPTER 21. Poincaré and Celestial Mechanics.

1. It is Hill's idea of the periodic orbit which, developed chiefly by Poincaré and G.H. Darwin, has given new life to the whole subject of celestial mechanics and has induced many mathematicians to investigate on these lines.

(E.W. Brown, 1916)

2. In regard to the question of the prize Weierstrass has promised me that he will write you his opinion on that in more detail as soon as he receives a letter from you. I did not inform him of what you wrote me in the letter before last with regard to the choice of jury, for I was sure in advance of his complete disapproval. Indeed I believe that in this way the thing presents many practical difficulties. Just consider how one could hope that four famous mathematicians, Weierstrass, Hermite, Cayley and Tschebychev would *ever* agree on the merits of a memoir. I believe it is certain that each of the four would refuse to become part of the jury as soon as he learned the names of the other three.

(from a letter of S. Kovalevskaya to G. Mittag-Leffler, 1884)

3. I consider three masses, the first very large, the second small but finite, the third infinitely small; I assume that the first two each describe a circle around their common centre of gravity and that third moves in the plane of these circles. An example would be the case of a small planet perturbed by Jupiter, if the eccentricity of Jupiter and the inclination of the orbits are disregarded.

(H. Poincaré, 'On the three-body problem and the equations of dynamics', 1890)

4. A new impetus was given to Dynamical Astronomy in 1890 by the publication of a memoir by Poincaré.

(E.T. Whittaker, 1899)

Exercises: Part B

Chapter-based essay questions.

1. Indicate some of the important developments in the problem of finding tangents to curves between 1600 and 1650. How did the work of Newton and Leibniz surpass what had gone before?

2. Indicate some of the important developments of the calculus of areas. How did the work of Newton and Leibniz surpass what had gone before?

3. Compare and contrast the contributions of Newton and Leibniz to the development of the calculus.

4. How was Leibniz's calculus disseminated in Europe between 1684 and the 1730s? To what extent did the Leibnizian calculus simplify mathematics?

5. Describe the circumstances surrounding the writing and publication of Newton's *Principia*. Why was this work so significant?

6. Describe the diversity of problems which Newton's *Principia* deals with. What concerns did Newton appear to have had in writing it?

7. Describe the reception of Newton's *Principia* between 1687 and 1750. What particular problems did it help to solve?

8. Why was the shape of the Earth of particular interest and concern in the early 18th century? What attempts were made to determine it?

9. Why were 18th-century mathematicians interested in the motion of the Moon? What hypotheses were investigated, and how was the matter finally resolved?

10. The question of where two curves intersect was surprisingly important in the 17th and 18th centuries. Describe some of the work on this problem. What light is cast upon the mathematical styles of the period by the kinds of solutions reached?

11. Give an account of the main discoveries in geometry in the 18th century. To what extent was there a decline in interest in the subject as the century wore on?

12. Why was the study of logarithms of interest in the 18th century? What related problems did Euler help to resolve?

13. Describe some aspects of the work of Euler. Why is he commonly regarded as the most important mathematician of the 18th century?

14. What is the significance of the problem of the vibrating string for mathematicians of the 18th century? How was the problem resolved?

15. Describe two important problems in celestial mechanics in the 18th century. To what extent were these problems solved during this period?

16. How was mathematics practised in France in the period 1790–1850? Indicate some of the mathematical advances made there during this period. To what extent can these advances be attributed to the institutional structure of early 19th-century French mathematics?

17. How was mathematics practised in Germany during the course of the 19th century? Indicate some of the mathematical advances made there during this period. To what extent can these advances be attributed to institutional support for mathematics in 19th-century Germany?

18. Describe the establishment of the École Polytechnique. What was the balance it struck between pure and applied mathematics, and how important was this in shaping the mathematics of the 19th century?

19. Given an account of the teaching of mathematics at universities in the 19th century. What differences were there between the French and German approaches, and what effect did these influences have?

20. What contribution to the development of mathematics in the 19th century was made by mathematical journals? How important was the role of the editor in such journals?

21. What were the key stages in the discovery of non-Euclidean geometry? How did it become accepted by mathematicians?

22. How did the work of Bolyai and Lobachevskii differ from earlier attempts on the parallel postulate? To what extent was their work on non-Euclidean geometry accepted by the mathematical community?

23. Describe the development of projective geometry in the first half of the 19th century. To what extent were there differences between French and German approaches to this subject, and for what reasons?

24. Describe some of the mathematical contributions of Felix Klein. How significant was Klein in the mathematical life of his day?

25. Compare the development of geometry in France and in Germany during the 19th century.

26. In what ways were mathematicians of the 19th century more interested in mathematical rigour than were their 18th-century predecessors? What factors may account for this interest?

27. What was unsatisfactory about the mid 19th-century definition of the real numbers? Why did it matter, and how was the problem resolved?

28. Why can Gottlob Frege be described as a logicist? How was Frege's work received by Bertrand Russell?

29. Describe Georg Cantor's contributions to the development of mathematics. Give an indication of how his work was received, and why.

30. Describe some of the disputes that arose in the late 19th and early 20th centuries around questions of the infinite and the use of sets in foundational questions. What solutions were offered to these disputes?

31. Describe the development of number theory in the 19th century. To what extent did it differ from earlier work in the subject?

32. What is significant about the theorem of quadratic reciprocity? How did French and German mathematicians vary in their approaches to number theory in the mid-19th century?

33. Give an account of the development of complex numbers. What was the philosophical barrier to their acceptance?

34. Describe the 19th-century development of vector methods. How important was Hamilton's discovery of quaternions in this development?

35. Why were mathematicians interested in solving polynomial equations of degree 5? What was Abel's contribution to the subject?

36. Describe the development of group theory in the 19th century. To what extent was it concerned with the solution of polynomial equations?

37. Give a brief account of Galois's work on the theory of equations. Why did it take several years to be recognised?

38. Describe the three classical problems in Greek geometry. What was Wantzel's contribution, and why was it ignored?

39. Give an account of two discoveries in applied mathematics in the 19th century. What physical problems did they help to solve?

40. Describe the mathematical contributions of Joseph Fourier. How was his work received?

41. Describe the significant features of Poincaré's work on the three-body problem. What were the circumstances of its publication in 1890?

42. What is significant about the International Congress of Mathematicians of 1900? To what extent could mathematics be described as an international discipline in 1900?

Exercises: Part C

General essay questions.

1. For what reasons has mathematics been studied? Give examples from three different centuries.

2. Describe three examples, drawn from different centuries, of major mathematical achievements. What do you consider their mathematical significance to have been?

3. Discuss the extent to which different mathematicians have struck a balance between geometry and algebra. Illustrate your answer with examples from different centuries.

4. How important has rigorous proof been to mathematicians? Illustrate your answer with at least three examples where mathematicians have made progress with, or without, proofs of varying levels of rigour.

5. 'What matters in mathematics is the ability to solve problems; proofs can wait.' How justified do you consider such an opinion to be? Illustrate your answer with examples from different centuries, of which at least one was where problem-solving was more important than giving proofs and one was where finding proofs was a stronger consideration.

6. Outline some of the ways in which mathematicians have communicated with each other. How important for the development of mathematics has been the growth of formal organisations such as colleges and learned societies?

7. How important has the role of applications been in the development of mathematics? Illustrate your answer with at least three examples showing differing degrees of importance.

8. How important has good notation been in the development of mathematics? Give examples where good notation has been important and where it has not.

9. Some mathematicians have worked in almost complete isolation, others in large mathematical communities. By discussing examples of particular mathematicians, consider in what ways being part of a wider community helps or hinders a mathematician to lead a creative life.

10. Non-mathematicians often wonder how there can be research in mathematics. Give examples, drawn from different centuries, to demonstrate how important new questions in mathematics get posed and answered, making clear what the importance of the questions was taken to be.

11. Describe a significantly useful discovery in the mathematical sciences made before 1900. Indicate some of the uses to which it has been put, and explain why it has been so useful.

12. Choose three mathematicians from different centuries, and explain what role your chosen individuals played in relation to their wider society. How far do any changes you can identify between the three centuries indicate any change in the status and role of mathematicians in general?

13. Histories of mathematics often concentrate on the achievements of the great mathematicians. What have been the contributions to mathematics of the less than great, and how important were these? Give examples from three different centuries.

14. Is there a difference between the history of mathematics and the history of great mathematicians? Illustrate your answer with examples from different centuries.

15. How important are biographies of mathematicians in the history of mathematics? Discuss examples drawn from different centuries that illustrate ways in which knowledge of the biography of a mathematician does, or does not, illuminate the history of mathematics.

16. It is sometimes said that the essential activity of mathematicians is proving theorems. Give examples, chosen from different centuries, to show that mathematics can involve more than proof.

17. Choose three locations in the past, and explain what it would have been like to practise mathematics there. Which would you have found most congenial, from the perspective of a practising mathematician, and why?

18. Describe three mathematical achievements that have a *cultural* significance reaching beyond the strict confines of the subject, and explain why they have been so important.

19. How true do you take the claim to be that 'Great mathematicians are born, not made'? Illustrate your answer with information about at least three well-chosen mathematicians.

20. To what extent is the history of mathematics the history of successes, from which failures have been omitted?

Bibliography

Abel, N.H. 1826. Recherches sur la série $1 + \frac{m}{1}x + \frac{m(m-1)}{1.2}x^2 + \cdots$, *Journal für Mathematik* 1, 311–339, in *Oeuvres Complètes* 1, 219–250.

Aiton, E.J. 1985. *Leibniz; A Biography*, Adam Hilger, Ltd., Bristol.

Ampère, A.-M. 1826. *Mémoire sur la Théorie Mathématique des Phénomènes Électrodynamiques, Uniquement Déduite de l'Expérience*, Paris.

Andersen, K. 1999. Wessel's work on complex numbers and its place in history, *Matematisk-fysiske Meddelelser* 46.1, 65–81.

Arago, D.F.J. 1855. *The History of my Youth*, Longman.

Archibald, R.C. 1936. Unpublished Letters of J.J. Sylvester, *Osiris* 1, 85–154.

Archimedes. *Archimedis Opera quae quidem exstant Omnia...*, T.G. Venatorius (ed.), Basel, 1544; *Archimedis opere nonnulla...*, Commandino (ed.), Venice, 1558; *The Works of Archimedes*, Sir T.L. Heath (ed.), Cambridge University Press.

Argand, J.R. 1813. Réflexions sur la nouvelle théorie des imaginaires, etc. *Annales de Mathématiques* 5, 197–209.

Arianrhod, R. 2012. *Seduced by Logic: Émilie du Châtelet, Mary Somerville and the Newtonian revolution*, Oxford University Press.

Aristotle, 2007. *Posterior Analytics*, transl. G.R.G. Mure, Adelaide University Press.

Arthur, R. 2013. Leibniz's syncategorematic infinitesimals, *Archive for History of Exact Sciences* 67, 553–593.

Arthur, R. and Rabouin, D., (to appear) Leibniz's syncategorematic infinitesimals II: Their existence, their use and their role in the justification of the differential calculus.

Aubin, D. 2009. Observatory mathematics in the nineteenth century, in *The Oxford Handbook of the History of Mathematics*, 273–298, E. Robson and J. Stedall (eds.), Oxford University Press.

Barrow, I. 1670. *Lectiones Geometriae*, Godlib.

Barrow-Green, J.E. 1997. *Poincaré and the Three-Body Problem*, American and London Mathematical Societies, HMath 11.

Barrow-Green, J.E. 2010. The dramatic episode of Sundman, *Historia Mathematica* 37, 164–203.

Bascelli, T. 2014. Galileo's *quanti*: understanding infinitesimal magnitudes, *Archive for History of Exact Sciences* 68, 121–136.

Beltrami, E. 1868. Saggio di interpretazione della geometria non-euclidea, *Giornale di Matematiche* 6, 285–315, in *Opere Matematiche* 1, 374–405, transl. in (Stillwell 1996, 7–34).

Ben-David, J. 1971. *The Scientist's Role in Society*, Prentice-Hall.

Berkeley, G. 1734. *The Analyst; or, a Discourse Addressed to an Infidel Mathematician*, London.

Bernoulli, D. 1738. *Hydrodynamica: Sive de Viribus et Motibus Fluidorum Commentarii*, Strasbourg.

Bernoulli, D. 1753. Réflexions et éclairicissements sur les nouvelles vibrations des cordes, etc., *Mémoires de l'Académie des Sciences de Berlin* 9, 147–172.

Bernoulli, Johann, 1691. Solutio Problematis Funicularii [etc.], *Acta Eruditorum*, 274–278.

Bernoulli, Johann, 1702. Solution d'un problème concernant le calcul intégral, avec quelques abrégés à ce calcul, *Mémoires de l'Académie Royale des Sciences*, 296–306, in *Opera Johannis Bernoulli* I, 393–400.

Bernoulli, Johann, 1742a. *Hydraulica Nunc Primum Detecta ac Demonstrata Directe ex Fundamentis Pure Mechanicis. Anno 1732*, in *Opera Omnia* 4, 387–493.

Bernoulli, Johann 1742b. Solution du problème inverse des forces centrales, in *Opera Johannis Bernoulli* I, 470–480, G. Cramer (ed.), Geneva.

Bertoloni Meli, D. 1993. *Equivalence and Priority: Newton versus Leibniz*, Oxford University Press.

Besterman, T. (ed.) 1958. *Les Lettres de la Marquise du Châtelet*, Vol. I, Institut et Musée Voltaire, Geneva.

Biermann, K.-R. 1959. *Johann Peter Gustav Lejeune Dirichlet: Dokumente für sein Leben und Werk*, Akademie-Verlag, Berlin.

Biggs, N.L., Lloyd, E.K., and Wilson R.J. 1998. *Graph Theory 1736–1936*, Oxford University Press.

Bioesmat-Martagon, L. 2010. *Eléments d'une Biographie de l'Espace Projectif*, Presses Universitaires de Nancy.

Birkhoff, G.D. 1913. Proof of Poincaré's Geometric Theorem, *Transactions of the American Mathematical Society* 14, 14–22

Birkhoff, G. and Merzbach, U. 1973. *A Source Book in Classical Analysis*, Harvard University Press.

Bjerknes, C.A. 1885. Niels Henrik Abel. Tableau de sa Vie et son Action Scientifique, *Bordeaux Mémoires* (3) 1, 1–365.

Blåsjö, V. 2017. On what has been called Leibniz's rigorous foundation of infinitesimal geometry by means of Riemannian sums, *Historia Mathematica* 44, 134–149.

Bolyai, J. 1832. Appendix scientiam spatii absolute veram exhibens, in Bolyai, W. and J. *Tentamen Juventutem Studiosam in Elementa Matheosis Purae, etc.*, Maros-Vásérhely, 2 vols. Transl. J. Hoüel, La Science Absolue de l'Espace, *Mémoires de la Société des Sciences Physiques et Naturelles de Bordeaux* 5, 1867, 189–248, transl. G.B. Halsted, *Science Absolute of Space*, Appendix in (Bonola 1912), ed. and repr. *Janos Bolyai, non-Euclidean Geometry and the Nature of Space*, J.J. Gray (ed.), Burndy Library, MIT. Press, 2004.

Bolzano, B. 1817. Rein analytischer Beweis, etc., Prague, in (Bolzano 2004, 251–278).

Bolzano, B. 1851. *Paradoxien des Unendlichen*, Reclam, Leipzig.

Bolzano, B. 2004. *The Mathematical Works of Bernard Bolzano*, S.B. Russ (transl. and ed.), Oxford University Press.

Boniface, J. and Schappacher, N. 2001. Sur le concept de nombre en mathématique: cours inédit de Leopold Kronecker à Berlin, *Revue d'Histoire des Mathématique* 7, 206–275.

Bonola, R. 1912. *Non-Euclidean Geometry*, transl. H.S. Carslaw, Open Court Publications, repr. Dover 1955, containing Halsted's translations of Bolyai's *Appendix*, and (Lobachevskii 1840).

Boutroux, P. 1914/1921. Lettre de M. Pierre Boutroux à M. Mittag-Leffler, *Acta Mathematica* 38, 197–201.

Bos, H.J.M 1980. Newton, Leibniz and the Leibnizian tradition, in *From the Calculus to Set Theory, 1640–1920: An Introductory History*, Grattan-Guinness, I. (ed.), Duckworth, 49–93.

Bottazzini, U. 1986. *The Higher Calculus, A History of Real and Complex Analysis from Euler to Weierstrass*, Springer.

Bottazzini, U. and Gray, J.J. 2013. *Hidden Harmonies — Geometric Fantasies: The Rise of Complex Function Theory*, Springer.

Boyer, C.B. 1959. *The History of the Calculus and its Conceptual Development*, Dover.

Bradley, R.E. and Sandifer, C.E. 2007. *Leonhard Euler: Life, Work, and Legacy*, Studies in the History and Philosophy of Mathematics, Vol. 5, Elsevier.

Braver, S. 2011. *Lobachevski Illuminated*, Mathematical Association of America.

Breitenberger, E. 1984. Gauss's Geodesy and the Axiom of Parallels, *Archive for History of Exact Sciences* 29, 273–289.

Brown, Ll. A. 1956. The Longitude, in *The World of Mathematics*, J.R. Newman (ed.), Allen and Unwin, many subsequent editions.

L. L. Bucciarelli and N. Dworsky, *Sophie Germain*, Studies in the History of Modern Science, vol. 6, D. Reidel Publishing Co., Dordrecht-Boston, Mass., 1980. An essay in the history of the theory of elasticity. MR619691

Buchwald, J.Z. and Feingold, M. 2012. *Newton and the Origin of Civilization*, Princeton University Press.

Bühler, W.K. 1981. *Gauss: a Biographical Study*, Springer. Corrected second printing 1987.

Burckhardt, J.J. 1976. Steiner, in *Dictionary of Scientific Biography*, Vol. XIII, C.C. Gillispie (ed.).

Burckhardt, J.J., Fellmann, E.A., and Habicht, W. (eds.) 1983. *Leonhard Euler 1707–1783, Beiträge zur Leben und Werk*, Birkhäuser.

Cajori, F. 1918. Pierre Laurent Wantzel, *Bulletin of the American Mathematical Society* 24, 339–347.

Calinger, R.S. 2016. *Leonhard Euler: Mathematical Genius in the Enlightenment*, Princeton University Press.

Cannell, M. 1993. *George Green, Mathematician and Physicist, 1793–1841. The Background to his Life and Work*, Athlone Press.

Cantor, G. 1872. Über die Ausdehnung eines Satzes aus der Theorie der trigonometrischen Reihen, *Mathematische Annalen* 5, in *Gesammelte Abhandlungen*, E. Zermelo (ed.), Berlin, 1932, 92–102.

Cantor, G. 1874a. Über eine Eigenschaften des Inbegriffs aller reellen algebraischen Zahlen, *Journal für Mathematik* 77, 258–263, in *Gesammelte Abhandlungen*, E. Zermelo (ed.), Berlin, 1932, 115–118.

Cantor, G. 1874b. Ein Beitrag zur Mannigfaltigkeitslehre, *Journal für Mathematik* 84, 242–259, in *Gesammelte Abhandlungen*, E. Zermelo (ed.), Berlin, 1932, 119–133.

Cantor, G. 1895. Beiträge zur Begründung der transfiniten Mengenlehre, *Mathematische Annalen* 46, 481–512, in *Gesammelte Abhandlungen*, E. Zermelo (ed.), Berlin, 1932, 282–289, transl. in P.E.B. Jourdain, *Contributions to the Founding of the Theory of Transfinite Numbers*, Dover, 1955.

Cantor, G. 1915. *Contributions to the Founding of the Theory of Transfinite Numbers*, transl. P.E.B. Jourdain, Open Court, repr. Dover, 1955.

Cantor, G. 1932. *Gesammelte Abhandlungen Mathematische und Philosophischen Inhalt*, E. Zermelo (ed.), Springer.

Cantor, G. 1991. *Briefe*, H. Meschkowski and W. Nilson (eds.), Springer.

Cantor, M. 1892–1908. *Vorlesungen über die Geschichte der Mathematik*, 4 vols., Teubner.

Cardano, G. 1968. *Ars Magna, The Great Art, or the Rules of Algebra*, T.R. Witmer (transl.), MIT Press.

Carlson, J., Jaffe, A., and Wiles, A. (eds.) 2006. *The Millennium Prize Problems*, Clay Mathematics Institute and American Mathematical Society.

Cauchy, A.-L. 1821. *Cours d'Analyse Algébrique*. Imprimerie Royale, Paris, in *Oeuvres Complètes* (2) 3, repr. U. Bottazzini (ed.), CLUEB, Bologna 1992.

Cauchy, A.-L. 1823. *Résumé des Leçons Donnés à l'École Royale Polytechnique sur le Calcul Infinitesimal. Tome Premier*, Paris, in *Oeuvres Complètes* (2) 4, 5–261.

Cauchy, A.-L. 1835. Mémoire sur l'intégration des équations différentielles (lith.), Prague, in *Oeuvres Complètes* (2) 11, 399–465.

Cauchy, A.-L. 1845. Sur le nombre des valeurs égales ou inégales que peut acquerir une fonction etc., *Comptes Rendus* (XXI) 779, in *Oeuvres Complètes* (1) 9, 277–292.

Cauchy, A.-L. 1981. *Équations Différentielles Ordinaires; Ordinary Differential Equations*, C. Gilain (ed.), Études vivantes and Johnson Reprint Corporation.

Cavalieri, B. 1653. *Geometria Indivisibilibus*, Bononiae.

Cayley, A. 1843. On the motion of rotation of a solid body, *Cambridge Mathematical Journal* 3, 224–232, in *Collected Mathematical Papers* I, 1897, no. 6, 28–35.

Cayley, A. 1845. On certain results relating to quaternions, *Philosophical Magazine* 26, 141–145, in *Collected Mathematical Papers* I, 1897, no. 20, 123–126.

Cayley, A. 1848. On the application of quaternions to the theory of rotation, *Philosophical Magazine* 33, 196–200, in *Collected Mathematical Papers* I, 1897, no. 68, 405–409.

Chasles, M. 1830. Note sur les propriétés genératés du systéme de deux corps semblables entre'eux, *Bulletin des Sciences Mathématiques* (1er. Section du *Bulletin Universel*) 14, 321–326.

Chasles, M. 1837. *Aperçu Historique sur l'Origine et le Développement des Méthodes en Géométrie*, Hayez, Brussels.

Child, J.M. 1920. *The Early Mathematical Manuscripts of Leibniz*, Open Court, Dover repr., 2005.

Clairaut, A.C. 1743. *Théorie de la Figure de la Terre*, Durand, Paris.

Clairaut, A.C. 1749. Du système du monde dans les principes de la gravitation universelle, *Mémoires de l' Académie des Sciences*, Paris, 329–364, 528.

Clebsch, R.F.A. 1872. Zum Gedächtnis an Julius Plücker, *Abhandlungen der Königlichen Gesellschaft der Wissenschaften zu Göttingen*, 15, 6.

Clebsch, R.F.A. 1895. *Gesammelte Mathematische Abhandlungen*, A. Schoenflies and F. Pockels (eds.), Leipzig.

Clifford, W.K. 1882. *Mathematical Papers*, R. Tucker (ed.), repr. Chelsea, 1968.

Cohen, P. 1963. The independence of the continuum hypothesis, 1 and 2, *Proceedings of the National Academy of Sciences of the United States of America* 50, 1143–1148, and 51, 105–110.

Cohen, I.B. 1980. *The Newtonian Revolution*, Cambridge University Press.

Cohen, I.B. and Smith, G.E. (eds.). 2002. *The Cambridge Companion to Newton*, Cambridge University Press.

Cohen, I.B. and Whitman, A. 1999. *Isaac Newton, The Principia, a New Translation*, University of California Press.

Cohen, M.R. and Drabkin, I.E. 1948. *A Source Book in Greek Science*, Harvard University Press.

Condorcet, N. 1802. Elogium of Euler, in *Letters of Euler on Different Subjects*, transl. H. Hunter, London.

Cooke, R. 1984. *The Mathematics of Sonya Kovalevskaya*, Springer.

Coolidge, J.L. 1934. The rise and fall of projective geometry, *American Mathematical Monthly* 34, 217–228.

Coolidge, J.L. 1940. *A History of Geometrical Methods*, Oxford University Press., Dover repr., 1963.

Cotes, R. 1722. *Harmonia Mensuarum*, Cambridge.

Courant, R. 1925. Felix Klein, *Jahresbericht den Deutschen Mathematiker Vereinigung* 34, 197–212.

Cowell, P.H. 1898. Periodic Orbits, *The Observatory* 21, 121–123.

Craig, J. 1685/1687. *Methodus Figuarum Lineis et Curvis Comprehensarum Quadraturas Determinandi*, London.

Cramer, G. 1750. *Introduction à l'Analyse des Lignes Courbes Algébriques*, Frères Cramer & Cl. Philbert, Geneva.

Crépel, P. 2005. Jean le Rond D'Alembert, *Traité de Dynamique* (1743, 1758), *Landmark Writings in Western Mathematics, 1640–1940*, I. Grattan-Guinness (ed.), 159–167.

Crowe, M.J. 1967. *A History of Vector Analysis*, University of Notre Dame Press, Dover repr., 1985.

D'Alembert, J. le Rond. 1743. *Traité de Dynamique*, Paris.

D'Alembert, J. le Rond. 1746. Recherches sur le calcul intégral. *Histoire de l'Académie de Berlin* 2 (1748), 182–224, in *Oeuvres Complètes* (1) 4a, 93–166.

D'Alembert, J. le Rond. 1747. Réflexions sur la Cause Générale des Vents, Paris.

D'Alembert, J. le Rond. 1749. Recherches sur la courbe que forme une corde tenduë mise en vibration, *Histoire de l'Académie de Berlin* 3, 214–219.

D'Alembert, J. le Rond. 1751. Discours Préliminaire, *Encyclopédie* 1, Paris, transl. *Preliminary Discourse to the Encyclopedia of 1751*, 1963.

D'Alembert, J. le Rond. 1754. Différentiels, Limites, *Encyclopédie* 4.

D'Alembert, J. le Rond. 1761. Sur les logarithmes de quantités negatives, *Opuscules Mathématiques* 1, 180–209.

Darrigol, O. 2000. *Electrodynamics from Ampère to Einstein*, Oxford University Press.

Darrigol, O. 2005. *Worlds of Flow*, Oxford University Press.

Darwin, G.H. 1900. Presentation of the medal of the Royal Astronomical Society to M. Henri Poincaré, *Monthly Notices of the Royal Astronomical Society* 60, 406–415.

Dauben, J.W. Jr. 1979. *Georg Cantor: His Mathematics and Philosophy of the Infinite*, Harvard University Press.

Dawson, J.W. Jr. 1997. *Logical Dilemmas: The Life and Work of Kurt Gödel*, A.K. Peters.

De Gua de Malves, J.P. 1740. *Usages de l'Analyse de Descartes, etc.*, Paris.

Dedekind, R. 1872. *Stetigkeit und irrationale Zahlen*, Vieweg, Braunschweig, in Dedekind, R. *Essays on the Theory of Numbers*, transl. W.W. Beman, Dover, 1963.

Dedekind, R. 1888. *Was Sind und was Sollen die Zahlen?*, Vieweg, Braunschweig, transl. in Dedekind, R. *Essays on the Theory of Numbers*, transl. W.W. Beman, Dover, 1963.

Dedekind, R. 1930–1932. *Gesammelte Mathematische Werke*, R. Fricke, E. Noether, and O. Ore (eds.), Vieweg.

Delambre, J. 1810. Rapport historique sur les progrès des sciences mathématiques depuis 1789, et sur leur état actual, Paris.

Descartes, R. 1637. *La Géométrie*, D.E. Smith and M.L. Latham (transl. and eds.), Dover repr. 1954.

Descartes, R. 1644. *Principia Philosophiae*, Amsterdam, in *Oeuvres de Descartes* 8, C. Adam and J. Tannery (eds.), Cerf, Paris, 1905.

Descartes, R. 1649. *Geometria*, transl. F. van Schooten, Leiden.

Dhombres, J.G. 1984. French textbooks in the sciences 1750–1850, *History of Education* 13, 153–161.

Diacu, F. and Holmes, P. 1996. *Celestial Encounters: The Origins of Chaos and Stability*, Princeton University Press.

Dick, O.L. (ed.) 1962. *Aubrey's Brief Lives*, Peregrine Books.

Diderot, D. 1981. Notice sur Clairaut, *Oeuvres Complètes*, Vol. 9, Hermann, Paris.

Diophantus, 1621. *Arithmetica*, G.B. de Méziriac Bachet (ed.).

Dirichlet, P.G.L. 1876. *Vorlesungen über die im ungekehrten Verhältniss des Quadrats der Entfernung wirkenden Kräfte*, F. Grube (ed.), Teubner.

Dittrich, W. 2018. *Reassessing Riemann's Paper: On the Number of Primes less than a given Magnitude*, Springer.

Domingues, J.C. 2008. *Lacroix and the Calculus*, Birkhäuser.

Dry, S. 2014. *The Newton Papers: The Strange and True Odyssey of Isaac Newton's Manuscripts*, Oxford University Press.

Dunham, W. 1999. *Euler: The Master of us All*, Mathematical Association of America.

Dunham, W. 2007. *The Genius of Euler*, Mathematical Association of America.

Dunnington, G.W. 2003. *Gauss: Titan of Science*, 2nd. edn., J.J. Gray (ed.), Mathematical Association of America.

Duren, P. 2009. Changing faces: the mistaken portrait of Legendre, *Notices of the American Mathematical Society* 56, 1440–1443.

Ebbinghaus, H.D. *et al.* 1990. *Numbers*, Springer.

Edwards, C.H. Jr. 1979. *The Historical Development of the Calculus*, Springer.

Edwards, H.M. 1977. *Fermat's Last Theorem: A Genetic Introduction to Algebraic Number Theory*, Springer.

Edwards, H.M. 1983. Euler and quadratic reciprocity, *Mathematics Magazine* 56, 285–291, repr. (Dunham 2007, 233–242).

Edwards, H.M. 1984. *Galois Theory*, Birkhäuser.

Eneström, G. 1905. Der Briefwechsel zwischen Leonhard Euler und Johann I Bernoulli, *Bibliotheca Mathematica* 6.

Engel, F. and Stäckel, P. (eds.) 1895. *Die Theorie der Parallellinien von Euklid bis auf Gauss*, Teubner.

Euler, L. 1731/1738. De summatione innumerabilium progressionum, *Commentarii Academiae Scientiarum Petropolitanae* 5, 91–105, in *Opera Omnia* (1) 14, 25–41 (E20).

Euler, L. 1732/1738. Observationes de theoremate quodam Fermatiano aliisque ad numeros primos spectantibus, *Commentarii Academiae Scientiarum Petropolitanae* 6, 103–107, in *Opera Omnia* (1) 2, 1–5 (E26).

Euler, L. 1734/1740a. De progressionibus harmonicis observationes, *Commentarii Academiae Scientiarum Petropolitanae* 7, 150–161, in *Opera Omnia* (1) 14, 87–100 (E43).

Euler, L. 1735/1740b. De summis serierum reciprocarum, *Commentarii Academiae Scientiarum Petropolitanae* 7, 123–134, *Opera Omnia* (1) 14, 73–86 (E41).

Euler, L. 1736. *Mechanica sive Motus Scientia Analytice Exposita*, 2 vols., St Petersburg, in *Opera Omnia* (2) 1, 2 (E15, E16).

Euler, L. 1738. Observationes de theoremate quodam Fermatiano aliisque ad numeros primos spectantibus, *Commentarii Academiae Scientiarum Petropolitanae* 6, 103–107, in *Opera Omnia* (1) 2, 1–5 (E26).

Euler, L. 1741. Solutio problematis ad geometriam situs pertinentis, *Commentarii Academiae Scientiarum Petropolitanae* 8, 128–140, in *Opera Omnia* (1) 7, 1–10 (E53), transl. in (Biggs, Lloyd, Wilson 1998, 3–8).

Euler, L. 1741. Theorematum quorundam ad numeros primos spectantium demonstratio, *Commentarii Academiae Scientiarum Petropolitanae* 8, 141–146, in *Opera Omnia* (1) 2, 33–37 (E54).

Euler, L. 1743. De integratione aequationum differentialium altiorum graduum, *Miscellanea Berolinensia* 7, in *Opera Omnia* (1) 22, 108–149 (E62).

Euler, L. 1744. *Methodus Inveniendi Lineas Curvas maximi minimivi Proprietate Gaudentes sive Solutio Problematis isoperimetrici latissimo sensu Accepti*, Lausanne and Geneva, in *Opera Omnia* (1) 24, (E65).

Euler, L. 1748a. *Introductio in Analysin Infinitorum*, two vols., *Opera Omnia* (1) Vols. 8 and 9, transl. *Introduction to Analysis of the Infinite*, Book I, Springer, 1988, Book II, Springer, 1990 (E101, E102).

Euler, L. 1748b. Demonstration sur le nombre des points, ou deux lignes des ordres quelconques peuvent se couper, *Mémoires de l'Académie des Sciences de Berlin* 4, 234–248, in *Opera Omnia* (1) 26, 46–59 (E148).

Euler, L. 1749. *Scientia Navalis*, 2 vols., St Petersburg, *Opera Omnia* (2) 18, 19 (E110, E111).

Euler, L. 1750. Sur la vibration des cordes, *Mémoires de l'Académie des Sciences de Berlin* 4, 69–85, in *Opera Omnia* (2), 10, 63–77 (E140).

Euler, L. 1751. De la controverse entre Mrs. Leibniz et Bernoulli sur les logarithmes des nombres negatifs et imaginaires, *Mémoires de l'Académie des Sciences de Berlin* 5, 139–179, in *Opera Omnia* (1), 17, 195–232 (E168).

Euler, L. 1751b. Recherches sur les racines imaginaires des equations, *Mémoires de l'Académie des Sciences de Berlin* 5, 222–288, in *Opera Omnia* (1) 6, 78–150 (E170).

Euler, L. 1752. Découverte d'un nouveau principe de Mécanique, *Mémoires de l'Académie des Sciences de Berlin* 6, 185–217, in *Opera Omnia* (2), 5, 81–108 (E177).

Euler, L. 1755a. Remarques sur les memoires precedens de M. Bernoulli, *Mémoires de l'Académie des Sciences de Berlin* 9, 196–222, in *Opera Omnia* (2) 10, 233–254 (E213).

Euler, L. 1755b. *Institutiones Calculi Differentialis cum eius usu in Analysi Finitorum ac Doctrina Serierum*, in *Opera Omnia* (1) 10 (E212).

Euler, L. 1758/1765. Du mouvement de rotation des corps solides autour d'un axe variable, *Mémoires de l'Académie Scientifiques Berlin* 14, in *Opera Omnia* (2) 8, 154–193 (E292).

Euler, L. 1761. Principia motus fluidorum, *Novi Commentarii Academiae Scientiarum Petropolitanae* 6, 271–311, in *Opera Omnia* (2) 12, 133–168 (E258).

Euler, L. 1763. Theoremata arithmetica nova methodo demonstrata, *Novi Commentarii Academiae Scientiarum Petropolitanae* 8, 74–104, in *Opera Omnia* (1) 2, 531–555 (E271).

Euler L. 1764/1766. Considerationes de motu corporum coelestium, *Novi Commentarii Academiae Scientiarum Imperialis Petropolitanae* 10, 544–558, in *Opera Omnia* (2) 25, 246–257 (E304).

Euler, L. 1765. *Theoria Motus Corporum Solidorum seu Rigidorum*, Rostock and Greifswald, in *Opera Omnia* (2) 3 (E289).

Euler, L. 1770a. *Institutionum Calculi Integralis* III, St Petersburg, in *Opera Omnia* (1), 13 (E385).

Euler, L. 1770b. *Vollständige Einleitung zur Algebra* in *Opera Omnia* (1) 1, transl. *Elements of Algebra*, Rev. J. Hewlett, London, 1840, repr. Springer, 1972 (E387).

Euler, L. 1770c. Considerations sur le problème de trois corps, *Mémoires de l'Académie des Sciences de Berlin*, 1770, 194–200, in *Opera Omnia* (2) (E400).

Euler L. 1764/1766. Considerationes de motu corporum coelestium, *Novi Commentarii Academiae Scientiarum Imperialis Petropolitanae* (15) 75–106, in *Opera Omnia* (1) 6, 286–315 (E407).

Euler, L. 1773. *Théorie complette de la construction et de la manoeuvre des vaisseaux*, St Petersburg. English transl. H. Watson, *A Compleat Theory of the Construction and Properties of Vessels*, London, 1776.

Euler, L. 1776. Formulae generates pro translatione quaecunque corporum rigidorum, *Novi Commentarii Academiae Scientiarum Imperialis Petropolitanae* 20, 189–217, in *Opera Omnia* (2) 9, 84–98 (E478).

Euler, L. 1775/1776. Nova methodus motum corporum rigidorum determinandi, *Novi Commentarii Academiae Scientiarum Imperialis Petropolitanae* 20, 208–238, in *Opera Omnia* (2) 9, 99–125 (E479).

Euler, L. 1849. De numeris amicibilibus, *Commentationes Arithmeticae* 2, 627–636, in *Opera Omnia*(1) 5, 353–365 (E798).

Euler, L. 1980. *Leonhard Euler Correspondence, Opera Omnia, Series Quarta A: Commercial Epistolicum*, Vol. V, A.P. Youschkevich and R. Taton (eds.), Birkhäuser.

Euler, L. 1998. *Leonhard Euler Correspondence: Briefwechsel mit Johann (I) Bernoulli und Nicolaus (I) Bernoulli, Opera Omnia: Series Quarta A: Commercium Epistolicum*, Vol. II, E.A. Fellmann and G.K. Mikhajlov (eds.), Springer.

Euler, L. 2015. *Leonhard Euler Correspondence, Opera Omnia, Series Quarta A: Commercium Epistolicum*, Vol. IV, F. Lemmermeyer and M. Mattmüller (eds.), Birkhäuser.

Ewald, W.B. 1996. *From Kant to Hilbert: A Source Book in the Foundations of Mathematics*, 2 vols., Oxford University Press.

Fara, P. 2002. *Newton, The Making of Genius*, Macmillan.

Fauvel, J. and Gray, J.J. (eds.) 1987. *The History of Mathematics: A Reader*, Macmillan, in association with the Open University (referred to as F&G in the text).

Fauvel, J., Flood, R., Shortland, M., and Wilson, R. (eds.) 1988. *Let Newton Be!: A New Perspective on his Life and Works*, Oxford University Press.

Fauvel, J., Flood R., and Wilson, R. (eds.) 1993. *Möbius and his Band*, Oxford University Press.

Fauvel, J., Flood R., and Wilson, R. (eds.) 2013. *Oxford Figures* (2nd edn.), Oxford University Press.

Feigenbaum, L. 1985. Brook Taylor and the method of increments, *Archive for History of Exact Sciences* 34, 1–140.

Feigenbaum, L. 1992. The fragmentation of the European mathematical community, in *The Investigation of Difficult Things*, P.M. Harman and A.E. Shapiro (eds.), Cambridge University Press, 383–398.

Feingold, M. (ed.) 1990. *Before Newton; The Life and Times of Isaac Barrow*, Cambridge University Press.

Fellmann, E.A. 2007. *Leonhard Euler*, Birkhäuser.

Fermat, P. de. 1891–1922. *Pierre de Fermat: Oeuvres*, 4 vols., P. Tannery and C. Henry (eds.), Paris.

Ferraro, G. 2007. Euler's treatises on infinitesimal analysis, in *Euler Reconsidered*, R. Baker (ed.), Kendrick Press, 39–101.

Ferraro, G. 2008. *The Rise and Development of the Theory of Series up to the Early 1820s*, Springer.

Ferraro, G. and Panza, M. 2012. Lagrange's Theory of Analytical Functions and his Ideal of Purity of Method, *Archive for History of Exact Sciences* 66, 95–197.

Flood, R., McCartney, M., and Whitaker, A. (eds.), 2008. *Kelvin: Life, Labours and Legacy*, Oxford University Press.

Föppl, A. 1904. *Einführung in die Maxwell'sche Theorie der Elektricität*, Leipzig.

Fourier, J. 1822. *Théorie Analytique de la Chaleur*, in *Oeuvres* 1, repr. Gabay, Paris, 1988, transl. *The Analytical Theory of Heat*, transl. A. Freeman, Cambridge,1878, Dover repr. 1950.

Franzen T. 2005. *Gödel's Theorem: An Incomplete Guide to Its Use and Abuse*, A.K. Peters.

Frederick the Great, *Oeuvres de Frédéric le Grand*, Vol. 7, Berlin, 1846–1857.

Frege, G. 1879. *Begriffsschrift*, Louis Nebert, Halle.

Fuss, N. 1783. Lobrede auf Herrn Leonhard Euler, in Euler, *Opera Omnia* (1), 1.

Fuss, N. 1843. *Correspondance Mathématique et Physique*, St Petersburg.

Galileo, G. 1914. *Dialogues Concerning Two New Sciences*, transl. H. Crew and A. de Salvio, Macmillan, Dover repr. n.d.

Galileo, G. 1968. *Le Opere*, Vol. 16, Firenze, G. Barbera.

Gauss, C.F. 1799. Demonstratio nova theorematis omnem functionem algebraicam ...resolvi posse, Helmstadt, in *Werke* III, 1–30.

Gauss, C.F. 1801. *Disquisitiones Arithmeticae*, G. Fleischer, in *Werke* I, transl. W.C. Waterhouse and A.A. Clarke, Springer, 1986.

Gauss, C.F. 1819. [Mutation des Raumes] *Werke* VIII, 1900, 357–362.

Gauss, C.F. 1828. *Disquisitiones Generales circa Superficies Curvas* and *Anzeige*, in *Werke* IV, 217–258, P. Dombrowski (ed.), in *Astérisque* 62, 1979, Latin original, with a reprint of the English translation by A. Hiltebeitel and J. Morehead, 1902, and as *General Investigations of Curved Surfaces*, P. Pesic (ed.), Dover, 2005.

Gauss, C.F. 1832. Theorie der biquadratischen Reste, *Commentiones Soc. Reg. Sci. Göttingensis*, 7, in *Werke* II, 93–1§48, and *Anzeige*, in *Werke* II, 169–178, transl. in (Ewald 1996), Vol. 1, 306–313.

Gauss, C.F. 1863–1929. *Werke*, 12 vols., Königlichen Gesellschaft der Wissenschaften zu Göttingen.

Gergonne, J.D. 1826. Philosophie mathématique. Considérations philosophiques sur les élémens de la science de l'étendue, *Annales de Mathématiques*, 209–231.

Gergonne, J.D. 1827–1828. Géométrie de situation. Rectification de quelques théorèmes énoncés dans les Annales, *Annales de Mathématiques* 18, 149–154.

Gibbs, J.W. 1881, 1884. *Elements of Vector Analysis*, 2 vols., New Haven.

Gilain, Ch. 1991. Sur l'histoire du théorème fondamental de l'algèbre: théorie des équations et calcul intégral, *Archive for History of Exact Sciences* 42, 91–136.

Gillispie, C.G. 1997. *Pierre-Simon Laplace 1749–1827. A Life in Exact Science*, Princeton University Press.

Gleick, J. 2003. *Isaac Newton*, Fourth Estate.

Golland, L.A. and Golland, R.W. 1993. Euler's troublesome series, *Historia Mathematica* 20, 54–67.

Grabiner, J.V. 1981. *The Origins of Cauchy's Rigorous Calculus*, MIT Press.

Granville, A. 2015. Primes in intervals of bounded length, *Bulletin of the American Mathematical Society* 52, 171–222.

Grassmann, H.G. 1844. *Die lineale Ausdehnungslehre, oder ein neue Zweig der Mathematik*, Wigand. *Linear Extension Theory*, transl. L.C. Kannenberg, in *A New Branch of Mathematics*, Open Court, 1995.

Grassmann, H.G. 1862. *Ausdehnungslehre*, Enslin, Berlin. *Extension Theory*, transl. L.C. Kannenberg, American and London Mathematical Societies, HMath 19, 2000.

Grassmann, H.G. 1894–1911. *Hermann Grassmann's Gesammelte Mathematische und Physikalische Werke*, 4 vols., F. Engel (ed.), Leipzig.

Grattan-Guinness, I. 1971. Towards a biography of Georg Cantor, *Annals of Science* 27, 345–391.

Grattan-Guinness, I. (ed.) 1980. *From the Calculus to Set Theory, 1630–1910*, Duckworth.

Grattan-Guinness, I. 1981. Essay review: Recent researches in French mathematical physics of the early 19th century, *Annals of Science* 38, 665–666.

Grattan-Guinness, I. 1990. *Convolutions in French Mathematics, 1800–1840*, 3 vols., Birkhäuser.

Gray, J.J. 1991. Did Poincaré say 'Set theory is a disease?', *The Mathematical Intelligencer* 13, 19–22.

Gray, J.J. 2000. *The Hilbert Challenge*, Oxford University Press.

Gray, J.J. 2011. *Worlds out of Nothing; A Course on the History of Geometry in the 19th Century*, 2nd edn., Springer.

Gray, J.J. 2013. *Henri Poincaré: A Scientific Biography*, Princeton University Press.

Green, G. 1871. *Mathematical Papers*, Cambridge University Press.

Greenberg, J.L. 1995. *The Problem of the Earth's Shape from Newton to Clairaut; The Rise of Mathematical Science in Eighteenth-Century Paris and the Fall of 'Normal' Science*, Cambridge University Press.

Gregorie, J. 1668. *Geometriae Pars Universalis*, Padua.

Grootendorst, A.W. and van Maanen, J.A. 1982. On the rectification of curves, *Nieuw Archiv voor Wiskunde* (3) 25, 101–105.

Guerlac, H. 1979. Some areas for further Newtonian studies, *History of Science* 17, 75–101.

Guerlac, H. 1981. *Newton on the Continent*, Cornell University Press.

Guicciardini, N. 1989. *The Development of the Newtonian Calculus in Britain, 1700–1800*, Cambridge University Press.

Guicciardini, N. 1999. *Reading the* Principia. *The Debate on Newton's Mathematical Methods for Natural Philosophy from 1687 to 1736*, Cambridge University Press.

Guicciardini, N. 2009. *Isaac Newton on Mathematical Certainty and Method*, MIT Press.

Hadamard, J. 1921. L'oeuvre de H. Poincaré, *Acta Mathematica* 38, 203–287.

Hahn, R. (ed.) 2013. *Correspondance de Pierre Simon Laplace (1749–1827)*, 2 vols., Brepols.

Hall, A.R. 1980. *Philosophers at War: The Quarrel between Newton and Gottfried Leibniz*, Cambridge University Press.

Hall, A. R. 1983. *The Revolution in Science, 1500–1750*, Longman.

Hamilton, Sir W. R. 1834. On conjugate functions, ..., British Association Report, Edinburgh 1834, 519–523, in *Mathematical Papers* III, Cambridge University Press, 1967, 97–100.

Hamilton, Sir W. R. 1843–1844. Letter to Graves on quaternions, *Philosophical Magazine* XXV, 489–495, in *Mathematical Papers* III, Cambridge University Press, 1967, 106–110.

Hamilton, Sir W. R. 1847. On a proof of Pascal's Theorem by means of Quaternions; and on some other connected Subjects, *Proceedings of the Royal Irish Academy* 3, 273–292, in *Mathematical Papers* III, Cambridge University Press, 1967, 367–386.

Hamilton, Sir W. R. 1853. *Lectures on Quaternions*, Dublin.

Hamilton, Sir W. R. 1854. *Elements of Quaternions*, Dublin.

Hamilton, Sir W. R. 1931–1967. *Mathematical Papers*, 4 vols., Royal Irish Academy.

Hankel, H. 1867. *Theorie der Complexen Zahlensysteme insbesondere der gemeinen imaginären Zahlen und der Hamilton'schen Quaternionen*, L. Voss, Leipzig.

Hankins, T.L. 1970. *Science and the Enlightenment*, Cambridge University Press.

Hankins, T.L. 1980. *Sir William Rowan Hamilton*, Johns Hopkins University.

Heaviside, O. 1876. On duplex telegraphy, *Philosophical Magazine* (5) 1, 32–43, in *Electrical Papers* I, Macmillan, 1892, 53–64.

Heaviside, O. 1893. *Electromagnetic Theory*, Vol. 1, The Electrician Printing and Publishing Co, London.

Hero, 1900. *Heronis Alexandrini Opera quae Supersunt Omnia, Vol. 2, Mechanica et Catoptrica*, L. Nix and W. Schmidt (eds.), Teubner.

Heilbron, J.L. 1982. *Elements of Early Modern Physics*, University of California Press.

Hilbert, D. 1899. *Grundlagen der Geometrie*, Teubner, many subsequent editions, transl. L. Unger as *Foundations of Geometry*, 10th English ed. of the second German ed., Illinois, 1971.

Hilbert, D. 1902. Mathematical Problems, *Bulletin of the American Mathematical Society* 8, 437–479.

Hilbert, D. 1926. Über das Unendliche, *Mathematische Annalen* 95, 161–190, in *Grundlagen der Geometrie*, Teubner, 1930, 262–289.

Hopkins, B. and Wilson, R.J. 2004. The Truth about Königsberg, *College Mathematics Journal* 35, 198–207, repr. in *The Genius of Euler*, W. Dunham (ed.), Mathematical Association of America 2007, 263– 272, and *Leonhard Euler: Life, Work, and Legacy*, R.E. Bradley and C.E. Sandifer (eds.), 2007, 409–420.

Jackson, P. N. W. 2008. William Thomson's determination of the age of the earth, in (Flood, McCartney, and Whitaker 2008), 160–174.

Jacobi, C. G. J. 1881. Bemerkungen zu einer Abhandlung Eulers über die orthogonale Substitution, *Mathematische Werke* III, 599–609, H. Kortum (ed.), Chelsea.

Jahnke, H.N. 2002. *A History of Analysis*, American and London Mathematical Societies, HMath 24.

Jarnik, V. 1981. *Bolzano and the Foundations of Mathematical Analysis*, Collets.

Jesseph, D. 2015. Leibniz on the Elimination of Infinitesimals, in *G. W. Leibniz: Interrelations Between Mathematics and Philosophy*, N. Goethe, P. Beeley, and D. Rabouin (eds.) (Archimedes, Vol. 41), Springer, 189–205.

Jones, W. 1706. *Synopsis Palmariorum Matheosos*, London.

Jordan, C. 1867. Sur les groupes de mouvement, *Comptes Rendus* LXV, 229–232, in *Oeuvres* IV, 1964, 113–115.

Jordan, C. 1868. Mémoire sur les groupes de mouvement, *Annali di Matematica Pura e Applicata* 11, 167–215, 322–345, in *Oeuvres* IV, 1964, 231–302.

Jordan, C. 1870. *Traité des Substitutions et des Équations Algébriques*, Gauthier-Villars, Paris.

Jungnickel, C. and MacCormach, R. 1986. *The Intellectual Mastery of Nature*, 2 vols., Chicago University Press.

Jurin, J. 1734. *Geometry no Friend to Infidelity: Or a Defence of Sir Isaac Newton and the British Mathematicians*, London.

Kant, I. 1789. *Critik der reinen Vernunft*, Riga, transl. N.K. Smith, *Critique of Pure Reason*, Macmillan, 1929.

Katz, V.J. and Parshall, K.H. 2014. *Taming the Unknown: A History of Algebra from Antiquity to the Early Twentieth Century*, Princeton University Press.

Kennedy, H.C. 1974. Peano, *Dictionary of Scientific Biography* 10, Scribners and Sons.

Kennedy, H.C. 1980. *Peano: Life and Works of Giuseppe Peano*, Reidel.

Kepler, J. 1615. *Nova Stereometria Doliorum Vinariorum*, Linz. *New Solid Geometry of Wine Barrels*, E. Knobloch (ed. and transl.), Sciences et Savoirs: Bibliothéque de Science, Tradition et Savoirs Humanistes 4, Paris, 2018.

Kirchhoff, G. 1857. Ueber die Bewegung der Elektricität in Drähten, *Annalen der Physik und Chemie* (4) 100, 193–217, in *Gesammelte Abhandlungen*, Barth, 131–155.

Klein, C.F. 1871. Ueber die sogenannte nicht-Euklidische Geometrie I, *Mathematische Annalen* 4, 573–615, in *Gesammelte Mathematische Abhandlungen* I, 254–305.

Klein, C.F. 1872. Vergleichende Betrachtungen über neuere geometrische Forschungen (Erlanger Program), in *Gesammelte Mathematische Abhandlungen* I, 460–497.

Klein, C.F. 1893. The present state of mathematics, *The Monist* 4, 1–4, and *Bulletin of the New York Mathematical Society* 3, 1–3, repr. in *Gesammelte Mathematische Abhandlungen* I, 2, 613–615.

Klein, C.F. 1894. *The Evanston Colloquium*, Macmillan.

Klein, C.F. 1923. Göttingen Professoren Lebensbilder von eigener hand, 4, Felix Klein. *Göttingen Mittheilungen*, 5.

Klein, C.F. 1926–1927. *Vorlesungen über die Entwicklung der Mathematik im 19. Jahrhundert*, 2 vols., R. Courant and O. Neugebauer (eds.), Springer, repr. Chelsea, 1967, transl. *Lectures on the Development of Mathematics in the Nineteenth Century*, MathSci Press, Brookline.

Kline, M. 1972. *Mathematical Thought from Ancient to Modern Times*, Oxford University Press.

Koblitz, A.H. 1983. *A Convergence of Lives. Sofia Kovalevskaia: Scientist, Writer, Revolutionary*, Birkhäuser.

Kollastrom, N. 2006. An hiatus in history: the British claim for Neptune's co-prediction, 1845–1846, *History of Science* 44, 1–28, 349–371.

Kollros, L. 1947. *Jakob Steiner*, Birkhäuser.

Kopelevich, Y.K. 1966. The Petersburg Academy contest in 1751, *Soviet Astronomy* 9, 653–661, and American Institute of Physics, NASA Astrophysics Data System.

Körner, T.W. 2002. *Fourier Analysis*, Cambridge University Press.

Kovalevskaya, S. 1978. *A Russian Childhood*, transl. B. Stillman, Springer.

L'Hôpital, G. de 1696. *Analyse des Infiniment Petits, pour l'Intelligence des Lignes Courbes*, Paris.

L'Huilier, S.-A.-J. 1786. *Exposition Élémentaire des Principes des Calculs Supérieurs*, Berlin.

Lagrange, J.-L. 1770. Réflexions sur la résolution algébriques des équations, *Mémoires de l'Académie Royale des Sciences de Paris*, in *Oeuvres* 3, J.-A. Serret (ed.), Paris, 205–421.

Lagrange, J.L. 1772a. Essai sur le problème des trois corps, *Nouvelles Mémoires de l'Académie des Sciences de Berlin*, in *Oeuvres* 6, J.-A. Serret (ed.), Paris, 229–324.

Lagrange, J.-L. 1772b. Sur la forme des racines imaginaires des équations, *Nouvelles Mémoires de l'Académie des Sciences de Berlin*, 222–261, in *Oeuvres* 3, J.-A. Serret (ed.), Paris, 479–518.

Lagrange, J.-L. 1773/1775. Recherches d'arithmétique, *Nouvelles Mémoires de l'Académie des Sciences de Berlin*, 265–365, in *Oeuvres* 3, J.-A. Serret (ed.) Paris, 695–795.

Lagrange, J.-L. 1774. Sur la forme des racines imaginaires des équations. *Nouvelles Mémoires de l'Académie des Sciences de Berlin*, 222–258, in *Oeuvres* 3, J.-A. Serret (ed.), Paris, 479–516.

Lagrange, J.-L. 1788. *Traité de Mécanique Analytique*, Paris. 2nd edn. 1811, in *Oeuvres* 11, J.-A. Serret (ed.), Paris.

Lagrange, J.-L. 1797. *Théorie des Fonctions Analytiques*. Imprimerie de la République, Paris. 2nd edn., Courcier, Paris 1813, in *Oeuvres* 9 J.-A. Serret (ed.), Paris.

Lambert, J.H. 1759. *Freye Perspektive*, Zürich.

Lambert, J.H. 1786. Theorie der Parallellinien, *Leipziger Magazin für die Reine und Angewandte Mathematik*, 137–164, 325–358.

Laplace, P.S. 1776. Recherches sur l'intégration des équations différentielles aux differences finies, et sur leur usage dans la théorie des hasards, *Oeuvres Complètes de Laplace* 8, 69–97, Gauthier Villars 1878–1912.

Laplace, P.S. 1784/1787. Mémoire sur les inégalités séculaires des planètes et des satellites, *Mémoires de l'Académie Royale des Sciences de Paris* 1–50, in *Oeuvres Complètes de Laplace* 11, 49–92.

Laplace, P.S. 1799–1805. *Traité de Mécanique Céleste*, also *Mécanique Céleste by the Marquis de Laplace, translated with commentary*, N. Bowditch (ed.), 4 vols., Boston, 1829–1839.

Laplace, P.S. 1814. *A Philosophical Essay on Probabilities*, transl. A.I. Dale, Springer, 1995.

Laubenbacher, R. and Pengelley, D. 2010 "Voici ce que j'ai trouvé": Sophie Germain's grand plan to prove Fermat's Last Theorem, *Historia Mathematica* 37, 641–692.

Lawrence, E. 1970. *The Origins and Growth of Modern Education*, Penguin.

Legendre, A.-M. 1788. Recherches d'analyse indéterminée, *Histoire de l'Académie Royale des Sciences*, 465–559.

Leibniz, G.W. 1858. *Mathematische Schriften* 5, C.I. Gerhardt (ed.), Halle.

Leibniz, G.W. 1906. Von dem Verhängnisse, in *Hauptschriften zur Grundlegung der Philosophie*, Vol. II, E. Cassirer (ed.), Meiner.

Leibniz, G.W. 1969. *Philosophical Papers and Letters*, 2nd. edn., L.E. Loemker (ed. and transl.), Kluwer.

Lemmermayer, F. 2000. *Reciprocity laws from Euler to Eisenstein*, Springer.

Linton, C.M. 2007. *From Eudoxus to Einstein: A History of Mathematical Astronomy*, Cambridge University Press.

Liouville, J. 1844. Nouvelle démonstration d'un théorème sur les irrationnelles algébriques, *Comptes Rendus* 18, 883–885.

Lobachevskii, N.I. 1837. Géométrie imaginaire, *Journal für die Reine und Angewandte Mathematik* 17, 295–320.

Lobachevskii, N.I. 1840. *Geometrische Untersuchungen zur Theorie der Parallellinien*, Berlin, repr. Mayer and Müller, 1887, transl. J. Hoüel as Études Géométriques sur la Théorie des Parallèles, *Mémoires de la Société des Sciences Physiques et Naturelles de Bordeaux*, 4, 1867, 83–128, repr. Paris 1866, Gauthier-Villars, Paris; transl. G.B. Halsted as Geometric Researches in the Theory of Parallels, Appendix in (Bonola 1912).

Lobachevskii, N.I. 1855. *Pangéométrie, Uchenye Zapiski Kazanskogo Imperatorskogo Universiteta*, Issue 1, 1–76. New edition with an English translation and a commentary by A. Papadopoulos, Heritage of European Mathematics, Vol. 4, European Mathematical Society, 2010.

Lützen, J. 1990. *Joseph Liouville, 1809–1882, Master of Pure and Applied Mathematics*, Springer.

Lützen, J. 2009. Why was Wantzel overlooked for a century? The changing importance of an impossibility result, *Historia Mathematica* 36, 374–394.

MacDonald Ross, G. 1984. *Leibniz*, Oxford University Press.

MacLaurin, C. 1720. *Geometria Organica, sive Descriptio Linearum Curvarum Universalis*, London.

MacLaurin, C. 1742. *A Treatise of Fluxions*, Edinburgh.

Mahoney, M.S. 1990. Barrow's mathematics: between ancients and moderns, in (Feingold 1990, 179–249).

Martinez, A. 2012. *The Cult of Pythagoras: Math and Myths*, University of Pittsburgh Press.

Maupertuis, P.L.M. 1738. *La Figure de la Terre déterminée par les observations de Messieurs De Maupertuis, Clairaut, Camus, Le Monnier … et Outhier … accompagnés de M. Celsius, … faites par ordre du Roy au Cercle Polaire*, Amsterdam, transl. *The Figure of the Earth, determined from observations made by order of the French King, at the Polar Circle: by Messrs. de Maupertuis, Camus, Clairaut …*, T. Cox, London, 1738.

Maxwell, J.C. 1873. *Treatise on Electricity and Magnetism*, 2 vols., Clarendon Press, Oxford.

Mazur, B. and Stein, W. 2018 *Prime Numbers and the Riemann Hypothesis*, Cambridge University Press.

McCartney, M. 2008. William Thomson: an introductory biography, in (Flood, McCartney, Whitaker 2008), 1–22.

McClelland, C. 1980. *State, Society, and University in Germany, 1700–1914*, Cambridge University Press.

Mehrtens, H. 1981. Mathematicians in Germany circa 1800, in H.N. Jahnke and M. Otto (eds.), *Epistemological and Social Problems of the Sciences in the Early Nineteenth Century*, Reidel, 410–411.

Mercator, N. 1668. *Logarithmotechnia sive Methodus Construendi Logarithmos Nova, Accurata, & Facilis*, London.

Mersenne, M. 1932. *Correspondance*, 8 vols., C. de Waard, R. Pintard, and B. Rochet (eds.), Presses Universitaires de France, Paris.

Merzbach, U. 2018. *Dirichlet: A Mathematical Biography*, Birkhäuser.

Merzbach, U. and Boyer, C.B. 2011. *A History of Mathematics*, 3rd ed., Wiley.

Meschkowski, H. 1964. *Non-Euclidean Geometry*, Academic Press.

Mittag-Leffler, G. 1902. Une page de la vie de Weierstrass, *Comptes Rendus de Deuxième Congrès International des Mathématiciens, tenu à Paris 6–12 août 1900*, Gauthier-Villars, Paris, 131–163.

Mittag-Leffler, G. 1912. Zur Biographie von Weierstrass, *Acta Mathematica* 36, 105–179.

Mittag-Leffler, G. 1923. Weierstrass et Sonja Kowalewsky, *Acta Mathematica* 39, 133–198.

Monk, R. 1997. *Bertrand Russell: The Spirit of Solitude, 1872–1920* Vol. I, Simon and Schuster.

Moore, E.H., Bolza, O., Maschke, H., and White, H.S. (eds.) 1896. *Mathematical Papers Read at the International Mathematical Congress held in Conjunction with the World's Columbian Exposition Chicago 1893*, Macmillan.

Moore, G.H. 2002. Hilbert on the infinite: The role of set theory in the evolution of Hilbert's thought, *Historia Mathematica* 29, 40–64.

Nagel, E. 1939. The formation of modern conceptions of formal logic in the development of geometry, *Osiris* 7, 142–223

Nagel, E. and Newman, J.R. 1959. *Gödel's Proof*, Routledge and Kegan Paul.

Neale, V. 2017. *Closing the Gap: The Quest to Understand Prime Numbers*, Oxford University Press.

Neeley, K.A. 2001. *Mary Somerville. Science, Illumination, and the Scientific Mind*, Cambridge University Press.

Neumann, P.M. 2011. *The Mathematical Writings of Évariste Galois*, Heritage of European Mathematics, European Mathematical Society, Zürich.

Newcomb, S. 1903. *The Reminiscences of an Astronomer*, Houghton, Mifflin and Company.

Newton, Sir I. 1676a. *Epistola Prior*, in (Turnbull 1960, 32–41).

Newton, Sir I. 1676b. *Epistola Posterior*, in (Turnbull 1960, 130–149).

Newton, Sir I. 1687. *Philosophiae Naturalis Principia Mathematica* London, 2nd edn. 1713, 3rd edn. 1726. English translations of the 3rd edn.: *Sir Isaac Newton's Mathematical Principles of Natural Philosophy and his System of the World*, Andrew Motte 1729, revised Florian Cajori, University of California Press, Berkeley, 1934, and *Isaac Newton's Philosophiae Naturalis Principia Mathematica. The third edition, 1726, with variant readings*, I.B. Cohen and A. Whitman (eds.), Cambridge University Press, 1972, 1999.

Newton, Sir I. 1704. *Opticks*, London, 2nd edn., 1718.

Newton, Sir I. 1707. *Arithmetica Universalis*, Cambridge University Press.

Ore, O. 1957. *Niels Henrik Abel*, University of Minnesota Press.

Oughtred, W. 1647. *The Key of the Mathematics Newly Forged and Filed*, transl. R. Wood, Tho. Harper for Rich. Whitaker, London.

Panza, M. 2007. Euler's *Introductio in Analysin Infinitorum*, in *Euler Reconsidered*, R. Baker (ed.), Kendrick Press, 119–166.

Parshall, K.H. and Rice, A.C. (eds.) 2002. *Mathematics Unbound: The Evolution of an International Mathematical Community, 1800–1945*, American and London Mathematical Societies, HMath 23.

Parshall, K.H. and Rowe, D.E. 1991. *The Emergence of the American Mathematical Community, 1876–1900: J.J. Sylvester, Felix Klein, and E.H. Moore*, American and London Mathematical Societies.

Pascal, B. 1914. *Oeuvres* VIII, L. Brunschvig and P. Boutroux (eds.), Hachette, Paris.

Pasch M. 1882. *Vorlesungen über neuere Geometrie*, Teubner.

Paul, M. 1980. *Gaspard Monges 'Géométrie Descriptive' und die École Polytechnique*, Institut für Didaktik der Mathematik, Bielefeld.

Peano, G. 1889a. *I Principi di Geometria Logicamente Esposití*, Turin.

Peano, G. 1889b. *Arithmetices Principia*, Frates Bocca, Rome.

Peckhaus, V. and Kahle, R. 2002. Hilbert's paradox, *Historia Mathematica* 29, 157–175.

Pedersen, K. 1980. Techniques of the calculus, 1630–1660, in I. Grattan-Guinness (ed.), *From the Calculus to Set Theory, 1630–1910*, Duckworth, 10–48.

Pesic, P. 2003. *Abel's Proof*, MIT Press.

Playfair, J. 1795. *Elements of Geometry; Containing the First Six Books of Euclid, with two Books on the Geometry of Solids. To Which are added, Elements of Plain and Spherical Geometry*, Edinburgh.

Plofker, K. 2009. *Mathematics in India*, Princeton University Press.

Poincaré, H. 1881. Mémoire sur les courbes définies par une équation différentielle, *Journal de Mathématiques* 7, 375–422.

Poincaré, H. 1882. Sur les fonctions uniformes qui se reproduisent par des substitutions linéaires, *Mathematische Annalen* 19, 553–564, in *Oeuvres* 2, 92–105.

Poincaré, H. 1891. Le Problème des trois corps, *Revue générale des Sciences* 2, 1?5, in *Oeuvres* 8, 532–534.

Poincaré, H. 1892, 1893, 1899. *Les Méthodes Nouvelles de la Mécanique Céleste*, 3 vols., Gauthier-Villars, Paris.

Poincaré, H. 1893a. Sur la propagation de l'électricité, *Comptes Rendus de l'Académie des Sciences* 117, 1027–1032, in *Oeuvres* 9, 278–283.

Poincaré, H. 1898. Sur la stabilité du système solaire, *Annuaire du Bureau des Longitudes*, B1–B16, in *Oeuvres* 8, 538–547, transl. On the stability of the solar system, *Nature* 58, 183–185.

Poincaré, H. 1902. *La Science et l'Hypothèse*, Flammarion, transl. W.G. Greenstreet, *Science and Hypothesis*, Walter Scott Publishing Co., 1905, and M. Frappier and D.J. Stump (eds.), M. Frappier, A. Smith, and D.J. Stump (transls.), *Science and Hypothesis; The Complete Text*, Bloomsbury, 2018.

Poincaré, H. 1905. *Le Valeur de la Science*, Flammarion, transl. G.B. Halsted, *The Value of Science*.

Poincaré, H. 1908. *Science et Méthode*, Flammarion, transl. F. Maitland, *Science and Method*, Nelson, 1914.

Poincaré, H. 1912. Sur un théorème de géométrie, *Rendiconti Circolo Matematico di Palermo* 33, 375–407, in *Oeuvres* 6, 499–538.

Poincaré, H. 1913. *Dernières Pensées,* Flammarion, transl. J.W. Bolduc, *Mathematics and Science: Last Essays*, Dover repr. 1963.

Poincaré, H. 1997. *Three Supplementary Essays on the Discovery of Fuchsian Functions*, J.J. Gray and S. Walter (eds.), Akademie Verlag, Berlin and Blanchard, Paris.

Poisson, S.-D. 1823. Mémoire sur la distribution de la chaleur dans les corps solide, *Journal de l'École Polytechnique*, Cahier 19, 1–144.

Poisson, S.-D. 1835. *Théorie Mathématique de la Chaleur*, Paris.

Poncelet, J.V. 1822. *Traité des Propriétés Projectives des Figures*, Gauthier-Villars.

Pourciau, B. 1998. The preliminary mathematical lemmas of Newton's *Principia*, *Archive for History of Exact Sciences* 52, 279–295.

Proclus, 1970. *A Commentary on the First Book of Euclid's* Elements, G. R. Morrow (ed. and transl.) , Princeton University Press.

Purkerts, W. and Ilgauds, H.J. 1985. *Georg Cantor*, Birkhäuser.

Pyenson, L. 1983. *Neo-humanism and the Persistence of Pure Mathematics in Wilhelmian Germany*, American Philosophical Society.

Rabouin, D. and Arthur, R.T.W. 2020. Leibniz's syncategorematic infinitesimals II: their existence, their use and their role in the justification of the differential calculus, *Archive for History of Exact Sciences*, 74, 401–443.

Raugh, M. and Probst, S. 2019. The Leibniz catenary and approximation of e – an analysis of his unpublished calculations, *Historia Mathematica* 49, 1–19.

Reichardt, H. 1976. *Gauss und die Nicht-Euklidische Geometrie*, Teubner.

Reid, C. 1970. *Hilbert*, Springer.

Richeson, D.S. 2008. *Euler's Gem: The Polyhedron Formula and the Birth of Topology*, Princeton University Press.

Richeson, D.S. 2019. *Tales of Impossibility: The 2000-Year Quest to Solve the Mathematical Problems of Antiquity*. Princeton University Press.

Rider, R. 1981. Poisson and algebra: against an eighteenth-century background, in *Siméon-Denis Poisson et la Science de son Temps*, M. Métivier, P. Costabel, and P. Dugac (eds.), École Polytechnique, 167–176.

Riemann, G.F.B. 1854. Über die Darstellbarkeit einer Function durch einer trigonometrische Reihe, *Königlichen Gesellschaft der Wissenschaften zu Göttingen* 13, 87–132, in *Bernhard Riemanns gesammelte mathematische Werke und wissenschaftliche Nachlass*, 3rd. edn., Narasimhan, R. (ed.). Springer, 259–303, transl. in Riemann, *Collected Papers*, Kendrick Press, 2004, 219–255.

Riemann, G.F.B. 1859. Ueber die Anzahl der Primzahlen unter einer gegebene Grösse, *Monatsberichte Berlin*, 671–680, in *Bernhard Riemanns gesammelte mathematische Werke und wissenschaftliche Nachlass*, 3rd. edn., Narasimhan, R. (ed.), Springer, 145–155, transl. in Riemann, *Collected Papers*, Kendrick Press, 2004, 135–145.

Riemann, G.F.B. 1867. Ueber die Hypothesen welche der Geometrie zu Grunde liegen, *Göttingen Abh.* 13, 133–152 in *Bernhard Riemanns Gesammelte Mathematische Werke und Wissenschaftliche Nachlass*, 3rd. edn., Narasimhan, R. (ed.), Springer, 304–319, transl. in Riemann, *Collected Papers*, Kendrick Press, 2004, 257–272,

Roberval, G.P. de. 1693. *Traité des Indivisibles*, Paris.

Romero, A. 2007. Physics and analysis: Euler and the search for fundamental principles of mechanics, in *Euler Reconsidered*, R. Baker (ed.), Kendrick Press, 232–281.

Rothman, T. 1989. *Science à la Mode*, Princeton University Press.

Rothman, P. 1996. Grace Chisholm Young and the Division of Laurels, *Notes and Records of the Royal Society* 50, 89–100.

Ruffini, P. 1799. *Teoria Generale delle Equazioni*, S. Tommaso d'Aquino, Bologna.

Russell, B. 1967. *The Autobiography of Bertrand Russell 1872–1914*, Routledge.

Saccheri, G. 1697. *Logica Demonstrativa*, Turin.

Saccheri, G. 1733. *Euclides ab Omne Naevo Vindicatus*, transl. as *Euclid Freed of every Flaw* by G.B. Halsted, Open Court, 1920; new edn., *Euclid Vindicated from every Blemish*, by V. De Risi (ed.), transl. and revd. by L. Allegri, Birkhäuser, 2014.

Saint-Vincent, G. de. 1647. *Opus Geometricum Quadraturae Circuli et Sectionum Coni*, Antwerp.

Sandifer, C.E. (ed.) 2007a. *The Early Mathematics of Euler*, Mathematical Association of America.

Sandifer, C.E. (ed.) 2007b. *How Euler Did it*, Mathematical Association of America.

Sandifer, C.E. (ed.) 2015. *How Euler Did Even More*, Mathematical Association of America.

Scharlau, W. and Opolka, H. 1985. *From Fermat to Minkowski*, Springer.

Schoenflies, A. 1927. Die Krisis in Cantor's mathematischen Schaffen, *Acta Mathematica* 50, 1–23.

Scholz, E. 1992. Gauss und die Begründung der 'höheren' Geodäsie, *Amphora*, S. Demidov, M. Folkerts, D.E. Rowe, and C.J. Scriba (eds.), Birkhäuser, 631–648.

Schubring, G. 1999. Argand and the early work on graphical representation: new sources and interpretations, *Matematisk-fysiske Meddelelser* 46.2, 125–146, repr. in *Around Caspar Wessel and the Geometric Representation of Complex Numbers*, J. Lützen (ed.), The Royal Danish Academy of Sciences and Letters, Copenhagen, 125–146.

Schubring, G. 2005. *Conflicts between Generalization, Rigor, and Intuition*, Springer.

Scriba, C.J. 1963. The inverse method of tangents: a dialogue between Leibniz and Newton, *Archive for History of Exact Sciences* 2, 113–137.

Shank, J.B. 2008. *The Newton Wars and the Beginning of the French Enlightenment*, Chicago University Press.

Shapiro, S. (ed.) 2005. *The Oxford Handbook of Philosophy of Mathematics and Logic*, Oxford University Press.

Smith, D.E. (ed.) 1929. *A Source Book in Mathematics*, McGraw-Hill Book Company, Dover repr. 1959.

Smith, G.E. 2002. The methodology of the *Principia*, in *The Cambridge Companion to Newton*, I.B. Cohen and G.E. Smith (eds.), Cambridge University Press, 138–173.

Smith, H.J.S. 1859. *Report on the Theory of Numbers*, British Association for the Advancement of Science, in *Collected Mathematical Papers* 1894, Vol. 1, 38–364, Chelsea repr., New York, 1965.

Sobel, D. 1995. *Longitude*, Fourth Estate.

Sohncke, L.A. 1838. Parallel, in *Allgemeine Encyclopaedie der Künste und Wissenschaften* III.11, 368–384.

Somerville, M. 1831. *The Mechanism of the Heavens*, John Murray.

Somerville, M. 1873. *Personal Recollections from Early Life to Old Age, of Mary Somerville. With selections of her correspondence, by her Daughter, Martha Somerville*, London.

Speziali, P. 1983. Leonhard Euler and Gabriel Cramer, in (Burckhardt, Fellmann, and Habicht (eds.), 1983), 421–434.

Spiess, O. (ed.) 1955. *Der Briefwechsel von Johann Bernoulli*, Birkhäuser.

Sprat, T. 1667. *History of the Royal Society*, London.

Stäckel. P. 1913. *Wolfgang und Johann Bolyai*, Teubner.

Stedall, J. 2004. *The Arithmetic of Infinitesimals: John Wallis 1656*, Springer.

Stedall, J. 2008. *Mathematics Emerging: A Sourcebook 1540–1900*, Oxford University Press.

Stedall, J. 2011. *From Cardano's* Great Art *to Lagrange's* Reflections*: Filling a Gap in the History of Algebra*, European Mathematical Society.

Steiner, J. 1832. *Systematischer Entwickelung der Abhängigkeit geometrischer Gestalten von einander*, Fincke.

Stenhouse, B. 2020. Mary Somerville's early contributions to the circulation of the differential calculus, *Historia Mathematica* 51, 1–25.

Stillwell, J. 1996. *Sources of Hyperbolic Geometry*, American and London Mathematical Societies, HMath 10.

Stone, E. 1730. *The Method of Fluxions Both Direct and Inverse*, London.

Struik, D. 1969. *A Source Book in Mathematics: 1200–1800*, Harvard University Press.

Sturm, C.-F. 1836. Mémoire sur une classe d'équations à differences partielles, *Journal de Mathématiques Pures et Appliquées* 1, 373–444.

Sundman, K.F. 1912. Mémoire sur le problème des trois corps, *Acta Mathematica* 36, 105–179.

Suzuki, J. 2006. Lagrange's proof of the Fundamental Theorem of Algebra, *American Mathematical Monthly*, 113, 705–714.

Takase, M. 2007. Euler's theory of numbers, in *Euler Reconsidered*, R. Baker (ed.), Kendrick Press, 377–421.

Taton, R. and Wilson, C. (eds.) 1995. *Planetary Astronomy from the Renaissance to the Rise of Astrophysics, Part B: The Eighteenth and Nineteenth Centuries*, Cambridge University Press.

Taylor, B. 1714. De motu nervi tensi, *Philosophical Transactions of the Royal Society* 28, 26–32.

Taylor, R. and Wiles, A. 1995. Ring-theoretic properties of certain Hecke algebras, *Annals of Mathematics* 141, 553–572.

Terrall, M. 2002. *The Man who Flattened the Earth*, Chicago University Press.

Thomas, I. 1939. *Greek Mathematical Works*, 2 vols., Loeb Classical Library, Harvard University Press.

Thomson, W. 1855. On the theory of the electric telegraph, *Proceedings of the Royal Society* 7, 382–399, in *Mathematical and Physical Papers* 2, Cambridge University Press, 61–75.

Thomson, W. and Tait, P.G. 1879. *Treatise on Natural Philosophy*, 2 vols., Cambridge University Press.

Thomson, W. 1882. The Tide Gauge, Tidal Harmonic Analyser, and Tide Predictor, *Minutes of the Proceedings of the Institution of Civil Engineers*, 1 March 1882, in Kelvin, *Mathematical and Physical Papers* 6, no. 271.

Tignol, J.-P. 2001. *Galois' Theory of Algebraic Equations*, World Scientific.

Torricelli, E. 1644. *Opera Geometrica*, Florence.

Trudeau, R.J. 1987. *The Non-Euclidean Revolution*, Birkhäuser, Boston.

Truesdell, C.A. 1955. The rational mechanics of flexible or elastic bodies 1638–1788, in Euler, *Opera Omnia* (2), 11.2.

Truesdell, C.A. 1960. A program toward rediscovering the rational mechanics of the Age of Reason, *Archive for History of Exact Sciences* 1, 1–36.

Truesdell, C.A. 1984. *An Idiot's Fugitive Essays on Science*, Springer.

Turnbull, H.W. (ed.) 1960. *The Mathematical Correspondence of Isaac Newton*, Vols. 1, 2, 1676–1687, Cambridge University Press.

Van Dalen, D. 1999, 2005. *Mystic, Geometer, and Intuitionist; The Life of L.E.J. Brouwer*, 2 vols., Clarendon Press, Oxford.

Van Schooten, F. 1657. *Exercitiones Mathematicae*, Elsevier.

Van Heijenoort, J. (ed.) 1967. *From Frege to Gödel: A Source Book in Mathematical Logic,1879–1931*, Harvard University Press.

Veblen, O. and Young, J.W. 1910, 1918. *Projective Geometry*, 2 vols., Ginn.

Verhulst, F. 2012. *Henri Poincaré: Impatient Genius*, Springer.

Vilenkin, N. Ya. 1968. *Stories about Sets*, transl. by Scripta Technica, Academic Press.

Voltaire, F. 1734. *Lettres Écrits de Londres sur les Anglais et Autres Sujets*, Basel, transl. *Letters on England*, Penguin, 1984.

Voltaire, F. 1753. *Diatribe du Docteur Akakia Medicin du Pape*, many editions.

Wallis, J. 1655. *De Sectionibus Conicis*, Oxford

Wallis, J. 1656. *Arithmetica Infinitorum*, Oxford, see also (Stedall 2004).

Wallis, J. 1663. De postulato quinto et definitione lib. 6, Euclidis deceptatio geometrica, *Operum Mathematicorum*, Vol. 2, 665–678.

Wallis, J. 1685. *A Treatise of Algebra*, London.

Waltershausen, W. S. von, 1856. *Gauss zum Gedächtnis*. 2nd edn., Hirzel, Stuttgart. repr. Sändig, Wiesbaden, 1965.

Wantzel, P.L. 1837. Recherches sur les moyens de reconnaître si un Problème de Géométrie peut se résoudre avec la règle et le compas, *Journal de Mathématiques Pures et Appliquées* 2, 366–372.

Weber, H.M. 1891. L. Kronecker, *Jahresbericht der Deutschen Mathematiker Vereinigung* 2, 5–23.

Weierstrass, K. 1911/1923. Eine Äusserung von Weierstrass an Mittag-Leffler über das Dreikörperproblem, *Acta Mathematica* 39, 257–258.

Weil, A. 1984. *Number Theory: An Approach through History from Hammurapi to Legendre*, Birkhäuser.

Westfall, R.S. 1983. *Never at Rest: A Biography of Isaac Newton*, Cambridge University Press.

Weyl, H. 1949. *Philosophy of Mathematics and Natural Science*, Princeton University Press.

Whiteside, D.T. 1961. Patterns of mathematical thought in the later 17th century, *Archive for History of Exact Sciences* 1, 179–388.

Whiteside, D.T. (ed.) 1964. *The Mathematical Works of Isaac Newton*, Johnson Reprint.

Whiteside, D.T. 1970. The mathematical principles underlying Newton's *Principia Mathematica*, *Journal of the History of Astronomy* 1, 116–138.

Whiteside, D.T. (ed.) 1967–1981. *The Mathematical Papers of Isaac Newton*, 8 vols., Cambridge University Press.

Whittaker, E.T. 1899. Report on the progress of the solution of the problem of three bodies, *Report of the British Association for the Advancement of Science*, 121–159.

Wiles, A. 1995. Modular elliptic curves and Fermat's Last Theorem, *Annals of Mathematics* 141, 443–551.

Wilson, C. 1985. The great inequality of Jupiter and Saturn, *Archive for History of Exact Sciences* 33, 15–290.

Wilson, C. 1995. The work of Lagrange in celestial mechanics, in (Taton and Wilson (eds.), 1995), 108–130, Cambridge University Press.

Wilson, C. 2002. Newton and celestial mechanics, *The Cambridge Companion to Newton*, I.B. Cohen and G.E. Smith (eds.), Cambridge University Press, 203–226.

Wilson, C. and Harper, W. 2014. The coming-to-be of Hansen's method, *Archive for History of Exact Sciences* 68, 409–497.

Wilson, R. 2020. *Number Theory: A Very Short Introduction*, Oxford University Press.

Wordsworth, W. 1850. *Prelude*, Book III.

Wussing, H. 1984. *The Genesis of the Abstract Group Concept*, transl. A. Shenitzer, MIT Press.

Yandell, B.H. 2002. *The Honors Class: Hilbert's Problems and their Solvers*, A.K. Peters.

Yavetz, I. 1995. *From Obscurity to Enigma: The Work of Oliver Heaviside, 1872–1889*, Birkhäuser.

Yoder, J.G. 2004. *Unrolling Time: Christiaan Huygens and the Mathematization of Nature*, Cambridge University Press.

Zagier, D. 1977. The first 50 million prime numbers, *Mathematical Intelligencer* 1, 7–19.

Zinsser, J.P. 2001. Translating Newton's *Principia*: the Marquise de Châtelet's revisions and additions for a French audience, *Notes and Records of the Royal Society* 55, 227–245.

Zweig, A. (ed.) 1967. *Kant's Philosophical Correspondence, 1759–1799*, Chicago University Press.

Index